CROSS-SCALE COUPLING AND ENERGY TRANSFER IN THE MAGNETOSPHERE-IONOSPHERE-THERMOSPHERE SYSTEM

CROSS-SCALE COUPLING AND ENERGY TRANSFER IN THE MAGNETOSPHERE-IONOSPHERE-THERMOSPHERE SYSTEM

Edited by

YUKITOSHI NISHIMURA

Research Associate Professor, Boston University, Boston, MA, United States

OLGA VERKHOGLYADOVA

Research Group Supervisor, Jet Propulsion Laboratory, California Institute of Technology, Pasadena, CA, United States

YUE DENG

Professor of Space Physics, Department of Physics, University of Texas at Arlington (UTA), Arlington, TX, United States

SHUN-RONG ZHANG

MIT Haystack Observatory, Westford, MA, United States

ELSEVIER

Elsevier
Radarweg 29, PO Box 211, 1000 AE Amsterdam, Netherlands
The Boulevard, Langford Lane, Kidlington, Oxford OX5 1GB, United Kingdom
50 Hampshire Street, 5th Floor, Cambridge, MA 02139, United States

Library of Congress Cataloging-in-Publication Data
A catalog record for this book is available from the Library of Congress

British Library Cataloguing-in-Publication Data
A catalogue record for this book is available from the British Library

ISBN: 978-0-12-821366-7

For information on all Elsevier publications
visit our website at https://www.elsevier.com/books-and-journals

Publisher: Candice Janco
Acquisitions Editor: Peter J. Llewellyn
Editorial Project Manager: Michelle Fisher
Production Project Manager: Vijayaraj Purushothaman
Cover Designer: Matthew Limbert

Typeset by STRAIVE, India

Contents

Contributors

Ercha Aa
MIT Haystack Observatory, Westford, MA, United States

Shane Coyle
Virginia Tech, Blacksburg, VA, United States

Tong Dang
CAS Key Laboratory of Geospace Environment, School of Earth and Space Sciences; Mengcheng National Geophysical Observatory, University of Science and Technology of China; CAS Center for Excellence in Comparative Planetology, Hefei, China

Yue Deng
Department of Physics, University of Texas at Arlington, TX, United States

Kshitija B. Deshpande
Department of Physical Sciences and Center for Space and Atmospheric Research (CSAR), Embry-Riddle Aeronautical University (ERAU), Daytona Beach, FL, United States

Yakov S. Dimant
Center for Space Physics, Boston University, Boston, MA, United States

Mark J. Engebretson
Augsburg University, Minneapolis, MN, United States

Scott England
Department of Aerospace and Ocean Engineering, Virginia Polytechnic Institute and State University, Blacksburg, VA, United States

Xiaohua Fang
Laboratory for Atmospheric and Space Physics, University of Colorado, Boulder, CO, United States

Evgeny N. Fedorov
Institute of Physics of the Earth, Moscow, Russia

Alex Glocer
NASA/GSFC, Greenbelt, MD, United States

Lindsay V. Goodwin
Center for Solar-Terrestrial Research, New Jersey Institute of Technology, Newark, NJ; Cooperative Programs for the Advancement of Earth System Science, University Corporation for Atmospheric Research, Boulder, CO, United States

Christine Gabrielse
The Aerospace Corporation, El Segundo, CA, United States

Bea Gallardo-Lacourt
Universities of Space Research Association, Columbia; NASA Goddard Space Flight Center, Greenbelt, MD, United States

Michael D. Hartinger
Virginia Tech, Blacksburg, VA, United States

Michael Hirsch
Boston University, Boston, MA, United States

Mingwu Jin
Department of Physics, University of Texas at Arlington, TX, United States

Stephen R. Kaeppler
Department of Physics and Astronomy, Clemson University, Clemson, SC, United States

Hyosub Kil
The Johns Hopkins University Applied Physics Laboratory, Laurel, MD, United States

Liam M. Kilcommons
University of Colorado, Boulder, CO, United States

Naritoshi Kitamura
Department of Earth and Planetary Science, Graduate School of Science, The University of Tokyo, Tokyo, Japan

Delores J. Knipp
Smead Aerospace Engineering Department, University of Colorado, Boulder, CO, United States

Leslie Lamarche
SRI International, Menlo Park, CA, United States

Woo Kyoung Lee
Korea Astronomy and Space Science Institute, Daejeon, Republic of Korea

Jiuhou Lei
CAS Key Laboratory of Geospace Environment, School of Earth and Space Sciences; Mengcheng National Geophysical Observatory, University of Science and Technology of China; CAS Center for Excellence in Comparative Planetology, Hefei, China

Cissi Y. Lin
Department of Physics, University of Texas at Arlington, TX, United States

Chaoqun Liu
Department of Mathematics, University of Texas at Arlington, TX, United States

Huixin Liu
Department of Earth and Planetary Science, Kyushu University, Fukuoka, Japan

William J. Longley
Department of Physics and Astronomy, Rice University, Houston, TX; University Corporation for Atmospheric Research, Boulder, CO, United States

Gang Lu
High Altitude Observatory, National Center for Atmospheric Research, Boulder, CO, United States

Larry R. Lyons
Department of Atmospheric and Oceanic Sciences, University of California, Los Angeles, CA, United States

Yukitoshi Nishimura
Department of Electrical and Computer Engineering and Center for Space Physics, Boston University, Boston, MA, United States

Meers M. Oppenheim
Center for Space Physics, Boston University, Boston, MA, United States

Larry J. Paxton
Korea Astronomy and Space Science Institute, Daejeon, Republic of Korea

Vyacheslav A. Pilipenko
Space Research Institute; Institute of Physics of the Earth, Moscow, Russia

Gareth W. Perry
Center for Solar-Terrestrial Research, New Jersey Institute of Technology, NJ, United States

Mark Redden
Department of Physical Sciences and Center for Space and Atmospheric Research (CSAR), Embry-Riddle Aeronautical University (ERAU), Daytona Beach, FL, United States

Cheng Sheng
Department of Physics, University of Texas at Arlington, TX, United States

Andres Spicher
Department of Physics and Technology, UiT the Arctic University of Norway, Tromsø, Norway

Olga P. Verkhoglyadova
Jet Propulsion Laboratory, California Institute of Technology, Pasadena, CA, United States

Chih-Ping Wang
UCLA, Los Angeles, CA, United States

Matthew A. Young
Space Science Center, University of New Hampshire, Durham, NH, United States

Yiqun Yu
School of Space and Environment, Beihang University; Key Laboratory of Space Environment Monitoring and Information Processing, Ministry of Industry and Information Technology, Beijing, China

Matthew D. Zettergren
Department of Physical Sciences and Center for Space and Atmospheric Research (CSAR), Embry-Riddle Aeronautical University (ERAU), Daytona Beach, FL, United States

Weijia Zhan
Department of Physics and Astronomy, Clemson University, Clemson, SC, United States

Shun-Rong Zhang
MIT Haystack Observatory, Westford, MA, United States

Qingyu Zhu
Department of Physics; Department of Mathematics, University of Texas at Arlington, TX, United States

Preface

The magnetosphere-ionosphere-thermosphere (M-I-T) coupling system evolves constantly due to variable energy inputs and complexity of exchange in mass, momentum, and energy. Determining the state of the M-I-T system is an important objective in the field of our community research because it is a fundamental science problem when nonlinear coupling occurs across regions and at different spatial and temporal scales. Coupling across multiple scales, among global (large-scale), regional (meso-scale), and turbulence (small-scale) domains, is a critical challenge because observational and modeling capabilities for bridging different scales are quite limited.

This book was motivated primarily by growing interests in the nature and challenges of the multiscale M-I-T coupling in the community. A dedicated session on multi-scale M-I-T coupling at the American Geophysical Union (AGU) Fall Meetings hosted every year since 2017. It has become one of the largest sessions in the Aeronomy subsection of the AGU Space Physics and Aeronomy (SPA) section, providing an active forum for discussing the current state of the topic with space physicists across the world. The Grand Challenge workshop on the Multi-Scale Ionosphere-Thermosphere System Dynamics in the Coupling, Energetics, and Dynamics of Atmospheric Regions (CEDAR) community started in 2018 for discussing challenges and collaborative projects among participants mainly from the United States. The international team on Multi-Scale Magnetosphere-Ionosphere-Thermosphere Interaction sponsored by the International Space Science Institute (ISSI) and ISSI-Beijing was formed in 2019 for focused research and collaboration among international experts. Multiscale processes in M-I-T have been addressed in several other research initiatives, including Multidisciplinary University Research Initiative (MURI), and are gaining broad attention from a system science perspective.

Through these community activities, we realized that this growing area of research does not have a reference manual that broadly describes key concepts. Thus, we decided to compile such a book responding to this need. The main theme of the book is to provide basic concepts, definitions, equations, and examples on M-I-T coupling processes across scales. The book is not intended for a collection of research papers or a thorough literature review. The book targets not only active research experts but also researchers and students who would like to learn M-I-T coupling topics. The book also includes original materials useful for classrooms and tutorials.

Chapter 1 provides an overview of the coupling system across scales. After a tutorial on large-scale dynamics, the chapter introduces multiscale processes in key regions, properties of meso-scale convection and impacts, and plasma upflows and outflows. Chapter 2 describes aurora and airglow and discusses their connection with magnetospheric processes and meso-scale flow dynamics. The topic of Chapter 3 is plasma density structures and irregularities. The chapter details instability mechanisms and impacts of density irregularities on space weather. Chapter 4 reviews multiscale convection, precipitation, and conductance, including magnetospheric plasma transport as well as modeling and data assimilation techniques of precipitation and conductance. Chapter 5 focuses on electromagnetic energy transport and dissipation in the form of Poynting flux and Joule heating on both observational and modeling aspects. In Chapter 6, kinetic physics as a regime beyond fluid scales is described. Treatment of kinetic effects, instabilities, particle acceleration, impact ionization, and ultralow-frequency (ULF) waves are detailed. Finally, Chapter 7 reviews ion-neutral interaction in terms of thermospheric structures and transport, and traveling atmospheric and ionospheric disturbances.

We thank all authors who have contributed to this book. We also appreciate the editorial project managers at Elsevier for their assistance and patience in developing the book. It was an extremely challenging time for everyone due to the pandemic, but we are pleased to complete and publish the book with high-quality materials.

Yukitoshi Nishimura
Yue Deng
Olga Verkhoglyadova
Shun-Rong Zhang

CHAPTER 1

Multiscale processes in the M-I-T system

Chapter outline

Cross-Scale Coupling and Energy Transfer in the Magnetosphere-Ionosphere-Thermosphere System
https://doi.org/10.1016/B978-0-12-821366-7.00007-X

Part 1 Basic concepts and overview of the coupling system

Yukitoshi Nishimura

Department of Electrical and Computer Engineering and Center for Space Physics, Boston University, Boston, MA, United States

1.1 Large-scale processes

1.1.1 Magnetospheric convection

The magnetosphere-ionosphere-thermosphere (M-I-T) system is driven by the Sun and lower atmosphere, and its system responses occur as a result of complex mass, momentum, and energy transfers among various regions within the system. The key regions and coupling schemes are illustrated in Fig. 1.1. The upper part of the diagram shows the circulation of the mass, momentum, and energy in the magnetosphere. The typical locations of these regions in the southward IMF are depicted in Fig. 1.2A. The plasma and electromagnetic energy from the solar wind enter the magnetosphere (via the bow shock and magnetosheath) through the magnetopause, and travel to the lobe, low latitude boundary layer (LLBL), and/or flanks. The magnetopause reconnection is the process where the Earth's closed magnetic field lines reconnect with the interplanetary magnetic field and become open magnetic

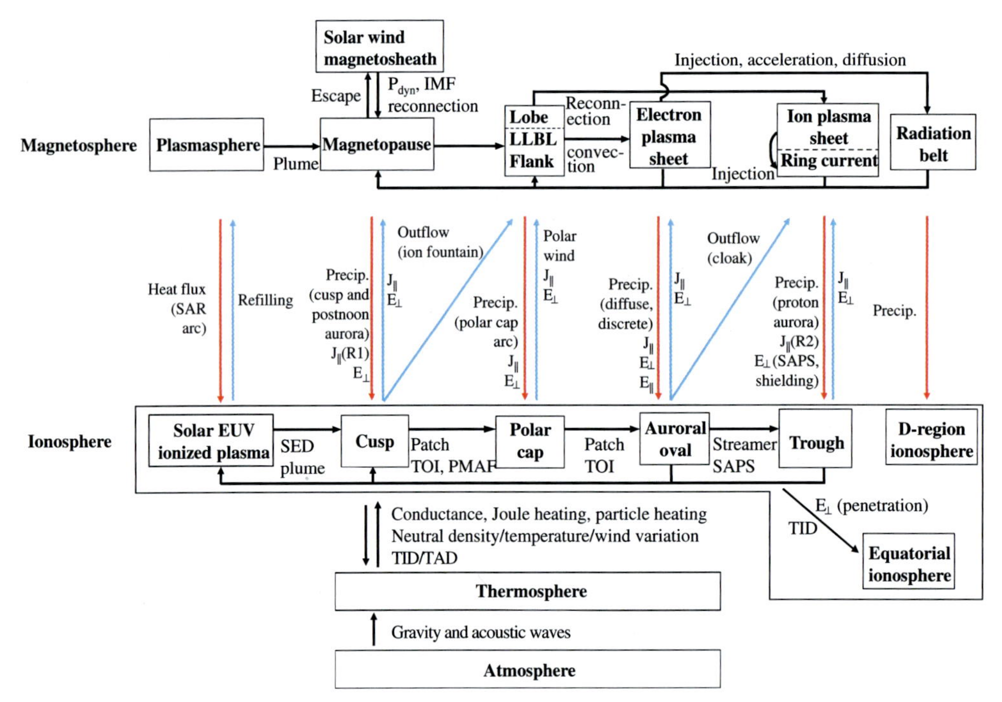

Fig. 1.1 Schematic diagram on how the key regions in the magnetosphere and ionosphere interact with the surrounding regions. The *arrows* indicate mass, momentum, and energy transport between regions. The *red and blue arrows* highlight the transport processes from the magnetosphere to the ionosphere, and from the ionosphere to the magnetosphere, respectively. The key parameters or phenomena of the transport are labeled next to the arrows.

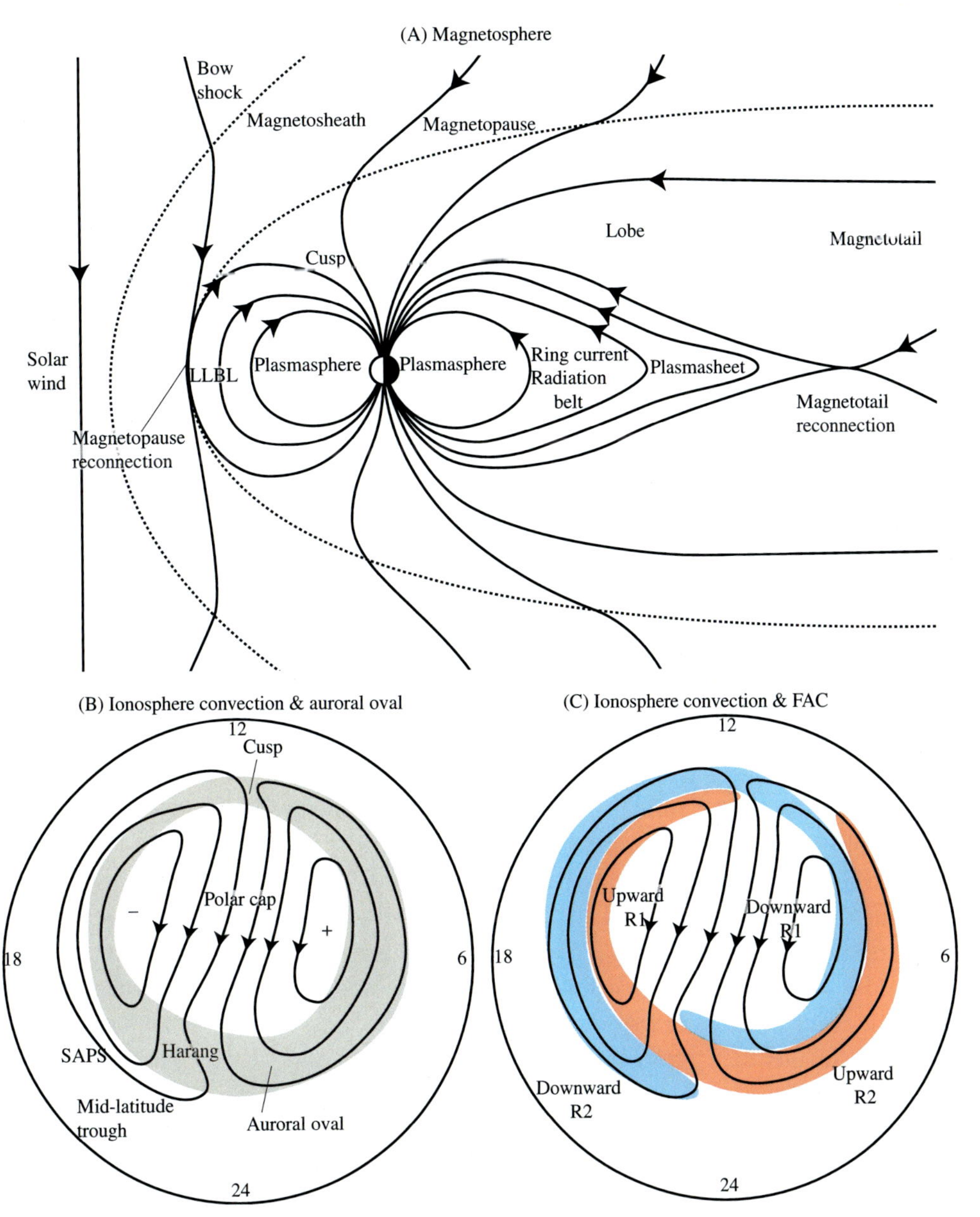

Fig. 1.2 Illustration of key regions in the (A) magnetosphere with representative magnetic field lines, and (B) ionosphere with representative convection streamlines and a typical auroral oval location. (C) The large-scale ionospheric current system.

field lines in the lobe. Magnetosheath plasma enters the magnetosphere and precipitates to the cusp through this process. Kelvin-Helmholtz instability and particle scattering also contribute to the plasma entry (see Section 4.1.3) (Wing et al., 2014).

The lobe magnetic fields are open to the interplanetary space, and the particle fluxes and energy are generally low with a mixture of magnetosheath plasma and outflowing plasma from the ionosphere. The lobe magnetic field lines near the nightside open-closed magnetic field boundary (outermost closed field line) reconnect in the magnetotail (magnetotail reconnection), and the closed magnetic field lines in the magnetotail form the plasma sheet (see Section 4.1). Plasma sheet plasma also comes from the flanks and the ionosphere. The plasma sheet plasma is transported earthward and forms the ring current and radiation belt. Then the plasma escapes out of the magnetosphere across the magnetopause, precipitates to the atmosphere, or repeats the circulation in the magnetosphere. The plasmasphere is the region of closed convection paths that is filled with cold plasma supplied from the ionosphere. The plasmaspheric plasma on the duskside forms the plasmaspheric plume and drifts toward the dayside magnetopause.

The plasma transport pathways and velocity in the magnetosphere are highly dependent on the energy. The cold plasma transport follows the $E \times B$ drift, and it is referred to as magnetospheric convection. The plasmaspheric plasma is cold, and the plume formation can be described by the $E \times B$ drift. The large-scale electric field consists of the convection electric field that arises from the solar wind-magnetosphere interaction (see Section 1.1.3) and the corotation electric field that arises from the Earth's rotation. In addition, the electric field is locally enhanced in association with the meso-scale processes discussed in Section 1.2 and with the waves in the magnetosphere.

Hot plasma motion is dominated by the magnetic drifts (∇B and curvature drifts). Ions and electrons drift to the opposite directions (ions: duskward, electrons: dawnward), and thus the motion of ions and electrons is often discussed separately (e.g., ion plasma sheet and electron plasma sheet). The magnetic drift motion of hot plasma along the electric field results in adiabatic acceleration. When electric and magnetic fields vary on a timescale comparable to or faster than gyro or drift periods, or on a spatial scale comparable to or smaller than gyroradii, the magnetic moments of the charged particles are not conserved, and the particles experience nonadiabatic acceleration, diffusion, and pitch angle scattering. Those include current sheet scattering, wave-particle interaction (such as with chorus, ultralow-frequency (ULF) waves). The large-scale and localized transport from the magnetotail contributes to developing the ring current and radiation belt.

1.1.2 Ionospheric convection and precipitation

Since the magnetosphere and ionosphere are connected by the Earth's magnetic field lines as illustrated in Fig. 1.2A, the ionosphere develops a convection pattern that is analogous to the magnetospheric convection. The ionospheric convection pattern under the southward

IMF is illustrated in Fig. 1.2B. The large-scale slowly varying convection can be described by the electrostatic potential field with a maximum on the dawnside and minimum on the duskside. The equipotential contours describe the $E \times B$ drift paths. The ionospheric plasma is created by the solar EUV radiation and impact ionization by particle precipitation from the magnetosphere. The ionospheric plasma is colder and follows the $E \times B$ drift. Precipitation from the magnetopause and its vicinity maps to the cusp in the polar ionosphere near noon, as a region of elevated plasma density and temperature in the F-region ionosphere. Plasma flows through the cusp cross the dayside open-closed magnetic field boundary and thus are the ionospheric signature of magnetopause reconnection. The plasma flows transport high-density plasma in the subauroral ionosphere on closed magnetic field lines into the polar cap. Enhanced plasma density in the cusp created by soft electron precipitation is also transported into the polar cap. The region of high density in the polar cap is called polar cap patches and tongue of ionization (TOI).

The flows across the nightside open-closed boundary are the ionospheric signature of the magnetotail reconnection. The flows crossing the boundary and subsequent equatorward flows in the auroral oval are associated with poleward boundary intensifications (PBIs) and auroral streamers. The equatorward flows turn duskward and dawnward, and the duskward turning often develops a large flow shear called the Harang flow reversal. While most of the dawnward flows stay in the auroral oval, the equatorward portion of the duskward flows extends into the subauroral ionosphere. The flow speed forms a peak in the subauroral ionosphere, and it is called subauroral polarization streams (SAPS). The SAPS flows contribute to the formation of the midlatitude trough via recombination, and transports subauroral ionospheric plasma toward the dayside, forming the plume. The convection electric field penetrating to the dayside subauroral ionosphere also transports the plasma upward and poleward, forming the storm enhanced density (SED). When the electric field stays enhanced, the SED plume plasma drifts into the cusp and becomes the source of the polar cap patches and TOIs. More quantitative large-scale convection and precipitation patterns can be found in studies by Thomas and Shepherd (2018) and Newell et al. (2014). The types of precipitation are described in Section 4.2.

1.1.3 Current system and M-I coupling

The magnetosphere and ionosphere interact through currents along the magnetic field lines (field-aligned currents or FACs, $J_{\|}$ in Fig. 1.1). Magnetic field distortions associated with large-scale FACs propagate at Alfvén speed ($v_A = B/\sqrt{\mu_0 \rho}$ where ρ is mass density). Alfvén waves are associated with electric and magnetic field distortions perpendicular to the background magnetic field, and Alfvén waves transfer electromagnetic energy between the magnetosphere and ionosphere as the Poynting flux (see Section 1.1.7 and Section 5.1) so that convection and currents in the magnetosphere and ionosphere evolve self-consistently. In the ionosphere, FACs can be expressed using the height-integrated current continuity equation,

$$J_{\|}\sin I = -(\nabla\cdot \boldsymbol{J}_{\perp}) = -\nabla\cdot(\Sigma \boldsymbol{E}) = -\Sigma_P(\nabla\cdot\boldsymbol{E}) - \boldsymbol{E}\cdot(\nabla\Sigma_P) + (\nabla\Sigma_H)\cdot(\mathbf{E}\times\mathbf{b}) \quad (1.1)$$

where E is the electric field in the ionosphere, $j_{\|}$ is the FAC density, $J_{\perp}$ is the height-integrated current perpendicular to the magnetic field and is the sum of the height-integrated Pedersen and Hall currents, I is the magnetic field inclination, and **b** is the unit vector of the background magnetic field. Eq. (1.1) describes how the FACs, ionospheric electric field, and conductances are related to one another. Here we used Ohm's law, where the currents perpendicular to the magnetic field are related to the conductance and electric field:

$$\mathbf{J}_{\perp} = \mathbf{J}_p + \mathbf{J}_H = \Sigma_P\mathbf{E} - \Sigma_H(\mathbf{E}\times\mathbf{b}) \quad (1.2)$$

FACs are related to the electric field divergence (or $E\times B$ flow shear) and gradients of the height-integrated Pedersen and Hall conductances. When the conductances are uniform, Eq. (1.1) is simplified as $J_{\|} = -\Sigma_P(\nabla\cdot\mathbf{E})$. In reality, conductance gradients exist due to the solar zenith angle dependence of the solar EUV radiation and structures of particle precipitation. When FACs, particle precipitation, and electric field from the magnetosphere change, the solution that satisfies Eq. (1.1) survives in the ionosphere. The modified FACs and electric field drive the ionospheric current system and convection, and map back to the magnetosphere as Alfvén waves. Magnetospheric plasma is redistributed.

The MHD momentum equation is

$$\rho\frac{d\boldsymbol{v}}{dt} = -\nabla p + \mathbf{j}\times\mathbf{B} \quad (1.3)$$

Using this equation and current continuity, the FACs in the magnetosphere in the MHD regime are given by

$$J_{\|} = -(\nabla\cdot\boldsymbol{J}_{\perp}) = B_I\int\left(\frac{\rho}{B}\frac{d}{dt}\left(\frac{\boldsymbol{\Omega}}{B}\right) + \frac{\mathbf{j}_{\perp}\cdot\nabla B}{B^2} - \frac{\mathbf{j}_{\mathbf{in}}\cdot\nabla n}{nB}\right)ds \quad (1.4)$$

where $\mathbf{j}_{\perp} = \mathbf{j}_{\nabla p} + \mathbf{j}_{in} = \frac{\mathbf{B}\times\nabla p}{\mathbf{B}^2} + \mathbf{B}\times\left(\frac{\rho}{\mathbf{B}^2}\frac{d\boldsymbol{v}}{dt}\right)$, and Ω is the parallel vorticity. The diamagnetic current $j_{\nabla p}$ represents the balance between the pressure gradient and $\boldsymbol{j}\times\boldsymbol{B}$ forces, the inertial current j_{in} represents the balance between the inertial and $\boldsymbol{j}\times\boldsymbol{B}$ forces, and the other term denotes the vortical current. The integral is performed along the magnetic field line. Force and pressure term analyses show that the dynamic pressure may only be important in the outflow near the magnetotail reconnection region and during solar wind dynamic pressure pulses, and that otherwise the thermal pressure dominates and the pressure gradient and $\boldsymbol{j}\times\boldsymbol{B}$ forces are nearly balanced (Tanaka et al., 2010; Fujita et al., 2003; Birn and Hesse, 2005). In the two major dynamos, icle precipitation and electric field map to the ionosphere. This feedback loop maintains the self-consistency between the ionosphere and magnetosphere.

The large-scale plasma convection in the M-I system described above is primarily driven by the dynamo of the R1 FAC system that emerges as a result of the solar wind-magnetosphere interaction (Tanaka, 2007). The dynamo is located in the high-latitude side of the cusp-mantle region and drives the magnetospheric convection and current system. Under the southward IMF, part of the currents close through the ionosphere (R1 FACs, Fig. 1.2C), and the associated electric field drives the ionospheric convection. The cross polar cap potential drop, the electrostatic potential difference between the maximum and minimum potentials of the two-cell convection pattern in Fig. 1.2B, characterizes the large-scale convection strength. It is also used as a measure of the strength of reconnection in the magnetosphere. The cross polar cap potential drop Φ is described as

$$\Phi = 2E_{sw}D\Sigma_A/(\Sigma_P + \Sigma_A) \tag{1.5}$$

where $E_{sw} = v_{sw}B_{sw}(\theta/2)$ is the solar wind electric field, D is the geoeffective length, Σ_A is the Alfvén conductance ($\Sigma_A = 1/(\mu_0 v_A)$), and Σ_P is the Pedersen conductance (Kivelson and Ridley, 2008). Here conductance gradients in Eq. (1.1) are assumed to be negligible for simplicity. The potential drop is not solely determined by the solar wind and magnetospheric parameters, but is also affected by the Pedersen conductance in the ionosphere. Φ is not linearly proportional to the southward IMF B_z but increases more slowly at the larger IMF $|B_z|$. This nonlinear behavior of Φ is called polar cap potential saturation (Siscoe et al., 2002) and is an important large-scale M-I interaction feature. Σ_P increases with solar and geomagnetic activities and also contributes to reducing Φ and magnetopause reconnection (Lopez, 2016; Jensen et al., 2017). The FAC and convection pattern is modified as the conductance and its distribution are altered (Ridley et al., 2004). The ionosphere conductance also influences the magnetotail flows (Ream et al., 2015) and storm strengths (Chen et al., 2015).

The R2 FAC system originates in the partial ring current in the nightside near-Earth magnetosphere. The ring current ion pressure peaks near midnight and decreases azimuthally. The ring current strength also decreases away from the pressure peak. For the current continuity ($\nabla \cdot \mathbf{J} = 0$), the excess current duskward of the pressure peak flows down to the ionosphere along the magnetic field lines (downward R2 FACs), and the current dawnward of the pressure peak flows up from the ionosphere (upward R2 FACs). While the upward R2 FACs are associated with electrons precipitating to the upper atmosphere and cause a dawnside diffuse aurora, the downward R2 FACs have much less electron precipitation though ion precipitation exists. The downward R2 FACs close partly through the upward R1 FACs poleward, where the convection electric field is directed poleward. The westward $E \times B$ drift enhances the recombination of the subauroral plasma (Schunk et al., 1975) and forms the low-conductance midlatitude trough. The decreasing conductance requires a stronger electric field, and this positive feedback process develops a region of the enhanced westward drift, which is the SAPS (Anderson et al., 1993). It is further discussed in Section 1.2.

It should be noted that this concept uses the electrostatic approximation and is only applicable when the system varies much slower than the Alfvén transit time between the magnetosphere and ionosphere ($\tau = l/v_A$ where l is the magnetic field line length). When the input from the magnetosphere changes faster, inductive coupling by the rotational electric field and currents becomes nonnegligible (Yoshikawa and Itonaga, 2000). Although inductive coupling is generally not considered, it is important to be considered for dynamically evolving phenomena. Dynamical evolution of the order of the Alfvén transit time occurs during the IMF and solar wind dynamic pressure changes, ULF waves, and meso-scale activities discussed in Section 1.2.

When the R1 and R2 FACs are nearly in equal strength, the convection electric field is shielded and does not penetrate to low and equatorial latitudes. When the R1 FACs are larger than R2 FACs, the electric field penetrates to low and equatorial latitudes (penetration electric field, dawn-to-dusk directed in the dayside equatorial ionosphere), while the overshielding electric field (dusk-to-dawn in the dayside equatorial ionosphere) develops in the opposite case.

More quantitative FAC distributions have been shown by Korth et al. (2014), and magnetospheric current systems are reviewed more in detail by Ganushkina et al. (2018).

1.1.4 Aurora

Upward FACs are generally associated with electrons precipitating to the atmosphere. When the amount of plasma is sufficient to carry the FACs, the electron distribution function in the loss cone in the magnetospheric source region represents the precipitating electron distribution function. Both electrons and ions are scattered into the loss cone due to the wave-particle interaction and curvature of the magnetic field being comparable to the particle gyroradius (current sheet scattering). The aurora caused by such scattering processes is called diffuse aurora (proton aurora for aurora caused by proton precipitation). When the amount of plasma is less than that needed to carry the current, a potential drop develops along the magnetic field (parallel potential drop) in the low-altitude magnetosphere (at ~1 R_E altitude) where the plasma density is low and accelerates the electrons downward (Hull et al., 2003). The accelerated electrons drive the discrete aurora. FACs often show a single or multiple sheets and the corresponding discrete aurora also forms an arc or arcs that are aligned in the east–west direction in the auroral oval or in the day-night direction in the polar cap. Electron acceleration and scattering also occur due to wave-particle interaction in the low-altitude magnetosphere and create structured and time-varying electron precipitation. Such aurora is called the Alfvénic aurora and shows vertically extended rays. The auroral oval includes a number of other types of auroral phenomena as reviewed in Chapter 2.

Precipitation of radiation belt electrons penetrates to the D-region ionosphere (~60–90 km altitude) and creates an additional layer of density and conductance enhancements near the equatorward boundary of the auroral oval. Particle precipitation and conductivity are discussed in Chapter 4.

1.1.5 Plasma outflow

The solar EUV, Joule heating, particle heating, and wave dissipation in the ionosphere accelerate ionospheric plasma upward along the magnetic field lines, and the population that exceeds the escape velocity (11.2 km/s at the ground level) flows out to the magnetosphere (plasma outflow). Ionospheric outflows are an important source of magnetospheric plasma. Representative types of outflows are indicated by blue arrows in Fig. 1.1. The solar EUV plasma on the closed magnetic field lines supplies plasmaspheric plasma (refilling). The cusp and its surrounding region are a major source of outflows known as the cleft ion fountain. The polar wind occurs on the open magnetic field lines in the polar cap. Outflows also occur in the nightside auroral oval. The outflows are reviewed more in detail in Section 1.4.

1.1.6 Ion-neutral interaction

The Joule heating (frictional heating) rate is the energy dissipation when the current flows through a resistive medium. Using the electric field in the neutral wind frame of reference $\mathbf{E}' = \mathbf{E} + \boldsymbol{u} \times \mathbf{B}$, the Joule heating rate is $q_J = \mathbf{j} \cdot \boldsymbol{E}'$. The Joule heating rate is the difference between the electromagnetic energy exchange rate $q_{EM} = \boldsymbol{j} \cdot \boldsymbol{E}$ and the mechanical work done on the wind $q_m = \boldsymbol{u} \cdot (\mathbf{j} \times \mathbf{B})$; $q_J = q_{EM} - q_m$ (Brekke and Rino, 1978). q_{EM} is the Joule heating when the neutral wind is absent. q_m can be positive (plasmas accelerate neutrals) or negative (neutrals accelerate plasma), and it is about 20% of the Joule heating (Aikio et al., 2012). Joule heating definitions, relevant observations, and modeling are further discussed in Chapter 5.

In addition to the Joule heating, particle heating (collision between precipitating particles and neutrals, and thermal conduction), solar EUV, and wave dissipation contribute to heating the thermosphere. The heating changes the pressure distribution, and the pressure gradient force changes the neutral wind. Collision between ionospheric plasma and neutrals drags the neutrals in the direction of the plasma drift (ion drag force). Eq. (1.6) is the momentum equation of neutrals for unit mass:

$$\frac{\partial \boldsymbol{u}}{\partial t} = \nu_{ni}(\mathbf{v} - \boldsymbol{u}) - \frac{1}{\rho}\nabla p - 2\boldsymbol{\Omega} \times \boldsymbol{u} - (\boldsymbol{u} \cdot \nabla)\boldsymbol{u} + \frac{1}{\rho}\nabla(\mu \nabla \boldsymbol{u}) \tag{1.6}$$

where u is the neutral wind velocity, v is the plasma velocity, ρ is the neutral mass density, p is the neutral thermal pressure, and μ is the viscosity. The right-hand side expresses the ion drag, pressure gradient, Coriolis, advection, and viscosity forces. During geomagnetically disturbed times, the ion drag force becomes larger due to enhanced plasma flows, and the neutral wind also shows a two-cell convection pattern. The acceleration timescale τ_{ni} by ion drag is $\tau_{ni}=1/\nu_{ni}=n_n/(n_i\nu_{in})$, and it is typically a few hours (Deng et al., 2009b). The solar EUV heating increases the thermal pressure in the dayside thermosphere, and the day-to-night pressure gradient force drives the antisunward neutral wind (Lühr et al., 2012).

The scale height of the thermosphere ($H=k_B T/mg$) is a function of the temperature and increases as the thermosphere is heated. Thermospheric heating is particularly intense in the cusp, and the neutral density forms a peak in the cusp region (Lühr et al., 2012). Heating in the auroral oval also increases thermospheric pressure, and abrupt heating due to substorms and cusp auroral brightenings creates neutral wind perturbations that propagate away from the auroral oval as gravity waves (traveling atmospheric disturbances, TADs). TADs are accompanied by plasma motion due to ion drag and create plasma density perturbations called traveling ionospheric disturbances (TIDs) (Shiokawa et al., 2002). Ionosphere-thermosphere interactions are reviewed in Chapter 7. While forcing from the atmosphere is outside the scope of this book, tropospheric disturbances such as tides also affect the thermosphere.

1.1.7 Large-scale energy budget

$$\frac{\partial W}{\partial t}=-\nabla\cdot\boldsymbol{S}-\boldsymbol{j}\cdot\boldsymbol{E} \tag{1.7}$$

Eq. (1.7) describes electromagnetic energy conservation. Changes in electromagnetic energy W are equal to the sum of convergence of the Poynting flux $\boldsymbol{S}=(\boldsymbol{E}\times\boldsymbol{\delta B})/\mu_0$ and energy exchange between particles and fields ($\boldsymbol{j}\cdot\boldsymbol{E}$). During geomagnetic storms, solar wind energy reaching the magnetosphere is some tens of TW, and ~1% of solar wind energy is supplied to the M-I system (Feldstein et al., 2003). Of the energy reaching the M-I system, ~50% of energy is supplied to the ionosphere as the Poynting flux or FACs and dissipates as Joule heating. Precipitation into the atmosphere carries ~25% of energy. The rest (~25%) goes into the ring current. Similar energy partition is seen during other storms (Knipp et al., 1998) and substorms (Østgaard et al., 2002a, 2002b). These numbers emphasize the importance of ionosphere processes for energy dissipation processes. Electromagnetic energy input and dissipation are reviewed in Chapter 5.

1.2 Multiscale processes

1.2.1 Three major localized and transient structures

The large-scale, quasi-steady system described in Section 1.1 represents the averaged picture of the M-I-T system. However, each region of the system also exhibits a localized structure and dynamic evolution that can be comparable to or more intense than the

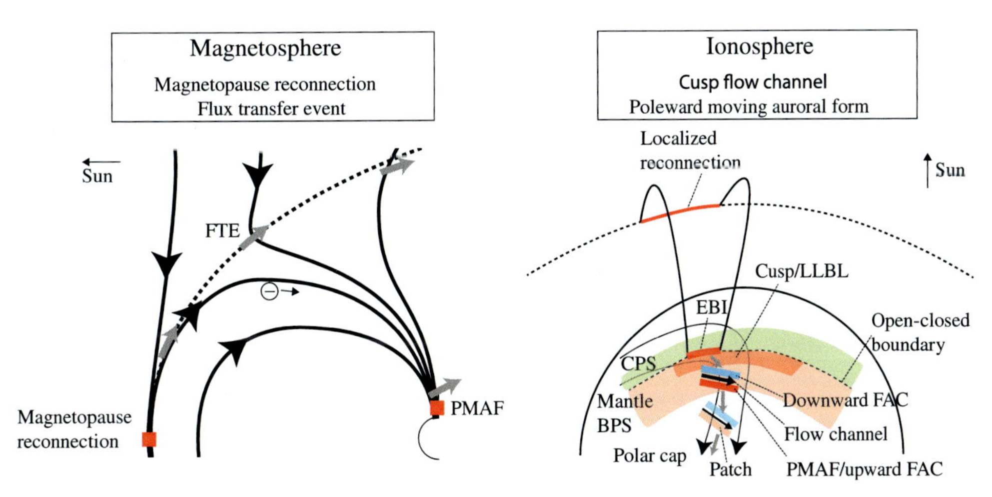

Fig. 1.3 Illustration of the meso-scale process in the dayside M-I system: connection among the magnetopause reconnection, flux transfer event, cusp flow channel, and poleward moving auroral form.

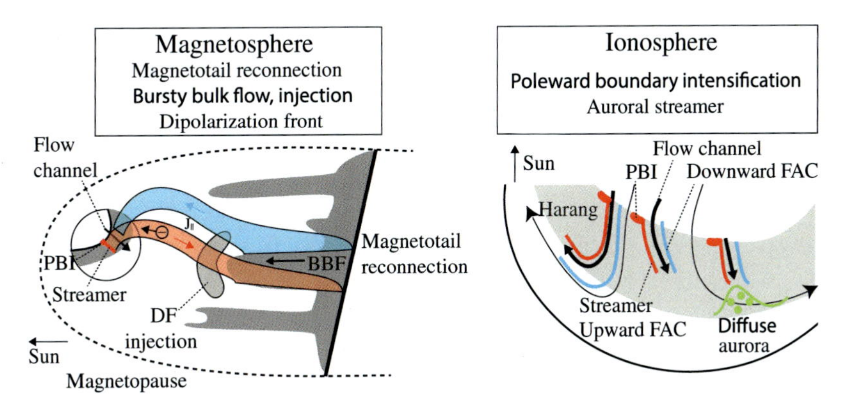

Fig. 1.4 Illustration of the meso-scale process in the nightside M-I system: connection among the magnetotail reconnection, bursty bulk flow, dipolarization front, poleward boundary intensification, and auroral streamer.

large-scale features. Figs. 1.3–1.6 schematically illustrate three key M-I coupling phenomena from the magnetosphere (left) and ionosphere (right) perspectives. They exhibit plasma flows and currents that are localized to a few hundred km widths and can be as intense as those on the large scale.

Fig. 1.3 shows the connection between magnetopause reconnection and poleward moving auroral forms (PMAFs) in the cusp. Magnetopause reconnection can be enhanced locally and transiently, and a burst of reconnected magnetic flux tubes moves away from the reconnection region (flux transfer events or FTEs). In the ionosphere, it is

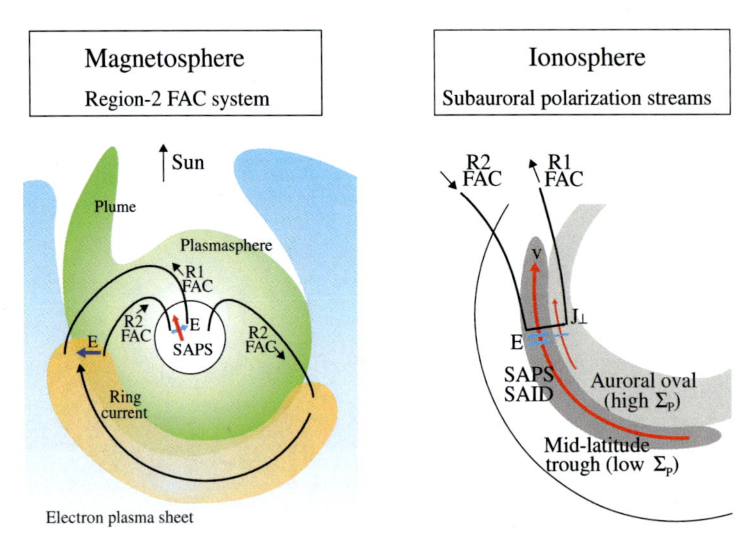

Fig. 1.5 Illustration of the meso-scale process in the duskside M-I system: Connection between the region-2 FAC system and subauroral polarization streams.

found to be a fast poleward flow channel across the cusp between a pair of upward and downward FACs. Electron precipitation is enhanced in the upward FAC region and shows a discrete auroral arc (PMAF). The PMAF begins with an equatorward boundary intensification (EBI), the moves poleward following the large-scale convection, and decays as the flux tube exists the cusp. The electron precipitation also increases the ionosphere density becomes one of the sources of the polar cap patch. The dayside cusp region also has meso-scale flow channels associated with dayside reconnection. Those are associated with PMAFs. They are associated with localized flow channels and a wedge-type FAC loop, where PMAFs correspond to the footprint of the upward FAC (Oksavik et al., 2004).

Similar localized and transient transport can be found in the magnetotail and nightside auroral ionosphere as illustrated in Fig. 1.4. Magnetotail reconnection creates channels of fast earthward flows (bursty bulk flows or BBFs) in the plasma sheet. The magnetic field ahead of a BBF is enhanced and forms a sharp front (dipolarization front, DF). Plasma sheet electrons and ions are accelerated by a strong electric field and injected into the inner magnetosphere. Pressure gradients and flow vortices associated with BBFs and DFs are linked to FACs as in Eq. (1.1), and electron precipitation along the magnetic field lines of the upward FACs creates PBIs and auroral streamers. A flow channel forms between the upward and downward FACs. The flow channels approximately follow large-scale convection; duskside streamers and flow channels move along the Harang flow shear and turn duskward, while dawnside streamers and flow channels turn dawnward. Streamers also interact with aurora near the equatorward boundary of the auroral

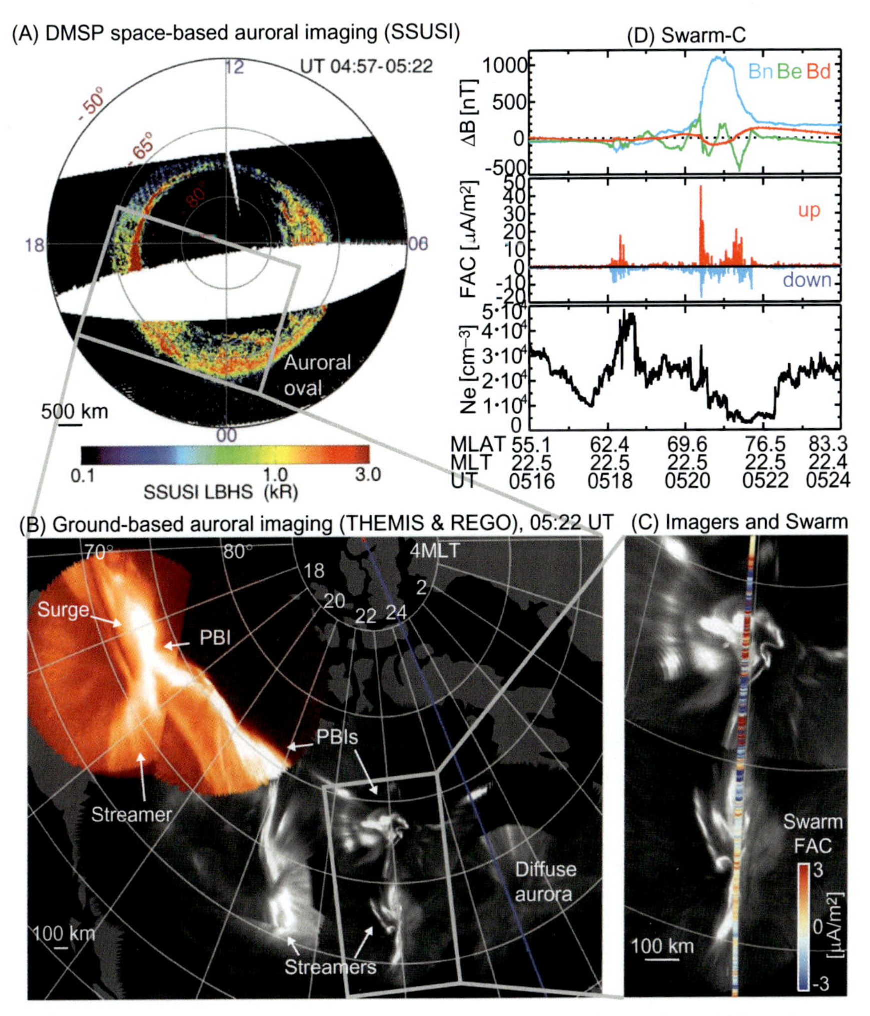

Fig. 1.6 Multiscale processes in the nightside auroral oval seen during the 5 UT, 14 January 2018 substorm. (A) Large-scale auroral structure by DMSP SSUSI, (B) a mosaic of nightside aurora by THEMIS ASIs and REGO, (C) the horizontal components of the magnetic field by Swarm-A overlaid on the THEMIS ASI data, and (D) Swarm-C measurements of the magnetic field, FAC, and plasma density.

oval such as the diffuse aurora and substorm onset arc. Enhanced flows and precipitation in the nightside auroral oval increase the neutral wind and excite TADs and TIDs (see Sections 2.4 and 4.2.2 for auroral precipitation. See Section 7.1.3 for TADs and TIDs).

Fig. 1.5 shows the connection between the R2 FAC system and SAPS. As discussed in Section 1.1.3, the R2 FAC system is driven by the partial ring current. On the duskside, the ring current ions extend more earthward than the plasma sheet electrons. The downward R2 FACs flow into the subauroral ionosphere, where the ionosphere conductance is lower than in the auroral oval. Part of the current flows poleward ($J_\perp$) and closes through the upward R1 FAC. Considering Ohm's law (Eq. 1.2), $J_p = \Sigma_P E$. The conductance in the midlatitude trough is lower than in the auroral oval, and the electric field in the trough is larger than in the auroral oval. The large subauroral electric field corresponds to the $E \times B$ SAPS westward flow. Neutral wind is enhanced in the SAPS region.

During the northward IMF, plasma convection in the polar cap is generally slower and develops two reversed cells near the cusp poleward of the two-cell convection pattern. Those are associated with a high-latitude reconnection poleward of the cusp. The polar cap density is lower due to slower transport from the dayside ionosphere, but discrete regions of particle precipitation occur along a roughly Sun-aligned auroral arc (polar cap arc), particularly during the large IMF B_y. Polar cap arcs are also associated with enhanced flow shears and FACs.

1.2.2 Multiscale structures in the nightside auroral oval

Fig. 1.6 shows an example of multiscale structures in the nightside auroral oval. This is a substorm event (-530 nT minimum AL index) at 5 UT on January 14, 2018, where the DMSP SSUSI provided space-based large-scale auroral imaging (Fig. 1.6A), the THEMIS and REGO ASIs provided ground-based imaging in the nightside auroral oval (Fig. 1.6B), and the Swarm satellite passed over the premidnight auroral oval (Fig. 1.6C–D). In the DMSP SSUSI data in Fig. 1.6A, the nightside aurora is characterized by the large-scale auroral oval with 1000s of km size, and the ~100-km scale aurora is superimposed. Fig. 1.6B shows a snapshot of the ground-based auroral imager mosaic near the end of the substorm expansion phase. The aurora in the premidnight sector is dominated by PBIs (auroral brightening near the oval poleward boundary) and streamers (approximately north–south aligned aurora extending equatorward from the PBIs). Each PBI and streamer have about a few 100 km east–west extent and consist of substructures of <100 km size. The Swarm-C satellite passed over the PBI and streamer at 22.5 MLT and detected multiple pairs of FAC enhancements and plasma density modulations (Fig. 1.6C). The satellite trajectory and horizontal magnetic field deflections are shown in Fig. 1.6D.

The 1000s of km size features in the ionosphere are large-scale structures that correspond to the global or a large portion of the magnetosphere when mapped along the magnetic field lines to the magnetosphere. A few 100 s of km size (more specifically 100–500 km size, see Fig. 1.7) features in the ionosphere are called meso-scale structures and map to about a few earth radii in the magnetosphere. Structures below 100 km size in the ionosphere are called small-scale structures and map to a fraction of earth radii in the magnetosphere.

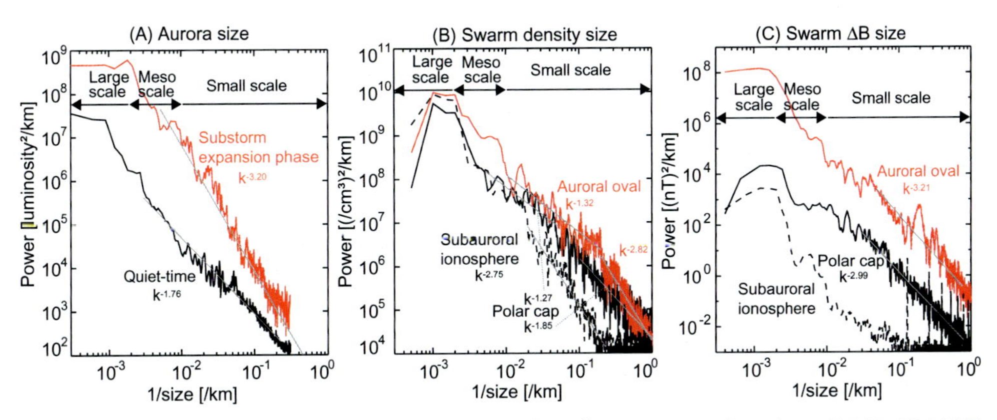

Fig. 1.7 Power spectra of (A) the auroral size during the substorm expansion phase (05:21–05:24 UT, *red*) and during a quiet time (06:30–07:00 UT, *black*), (B) plasma density measured by Swarm-C, and (C) magnetic field measured by Swarm-C. The Swarm data are split in three regions: auroral oval (5:17:40–5:22:00 UT, *red*), polar cap (5:22:40–5:27:00 UT, *black solid*), and subauroral ionosphere (5:13:00–5:17:20 UT, *black dashed*). Major power law slopes are indicated in each panel.

To quantify the multiscale nature, power spectra of the aurora from the ASIs, and density and magnetic field from the Swarm-C are shown in Fig. 1.7. The ASI data were sliced in the east–west direction at 68 degree MLAT. The FFT analysis was performed as functions of space at each time of data and the frequency power spectra were averaged over 05:21–05:23 UT (during the Swarm-C crossing in the substorm expansion phase, red line in Fig. 1.7A) and over 06:30–07:00 UT (quiet time, black line in Fig. 1.7A). North–south spectra are not shown but have similar profiles. The spectral power during the substorm expansion phase is larger than during the quiet time at all scales. A spectral power peak is found at the ~1000 km size, which corresponds to large-scale auroral structures in the oval. >1000-km auroral structures exist but are not covered by the available ASIs. The enhanced spectral power at ~100–500 km corresponds to the PBIs and streamers, and we refer to them as meso-scale structures. Auroral structures at <~100 km represent substructures embedded in the meso-scale structures and individual auroral arcs, which we refer to as small-scale structures. The power spectra at <~100 km roughly follow a power law with a slope of −3.2. It is close to the energy spectra of two-dimensional turbulence, where the spectral slope is −3 for enstrophy (integrated square of vorticity) cascades (Kraichnan and Montgomery, 1980). Kinetic energy cascades in the magnetosphere may create the spectra of the small-scale aurora. However, the spectrum has several peaks that correspond to coherent auroral structures, and the spectrum varies in time due to the dynamical evolution of auroral structures. The quiet-time power spectrum (black) shows a power-law profile with a slope of −1.76, which indicates the ~−5/3 slope energy cascade (Kozelov et al., 2004).

An FFT analysis was also performed for the Swarm-C plasma density data (Fig. 1.7B). The red line shows the spectrum in the auroral oval, and the black solid and dashed lines show the spectra in the polar cap and subauroral ionosphere. Similar to the auroral intensity spectra, the density spectra peak near the 1000 km size (large-scale) in all regions and the spectra in the auroral oval are enhanced at all scales. The enhanced spectral power at the ~100–500-km size corresponds to density structures associated with the meso-scale auroral forms. The density spectrum in the auroral oval below the 100 km size is comparable to that in the polar cap, while the spectrum in the subauroral ionosphere is much weaker. The density spectra at the 10–100 km size in the auroral oval and polar cap have a slope close to -1, which is again consistent with the two-dimensional turbulent enstrophy cascade (Kintner and Seyler, 1985). The density spectral power in the subauroral ionosphere is much smaller.

The magnetic field also shows clearly different spectral profiles in the three regions, and the meso- and small-scale magnetic fields show a pronounced spectral peak at the ~300-km size, which corresponds to FAC structures associated with the meso-scale auroral forms. The small-scale spectrum in the auroral oval shows a power law index of -3.21, similar to earlier results by Kintner (1976).

As in Fig. 1.7B, the density below the 10 km size in the auroral oval also has a power law but has a steeper slope. The enhanced power of the plasma density near the 1-km scale is the source of density irregularities that causes radio signal scintillation. When the plasma density along a radio wave propagation path has a transverse size less than the Fresnel scale ($\sqrt{2\lambda z}$, λ: wavelength, z: altitude, ~360 m at 350 km altitude for GPS L1 1.6 GHz), the density structure diffractively scatters the radio wave and causes amplitude and phase scintillation (Rino, 1979). When transverse density perturbations larger than the Fresnel scale are present, the density perturbations modulate the phase velocity of radio signals due to different refractive indices and produce phase scintillation. In the mid- and high-latitude ionosphere, the phase scintillation occurs mainly in the auroral oval and is most intense in the cusp (Prikryl et al., 2015), while the amplitude scintillation is less often but can still be detected in the cusp (Jin et al., 2019).

Meso- and small-scale structures also exist in the polar cap. Similar to arcs in the auroral oval, polar cap arcs are associated with a wedge-type current system where flow channels with a few kV potential drop are located between upward and downward FACs (Robinson et al., 1987). Patches are not always associated with flow structures but a type of patches is associated with flow channels (Kivanç and Heelis, 1997). Small-scale flow and density structures of less than 100 km can occur as shown in Fig. 1.7C.

While it is difficult to measure kinetic scale structures (ion gyrofrequency in the ionosphere ~1 m), recent particle-in-cell simulations have predicted the existence of meter-scale turbulent density structures created by the Farley-Buneman instability (Oppenheim and Dimant, 2013). The process is nonlinear, and there are net changes in conductance due to the density structures. While studies on how global processes are affected by small-scale processes are limited, Wiltberger et al. (2017) incorporated the kinetic turbulence conductance into the coupled MHD simulation and evaluated the effects of kinetic turbulence. They showed that the inclusion of kinetic turbulence processes led to a 20% increase in conductance which is in better agreement with the measured D_{st} index. The cross polar cap potential decreased by 13%, and this may substantially improve estimation of the cross polar cap potential, which tends to be overestimated in MHD simulations. Thus, the multiscale processes may play a substantial role in understanding ionosphere processes.

Part 2 Electric field variability and impact on the ionosphere-thermosphere

Yue Deng[1], Qingyu Zhu[1,3], Cissi Y. Lin[1], Mingwu Jin[1], Chaoqun Liu[2], Cheng Sheng[1]

[1] Department of Physics, University of Texas at Arlington, TX, United States
[2] Department of Mathematics, University of Texas at Arlington, TX, United States
[3] National Center for Atmospheric Research, CO, United States

1.3 Electric field variability and impact on the ionosphere-thermosphere

1.3.1 Introduction

As mentioned in the previous section, the energy input from the magnetosphere to the ionosphere and thermosphere is highly variable with the geomagnetic conditions and can cause global-scale disturbances in the I-T system during a storm period. In general circulation models (GCMs), the high-latitude electric field and particle precipitation are typically specified by empirical models (Weimer, 2005; Heelis et al., 1981; Fuller-Rowell and Evans, 1987; Newell et al., 2009; Zhang and Paxton, 2008) or through data assimilation techniques, such as the Assimilative Mapping of Ionospheric Electrodynamics (AMIE) procedure (Richmond and Kamide, 1988). Fig. 1.8 shows simulated and observed plasma and neutral mass densities during the March 17, 2013 storm. The Global Ionosphere-Thermosphere Model (GITM) simulation (Ridley et al., 2006) as one of the GCMs can reasonably reproduce the large-scale trend. However, the simulated densities are much smoother and the observations show substantial strength of meso-scale structures or fast variations.

A major reason is that the empirical model drivers only describe large-scale and slowly varying input and lack localized and fast-varying components. In addition, Joule heating is usually found to be underestimated in GCMs when using those models to specify the high-latitude electrodynamics in GCMs (Emery et al., 1999), and this insufficient energy is attributed to the neglect of the contribution of electric field variability to the Joule heating (Codrescu et al., 1995). Joule heating and its variabilities are described in detail in

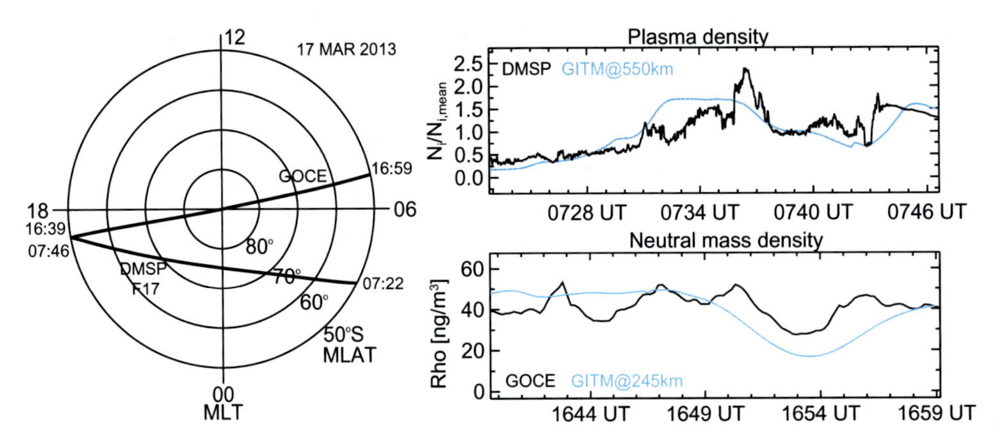

Fig. 1.8 (*Left*) Orbits of the DMSP F17 and GOCE satellites in the magnetic coordinates in the southern hemisphere. (*Right*) The plasma density and neutral mass density measured by DMSP and GOCE (*black*) and simulated by GITM (*blue*). The GITM neutral density is obtained at the GOCE altitude (~245 km). The DMSP is at ~830 km, while the GITM plasma density is obtained at the top boundary of the simulation domain (550 km). The plasma density is normalized by the mean density to remove the offset due to the altitude difference.

Chapter 5 and section 5.1.6. Empirical models of the electric potential used to drive GCMs represent only the statistical average of the vector field $\langle E \rangle$ and the difference between $\mathbf{E}$ and $\langle E \rangle$, called the "residual electric field" or "electric field variability," has been ignored. Previous studies (Codrescu et al., 1995; Matsuo et al., 2003; Matsuo and Richmond, 2008; Golovchanskaya, 2008; Deng et al., 2009a) showed that the electric field variability can be comparable to or even larger than the average electric field, with the consequence that the variable component can contribute as much to Joule heating as the average field. Codrescu et al. (1995) first pointed out that the neglect of the electric field variability may lead to a significant underestimation of Joule heating. Since then, great efforts have been made to investigate the patterns of electric field variabilities on different scales under different conditions utilizing both satellite and ground-based radar observations (Codrescu et al., 2000; Golovchanskaya et al., 2002; Matsuo et al., 2002, 2003; Johnson and Heelis, 2005; Cosgrove and Thayer, 2006; Matsuo and Richmond, 2008; Golovchanskaya, 2008; Abel et al., 2009; Cousins and Shepherd, 2012a, 2012b; Cousins et al., 2013; Zhu et al., 2018). To quantify the Joule heating associated with the residual electric field in a way consistent with the empirical model of electric potential used as GCM inputs, statistical characterization and empirical models with an electric field variability component have been developed, which supply a more realistic way to include electric field variability in GCMs for the energy estimation than through ad hoc increases of the Joule heating. The impacts on the I-T system associated with the electric field variability have been further studied by GCM simulations (Codrescu et al., 2008; Matsuo and Richmond, 2008; Deng et al., 2009a; Zhu et al.,

2018). In this section, the estimation of electric field variability is described and the impact of electric field variability on Joule heating and the I-T system is reviewed.

1.3.2 Description of electric field variability

The electric field variability is typically defined as the "residual electric field," which represents the difference between the observations and the statistical average. The statistical characteristics of those residuals have been examined (Matsuo et al., 2003; Matsuo and Richmond, 2008; Cosgrove and Thayer, 2006; Cousins and Shepherd, 2012a, b) and empirical models have been developed based on statistical results (Deng et al., 2009a; Cousins et al., 2013). Since the electric field variability naturally includes multiscale components, the further separation between large-scale and mesoscale has been conducted. Conductance, which is strongly under the influence of particle precipitation in the high latitudes, is closely related to the electric field through Ohm's law ($J=\sigma \cdot E$). Therefore, the correlation between electric field variability and particle precipitation variability is important for the magnetosphere-ionosphere (M-I) coupling. The detailed description of these points has been included below.

1.3.2.1 Statistical characteristics of electric field variability

Different datasets have been utilized to characterize the statistical features of electric field variability, including measurements from the Dynamic Explorer 2 (DE-2) satellite (Matsuo et al., 2003; Matsuo and Richmond, 2008; Maute et al., 2005; Deng et al., 2009a; Zhu et al., 2018), Super Dual Auroral Radar Network (SuperDARN) radars (Cousins and Shepherd, 2012a, b), and Sonderstrom incoherent scatter radar (Cosgrove and Thayer, 2006; Cosgrove et al., 2011). The analysis in Matsuo et al. (2003) reveals that electric field variability varies with the magnetic latitude, magnetic local time (MLT), interplanetary magnetic field (IMF), and season in a manner distinct from that of the climatological electric field. This indicates that empirical models and data assimilation models designed to reproduce the average electric fields correctly are not necessarily well suited to represent the electric field variability correctly. It is also found that the standard deviation of electric field variability scales linearly with the K_p index (Cosgrove and Thayer, 2006) and the characteristics of electric field variability is consistent with the expected properties of a turbulent flow (Cousins and Shepherd, 2012b). Fitting (Maute et al., 2005) or empirical orthogonal function (EOF) analysis (Cousins et al., 2013) has been applied to the observational data to identify the dominant modes of electric field variability. In the following paragraphs, a specific example has been shown about how to use DE-2 data conducting the statistical study and developing an empirical model.

By analyzing observations from the DE-2 spacecraft, a comprehensive, mutually consistent model of high-latitude thermospheric forcing has been developed at the National Center for Atmospheric Research (NCAR), and a brief description has been included in the study by Maute et al. (2005) and Deng et al. (2009a, b). In total, 2895 satellite passes during August 1981–March 1983 have been used in the process. DE-2 was a polar-orbiting satellite at the altitudes roughly between 300 and 1000 km with an orbital period of 98 min (i.e., the velocity of DE-2 is ~8 km/s). DE-2 precessed through all local time sectors once per year, producing a dataset with a relatively good local time coverage but limited separation of local time and seasonal variations. The temporal resolution of the dataset used in this study is 1 s, and therefore the corresponding spatial resolution of the DE-2 measurement is ~8 km. DE-2 measured the along-track and cross-track ion drifts, which were taken by the Retarding Potential Analyzer (RPA) (Hanson et al., 1982) and the Ion Drift Meter (IDM) (Heelis et al., 1981), respectively. The bulk ion drift vector (**V**) is the combination of those two components and provides the electric field according to $E = -V \times B_0$. Here $\mathbf{B_0}$ is the geomagnetic main field from the International Geomagnetic Reference Field (IGRF) model. The electric field is decomposed in modified apex coordinates (Richmond, 1995) according to the following formula: $E = E_{d1} \cdot d_1 + E_{d2} \cdot d_2$. Here E_{d1} and E_{d2} represent the magnetic eastward and northward components of the electric field, respectively, while $\mathbf{d_1}$ and $\mathbf{d_2}$ are the base vectors at the magnetic eastward and northward directions, respectively, at the satellite height. The electric field is then mapped to 110 km altitude along the same geomagnetic field line. More details involving the modified apex coordinates and the decomposition procedure can be found in Richmond (1995).

At each magnetic latitude, the observations are binned and fitted to analytical functions of MLT, dipole tilt angle with respect to the plane normal to the Sun-Earth line, and strength and clock angle of the IMF obtained from the IMP 8 and ISEE 3 satellite measurements. The climatological patterns of electric potential are then produced, which are generally consistent with those in other empirical models such as Weimer (2005). The electric field variabilities are then calculated by subtracting the averaged electric field pattern from the electric field observations ($E - \langle E \rangle$). Similar binning and fitting processes have been conducted to generate the electric field variability patterns in the empirical model (Maute et al., 2005). It should be noted that the values of electric field variability depend on the manner in which the statistical averages of **E** are defined. In general, the better the statistical models of $\langle \mathbf{E} \rangle$ manage to represent the actual fields for the given geophysical and interplanetary conditions, the smaller the variability components will be. The model of electric field variability represents the standard deviations of the magnetic-northward and eastward components of **E**, or $\sqrt{\langle (E_n - \langle E_n \rangle)^2 \rangle}$ and $\sqrt{\langle (E_e - \langle E_e \rangle)^2 \rangle}$, as shown in Fig. 1.9. It includes both small- and large-scale spatial variations, as well as temporal variations. The patterns and magnitudes of the electric field variability in Fig. 1.9 are comparable with those shown by Matsuo et al. (2003), which are different from those of the average electric field. For example, the persistent peak on the dayside polar cap boundary region (such as the cusp) is a distinct feature in the electric

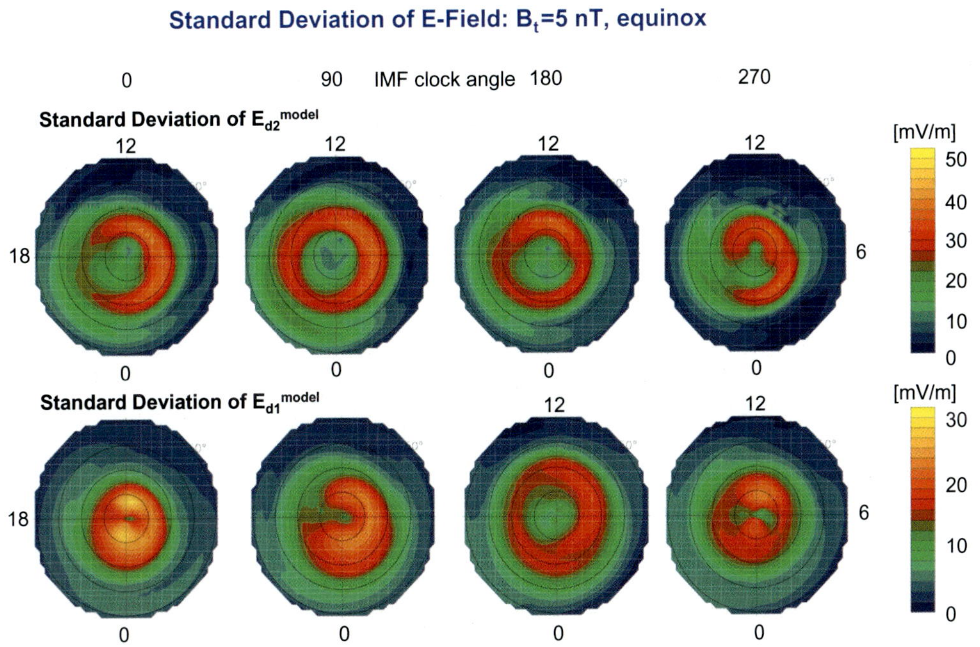

Fig. 1.9 The distributions of standard deviation of electric field in two directions from an empirical model. *Courtesy of Astrid Maute.*

field variability patterns. Maute et al. (2005) show the first empirical model in the community which includes an electric field variability component consistent with the average electric field.

1.3.2.2 Electric field variability on different scales

The electric field variability, which is calculated from the difference between the observations and the climatological average electric field, includes multiscale components in nature. The distinction between large-scale electric field variability and small- and meso-scale electric field variability has been made previously (Matsuo and Richmond, 2008; Cosgrove et al., 2011; Cousins et al., 2013; Zhu et al., 2018). In a study by Matsuo and Richmond (2008), the residual electric field variability is decomposed into resolved-scale and subgrid-scale corresponding to the TIEGCM grid size (5 degree). The analysis reveals that the subgrid-scale electric field variability varies with the magnetic latitude, MLT, IMF, and season in a manner distinct from that of the resolved-scale electric field variability and of the climatological electric field. Cosgrove et al. (2011) separately analyze the small- and meso-scale variability and the resolved-scale model uncertainty using the Sondrestrom radar database, and it is found that small-scale variability, being the result of actual physical fluctuations on small scales, is a more general quantity than

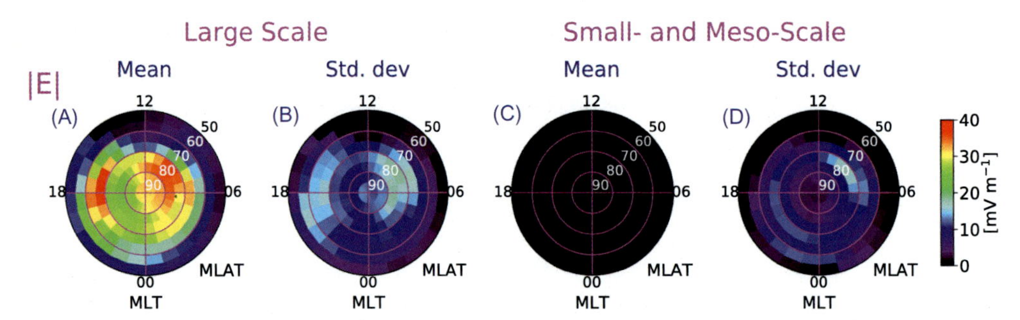

Fig. 1.10 The distributions of the mean and standard deviations of the (A and B) large-scale electric field intensity and (C and D) small- and meso-scale variabilities of electric field intensity under the condition when the IMF clock angle is between 135 and 225 degrees, and IMF-B- ranges from 4 to 10 nT. All plots are presented as a function of MLAT and MLT (Zhu et al., 2018).

resolved-scale model uncertainty since the latter is only defined with respect to a particular model. Statistical maps of electric field variability at small- and mesoscales (smaller than ~500 km) have been produced using both ground-based (Cousins et al., 2013) and satellite (Zhu et al., 2018) observations. In a study by Zhu et al. (2018), a 500-km moving window is applied to the DE-2 data along each polar pass in order to extract the small- and meso-scale variabilities from the electric field observations. It is worth noting that the satellite measurements involve both spatial and temporal variations. Since DE-2 travels 500 km in ~63 s, most of the variabilities extracted by the 500-km moving window can be treated as spatial variabilities below 500 km if the structures encountered by the satellite are assumed to be stationary for a short period (~1–2 min). One example using the moving window is shown in a study by Zhu et al. (2018), and we emphasize that the small- and meso-scale variabilities shown in that paper represent the variabilities at a scale size smaller than 500 km.

The separated large-scale, and small- and meso-scale quantities are then binned as a function of MLT and magnetic latitude (MLAT). As shown in Fig. 1.10, the bin size is chosen to be 5 degree in MLAT and variable size in MLT (0.64 h at 62.5 degree MLAT and 2.25 h at 82.5 degree MLAT, i.e., 500 km along MLT) to keep a roughly constant area at different latitudes. Zhu et al. (2018) focused on the cases when the IMF |B| is between 4 and 10 nT and the IMF clock angle is between 135 and 225 degrees. It is worth mentioning that by including all data under such IMF conditions, the variability due to the IMF By component, which is a major driver of variability (Cousins et al., 2013), has been included. The data from both hemispheres and all seasons are combined in order to have a reasonable data coverage. Fig. 1.10 displays the binning results for the electric field intensity. The averages of the large-scale electric field intensity are calculated in each bin and the distribution is shown in the first column of Fig. 1.10. Clearly, the pattern shown in the first column represents the climatological average very well and is similar to the patterns from empirical models (Weimer, 2005; Fuller-Rowell and Evans, 1987;

Newell et al., 2009). The second column exhibits the distributions of the standard deviation of the large-scale electric field intensity in each bin, which is similar to the resolved-scale electric field variability shown in Matsuo and Richmond (2008). The standard deviation of large-scale quantity primarily reflects variations in the solar wind and IMF conditions. In addition, seasonal variations and hemispherical asymmetry may also contribute to the large-scale standard deviations. The standard deviations of large-scale electric field intensity are clearly not negligible, yet they are generally smaller than the mean fields. The standard deviations of the large-scale quantities may be reduced in the future studies if there are sufficient data for subdividing the case into more specific conditions with respect to the IMF, season, and hemisphere, since the variability represented by these conditions adds to the large-scale standard deviations shown in the second column. The third and fourth columns show the averages and the standard deviations of small- and meso-scale electric field variabilities, respectively. Unlike what is shown on a large scale, the averages of the small- and meso-scale variabilities are close to zero since the subtraction of the moving average tends to leave residuals with zero means. On the other hand, the standard deviations of the small- and meso-scale electric field variabilities are generally 10–15 $\mathrm{mV\,m^{-1}}$ at 60–75 degrees MLAT, which are comparable to the standard deviation of the large-scale electric field intensity.

1.3.2.3 Correlation between electric field variability and particle precipitation variability

Conductivity, which is strongly under the influence of particle precipitation at the high latitudes, also displays variabilities on different scales (McGranaghan et al., 2016a, 2016b). It has been found that the electric field is anticorrelated with particle precipitation in small- and meso-scale structures, which can cause a clear reduction in Joule heating (Evans et al., 1977; Baker et al., 2004; Zhu et al., 2018). For example, Zhu et al. (2018) investigated the distribution of particle precipitation variabilities and quantified their correlations with electric field variabilities at different scales in a statistical sense. Specifically, the simultaneous electric field and particle precipitation measurements from the DE-2 satellite have been utilized to quantify the distributions of the high-latitude electric field and particle precipitation variabilities and their correlation. The global distributions of the correlation between electric field and particle precipitation at different scale sizes are quantified for the first time in a study by Zhu et al. (2018).

The particle precipitation measurements on DE-2 were taken by the Low-Altitude Plasma Instrument (LAPI) (Winningham et al., 1981). LAPI provides the differential particle energy flux on different energy channels for both the electron and the ion ranging from 5 eV to 32 keV. The total particle energy flux (Φ_E) is determined by integrating the differential particle energy flux over different energy channels and multiplying by the factor of π when assuming the downward differential particle energy flux is isotropic. Zhu et al. (2018) specifically focused on the total energy flux of electron precipitation.

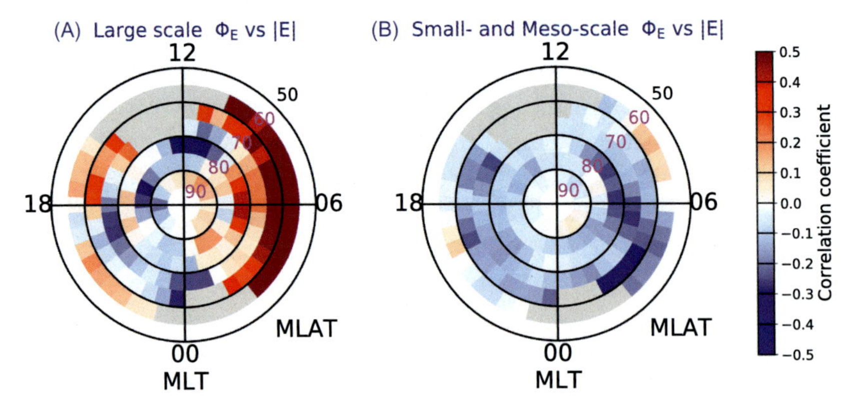

Fig. 1.11 The distributions of the linear correlation coefficient (A) between the large-scale electric field intensity and particle energy flux and (B) between small- and meso-scale variabilities of electric field intensity and particle energy flux when the IMF clock angle is between 135 and 225 degrees, and IMF |B| ranges from 4 to 10nT. All plots are presented as a function of MLAT and MLT. The *gray shaded areas* represent bins without sufficient data (Zhu et al., 2018).

The linear correlations between the large-scale electric field intensity and particle energy flux and between small- and meso-scale variabilities of electric field intensity and particle energy flux are calculated in each bin, and distributions of the correlation coefficient are presented in Fig. 1.11. Here only the correlation coefficients in the bins where the number of trajectories passing through that bin is greater than four and the number of data points is larger than 200 are kept, otherwise they are set to be zero (gray shaded areas, which indicate the data are not sufficient). It is clear that the pattern shown in Fig. 1.11A is more complicated than that shown in Fig. 1.11B. More specifically, for the large-scale electric field and particle precipitation, Fig. 1.11A shows that a positive correlation occurs mostly on the dawnside, whereas an anticorrelation occurs mostly in the early evening sector as well as around noon and midnight. In contrast, the electric field variability tends to be anticorrelated with the particle energy flux variability in general at small- and meso-scales, as shown in Fig. 1.11B.

The FACs are closed by the ionospheric currents which are related to the electric field and the conductance in the ionosphere. The correlation between the electric field intensity and the particle energy flux shown in Fig. 1.11 is helpful to address the question whether the magnetosphere tends to act as a current generator or a voltage generator in the magnetosphere-ionosphere coupled system at different scales. Evidently, Fig. 1.11B shows a consistent anticorrelation between the electric field intensity and particle energy flux variabilities in the aurora region at small- and mesoscales, indicating that the magnetosphere tends to behave as a current generator at those scales, which is consistent with previous findings (Lysak, 1985; Vickrey et al., 1986). On large scales, the electric field intensity clearly tends to be positively correlated with the particle energy flux on the morning side. This may result from expansion of the auroral oval low-latitude boundary, accompanied by the enhancement of the electric field at a given location near the boundary as the boundary moves equatorward. However, the electric field intensity appears to be anticorrelated with the particle energy flux on the evening side, which is different from the theoretical prediction in a study by Lysak (1985). It has been proposed that the type of magnetospheric generator depends on the solar wind conditions. Recently, Weimer et al. (2017) found that the magnetosphere probably acts as a current source on large scales when the interplanetary electric field is large. Since the interplanetary electric field intensity has not been utilized as a binning parameter in a study by Zhu et al. (2018), the data trend shown in Fig. 1.11 may have been strongly influenced by the observations when the interplanetary electric field was fairly large, which may contribute to the anticorrelation at a large scale. In the future, larger data sets for the ionospheric electric field and particle precipitation as well as better specification of the solar wind conditions may help to extend the analysis.

1.3.3 Impact of electric field variability

The previous section shows that in general the large-scale electric potential patterns overlook the considerable electric field variability, which tends to be anticorrelated with the particle energy flux variability at small- and meso-scales. The question is to what degree the electric field variability can contribute to the Joule heating and impact the density and dynamics in the I-T system, which has been addressed by implementing the distributions of electric field variability from statistical analysis of observational data into modeling.

1.3.3.1 Impact on joule heating

If we focus on the contribution of electric field variability to the Joule heating and ignore the influence of conductance variability, the Joule heating over a certain domain can be written as $Q_J = \Sigma_p \langle E^2 \rangle = \Sigma_p (\langle E \rangle^2 + \sigma_E^2)$. Here, Σ_p is the Pedersen conductance, which is dominated by the solar irradiation on the dayside and the particle precipitation on the nightside (Wallis and Budzinski, 1981). $\langle E^2 \rangle$ and $\langle E \rangle^2$ are the average of the electric field square and the square of the average electric field, respectively. σ_E is the standard deviation of the electric field, which represents the electric field variability. Clearly, the electric field variability plays an equally important role as the average electric field to the estimation of Joule heating. The significance of electric field variability to the Joule heating has been pointed out by Codrescu et al. (1995) and the effects of electric field variability on the Joule heating are investigated by incorporating the electric field variability into GCMs (Matsuo and Richmond, 2008; Deng et al., 2009a; Zhu et al., 2018).

To investigate the importance of electric field variability to the Joule heating, Deng et al. (2009a, b) implemented both the large-scale convection pattern and the electric field variability from the empirical model based on the DE-2 satellite measurements

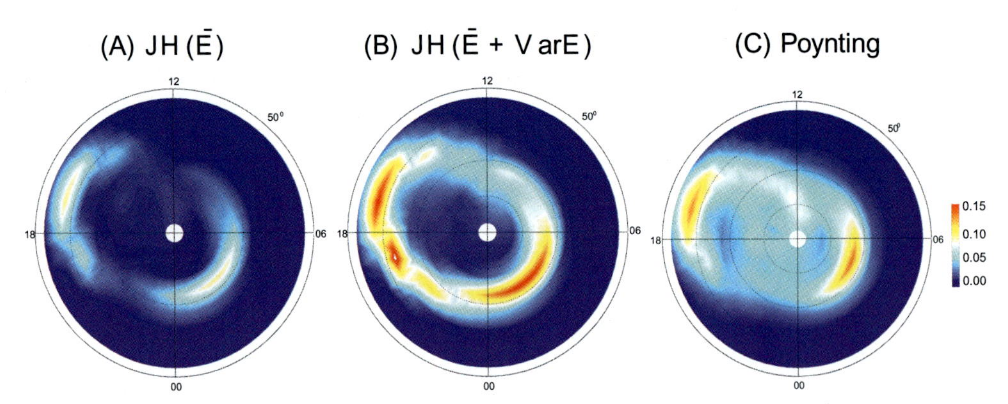

Fig. 1.12 (A) Distribution of the altitude-integrated Joule heating (W/m^2) in the northern hemisphere in equinox from a TIEGCM simulation, when the average electric field is used in the Joule heating calculation. The IMF conditions are $B_y=0$ and $B_z=5\,\text{nT}$. The hemispheric power is 30 GW and $F_{10.7}$ is $150\times10^{-22}\,\text{W/m}^2\text{Hz}$. Geographic coordinates are used in this figure. (B) Same as (A), but the electric field variability from the empirical model is also included in the calculation. (C) Poynting flux at the top of the thermosphere from the empirical model (Deng et al., 2009a).

(Maute et al., 2005) into the National Center for Atmosphere Research Thermosphere Ionosphere Electrodynamics General Circulation Model (NCAR-TIEGCM). Fig. 1.12A shows the distribution of the altitude-integrated Joule heating from an equinox simulation in the northern hemisphere, when the empirical large-scale electric potential has been used to drive the ion drift under moderate geomagnetic condition in solar maximum (IMF $B_y=0$ and $B_z=-5$ nT, HP$=$30 GW, and $F_{10.7}=150\times10^{-22}\,\text{W/m}^2\text{Hz}$). Fig. 1.12B is the same as Fig. 1.12A, except that the electric field variability from the empirical forcing model has also been implemented in the TIEGCM. The electric field variation is used by alternating the sign of the electric field standard deviation from the model for a given point at every time step (2 min) in both the north–south and east–west directions. This methodology ensures that the mean squared electric field variability matches that of the empirical model. It effectively assumes that there is no temporal coherence of the variability, which is an oversimplification. However, the methodology does capture the dominant effect of electric field variability on the Joule heating. A comparison between Fig. 1.12A and B shows that the electric field variability increases the Joule heating significantly. For example, the maximum at dawn and dusk increases from 0.009 to 0.018 W/m^2. The estimates of height-integrated Joule heating have been calibrated against the estimation of the Poynting flux, which is obtained from the vector cross product of the electric field and perturbation magnetic field ($S=\frac{<E\times\Delta B>}{\mu_0}$) (e.g., Kelley et al., 1991; Gary et al., 1995) and represents the total electromagnetic energy input from the magnetosphere into the I-T system. Fig. 1.12C shows the Poynting flux at the top of the thermosphere from the empirical model (Maute et al., 2005). The results indicate that the dayside can have large electromagnetic energy inputs due to some mechanisms, such as electric field variability. Both Fig. 1.12B and C show dawn and dusk peaks with similar magnitudes, and a large amount of energy flux in the dayside cusp region. But the Poynting flux is larger in the polar cap and smaller on the night side than the Joule heating calculated with the average electric field and electric field variability. When integrating the Joule heating in the northern hemisphere as shown in Fig. 1.12, the total Joule heating increases by more than 100% after including the electric field variability, which indicates the electric field variability has a comparable contribution to the Joule heating as the average electric field. Clearly, the calculated Joule heating with the average electric field and electric field variability is much closer to the Poynting flux than that using only the average electric field. Generally, after adding the electric field variability, the Joule heating undergoes a substantial enhancement. The electric field variability strongly improves the agreement between the Joule heating and the Poynting flux, while their horizontal distributions have some detailed differences in the polar cap and nightside regions.

1.3.3.2 Impact of flow bursts on the ionosphere-thermosphere

Flow bursts are a specific meso-scale phenomenon in the high latitudes, which significantly contributes to the electric field variability described in the previous section. Various properties of meso-scale flow bursts have been recognized in different data sets. The statistical features of meso-scale plasma flows in the nightside high-latitude ionosphere have been characterized using the SuperDARN line-of-sight ion velocity data (Gabrielse et al., 2018). Results showed that the typical width of a characteristic equatorward flow perturbation is ~180 km in the polar cap and ~140–150 km in the auroral oval, and that the meso-scale flows tend to follow the large-scale background convection. It was also found that flow bursts and auroral streamers in the nightside auroral zone are strongly correlated, and the flow bursts are often directed equatorward and appear simultaneously with the streamers (Gallardo-Lacourt et al., 2014). Meanwhile, the ion drift measurements from the Defense Meteorological Satellite Program (DMSP) F17 satellite were utilized to identify meso-scale flow perturbations with spatial scale sizes between 100 and 500 km in the high-latitude ionosphere (Chen and Heelis, 2018) and the data analysis suggests that flow perturbation locations strongly depend on the IMF orientation as does their occurrence frequency. Such localized momentum and energy inputs related to flow burst can significantly impact on the thermospheric density and temperature (Carlson et al., 2012; Deng et al., 2013). For example, the all-sky imaging and GPS TEC observations show traveling ionospheric disturbances (TIDs) with a magnitude of 0.3 TECu driven by auroral oval flow bursts (Lyons et al., 2019).

To investigate the influence of localized and bursty ionospheric signatures on the upper atmosphere, Deng et al. (2019) have specified meso-scale flow bursts in the nightside auroral zone in GITM simulations according to the ground-based and satellite observations (Gallardo-Lacourt et al., 2014; Chen and Heelis, 2018; Gabrielse et al., 2018) to quantify their roles in energy and momentum exchange between the ionosphere and the thermosphere. Specifically, one flow burst has been added in the aurora zone close to local midnight. The central location of the flow burst is (2330 LT, 65 degree Lat) in the geographic coordinates. The spatial extent of the equatorward flow in longitude is 100 km and there are poleward returning flows each of 50 km in spatial extent on each side. In the center of the flow burst, the equatorward ion drift increases by more than 900 m/s and the returning poleward flow on two sides is 400 m/s, which is comparable to the background large-scale ion drift convection speed. The flow burst lasts for 15 min, which represent the upper bound of lifetime of a flow burst according to the statistical studies. After 15 min, the enhancement of both convection and particle precipitation is turned off and the electrodynamic forcing goes back to the background large-scale specification. The 900 m/s equatorward flow speed and 15-min lifetime are larger than the average values for a flow burst, but are still within the range of commonly observed values (2–15 min) (Sergeev et al., 2004; Chen and Heelis, 2018; Gabrielse et al., 2018). The particle precipitation flux in the center of the flow burst (overlapping with the equatorward ion drift flow) has been doubled without changing the average energy of the precipitating particles. In order to resolve the narrow longitudinal structure of plasma flow, the resolution of GITM has been set up as 0.5 degree in Lon 2 degree in Lat. To emphasize the influence of a flow burst on the ionosphere/thermosphere, the difference fields between the cases with and without a flow burst are shown in Fig. 1.13.

The top two rows in Fig. 1.13 show the perturbation in the neutral wind and the neutral density at 5, 15, and 30 min. Those three times are picked since 5 min are roughly the length of time needed for an acoustic wave to propagate to 300 km altitude (Deng et al., 2008), 15 min are the lifetime of the flow burst we add in, and 30 min represent the time without flow burst when the perturbation is mainly related to TADs propagating out of the source region after the forcing has been turned off. In the top panel, the vector shows the difference in the horizontal wind and the color contour shows the difference in the vertical wind. The horizontal wind perturbation can reach 30 m/s and propagates out as a traveling atmospheric disturbances (TADs) even after the flow burst has been turned off at 15 min. The phase speed is 800 m/s, which is close to the sound speed at the altitudes displayed. A clear dawn-dusk asymmetry can be identified in the horizontal wind disturbance. In general, the duskside has a larger perturbation with a more complex structure than the dawnside, which may be related to the difference in the Coriolis force and preconditioning at these locations. The vertical wind is upward with a magnitude of 20 m/s at 300 km altitude at 5 min. However, after the flow burst has been turned off at 15 min, the vertical wind is mainly downward and the perturbation also propagates outward in concert with the perturbation in the horizontal wind. In the second row of Fig. 1.13, the color contour shows the percentage difference in the

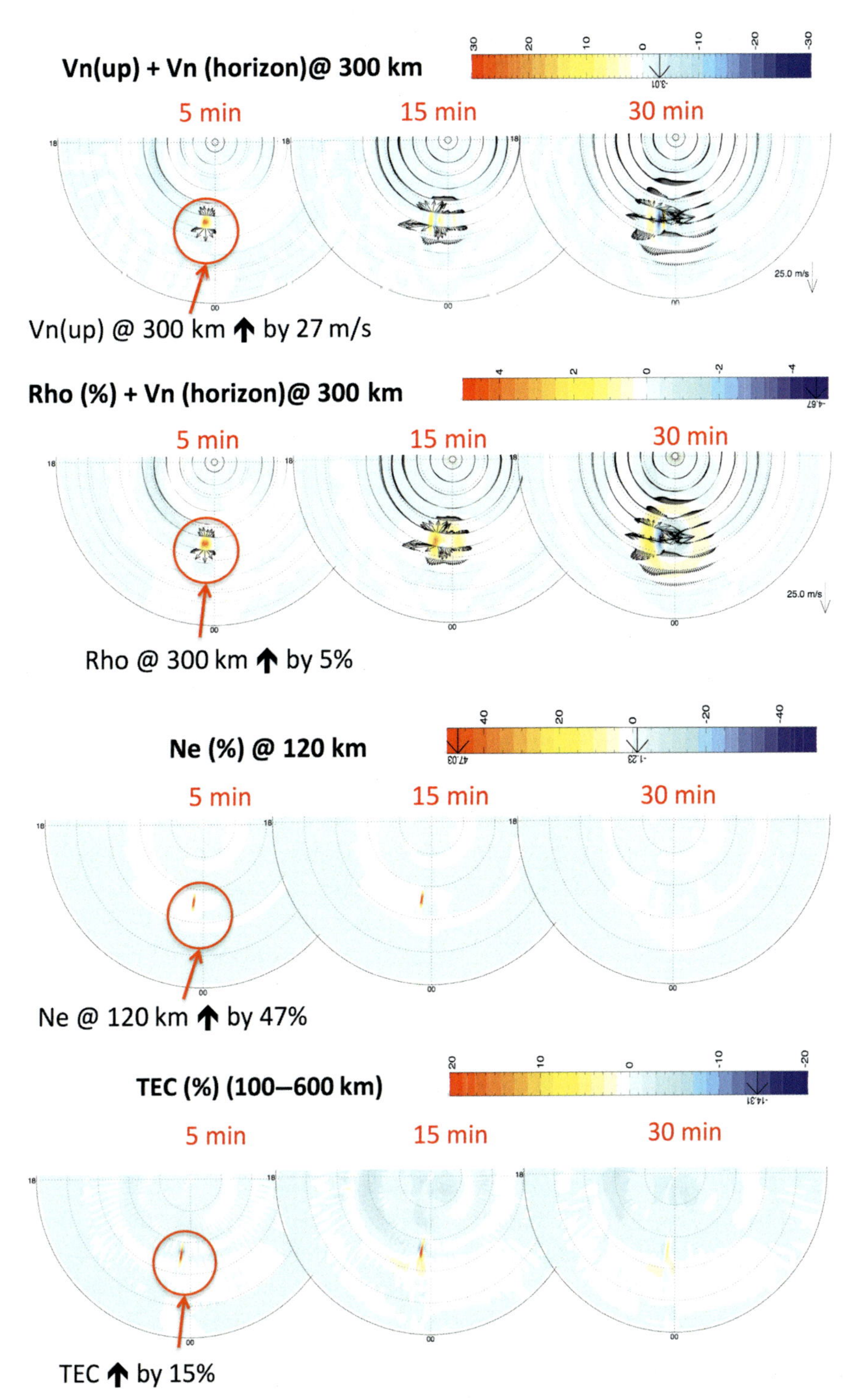

Fig. 1.13 (*First row*) Difference fields for vertical neutral wind (color contour) and horizontal neutral wind (vector) between the simulations with and without flow burst. (*Second row*) The same as top, except the color contour for neutral density (Rho). $T=0$ is the onset of flow burst. (*Third row*) Percentage difference of electron density at 120 km altitude between the simulations with and without flow burst. (*Fourth row*) Percentage difference of total electron content (TEC) integrated from 100 to 600 km (Deng et al., 2019).

neutral density at 300 km altitude. The neutral density enhancement can reach 5% at 5 min and the neutral density variation remains positive until the flow burst is turned off at 15 min. A strong negative neutral density perturbation with a magnitude of 4%–5% appears at 30 min. Actually, the Joule heating is more than doubled at the center of the flow burst at 15 min due to the enhancement of V_i. The neutral density perturbation is primarily driven by the atmosphere upwelling associated with the Joule heating increase and the corresponding gravity waves propagating out of the source region.

The bottom two rows in Fig. 1.13 show the perturbation in the electron density at 120 km altitude and in the total electron content (TEC). As shown in the third row, the electron density at 120 km increases by almost 50% at 5 min due to the enhanced particle precipitation flux. However, the electron density disturbance quickly decays when the flow burst and particle precipitation enhancement are turned off at 15 min. It is difficult to identify a noticeable variation in electron density at 30 and 50 min. The TEC perturbation at 5 min is close to 15% (1 TECU), and goes down to 1%–2% (0.1 TECU) at 30 and 50 min. Fig. 1.13 illustrates that the electron density responds quite differently before 15 min and after, which can be explained as the consequence of the particle precipitation enhancement and the interaction between TADs and TIDs. Before 15 min, when there is a significant particle precipitation enhancement in the center of the flow burst, the E-region ion density increases significantly due to the augmented ionization. But the E-region ionosphere is close to a chemical equilibrium and the electron density enhancement quickly decays when the enhanced particle precipitation is removed after 15 min. Following the period of the flow burst, the ion density can be strongly influenced by TADs through ion-neutral coupling and exhibits a wave structure, called a TID (Hocke and Schlegel, 1996; Balthazor and Moffett, 1997). While a TAD can be clearly identified in the top two panels of Fig. 1.13, the accompanying TID signal with a magnitude of 2% can be barely visualized after 15 min in the bottom row of Fig. 1.13, which may be due to the relatively small neutral density perturbation (5%). However, the signal of TIDs in TEC becomes much clearer when the size of the flow burst is doubled, as shown in a study by Deng et al. (2019).

1.3.3.3 Model development for further studies: GITM-R

As described in Section 1.3.2, using the standard deviation to represent the electric field variability is an important and practical way to statistically characterize meso-scale structures in the electrodynamic forcing. To better capture the meso-scale variations in the upper atmosphere dynamics and further improve our capability to resolve the meso-scale I-T disturbances in GCMs, the multiscale numerical representation is critical. While having a self-consistent atmosphere, spatial resolution is often limited to studying large-scale

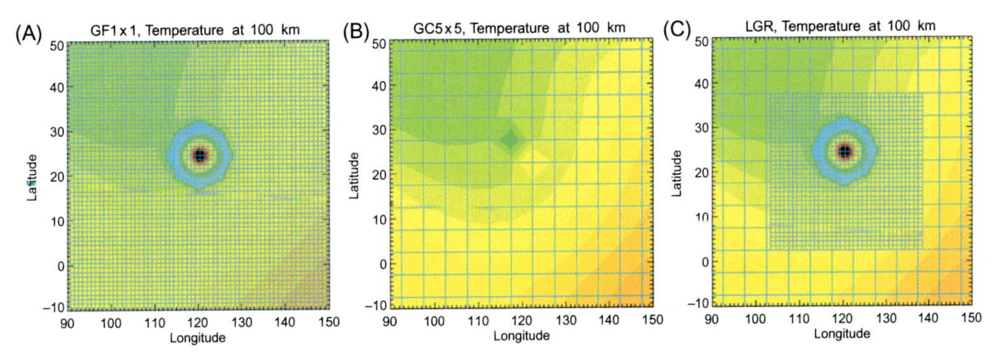

Fig. 1.14 Comparison of simulations with (A) global fine (1° × 1°), (B) global coarse (5° × 5°), and (C) local-mesh refinement (1° × 1° in regional domain) grid structures.

phenomena owing to the global nature of GCMs. Numerical simulation with the capability of dealing with multiscale problems is critical for improving the understanding of ion-neutral coupling in the upper atmosphere.

Recently, the Global Ionosphere-Thermosphere Model with the local-mesh refinement technique (GITM-R) has been developed. Using the local-mesh refinement technique, a high-resolution grid in the regional domain of interest can be nested within the coarse-resolution grid of the global domain, as shown in Fig. 1.14. Specifically, a user-defined high-resolution mesh grid is embedded in a uniform coarse-resolution grid of GITM and these two grid layers exchange information at the boundary of the regional domain to fulfill a two-way coupling between them. Through this approach, the regional domain achieves a higher resolution; meanwhile, the global domain supplies more realistic boundary values for the regional domain than the simple open boundary conditions, which are typically used for localized simulations. Multiple-level refinement can be applied to achieve an even higher resolution in the nested area. Mesh refinement techniques offer a flexible framework for variable-resolution models since they can focus their computational mesh on certain geographical areas.

To assess the efficiency and effectiveness of GITM-R in three-dimensional simulations, nested grids have been placed in a region of 35 degree in latitude and 35 degree in longitude centered at [120°E, 25°N] as shown in Fig. 1.14, where a meso-scale perturbation has been imposed at the lower boundary of GITM. This boundary perturbation triggers gravity waves propagating into the upper atmosphere and the scientific study of gravity wave propagation using GITM-R has been included in a study by Zhao et al. (2020). The comparison between different resolutions and grid structures has been conducted to illustrate the capability of the local-mesh refinement technique to resolve the meso-scale structures caused by different forcing. Fig. 1.14A represents the GITM simulation results with a global uniform high resolution ($1^\circ \times 1^\circ$), which shows meso-scale perturbations on top of the large-scale temperature variations and serves as a reference for the other cases. Fig. 1.14B represents the GITM simulation results with a global uniform low resolution ($5^\circ \times 5^\circ$). Clearly, the meso-scale perturbation has not been resolved under the coarse resolution. Fig. 1.14C represents the simulation results with a local-mesh refinement grid structure, in which the global resolution is $5^\circ \times 5^\circ$ and the resolution of the local domain of interest is refined to $1^\circ \times 1$. Both the meso-scale and large-scale structures are well represented in this case. Overall, the run with local-mesh refinement matches the uniform high resolution run very closely (the average difference between them in the localized region is less than 1%), but the needed computational resource for the local-mesh refinement case is only 10% of what is needed in the uniform high resolution case. Fig. 1.14 demonstrates that the local-mesh refinement technique is practical for simulating the meso-scale phenomena with GCMs. Certainly a higher resolution of GITM-R, such as a subdegree, can be achieved as well and has been applied for the studies of a tropical cyclone (Zhao et al., 2020) and volcanic activity (Tyska et al., 2019) induced ionospheric disturbance. The development of GITM-R makes it possible to combine a large-scale simulation with meso-scale features in a region of interest and will greatly improve the simulation capability for multiscale I-T perturbations caused by different forcing. In Chapter 5, Sections 5.6 and 5.7 discuss additional modeling approaches and challenges to characterization of Joule heating across different scales.

1.3.4 Summary

The large-scale ion convection flows in the high latitudes may often contain localized velocity enhancements, especially in the mesoscales (100–500 km). Studies with both ground-based and satellite observations illustrate that the electric field variability is not negligible, as compared with the large-scale average electric field, and has quite different horizontal distributions and different dependence on the geomagnetic conditions than the average electric field. The electric field variability tends to be anticorrelated with the particle precipitation variability at small- and mesoscales under the southward IMF conditions, indicating that the magnetosphere is likely to behave as a current generator at those scales.

The GCM simulations reveal that the electric field variability can increase the Joule heating by more than 100%, and significantly improves the consistency between the Joule heating and the Poynting flux. Consequently, including the electric field variability into the energy calculation results in significant changes in the neutral and ion density and dynamics. Specifically, the influence of meso-scale flow bursts associated with enhanced electron precipitation on the I-T has been simulated. The results show that a single flow burst with an equatorward flow speed of ~900 m/s can cause the neutral density and the horizontal wind at 300 km altitude to increase by 5% and 30 m/s, respectively. TADs triggered by the flow burst produce significant wave structures after the flow event ceases. Meanwhile, TEC increases up to ~15% (~1 TECU) due to the enhancement of electron precipitation associated with the flow burst.

In the future, systematic studies with improved data coverage are urgently needed to address the research goals of characterizing the multiscale electric field variabilities and their impact on the I-T system. The statistical analysis of data with a complete and even data coverage will strongly help to build up robust models, which are capable of accurately describing the electrodynamic forcing, including the average electric field, electric field variability, and consistent particle precipitation patterns.

Part 3 A review of ion outflow

Alex Glocer[1] and Naritoshi Kitamura[2]

[1] NASA/GSFC, Greenbelt, MD, United States
[2] Department of Earth and Planetary Science, Graduate School of Science, The University of Tokyo, Tokyo, Japan

1.4 A review of ion outflow

1.4.1 History and basic concepts

The origin of near-Earth plasma has been a topic of intense scientific study since the beginning of the space age. Earth's magnetosphere, the magnetic cavity carved out of the solar wind by the planet's intrinsic magnetic field, is populated by two sources. One source is the solar wind, the tenuous plasma comprised primarily of protons that are constantly blowing outward from the sun at super Alfvénic speeds. The second source is Earth itself, where plasma flows from the ionosphere to the magnetosphere supplying protons as well as heavier ion species. The plasmasphere, referring to the cold dense plasma at the low L-shell, was largely considered to be of ionospheric origin since its discovery (Carpenter, 1966). With this exception, during the early portion of the space age, the solar wind was considered the primary source of plasma for much of the magnetosphere. Thinking regarding the relative importance of the solar wind and ionospheric source of near-Earth plasma was called into question in the early 1970s with the surprising observations of heavy O^+ ions raining down into the atmosphere from the magnetosphere (Shelley et al., 1972). As O^+ in the magnetosphere can only originate from the ionosphere, its presence constitutes incontrovertible proof that the ionosphere acts as a source of plasma for the magnetosphere. Moreover, the observed O^+ was found to be at multiple keV energies, much higher than ionospheric energies, thereby indicating that the ionosphere can supply the hot magnetospheric plasma as well as the cold plasmasphere.

Much attention in the literature regarding the ionospheric supply of magnetospheric plasma concerns O^+ as it is indisputably from the ionosphere. However, there are also other ionospheric contributions which can contribute significantly to the magnetospheric composition. For instance, protons can come from either the solar wind or the magnetosphere, making it difficult to discern what is the ultimate source of any particular proton in the magnetosphere. The question of what fraction of protons in the magnetosphere originate in the ionosphere has been evaluated statistically in a number of studies. One such approach that has been used is to assume that protons originating in the solar wind will maintain a similar ratio to minor species in the magnetosphere as they had in the solar wind prior to entry. This technique has been used to estimate that during storms, up to 65% of protons in the magnetosphere may originate in the ionosphere (Gloeckler and Hamilton, 1987). Another approach assumes that you can estimate the fraction of protons of ionospheric origin in the magnetosphere by assuming their ratio to O^+ is the same as the ratio typically observed in those ions as they escape from the ionosphere. This approach shows that the fraction of geogenic protons can vary from 10%–50% depending on the conditions of magnetic activity (Peterson, 2002; Shelley et al., 1986; Shelley, 1986). Beyond protons, other heavy species have been observed outflowing from the ionosphere including He^+, N^+, and heavy molecular ions such as NO^+ (Lin et al., 2020).

As of today, it is now generally accepted that the ionosphere can indeed provide a significant quantity of magnetospheric plasma (Chappell et al., 1987), but the relative importance of different processes driving ionospheric outflow remains intensely debated. There are a multitude of processes at play in the aurora, cusp, polar cap, and plasmasphere refilling region that all contribute to ion outflow at high- and mid-latitudes. Likewise, there are different processes driving outflows of light and heavy ions, as well as cold and energized outflows. In the following subsections, we will review a number of the key mechanisms involved in generating ionospheric outflow. After this review of processes, we will discuss how the outflow is modeled, how it is tracked in the magnetosphere, and what are some of its impacts in the magnetosphere.

1.4.1.1 The polar wind

The "polar wind" is one of the longest studied types of ionospheric outflow and refers to the supersonic escape of protons from the high-latitude region. The classical polar wind acceleration mechanism is driven by a field-aligned, or "ambipolar," electric field, caused by the charge separation between heavy ions and electrons. A mental picture is quite helpful when trying to understand the origin of this electric field. Imagine the electrons, which are very light compared to the much heavier ions, feel very little influence due to gravity and on their own would fly off unrestrained. The ions in contrast are sluggish and more constrained by gravity. As the electrons try to separate from the ions, an electric field arises which holds them back, but also pulls on the ions. You can liken this picture to a large individual walking an energetic dog on a leash. As the dog starts to run, the separation is limited by the leash which restrains the dog, while simultaneously pulling the human walker forward. In this analogy, the tension in the leash acts as the electric field in the ambipolar polar wind. As we will discuss below, the magnitude of this electric field is such that lighter ion species can be accelerated to supersonic velocities. This mechanism was originally put forward by Axford (1968) and Banks and Holzer (1968), and the first observations of the polar wind were obtained by Explorer 31 and ISIS2 (Hoffman, 1970; Brinton et al., 1971; Hoffman et al., 1974).

An expression for the ambipolar electric field can be derived from the electron momentum equation along an open magnetic field line with an expanding cross section given by (e.g., Gombosi and Nagy, 1989; Varney et al., 2014):

$$\frac{\partial \rho_e u_e}{\partial t} + \frac{1}{A}\frac{\partial}{\partial s}\left(A\rho_e u_e^2\right) + \frac{\partial p_e}{\partial s} = -en_e E_{\parallel} + \frac{\delta M_e}{\delta t} \tag{1.8}$$

where s represents the distance along the field line, ρ_e and n_e represent the mass and number density of electrons respectively, u_e is the electron velocity along the magnetic field, p_e is the electron pressure, e is the electron charge, and $\frac{\delta M_e}{\delta t}$ represents collisional terms. We assume a steady state (eliminating the first term on the left), assume that u_e is less than the thermal speed (making the second term on the left negligible), and assume that collisional terms can be neglected (eliminating the last term on the right). It has been noted in some studies, however, that some of the collisional terms may be significant including the contribution of collisions with superthermal electrons and the "thermal diffusion effect" (Liemohn et al., 1997; Varney et al., 2014), and so collisional terms should be neglected with caution. The remaining terms can be rearranged to find the following expression for the ambipolar field.

$$E_{\parallel} = \frac{-1}{en_e}\frac{\partial p_e}{\partial s} \tag{1.9}$$

A very crude estimate of the magnitude of the field can be found by assuming that the major ion species in the topside ionosphere, O^+, is in hydrostatic equilibrium and the temperature profile is approximately isothermal. In this case, it can be shown that the magnitude of the electric field is approximately (Gombosi et al., 2004; Cohen and Glocer, 2012):

$$E_{\parallel} \approx \frac{m_O g}{2e} \tag{1.10}$$

In this simple approximation, the force associated with the ambipolar field is approximately half the strength of the gravitational force exerted on the major ion species.

For lighter ion species, such as H^+ and He^+, this force exceeds the gravitational force and leads to a supersonic escape of lighter plasma constituents.

1.4.1.2 Transverse heating by resonant wave-particle interactions

The polar wind of the previous section is primarily accelerated by the electric field without significant heating. Such a mechanism can produce significant outflows that are relatively cold with temperatures of a fraction of an eV. In contrast, heated outflows, with perpendicular temperatures ranging from a few to 100 s of eV, are frequently observed above the cusp and auroral regions. This population is strongly heated perpendicular to the magnetic field line through wave-particle interactions. The result is an ion distribution function that spreads out more transversely to the local magnetic field direction than parallelly. The magnetic mirror force, given by $\mu_i \frac{\partial B}{\partial s}$ where μ_i is the first adiabatic invariant and B is the magnetic field, works to convert this excess perpendicular energy into field-aligned transport. As this force is stronger for stronger perpendicular velocity, the heated edges of the distribution function experience stronger mirror force acceleration, thus bending up the edges of the distribution function. The result is a "V" or cone-shaped distribution function known as an "ion conic." An example of such a conic distribution is presented in Fig. 1.15.

There are a number of different wave-particle interactions proposed that can lead to transverse heating of ion distributions. They will not all be reviewed here due to space limitation, but the most prominent proposed mechanisms will be briefly summarized. These include transverse heating by the interaction of particles with broadband electromagnetic Ion Cyclotron Resonance Heating (ICRH) and Lower Hybrid Waves (LHW).

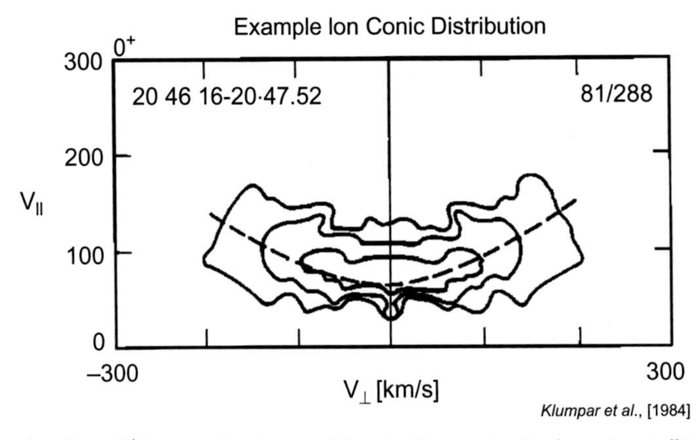

Fig. 1.15 Example of an O^+ ion conic observed by the Dynamics Explorer 1 satellite. *Adapted from Klumpar, D.M., Peterson, W.K., Shelley, E.G., 1984. Direct evidence for two-stage (bimodal) acceleration of ionospheric ions. J. Geophys. Res. Space Phys. 89 (A12), 10779–10787, https://doi.org/10.1029/JA089iA12p10779.*

Heating by Dispersive Alfvén Waves (DAW) can have a similar effect, but will be discussed in the next subsection.

ICRH resonant heating refers to the resonant interaction of ions with the broadband electromagnetic wave turbulence observed above the cusp and auroral regions. The application of ICRH theory to explaining ion conics was first introduced by Chang et al. (1986). This heating mechanism can be represented as a stochastic diffusion process with a quasi-linear diffusion coefficient which can be approximated by Crew et al. (1990). This form of the heating terms has been directly applied in kinetic models for specific parameterization of the wave inputs (Retterer et al., 1987; Crew et al., 1990; Barakat and Barghouthi, 1994; Glocer et al., 2018).

Another wave-particle interaction mechanism that can lead to the generation of energized ion conic distributions is the heating associated with LHW. A broad spectrum of these waves, near the lower hybrid frequency, is thought to be excited in the auroral region by accelerated electrons (Chang and Coppi, 1981). These waves are mostly effective at scattering particles well above the thermal velocity where they can generate high energy tails. They are not effective at directly heating the core plasma population, however, as those particles do not have enough energy to experience significant wave-particle interactions with LHW.

Given the different mechanisms for generating ion conics, it is interesting to ask which energization mechanism seems to be the most important. A survey of 200 events observed by the Freja satellite was conducted by André et al. (1998) where each event was classified by heating type. They found that the most common type of ion heating was of the ICRH variety, although other types of heating were also observed. This implies that ICRH heating is very important for understanding energized ion outflow. However, there are a number of important questions about the mechanism. For instance, how can these waves be predicted from first principles? What explains the saturation of the observed ion energies? What is the altitudinal distribution of these waves? These open questions require new observations and investigations.

1.4.1.3 Acceleration by interactions with low-frequency Alfvén waves

Low-frequency Alfvén waves, whose frequency is much less than the ion gyrofrequency, have also been argued to play an important role in the acceleration of plasma in the cusp and auroral regions. There are multiple pathways through which Alfvén waves can exert influence on plasmas, but here we group them into two general categories. First, when the Alfvén waves have a perpendicular wavelength much larger than the ion gyroradius, a number of so-called "ponderomotive" forces can be exerted on the plasma that can lift charged particles to higher altitudes. The second case deals with the case of scattering of ions by DAW, which are obliquely propagating with perpendicular wavelength comparable to the ion gyroradius. Both of these mechanisms are nonresonant but can significantly accelerate ions as will be described in the following paragraphs.

Ponderomotive forces are obtained by considering the net force by the electric and magnetic fields of an Alfvén wave acting on a particle over its gyration. For example, averaging the force acting on an ion over the course of its gyration in a wave field with a spatial gradient yields a force oppositely directed to the spatial gradient in the wave amplitude. This force is often referred to as the "Miller force" (Miller, 1958). The application of this type of force to low frequencies above the aurora was investigated by Li et al. (1993) for plasmas above the auroral region. They demonstrated numerically that this force could lead to significant acceleration of O^+ to escape energies. Another commonly applied ponderomotive force deals with the situation of interaction with a wave in an inhomogeneous magnetic field. In this case, changes in the perpendicular velocity of the particle result in upward or downward forces as that particle tries to conserve its first adiabatic invariant. This "magnetic pumping force," described in detail by Lundin and Hultqvist (1989) and by Hultqvist (1996), preferentially pushes particles to regions of a weaker background magnetic field, which above the auroral regions would be toward higher altitudes. A comprehensive review of these and other ponderomotive forces is provided by Lundin and Guglielmi (2006).

A particularly useful and simple form of the acceleration provided by the ponderomotive force, a_{PM}, is found when combining the Miller force and the magnetic pumping force into a single expression (Guglielmi et al., 1996):

$$a_{PM} = \frac{b^2}{8\mu_o\rho}\frac{\partial \ln\rho}{\partial s} \tag{1.11}$$

where ρ is the mass density and b is the wave amplitude. This formulation shows that ponderomotive force is directed toward decreasing densities and thus shows how Alfvén waves can lift plasma in the topside ionosphere. This force can be easily added to the equation of motion for a particle or as a source term for a fluid model.

The effect of ponderomotive forces on ion outflow can be quite substantial. Fig. 1.16 from Miller et al. (1995), shows an example of a numerical simulation of the high-latitude outflow problem with these forces included. The top part of the figure shows the effect on the density and the bottom part shows the effect on the velocity. Both H^+ and O^+ are considered. The results of the study show that ponderomotive forces not only lift plasma to higher altitudes, but they can accelerate O^+ to significant velocities which can exceed escape velocity. Interestingly, this acceleration happens with minimal heating and so outflow accelerated by this mechanism gives strong flow, but not a heated outflow distribution.

Another mechanism through which Alfvén waves can accelerate ions is through heating of the ion as it traverses the perpendicular field of an oblique wave whose perpendicular wavelength approaches the ion gyroradius (Chaston et al., 2004). This is often referred to in the literature as heating by DAW. In this mechanism, the particle perpendicular velocity is perturbed as the particle gyrates through the wave field, causing a

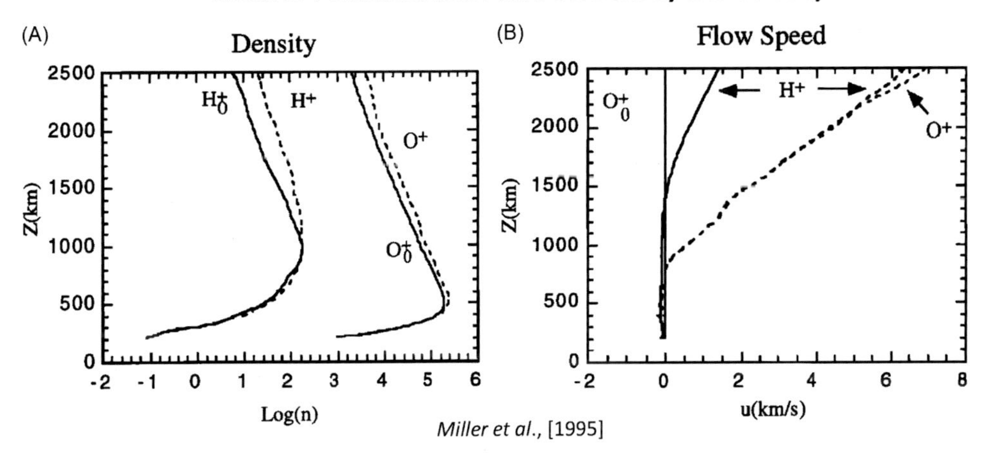

Fig. 1.16 A simulation result showing the effect of ponderomotive forces on the outflow solution. The force is effective at lifting an accelerating plasma, particularly O^+ to escape velocities. *Adapted from Miller, R.H., Rasmussen, C.E., Combi, M.R., Gombosi, T.I., Winske, D., 1995. Ponderomotive acceleration in the auroral region: a kinetic simulation. J. Geophys. Res. Space Phys. 100 (A12), 23901–23916, https://doi.org/10.1029/95JA01908.*

stochastic heating of the particle distribution perpendicular to the magnetic field. An explanation of how this heating works was introduced for laboratory plasmas initially (McChesney et al., 1987) and later applied to explain heating events seen by the Freja satellite (Stasiewicz et al., 2000). Much like the ICRH and LHW heating mechanisms, the heating by DAW heats ions transverse to the field line, and then the mirror force takes over to accelerate the plasma parallel to the field. It is interesting to note that unlike the ponderomotive interaction which accelerates but does not heat the ions significantly, this mechanism relies on transverse heating.

A study using slightly more than 3 months of observations from the FAST spacecraft found that only 15%–34% of energetic ion outflow events were associated with the presence of DAWs (Chaston et al., 2007). In some regions, however, the fraction of outflow events during which DAWs are observed can be higher. Some O+ outflows at lower L-shells observed by the Van Allen Probes have also been found to be associated with DAWs (Hull et al., 2019). Indeed, a recent statistical study for a storm event found a significant positive correlation of outflow events with the presence of DAW (Hull et al., 2020).

1.4.1.4 Superthermal electrons

Superthermal electrons, sometimes also called suprathermal electrons, refer to electrons with energies much greater than the thermal electron energy. In the topside ionosphere, thermal electrons have typical temperatures of around 2000–6000 K, with some variation. This corresponds to a thermal energy on the order of half an electronvolt (eV), and therefore superthermal electrons in the topside ionosphere are typically above 1 eV. There are three sources of these electrons including photoionization of the atmosphere, precipitating electrons of magnetospheric origin (polar rain, cusp, and auroral electrons), and secondary electrons produced by impact ionization of the neutral atmosphere. They affect the outflowing solution through increasing the ionization rate, enhancing the ambipolar electric field, as well as the energy deposition from the superthermal population to the thermal population.

Photoelectrons whose origin lies in the ionization of the neutral atmosphere from solar Extreme Ultraviolet (EUV) radiation have long been suggested to enhance ionospheric outflows (Axford, 1968; Lemaire, 1972). In these early studies, it was argued that the more energetic and lighter photoelectrons would separate from the heavier and less energetic ions, thus enhancing the ambipolar electric field. A later modeling study demonstrated a strong connection between photoelectrons and the ambipolar electric field (Khazanov et al., 1997). That study found that even a small amount of photoelectrons relative to the thermal electrons can have a major impact on the polar wind solution. The electric field is not the only channel of influence, these electrons can also transfer energy to the thermal electron population through Coulomb collisions, thereby heating the thermal population (e.g., Yau et al., 1995). This is an elastic collisional process which conserves the total amount of energy, but transfers it from one population to the other. These photoelectrons can also contribute to the formation of a high-altitude reflection potential (Khazanov et al., 2019a) which is frequently observed above the polar cap (Kitamura et al., 2012). This reflection potential can effectively reflect a portion of the photoelectrons leaving the atmosphere back, thus effectively bottling them up where they can thus heat the topside ionosphere (Varney et al., 2014). The trapping of photoelectrons at high altitudes between the reflection potential and the magnetic mirror point has also been argued as a source of the topside electron heat flux over the sunlit polar cap (Khazanov et al., 2019b). Electron heat flux is particularly important to defining the high-altitude ionosphere density and temperature profile and thus has an important impact on ionospheric outflow.

Both observations and numerical simulations support the idea that photoelectrons influence the quiet-time polar wind solution. As an example, the polar wind velocity is observed to be larger under sunlit conditions than under dark conditions (Abe et al., 1993). This is coincident with the observed photoelectron flux (Lee et al., 1980; Peterson et al., 2008), and the electron density observed by the Akebono satellite as well as temperatures observed by EISCAT Svalbard Radar (ESR) (Kitamura et al., 2011). Numerical simulations demonstrate that photoelectron effects likely explain this dependence (Glocer et al., 2012). Indeed, subsequent studies using a kinetic electron

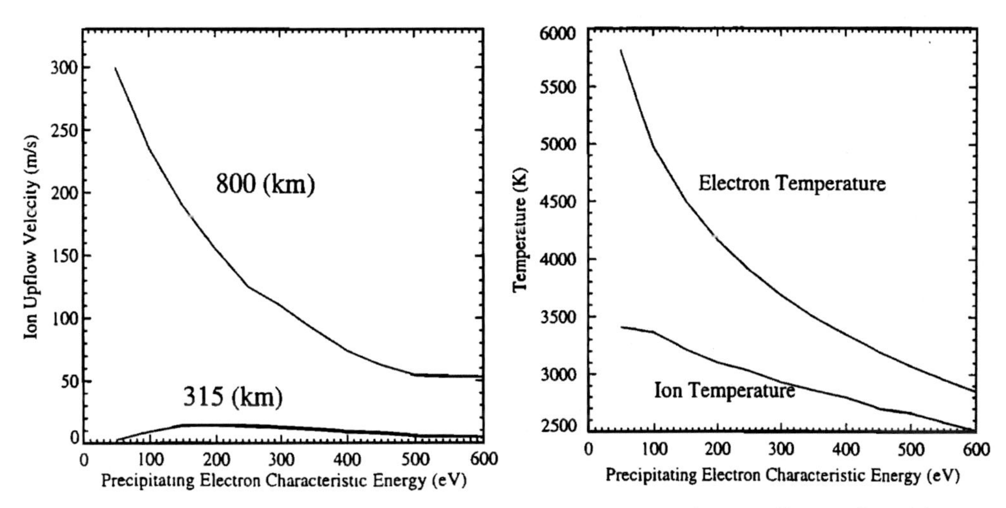

Fig. 1.17 The relationship between soft electron precipitation and ion upflow and ion/electron temperatures. The soft electron precipitation is assumed to have a fixed energy flux of 1.75 ergs cm^{-2} s^{-1}. *Adapted from Su, Y.-J., Caton, R.G., Horwitz, J.L., Richards, P.G., 1999. Systematic modeling of soft-electron precipitation effects on high-latitude f region and topside ionospheric upflows. J. Geophys. Res. Space Phys. 104 (A1), 153–163. https://doi.org/10.1029/1998JA900068.*

model coupled to a fluid ion model of ion outflow demonstrate that photoelectrons are essential to the global quiet-time polar wind outflow solution.

Beyond photoelectrons, there is a strong association observed between soft (<1 keV) electron precipitation and ion outflow. This relationship was found from detailed studies of data from the Polar spacecraft (Zheng et al., 2005) as well as from the FAST spacecraft (Strangeway et al., 2005). These electrons deposit most of their energy in the topside F-region of the ionosphere where they can exert a significant effect on ion outflow. The lower characteristic energy of this population can enhance the ambipolar field, ionization, and heating, thus enhancing upward $O+$ flows (Su et al., 1999). An example of this is shown in Fig. 1.17.

1.4.1.5 Other outflow processes

In addition to the mechanisms above, ion neutral frictional heating, also known as Joule heating, can contribute to ion heating and upwelling in the high-latitude region. The basic idea is that the differential motion between the ion and neutral velocity perpendicular to the magnetic field can lead to frictional heating that results in an increase in the vertical velocity of ions. Much of the Joule heating is confined to the E-region of the ionosphere where it is not active in the generation of ion outflow. However, measurements from Dynamics Explorer show a large and localized Joule heating event in the F-region which can be much more effective in creating ion upwelling in the cusp

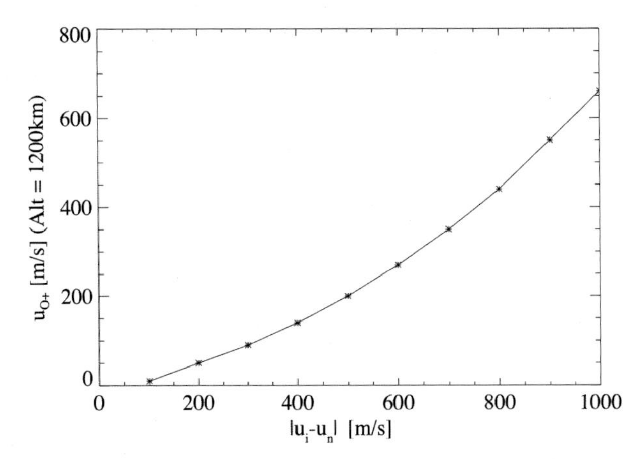

Fig. 1.18 An illustration of the effects of Joule heating associated with ion convection on ion upflow using the model of Glocer et al. (2009a, b).

and auroral regions (Killeen and Roble, 1984). Including such Joule heating into a fluid model, it was found that significant O+ upwelling can be generated with upward velocities of about 2 km/s (Gombosi and Killeen, 1987).

Fig. 1.18 provides a demonstration of the effectiveness of Joule heating in the formation of ion upflow events. This figure shows the peak upflow velocity at 1200 km altitude as a function of the difference between the ion convection and neutral velocity. To create this figure, we consider a steady-state field line on the dayside that is suddenly picked up by convection in the high-latitude region. The calculation is done using a polar wind model described by Glocer et al. (2009a). A source term representing the Joule heating is added to the energy equation in that model as (Schunk and Nagy, 2000):

$$\frac{\delta E_{Joule}}{\delta t} = \sum_n \frac{\rho_i \nu_{in}}{m_i + m_n} \left[m_n (u_{i\perp} - u_{n\perp})^2 \right] \tag{1.12}$$

where $u_{i\perp}$ and $u_{n\perp}$ are the ion and neutral velocities perpendicular to the local magnetic field.

Another way that convection across the polar cap can affect outflow is though centrifugal acceleration (Cladis, 1986). In the rest frame of a convecting ion, there is an outward centrifugal acceleration given by Horwitz et al. (1994)

$$a_{cent,\theta} = 1.5(E_i/B_i)^2 (r^2/r_i^3) \tag{1.13}$$

for meridional convection near the magnetic pole. In this equation, E_i is the convection electric field in the ionosphere, B_i is the magnetic field in the ionosphere, r is the radial distance along the field, and r_i is the radial distance at the foot point of the field line in the ionosphere. As you can see, this effect becomes more pronounced at high altitudes and is only a very modest factor at low altitudes. In fact, a detailed modeling study was undertaken by Demars et al. (1996) in which they found that centrifugal acceleration has only a mostly negligible effect on outflow. In that study, the only significant effect was found for O+ during periods of enhanced convection and electron temperature. Such an effect can be important to understand the extremely high-density plasma corresponding to the low-energy portion of the cleft ion fountain (Kitamura et al., 2010).

1.4.2 Approaches of modeling outflow

All techniques to model ion outflow start with the well-known Boltzmann equation (Boltzmann, 1872) given by

$$\frac{\partial f}{\partial t} + v_i \frac{\partial f}{\partial x_i} + a_i \frac{\partial f}{\partial v_i} = \frac{\delta f}{\delta t} \tag{1.14}$$

In this equation, f is the distribution function, the subscript i refers to the dimension index, the velocity is given by v_i, the acceleration is given by a_i, and x_i is the position. Note that this equation is presented in index notation with repeated index equal to the sum. For a space plasma, the acceleration a_i is given by the Lorenz force divided by the mass. The right-hand side, $\frac{\delta f}{\delta t}$, represents collision terms which can include wave particle interaction terms as well. This equation describes the evolution of the particle distribution function which is in general a six-dimensional function of velocity and location.

The approaches to solving Eq. (1.14) can be loosely grouped into two categories: kinetic and hydrodynamic. In the kinetic approach, one solution is to directly solve the equation (e.g., Khazanov et al., 1994). The advantage to the direct solution is the numerical noise and diffusion can be suppressed with an appropriate choice of numerical scheme and transformation of variables. This can be quite complex, particularly as the collision terms are nonlinear, as their evolution depends on the shape of the distribution function itself (e.g., Khazanov et al., 2012).

Another commonly used kinetic approach is to solve the Boltzmann equation statistically using a particle-in-cell or PIC method. In this case, macro-particles, each representing some number of real particles, are randomly sampled from the lower altitude boundary of the outflow simulation. These macro-particles are evolved according to their equation of motion. For a macro-particle moving along an expanding magnetic flux tube, the equation of motion is given by

$$m_i \frac{\partial v_{i\|}}{\partial t} - q_i E_\| + \frac{G m_i M_{planet}}{r^2} + \mu_i \frac{\partial B}{\partial s} = 0 \tag{1.15}$$

where i provides the index of the ion species, the mass is given by m_i, the velocity is given by v_i, the time is given by t, the charge is given by q_i, the parallel electric field is given by $E_\|$, the gravitational constant is given by G, the mass of the planet is given by M_{planet}, the radial location of the particle is given by r, the magnetic field magnitude is given by B, and

s is the distance along the magnetic field. The PIC method of solution can be thought of as akin to understanding public opinion on a topic by polling a representative sample of the general population. In this case, the collection of macro-particles followed are our sample, and by solving for millions of macro-particles the kinetic solution is recovered statistically. The main drawback of this approach, as compared to the direct solution, is that it is inherently subject to sampling noise that can only be controlled by following an extensive number of particles.

An alternative to the kinetic approach is the hydrodynamic approach. In this case, velocity moments of the Boltzmann equation are taken in order to obtain fluid equations. While there are many forms that these fluid equations have been presented in, a useful form for the study of outflow problems is the field-aligned gyrotropic transport equations which describe the motion of a low beta plasma along a magnetic field line (Gombosi and Nagy, 1989):

$$\frac{\partial}{\partial t}(A\rho_i) + \frac{\partial}{\partial r}(A\rho_i u_i) = AS_i \tag{1.16}$$

$$\frac{\partial}{\partial t}(A\rho_i u_i) + \frac{\partial}{\partial r}(A\rho_i u_i^2) + A\frac{\partial p_i}{\partial r} = A\rho_i\left(\frac{e}{m_i}E_\| - g\right) + A\frac{\delta M_i}{\delta t} + Au_i S_i \tag{1.17}$$

$$\frac{\partial}{\partial t}\left(\frac{1}{2}A\rho_i u_i^2 + \frac{1}{\gamma_i - 1}Ap_i\right) + \frac{\partial}{\partial r}\left(\frac{1}{2}A\rho_i u_i^3 + \frac{\gamma_i}{\gamma_i - 1}Au_i p_i\right) = A\rho_i u_i\left(\frac{e}{m_i}E_\| - g\right) + \frac{\partial}{\partial r}\left(A\kappa_i\frac{\partial T_i}{\partial r}\right) + A\frac{\delta E_i}{\delta t} + Au_i\frac{\delta M_i}{\delta t} + \frac{1}{2}Au_i^2 S_i \tag{1.18}$$

where A is the cross section of the expanding magnetic field, and A' is the derivative of the cross section along the field line. These equations are effectively conservation laws for each ion fluid and reflect the conservation of particles, momentum, and energy. The fluid equations also include various source terms. For instance, the mass source term, S_i, represents the net production rate minus loss rate of a given species. The production of a species i could be due to photoionization, charge exchange, or some other reaction. $\frac{\delta M_i}{\delta t}$ is the momentum exchange rate and $\frac{\delta E_i}{\delta t}$ is the energy exchange rate given by Schunk and Nagy (2000).

It is interesting to note that there is a long running historical controversy in polar wind modeling about the appropriate way to model ion outflow (Donahue, 1971; Lemaire and Scherer, 1973). In fact, after the first hydrodynamic model of polar wind was introduced (Banks and Holzer, 1968), it was immediately criticized as being inappropriate to the problem (Dessler and Cloutier, 1969). The crux of many of the criticisms of the hydrodynamic approach boils down to the assumption of collisionality inherent in the derivation of the fluid equations. For the ion outflow problem, the ratio of the mean free path between collisions to the characteristic length scale of the problem, known as the Knudsen number, starts to approach one at about 2000 km for longer-range Coulomb collisions (Lemaire and Scherer, 1970). Therefore, the perturbation theory that underlies the hydrodynamic approach starts to lose validity above this altitude. Additionally, there are many non-Maxwellian outflow distributions including ion conics, beams, and double hump distributions that can be difficult to represent in a fluid model (e.g., Barakat et al., 1995).

Despite these difficulties, there are many good reasons to use the hydrodynamic approach. First and foremost, the approach is applicable in the low altitude collision dominated regime and in the transition region. In this regime, the approach is not only applicable, but much more computationally efficient than the kinetic approach which must resolve the collision time. The low altitude region below the exobase is also what sets the limit on the mass flux. So while additional acceleration may occur at high altitudes, it is the low altitude gateway that determines how much mass can escape. Indeed, many comparisons between kinetic and hydrodynamic models show the results can be very similar (Marubashi, 1970; Holzer et al., 1971; Glocer et al., 2018).

1.4.3 Tracking outflow in the magnetosphere

Once plasma leaves the ionosphere and enters the magnetosphere, it can have many effects. The enhanced O^+ affects the reconnection rate (Shay et al., 2004), contributes a major portion of the energy density of the ring current during active times (Daglis et al., 1999), among other effects. To study these effects, researchers often try to follow the path that a particle takes during its traverse from the ionosphere through the magnetosphere. Tracking plasma of ionospheric origin as it moves through the magnetosphere takes many forms but can be loosely grouped into two categories: (1) Tracing test particles through empirical and MHD fields or (2) the use of multifluid MHD.

Tracking test particles through empirical or MHD fields is often used as a part of data analysis to understand the fate of particles leaving the ionosphere, or the origin of particles entering the magnetosphere. An excellent example of the former approach is the combination of Polar satellite observations of escaping polar wind ions with test particle data presented by Huddleston et al. (2005). An example of a trajectory from this study and the particle energization along the trajectory is given in Fig. 1.19. This example demonstrates that a polar wind proton starting at less than 1 eV is able to reach multiple keV of energy in the magnetosphere.

The test particle approach has also been used on a massive scale to examine the large-scale impacts of outflow on the magnetosphere. This approach can involve tracking thousands or millions of particles (e.g., Peroomian et al., 2007; Moore et al., 2005). The advantages of this technique are that it allows for kinetic effects and for the many exotic velocity space distribution functions observed throughout the magnetosphere. The disadvantage is that particles and fields are not evolved in a self-consistent manner as the movement of particles does not feed back into the electromagnetic fields.

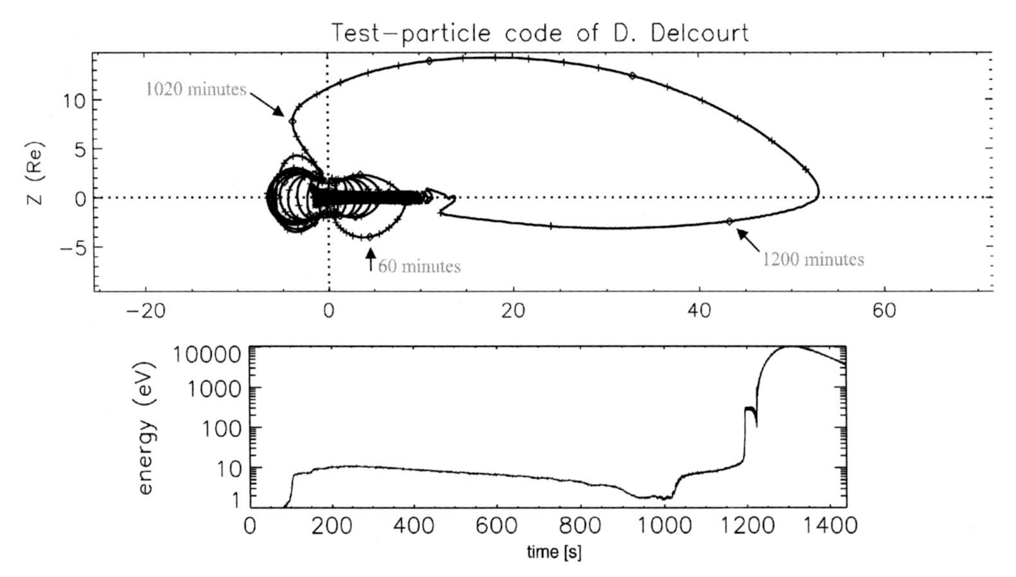

Fig. 1.19 The figure presents the result of a test particle simulation following the trajectory of an escaping cold ion and its subsequent energization in the magnetosphere. The result shows how cold ionospheric particles can contribute to the hot magnetospheric population. *Adapted from Huddleston, M.M., Chappell, C.R., Delcourt, D.C., Moore, T.E., Giles, B.L., Chandler, M.O., 2005.* An examination of the process and magnitude of ionospheric plasma supply to the magnetosphere. J. Geophys. Res. 110, 12202, *https://doi.org/10.1029/2004JA010401.*

An alternative to the test particle approach is the multifluid MHD method. In this approach, each species in the magnetosphere is followed with a separate fluid. These equations are given as follows (Glocer et al., 2009a, b):

$$\frac{\partial \rho_s}{\partial t} + \nabla \cdot (\rho_s \boldsymbol{u}_\mathbf{s}) = S_{\rho_s} \tag{1.19}$$

$$\frac{\partial \rho_s \boldsymbol{u}_\mathbf{s}}{\partial t} + \nabla \cdot (\rho_s \boldsymbol{u}_\mathbf{s} \boldsymbol{u}_\mathbf{s} + I p_s) = n_s q_s (\boldsymbol{u}_\mathbf{s} - \boldsymbol{u}_+) \times \mathbf{B} + \frac{n_s q_s}{n_e e} (\mathbf{J} \times \mathbf{B} - \nabla p_e) + S_{\rho_s \boldsymbol{u}_\mathbf{s}} \tag{1.20}$$

$$\frac{\partial p_s}{\partial t} + \nabla \cdot (p_s \boldsymbol{u}_\mathbf{s}) = -(\gamma_s - 1) p_s \nabla \cdot \boldsymbol{u}_\mathbf{s} + S_{p_s} \tag{1.21}$$

$$\frac{\partial \mathbf{B}}{\partial t} - \nabla \times (\boldsymbol{u}_+ \times \mathbf{B}) = 0 \tag{1.22}$$

$$\frac{\partial p_e}{\partial t} + \nabla \cdot (p_e \boldsymbol{u}_\mathbf{e}) = -(\gamma_e - 1) p_e \nabla \cdot \boldsymbol{u}_\mathbf{e} + S_{p_e} \tag{1.23}$$

in this case, γ_s and γ_e are the adiabatic index for species s and electrons. Note that *s* and *e* subscripts represent the ion or electron fluids, respectively. The charges for ion species s and electrons are given by q_s and e. Additionally, n, and ρ are number and mass density, respectively, $\boldsymbol{u}$ describes the velocity, p is the pressure, $\mathbf{J}$ is the electrical current density, $\mathbf{B}$ describes the magnetic field, and source terms are given by S whose type is provided by the corresponding subscript. Additionally, $\boldsymbol{u}_+$ represents the charge averaged ion velocity:

$$\boldsymbol{u}_+ = \frac{\sum_s n_s q_s \boldsymbol{u}_\mathbf{s}}{e n_e} \tag{1.24}$$

In contrast to the test particle approach, multifluid MHD allows for the self-consistent evolution of plasma and fields, but at the expense of limiting the complexity of the velocity space distribution function. This approach was first championed by Winglee (1998) who used the multifluid approach to study the "geopause." The geopause is a concept first introduced by Moore and Delcourt (1995), and refers to the region of near-Earth space whose plasma is primarily of ionospheric origin. It was another 10 years after these first multifluid MHD studies for this approach to gain traction in other modeling efforts, but has since become more widely used (Glocer et al., 2009a, b; Wiltberger et al., 2010). Fig. 1.20 provides an example of the application of such a model to the study of sources of near-Earth plasma (Glocer et al., 2020). This study used separate fluids for solar wind H+ and ionospheric H+ and O+ to look at how each source of plasma contributes to the magnetosphere at different phases of a particular geomagnetic storm.

1.4.4 Impacts of ionospheric outflow on the magnetosphere

In recent years, there has been a growing number of studies demonstrating the impact that outflow has on the magnetosphere. It is found to affect many aspects of magnetospheric processes including reconnection, magnetotail dynamics, composition, and more. In the following subsection, we walk through some examples of the larger impacts that outflowing plasma is found to have on the magnetosphere.

1.4.4.1 The role of outflow on reconnection

Magnetic reconnection is one of the fundamental processes through which we understand magnetospheric dynamics (Dungey, 1961). Reconnection governs the transfer of solar wind energy into the magnetosphere, and is also implicated in many dynamical processes in the magnetotail as we will discuss further in the next subsection. The presence of ionospheric plasma in the magnetosphere has been argued to affect the reconnection process in various ways.

Mass loading of the reconnection region can directly affect the reconnection rate. In this process, if the mass near the reconnection *x*-line goes up, either through the addition

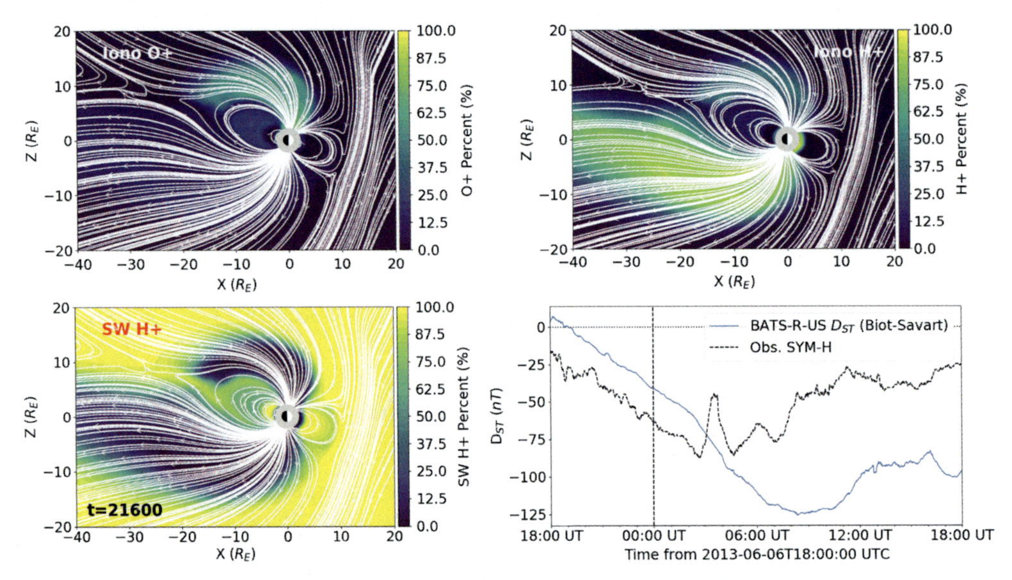

Fig. 1.20 Example of the application of the multifluid MHD to understand the various sources of near-Earth plasma during a geomagnetic storm during the Early main phase. *Adapted from Glocer, A., Welling, D., Chappell, C.R., Toth, G., Fok, M.C., Komar, C., Kang, S.B., Buzulukova, N., Ferradas, C., Bingham, S., Mouikis, C., 2020. A case study on the origin of near-earth plasma. J. Geophys. Res. Space Phys. 125 (11), e28205, https://doi.org/10.1029/2020JA028205.*

of ionospheric H^+ or the inclusion of a heavy O^+ population, the local Alfvén speed can go down and thus reduce the reconnection rate (Shay et al., 2004). There are many places this process is likely to happen. For example, when the magnetosphere experiences enhanced convection, such as during a geomagnetic storm, the plasmasphere can drain to the dayside magnetopause where it can encounter the dayside reconnection separator or *x*-line. This dense plasma, whose origin is in mid to low latitude ionospheric outflow, then mass loads the reconnection and suppresses the local reconnection rate (Borovsky et al., 2013). The left portion of Fig. 1.21 shows a schematic illustration of the geometric configuration, while the right side of the figure presents the result of a simulation with the plume included. It is seen that the arrival of the plume suppresses the cross polar cap potential which is often used as a proxy for the global reconnection rate. Another example of mass loading reconnection sites with ionospheric plasma is found in the magnetotail. During geomagnetically active times, the magnetotail can be dominated by O^+ which can only come from the ionosphere Kistler et al. (2005). The addition of such heavy ions similarly can reduce the reconnection rate in the tail.

The presence of outflow has many effects beyond just mass loading. For instance, O^+ is much heavier than H^+ and thus has a larger and slower gyroperiod. It is therefore more easily demagnetized, losing its frozen to the magnetic field nature more quickly. They can therefore be accelerated by the Hall fields near the reconnection site affecting the

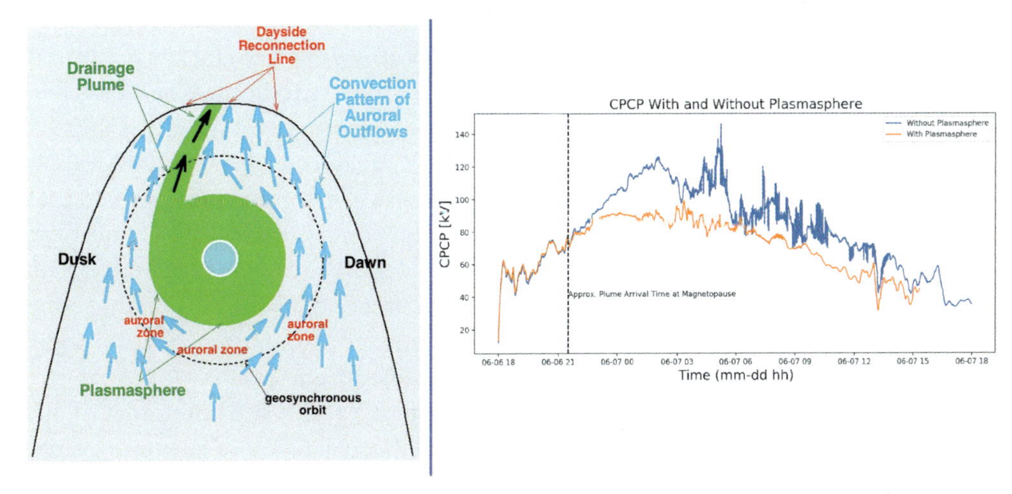

Fig. 1.21 (*Left*) Illustration of the plasmasphere plume reaching the magnetopause where it can disrupt reconnection. (*Right*) Effect of a plume on the cross polar cap potential which is a proxy of the dayside reconnection rate. *Left: Adapted from Borovsky, J.E., Denton, M.H., Denton, R.E., Jordanova, V.K., Krall, J., 2013. Estimating the effects of ionospheric plasma on solar wind/magnetosphere coupling via mass loading of dayside reconnection: ion-plasma-sheet oxygen, plasmaspheric drainage plumes, and the plasma cloak. J. Geophys. Res. Space Phys. 118 (9), 5695–5719. https://doi.org/10.1002/jgra.50527.); right: adapted from Glocer, A., Welling, D., Chappell, C.R., Toth, G., Fok, M.C., Komar, C., Kang, S.B., Buzulukova, N., Ferradas, C., Bingham, S., Mouikis, C., 2020. A case study on the origin of near-earth plasma. J. Geophys. Res. Space Phys. 125 (11), e28205. https://doi.org/10.1029/2020JA028205.*

evolution of bursty flows in the magnetotail (Liang et al., 2017; Tenfjord et al., 2018). Additionally, PIC simulations have shown that demagnetized O^+ ions in the reconnection region can act as a sponge effectively absorbing energy from the fields and reducing the reconnection rate and resulting acceleration (Tenfjord et al., 2019). The outflow rate and composition are also found to be asymmetric due to seasonal and other effects, and simulations show that an asymmetric composition can result in motion of the reconnection location (Kolstø et al., 2020). Another issue is that outflowing populations are often streaming along the field and retain a significant parallel motion when they encounter the reconnection site. This streaming can affect the reconnection as well causing a tailward propagation of the x-line (Tenfjord et al., 2020). Note that many of these effects outside of mass loading are due to kinetic processes and are therefore difficult to be included in current global models which tend to rely on fluid approaches.

1.4.4.2 Outflow and magnetotail dynamics

In addition to affecting reconnection, it has also been argued that ion outflow can affect the stability of the magnetotail. This possibility was initially suggested when considering the effect of the introduction of O^+ to the magnetotail. It was suggested by Baker et al.

(1982) that the presence of O^+ would make the plasmasheet unstable to the tearing instability and thus could be a trigger for substorms. Observational evidence for this prediction has been mixed. Kistler et al. (2005) found that the O^+:H^+ ratio in the Cluster satellite data can reach 10:1 during a substorm. Similarly, Geotail observations show that the O^+: H^+ energy density ratio is enhanced during dipolarization events (Nosé et al., 2009a). In contrast, long-term studies showed little connection between O + and substorms. In particular, Nosé et al. (2009b) compared plasma sheet ion composition with Pi2 pulsation, which was used as a proxy for the occurrence of substorms. They found that there was no correlation between these two parameters, which indicates that the O^+ triggering hypothesis for substorms may not be correct, or at least not complete. They also argue that O^+ may actually have a stabilizing influence on the magnetotail.

Beyond the controversy regarding the connection of O^+ and substorms, it has recently been argued that O^+ may be responsible for the generation of sawtooth events. Sawtooth events are loosely defined as a type of injection that is observable over a wide range of local times and recurs periodically. As such this represents a global-scale phenomenon for the magnetosphere. The connection between sawtooth events and the outflow of plasma from the ionosphere was first put forward by Brambles et al. [2011] who used numerical multifluid simulations of the magnetosphere with an empirical specification of outflow driven causally by energy inputs from the magnetosphere. They found that under certain conditions magnetospheric field lines can get overloaded with ionospheric plasma causing them to stretch out, release a plasmoid down tail, and snap back. This pattern was found to repeat periodically. Sawtooth period changes in the inclination angle represent the stretching and snapping back of the field during such events. More advanced simulations confirmed that these sawtooth events are connected with ion outflow close to midnight (Varney et al., 2016a, b). Observational evidence for this connection, however, is mixed. In particular, a study from Lund et al. (2018) found that outflow during sawtooth events tends to originate preferentially from the dayside cusp, not the nightside as argued for in simulations. Such an observation would seemingly contradict the idea from simulations that right side outflow is critical for sawtooth events, but it should be noted that it is possible that dayside outflow could also play this role (Zhang et al., 2020). Future study is clearly required to further understand this possible connection.

1.4.4.3 Asymmetric ion outflow affecting the magnetosphere

It is well known from observations and simulations that ion outflow demonstrates significant asymmetries. Differences in illumination and energy inputs going into each hemisphere translate into hemispheric asymmetries in ionospheric outflow. Polar observations, for example found that the O^+ flux changes more strongly with season than the H^+ flux (Lennartsson et al., 2004). Such seasonal asymmetries are likewise found in simulation studies of ion outflow (Schunk and Sojka, 1997; Maes et al., 2016), the offset between the magnetic dipole axis and the rotation axis can also induce diurnal variations in the outflow that can lead to oscillations in the magnetotail (Barakat et al., 2015).

1.4.4.4 Outflow and the inner magnetosphere

Ionospheric outflow is particularly important to the inner magnetosphere. During storm times in particular, O^+ is found to provide a major portion of the ring current pressure (Moore et al., 2001; Daglis et al., 1993, 1999). As the pressure in the ring current exerts a strong influence on the fields in the inner magnetosphere, ion outflow can be important for understanding the magnetic field in this region (Glocer et al., 2009a, b). Additionally, the $O+$ in the ring current has a more favorable cross-section for charge exchange (Smith et al., 1981). It has therefore been argued that enhanced O^+ in the storm time ring current may contribute to faster ring current decay (Hamilton et al., 1988). Interestingly, much of the $H+$ in the ring current may also be from the ionosphere at least in some circumstances. Fig. 1.22 presents a simulation by Glocer et al. (2020) showing the integral of the ring current energy contributed by each species over an event. At least in the event shown, ionospheric sources dominate over the solar wind source in the ring current.

The significant presence of ionospheric plasma in the inner magnetosphere has significant implications for the wave environment which can impact radiation belt particles. The existence of heavy ions in the ring current can alter the growth rate of EMIC waves (Kozyra et al., 1984; Jordanova et al., 1996). Using a series of hybrid simulations, Ofman et al. (2017) found that a large amount of O^+ can reduce the linear growth of EMIC waves as well as their saturation wave amplitude. Note that, while we often refer to heavy ions as synonymous with O^+ it is possible that similar effects on EMIC waves are created by N^+ (Bashir and Ilie, 2018). This is significant as N^+ may be as much as 10% of O^+ in the outflow solution (Lin et al., 2020). Beyond EMIC waves, cold plasma of ionospheric origin also affects the ultralow-frequency waves which affect radiation belt particles through drift resonant interactions. In particular, the plasmasphere, whose particles originate in the ionosphere, affects how deeply these waves can penetrate into the inner magnetosphere (Claudepierre et al., 2016).

1.4.5 Concluding thoughts

Ionospheric outflow has myriad impacts on the magnetosphere, touching upon reconnection, ring current evolution, and more. In recent years, there has been a growing trend toward the incorporation of this source of plasma into global simulations. These models seek to include global models incorporating the many competing mechanisms underlying ionospheric outflow. However, the ability to predict ionospheric outflow remains elusive. Much of this uncertainty is due to the inability to predict key mechanisms. For example, waves that drive the ion conic generation are not currently predicted in global models, meaning that empirical specifications of these waves are required.

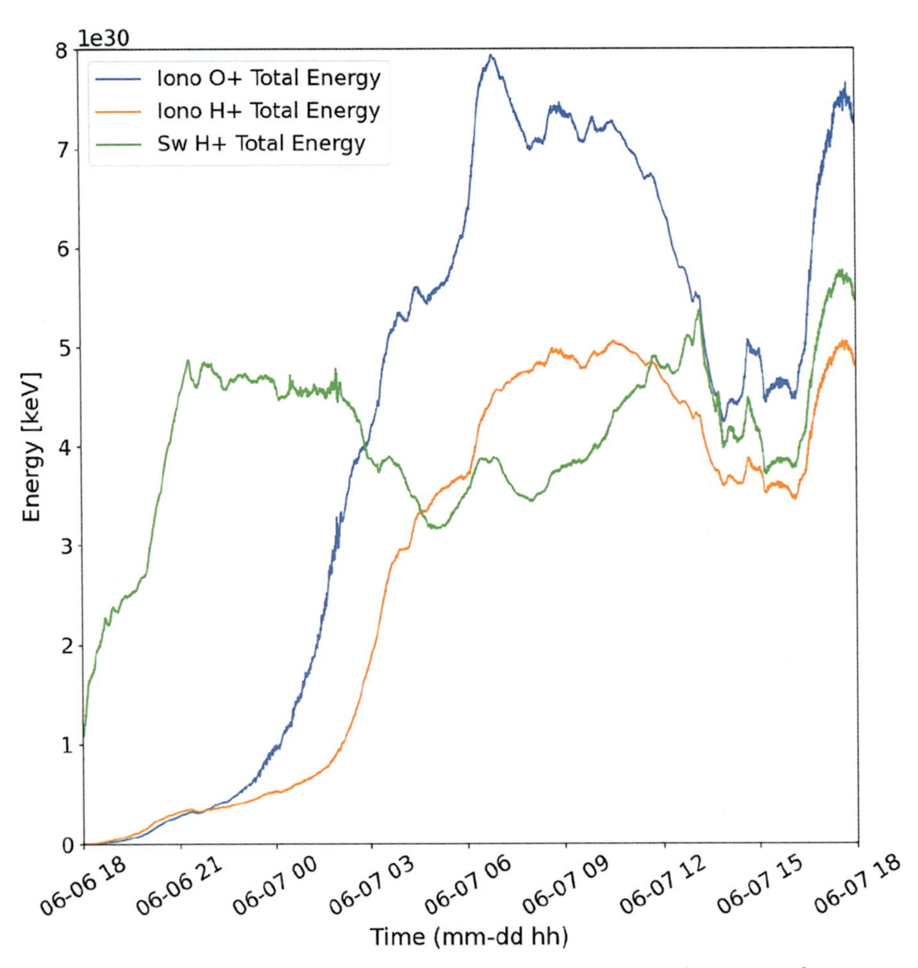

Fig. 1.22 Simulations of the total energy content in the ring current by species for a particular geomagnetic storm. *Adapted from Glocer, A., Welling, D., Chappell, C.R., Toth, G., Fok, M.C., Komar, C., Kang, S.B., Buzulukova, N., Ferradas, C., Bingham, S., Mouikis, C., 2020. A case study on the origin of near-earth plasma. J. Geophys. Res. Space Phys. 125 (11), e28205, https://doi.org/10.1029/2020JA028205.*

However, even these empirical specifications are insufficient as the altitude variation of the waves and the generation mechanism are often unknown. Further complicating this picture is that outflow can often be a multistep process involving multiple mechanisms simultaneously. Disentangling the multitude of simultaneous processes will require future observational and modeling investigations.

Acknowledgments

The work of YN was supported by National Aeronautics and Space Administration grants 80NSSC18K0657, 80NSSC20K0604, 80NSSC21K1321, and 80NSSC20K0725, NSF grant AGS-1907698 and AGS-2100975, and AFOSR grant FA9559-16-1-0364. The THEMIS and REGO all-sky imagers are maintained by University of Calgary under a support from Canadian Foundation of Innovation. AGS-1840962 to SRI International. GPS TEC data processing is supported by NSF grant AGS-1952737. The work of YD and QZ was supported by National Aeronautics and Space Administration through grants 80NSSC20K0195, 80NSSC20K0606, and 80NSSC20K1786, and AFOSR through award FA9559-16-1-0364. We acknowledge the Texas Advanced Computing Center (TACC) at the University of Texas at Austin for providing High performance computing (HPC) and visualization resources that have contributed to the research results reported within this paper (http://www.tacc.utexas.edu). We thank support from the CEDAR workshop "Grand Challenge: Multiscale I-T system dynamics" and ISSI/ISSI-BJ team "Multiscale Magnetosphere-Ionosphere-Thermosphere Interaction."

References

Abe, T., Whalen, B.A., Yau, A.W., Watanabe, S., Sagawa, E., Oyama, K.I., 1993. Altitude profile of the polar wind velocity and its relationship to ionospheric conditions. Geophys. Res. Lett. 20, 2825–2828. https://doi.org/10.1029/93GL02837.

Abel, G., Freeman, M., Chisham, G., 2009. Imf clock angle control of multifractality in ionospheric velocity fluctuations. Geophys. Res. Lett. 36 (19).

Aikio, A.T., Cai, L., Nygrén, T., 2012. Statistical distribution of height-integrated energy exchange rates in the ionosphere. J. Geophys. Res. 117. https://doi.org/10.1029/2012JA018078, A10325.

Anderson, P.C., Hanson, W.B., Heelis, R.A., Craven, J.D., Baker, D.N., Frank, L.A., 1993. A proposed production model of rapid subauroral ion drifts and their relationship to substorm evolution. J. Geophys. Res. 98 (A4), 6069–6078. https://doi.org/10.1029/92JA01975.

André, M., Norqvist, P., Andersson, L., Eliasson, L., Eriksson, A.I., Blomberg, L., Erlandson, R.E., Waldemark, J., 1998. Ion energization mechanisms at 1700 km in the auroral region. J. Geophys. Res. Space Phys. 103 (A3), 4199–4222. https://doi.org/10.1029/97JA00855.

Axford, W.I., 1968. The polar wind and the terrestrial helium budget. J. Geophys. Res. 73, 6855.

Baker, D.N., Hones Jr., E.W., Young, D.T., Birn, J., 1982. The possible role of ionospheric oxygen in the initiation and development of plasma sheet instabilities. Geophys. Res. Lett. 9, 1337–1340. https://doi.org/10.1029/GL009i012p01337.

Baker, J.B.H., Zhang, Y., Greenwald, R.A., Paxton, L.J., Morrison, D., 2004. Height-integrated Joule and auroral particle heating in the night side high latitude thermosphere. Geophys. Res. Lett. 31. https://doi.org/10.1029/2004GL019,535.

Balthazor, R.L., Moffett, R.J., 1997. A study of atmospheric gravity waves and travelling ionospheric disturbances at equatorial latitudes. Ann. Geophys. 15, 1048–1056. https://doi.org/10.1007/s00585-997-1048-4.

Banks, P.M., Holzer, T.E., 1968. The polar wind. J. Geophys. Res. 73, 6846–6854. https://doi.org/10.1029/JA073i021p06846.

Barakat, A.R., Barghouthi, I.A., 1994. The effect of wave-particle interactions on the polar winds O(+). Geophys. Res. Lett. 21, 2279–2282.

Barakat, A.R., Barghouthi, I.A., Schunk, R.W., 1995. Double-hump H^+ velocity distribution in the polar wind. Geophys. Res. Lett. 22, 1857–1860. https://doi.org/10.1029/95GL01519.

Barakat, A.R., Eccles, J.V., Schunk, R.W., 2015. Effects of geographic-geomagnetic pole offset on ionospheric outflow: can the ionosphere wag the magnetospheric tail? Geophys. Res. Lett. 42 (20), 8288–8293. https://doi.org/10.1002/2015GL065736.

Bashir, M.F., Ilie, R., 2018. A new n+ band of electromagnetic ion cyclotron waves in multi-ion cold plasmas. Geophys. Res. Lett. 45 (19), 10150–10159. https://doi.org/10.1029/2018GL080280.

Birn, J., Hesse, M., 2005. Energy release and conversion by reconnection in the magnetotail. Ann. Geophys. 23, 3365–3373.

Boltzmann, L., 1872. Weitere studien über das wärmegleichgewicht unter gasmolekülen. Sitz. Math. Naturwiss. Cl. Akad. Wiss. Wien 66, 275–370.

Borovsky, J.E., Denton, M.H., Denton, R.E., Jordanova, V.K., Krall, J., 2013. Estimating the effects of ionospheric plasma on solar wind/magnetosphere coupling via mass loading of dayside reconnection:

ion-plasma-sheet oxygen, plasmaspheric drainage plumes, and the plasma cloak. J. Geophys. Res. Space Phys. 118 (9), 5695–5719. https://doi.org/10.1002/jgra.50527.

Brambles, O.J., Lotko, W., Zhang, B., Wiltberger, M., Lyon, J., Strangeway, R.J., 2011. Magnetosphere sawtooth oscillations induced by ionospheric outflow. Science 332 (6034), 1183–1186.

Brekke, A., Rino, C.L., 1978. High-resolution altitude profiles of the auroral zone energy dissipation due to ionospheric currents. J. Geophys. Res. 83 (A6), 2517–2524. https://doi.org/10.1029/JA083iA06p02517.

Brinton, H.C., Grebowsky, J.M., Mayr, H.G., 1971. Altitude variation of ion composition in the midlatitude trough region: evidence for upward plasma flow. J. Geophys. Res. 76, 3738–3745. https://doi.org/10.1029/JA076i016p03738.

Carlson, H.C., Spain, T., Aruliah, A., Skjaeveland, A., Moen, J., 2012. First-principles physics of cusp/polar cap thermospheric disturbances. Geophys. Res. Lett. 39. https://doi.org/10.1029/2012GL053034, L19103.

Carpenter, D., 1966. Whistler studies of the plasma-pause in the magnetosphere, 1. Temporal variations in the position of the knee and some evidence of plasma motions near the knee. J. Geophys. Res. 71, 693.

Chang, T., Coppi, B., 1981. Lower hybrid acceleration and ion evolution in the suprauroral region. Geophys. Res. Lett. 8 (12), 1253–1256. https://doi.org/10.1029/GL008i012p01253.

Chang, T., Crew, G.B., Hershkowitz, N., Jasperse, J.R., Retterer, J.M., Winningham, J.D., 1986. Transverse acceleration of oxygen ions by electromagnetic ion cyclotron resonance with broad band left-hand polarized waves. Geophys. Res. Lett. 13 (7), 636–639. https://doi.org/10.1029/GL013i007p00636.

Chappell, C.R., Moore, T.E., Waite, J.H., 1987. The ionosphere as a fully adequate source of plasma for the earth's magnetosphere. J. Geophys. Res. 92, 5896–5910.

Chaston, C.C., Bonnell, J.W., Carlson, C.W., McFadden, J.P., Ergun, R.E., Strangeway, R.J., Lund, E.J., 2004. Auroral ion acceleration in dispersive Alfvén waves. J. Geophys. Res. Space Phys. 109 (A4). https://doi.org/10.1029/2003JA010053, A04205.

Chaston, C.C., Carlson, C.W., McFadden, J.P., Ergun, R.E., Strangeway, R.J., 2007. How important are dispersive alfvén waves for auroral particle acceleration? Geophys. Res. Lett. 34 (7). https://doi.org/10.1029/2006GL029144.

Chen, Y.-J., Heelis, R.A., 2018. Mesoscale plasma convection perturbations in the high-latitude ionosphere. J. Geophys. Res. Space Phys. 123 (9), 7609–7620.

Chen, M.W., Lemon, C.L., Guild, T.B., Keesee, A.M., Lui, A., Goldstein, J., Rodriguez, J.V., Anderson, P. C., 2015. Effects of modeled ionospheric conductance and electron loss on self-consistent ring current simulations during the 5–7 April 2010 storm. J. Geophys. Res. Space Phys. 120, 5355–5376. https://doi.org/10.1002/2015JA021285.

Cladis, J.B., 1986. Parallel acceleration and transport of ions from polar ionosphere to plasma sheet. Geophys. Res. Lett. 13, 893–896.

Claudepierre, S.G., Toffoletto, F.R., Wiltberger, M., 2016. Global mhd modeling of resonant ulf waves: simulations with and without a plasmasphere. J. Geophys. Res. Space Phys. 121 (1), 227–244. https://doi.org/10.1002/2015JA022048.

Codrescu, M.V., Fuller-Rowell, T.J., Foster, J.C., 1995. On the importance of E-field variability for joule heating in the high-latitude thermosphere. Geophys. Res. Lett. 22, 2393.

Codrescu, M.V., Fuller-Rowell, T.J., Foster, J.C., Holt, J.M., Cariglia, S.J., 2000. Electric field variability associated with the Millstone Hill electric field model. J. Geophys. Res. 105, 5265.

Codrescu, M.V., Fuller-Rowell, T.J., Munteanu, V., Minter, C.F., Millward, G.H., 2008. Validation of the coupled thermosphere ionosphere plasmasphere electrodynamics model: CTIPE-mass spectrometer incoherent scatter temperature comparison. Space Weather 6, 9005. https://doi.org/10.1029/2007SW000364.

Cohen, O., Glocer, A., 2012. Ambipolar electric field, photoelectrons, and their role in atmospheric escape from hot jupiters. Astrophys. J. 753 (1), L4. https://doi.org/10.1088/2041-8205/753/1/l4.

Cosgrove, R.B., Thayer, J.P., 2006. Parametric dependence of electric field variability in the sondrestrom database: a linear relation with kp. J. Geophys. Res. Space Phys. 111 (A10).

Cosgrove, R., McCready, M., Tsunoda, R., Stromme, A., 2011. The bias on the joule heating estimate: small-scale variability versus resolved-scale model uncertainty and the correlation of electric field and conductance. J. Geophys. Res. Space Phys. 116 (A9).

Cousins, E., Shepherd, S., 2012a. Statistical characteristics of small-scale spatial and temporal electric field variability in the high-latitude ionosphere. J. Geophys. Res. Space Phys. 117 (A3).

Cousins, E., Shepherd, S., 2012b. Statistical maps of small-scale electric field variability in the high-latitude ionosphere. J. Geophys. Res. Space Phys. 117 (A12).

Cousins, E., Matsuo, T., Richmond, A., 2013. Mesoscale and large-scale variability in high-latitude ionospheric convection: dominant modes and spatial/temporal coherence. J. Geophys. Res. Space Phys. 118 (12), 7895–7904.

Crew, G.B., Chang, T., Retterer, J.M., Peterson, W.K., Gurnett, D.A., 1990. Ion cyclotron resonance heated conics—theory and observations. J. Geophys. Res. 95, 3959.

Daglis, I.A., Sarris, E.T., Wilken, B., 1993. AMPTE/CCE CHEM observations of the energetic ion population at geosynchronous altitudes. Ann. Geophys. 11, 685–696.

Daglis, I.A., Thorne, R.M., Baumjohann, W., Orsini, S., 1999. The terrestrial ring current: origin, formation, and decay. Rev. Geophys. 37, 407. https://doi.org/10.1029/1999RG900009. (c) 1999: American Geophysical Union.

Demars, H.G., Barakat, A.R., Schunk, R.W., 1996. Effect of centrifugal acceleration on the polar wind. J. Geophys. Res. Space Phys. 101 (A11), 24565–24571. https://doi.org/10.1029/96JA02234.

Deng, Y., Richmond, A.D., Ridley, A.J., Liu, H.-L., 2008. Assessment of the non-hydrostatic effect on the upper atmosphere using a general circulation model (GCM). Geophys. Res. Lett. 35, 1104. https://doi.org/10.1029/2007GL032182.

Deng, Y., Maute, A., Richmond, A.D., Roble, R.G., 2009a. Impact of electric field variability on Joule heating and thermospheric temperature and density. Geophys. Res. Lett. 36, 8105. https://doi.org/10.1029/2008GL036916.

Deng, Y., Lu, G., Kwak, Y.-.S., Sutton, E., Forbes, J., Solomon, S., 2009b. Reversed ionospheric convections during the November 2004 storm: impact on the upper atmosphere. J. Geophys. Res. 114. https://doi.org/10.1029/2008JA013793, A07313.

Deng, Y., Fuller-Rowell, T.J., Ridley, A.J., Knipp, D., Lopez, R.E., 2013. Theoretical study: influence of different energy sources on the cusp neutral density enhancement. J. Geophys. Res. Space Phys. 118, 2340–2349. https://doi.org/10.1002/jgra.50197.

Deng, Y., Heelis, R., Lyons, L.R., Nishimura, Y., Gabrielse, C., 2019. Impact of flow bursts in the auroral zone on the ionosphere and thermosphere. J. Geophys. Res. Space Phys. 124 (12), 10459–10467.

Dessler, A.J., Cloutier, P.A., 1969. Discussion of letter by peter m. banks and Thomas e. holzer, 'the polar wind'. J. Geophys. Res. 74 (14), 3730–3733. https://doi.org/10.1029/JA074i014p03730.

Donahue, T.M., 1971. Polar ion flow: wind or breeze? Rev. Geophys. 9 (1), 1–9. https://doi.org/10.1029/RG009i001p00001.

Dungey, J., 1961. Interplanetary magnetic field and the auroral zones. Phys. Rev. Lett. 93, 47.

Emery, B.A., Lathuillere, C., Richards, P.G., Roble, R.G., Buonsanto, M.J., Knipp, D.J., Wilkinson, P., Sipler, D.P., Niciejewski, R., 1999. Time dependent thermospheric neutral response to the 2–11 November 1993 storm period. J. Atmos. Sol. Terr. Phys. 61 (3–4), 329–350.

Evans, D., Maynard, N., Trøim, J., Jacobsen, T., Egeland, A., 1977. Auroral vector electric field and particle comparisons, 2, electrodynamics of an arc. J. Geophys. Res. 82 (16), 2235–2249.

Feldstein, Y.I., Dremukhina, L.A., Levitin, A.E., Mall, U., Alexeev, I.I., Kalegaev, V.V., 2003. Energetics of the magnetosphere during the magnetic storm. J. Atmos. Sol. Terr. Phys. 65, 429–446.

Fujita, S., Tanaka, T., Kikuchi, T., Fujimoto, K., Hosokawa, K., Itonaga, M., 2003. A numerical simulation of the geomagnetic sudden commencement: 1. Generation of the field-aligned current associated with the preliminary impulse. J. Geophys. Res. 108, 1416. https://doi.org/10.1029/2002JA009407. A12.

Fuller-Rowell, T.J., Evans, D., 1987. Height-integrated Pedersen and Hall conductivity patterns inferred from TIROS–NOAA satellite data. J. Geophys. Res. 92, 7606.

Gabrielse, C., Nishimura, Y., Lyons, L., Gallardo-Lacourt, B., Deng, Y., Donovan, E., 2018. Statistical properties of mesoscale plasma flows in the nightside high-latitude ionosphere. J. Geophys. Res. Space Phys. 123, 6798–6820. https://doi.org/10.1002/2018JA025440.

Gallardo-Lacourt, B., Nishimura, Y., Lyons, L.R., Zou, S., Angelopoulos, V., Donovan, E., McWilliams, K.A., Ruohoniemi, J.M., Nishitani, N., 2014. Coordinated SuperDARN THEMIS ASI observations of mesoscale flow bursts associated with auroral streamers. J. Geophys. Res. Space Phys. 119, 142–150. https://doi.org/10.1002/2013JA019245.

Ganushkina, N.Y., Liemohn, M.W., Dubyagin, S., 2018. Current systems in the Earth's magnetosphere. Rev. Geophys. 56, 309–332. https://doi.org/10.1002/2017RG000590.

Gary, J.B., Heelis, R.A., Thayer, J.P., 1995. Summary of field-aligned poynting flux observations from DE 2. Geophys. Res. Lett. 22, 1861.
Glocer, A., Toth, G., Gombosi, T., Welling, D., 2009a. Modeling ionospheric outflows and their impact on the magnetosphere, initial results. J. Geophys. Res. 114 (A05216). https://doi.org/10.1029/2009JA014053.
Glocer, A., Tóth, G., Ma, Y., Gombosi, T., Zhang, J.-C., Kistler, L.M., 2009b. Multifluid block-adaptive-tree solar wind roe-type upwind scheme: magnetospheric composition and dynamics during geomagnetic storms—initial results. J. Geophys. Res. Space Phys. 114 (A13). https://doi.org/10.1029/2009JA014418, A12203.
Glocer, A., Kitamura, N., Toth, G., Gombosi, T., 2012. Modeling solar zenith angle effects on the polar wind. J. Geophys. Res. Space Phys. 117 (A4), a04318. https://doi.org/10.1029/2011JA017136.
Glocer, A., Toth, G., Fok, M.-C., 2018. Including kinetic ion effects in the coupled global ionospheric outflow solution. J. Geophys. Res. Space Phys. 123 (4), 2851–2871. https://doi.org/10.1002/2018JA025241.
Glocer, A., Welling, D., Chappell, C.R., Toth, G., Fok, M.C., Komar, C., Kang, S.B., Buzulukova, N., Ferradas, C., Bingham, S., Mouikis, C., 2020. A case study on the origin of near-earth plasma. J. Geophys. Res. Space Phys. 125 (11). https://doi.org/10.1029/2020JA028205, e28205.
Gloeckler, G., Hamilton, D.C., 1987. AMPTE ion composition results. Phys. Scr. T18, 73–84. https://doi.org/10.1088/0031-8949/1987/t18/009.
Golovchanskaya, I.V., 2008. Assessment of Joule heating for the observed distributions of high-latitude electric fields. Geophys. Res. Lett. 35, 16102. https://doi.org/10.1029/2008GL034413.
Golovchanskaya, I., Maltsev, Y., Ostapenko, A., 2002. High-latitude irregularities of the magneto-spheric electric field and their relation to solar wind and geomagnetic conditions. J. Geophys. Res. Space Phys. 107 (A1). SMP–1.
Gombosi, T.I., Killeen, T.L., 1987. Effects of thermospheric motions on the polar wind: a time-dependent numerical study. J. Geophys. Res. 92 (A5), 4725–4729.
Gombosi, T.I., Nagy, A., 1989. Time-dependent modeling of field aligned current-generated ion transients in the polar wind. J. Geophys. Res. 94, 359–369.
Gombosi, T., Powell, K.G., Zeeuw, D.L.D., Clauer, R., Hansen, K.C., Manchester, W.B., Ridley, A., Roussev, I.I., Sokolov, I.V., Stout, Q.F., Tóth, G., 2004. Solution-adaptive magnetohydro-dynamics for space plasmas: sun-to-earth simulations, computing in science and engineering. Front. Simul., 14.
Guglielmi, A., Kangas, J., Mursula, K., Pikkarainen, T., Pokhotelov, O., Potapov, A., 1996. Pc 1 induced electromagnetic lift of background plasma in the magnetosphere. J. Geophys. Res. 101, 21493–21500. https://doi.org/10.1029/96JA01750.
Hamilton, D.C., Gloeckler, G., Ipavich, F.M., Wilken, B., Stuedemann, W., 1988. Ring current development during the great geomagnetic storm of February 1986. J. Geophys. Res. 93, 14343–14355. https://doi.org/10.1029/JA093iA12p14343.
Hanson, W., Heelis, R., Power, R., Lippincott, C., Zuccaro, D., Holt, B., Harmon, L., Heelis, R.A., Lowell, J.K., Spiro, R.W., 1982. A model of the high-latitude ionospheric convection pattern. J. Geophys. Res. 87, 6339.
Heelis, R., Hanson, W., Lippincott, C., Zuccaro, D., Harmon, L., Holt, B., Doherty, J., Power, R., 1981. The ion drift meter for dynamics explorer-b. Space Sci. Instrum. 5, 511–521.
Hocke, K., Schlegel, K., 1996. A review of atmospheric gravity waves and travelling ionospheric disturbances: 1982–1995. Ann. Geophys. 14, 917–940. https://doi.org/10.1007/s00585-996-0917-6.
Hoffman, J.H., 1970. Studies of the composition of the ionosphere with a magnetic deflection mass spectrometer. Int. J. Mass Spectrom. 4 (315).
Hoffman, J.H., Dodson, W.H., Lippincott, C.R., Hammack, H.D., 1974. Initial ion composition results from the Isis 2 satellite. J. Geophys. Res. 79, 4246–4251. https://doi.org/10.1029/JA079i028p04246.
Holzer, T.E., Fedder, J.A., Banks, P.M., 1971. A comparison of kinetic and hydrodynamic models of an expanding ion-exosphere. J. Geophys. Res. 76 (10), 2453–2468. https://doi.org/10.1029/JA076i010p02453.
Horwitz, J.L., Ho, C.W., Scarbro, H.D., Wilson, G.R., Moore, T.E., 1994. Centrifugal acceleration of the polar wind. J. Geophys. Res. 99, 15051–15064. https://doi.org/10.1029/94JA00924.

Huddleston, M.M., Chappell, C.R., Delcourt, D.C., Moore, T.E., Giles, B.L., Chandler, M.O., 2005. An examination of the process and magnitude of ionospheric plasma supply to the magnetosphere. J. Geophys. Res. 110, 12202. https://doi.org/10.1029/2004JA010401.

Hull, A.J., Bonnell, J.W., Mozer, F.S., Scudder, J.D., 2003. A statistical study of large-amplitude parallel electric fields in the upward current region of the auroral acceleration region. J. Geophys. Res. 108 (A1), 1007. https://doi.org/10.1029/2001JA007540.

Hull, A.J., Chaston, C.C., Bonnell, J.W., Wygant, J.R., Kletzing, C.A., Reeves, G.D., Gerrard, A., 2019. Dispersive alfvén wave control of o+ ion outflow and energy densities in the inner magnetosphere. Geophys. Res. Lett. 46 (15), 8597–8606. https://doi.org/10.1029/2019GL083808.

Hull, A.J., Chaston, C.C., Bonnell, J.W., Damiano, P.A., Wygant, J.R., Reeves, G.D., 2020. Correlations between dispersive alfvén wave activity, electron energization, and ion outflow in the inner magnetosphere. Geophys. Res. Lett. 47 (17). https://doi.org/10.1029/2020GL088985. e2020GL088985 .

Hultqvist, B., 1996. On the acceleration of positive ions by high-latitude, large-amplitude electric field fluctuations. J. Geophys. Res. 101 (A12), 27111–27122. https://doi.org/10.1029/96JA02435.

Jensen, J.B., Raeder, J., Maynard, K., Cramer, W.D., 2017. Particle precipitation effects on convection and the magnetic reconnection rate in Earth's magnetosphere. J. Geophys. Res. Space Phys. 122, 11413–11427. https://doi.org/10.1002/2017JA024030.

Jin, Y., Moen, J.I., Spicher, A., Oksavik, K., Miloch, W.J., Clausen, L.B.N., et al., 2019. Simultaneous rocket and scintillation observations of plasma irregularities associated with a reversed flow event in the cusp ionosphere. J. Geophys. Res. Space Phys. 124, 7098–7111. https://doi.org/10.1029/2019JA026942.

Johnson, E.S., Heelis, R.A., 2005. Characteristics of ion velocity structure at high latitudes during steady southward interplanetary magnetic field conditions. J. Geophys. Res. 110. https://doi.org/10.1029/2005JA011130.

Jordanova, V., Kistler, L., Kozyra, J., Khazanov, G., Nagy, A., 1996. Collisional losses of ring current ions. J. Geophys. Res. 101, 111.

Kelley, M.C., Knudsen, D.J., Vickrey, J.F., 1991. Poynting flux measurements on a satellite—a diagnostic tool for space research. J. Geophys. Res. 96, 201.

Khazanov, G.V., Neubert, T., Gefan, G.D., 1994. Unified theory of ionosphere-plasmasphere transport of suprathermal electrons. IEEE Trans. Plasma Sci. 22, 187–198. https://doi.org/10.1109/27.279022.

Khazanov, G.V., Liemohn, M.W., Moore, T.E., 1997. Photoelectron effects on the self-consistent potential in the collisionless polar wind. J. Geophys. Res. 102, 7509–7522. https://doi.org/10.1029/96JA03343.

Khazanov, G.V., Khabibrakhmanov, I.K., Glocer, A., 2012. Kinetic description of ionospheric outflows based on the exact form of fokker-planck collision operator: electrons. J. Geophys. Res. Space Phys. 117 (A11). https://doi.org/10.1029/2012JA018082.

Khazanov, G.V., Krivorutsky, E.N., Sibeck, D.G., 2019a. Formation of the potential jump over the geomagnetically quiet sunlit polar cap region. J. Geophys. Res. Space Phys. 124 (6), 4384–4401. https://doi.org/10.1029/2019JA026576.

Khazanov, G.V., Sibeck, D.G., Zesta, E., 2019b. The formation of electron heat flux over the sunlit quiet polar cap ionosphere. Geophys. Res. Lett. 46 (17–18), 10201–10208. https://doi.org/10.1029/2019GL084522.

Killeen, T., Roble, R., 1984. An analysis of the high-latitude thermospheric wind pattern calculated by a thermospheric general circulation model 1. Momentum forcing. J. Geophys. Res. 89, 7509.

Kintner, P.M., 1976. Observations of velocity shear driven plasma turbulence. J. Geophys. Res. 81 (28), 5114–5122. https://doi.org/10.1029/JA081i028p05114.

Kintner, P.M., Seyler, C.E., 1985. The status of observations and theory of high latitude ionospheric and magnetospheric plasma turbulence. Space Sci. Rev. 41 (1), 91–129. https://doi.org/10.1007/BF00241347.

Kistler, L.M., Mouikis, C., Möbius, E., Klecker, B., Sauvaud, J.A., Réme, H., Korth, A., Marcucci, M.F., Lundin, R., Parks, G.K., Balogh, A., 2005. Contribution of nonadiabatic ions to the cross-tail current in an Ô+ dominated thin current sheet. J. Geophys. Res. Space Phys. 110, 6213. https://doi.org/10.1029/2004JA010653.

Kitamura, N., Nishimura, Y., Ono, T., Ebihara, Y., Terada, N., Shinbori, A., Kumamoto, A., Abe, T., Yamada, M., Watanabe, S., Matsuoka, A., Yau, A.W., 2010. Observations of very-low-energy (< 10 ev) ion outflows dominated by o + ions in the region of enhanced electron density in the polar cap magnetosphere during geomagnetic storms. J. Geophys. Res. Space Phys. 115 (A11). https://doi.org/10.1029/2010JA015601.

Kitamura, N., Ogawa, Y., Nishimura, Y., Terada, N., Ono, T., Shinbori, A., Kumamoto, A., Truhlik, V., Smilauer, J., 2011. Solar zenith angle dependence of plasma density and temperature in the polar cap ionosphere and low-altitude magnetosphere during geomagnetically quiet periods at solar maximum. J. Geophys. Res. 116. https://doi.org/10.1029/2011JA016631, AO8227.

Kitamura, N., Seki, K., Nishimura, Y., Terada, N., Ono, T., Hori, T., Strangeway, R.J., 2012. Photoelectron flows in the polar wind during geomagnetically quiet periods. J. Geophys. Res. Space Phys. 117 (A7). https://doi.org/10.1029/2011JA017459.

Kivanç, Ö., Heelis, R.A., 1997. Structures in ionospheric number density and velocity associated with polar cap ionization patches. J. Geophys. Res. 102 (A1), 307–318, https://doi.org/10.1029/96JA03141.

Kivelson, M.G., Ridley, A.J., 2008. Saturation of the polar cap potential: inference from Alfvén wing arguments. J. Geophys. Res. 113. https://doi.org/10.1029/2007JA012302, A05214.

Klumpar, D.M., Peterson, W.K., Shelley, E.G., 1984. Direct evidence for two-stage (bimodal) acceleration of ionospheric ions. J. Geophys. Res. Space Phys. 89 (A12), 10779–10787. https://doi.org/10.1029/JA089iA12p10779.

Knipp, D.J., et al., 1998. An overview of the early November 1993 geomagnetic storm. J. Geophys. Res. 103 (A11), 26197–26220. https://doi.org/10.1029/98JA00762.

Kolstø, H.M., Hesse, M., Norgren, C., Tenfjord, P., Spinnangr, S.F., Kwagala, N., 2020. Collisionless magnetic reconnection in an asymmetric oxygen density configuration. Geophys. Res. Lett. 47 (1). https://doi.org/10.1029/2019GL085359. e2019GL085359.

Korth, H., Zhang, Y., Anderson, B.J., Sotirelis, T., Waters, C.L., 2014. Statistical relationship between large-scale upward field-aligned currents and electron precipitation. J. Geophys. Res. Space Phys. 119, 6715–6731. https://doi.org/10.1002/2014JA01996.

Kozelov, B.V., Uritsky, V.M., Klimas, A.J., 2004. Power law probability distributions of multiscale auroral dynamics from ground-based TV observations. Geophys. Res. Lett. 31. https://doi.org/10.1029/2004GL020962, L20804.

Kozyra, J.U., Cravens, T.E., Nagy, A.F., Fontheim, E.G., Ong, R.S.B., 1984. Effects of energetic heavy ions on electromagnetic ion cyclotron wave generation in the plasmapause region. J. Geophys. Res. 89, 2217–2233. https://doi.org/10.1029/JA089iA04p02217.

Kraichnan, R.H., Montgomery, D., 1980. Two-dimensional turbulence. Rep. Prog. Phys. 43, 547–619.

Lee, J.S., Doering, J.P., Potemra, T.A., Brace, L.H., 1980. Measurements of the ambient photoelectron spectrum from atmosphere explorer: II. AE-E measurements from 300 to 1000 km during solar minimum conditions. Planet. Space Sci. 28, 973–996. https://doi.org/10.1016/0032-0633(80)90059-8.

Lemaire, J., 1972. Effect of escaping photoelectrons in a polar exospheric model. Space Res. 12, 1413–1416.

Lemaire, J., Scherer, M., 1970. Model of the polar ion-exosphere. Planet. Space Sci. 18 (1), 103–120. https://doi.org/10.1016/0032-0633(70)90070-X.

Lemaire, J., Scherer, M., 1973. Kinetic models of the solar and polar winds. Rev. Geophys. 11 (2), 427–468. https://doi.org/10.1029/RG011i002p00427.

Lennartsson, O.W., Collin, H.L., Peterson, W.K., 2004. Solar wind control of earth's h + and o + outflow rates in the 15-ev to 33-kev energy range. J. Geophys. Res. Space Phys. 109 (A12). https://doi.org/10.1029/2004JA010690.

Li, X., Roth, I., Temerin, M., Wygant, J.G., Hudson, M.K., et al., 1993. Simulation of the prompt energization and transport of radiation belt particles during the March 24, 1991 SSC. Geophys. Res. Lett. 20, 2423.

Liang, H., Lapenta, G., Walker, R.J., Schriver, D., El-Alaoui, M., Berchem, J., 2017. Oxygen acceleration in magnetotail reconnection. J. Geophys. Res. Space Phys. 122 (1), 618–639. https://doi.org/10.1002/2016JA023060.

Liemohn, M.W., Khazanov, G.V., Moore, T.E., Guiter, S.M., 1997. Self-consistent superthermal electron effects on plasmaspheric refilling. J. Geophys. Res. 102, 7523–7536. https://doi.org/10.1029/96JA03962.

Lin, M.-Y., Ilie, R., Glocer, A., 2020. The contribution of n+ ions to earth's polar wind. Geophys. Res. Lett. 47 (18). https://doi.org/10.1029/2020GL089321. e2020GL089321.

Lopez, R.E., 2016. The integrated dayside merging rate is controlled primarily by the solar wind. J. Geophys. Res. Space Phys. 121, 4435–4445. https://doi.org/10.1002/2016JA022556.

Lühr, H., Park, J., Ritter, P., Liu, H., 2012. In-situ CHAMP observation of ionosphere-thermosphere coupling. Space Sci. Rev. 168 (1–4), 237–260. https://doi.org/10.1007/s11214-011-9798-4.

Lund, E.J., Nowrouzi, N., Kistler, L.M., Cai, X., Frey, H.U., 2018. On the role of ionospheric ions in sawtooth events. J. Geophys. Res. Space Phys. 123 (1), 665–684. https://doi.org/10.1002/2017JA024378.

Lundin, R., Guglielmi, A., 2006. Ponderomotive forces in cosmos. Space Sci. Res. 127 (1–4), 1–116. https://doi.org/10.1007/s11214-006-8314-8.

Lundin, R., Hultqvist, B., 1989. Ionospheric plasma escape by high-altitude electric fields: magnetic moment "pumping". J. Geophys. Res. 94 (A6), 6665–6680. https://doi.org/10.1029/JA094iA06p06665.

Lyons, L., Nishimura, Y., Zhang, S.-R., Coster, A., Bhatt, A., Kendall, E., Deng, Y., 2019. Identification of auroral zone activity driving large? Scale traveling ionospheric disturbances. J. Geophys. Res. Space Phys. 124. https://doi.org/10.1029/2018JA025980.

Lysak, R.L., 1985. Auroral electrodynamics with current and voltage generators. J. Geophys. Res. Space Phys. 90 (A5), 4178–4190.

Maes, L., Maggiolo, R., De Keyser, J., 2016. Seasonal variations and north–south asymmetries in polar wind outflow due to solar illumination. Ann. Geophys. 34 (11), 961–974. https://doi.org/10.5194/angeo-34-961-2016.

Marubashi, K., 1970. Escape of the Polar-Ionospheric Plasma into the Magnetospheric Tail. Tech. Rep. Tokyo University.

Matsuo, T., Richmond, A.D., 2008. Effects of high-latitude ionospheric electric field variability on global thermospheric joule heating and mechanical energy transfer rate. J. Geophys. Res. 113, 7309. https://doi.org/10.1029/2007JA012993.

Matsuo, T., Richmond, A.D., Nychka, D.W., 2002. Modes of high-latitude electric field variability derived from de-2 measurements: empirical orthogonal function (eof) analysis. Geophys. Res. Lett. 29, 11–1.

Matsuo, T., Richmond, A.D., Hensel, K., 2003. High-latitude ionospheric electric field variability and electric potential derived from DE-2 plasma drift measurements: dependence on IMF and dipole tilt. J. Geophys. Res. Space Phys. 108, 1005. https://doi.org/10.1029/2002JA009429.

Maute, A., Richmond, A., France, J., St-Jean Leblanc, C., 2005. Empirical Model of High-Latitude Magnetosphere-Ionosphere Energy Transfer Based on Dynamic Explorer-2 Data. AGUFM. SA31A–0334.

McChesney, J.M., Stern, R.A., Bellan, P.M., 1987. Observation of fast stochastic ion heating by drift waves. Phys. Rev. Lett. 59, 1436–1439. https://doi.org/10.1103/PhysRevLett.59.1436.

McGranaghan, R., Knipp, D.J., Matsuo, T., 2016a. High-latitude ionospheric conductivity variability in three dimensions. Geophys. Res. Lett. 43 (15), 7867–7877.

McGranaghan, R., Knipp, D.J., Matsuo, T., Cousins, E., 2016b. Optimal interpolation analysis of high-latitude ionospheric hall and pedersen conductivities: application to assimilative ionospheric electrodynamics reconstruction. J. Geophys. Res. Space Phys. 121 (5), 4898–4923.

Miller, M.A., 1958. Motion of charged particles in the high-frequency electromagnetic fields. Radiophysics 1(3), 110–123.

Miller, R.H., Rasmussen, C.E., Combi, M.R., Gombosi, T.I., Winske, D., 1995. Ponderomotive acceleration in the auroral region: a kinetic simulation. J. Geophys. Res. Space Phys. 100 (A12), 23901–23916. https://doi.org/10.1029/95JA01908.

Moore, T.E., Delcourt, D.C., 1995. The geopause. Rev. Geophys. 33, 175–210. https://doi.org/10.1029/95RG00872.

Moore, T.E., Chandler, M.O., Fok, M.-C., Giles, B.L., Delcourt, D.C., Horwitz, J.L., Pollock, C.J., 2001. Ring currents and internal plasma sources. Space Sci. Rev. 95, 555–568.

Moore, T.E., Fok, M.-C., Chandler, M.O., Chappell, C.R., Christon, S.P., Delcourt, D.C., Fedder, J., Huddleston, M., Liemohn, M., Peterson, W.K., Slinker, S., 2005. Plasma sheet and (nonstorm) ring current formation from solar and polar wind sources. J. Geophys. Res. Space Phys. 110 (A2). https://doi.org/10.1029/2004JA010563.

Newell, P.T., Liou, K., Wilson, G.R., 2009. Polar cap particle precipitation and aurora: review and commentary. J. Atmos. Sol. Terr. Phys. 71, 199–215. https://doi.org/10.1016/j.jastp.2008.11.004.

Newell, P.T., Liou, K., Zhang, Y., Sotirelis, T., Paxton, L.J., Mitchell, E.J., 2014. OVATION prime-2013: extension of auroral precipitation model to higher disturbance levels. Space Weather 12, 368–379. https://doi.org/10.1002/2014SW001056.

Nosé, M., Ieda, A., Christon, S.P., 2009a. Geotail observations of plasma sheet ion composition over 16 years: on variations of average plasma ion mass and o+ triggering substorm model. J. Geophys. Res. Space Phys. 114 (A7). https://doi.org/10.1029/2009JA014203.

Nosé, M., Taguchi, S., Christon, S.P., Collier, M.R., Moore, T.E., Carlson, C.W., McFadden, J.P., 2009b. Response of ions of ionospheric origin to storm time substorms: coordinated observations over the ionosphere and in the plasma sheet. J. Geophys. Res. Space Phys. 114, 5207. https://doi.org/10.1029/2009JA014048.

Ofman, L., Denton, R.E., Bortnik, J., An, X., Glocer, A., Komar, C., 2017. Growth and nonlinear saturation of electromagnetic ion cyclotron waves in multi-ion species magnetospheric plasma. J. Geophys. Res. Space Phys. 122 (6), 6469–6484. https://doi.org/10.1002/2017JA024172.

Oksavik, K., Søraas, F., Moen, J., Pfaff, R., Davies, J.A., Lester, M., 2004. Simultaneous optical, CUTLASS HF radar, and FAST spacecraft observations: signatures of boundary layer processes in the cusp. Ann. Geophys. 22 (2), 511–525. https://doi.org/10.5194/angeo-22-511-2004.

Oppenheim, M.M., Dimant, Y.S., 2013. Kinetic simulations of 3-D Farley-Buneman turbulence and anomalous electron heating. J. Geophys. Res. Space Phys. 118, 1306–1318. https://doi.org/10.1002/jgra.50196.

Østgaard, N., Germany, G., Stadsnes, J., Vondrak, R.R., 2002a. Energy analysis of substorms based on remote sensing techniques, solar wind measurements, and geomagnetic indices. J. Geophys. Res. 107 (A9), 1233. https://doi.org/10.1029/2001JA002002.

Østgaard, N., Vondrak, R.R., Gjerloev, J.W., Germany, G., 2002b. A relation between the energy deposition by electron precipitation and geomagnetic indices during substorms. J. Geophys. Res. 107 (A9), 1246. https://doi.org/10.1029/2001JA002003.

Peroomian, V., El-Alaoui, M., Abdalla, M.A., Zelenyi, L.M., 2007. A comparison of solar wind and ionospheric plasma contributions to the september 24–25, 1998 magnetic storm. J. Atmos. Sol. Terr. Phys. 69 (3), 212–222. https://doi.org/10.1016/j.jastp.2006.07.025. Global Aspects of Magnetosphere-Ionosphere Coupling.

Peterson, W., 2002. Ionospheric influence on substorm development. In: Proceedings of the Sixth International Conference on Substorms. University of Washington, p. 143.

Peterson, W.K., Woods, T.N., Chamberlin, P.C., Richards, P.G., 2008. Photoelectron flux variations observed from the FAST satellite. Adv. Space Res. 42, 947–956. https://doi.org/10.1016/j.asr.2007.08.038.

Prikryl, P., Jayachandran, P.T., Chadwick, R., Kelly, T.D., 2015. Climatology of GPS phase scintillation at northern high latitudes for the period from 2008 to 2013. Ann. Geophys. 33, 531–545. https://doi.org/10.5194/angeo-33-531-2015.

Ream, J.B., Walker, R.J., Ashour-Abdalla, M., El-Alaoui, M., Wiltberger, M., Kivelson, M.G., Goldstein, M.L., 2015. Propagation of Pi2 pulsations through the braking region in global MHD simulations. J. Geophys. Res. Space Phys. 120, 10574–10591. https://doi.org/10.1002/2015JA021572.

Retterer, J.M., Chang, T., Crew, G.B., Jasperse, J.R., Winningham, J.D., 1987. Monte Carlo modeling of ionospheric oxygen acceleration by cyclotron resonance with broad-band electromagnetic turbulence. Phys. Rev. Lett. 59, 148–151. https://doi.org/10.1103/PhysRevLett.59.148.

Richmond, A.D., 1995. Ionospheric electrodynamics using magnetic apex coordinates. J. Geomagn. Geoelectr. 47, 191.

Richmond, A.D., Kamide, Y., 1988. Mapping electrodynamic features of the high-latitude ionosphere from localized observations: technique. J. Geophys. Res. 93, 5741.
Ridley, A.J., Gombosi, T.I., DeZeeuw, D.L., 2004. Ionospheric control of the magnetosphere: conductance. Ann. Geophys. 22 (2), 567–584. https://doi.org/10.5194/angeo-22-567-2004.
Ridley, A.J., Deng, Y., To'th, G., 2006. The global ionosphere thermosphere model. J. Atmos. Sol. Terr. Phys. 68, 839–864. https://doi.org/10.1016/j.jastp.2006.01.008.
Rino, C.L., 1979. A power law phase screen model for ionospheric scintillation: 1. Weak scatter. Radio Sci. 14 (6), 1135–1145. https://doi.org/10.1029/RS014i006p01135.
Robinson, R.M., Vondrak, R.R., Friis-Christensen, E., 1987. Ionospheric currents associated with a sun-aligned arc connected to the auroral oval. Geophys. Res. Lett. 14, 656–659. https://doi.org/10.1029/GL014i006p00656.
Schunk, R.W., Nagy, A.F., 2000. Ionospheres: Physics, Plasma Physics, and Chemistry. Cambridge University Press, Cambridge, UK.
Schunk, R.W., Sojka, J.J., 1997. Global ionosphere-polar wind system during changing magnetic activity. J. Geophys. Res. 102, 11625. https://doi.org/10.1029/97JA00292.
Schunk, R.W., Raitt, W.J., Banks, P.M., 1975. Effect of electric fields on the daytime high-latitude E and F regions. J. Geophys. Res. 80 (22), 3121–3130. https://doi.org/10.1029/JA080i022p03121.
Sergeev, V., Liou, K., Newell, P., Ohtani, S., Hairston, M., Rich, F., 2004. Auroral streamers: characteristics of associated precipitation,convection and field-aligned currents. Ann. Geophys. 22, 537–548. https://doi.org/10.5194/angeo-22-537-2004.
Shay, M.A., Drake, J.F., Swisdak, M., Rogers, B.N., 2004. The scaling of embedded collisionless reconnection. Phys. Plasmas 11, 2199. https://doi.org/10.1063/1.1705650.
Shelley, E., 1986. Magnetospheric energetic ions from the earth's ionosphere. Adv. Space Res. 6 (3), 121–132. https://doi.org/10.1016/0273-1177(86)90325-X.
Shelley, E.G., Johnson, R.G., Sharp, R.D., 1972. Satellite observations of energetic heavy ions during a geomagnetic storm. J. Geophys. Res. 77, 6104–6110. https://doi.org/10.1029/JA077i031p06104.
Shelley, E., Collin, H.I., Drake, J.K., Lennartsson, W., Yau, A., 1986. Origin of plasma sheet ions: substorm and solar cycle dependence, Eos. Trans. Am. Geophys. Union 67 (44), 1133. https://doi.org/10.1029/EO067i044p00867.
Shiokawa, K., Otsuka, Y., Ogawa, T., Balan, N., Igarashi, K., Ridley, A.J., Knipp, D.J., Saito, A., Yumoto, K., 2002. A large-scale traveling ionospheric distrubance during the magnetic storm of 15 September 1999. J. Geophys. Res. 107 (A6). https://doi.org/10.1029/2001JA000245.
Siscoe, G.L., Crooker, N.U., Siebert, K.D., 2002. Transpolar potential saturation: roles of region 1 current system and solar wind ram pressure. J. Geophys. Res. 107 (A10), 1321. https://doi.org/10.1029/2001JA009176.
Smith, P.H., Hoffman, R.A., Bewtra, N.K., 1981. Inference of the ring current ion composition by means of charge exchange decay. J. Geophys. Res. 86, 3470–3480. https://doi.org/10.1029/JA086iA05p03470.
Stasiewicz, K., Lundin, R., Marklund, G., 2000. Stochastic ion heating by orbit chaotization on electrostatic waves and nonlinear structures. Phys. Scr. T84 (1), 60. https://doi.org/10.1238/physica.topical.084a00060.
Strangeway, R.J., Ergun, R.E., Su, Y.-J., Carlson, C.W., Elphic, R.C., 2005. Factors controlling ionospheric outflows as observed at intermediate altitudes. J. Geophys. Res. Space Phys. 110, 3221. https://doi.org/10.1029/2004JA010829.
Su, Y.-J., Caton, R.G., Horwitz, J.L., Richards, P.G., 1999. Systematic modeling of soft-electron precipitation effects on high-latitude f region and topside ionospheric upflows. J. Geophys. Res. Space Phys. 104 (A1), 153–163. https://doi.org/10.1029/1998JA900068.
Tanaka, T., 2007. Magnetosphere–ionosphere convection as a compound system. Space Sci. Rev. 133, 1–72. https://doi.org/10.1007/s11214-007-9168-4.
Tanaka, T., Nakamizo, A., Yoshikawa, A., Fujita, S., Shinagawa, H., Shimazu, H., Kikuchi, T., Hashimoto, K.K., 2010. Substorm convection and current system deduced from the global simulation. J. Geophys. Res. 115. https://doi.org/10.1029/2009JA014676, A05220.
Tenfjord, P., Hesse, M., Norgren, C., 2018. The formation of an oxygen wave by magnetic reconnection. J. Geophys. Res. Space Phys. 123 (11), 9370–9380. https://doi.org/10.1029/2018JA026026.

Tenfjord, P., Hesse, M., Norgren, C., Spinnangr, S.F., Kolstø, H., 2019. The impact of oxygen on the reconnection rate. Geophys. Res. Lett. 46 (12), 6195–6203. https://doi.org/10.1029/2019GL082175.

Tenfjord, P., Hesse, M., Norgren, C., Spinnangr, S.F., Kolstø, H., Kwagala, N., 2020. Interaction of cold streaming protons with the reconnection process. J. Geophys. Res. Space Phys. 125 (6). https://doi.org/10.1029/2019JA027619. e2019JA027619.

Thomas, E.G., Shepherd, S.G., 2018. Statistical patterns of ionospheric convection derived from mid-latitude, high-latitude, and polar SuperDARN HF radar observations. J. Geophys. Res. Space Phys. 123, 3196–3216. https://doi.org/10.1002/2018JA025280.

Tyska, J., Lin, C.Y.-T., Deng, Y., Zhang, S., 2019. Volcano-generated ionospheric disturbances: comparison of GITM-R simulations with GNSS observations. In: AGU Fall Meeting 2019. AGU.

Varney, R.H., Solomon, S.C., Nicolls, M.J., 2014. Heating of the sunlit polar cap ionosphere by reflected photoelectrons. J. Geophys. Res. Space Phys. 119 (10), 8660–8684. https://doi.org/10.1002/2013JA019378.

Varney, R.H., Wiltberger, M., Zhang, B., Lotko, W., Lyon, J., 2016a. Influence of ion outflow in coupled geospace simulations: 1. Physics-based ion outflow model development and sensitivity study. J. Geophys. Res. Space Phys. 121 (10), 9671–9687. https://doi.org/10.1002/2016JA022777.

Varney, R.H., Wiltberger, M., Zhang, B., Lotko, W., Lyon, J., 2016b. Influence of ion outflow in coupled geospace simulations: 2. Sawtooth oscillations driven by physics-based ion outflow. J. Geophys. Res. Space Phys. 121 (10), 9688–9700. https://doi.org/10.1002/2016JA022778.

Vickrey, J., Livingston, R., Walker, N., Potemra, T., Heelis, R., Kelley, M., Rich, F., 1986. On the current-voltage relationship of the magnetospheric generator at intermediate spatial scales. Geophys. Res. Lett. 13 (6), 495–498.

Wallis, D.D., Budzinski, E.E., 1981. Empirical models of height-integrated conductivities. J. Geophys. Res. 86, 125.

Weimer, D.R., 2005. Improved ionospheric electrodynamic models and application to calculating joule heating rates. J. Geophys. Res. 110, 5306. https://doi.org/10.1029/2004JA010884.

Weimer, D., Edwards, T., Olsen, N., 2017. Linear response of field-aligned currents to the interplanetary electric field. J. Geophys. Res. Space Phys. 122 (8), 8502–8515.

Wiltberger, M., Lotko, W., Lyon, J.G., Damiano, P., Merkin, V., 2010. Influence of cusp o+ outflow on magnetotail dynamics in a multifluid mhd model of the magnetosphere. J. Geophys. Res. Space Phys. 115 (A10). https://doi.org/10.1029/2010JA015579, a00J05.

Wiltberger, M., et al., 2017. Effects of electrojet turbulence on a magnetosphere-ionosphere simulation of a geomagnetic storm. J. Geophys. Res. Space Phys. 122, 5008–5027. https://doi.org/10.1002/2016JA023700.

Wing, S., Johnson, J.R., Chaston, C.C., Echim, M., Escoubet, C.P., Lavraud, B., Lemon, C., Nykyri, K., Otto, A., Raeder, J., Wang, C.-.P., 2014. Review of solar wind entry into and transport within the plasma sheet. Space Sci. Rev. 184 (1), 33–86. https://doi.org/10.1007/s11214-014-0108-9.

Winglee, R.M., 1998. Multi-fluid simulations of the magnetosphere: the identification of the geopause and its variation with IMF. Geophys. Res. Lett. 25, 4441–4444.

Winningham, J., Burch, J., Eaker, N., Blevins, V., Hoffman, R., 1981. The low altitude plasma instrument (lapi). Space Sci. Instrum. 5, 465–475.

Yau, A.W., Whalen, B.A., Abe, T., Mukai, T., Oyama, K.I., Chang, T., 1995. Akebono observations of electron temperature anisotropy in the polar wind. J. Geophys. Res. 100 (A9), 17451–17464. https://doi.org/10.1029/95JA00855.

Yoshikawa, A., Itonaga, M., 2000. The nature of reflection and mode conversion of MHD waves in the inductive ionosphere: multistep mode conversion between divergent and rotational electric fields. J. Geophys. Res. 105 (A5), 10565–10584. https://doi.org/10.1029/1999JA000159.

Zhang, Y., Paxton, L., 2008. An empirical Kp-dependent global auroral model based on TIMED/GUVI FUV data. J. Atmos. Sol. Terr. Phys. 70, 1231–1242. https://doi.org/10.1016/j.jastp.2008.03.008.

Zhang, B., Brambles, O.J., Lotko, W., Lyon, J.G., 2020. Is nightside outflow required to induce magnetospheric sawtooth oscillations. Geophys. Res. Lett. 47 (6). https://doi.org/10.1029/2019GL086419. e2019GL086419.

Zhao, Y., Deng, Y., Wang, J.-S., Zhang, S.-R., Lin, C.Y., 2020. Tropical cyclone-induced gravity wave perturbations in the upper atmosphere: GITM-R simulation. J. Geophys. Res. Space Phys. 125 (12). e2019JA027675.

Zheng, Y., Moore, T.E., Mozer, F.S., Russell, C.T., Strangeway, R.J., 2005. Polar study of ionospheric ion outflow versus energy input. J. Geophys. Res. Space Phys. 110, 7210. https://doi.org/10.1029/2004JA010995.

Zhu, Q., Deng, Y., Richmond, A., Maute, A., 2018. Small-scale and mesoscale variabilities in the electric field and particle precipitation and their impacts on joule heating. J. Geophys. Res. Space Phys. 123 (11), 9862–9872.

CHAPTER 2

Auroral structures: Revealing the importance of meso-scale M-I coupling

Larry R. Lyons[a], **Bea Gallardo-Lacourt**[b,c], **and Yukitoshi Nishimura**[d]
[a]Department of Atmospheric and Oceanic Sciences, University of California, Los Angeles, CA, United States
[b]Universities of Space Research Association, Columbia, MD, United States
[c]NASA Goddard Space Flight Center, Greenbelt, MD, United States
[d]Department of Electrical and Computer Engineering and Center for Space Physics, Boston University, Boston, MA, United States

Chapter outline

2.1 Introduction

On a large scale, plasma convection has a relatively smooth two-cell configuration. Within the ionosphere, the flow is generally antisunward across the polar caps and returns to the dayside at auroral and subauroral latitudes. The configuration's shape and the flow strength are primarily determined by the interplanetary magnetic field (IMF) (Ruohoniemi and Greenwald, 2005, and references therein). Large-scale convection responds globally to IMF changes quite rapidly, delays for an initial response being only a few minutes at all magnetic local times (MLTs). The time until full reconfiguration of the shape to that of a new IMF is ~10 min, being somewhat shorter on the dayside than on the nightside (Murr and Hughes, 2001; Ruohoniemi et al., 2002; Xu et al., 2007).

Cross-Scale Coupling and Energy Transfer in the Magnetosphere-Ionosphere-Thermosphere System
https://doi.org/10.1016/B978-0-12-821366-7.00004-4

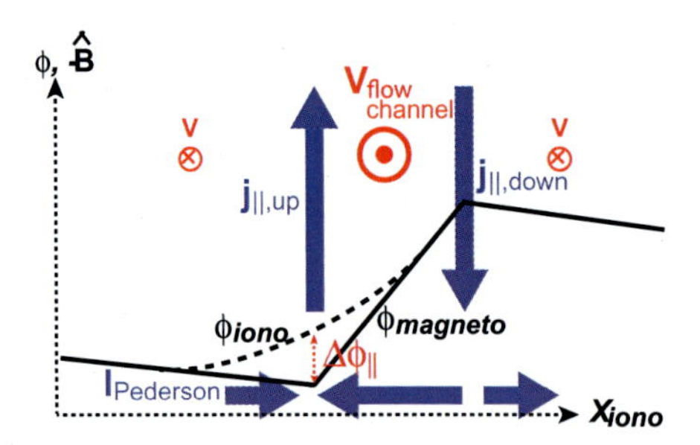

Fig. 2.1 Schematic illustration of the formation of auroral electron acceleration by a $\Delta\phi_{||}$. The *solid line* gives the electric potential as mapped along magnetic field lines to the ionosphere ($\phi_{magneto}$) of a magnetospheric flow channel flowing perpendicular to the *x*-axis. For illustration, the figure shows the flow (electric field) direction outside the flow channel being opposite to that within the flow channel. Such a potential distribution gives converging (diverging) height-integrated Pederson currents to the *right* (*left*) of the flow channel direction that connect to upward (downward) FACs $j_{||}$).

This response time is very much shorter than for an IMF change to propagate across the open field line region from the dayside to the nightside, that time being ~1.5 h for a ~400-km/s solar wind and 300-R_E length of the open field line region as mapped along magnetic field lines into the solar wind (length estimated for a 100 kV cross-polar cap potential ($\Delta\phi_{pc}$) following Dungey, 1965; Stern, 1973).

In addition to this large-scale structure, strong meso-scale flows (~1–2 R_E width in the equatorial plane) are known to be a major contributor to earthward plasma transport in the plasma sheet (Angelopoulos et al., 1992, 1994; Baumjohann et al., 1989). These flow structures extend along magnetic field lines from the magnetosphere to the ionosphere, and their changes occur nearly simultaneously in the magnetosphere to the ionosphere (Sergeev et al., 1990). In the ionosphere, the flows are associated with a nonzero divergence of ionospheric Pedersen currents driven by the electric field as illustrated in Fig. 2.1 for the northern hemisphere (with uniform Hall conductance for simplicity). Based on Lyons (1980), the solid line gives the electric potential as mapped along magnetic field lines to the ionosphere ($\phi_{magneto}$) of a magnetospheric flow channel flowing perpendicular to the *x*-axis. For illustration, the figure shows the flow (electric field) direction outside the flow channel being opposite to that within the flow channel, but the illustrated principal applies if the outside flow speed (electric field) is lower than within the flow channel but in the same direction. Such a potential distribution gives converging (diverging) height-integrated Pederson currents to the right (left) of the flow channel direction that connects to upward (downward) magnetic field-aligned currents (FACs) as illustrated in the figure. The $\Delta\phi_{||}$ forces more plasma sheet electrons into the loss cone and thus increases the upward FAC and also smooths the potential distribution in the ionosphere (ϕ_{iono}), as illustrated by the dashed curve in Fig. 2.1. This results in a horizontal region of nonzero $\Delta\phi_{||}$ that energizes electrons downward toward the atmosphere. The flow speed in the ionosphere is reduced, while the ionospheric conductance increases as a result of the electron precipitation, and both together contribute to maintain current continuity.

The electrons energized by the $\Delta\phi_{||}$ follow the magnetic field of the Earth down to the upper atmosphere where they collide with neutral oxygen and nitrogen atoms and molecules. During these collisions, the electrons transfer energy to the atmospheric particles upon which they collide, causing the atoms and molecules to be excited to higher energy states. When the atoms and molecules relax back down to a lower energy state, they release their energy in the form of light that, when bright enough, we can see with naked eyes as the aurora. The aurora extends from ~80 to ~500 km, the brightest auroral features typically occurring at altitudes of ~100–150 km.

The precipitating electrons that undergo energization from $\Delta\phi_{||}$ give rise to the brightest and most dynamic auroral features that are referred to as discrete auroral arcs, which have well-defined arc or ray structures. In contrast, diffuse auroras, which are caused by precipitating electrons without the further energization from $\Delta\phi_{||}$, are generally dimer and less structured.

As a result, discrete auroral arcs give a two-dimensional (2D) picture of the region of upward FACs that lies adjacent to meso-scale flow channels as mapped to the ionosphere, and give a time-dependent picture of the currents and adjacent flow channels. While not all flow channels are strong enough and have a sufficiently strong shear on their edge to form a discrete arc, those that do reveal important meso-scale flow structures occurring within the dayside cusp region, within the polar caps, and throughout the auroral oval.

2.2 Dayside cusp region

Heated solar wind electrons and protons from the dayside magnetosheath interact with the magnetosphere and precipitate into the upper atmosphere just poleward of the dayside boundary between open and closed magnetic field lines (Heikkila and Winningham, 1971). During southward IMF, antisunward convection causes the protons to precipitate with a characteristic energy that decreases with increasing latitudinal distance from the last closed field line. This is because protons with lower energies take longer time to reach the atmosphere, and thus convect further poleward during their transit to the atmosphere (Reiff et al., 1977). The high higher energy protons lead to a broad region of intense proton auroral emissions centered near local noon (Fuselier et al., 2002) that lies adjacent to the electron aurora. When the IMF is northward, the magnetosphere protons form a spot poleward of the main auroral oval where they first enter open polar cap fields (Frey et al., 2002).

However, convection is typically not uniform and steady. On the dayside, meso-scale flows of ≥500 m/s lasting ~10–15 min are often observed associated with localized enhancements in dayside reconnection (Carlson, 2012; Fear et al., 2017; Frey et al., 2019 and references therein; Kim et al., 2009; Lyons et al., 2009; Sandholt et al.,

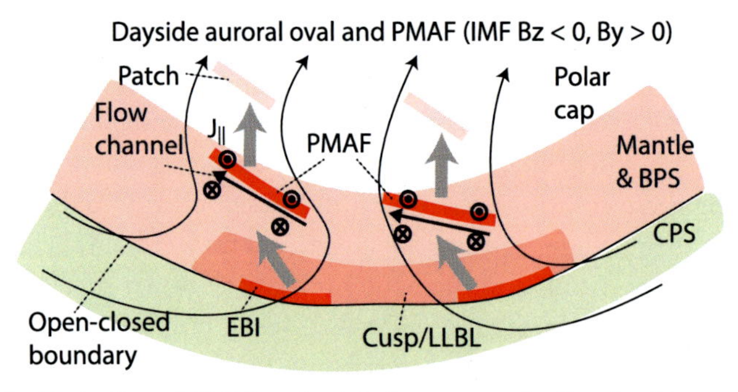

Fig. 2.2 Schematic illustration of PMAF and related processes in the dayside auroral oval under a typical IMF condition for PMAF occurrence ($B_z < 0$ and $B_y > 0$). The *gray lines* depict the sequence of poleward propagation after a brightening along the dayside open-closed magnetic field line boundary (EBI) to form a PMAF and then a patch. *From Frey, H.U., Han, D., Kataoka, R., Lessard, M.R., Milan, S.E., Nishimura, Y., Strangeway, R.J., Zou, Y., 2019. Dayside aurora. Space Sci. Rev. 215, 51. https://doi.org/10.1007/s11214-019-0617-7.*

2002; Sandholt and Farrugia, 2007; Wang et al., 2016). Upward FACs along the afternoon edge of these enhancements often give auroral intensifications starting at the equatorward boundary of the dayside 630.0 nm aurora. These discrete auroral structures typically move poleward and are referred to as "poleward-moving auroral forms" (PMAFs) (Oksavik et al., 2004, 2005; Sandholt and Farrugia, 2007).

That PMAFs move poleward into the polar cap, which implies that the associated enhanced flows do so as well (Lockwood et al., 2000; Oksavik et al., 2004, 2005). Fig. 2.2 (from Frey et al., 2019) illustrates the relationship between FACs, flow channels, and aurora as they propagate into the dayside polar cap after initiation as a brightening (EBI in the figure) along the dayside open-closed magnetic field line boundary.

Furthermore, PMAFs have been observed to decay into airglow patches that continue propagating into the polar cap (Lorentzen et al., 2010), and observations over polar cap patches that have moved poleward into the dayside polar cap (Wang et al., 2016) show meso-scale flow enhancements. These observations have verified that such flows do indeed move into the dayside polar caps. Patches are high-density, meso-scale regions of F-region plasma. Patches arise from PMAFs since the electron precipitation that creates PMAFs enhances F-region densities regions. In addition, in the absence of a PMAF, high-density F-region plasma from lower latitudes often enters the dayside polar cap near the cusp in association with dayside meso-scale flow enhancements (Carlson, 2012; Carlson et al., 2006; Oksavik et al., 2004, 2005).

Fig. 2.2 illustrates the dayside open-closed field line boundary as a smooth curved surface with equatorward protrusions at the locations of auroral brightenings. Strong meso-scale poleward directed plasma flows can lie within such protrusions, giving a pair of approximately north-south oriented FACs. The upward FACs on the afternoon side of the flow enhancements give rise to an approximately north-south oriented auroral arc referred to as "throat aurora," and the protrusions associated with throat aurora can become quite substantial and localized (Han et al., 2016, 2017).

2.3 Polar cap

Polar cap patches are islands of high-density ionospheric plasma in the F-region ionosphere surrounded by plasma that is half or less than half as dense as the patch (Weber et al., 1986; Crowley, 1996). Chapter 3 describes plasma density aspects of patches. Patches are also observable as enhanced 630.0nm airglow emissions that are traceable over long distances within the polar caps due to slow F-region recombination. The emissions arise as O_2^+ ions recombine with electrons, yielding O atoms in an excited state that relaxes by emitting 630.0nm emissions (e.g., Hosokawa et al., 2011a, b; Link and Cogger, 1988). Patch properties within the ionosphere can be found in Ren et al. (2018). Some patches have been observed to propagate deep into the polar cap and reach the nightside polar cap (Lorentzen et al., 2004; Moen et al., 2007; Nishimura et al., 2013; Oksavik et al., 2010), and they can exit the polar cap through magnetic reconnection and enter the nightside auroral oval (Zhang et al., 2013); see also Section 4.2.2 for airglow patches.

To propagate far across the polar cap without disappearing from view due to recombination, patches should be carried antisunward within the polar cap by flows that are larger than the average background. This implies a patch connection to meso-scale flow enhancements. This possibility has been examined deep within the polar cap by Zou et al. (2015) who found that flow enhancements are indeed collocated with airglow patches, an example of which is shown in Fig. 2.3. Fig. 2.3 shows three patches seen by 630.0nm all-sky imager (ASI) as the patches swept eastward within the polar cap. They can be seen extending from the dayside and approaching the auroral oval, and show the typical tendency to be elongated in the dayside to nightside direction during large IMF By. Channels of enhanced flow can be seen in the SuperDARN radar measurement as each patch moves into the region of detected flow vectors (obtained from line-of-sight measurements from two radars). Also, flow directions and speeds were found by Zou et al. (2015) to be consistent with patch propagation directions and speeds during their evolution across the polar cap. These correspondences indicate that patches that move to deep within the polar cap, including those that enter the nightside auroral oval, optically trace localized flow enhancements. In fact, polar cap patches have been tracked all the way from the near the dayside polar cap boundary to the nightside auroral oval (Nishimura et al., 2014a). This indicates that meso-scale flow enhancements can traverse the polar cap. Zou et al. (2015) emphasized the overall importance of these flow enhancements by showing that even one polar cap flow channel can account for a substantial portion ($\sim$30%) of the cross-polar cap potential drop. Also, as expected from patches being

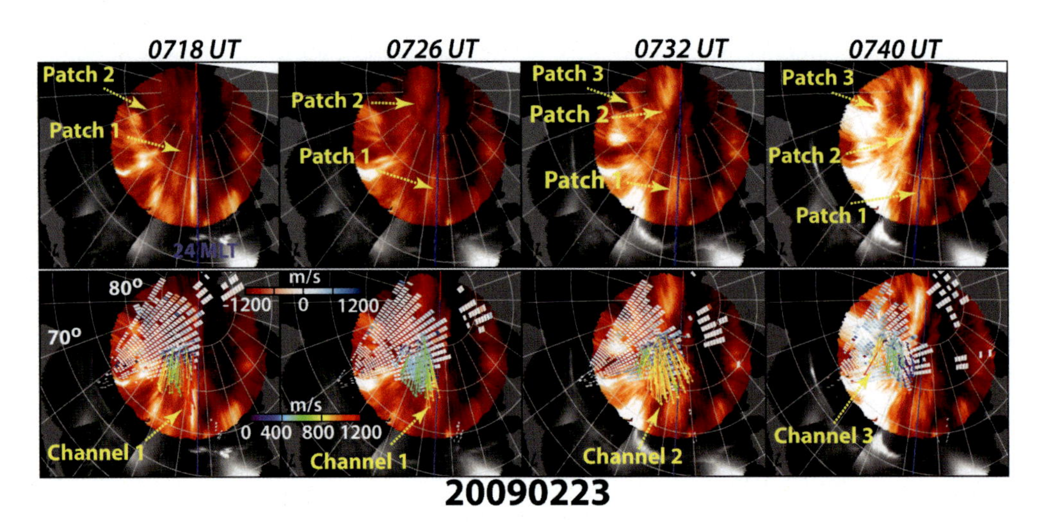

Fig. 2.3 ASI images showing three patches in 630.0 nm emission as they swept eastward within the polar cap. Images from the THEMIS all-sky imager array over North America (Mende et al., 2008) are shown in regions equatorward of the 630.0 nm emission image. SuperDARN radar LOS flows are shown by dots, and vector flows are shown where LOS echoes occur at the same measurement location from two radars. Channels of enhanced flow can be seen in the SuperDARN radar measurement as each of three patches moves into the region of detected flow vectors.

associated with flow channels, which have shears along their edges, they are also associated with a pair of weak upward and downwards FACs (Zou et al., 2016), which indicate coupling to driving processes along magnetospheric lobe magnetic field lines (Goodwin et al., 2019).

In addition to patches, polar cap arcs are known to be associated with localized and enhanced antisunward sheared flows embedded in large-scale convection (e.g., Koustov et al., 2009; Robinson et al., 1987; Valladares and Carlson, 1991). The arcs can extend across the polar cap from the dayside to the nightside of the auroral oval, are Sun-aligned (Meng and Akasofu, 1976; Murphree and Cogger, 1981; Frank et al., 1982, 1986; Hones Jr. et al., 1989), with the optical emission pattern sometimes resembling the Greek letter "theta" (Frank et al., 1982, 1986; Nielsen et al., 1990). To produce an arc, the flow shear at the edge of the flow channel must be strong enough to generate upward FACs sufficiently large to require a $\Delta\phi_{||}$ as illustrated in Fig. 2.1. Many polar cap arcs are localized structures, while some traverse the entire polar cap, and there can be several across the polar cap at any given time as shown by the example in Fig. 2.4 (from Zhang et al., 2016). Many polar cap arcs move azimuthally across the polar cap (Hosokawa et al., 2011b; Valladares et al., 1994), which allows us to infer that flow channels can have significant azimuthal motion across the polar as the plasma flows from the dayside to the nightside.

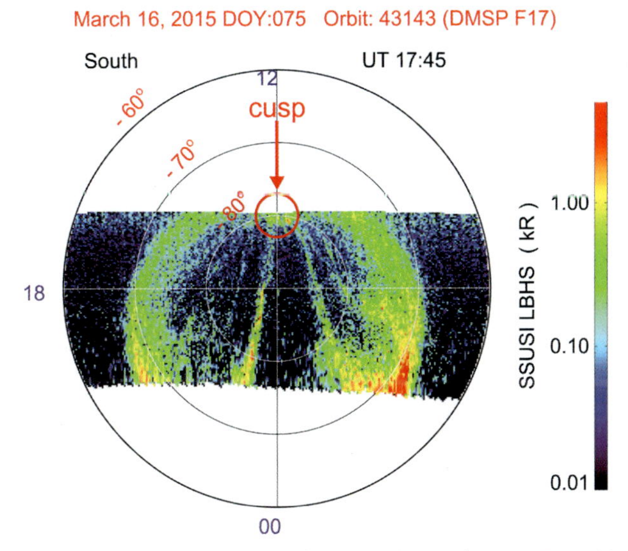

Fig. 2.4 SSUSI LBHS (Paxton et al., 1993) auroral images (around 17:45UT on March 16, 2015) (F17 DMSP). The associated IMF B_x, B_y, and B_z is (−7.1, −4.0, 9.6) nT. *From Zhang, Y., Paxton, L.J., Zhang, Q., Xing, Z., 2016. Polar cap arcs: Sun-aligned or cusp-aligned? J. Atmos. Sol. Terr. Phys. 146, 123–128. https://doi.org/10.1016/j.jastp.2016.06.001.*

2.4 Auroral oval

2.4.1 Poleward boundary intensifications

Meso-scale flow enhancements that approach the nightside auroral oval are closely followed by auroral poleward boundary intensifications (PBIs), which appear near the nightside auroral oval poleward boundary with a few minutes delay. They are near the same longitude as the flow enhancements and are associated with bursty bulk flows (BBFs) in the plasma sheet (de la Beaujardière et al., 1994; Lyons et al., 2011; Nishimura et al., 2010c; Pitkänen et al., 2013; Shi et al., 2012; Zesta et al., 2000). Three PBIs are identified in the mosaic of images from the THEMIS ASI array over North America (Mende et al., 2008) in Fig. 2.5. Such observations imply that flow enhancements traverse the open-closed field line boundary through enhanced nightside reconnection and couple to BBFs.

Observations of drifting 630.0 nm airglow patches in the polar ionospheric F-layer interacting with PBIs measured by the meridian scanning photometers (MSP) at magnetic latitude $\Lambda = 75.3$ degree near Longyearbyen, Svalbard (Lorentzen et al., 2004; Moen et al., 2007) support the hypothesis that PBIs can be triggered by flow channels, and indicate that the flow channels come from deep within the polar cap. As seen in Fig. 2.6, such patches were often observed within the nightside polar cap drifting from the poleward boundary of the MSP field of view (FOV), which is above $\Lambda = 85$ degree,

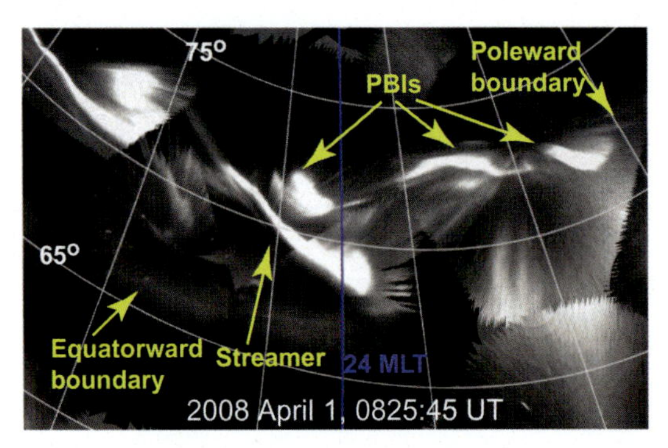

Fig. 2.5 Mosaic of images from the THEMIS ASI array over North America showing three PBIs and an auroral streamer.

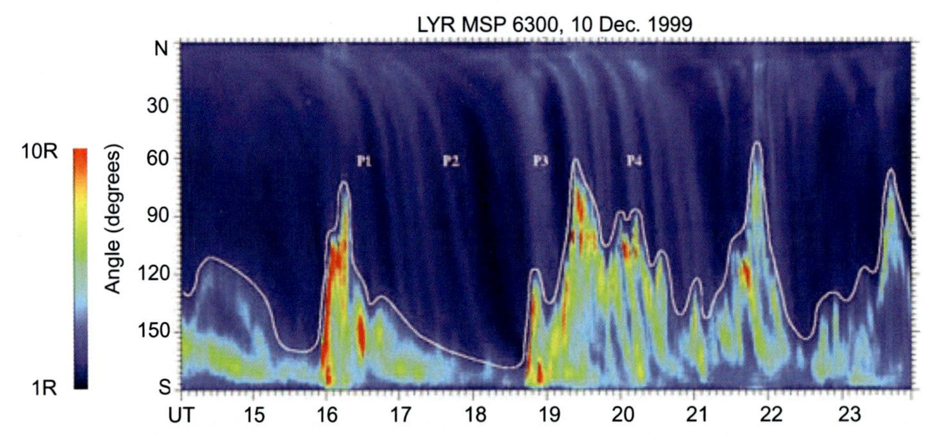

Fig. 2.6 Nighttime meridian scanning photometer observations from LYR, December 10, 1999, from 1400 to 2400 UT, as a function of meridian scan angle from 0° north to 180° south. Intensity is color coded in Rayleighs. *From Lorentzen, D.A., Shumilov, N., Moen, J., 2004. Drifting airglow patches in relation to tail reconnection. Geophys. Res. Lett. 31. https://doi.org/10.1029/2003GL017785.*

toward the polar cap boundary with meridional drift speeds from 350 to 1000 m/s. When the poleward boundary of the auroral oval (the open/closed field line boundary) was within the MSP FOV, all patches were observed to drift into that boundary with subsequent brightenings of PBIs (Lorentzen et al., 2004). These brightenings are a direct signature of tail reconnection bursts driven by flow channels from the polar cap which carry plasma from open polar cap field lines into the plasma sheet.

As discussed by Lorentzen et al. (2004), polar cap patches are believed to originate near the dayside polar boundary, indicating that the polar cap flow channels that traverse the nightside polar cap and enter the auroral oval could originate with the flow channels near the dayside cusp as discussed in Section 2.2. This was directly observed by flow and all-sky imager (ASI) observations by Nishimura et al. (2014a), who estimated that it took ~90 min for a flow channel to move the dayside to the nightside auroral oval. These observations are consistent with a polar cap flow channels being magnetically connected to the solar wind as they flow across the open field line region.

2.4.2 Streamers

ASI images show that PBIs can extend equatorward from the poleward boundary of the auroral zone and become elongated in roughly the north-south direction as seen in the example in Fig. 2.5. Such north-south auroral structures, generally referred to as auroral streamers, have been reported in images of the aurora for many years (e.g., Nakamura et al., 1993; Rostoker et al., 1987), and they have been suggested to be related to flow bursts in the tail (Henderson et al., 1998). Sergeev et al. (2000) found direct correspondence between an individual streamer and an earthward going flow burst observed in the tail (see Fig. 4.12), an association that Zesta et al. (2000, 2002) found to be common. More recently, the direct connection has been made directly between an auroral streamer and a flow burst seen simultaneously in the ionosphere and in the plasma sheet (Pitkänen et al., 2011). Auroral streamers are relatively common as they tend to occur during active and quiet conditions, with an increase in number of observations during disturbed periods (Gabrielse et al., 2018; Paschmann et al., 2012).

The connection with PBIs implies that streamers and PBIs are driven by the same plasma sheet flow bursts, so that both should be connected to flow channels from the polar cap. Some streamers reach the equatorward portion of the auroral oval, such streamers therefore being connected to flow bursts that move earthward from the distant plasma sheet to near the inner boundary of the plasma sheet, which can be at equatorial radial distances from a few to ~10 R_E. Plasma sheet flow bursts were proposed to consist of depleted magnetic flux tubes (i.e., flux tubes with lower total entropy than the surroundings, leading to earthward interchange motion) (e.g., Pontius and Wolf, 1990; Yang et al., 2011), and this has now been strongly supported by observations (Dubyagin et al., 2010; Panov et al., 2010; Sergeev et al., 1996, 2012; Xing et al., 2010). It is likely that the amount of earthward penetration of a depleted flux tube depends upon its total entropy relative to that of the surrounding plasma (Wolf et al., 2012).

Taking advantage of the extensive THEMIS array of ASIs over the American sector auroral oval and the concurrently available SuperDARN radar observations with fields-of-view (FOVs) overlapping that of the ASIs, Gallardo-Lacourt et al. (2014a, b) examined the characteristics of flow bursts associated with streamers. Examples from the 135 isolated examples included in their study are shown in Fig. 2.7. In Fig. 2.7, line-of-sight (LOS) flow velocities along the 16 SuperDARN Rankin Inlet radar beams are shown as small rectangles that are overlaid on a merger of images from the ASIs. The LOS

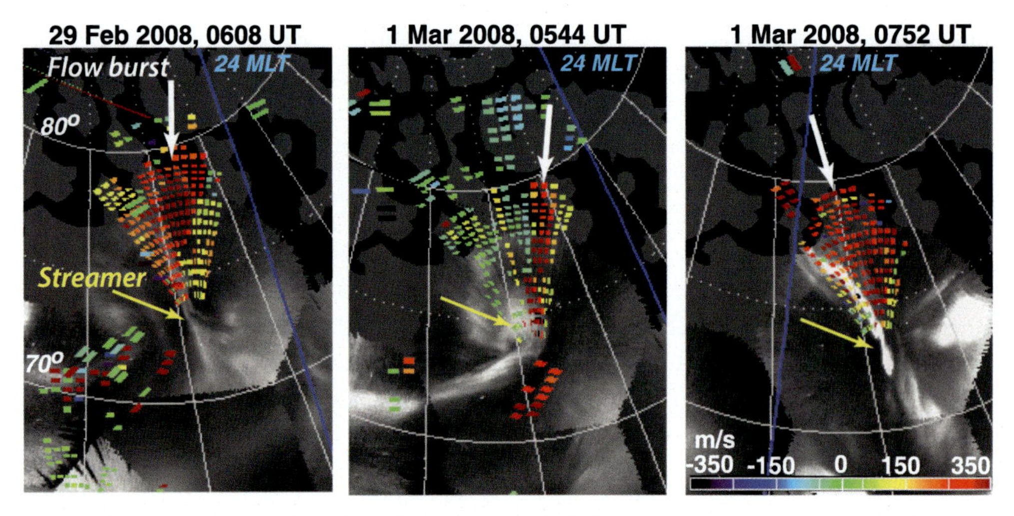

Fig. 2.7 Three examples of LOS flows from the SuperDARN radars, primarily the one at Rankin Inlet, overlaid on auroral images from several THEMIS ASIs. LOS flows toward the radar are shown as positive, and longitude lines are separated by 1 h of MLT. *Yellow arrows* point to the streamers of interest, and *white arrows* point to the corresponding flow channels.

velocities are color coded with reddish colors showing flows toward the radar. Gallardo-Lacourt et al. (2014a, b) found that all streamers were correlated with fast flows around the streamers, most of them being directed equatorward and being elongated so as to appear as channels. As can be seen from the examples in Fig. 2.7, the equatorward flows (identified by white arrows) lie just to the east of their respective streamer (identified by yellow arrows), with their western edge lying almost directly on the streamer. The width of the flow channels in the ionosphere was found to average ~75 km, which maps to ~1.5–2.5 R_E in the equatorial plasma sheet, consistent with typical widths of flow burst seen by satellites (Angelopoulos et al., 1997; Nakamura et al., 2004). The speed of the fast flows varies considerably, the LOS speeds in the examples in Fig. 2.7 being ~350–400 km/s, and, as seen in the ionosphere, the fast flows have substantial temporal variations and spatial motion. Because of their variability and motion, they typically appear as bursts of enhanced flow when seen by spacecraft within the plasma sheet.

Large, abrupt nighttime magnetic perturbations are often been associated with substorms (Akasofu and Meng, 1969; Nishida and Kokubun, 1971). However, more recently, ground-based ASIs have been used to demonstrate that streamers and PBIs can also drive these events (Lyons, 2000; Nishimura et al., 2020), including during quite quiet conditions (Sutcliffe and Lyons, 2002). Engebretson et al. (2019a, b) used ground-based imagers to show intense magnetic perturbations events well after substorm onsets that were associated with PBIs and streamers.

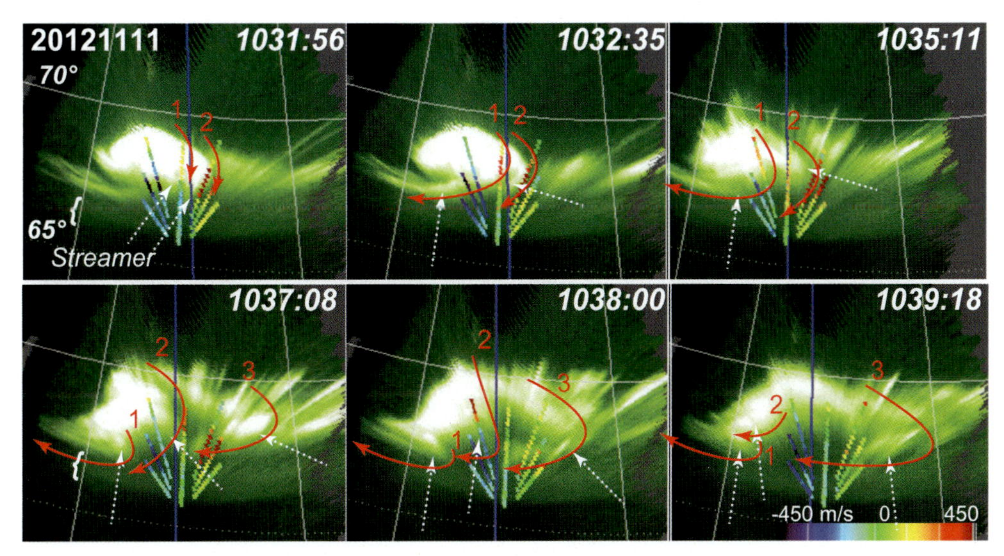

Fig. 2.8 Time series of LOS flows from PFISR overlaid on *green line images* from the multispectral ASI at Poker Flat on November 11, 2012. *Curved red arrows* illustrate the ionospheric paths of the plasma sheet flow bursts estimated using paths of the auroral streamers and the LOS flows observed by PFISR. *Dashed white arrows* identify streamers used for estimating these paths, *white bracket in the images along the left edge* identifies the equatorward diffuse auroral band attributable to proton precipitation. *From Lyons, L.R., Nishimura, Y., Gallardo-Lacourt, B., Nicolls, M.J., Chen, S., Hampton, D. L., Bristow, W.A., Ruohoniemi, J.M., Nishitani, N., Donovan, E.F., Angelopoulos, V., 2015. Azimuthal flow bursts in the inner plasma sheet and possible connection with SAPS and plasma sheet earthward flow bursts. J. Geophys. Res. Space Physics 120, 2015JA021023. https://doi.org/10.1002/2015JA021023.*

Within the auroral oval, there is a boundary between the eastward and the westward electrojet on the nightside (Harang, 1946), which corresponds to a reversal of the convection electric field where the dusk convection protrudes toward the dawnside and then turns duskward at aurora latitudes (referred to as the Harang discontinuity or, more appropriately, the Harang reversal). An auroral streamer often can be seen following this convection feature that has been termed as loops (Akasofu et al., 1965) or the "Harang aurora" (Nishimura et al., 2010a), which has the appearance of a hook (Zou et al., 2012). An example of a hook-shaped Harang aurora is shown in Fig. 2.8. As revealed by the LOS measurement from the Poker Flat Incoherent Scatter Radar (PFISR), and indicated by red arrows, the plasma flow of the Harang reversal is quite dynamic, often consisting of more than one flow channel and hook-shaped auroral features that move and evolve over times scales of minutes. As discussed later, these time-dependent flow structures reveal themselves as bursts of westward flow in the equatorial portion of the auroral oval and further equatorward into the subauroral region.

2.4.3 Omega bands and DAPS

The flow channels associated with the Harang aurora become part of the dusk convection cell. Flow channels further to the east often end up within the morning-side convection cell, which then form strong eastward flows recently identified as dawnside auroral polarization streams (or DAPS) (Liu et al., 2020). There they flow eastward while staying poleward of the most intense morning-side aurora. More intense aurora occurs within the region of upward Region 2 currents, whereas DAPS lie within the region of downward Region 1 currents and often display a steep gradient at the boundary between Regions 1 and 2 currents.

DAPS are associated with a major disturbance that only occurs in the midnight-to-morning sector known as omega bands. Omega bands are large-scale, auroral folds that move eastward in the morning sector with velocities of a few to several hundred m/s (Akasofu, 1974; Andre and Baumjohann, 1982; Opgenoorth et al., 1983). They have a tendency to occur during periods of enhanced convection (e.g., Solovyev et al., 1999), and they typically occur along the poleward edge of the diffuse auroral region that lies within the upward Region 2 currents, and have wavelengths of ~500–1000 km (Henderson, 2012; Lyons and Walterscheid, 1985). They protrude poleward from the equatorward portion of the auroral oval toward the higher latitude portion of the oval and are separated by dark regions (Henderson, 2012). Auroral torches protrude more strongly poleward (Akasofu and Kimball, 1964), though they can be seen to evolve into omega bands as they move eastward. Omega bands are associated with substantial magnetic field perturbations on the ground that can be as large as several hundred nT with periods of several to 10s of minutes (Kawasaki and Rostoker, 1979; Rostoker and Barichello, 1980), consistent with them being a major disturbance.

Henderson (2012) took advantage of global imaging from the NASA Polar spacecraft and showed examples where auroral streamers evolved into torch-like structures when they reached the diffuse auroral region, and soon evolve into clear, eastward moving, omega bands. These observations demonstrate that flow channels on the morning side, appearing to be within the morning convection cell, can evolve into omega bands when they extend sufficiently earthward in the plasma sheet. We have been able to see this connection in the THEMIS ASI images, since they cover a large-enough region of the post-midnight auroral region when sky conditions are clear. While a detailed study of such events has not yet been performed using the THEMIS ASI observations, an example was shown by Nishimura et al. (2010b) and is given in Fig. 2.9. In Fig. 2.9, the first panel shows a PBI near the auroral poleward boundary that extended equatorward to near the equatorward auroral region as a streamer as seen in the second panel. Connection to the newly formed torch structure is seen in the third, and its eastward propagation as can be seen in the fourth panel. The same process is seen in Fig. 2.10, where high-altitude measurements of the aurora from the Polar spacecraft VIS low resolution camera show

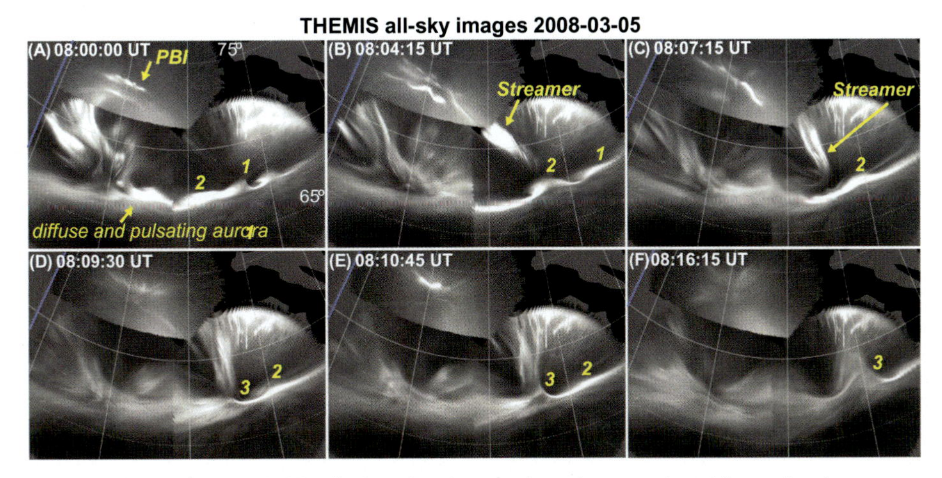

Fig. 2.9 Snapshots of THEMIS ASIs during showing torch and omega band formation by an auroral streamer on February 4, 2008. *From Lyons, L.R., Nishimura, Y., Gallardo-Lacourt, B., Zou, Y., Donovan, E.F., Mende, S., Angelopoulos, V., Ruohoniemi, J.M., McWilliams, K.A., Hampton, D.L., Nicolls, M.J., 2015. Dynamics related to plasmasheet flow bursts as revealed from the aurora, in: Zhang, Y., Paxton, L.J. (Eds.), Auroral Dynamics and Space Weather. John Wiley & Sons, Inc, pp. 95–113.*

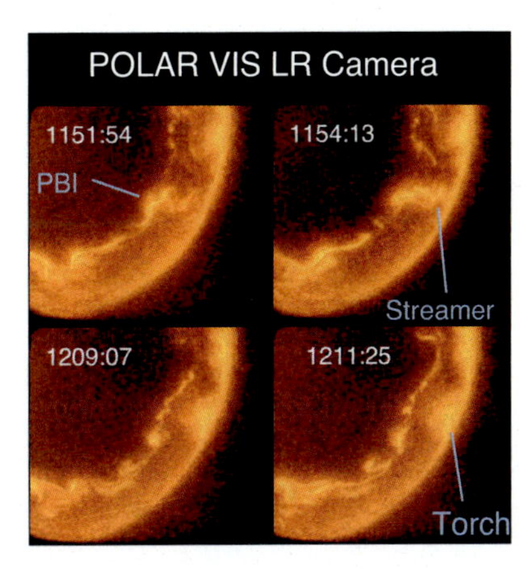

Fig. 2.10 High-altitude measurements of the aurora from the polar spacecraft VIS low resolution camera show the sequence of an omega band formation from a PBI evolving into an auroral streamer that subsequently gives rise to the torch formation. *Adapted from Henderson, M.G. (2012) Auroral substorms, poleward boundary activations, auroral streamers, omega bands, and onset precursor activity, in Auroral Phenomenology and Magnetospheric Processes: Earth and Other Planets (eds A. Keiling, E. Donovan, F. Bagenal and T. Karlsson), American Geophysical Union, Washington, DC. https://doi.org/10.1029/2011GM001165.*

the sequence of an omega band formation from a PBI evolving into an auroral streamer that subsequently gives rise to the torch formation (figure adapted from Henderson, 2012).

Liu et al. (2018) showed clearly that the poleward edge of omega bands corresponds to the equatorward boundary of DAPS, so that it is reasonable to propose that the omega bands are associated with the strong shear flow at this boundary and with the reduced entropy of the plasma within flow channels.

The physical process by which flow channels lead to strong SAPS and DAPS flows is discussed later.

2.4.4 Substorms

One of the most fascinating phenomena in auroral physics is the substorm. Substorms were first identified in auroral observation by Akasofu (1964), who described the auroral evolution of the substorm. However, substorms are a disturbance of the global magnetosphere-ionosphere system. Its expansion-phase onset is characterized in the aurora by a brightening along a preexisting, growth-phase arc or new arc that emerges near the equatorward boundary of the auroral oval (Akasofu, 1964; Deehr and Lummerzheim, 2001; Samson et al., 1992). The brightening first occurs along a section of the growth-phase arc and then expands in longitude along that arc (Lyons et al., 2013; Sakaguchi et al., 2009; Shiokawa et al., 2009), and the brightening appears as beads along the growth-phase arc. Such beading is a general feature of substorm auroral onset, being seen nearly whenever there is good auroral viewing (Kalmoni et al., 2017; Nishimura et al., 2016). The auroral beads propagate azimuthally and evolve into wavy structures, as expected for the beads being an auroral manifestation of waves associated with an instability responsible for substorm onset.

Two examples of the evolution of beading at and following a substorm auroral onset are shown in Fig. 2.11. A sequence of combined auroral images from the THEMIS ASI is shown for each event. The onset beads initially appear as faint rays due to the convergence of auroral brightenings toward the center of each imager's FOV (identified by yellow arrows in Fig. 2.11Aa and Ba). The beads are initially regularly spaced like waves. They then grow in intensity (Kalmoni et al., 2017; Nishimura et al., 2016) as the region of detectable wave structure spreads longitudinally (indicated by orange arrows in Fig. 2.11Ab–c and Bb–d). The wave-like structure grows in amplitude (Kalmoni et al., 2017), and then develops nonlinearities that become the active substorm expansion-phase aurora (identified by yellow arrows in Fig. 2.11Ad and Bd). Such growing waves are the signature of an instability that grows until it becomes nonlinear. East-west keograms of differential intensity along the onset arc in the lower panels of Fig. 2.12 give a different view of the growth and longitudinal expansion of the waves, and their evolution into the bright substorm expansion-phase aurora. The middle panels of Fig. 2.12 give the maximum emission intensity along the onset arc, which gradually

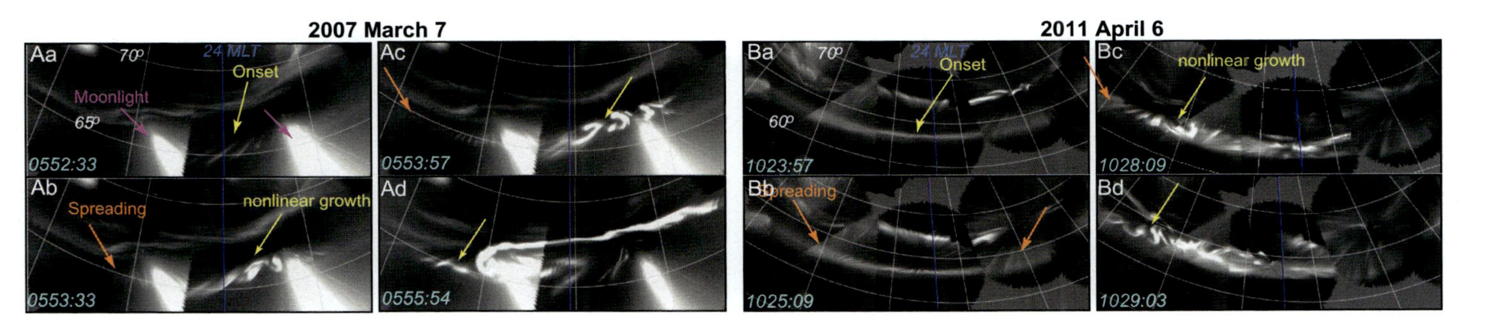

Fig. 2.11 Selected mergers of the auroral images from the THEMIS ASIs for the time interval of substorms on (A) March 7, 2007 and (B) April 6, 2011. The *dark blue line marks* magnetic midnight, and longitude lines are space 1 h in MLT apart. Moonlight contamination is identified in the first panel of (A). *Adapted from Lyons, L.R., Nishimura, Y., 2020. Substorm onset and development: the crucial role of flow channels. J. Atmos. Sol. Terr. Phys. 211, 105474. https://doi.org/10.1016/j.jastp.2020.105474.*

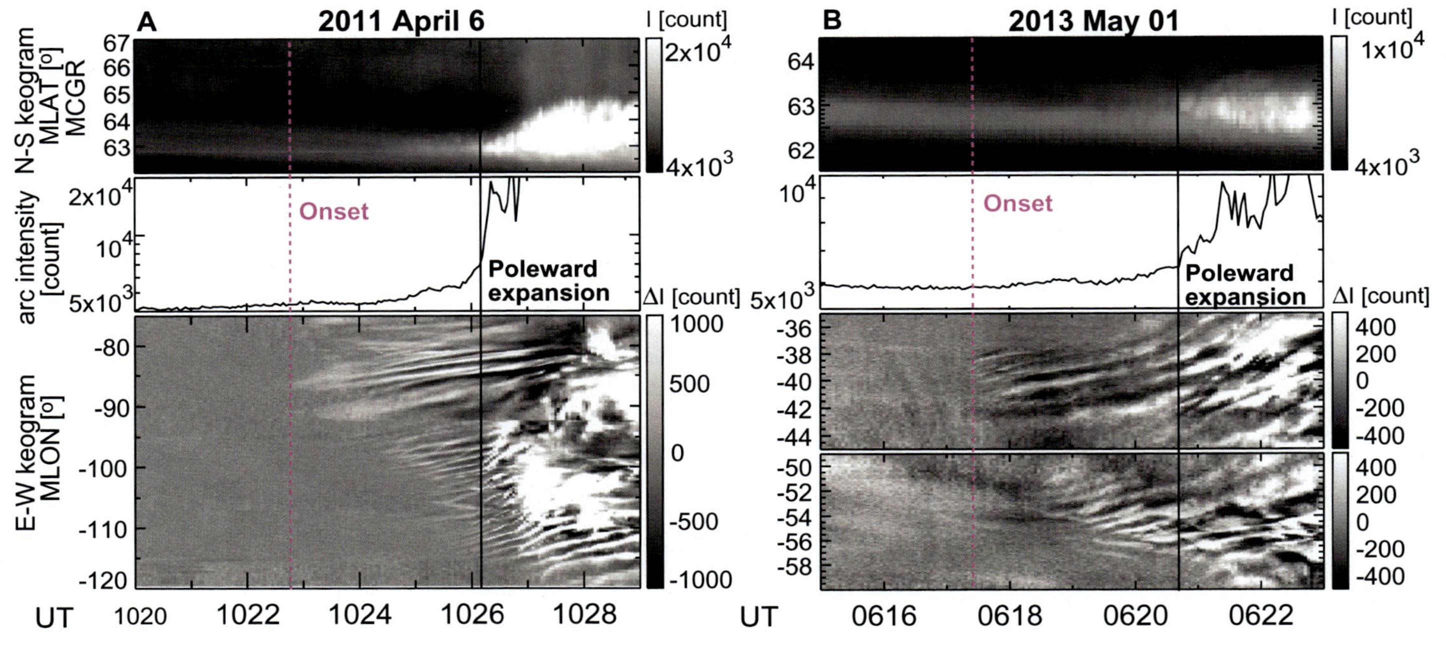

Fig. 2.12 From top to bottom, north-south auroral keograms, maximum intensity along the onset arc, and east-west auroral keograms in 5-min detrended intensity scales around onset times for substorms on (A) April 6, 2011 and (B) May 1, 2013. The *dashed magenta and black vertical lines mark*, respectively, the initial auroral brightening and the initiation of poleward expansion. North-south keograms use maximum intensity within ±15° longitude from imager zenith longitude at each latitude. East-west keograms are sliced along the initial brightening arcs. *From Lyons, L.R., Nishimura, Y., 2020. Substorm onset and development: the crucial role of flow channels. J. Atmos. Sol. Terr. Phys. 211, 105474. https://doi.org/10.1016/j.jastp.2020.105474.*

increases until more explosively increasing when the nonlinear development leads to the poleward expansion of the region of active aurora seen in the upper panels.

The beads occur simultaneously in both hemispheres with similar wavelengths and periodicities (Motoba et al., 2012), indicating that the instability giving the beads is generated in the magnetosphere rather than in the auroral acceleration region or in the ionosphere. Otherwise, different ionospheric conditions would create different characteristics of waves in the different hemispheres. Highly fluctuating magnetic fields in the near-Earth plasma sheet and have been suggested to be linked to a magnetospheric substorm onset instability (Cheng and Lui, 1998; Lui, 1996, p. 196; Ohtani, 1998; Park et al., 2010; Roux et al., 1991; Shiokawa et al., 2005; Takahashi et al., 1987). The longer period component of those fluctuations (several tens of seconds) corresponds to periods seen in the aurora as the beads pass overhead and in the ground magnetic field.

A dramatic feature of the instability is its electric field signatures, which have been seen by ground radars. Fast oscillating flows (~1000 m/s) are correlated with the onset beads as they propagate across the meridian of a poleward looking radar beam (Gallardo-Lacourt et al., 2014a; Hosokawa et al., 2013). The association between the flow oscillations and the beads is shown in a time-series format in Fig. 2.13. The vertical gray dashed lines give the times when each bead crosses the radar beam longitude (white horizontal line in the east-west keogram), and each auroral bead is marked by a white dashed line in the east-west keogram. As each auroral bead crosses the radar beam longitude, we see an equatorward flow enhancement followed by a poleward flow enhancement. As sketched in the right side of Fig. 2.13, Gallardo-Lacourt et al. (2014a) suggested that a clockwise flow shear was associated with the formation of an upward FAC at the center of this flow, which appears as an auroral bead, and a counterclockwise flow shear occurred between the auroral beads as expected to associated with a downward FAC.

For understanding substorms, it is necessary to determine what causes the transition from a stable system to an unstable system, and what is the resulting instability. Fundamental for this understanding is that, in the aurora, the instability is along a nearly east-west oriented auroral arc that maps to the inner plasma sheet, and that the instability spreads longitudinally along this arc. Thus, a process that is aligned and expands longitudinally is required.

Nishimura et al. (2010a, b, c) reported a repeatable sequence of events leading to substorm onset in observations from the THEMIS ASI array that should be associated with an abrupt change in conditions in the inner plasma sheet. The preonset sequence starts with the formation of a poleward boundary intensification (PBI) along the auroral poleward boundary. An auroral streamer then extends equatorward from the PBI toward the equatorward boundary of the aurora oval, which may turn into enhanced auroral brightness that drifts azimuthally (westward or eastward for auroral onsets in the dusk cell or dawn cell, respectively). Onset occurs when the enhanced auroral luminosity region reaches the onset location, or substorm onset can occur near the location where the

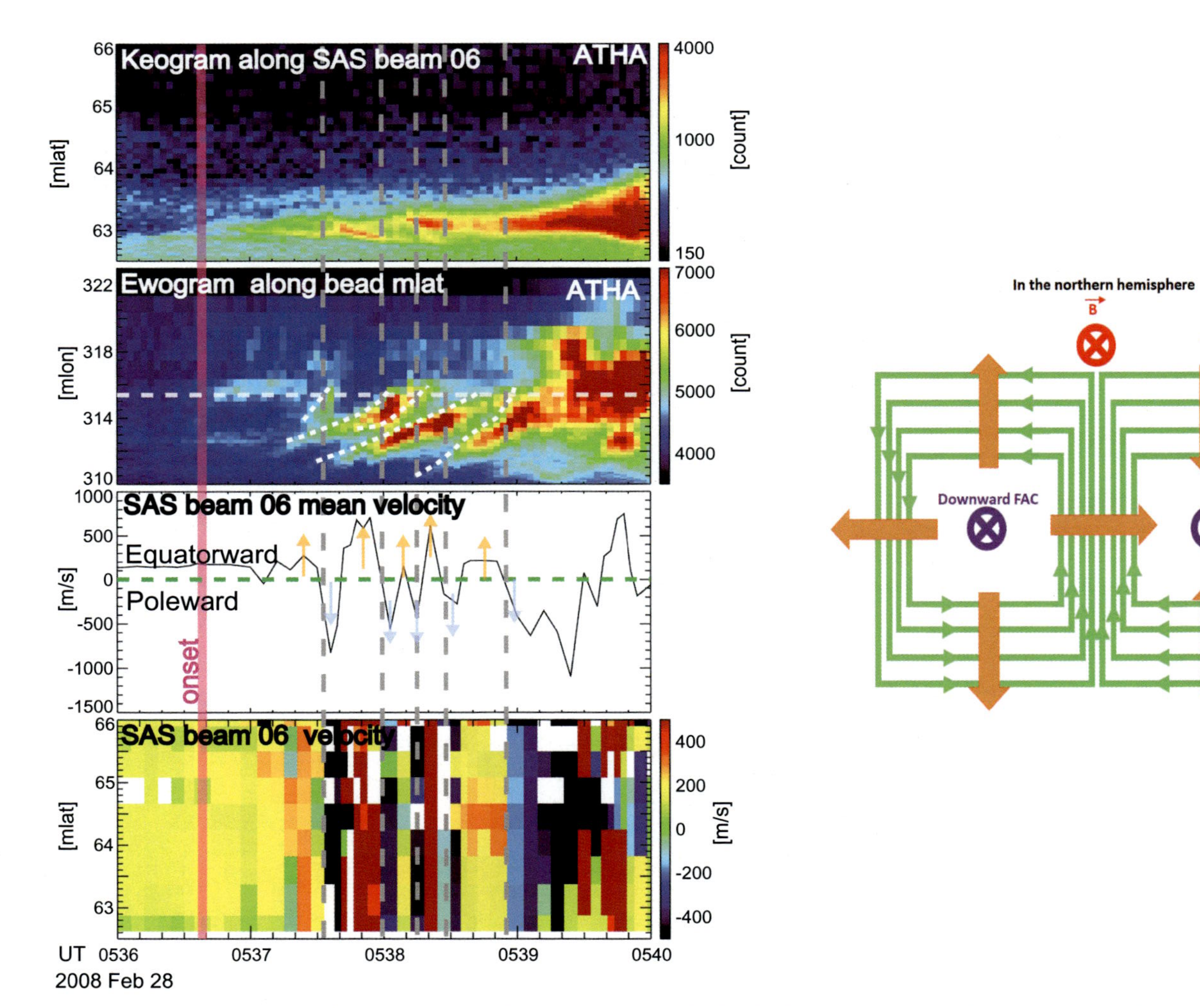

Fig. 2.13 Time-series evolution of the auroral beads observed using THEMIS ASI (first and second panel) and the flow velocities measured by SuperDARN radars (third and fourth panel). (*Right*) Schematic illustration of the flow shears reported by Gallardo-Lacourt et al. (2014a).

streamer first reaches a growth-phase arc located near the equatorward boundary of the auroral oval.

Using the known relation between auroral enhancements and plasma sheet flow as mapped to the ionosphere (de la Beaujardière et al., 1994; Gallardo-Lacourt et al., 2014a, b; Haerendel, 2011; Henderson et al., 1998; Lyons et al., 1999; Nakamura et al., 2001; Pitkänen et al., 2011; Sergeev et al., 1999, 2000; Zesta et al., 2000), the observed auroral sequence implies that new plasma crosses the polar cap boundary onto closed auroral oval/plasma sheet field lines (corresponding to localized reconnection in the distant tail) and then intrudes as a longitudinally localized flow channel to the equatorward/near-Earth region of the auroral oval/plasma sheet, leading to the onset instability. Since flow bursts in the plasma sheet consist of depleted magnetic flux tubes, it is thus reasonable that the plasma sheet flow channels trigger the onset instability by bringing reduced entropy plasma to the inner plasma sheet where it abruptly changes the entropy distribution. In particular, it should lead to decrease in the tailward gradient of entropy, a gradient change that could be important (e.g., Xing and Wolf, 2007) for an abrupt transition to instability. This instability must be very different from just the earthward (equatorward as mapped to the ionosphere) interchange motion, since onset extends longitudinally along an east-west oriented auroral arc near the equatorward boundary of the auroral oval.

While there has been debate about how often auroral streamers are seen leading to substorms (e.g., Frey, 2010), they are very commonly seen when viewing conditions are good and the auroral oval is not very thin so that streamer identification from ASIs is difficult (Lyons et al., 2018; Nishimura et al., 2010a, 2011). However, direct observation of flows in the ionosphere using radars allows a test of the flows to the onset location scenario without the ambiguities that can occasionally occur with auroral observations. An example from an initial study of such flows (Lyons et al., 2010) is shown in Fig. 2.14. The flow direction changed and speed increased several minutes before the substorm onset, leading to a southeastward-directed flow enhancement that moved equatorward and reached the onset latitude at approximately the onset time. The densities prior to the onset peak at ~130 km and fall off above ~150 km, indicating the pure proton precipitation (Zou et al., 2009, and references therein) that typically extends equatorward of the electron auroral oval and lies equatorward of the onset. The density reduction a few minutes before onset indicates an arrival of a flux tube with reduced entropy. Similar events with preonset equatorward flow enhancements have been presented by Lyons and Nishimura (2020) and Nishimura et al. (2014b).

As seen in MHD simulations (e.g., Birn et al., 2004), flow bursts in the plasma sheet are normally thought of extending radially but being azimuthally narrow. If this were true when flow channels reached the inner plasma sheet, they could not account for the azimuthal alignment and spreading of the substorm onset instability. However, Rice Convection Model (RCM) modeling shows a very clear longitudinal expansion of the

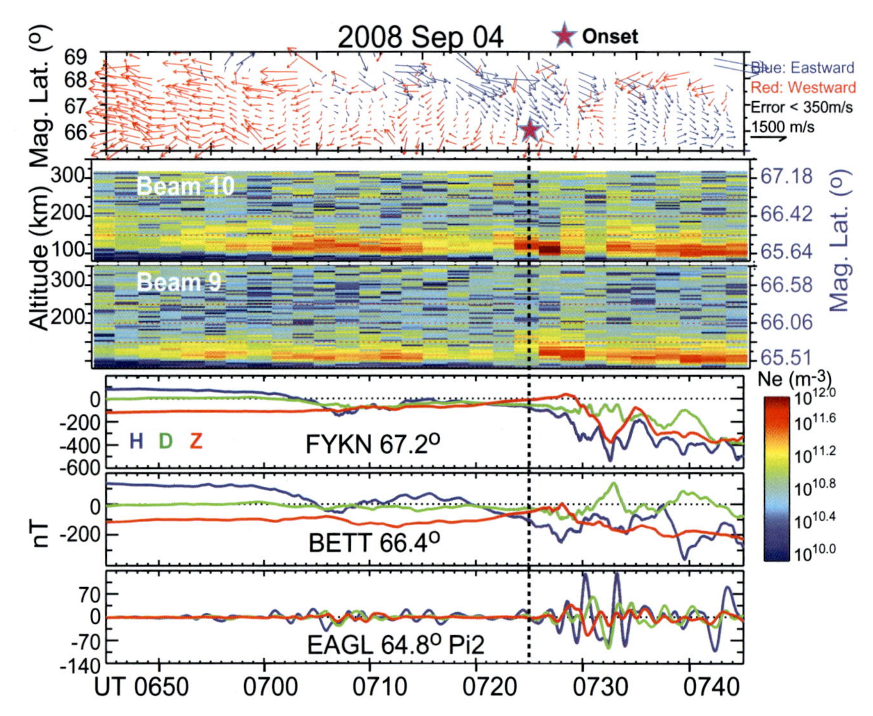

Fig. 2.14 PFISR observations from 0640 to 0750 UT on September 4, 2008. The top three panel shows F-region velocity vectors. Electron densities along two of the radar beams directed poleward along the magnetic meridian are shown in the second and third panels, altitude being shown along the left axis and magnetic latitude along the right axis. Ground magnetic observations are shown from BETT and FYKN approximately along the PFISR meridian. Pi2 pulsations from EAGLE are shown in the bottom panel. The time of the identified substorm onset is indicated by a *dashed vertical line*; the latitude of initial substorm brightening is identified by a star in the flow vector panel. *Adapted from Lyons, L.R., Nishimura, Y., Shi, Y., Zou, S., Kim, H.-J., Angelopoulos, V., Heinselman, C., Nicolls, M.J., Fornacon, K.-H., 2010. Substorm triggering by new plasma intrusion: Incoherent-scatter radar observations. J. Geophys. Res. 115. https://doi.org/10.1029/2009JA015168.*

reduced entropy plasma that comprises a plasma sheet flow channel as it extends to the inner plasma sheet (Wang et al., 2018; Yang et al., 2014). The expansion results from a combination of lower energy electrons following electric drift contours to the dawnside, and the energy-dependent magnetic drift causing more energetic electrons to drift duskward. This would lead to a longitudinal radial of the region of decrease in the tailward gradient of entropy, and offers a viable explanation for why the onset instability initially extends, and then expands, longitudinally.

2.5 Subauroral response to flow channels

At latitudes equatorward of the electron auroral oval, we encounter the subauroral region. In the afternoon to midnight sector, this region is characterized by westward ionospheric flows, which include latitudinally narrow (~1–2 degree) flows in the premidnight sector known as polarization jets (Galperin et al., 1974) or subauroral ion drifts (SAIDs) (Anderson et al., 1991; Spiro et al., 1979). These are often imbedded in a broader region of westward flow referred to earlier and named as subauroral polarization streams (SAPS) (Foster and Burke, 2002; Foster and Vo, 2002). SAPS are part of the duskside convection cell and form in the region of downward Region 2 FACs, which lie adjacent to, but equatorward of the electron auroral oval (Anderson et al., 2001). SAPS form in downward FAC regions because of the substantially lower conductivity than in the regions of upward FACs, which support the formation of discrete aurora.

Initial studies established a relationship between intensifications of SAPS flows and substorm dynamics (Anderson et al., 1993; Erickson et al., 2002; Foster et al., 2004; Makarevich and Dyson, 2007; Nishimura et al., 2008). More recently, the transient aspect of SAPS flow enhancements has become increasingly recognized. In particular, Gallardo-Lacourt et al. (2017) and Makarevich et al. (2011) found strong evidence that some plasma sheet flow bursts can turn azimuthally after reaching the inner plasma sheet leading to strong bursts of westward flow in the SAPS region, and such bursts can be accompanied by an auroral enhancement along their poleward edge (Lyons et al., 2015). The result is a flow enhancement that tends to follow the dusk cell of the large-scale convection pattern. Fig. 2.15 shows a schematic illustration of the flow and streamer motion.

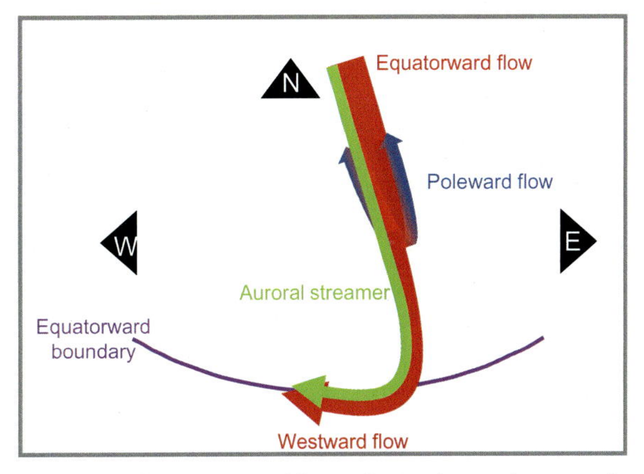

Fig. 2.15 Schematic illustration of streamer and flow relationship in the auroral and subauroral region. *Adapted from Gallardo-Lacourt, B., Nishimura, Y., Lyons, L.R., Mishin, E.V., Ruohoniemi, J.M., Donovan, E.F., Angelopoulos, V., Nishitani, N., 2017. Influence of auroral streamers on rapid evolution of ionospheric SAPS flows. J. Geophys. Res. Space Physics 122, 12406-12420. https://doi.org/10.1002/2017JA024198.*

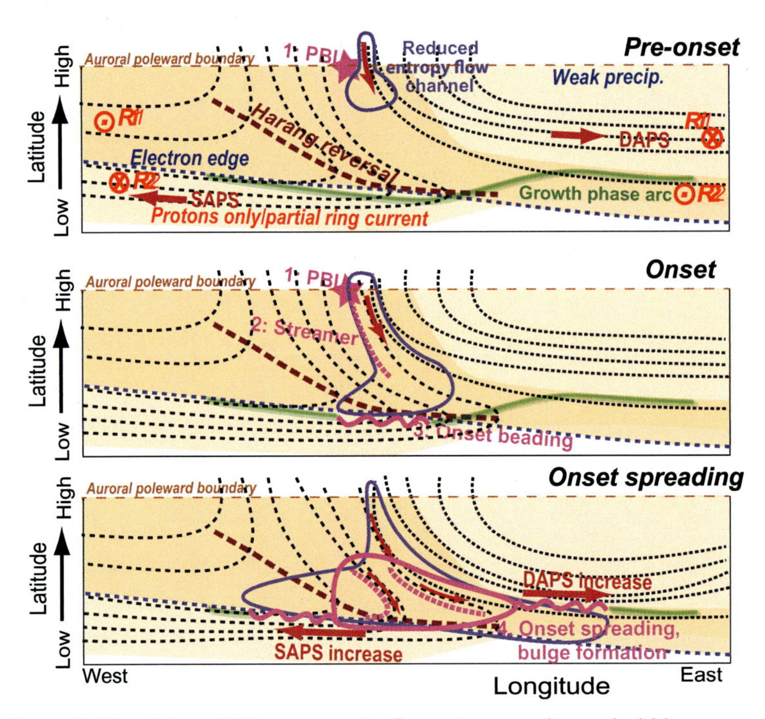

Fig. 2.16 Schematic illustration of the sequence of events as a plasma bubble moves earthward toward the inner plasma sheet. This figure applies to the flows that lead to a substorm. It also applies to cases without a substorm if the substorm phenomena in magenta and the growth-phase arc in green are not considered. *From Lyons, L.R., Nishimura, Y., 2020. Substorm onset and development: the crucial role of flow channels. J. Atmos. Sol. Terr. Phys. 211, 105474. https://doi.org/10.1016/j.jastp.2020.105474.*

The use of ionospheric flows relative to the aurora allows testing of whether the longitudinal expansion of the reduced entropy plasma that comprises a plasma sheet flow channel offers a viable explanation for why the substorm onset instability initially extends, and then expands, azimuthally. As found by Zou et al. (2009) and Lyons and Nishimura, 2020), observations following a substorm onset as seen in the aurora show a flow increase that is equatorward of detectable aurora so that is likely SAPS, and this flow enhancement expands westward as the auroral brightening of substorm onset expands westward. In addition, an increase is seen in the DAPS flows as auroral onset spread eastward. These are just as expected from an azimuthally expanding bubble as modeled by Wang et al. (2018).

Fig. 2.16 schematically summarizes the scenario for how flow channels drive substorm onset as seen in the aurora that is supported by the RCM modeling and the data analysis. The top panel shows a flow channel entering the auroral oval (plasma sheet) from the polar cap (magnetotail lobes), where it gives rise to an auroral poleward boundary intensification (PBI). It starts to expand azimuthally as it moves equatorward (earthward in the plasma sheet) due to the combination of electric drift along dawnward tilted equipotentials and westward magnetic drift of the more energetic ions. As shown in the middle panel, when the azimuthally expanding bubble reaches to near the equatorward boundary of the electron auroral oval, the instability indicated by the auroral beading and its associated flows is initiated if conditions are appropriate. As the instability grows (lower panel), auroral activity expands poleward forming the auroral bulge, which has imbedded flow channels and streamers, and the onset instability extends longitudinally with the continued azimuthal expansion of the bubble within the plasma sheet. Thus, by bringing together observations of aurora relative to observed flows within both the auroral oval and subauroral region, we obtain a consistent picture of substorm onset via an instability triggered by an intruding low entropy flow channels and the azimuthal evolution of the instability.

A large variety of mesoscale structures has been found in the subauroral region. For example, the subauroral proton aurora, which forms equatorward of the proton auroral oval, has typical scale sizes of mesoscale structures. These structures appear as elongated arcs, spots, or patches (Frey, 2007; Immel et al., 2002). In particular, Nishimura et al. (2014b) analyzed the rapid temporal variations of subauroral proton aurora in the premidnight sector by using coordinate ground-based optical measurements and satellite data. In their analysis, the authors revealed a close connection between the initiation of proton aurora and an auroral streamer approaching the equatorward boundary. Satellite measurements demonstrated that plasma sheet injections are able to supply energetic ions into the plasmasphere and excite EMIC waves, which subsequently enhance the ion precipitation into the atmosphere. This sequence indicates that flow burst plasma sheet transport is a fundamental process for injecting plasma toward the inner magnetosphere as well as for subauroral proton precipitation. For more information on proton aurora and subauroral optical structures, see Gallardo-Lacourt et al. (2021).

2.5.1 STEVE

Ground-based all-sky cameras have recently played an important role in the discovery of a new subauroral optical structure. Amateur night sky watchers and auroral photographers have documented this upper atmospheric phenomenon for decades, but the scientific community only recently began analyzing it in detail. The story behind this discovery can be found in Gallardo-Lacourt et al. (2019). This new optical structure appears as a narrow luminous mauve structure extending across the night sky over thousands of kilometers in the east-west direction. Initially, observers named the structure as "Steve," a denomination without physical implications. Currently, the phenomenon is known as Strong Thermal Emission Velocity Enhancement or STEVE, based on its observed characteristics (MacDonald et al., 2018). STEVE is sometimes observed

Fig. 2.17 STEVE and picket Fence observation on September 16, 2016 at Berg Lake, British Columbia, Canada. *Photograph courtesy of citizen scientist Robert Downie.*

together with a green feature resembling a picket fence (Archer et al., 2019a; MacDonald et al., 2018; Nishimura et al., 2019). Fig. 2.17 shows STEVE with an associated green picket fence observed on September 16, 2017 over Berg Lake in British Columbia, Canada.

Observations using ASIs in combination with Swarm satellite data revealed that STEVE is a subauroral phenomenon that lies equatorward of the auroral oval (MacDonald et al., 2018), and it corresponds to an optical signature of unusually strong subauroral ion drifts (SAID) (Archer et al., 2019a; MacDonald et al., 2018; Nishimura et al., 2019). Since SAID are westward flow enhancements driven by plasma sheet flow channels, this implies that STEVE is another manifestation of the azimuthal deflection of flow channels that penetrate equatorward of the electron auroral oval. As found by Gallardo-Lacourt et al. (2018), STEVE typically occurs from 22 to 01 MLT, lasts for about 1 h, and has a latitudinal width of ~20 km. Its maximum observed longitudinal extent is ~2145 km, limited by the ASI FOVs. In addition, a superposed epoch analysis shows that STEVE typically occurs after a prolonged of auroral oval activity as indicated by the AL index (~ twice the typical AL enhancement duration of ~25 min). This is consistent with STEVE being related to plasma sheet flow channels and associated auroral streamers, and indicates that the emergence of STEVE requires a prolonged period of such activity.

Fig. 2.18A presents an example of a STEVE event showing the evolution of a narrow STEVE structure observed by the Athabasca (ATHA) THEMIS ASI. Panels (B) and (C) show the AL index for 16,000 substorms from the SuperMAG database and for the 28 STEVE events analyzed in the statistical study.

In addition, Swarm measurements show only small magnetic field perturbations in association with STEVE, which imply small downward FACs. This indicates that STEVE is not generated like auroral arcs, which are associated with upward FACs. Consistent with this indication, Gallardo-Lacourt et al. (2018) analyzed low-altitude satellite data in conjunction with ASI observations and found that the precipitating particle energy flux was at least two orders of magnitude smaller than the average energy fluxes reported for visible auroras, concluding that these precipitating particles cannot be responsible for the luminosity of STEVE. This indicates that STEVE is not produced by particle precipitation like the aurora, but generated locally in the ionosphere.

Nishimura et al. (2019) evaluated the magnetospheric drivers of several STEVE events and the sometimes associated "picket fence" structures using a combination of citizen scientists' photographs, ground and space-based imagers, and satellite data. They found an absence of proton precipitation, thereby confirming that STEVE does not correspond to proton precipitation. Moreover, electron precipitation was only observed for events where the picket fence was present, indicating that, in the absence of the green-rayed structure, there is no significant electron precipitation. In contrast, the picket fence structure was associated with electron precipitation at energies of the order of 10 keV and weak upward FAC. Mishin and Streltsov (2019) demonstrated that these energetic electrons are critical for the structuring observed in the picket fence. Nishimura et al. (2019) also reported observations of waves and strong electric fields that could be responsible for producing heating in the F region ionosphere, as well as a structured electron boundary that may drive precipitation for the picket fence structure.

Research on the topic of STEVE is currently rapidly increasing. For example, Gillies et al. (2019) showed the first spectrographic measurements of STEVE and the picket fence. For STEVE, they found a continuous emission spectrum between 400 and 800 emission, and suggested that nitric oxide may be responsible for the continuum. In contrast, the picket fence presented a singular discrete emission at 557.7 nm, corresponding to the green line emission. Harding et al. (2020) described a physical mechanism for the formation of STEVE continuum emission by using a simple photochemical model. In this model, molecular nitrogen colliding with fast-moving ions observed in extreme SAIDs undergo chemical reactions to produce a continuous spectrum. The results of this work explain the necessity for extreme SAIDs at the time of STEVE formation and predict emission altitudes comparable to those previously estimated by Archer et al. (2019b).

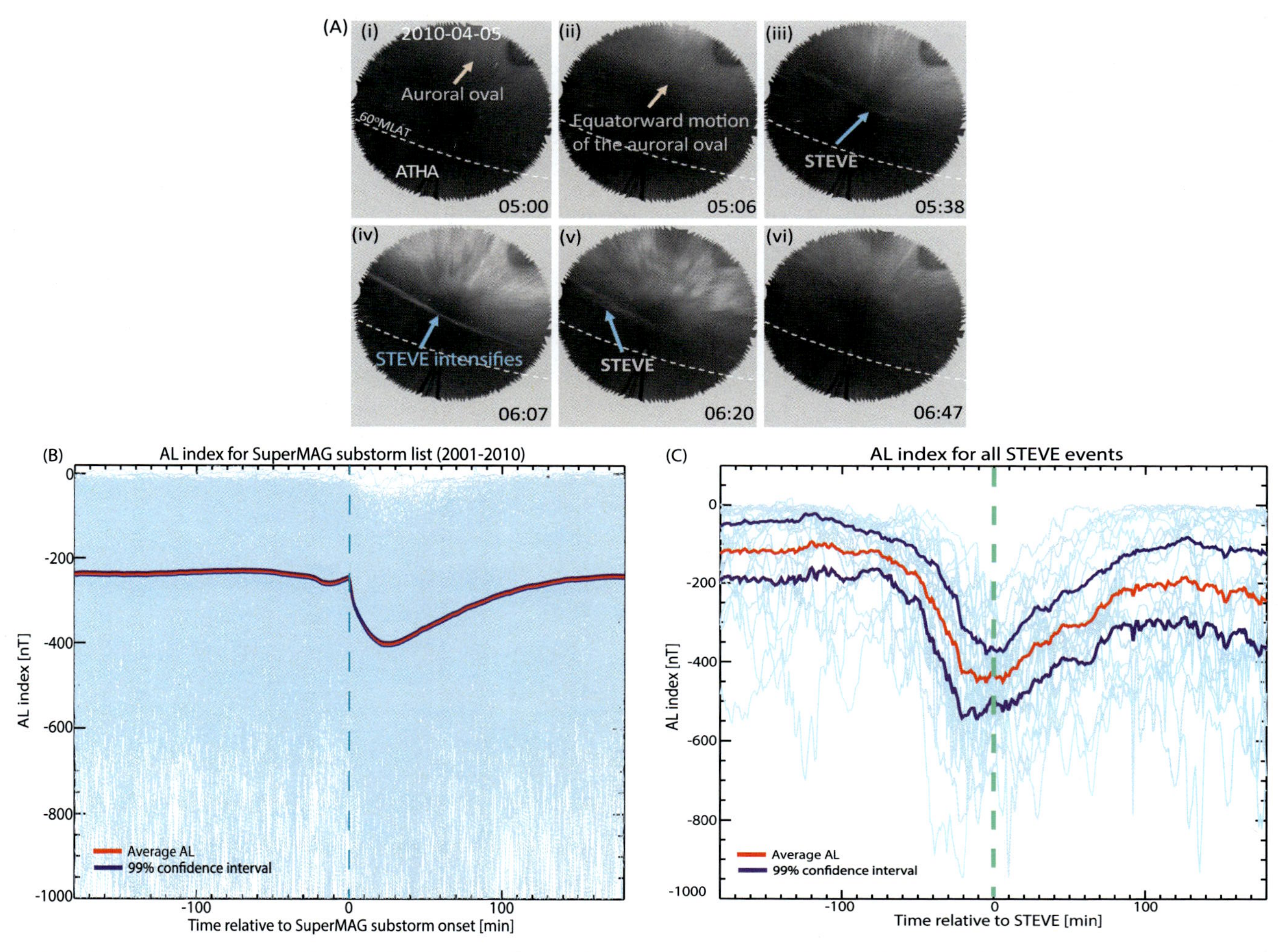

Fig. 2.18 (A) Time-series evolution of a STEVE event analyzed by Gallardo-Lacourt et al. (2018), (B) average AL index for the 16k substorms, and (C) AL index for the 28 STEVE events analyzed by Gallardo-Lacourt et al. (2018).

The potential involvement of nitric oxide or molecular nitrogen suggest that unusually strong heating may transport molecular neutrals from lower altitudes and change the composition in the thermosphere. The composition change could result in the emergence of the characteristic wavelength spectrum of STEVE.

2.6 Concluding statement

Auroral observations in combination with radar observations of ionospheric flows have shown that the structure and evolution of the aurora are features of, and reveal much about, the critical roles played by meso-scale structuring of the coupled magnetosphere-ionosphere. Auroral forms emanate from the dayside boundary of the polar cap and reflect flow channels that propagate across the polar cap. As the flows impinge upon the poleward boundary of the auroral oval, they penetrate onto nightside, plasma sheet magnetic field lines as localized regions of enhanced reconnection. They then become important for major auroral and associated geomagnetic disturbance phenomena within the nightside magnetosphere-ionosphere system. These include the disturbances associated with PBIs and streamers themselves. They also lead to auroral arc formation during the substorm growth phase and to the substorm onset instability in the near-Earth plasma sheet, and to omega bands in the midnight-to-dawn sector. Furthermore, auroral streamers and their associated flows appear to give the major magnetic signatures that occur during the substorm expansion-phase substorms.

As modeled by the RCM, flow bursts can be guided to the SAPS (DAPS) region forming strong westward (eastward) flows in the region of downward Region 2 (Region 1) that lie equatorward (poleward) and the duskside electron auroral oval (dawnside region of strongest aurora). This guiding results from the plasma sheet bubble expansion and growth that results from the combination of electric and magnetic drift. Some auroral streamers approach the equatorward boundary of the auroral oval, indicating flow channels that can sometimes lead to optical emissions equatorward of the nominal auroral oval as well as the dramatic and highly publicized STEVE emissions.

2.6.1 Areas for future research

We call attention to:

1. What processes cause poleward-moving auroral forms and their associated flow channels to arise on the dayside and what determines their size, intensity, and duration?
2. How do polar cap aurora and patches propagate across the polar cap, what is the structure of their associated flow channels as they propagate, and do these structures lie on magnetic field lines that remain connected to the solar wind flow?
3. Upon impinging on the poleward boundary of the nightside auroral oval, how do flow channels drive the PBIs and the associated magnetotail reconnection?

4. How, and how often, does this reconnection lead to the reduced entropy bubbles and associated auroral streamers, when do these streamers extend a substantial distance earthward in the plasma sheet, and when to they impact substorm and subauroral processes?
5. What are the full effects of the spreading and growth of the bubbles within the plasma sheet on substorm and other auroral development?
6. How does the interaction of multiple bubbles affects the evolution of substorm aurora.
7. What is the physics of the substorm onset instability, including its nonlinear evolution into substorm expansion-phase dynamics as revealed by substorm auroral evolution.
8. What produces the continuum spectrum of STEVE at ~300 km altitude?
9. What's the physical mechanism that allows for the formation of subauroral proton aurora?

Acknowledgments

This work at UCLA was supported by the National Science Foundation grant 1907483 and AGS-2100975, NASA grant 80NSSC20K1314. The work at the Boston University was supported by NASA 80NSSC18K0657 and 80NSSC20K0604, 80NSSC21K1321, and 80NSSC20K0725, NSF AGS-1907698, AGS-2100975, and AGS-1762141, and AFOSR FA9559-16-1-0364. We thank Nozomu Nishitani, Xueling Shi, and Bill Bristow for assistance and access to the SuperDARN data, Don Hampton for assistance and access to the Poker Flat ASI data, and Roger Varney and Ashton S. Reimer for assistance and access to the PFISR data, Kazuo Shiokawa for the Resokute Bay ASI, and Vassilis Angelopoulos and Eric Donovan for access to the THEMIS ASI data. The plots combining the above data sets are part of ongoing collaboration with Jiang Liu. We thank Robert Downie for his beautiful picture of STEVE.

References

Akasofu, S.-I., 1964. The development of the auroral substorm. Planet. Space Sci. 12, 273–282. https://doi.org/10.1016/0032-0633(64)90151-5.

Akasofu, S.-I., 1974. A study of auroral displays photographed from the DMSP-2 satellite and from the Alaska meridian chain of stations. Space Sci. Rev. 16, 617–725. https://doi.org/10.1007/BF00182598.

Akasofu, S.-I., Kimball, D.S., 1964. The dynamics of the aurora—I: instabilities of the aurora. J. Atmos. Terr. Phys. 26, 205–211. https://doi.org/10.1016/0021-9169(64)90147-3.

Akasofu, S.-I., Meng, C.-I., 1969. A study of polar magnetic substorms. J. Geophys. Res. 74, 293–313. https://doi.org/10.1029/JA074i001p00293.

Akasofu, S.-I., Kimball, D.S., Meng, C.-I., 1965. The dynamics of the aurora—III westward drifting loops. J. Atmos. Terr. Phys. 27, 189–196. https://doi.org/10.1016/0021-9169(65)90115-7.

Anderson, P.C., Heelis, R.A., Hanson, W.B., 1991. The ionospheric signatures of rapid subauroral ion drifts. J. Geophys. Res. Space Physics 96, 5785–5792. https://doi.org/10.1029/90JA02651.

Anderson, P.C., Hanson, W.B., Heelis, R.A., Craven, J.D., Baker, D.N., Frank, L.A., 1993. A proposed production model of rapid subauroral ion drifts and their relationship to substorm evolution. J. Geophys. Res. Space Physics 98, 6069–6078. https://doi.org/10.1029/92JA01975.

Anderson, P.C., Carpenter, D.L., Tsuruda, K., Mukai, T., Rich, F.J., 2001. Multisatellite observations of rapid subauroral ion drifts (SAID). J. Geophys. Res. Space Physics 106, 29585–29599. https://doi.org/10.1029/2001JA000128.

Andre, D., Baumjohann, W., 1982. Joint two-dimensional observations of ground magnetic and ionospheric electric-fields associated with auroral currents. 5. Current system associated with eastward drifting omega bands. J. Geophys. Z. Geophys. 50, 194–201.
Angelopoulos, V., Baumjohann, W., Kennel, C.F., Coroniti, F.V., Kivelson, M.G., Pellat, R., Walker, R.J., Lühr, H., Paschmann, G., 1992. Bursty bulk flows in the inner central plasma sheet. J. Geophys. Res. 97, 4027–4039. https://doi.org/10.1029/91JA02701.
Angelopoulos, V., Kennel, C.F., Coroniti, F.V., Pellat, R., Kivelson, M.G., Walker, R.J., Russell, C.T., Baumjohann, W., Feldman, W.C., Gosling, J.T., 1994. Statistical characteristics of bursty bulk flow events. J. Geophys. Res. 99, 280.
Angelopoulos, V., Phan, T.D., Larson, D.E., Mozer, F.S., Lin, R.P., Tsuruda, K., Hayakawa, H., Mukai, T., Kokubun, S., Yamamoto, T., Williams, D.J., McEntire, R.W., Lepping, R.P., Parks, G.K., Brittnacher, M., Germany, G., Spann, J., Singer, H.J., Yumoto, K., 1997. Magnetotail flow bursts: association to global magnetospheric circulation, relationship to ionospheric activity and direct evidence for localization. Geophys. Res. Lett. 24, 2271–2274. https://doi.org/10.1029/97GL02355.
Archer, W.E., Gallardo-Lacourt, B., Perry, G.W., St. Maurice, J.P., Buchert, S.C., Donovan, E., 2019a. Steve: the optical signature of intense subauroral ion drifts. Geophys. Res. Lett. 46, 6279–6286. https://doi.org/10.1029/2019GL082687.
Archer, W.E., Maurice, J.-P.S., Gallardo-Lacourt, B., Perry, G.W., Cully, C.M., Donovan, E., Gillies, D. M., Downie, R., Smith, J., Eurich, D., 2019b. The vertical distribution of the optical emissions of a Steve and picket fence event. Geophys. Res. Lett. 46, 10719–10725. https://doi.org/10.1029/2019GL084473.
Baumjohann, W., Paschmann, G., Cattell, C.A., 1989. Average plasma properties in the central plasma sheet. J. Geophys. Res. Space Physics 94, 6597–6606. https://doi.org/10.1029/JA094iA06p06597.
Birn, J., Raeder, J., Wang, Y.L., Wolf, R.A., Hesse, M., 2004. On the propagation of bubbles in the geomagnetic tail. Ann Geophys 22, 1773–1786. https://doi.org/10.5194/angeo-22-1773-2004.
Carlson, H.C., 2012. Sharpening our thinking about polar cap ionospheric patch morphology, research, and mitigation techniques. Radio Sci. 47. https://doi.org/10.1029/2011RS004946, RS0L21.
Carlson, H.C., Moen, J., Oksavik, K., Nielsen, C.P., McCrea, I.W., Pedersen, T.R., Gallop, P., 2006. Direct observations of injection events of subauroral plasma into the polar cap. Geophys. Res. Lett. 33. https://doi.org/10.1029/2005GL025230, L05103.
Cheng, C.Z., Lui, A.T.Y., 1998. Kinetic ballooning instability for substorm onset and current disruption observed by AMPTE/CCE. Geophys. Res. Lett. 25, 4091–4094. https://doi.org/10.1029/1998GL900093.
Crowley, G., 1996. Critical review of ionospheric patches and blobs. In: Stone, W.R. (Ed.), Review of Radio Science 1993–1996, Chapter 27. Oxford Science Publication, UK, pp. 619–648.
de la Beaujardière, O., Lyons, L.R., Ruohoniemi, J.M., Friis-Christensen, E., Danielsen, C., Rich, F.J., Newell, P.T., 1994. Quiet-time intensifications along the poleward auroral boundary near midnight. J. Geophys. Res. 99, 287–298. https://doi.org/10.1029/93JA01947.
Deehr, C., Lummerzheim, D., 2001. Ground-based optical observations of hydrogen emission in the auroral substorm. J. Geophys. Res. Space Physics 106, 33–44. https://doi.org/10.1029/2000JA002010.
Dubyagin, S., Sergeev, V., Apatenkov, S., Angelopoulos, V., Nakamura, R., McFadden, J., Larson, D., Bonnell, J., 2010. Pressure and entropy changes in the flow-braking region during magnetic field dipolarization. J. Geophys. Res. 115. https://doi.org/10.1029/2010JA015625.
Dungey, J.W., 1965. The length of the magnetospheric tail. J. Geophys. Res. 70, 1753. https://doi.org/10.1029/JZ070i007p01753.
Engebretson, M.J., Pilipenko, V.A., Ahmed, L.Y., Posch, J.L., Steinmetz, E.S., Moldwin, M.B., Connors, M.G., Weygand, J.M., Mann, I.R., Boteler, D.H., Russell, C.T., Vorobev, A.V., 2019a. Nighttime magnetic perturbation events observed in arctic Canada: 1. Survey and statistical analysis. J. Geophys. Res. Space Physics 124, 7442–7458. https://doi.org/10.1029/2019JA026794.
Engebretson, M.J., Steinmetz, E.S., Posch, J.L., Pilipenko, V.A., Moldwin, M.B., Connors, M.G., Boteler, D.H., Mann, I.R., Hartinger, M.D., Weygand, J.M., Lyons, L.R., Nishimura, Y., Singer, H.J., Ohtani, S., Russell, C.T., Fazakerley, A., Kistler, L.M., 2019b. Nighttime magnetic perturbation events observed in arctic Canada: 2. Multiple-instrument observations. J. Geophys. Res. Space Physics 124, 7459–7476. https://doi.org/10.1029/2019JA026797.

Erickson, P.J., Foster, J.C., Holt, J.M., 2002. Inferred electric field variability in the polarization jet from Millstone Hill E region coherent scatter observations. Radio Sci. 37, 11-1–11-14. https://doi.org/10.1029/2000RS002531.

Fear, R.C., Trenchi, L., Coxon, J.C., Milan, S.E., 2017. How much flux does a flux transfer event transfer? J. Geophys. Res. Space Physics 122 (12), 310–12,327. https://doi.org/10.1002/2017JA024730.

Foster, J.C., Burke, W.J., 2002. SAPS: a new categorization for sub-auroral electric fields. Eos Trans. Am. Geophys. Union 83, 393–394. https://doi.org/10.1029/2002EO000289.

Foster, J.C., Vo, H.B., 2002. Average characteristics and activity dependence of the subauroral polarization stream. J. Geophys. Res. Space Physics 107, 1475. https://doi.org/10.1029/2002JA009409.

Foster, J.C., Erickson, P.J., Lind, F.D., Rideout, W., 2004. Millstone Hill coherent-scatter radar observations of electric field variability in the sub-auroral polarization stream. Geophys. Res. Lett. 31. https://doi.org/10.1029/2004GL021271.

Frank, L.A., Craven, J.D., Burch, J.L., Winningham, J.D., 1982. Polar views of the Earth's aurora with dynamics explorer. Geophys. Res. Lett. 9, 1001–1004. https://doi.org/10.1029/GL009i009p01001.

Frank, L.A., Craven, J.D., Gurnett, D.A., Shawhan, S.D., Burch, J.L., Winningham, J.D., et al., 1986. The theta aurora. J. Geophys. Res. 91, 3177–3224. https://doi.org/10.1029/JA091iA03p03177.

Frey, H.U., 2007. Localized aurora beyond the auroral oval. Rev. Geophys. 45. https://doi.org/10.1029/2005RG000174.

Frey, H.U., 2010. Comment on "Substorm triggering by new plasma intrusion: THEMIS all-sky imager observations" by Y. Nishimura et al. J. Geophys. Res. Space Physics 115. https://doi.org/10.1029/2010JA016113, A12232.

Frey, H.U., Mende, S.B., Immel, T.J., Fuselier, S.A., Claflin, E.S., Gérard, J.-C., Hubert, B., 2002. Proton aurora in the cusp. J. Geophys. Res. Space Physics 107, SMP 2-1–SMP 2-17. https://doi.org/10.1029/2001JA900161.

Frey, H.U., Han, D., Kataoka, R., Lessard, M.R., Milan, S.E., Nishimura, Y., Strangeway, R.J., Zou, Y., 2019. Dayside aurora. Space Sci. Rev. 215, 51. https://doi.org/10.1007/s11214-019-0617-7.

Fuselier, S.A., Frey, H.U., Trattner, K.J., Mende, S.B., Burch, J.L., 2002. Cusp aurora dependence on interplanetary magnetic field Bz. J. Geophys. Res. Space Physics 107, SIA 6-1–SIA 6-10. https://doi.org/10.1029/2001JA900165.

Gabrielse, C., Nishimura, Y., Lyons, L., Gallardo-Lacourt, B., Deng, Y., Donovan, E., 2018. Statistical properties of mesoscale plasma flows in the nightside high-latitude ionosphere. J. Geophys. Res. Space Physics 123, 6798–6820. https://doi.org/10.1029/2018JA025440.

Gallardo-Lacourt, B., Nishimura, Y., Lyons, L.R., Ruohoniemi, J.M., Donovan, E., Angelopoulos, V., McWilliams, K.A., Nishitani, N., 2014a. Ionospheric flow structures associated with auroral beading at substorm auroral onset. J. Geophys. Res. Space Physics 119, 9150–9159. https://doi.org/10.1002/2014JA020298.

Gallardo-Lacourt, B., Nishimura, Y., Lyons, L.R., Zou, S., Angelopoulos, V., Donovan, E., McWilliams, K.A., Ruohoniemi, J.M., Nishitani, N., 2014b. Coordinated SuperDARN THEMIS ASI observations of mesoscale flow bursts associated with auroral streamers. J. Geophys. Res. Space Physics 119, 142–150. https://doi.org/10.1002/2013JA019245.

Gallardo-Lacourt, B., Perry, G., Archer, W., Donovan, E., 2019. How did we miss this? An upper atmospheric discovery named STEVE [WWW document]. Eos. https://eos.org/features/how-did-we-miss-this-an-upper-atmospheric-discovery-named-steve. accessed 10.8.20.

Gallardo-Lacourt, B., Nishimura, Y., Lyons, L.R., Mishin, E.V., Ruohoniemi, J.M., Donovan, E.F., Angelopoulos, V., Nishitani, N., 2017. Influence of auroral streamers on rapid evolution of ionospheric SAPS flows. J. Geophys. Res. Space Physics 122, 12406–12420. https://doi.org/10.1002/2017JA024198.

Gallardo-Lacourt, B., Nishimura, Y., Donovan, E., Gillies, D.M., Perry, G.W., Archer, W.E., Nava, O.A., Spanswick, E.L., 2018. A statistical analysis of STEVE. J. Geophys. Res. Space Physics 123, 9893–9905. https://doi.org/10.1029/2018JA025368.

Gallardo-Lacourt, B., Frey, H.U., Martinis, C., 2021. Proton aurora and optical emissions in the subauroral region. Space Sci. Rev. 217, 10. https://doi.org/10.1007/s11214-020-00776-6.

Galperin, Y., Ponomarev, V.N., Zosimova, A.G., 1974. Plasma convection in polar ionosphere. Ann. Geophys. 30, 1.

Gillies, D.M., Donovan, E., Hampton, D., Liang, J., Connors, M., Nishimura, Y., Gallardo-Lacourt, B., Spanswick, E., 2019. First observations from the TREx spectrograph: the optical spectrum of STEVE and the picket fence phenomena. Geophys. Res. Lett. 46, 7207–7213. https://doi.org/10.1029/2019GL083272.

Goodwin, L.V., Nishimura, Y., Zou, Y., Shiokawa, K., Jayachandran, P.T., 2019. Mesoscale convection structures associated with airglow patches characterized using cluster-imager conjunctions. J. Geophys. Res. Space Physics 124, 7513–7532. https://doi.org/10.1029/2019JA026611.

Haerendel, G., 2011. Six auroral generators: a review. J. Geophys. Res. Space Physics 116. https://doi.org/10.1029/2010JA016425.

Han, D.-S., Nishimura, Y., Lyons, L.R., Hu, H.-Q., Yang, H.-G., 2016. Throat aurora: the ionospheric signature of magnetosheath particles penetrating into the magnetosphere. Geophys. Res. Lett. 43, 1819–1827. https://doi.org/10.1002/2016GL068181.

Han, D.-S., Li, J.-X., Nishimura, Y., Lyons, L.R., Bortnik, J., Zhou, M., Liu, J.-J., Hu, Z.-J., Hu, H.-Q., Yang, H.-G., Fuselier, S.A., Contel, O.L., Ergun, R.E., Malaspina, D., Lindqvist, P.-A., Pollock, C.J., 2017. Coordinated observations of two types of diffuse auroras near magnetic local noon by magnetospheric multiscale mission and ground all-sky camera. Geophys. Res. Lett. 44, 8130–8139. https://doi.org/10.1002/2017GL074447.

Harang, L., 1946. The mean field of disturbance of polar geomagneticstorms. Terr. Magn. Atmos. Elec. 51, 353–371.

Harding, B.J., Mende, S.B., Triplett, C.C., Wu, Y.-J.J., 2020. A mechanism for the STEVE continuum emission. Geophys. Res. Lett. 47. https://doi.org/10.1029/2020GL087102, e2020GL087102.

Heikkila, W.J., Winningham, J.D., 1971. Penetration of magnetosheath plasma to low altitudes through the dayside magnetospheric cusps. J. Geophys. Res. 1896-1977 (76), 883–891. https://doi.org/10.1029/JA076i004p00883.

Henderson, M.G., 2012. Auroral substorms, poleward boundary activations, auroral streamers, omega bands, and onset precursor activity. In: Keiling, A., Donovan, E., Bagenal, F., Karlsson, T. (Eds.), Auroral Phenomenology and Magnetospheric Processes: Earth and Other Planets. American Geophysical Union, Washington, DC, https://doi.org/10.1029/2011GM001165.

Henderson, M.G., Reeves, G.D., Murphree, J.S., 1998. Are north-south aligned auroral structures an ionospheric manifestation of bursty bulk flows? Geophys. Res. Lett. 25, 3737–3740.

Hones Jr., E.W., Craven, J.D., Frank, L.A., Evans, D.S., Newell, P.T., 1989. The horse-collar aurora: a frequent pattern of the aurora in quiet times. Geophys. Res. Lett. 16, 37.

Hosokawa, K., Moen, J.I., Shiokawa, K., Otsuka, Y., 2011a. Decay of polar cap patch. J. Geophys. Res. Space Physics 116. https://doi.org/10.1029/2010JA016297.

Hosokawa, K., Moen, J.I., Shiokawa, K., Otsuka, Y., 2011b. Motion of polar cap arcs. J. Geophys. Res. Space Physics 116. https://doi.org/10.1029/2010JA015906.

Hosokawa, K., Milan, S.E., Lester, M., Kadokura, A., Sato, N., Bjornsson, G., 2013. Large flow shears around auroral beads at substorm onset. Geophys. Res. Lett. 40, 4987–4991. https://doi.org/10.1002/grl.50958.

Immel, T.J., Mende, S.B., Frey, H.U., Peticolas, L.M., Carlson, C.W., Gérard, J.-C., Hubert, B., Fuselier, S.A., Burch, J.L., 2002. Precipitation of auroral protons in detached arcs. Geophys. Res. Lett. 29, 14-1–14-4. https://doi.org/10.1029/2001GL013847.

Kalmoni, N.M.E., Rae, I.J., Murphy, K.R., Forsyth, C., Watt, C.E.J., Owen, C.J., 2017. Statistical azimuthal structuring of the substorm onset arc: implications for the onset mechanism. Geophys. Res. Lett. 44, 2078–2087. https://doi.org/10.1002/2016GL071826.

Kawasaki, K., Rostoker, G., 1979. Auroral motions and magnetic variations associated with the onset of auroral substorms. J. Geophys. Res. Space Physics 84, 7113–7122. https://doi.org/10.1029/JA084iA12p07113.

Kim, H.-J., Lyons, L.R., Zou, S., Boudouridis, A., Lee, D.-Y., Heinselman, C., McCready, M., 2009. Evidence that solar wind fluctuations substantially affect the strength of dayside ionospheric convection. J. Geophys. Res. 114. https://doi.org/10.1029/2009JA014280.

Koustov, A.V., St. Maurice, J.-P., Sofko, G.J., Andre, D., MacDougall, J.W., Hairston, M.R., Fiori, R.A., Kadochnikov, E.E., 2009. Three-way validation of the Rankin inlet PolarDARN radar velocity measurements. Radio Sci. 44. https://doi.org/10.1029/2008RS004045.

Link, R., Cogger, L.L., 1988. A reexamination of the O I 6300-Å nightglow. J. Geophys. Res. Space Physics 93, 9883–9892. https://doi.org/10.1029/JA093iA09p09883.

Liu, J., Lyons, L.R., Archer, W.E., Gallardo-Lacourt, B., Nishimura, Y., Zou, Y., Gabrielse, C., Weygand, J.M., 2018. Flow shears at the poleward boundary of omega bands observed during conjunctions of swarm and THEMIS ASI. Geophys. Res. Lett. 45, 1218–1227. https://doi.org/10.1002/2017GL076485.

Liu, J., Lyons, L.R., Wang, C.-P., Hairston, M.R., Zhang, Y., Zou, Y., 2020. Dawnside auroral polarization streams. J. Geophys. Res. Space Physics 125. https://doi.org/10.1029/2019JA027742, e2019JA027742.

Lockwood, M., McCrea, I.W., Milan, S.E., Moen, J., Cerisier, J.C., Thorolfsson, A., 2000. Plasma structure within poleward-moving cusp/cleft auroral transients: EISCAT Svalbard radar observations and an explanation in terms of large local time extent of events. Ann. Geophys. 18, 1027–1042. https://doi.org/10.1007/s00585-000-1027-5.

Lorentzen, D.A., Shumilov, N., Moen, J., 2004. Drifting airglow patches in relation to tail reconnection. Geophys. Res. Lett. 31. https://doi.org/10.1029/2003GL017785.

Lorentzen, D.A., Moen, J., Oksavik, K., Sigernes, F., Saito, Y., Johnsen, M.G., 2010. In situ measurement of a newly created polar cap patch. J. Geophys. Res. Space Physics 115. https://doi.org/10.1029/2010JA015710, A12323.

Lui, A.T.Y., 1996. Current disruption in the Earth's magnetosphere: observations and models. J. Geophys. Res. Space Physics 101, 13067–13088. https://doi.org/10.1029/96JA00079.

Lyons, L.R., 1980. Generation of large-scale regions of auroral currents, electric potentials, and precipitation by the divergence of the convection electric field. J. Geophys. Res. 85, 17–24. https://doi.org/10.1029/JA085iA01p00017.

Lyons, L., 2000. Geomagnetic disturbances: characteristics of, distinction between types, and relations to interplanetary conditions. J. Atmos. Sol. Terr. Phys. 62, 1087–1114.

Lyons, L.R., Nishimura, Y., 2020. Substorm onset and development: the crucial role of flow channels. J. Atmos. Sol. Terr. Phys. 211, 105474. https://doi.org/10.1016/j.jastp.2020.105474.

Lyons, L.R., Walterscheid, R.L., 1985. Generation of auroral omega bands by shear instability of the neutral winds. J. Geophys. Res. 90, 12321–12329. https://doi.org/10.1029/JA090iA12p12321.

Lyons, L.R., Nagai, T., Blanchard, G.T., Samson, J.C., Yamamoto, T., Mukai, T., Nishida, A., Kokubun, S., 1999. Association between Geotail plasma flows and auroral poleward boundary intensifications observed by CANOPUS photometers. J. Geophys. Res. 104, 4485–4500. https://doi.org/10.1029/1998JA900140.

Lyons, L.R., Kim, H.-J., Xing, X., Zou, S., Lee, D.-Y., Heinselman, C., Nicolls, M.J., Angelopoulos, V., Larson, D., McFadden, J., Runov, A., Fornacon, K.-H., 2009. Evidence that solar wind fluctuations substantially affect global convection and substorm occurrence. J. Geophys. Res. 114. https://doi.org/10.1029/2009JA014281.

Lyons, L.R., Nishimura, Y., Shi, Y., Zou, S., Kim, H.-J., Angelopoulos, V., Heinselman, C., Nicolls, M.J., Fornacon, K.-H., 2010. Substorm triggering by new plasma intrusion: Incoherent-scatter radar observations. J. Geophys. Res. 115. https://doi.org/10.1029/2009JA015168.

Lyons, L.R., Nishimura, Y., Kim, H.-J., Donovan, E., Angelopoulos, V., Sofko, G., Nicolls, M., Heinselman, C., Ruohoniemi, J.M., Nishitani, N., 2011. Possible connection of polar cap flows to pre- and post-substorm onset PBIs and streamers. J. Geophys. Res. 116, 14. https://doi.org/10.1029/2011JA016850.

Lyons, L.R., Nishimura, Y., Gallardo-Lacourt, B., Zou, Y., Donovan, E., Mende, S., Angelopoulos, V., Ruohoniemi, J., McWilliams, K., 2013. Westward traveling surges: sliding along boundary arcs and distinction from onset arc brightening. J. Geophys. Res. Space Physics 118, 7643–7653. https://doi.org/10.1002/2013JA019334.

Lyons, L.R., Zou, Y., Nishimura, Y., Gallardo-Lacourt, B., Angelopulos, V., Donovan, E.F., 2018. Storm-time substorm onsets: occurrence and flow channel triggering. Earth Planets Space 70, 81. https://doi.org/10.1186/s40623-018-0857-x.

Lyons, L.R., Nishimura, Y., Gallardo-Lacourt, B., Nicolls, M.J., Chen, S., Hampton, D.L., Bristow, W.A., Ruohoniemi, J.M., Nishitani, N., Donovan, E.F., Angelopoulos, V., 2015. Azimuthal flow bursts in the inner plasma sheet and possible connection with SAPS and plasma sheet earthward flow bursts. J. Geophys. Res. Space Physics 120. https://doi.org/10.1002/2015JA021023. 2015JA021023.

MacDonald, E.A., Donovan, E., Nishimura, Y., Case, N.A., Gillies, D.M., Gallardo-Lacourt, B., Archer, W.E., Spanswick, E.L., Bourassa, N., Connors, M., Heavner, M., Jackel, B., Kosar, B., Knudsen, D.J., Ratzlaff, C., Schofield, I., 2018. New science in plain sight: citizen scientists lead to the discovery of optical structure in the upper atmosphere. Sci. Adv. 4. https://doi.org/10.1126/sciadv.aaq0030, eaaq0030.

Makarevich, R.A., Dyson, P.L., 2007. Dual HF radar study of the subauroral polarization stream. Ann. Geophys. 25, 2579–2591. https://doi.org/10.5194/angeo-25-2579-2007.

Makarevich, R.A., Kellerman, A.C., Devlin, J.C., Ye, H., Lyons, L.R., Nishimura, Y., 2011. SAPS intensification during substorm recovery: a multi-instrument case study. J. Geophys. Res. 116. https://doi.org/10.1029/2011JA016916.

Mende, S.B., Harris, S.E., Frey, H.U., Angelopoulos, V., Russell, C.T., Donovan, E., Jackel, B., Greffen, M., Peticolas, L.M., 2008. The THEMIS array of ground-based observatories for the study of auroral substorms. Space Sci. Rev. 141, 357–387. https://doi.org/10.1007/s11214-008-9380-x.

Meng, C.-I., Akasofu, S.-I., 1976. The relation between the polar cap auroral arc and the auroral oval arc. J. Geophys. Res. 81 (22). https://doi.org/10.1029/JA081i022p04004.

Mishin, E., Streltsov, A., 2019. STEVE and the picket fence: evidence of feedback-unstable magnetosphere-ionosphere interaction. Geophys. Res. Lett. 46, 14247–14255. https://doi.org/10.1029/2019GL085446.

Moen, J., Gulbrandsen, N., Lorentzen, D.A., Carlson, H.C., 2007. On the MLT distribution of *F* region polar cap patches at night. Geophys. Res. Lett. 34. https://doi.org/10.1029/2007GL029632.

Motoba, T., Hosokawa, K., Kadokura, A., Sato, N., 2012. Magnetic conjugacy of northern and southern auroral beads. Geophys. Res. Lett. 39, 5. https://doi.org/10.1029/2012GL051599.

Murphree, J.S., Cogger, L.L., 1981. Observed connections between apparent polar cap features and the instantaneous diffuse auroral oval. Planet. Space Sci. 29, 1143.

Murr, D.L., Hughes, W.J., 2001. Reconfiguration timescales of ionospheric convection. Geophys. Res. Lett. 28, 2145–2148. https://doi.org/10.1029/2000GL012765.

Nakamura, R., Oguti, T., Yamamoto, T., Kokubun, S., 1993. Equatorward and poleward expansion of the auroras during auroral substorms. J. Geophys. Res. 98, 5743–5759. https://doi.org/10.1029/92JA02230.

Nakamura, R., Baumjohann, W., Schödel, R., Brittnacher, M., Sergeev, V.A., Kubyshkina, M., Mukai, T., Liou, K., 2001. Earthward flow bursts, auroral streamers, and small expansions. J. Geophys. Res. 106, 10791–10802. https://doi.org/10.1029/2000JA000306.

Nakamura, R., Baumjohann, W., Mouikis, C., Kistler, L.M., Runov, A., Volwerk, M., Asano, Y., Vörös, Z., Zhang, T.L., Klecker, B., Rème, H., Balogh, A., 2004. Spatial scale of high-speed flows in the plasma sheet observed by cluster. Geophys. Res. Lett. 31, L09804. https://doi.org/10.1029/2004GL019558.

Nielsen, E., Craven, J.D., Frank, L.A., Heelis, R.A., 1990. Ionospheric flows associated with a transpolar arc. J. Geophys. Res. 95 (A12), 21,169–21,178. https://doi.org/10.1029/JA095iA12p21169.

Nishida, A., Kokubun, S., 1971. New polar magnetic disturbances: Sqp, SP, DPC, and DP2. Rev. Geophys. 9, 417. https://doi.org/10.1029/RG009i002p00417.

Nishimura, Y., Wygant, J., Ono, T., Iizima, M., Kumamoto, A., Brautigam, D., Friedel, R., 2008. SAPS measurements around the magnetic equator by CRRES. Geophys. Res. Lett. 35. https://doi.org/10.1029/2008GL033970.

Nishimura, Y., Lyons, L., Zou, S., Angelopoulos, V., Mende, S., 2010a. Substorm triggering by new plasma intrusion: THEMIS all-sky imager observations. J. Geophys. Res. 115. https://doi.org/10.1029/2009JA015166.

Nishimura, Y., Lyons, L.R., Zou, S., Angelopoulos, V., Mende, S.B., 2010b. Reply to comment by Harald U. Frey on "Substorm triggering by new plasma intrusion: THEMIS all-sky imager observations.". J. Geophys. Res. 115. https://doi.org/10.1029/2010JA016182.

Nishimura, Y., Lyons, L.R., Zou, S., Xing, X., Angelopoulos, V., Mende, S.B., Bonnell, J.W., Larson, D., Auster, U., Hori, T., Nishitani, N., Hosokawa, K., Sofko, G., Nicolls, M., Heinselman, C., 2010c. Preonset time sequence of auroral substorms: coordinated observations by all-sky imagers, satellites, and radars. J. Geophys. Res. 115. https://doi.org/10.1029/2010JA015832.

Nishimura, Y., Lyons, L.R., Angelopoulos, V., Kikuchi, T., Zou, S., Mende, S.B., 2011. Relations between multiple auroral streamers, pre-onset thin arc formation, and substorm auroral onset. J. Geophys. Res. Space Physics 116. https://doi.org/10.1029/2011JA016768, A09214.

Nishimura, Y., Lyons, L.R., Shiokawa, K., Angelopoulos, V., Donovan, E.F., Mende, S.B., 2013. Substorm onset and expansion phase intensification precursors seen in polar cap patches and arcs. J. Geophys. Res. Space Physics 118, 2034–2042. https://doi.org/10.1002/jgra.50279.

Nishimura, Y., Lyons, L.R., Zou, Y., Oksavik, K., Moen, J.I., Clausen, L.B., Donovan, E.F., Angelopoulos, V., Shiokawa, K., Ruohoniemi, J.M., Nishitani, N., McWilliams, K.A., Lester, M., 2014a. Day-night coupling by a localized flow channel visualized by polar cap patch propagation. Geophys. Res. Lett. 41. https://doi.org/10.1002/2014GL060301, 2014GL060301.

Nishimura, Y., Bortnik, J., Li, W., Lyons, L.R., Donovan, E.F., Angelopoulos, V., Mende, S.B., 2014b. Evolution of nightside subauroral proton aurora caused by transient plasma sheet flows. J. Geophys. Res. Space Physics 119. https://doi.org/10.1002/2014JA020029, 2014JA020029.

Nishimura, Y., Yang, J., Pritchett, P.L., Coroniti, F.V., Donovan, E.F., Lyons, L.R., Wolf, R.A., Angelopoulos, V., Mende, S.B., 2016. Statistical properties of substorm auroral onset beads/rays. J. Geophys. Res. Space Physics 121, 8661–8676. https://doi.org/10.1002/2016JA022801.

Nishimura, Y., Gallardo-Lacourt, B., Zou, Y., Mishin, E., Knudsen, D.J., Donovan, E.F., Angelopoulos, V., Raybell, R., 2019. Magnetospheric signatures of STEVE: implications for the magnetospheric energy source and interhemispheric conjugacy. Geophys. Res. Lett. 46, 5637–5644. https://doi.org/10.1029/2019GL082460.

Nishimura, Y., Lyons, L.R., Gabrielse, C., Sivadas, N., Donovan, E.F., Varney, R.H., Angelopoulos, V., Weygand, J.M., Conde, M.G., Zhang, S.R., 2020. Extreme magnetosphere-ionosphere-thermosphere responses to the 5 April 2010 supersubstorm. J. Geophys. Res. Space Physics 125. https://doi.org/10.1029/2019JA027654, e2019JA027654.

Ohtani, S., 1998. Earthward expansion of tail current disruption: dual-satellite study. J. Geophys. Res. 103, 6815–6825.

Oksavik, K., Moen, J., Carlson, H.C., 2004. High-resolution observations of the small-scale flow pattern associated with a poleward moving auroral form in the cusp. Geophys. Res. Lett. 31. https://doi.org/10.1029/2004GL019838, L11807.

Oksavik, K., Moen, J., Carlson, H.C., Greenwald, R.A., Milan, S.E., Lester, M., Denig, W.F., Barnes, R.J., 2005. Multi-instrument mapping of the small-scale flow dynamics related to a cusp auroral transient. Ann Geophys 23, 2657–2670. https://doi.org/10.5194/angeo-23-2657-2005.

Oksavik, K., Barth, V.L., Moen, J., Lester, M., 2010. On the entry and transit of high-density plasma across the polar cap. J. Geophys. Res. Space Physics 115. https://doi.org/10.1029/2010JA015817, A12308.

Opgenoorth, H.J., Oksman, J., Kaila, K.U., Nielsen, E., Baumjohann, W., 1983. Characteristics of eastward drifting omega bands in the morning sector of the auroral oval. J. Geophys. Res. Space Physics 88, 9171–9185. https://doi.org/10.1029/JA088iA11p09171.

Panov, E.V., Nakamura, R., Baumjohann, W., Angelopoulos, V., Petrukovich, A.A., Retinò, A., Volwerk, M., Takada, T., Glassmeier, K.-H., McFadden, J.P., Larson, D., 2010. Multiple overshoot and rebound of a bursty bulk flow. Geophys. Res. Lett. 37. https://doi.org/10.1029/2009GL041971.

Park, M.Y., Lee, D.-Y., Ohtani, S., Kim, K.C., 2010. Statistical characteristics and significance of low-frequency instability associated with magnetic dipolarizations in the near-Earth plasma sheet. J. Geophys. Res. 115. https://doi.org/10.1029/2010JA015566, A11203.

Paschmann, G., Haaland, S., Treumann, R., 2012. Auroral Plasma Physics. Springer Science & Business Media.

Paxton, L.J., Meng, C.-I., Fountain, G.H., Ogorzalek, B.S., Darlington, E.H., Gary, S.A., Goldsten, J.A., Kusnierkiewicz, D.Y., Lee, S.C., Linstrom, L.A., Maynard, J.J., Peacock, K., Persons, D.F., Smith, B.E., Strickland, D.G., Daniell Jr., R.E., 1993. SSUSI—Horizon-to-horizon and limb viewing spectrographic imager for remote sensing of environmental parameters. In: Proc. SPIE 1764. Ultraviolet Technology IV. https://doi.org/10.1117/12.140846.

Pitkänen, T., Aikio, A.T., Amm, O., Kauristie, K., Nilsson, H., Kaila, K.U., 2011. EISCAT-cluster observations of quiet-time near-Earth magnetotail fast flows and their signatures in the ionosphere. Ann. Geophys. 29, 299–319. https://doi.org/10.5194/angeo-29-299-2011.

Pitkänen, T., Aikio, A.T., Juusola, L., 2013. Observations of polar cap flow channel and plasma sheet flow bursts during substorm expansion. J. Geophys. Res. Space Physics 118, 774–784. https://doi.org/10.1002/jgra.50119.

Pontius, D.H., Wolf, R.A., 1990. Transient flux tubes in the terrestrial magnetosphere. Geophys. Res. Lett. 17, 49. https://doi.org/10.1029/GL017i001p00049.

Reiff, P.H., Hill, T.W., Burch, J.L., 1977. Solar wind plasma injection at the dayside magnetospheric cusp. J. Geophys. Res. 1896-1977 (82), 479–491. https://doi.org/10.1029/JA082i004p00479.

Ren, J., Zou, S., Gillies, R.G., Donovan, E., Varney, R.H., 2018. Statistical characteristics of polar cap patches observed by RISR-C. J. Geophys. Res. Space Physics 123, 6981–6995. https://doi.org/10.1029/2018JA025621.

Robinson, R.M., Vondrak, R.R., Friis-Christensen, E., 1987. Ionospheric currents associated with a Sun-aligned arc connected to the auroral oval. Geophys. Res. Lett. 14, 656–659. https://doi.org/10.1029/GL014i006p00656.

Rostoker, G., Barichello, J.C., 1980. Seasonal and diurnal variation of Ps 6 magnetic disturbances. J. Geophys. Res. Space Physics 85, 161–163. https://doi.org/10.1029/JA085iA01p00161.

Rostoker, G., Lui, A.T.Y., Anger, C.D., Murphree, J.S., 1987. North-south structures in the midnight sector auroras as viewed by the Viking imager. Geophys. Res. Lett. 14, 407. https://doi.org/10.1029/GL014i004p00407.

Roux, A., Perraut, S., Robert, P., Morane, A., Pedersen, A., Korth, A., Kremser, G., Aparicio, B., Rodgers, D., Pellinen, R., 1991. Plasma sheet instability related to the westward traveling surge. J. Geophys. Res. 96, 17697–17,714. https://doi.org/10.1029/91JA01106.

Ruohoniemi, J.M., Greenwald, R.A., 2005. Dependencies of high-latitude plasma convection: consideration of interplanetary magnetic field, seasonal, and universal time factors in statistical patterns. J. Geophys. Res. Space Physics 110. https://doi.org/10.1029/2004JA010815, A09204.

Ruohoniemi, J.M., Shepherd, S.G., Greenwald, R.A., 2002. The response of the high-latitude ionosphere to IMF variations. J. Atmos. Sol. Terr. Phys. 64, 159–171. https://doi.org/10.1016/S1364-6826(01)00081-5.

Sakaguchi, K., Shiokawa, K., Ieda, A., Nomura, R., Nakajima, A., Greffen, M., Donovan, E., Mann, I.R., Kim, H., Lessard, M., 2009. Fine structures and dynamics in auroral initial brightening at substorm onsets. Ann Geophys 27, 623–630. https://doi.org/10.5194/angeo-27-623-2009.

Samson, J.C., Lyons, L.R., Newell, P.T., Creutzberg, F., Xu, B., 1992. Proton aurora and substorm intensifications. Geophys. Res. Lett. 19, 2167. https://doi.org/10.1029/92GL02184.

Sandholt, P.R., Farrugia, C.J., 2007. Poleward moving auroral forms (PMAFs) revisited: responses of aurorae, plasma convection and Birkeland currents in the pre- and postnoon sectors under positive and negative IMF BY conditions. Ann. Geophys. 25. https://doi.org/10.5194/angeo-25-1629-2007.

Sandholt, P.E., Carlson, H.C., Egeland, A., 2002. Dayside and polar cap aurora. In: Astrophysics and Space Science Library. Springer Netherlands, https://doi.org/10.1007/0-306-47969-9.

Sergeev, V.A., Aulamo, O.A., Pellinen, R.J., Vallinkoski, M.K., Bösinger, T., Cattell, C.A., Elphic, R.C., Williams, D.J., 1990. Non-substorm transient injection events in the ionosphere and magnetosphere. Planet. Space Sci. 38, 231–239. https://doi.org/10.1016/0032-0633(90)90087-7.

Sergeev, V.A., Angelopoulos, V., Gosling, J.T., Cattell, C.A., Russell, C.T., 1996. Detection of localized, plasma-depleted flux tubes or bubbles in the midtail plasma sheet. J. Geophys. Res. Space Physics 101, 10817–10826. https://doi.org/10.1029/96JA00460.

Sergeev, V.A., Liou, K., Meng, C.-I., Newell, P.T., Brittnacher, M., Parks, G., Reeves, G.D., 1999. Development of auroral streamers in association with localized impulsive injections to the inner magnetotail. Geophys. Res. Lett. 26, 417–420. https://doi.org/10.1029/1998GL900311.

Sergeev, V.A., Sauvaud, J.-A., Popescu, D., Kovrazhkin, R.A., Liou, K., Newell, P.T., Brittnacher, M., Parks, G., Nakamura, R., Mukai, T., Reeves, G.D., 2000. Multiple-spacecraft observation of a narrow transient plasma jet in the Earth's plasma sheet. Geophys. Res. Lett. 27, 851–854. https://doi.org/10.1029/1999GL010729.

Sergeev, V., Nishimura, Y., Kubyshkina, M., Angelopoulos, V., Nakamura, R., Singer, H., 2012. Magnetospheric location of the equatorward prebreakup arc. J. Geophys. Res. 117. https://doi.org/10.1029/2011JA017154.

Shi, Y., Zesta, E., Lyons, L.R., Yang, J., Boudouridis, A., Ge, Y.S., Ruohoniemi, J.M., Mende, S., 2012. Two-dimensional ionospheric flow pattern associated with auroral streamers. J. Geophys. Res. 117. https://doi.org/10.1029/2011JA017110.

Shiokawa, K., Miyashita, Y., Shinohara, I., Matsuoka, A., 2005. Decrease in Bz prior to the dipolarization in the near-Earth plasma sheet. J. Geophys. Res. Space Physics 110. https://doi.org/10.1029/2005JA011144.

Shiokawa, K., Ieda, A., Nakajima, A., Sakaguchi, K., Nomura, R., Aslaksen, T., Greffen, M., Spanswick, E., Donovan, E., Mende, S.B., McFadden, J., Glassmeier, K.-H., Angelopoulos, V., Miyashita, Y., 2009. Longitudinal development of a substorm brightening arc. Ann. Geophys. 27, 1935–1940. https://doi.org/10.5194/angeo-27-1935-2009.

Solovyev, S.I., Baishev, D.G., Barkova, E.S., Engebretson, M.J., Posch, J.L., Hughes, W.J., Yumoto, K., Pilipenko, V.A., 1999. Structure of disturbances in the dayside and nightside ionosphere during periods of negative interplanetary magnetic field Bz. J. Geophys. Res. 104, 039.

Spiro, R.W., Heelis, R.A., Hanson, W.B., 1979. Rapid subauroral ion drifts observed by atmosphere explorer C. Geophys. Res. Lett. 6, 657–660. https://doi.org/10.1029/GL006i008p00657.

Stern, D.P., 1973. A study of the electric field in an open magnetospheric model. J. Geophys. Res. 78, 7292–7305. https://doi.org/10.1029/JA078i031p07292.

Sutcliffe, P., Lyons, L., 2002. Association between quiet-time Pi2 pulsations, poleward boundary intensifications, and plasma sheet particle fluxes. Geophys. Res. Lett. 29. https://doi.org/10.1029/2001GL014430.

Takahashi, K., Zanetti, L.J., Lopez, R.E., McEntire, R.W., Potemra, T.A., Yumoto, K., 1987. Disruption of the magnetotail current sheet observed by AMPTE/CCE. Geophys. Res. Lett. 14, 1019–1022.

Valladares, C.E., Carlson, H.C., 1991. The electrodynamic, thermal, and energetic character of intense Sun-aligned arcs in the polar cap. J. Geophys. Res. Space Physics 96, 1379–1400. https://doi.org/10.1029/90JA01765.

Valladares, C.E., Carlson, H.C., Fukui, K., 1994. Interplanetary magnetic field dependency of stable sun-aligned polar cap arcs. J. Geophys. Res. Space Physics 99, 6247–6272. https://doi.org/10.1029/93JA03255.

Wang, B., Nishimura, Y., Lyons, L.R., Zou, Y., Carlson, H.C., Frey, H.U., Mende, S.B., 2016. Analysis of close conjunctions between dayside polar cap airglow patches and flow channels by all-sky imager and DMSP. Earth Planets Space 68, 150. https://doi.org/10.1186/s40623-016-0524-z.

Wang, C.-P., Gkioulidou, M., Lyons, L.R., Wolf, R.A., 2018. Spatial distribution of plasma sheet entropy reduction caused by a plasma bubble: rice convection model simulations. J. Geophys. Res. Space Physics 123, 3380–3397. https://doi.org/10.1029/2018JA025347.

Weber, E.J., Klobuchar, J.A., Buchau, J., Carlson, H.C., Livingston, R.C., de la Beaujardiere, O., McCready, M., Moore, J.G., Bishop, G.J., 1986. Polar cap F layer patches: structure and dynamics. J. Geophys. Res. Space Phys. 91, 12121–12129. https://doi.org/10.1029/JA091iA11p1212.

Wolf, R.A., Chen, C.X., Toffoletto, F.R., 2012. Thin filament simulations for Earth's plasma sheet: interchange oscillations. J. Geophys. Res. Space Physics 117. https://doi.org/10.1029/2011JA016971.

Xing, X., Wolf, R.A., 2007. Criterion for interchange instability in a plasma connected to a conducting ionosphere. J. Geophys. Res. Space Physics 112. https://doi.org/10.1029/2007JA012535, A12209.

Xing, X., Lyons, L., Nishimura, Y., Angelopoulos, V., Larson, D., Carlson, C., Bonnell, J., Auster, U., 2010. Substorm onset by new plasma intrusion: THEMIS spacecraft observations. J. Geophys. Res. Space Phys. 115. https://doi.org/10.1029/2010JA015528.

Xu, L., Xu, J.-S., Kaustov, A.V., 2007. The response of high-latitude ionospheric convection during a southward IMF turning event. Chin. J. Geophys. 50. https://doi.org/10.1002/cjg2.1159.

Yang, J., Toffoletto, F.R., Wolf, R.A., Sazykin, S., 2011. RCM-E simulation of ion acceleration during an idealized plasma sheet bubble injection. J. Geophys. Res. 116. https://doi.org/10.1029/2010JA016346.

Yang, J., Toffoletto, F.R., Wolf, R.A., 2014. RCM-E simulation of a thin arc preceded by a north-south-aligned auroral streamer. Geophys. Res. Lett. 41, 2695–2701. https://doi.org/10.1002/2014GL059840.

Zesta, E., Lyons, L., Donovan, E., 2000. The auroral signature of Earthward flow bursts observed in the magnetotail. Geophys. Res. Lett. 27, 3241–3244.

Zesta, E., Donovan, E., Lyons, L., Enno, G., Murphree, J., Cogger, L., 2002. Two-dimensional structure of auroral poleward boundary intensifications. J. Geophys. Res. Space Phys. 107. https://doi.org/10.1029/2001JA000260.

Zhang, Q.-H., Zhang, B.-C., Lockwood, M., Hu, H.-Q., Moen, J., Ruohoniemi, J.M., Thomas, E.G., Zhang, S.-R., Yang, H.-G., Liu, R.-Y., McWilliams, K.A., Baker, J.B.H., 2013. Direct observations of the evolution of polar cap ionization patches. Science 339, 1597–1600. https://doi.org/10.1126/science.1231487.

Zhang, Y., Paxton, L.J., Zhang, Q., Xing, Z., 2016. Polar cap arcs: Sun-aligned or cusp-aligned? J. Atmos. Sol. Terr. Phys. 146, 123–128. https://doi.org/10.1016/j.jastp.2016.06.001.

Zou, S., Lyons, L.R., Nicolls, M.J., Heinselman, C.J., Mende, S.B., 2009. Nightside ionospheric electrodynamics associated with substorms: PFISR and THEMIS ASI observations. J. Geophys. Res. Space Physics 114. https://doi.org/10.1029/2009JA014259.

Zou, S., Lyons, L.R., Nishimura, Y., 2012. Mutual evolution of aurora and ionospheric electrodynamic features near the Harang reversal during substorms. Geophys. Monogr. Ser. 197, 159–169. https://doi.org/10.1029/2011GM001163.

Zou, Y., Nishimura, Y., Lyons, L.R., Shiokawa, K., Donovan, E.F., Ruohoniemi, J.M., McWilliams, K.A., Nishitani, N., 2015. Localized polar cap flow enhancement tracing using airglow patches: statistical properties, IMF dependence, and contribution to polar cap convection. J. Geophys. Res. Space Physics. https://doi.org/10.1002/2014JA020946, 2014JA020946.

Zou, Y., Nishimura, Y., Burchill, J.K., Knudsen, D.J., Lyons, L.R., Shiokawa, K., Buchert, S., Chen, S., Nicolls, M.J., Ruohoniemi, J.M., McWilliams, K.A., Nishitani, N., 2016. Localized field-aligned currents in the polar cap associated with airglow patches. J. Geophys. Res. Space Physics 121 (10), 172–10,189. https://doi.org/10.1002/2016JA022665.

CHAPTER 3

Density, irregularity, and instability

Chapter Outline

Cross-Scale Coupling and Energy Transfer in the Magnetosphere-Ionosphere-Thermosphere System
https://doi.org/10.1016/B978-0-12-821366-7.00001-9

3.1 High-latitude F-region plasma irregularities

Gareth W. Perry[a] *and Lindsay V. Goodwin*[a,b]

[a]Center for Solar-Terrestrial Research, New Jersey Institute of Technology, Newark, NJ, United States
[b]Cooperative Programs for the Advancement of Earth System Science, University Corporation for Atmospheric Research, Boulder, CO, United States

3.1.1 Introduction

Some of the most striking evidence of the complex, cross-scale interactions and interconnections present and underway in the terrestrial magnetosphere-ionosphere-thermosphere (M-I-T) system can be found at high latitudes. The region is home to density irregularities, auroral dynamics, and extremely fast plasma flows that set it apart from any other geomagnetic regions in terms of scale size and intensity. In the high-latitude ionosphere, the geomagnetic field lines are nearly vertical, with a magnetic dip angle $I \gtrsim 80$ degrees. Furthermore, the magnetic topology can be either "open" or "closed" (although the former is more likely), that is, the magnetic field lines are either connected to the interplanetary magnetic field (IMF)—"open" field lines, or they are connected to the opposite geomagnetic pole, located in the opposite hemisphere—"closed" field lines.

The first radio instruments used to study the ionosphere in detail were ionosondes. Also known as a "vertical sounder," an ionosonde provides a vertical plasma density profile of the bottomside ionosphere up to the altitude, *hmF*2, of the ionosphere's critical frequency, *foF*2. This is accomplished by increasing the ionosonde's operating frequency in successively transmitted pulses. The pulses are reflected from the altitude where the ionosphere's plasma frequency matches the frequency of the transmitted ionosonde pulse; higher frequencies continue to propagate vertically. A vertical profile of the ionosphere's plasma density below *hmF*2, a region commonly referred to as the "bottomside" ionosphere, can be deduced from the reflected transmissions. The ionosonde technique cannot measure the "topside" ionosphere, above *hmF*2, where the plasma density decreases with increasing altitude. This is because the ionosonde pulses that would have been reflected in the topside cannot reach those altitudes as they would have been already reflected (back to the ground) in the bottomside.

The high-latitude ionosphere's extreme variability in terms of ionospheric plasma density and dynamics was readily apparent early in the history of the studyed region. Ionosonde measurements of the region were peculiar; they frequently showed vertical plasma density profile traces with ionospheric peak densities at altitudes that were several factors higher than what was normally expected for the F region. The traces were also dynamic; they were seen to quickly descend before settling to an altitude that was more consistent with an F-region profile. It was quickly realized that, in fact, an anomalously high altitude F-region ionosphere was not responsible for the signature. Rather, "ionized clouds," localized enhancements in ionospheric plasma density, traveling horizontally at 300–400 m/s were responsible for the curious traces (Meek, 1949). In general, ionosondes transmit toward the local zenith. But, like all radio systems, their transmitted power also leaks in all directions, albeit at a lower intensity, which is how the clouds were detected. The apparent vertical movement of the plasma profile was a manifestation of clouds' horizontal motion with respect to the ionosonde.

These ionosonde measurements gave the first glimpse of the highly dynamic and variable high-latitude ionosphere, and among the first reports of a phenomenon that would come to be known as a "polar cap patch" (see also Section 2.3). In this chapter, we will pay particularly close attention to this phenomenon as it has garnered a great deal of attention in the decades since it was first detected. We will also examine its counterpart, the plasma density depletion, sometimes referred to as a "trough" or "hole."

Before we proceed any further, we would like to point out that the terms "patch," "trough," "hole," etc. are colloquial. For a more meticulous treatment and description of these phenomena, we will avoid using such terminology. It is our hope that our colleagues will follow suit in the future. The first term, "patch" is operationally defined by Crowley (1996) as a plasma density enhancement of a magnitude that is at least twice that of the background ionosphere, with a scale size that is of the order of 100 km, i.e., a mesoscale density structure. However, this definition is several decades old and is awkward because (1) it requires one to define the "background" plasma density, which is variable in space and time on both short-term and long-term scales in the high-latitude region, (2) it is arbitrary and excludes enhancements slightly less dense than twice the background, and (3) it is too general and does not reflect the diversity of plasma density enhancement generators present at high latitudes. In the light of new measurements and analysis of the high-latitude ionosphere that have taken place since its inception, it should be revisited, if not discontinued altogether. We do not want to develop or propagate terminology that may become redundant or ambiguous in the long term. Accordingly, we choose to classify the aforementioned phenomena simply as "plasma density irregularities," or "irregularities" for brevity.

3.1.1.1 The continuity equation

Even during the winter, when the region is completely devoid of photoionization, the high-latitude F region is swarming with plasma density irregularities of a variety of scale sizes. The irregularities include density enhancements with magnitudes that are close to that of the lower-latitude dayside ionosphere. How can this be if the main source of ionization, solar EUV flux, is absent, and even if another source of ionization, particle precipitation, is also absent? Such unique conditions are what set the high-latitude ionosphere apart from other geographic regions. As it turns out, the irregularities are a byproduct of complex and multiscale M-I-T coupling processes. To gain better insight into the sources and sinks of these irregularities, let us first consider the continuity equation.

The continuity equation is expressed as (Schunk and Nagy, 2000):

$$\frac{\partial n_i}{\partial t} + n_i \nabla \cdot \vec{v}_i + \vec{v}_i \cdot \nabla n_i = P_i - L_i, \tag{3.1}$$

where n_i is the plasma density, v_i is the ion plasma flow, P_i is the production term, and L_i is the loss term. While irregularities spread over a wide-scale range between 100 km and <1 m (the kinetic scale), we pay particular attention to Eq. (3.1) at spatial scales $\lambda > 10$ km and temporal scales $60 \leq t \leq 3600$ s. These scales are largely determined and limited by our ability to measure and study polar F-region phenomena.

Let us first deal with the second term on the left-hand side of Eq. (3.1). In the polar F region, plasma flows at the convection velocity, $\vec{v}_i = (\vec{E} \times \vec{B})/B^2$, in which $\vec{E}$ is the electric field and $\vec{B}$ is the local magnetic field. Thus, according to Rishbeth and Hanson,

$$\nabla \cdot \vec{v}_i = \nabla \cdot (\vec{E} \times \vec{B}), \tag{3.2}$$

which, after some algebraic manipulations, can be rewritten as

$$\nabla \cdot \vec{v}_i = \frac{\vec{B} \cdot (\nabla \times \vec{E}) - \vec{E} \cdot (\nabla \times \vec{B})}{B^2} + \vec{E} \times \vec{B} \cdot \nabla \frac{1}{B^2}. \tag{3.3}$$

The first two terms are zero because $\nabla \times \vec{E} = \nabla \times (-\nabla \phi) = 0$, where ϕ is an electric potential, and the F region is nearly void of strong electric currents. Furthermore, at the scales we are interested in, gradients in the magnetic field are considered to be negligible. Therefore, the F-region plasma is considered to be incompressible; $n_i \nabla \cdot \vec{v}_i \simeq 0$ (Rishbeth and Hanson, 1974).

F-region plasma production is dominated by the ionization of O via solar photons or collisions with energetic particles. However, we will consider Eq. (3.1) in the case of the F-region polar ionosphere during winter. Here we can assume that $n_i \simeq [O^+]$ (square brackets denote concentration), and $P_i \simeq 0$ (or the equivalent scenario in which protoproduction is balanced by chemical recombination). For the moment, we are also assuming that plasma production by precipitation, a second-order effect, is also negligible.

Plasma loss, L_i, is governed by chemistry, i.e., the chemical recombination of O^+, and diffusion. The two rate determining reactions in the depletion of O^+ are

$$O^+ + O_2 \xrightarrow{k_1(T)} O_2^+ + O \tag{3.4}$$

and

$$O^+ + N_2 \xrightarrow{k_2(T)} NO^+ + N \tag{3.5}$$

in which $k_1(T)$ and $k_2(T)$ are the temperature-dependent reaction rates. Note that the resulting molecular ions are much more quickly recombined with the ambient molecular neutrals than O^+. The reduction in $[O^+]$ in a reference frame moving with the plasma can be expressed as (Perry et al., 2013):

$$\frac{\partial[O^+]}{\partial t} = -[O^+](k_1(T)[O_2] + k_2(T)[N_2]). \tag{3.6}$$

Eq. (3.6) shows the sensitivity of the temperature of the plasma and the neutrals. The reaction rates can increase substantially with temperature (St.-Maurice and Torr, 1978). For example, if the effective temperature of the gas doubles from 1700 to 3400 K, $k_1(T)$ doubles and $k_2(T)$ increases eightfold. Such temperature spikes are not unheard of in the high-latitude region (Goodwin et al., 2014; Perry et al., 2015) and are predominately a byproduct of frictional heating between fast flowing ions and the neutral gas.

Under typical conditions, the reaction rates are, $k_1 = 1.5 \times 10^{-5} m^3 s^{-1}$ and $k_2 = 6.5 \times 10^{-7} m^3 s^{-1}$. Therefore, a volume of plasma of the order of $2 \times 10^{11} m^{-3}$ takes approximately 2 h to recombine and decrease by a factor of e at 300 km altitude. Under these conditions, a F-region irregularity, such as a plasma density enhancement, can be transported over large distances of the high-latitude ionosphere via $\vec{E} \times \vec{B}$ plasma convection, before it is chemically recombined.

Diffusion perpendicular to the magnetic field (i.e., "cross-field" diffusion) is strongly dependent on spatial scales. The cross-field diffusion coefficient for ions can be expressed as $D_{i\perp} = r_g^2 \nu_{in}$ (Kelley, 2009), where r_g is the ion gyroradius and ν_{in} is the ion-neutral collision frequency. For typical F-region values and scale sizes of interest here ($\lambda > 10$km), it can be shown that the contribution of diffusion to L_i is orders of magnitudes lower than chemical recombination (Tsunoda, 1988). At smaller spatial scales, cross-scale diffusion does become more important and must be considered when interpreting the presence of irregularities.

We have now dealt with most of the terms in Eq. (3.1). In the wintertime polar ionosphere (or when photoproduction and chemical recombination are balanced), the equation can be reduced to one that describes temporal variations in plasma density as a function of the transport of plasma density gradients and plasma loss (Eq. 3.6), that is:

$$\frac{\partial n_i}{\partial t} = -\vec{v}_i \cdot \nabla n_i \tag{3.7}$$

It is important to note that this equation is in the local rest frame (the Eulerian frame). That is, Eq. (3.7) describes what a ground-based instrument would measure in the presence of a plasma density irregularity. In the case of a density enhancement moving toward the instrument, $\partial n_i/\partial t > 0$ on the leading edge of the enhancement since $\vec{v}_i \cdot \nabla n_i < 0$ there, and vice versa for its trailing edge.

3.1.2 Plasma density enhancements

High-latitude F-region plasma density irregularities can be separated into two categories: enhancements and depletions. Plasma density enhancements are, simply put, structures with an enhanced plasma density relative to the ambient ionosphere. In this section, we will focus on the detection of these density enhancements and their source.

3.1.2.1 Detecting plasma density enhancements

In situ techniques

Arguably the most straightforward way of detecting plasma density enhancements is through in situ rocket or spacecraft observations. Rockets provide flexibility and high spatiotemporal resolution, making it easier to view a given region at a specific time, or perform common volume measurements with other instruments. Rocket missions (Moen et al., 2012) were able to identify enhancements which contained small scale ($\lambda < 10$ km)—decameter—variations as well. They used Langmuir probes to do so, a popular technique given its relative simplicity. Satellite missions use a variety of techniques (e.g., Langmuir probe measurements, retarding potential analyzers) and have contributed significantly to the study of plasma density enhancements and other plasma density structuring at high latitudes (Basu et al., 1990; Goodwin et al., 2015; Kivanç and Heelis, 1997; Zhang et al., 2017a). Rocket measurements can detect density variations down to the decameter scale, but only for several minutes over a confined geographic region, whereas spacecraft measurements can rarely achieve such a fine resolution, but they can collect data continuously over large swaths of the high-latitude ionosphere (Goodwin et al., 2015).

Radio techniques

Ionospheric plasma density enhancements can also be detected from the ground using remote sensing techniques. In Section 3.1.1, we discussed how enhancements can be detected using the ionosonde technique. However, more sophisticated radar systems can also be employed, as outlined here.

Incoherent scatter radar A type of radar system that is commonly used to resolve plasma density enhancements is the incoherent scatter radar (ISR). The ISR technique uses the backscatter power generated by ionospheric plasma to deduce the ionosphere's state parameters including plasma density, ion and electron, and bulk plasma velocity. The spatial resolution of ISR measurements varies depending on the system, but is typically of the order of tens of kilometers, with a temporal resolution of the order of 1–5 min. Unlike ionosondes, ISR measurements are not restricted to the bottomside ionosphere and can provide observations in multiple directions well into the topside ionosphere. However, ISRs are extremely sophisticated and, therefore, very expensive instruments. Accordingly, only a few ISRs have been built and are operating in the world at the moment.

An example of ISR measurements of plasma density irregularities is provided in Fig. 3.1, which shows data spanning approximately 18 h collected by the Resolute Bay Incoherent Scatter Radar—North (RISR-N) (Bahcivan et al., 2010), which faces northwards, and RISR-Canada (RISR-C), which faces southwards (Gillies et al., 2016). Fig. 3.1 shows RISR-C data collected by a beam directed at an azimuth of −157.0 degrees of elevation angle of 55 degrees, and RISR-N data from a beam directed at an azimuth of 26.0 degrees of elevation angle of 55 degrees. Note that these beams were directed in opposite directions—their azimuths were separated by approximately 180 degrees. These radars operate at radio frequencies well above the ionosphere's critical frequency, so their narrow radar beams do not suffer from refractive effects.

Plasma density is plotted in panels i and ii. Panels iii and iv show the median plasma density along two beams from each radar, covering 200 and 400 km altitude. The first beam has an elevation angle of 55 degrees (orange), while the second has an elevation angle of 75 degrees (blue). Both beams are directed along the same azimuth. Panels v and vi show the line-of-sight plasma velocity at a 55 degrees elevation angle, and panels vii and viii show the electron temperature at a 55 degrees elevation angle. In panels i and ii, a variety of relative density enhancements and minimums are apparent. In general, the plasma density is enhanced on the dayside (between 18:00 and 00:00 universal time, UT) relative to the nightside (after 00:00 UT).

The traces in panels iii and iv show the same trends as in panels i and ii, respectively. Variations in the median plasma density in both traces, especially after 03:00 UT, are readily apparent. The sudden drop in electron temperature at approximately 00:00 UT marks local sunset; the lack of electron temperature enhancements after that show there is negligible particle precipitation present in either radar's field of view at night. We can, therefore, apply the same logic used in deriving Eq. (3.7) to conclude that the variations detected in panels ii are the result of F-region plasma density irregularities being transported into the region.

Both radars measure $\partial n_i/\partial t > 0$ just before 04:00 UT (panels iii and iv), followed by a $\partial n_i/\partial t < 0$ signature, consistent with the signature of an enhancement moving toward,

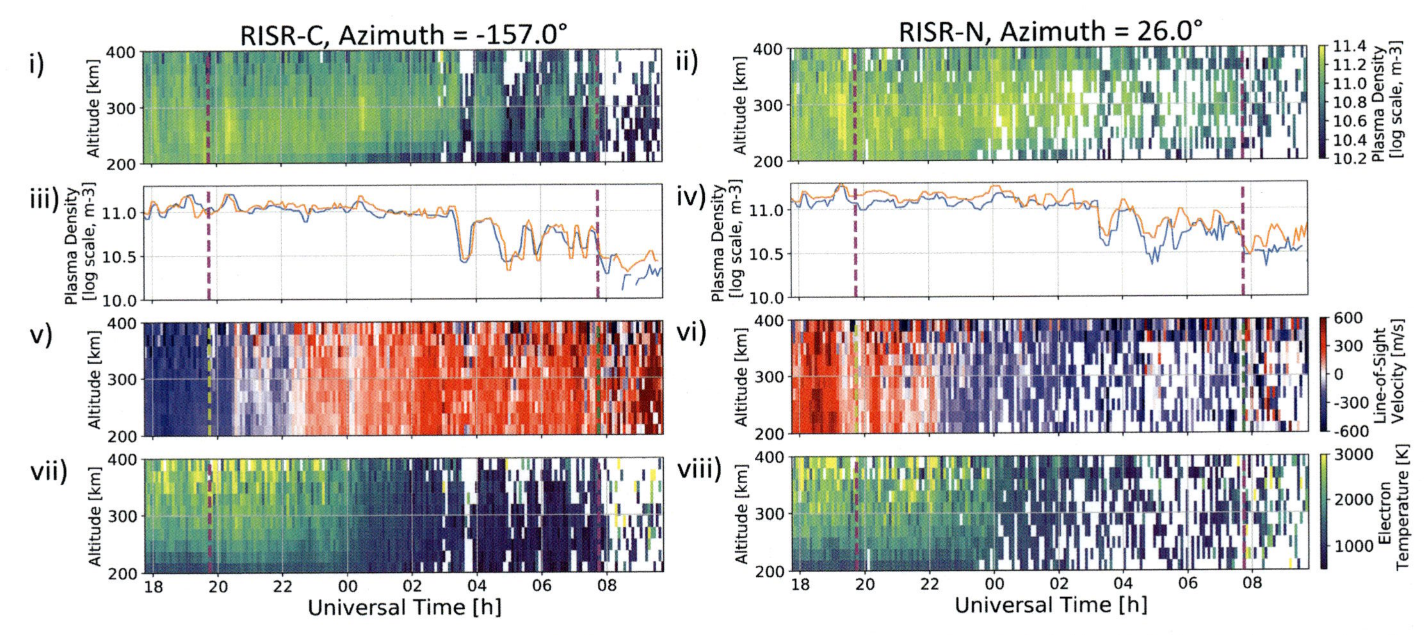

Fig. 3.1 September 29–30, 2019 observations from RISR-C and -N as a function of Universal Time (UT). (left) RISR-C observations at azimuth = −157.0 degrees and elevation = 55.0 degrees. (right) RISR-N observations at azimuth = 26.0 degrees and elevation = 55.0 degrees. In each subfigure: (i + ii) plasma density (iii + iv) the median plasma density between 200 and 400 km (a 15 min median filter has been applied), where *orange* corresponds to the beam of interest and *blue* is a similarly directed beam at an elevation angle of 75.0 degrees (v + vi) line-of-sight plasma velocity, where positive is away *(red)* from the radar and negative is toward *(blue)* the radar (vii + viii) electron temperature. *Dashed lines* indicate the locations of local noon (19:45 UT) and local midnight (7:45 UT). A Python program used to generate panels (i), (ii), and (v)–(viii) has been supplied in the supplementary information available at https://doi.org/10.1016/B978-0-12-821366-7.00001-9.

then away from each radar. A comparison of this feature in the higher (blue) and lower (orange) elevation beams of each radar shows that the enhancement is moving away from RISR-C and toward RISR-N, that is, in a southerly direction. This is consistent with the direction of the plasma flows measured by each radar (panels v and vi). This movement is consistent with the southerly global convection flow expected in that region at that time of day (Ruohoniemi and Greenwald, 2005). A similar effect can be seen on the dayside, such as near 19:00 UT, with the exception that the flows are $v_i > 0$ for RISR-N and $v_i < 0$ for RISR-C, suggesting a northerly motion. This movement is also consistent with the global convection flow expected in that region at that time of day.

We feel that it is necessary to point out that, in accordance with contemporary jargon, these enhancements would be referred to as "patches" depending on our definition of the ambient ionosphere. This is one of the several reasons why we feel that the term "patch" is outdated. Instead, language such as "large-scale plasma density irregularities" is, arguably, more descriptive, without implying that a particular or exceptional process is associated with enhancements that are twice as dense as the background. Furthermore, as Fig. 3.1 illustrates, quantifying the "background or ambient ionosphere" in the polar cap is exceedingly difficult.

Finally, panels vii and viii show that overall the dayside plasma has an elevated electron temperature, which is an expected result of photoionization. However, note that the plasma density enhancements are coincident with slight decreases in electron temperature. This is because the thermal capacity of the enhancements is larger than the ambient ionosphere. Under a constant supply of heat to the plasma, the temperature of the enhancements will be less than the ambient ionosphere, owing to their increased density with respect to the ambient ionosphere. This effect is not as readily apparent under nighttime conditions.

ISRs have, by far, been the subject of the most irregularity studies, no doubt given their capacity to estimate plasma density and plasma density enhancements. Some of the first high-latitude ISR measurements were performed with an ISR located in Chatanika, Alaska (which was subsequently moved to Greenland and contributed even further to the study of enhancements) (Kelley et al., 1982; Vickrey et al., 1980), which demonstrated that, indeed, the nighttime high-latitude ionosphere is replete with plasma density enhancements. Several ISR-based studies (we only provide a few here and suggest to the reader to consider the references contained therein) have followed since and became more common as more ISR facilities were installed in Scandinavia (Carlson et al., 2004; de la Beaujardière et al., 1986), Nunavut, Canada (Dahlgren et al., 2012; Perry and St.-Maurice, 2018; Ren et al., 2018), and Poker Flat, Alaska (Liang et al., 2018; Nicolls and Heinselman, 2007). Observations by these systems have been pivotal in characterizing plasma irregularities, including enhancements, their generation mechanisms, their dynamic properties, and their connection to other M-I-T coupling processes.

Global navigation satellite system Global navigation satellite system (GNSS) transmissions serve as another favored way of detecting and studying irregularities. Plasma densities are inferred by comparing the phase delay multifrequency signals propagating between transmitters (usually located in orbit) to a receiver (usually located on the ground) (Mannucci et al., 1998). With this technique, one can estimate the total electron content (TEC) along the path length of the transmissions and, therefore, irregularities in TEC. Since the frequency of the transmissions is well above the critical frequency of the ionosphere, refractive effects are negligible and the propagation path of a transmission is essentially line-of-sight.

The temporal resolution of TEC measurements is typically of the order of five minutes with a 1 degree by 1 degree geographic latitude and longitude resolution. The technique provides a large-scale perspective. Thanks to the proliferation of GNSS receivers in recent years, we are now able to detect and track enhancements over several hours. A notable example of this is shown in Fig. 3.2, where GNSS techniques were used to track plasma density enhancements as they circulated the polar cap ionosphere in the high-latitude convection flow (Zhang et al., 2013).

In Fig. 3.2, the data are presented in polar magnetic coordinates, with longitudinal lines of magnetic local time (MLT). In Fig. 3.2B, a grouping of TEC enhancements (circled) is seen entering the polar cap, in the late-morning sector. The higher TEC values at lower latitudes, in the morning, noon, and afternoon sectors is the sunlit ionosphere. Proceeding through Fig. 3.2C–I, the enhancements move across the polar cap, in an anti-sunward direction, along plasma convection streamlines, which are parallel to the equipotential contour line estimates derived from ground-based radar measurements (Ruohoniemi and Baker, 1998). It takes approximately 3–4 h for the grouping of TEC enhancements to complete the circuit—entering (B) and exiting (F) the polar cap region, and returning to the approximate entry location (I). Interestingly, the enhancements move along both open and closed magnetic field lines; the latter segment as the enhancements move sunward along the evening sector of the polar cap region.

One of the major goals of polar cap studies has been the ability to detect and track plasma density irregularities in the region, from their sources to their sinks. Fig. 3.2 was the first to demonstrate the "life cycle" of these irregularities. A GNSS-based technique is the only capable of achieving such a large-scale picture of the coupled M-I-T system. A rich variety of other notable works has demonstrated the utility of GNSS-based techniques for studying plasma density irregularities at high latitudes (Chartier et al., 2018, 2019; David et al., 2019; Noja et al., 2013; Watson et al., 2016).

Optical techniques

Not all observations of plasma density enhancements require radio waves. Optical measurements of F-region "red line" emissions (630 nm), reveal plasma undergoing dissociative recombination of O_2^+ (Wickwar et al., 1974), expressed by

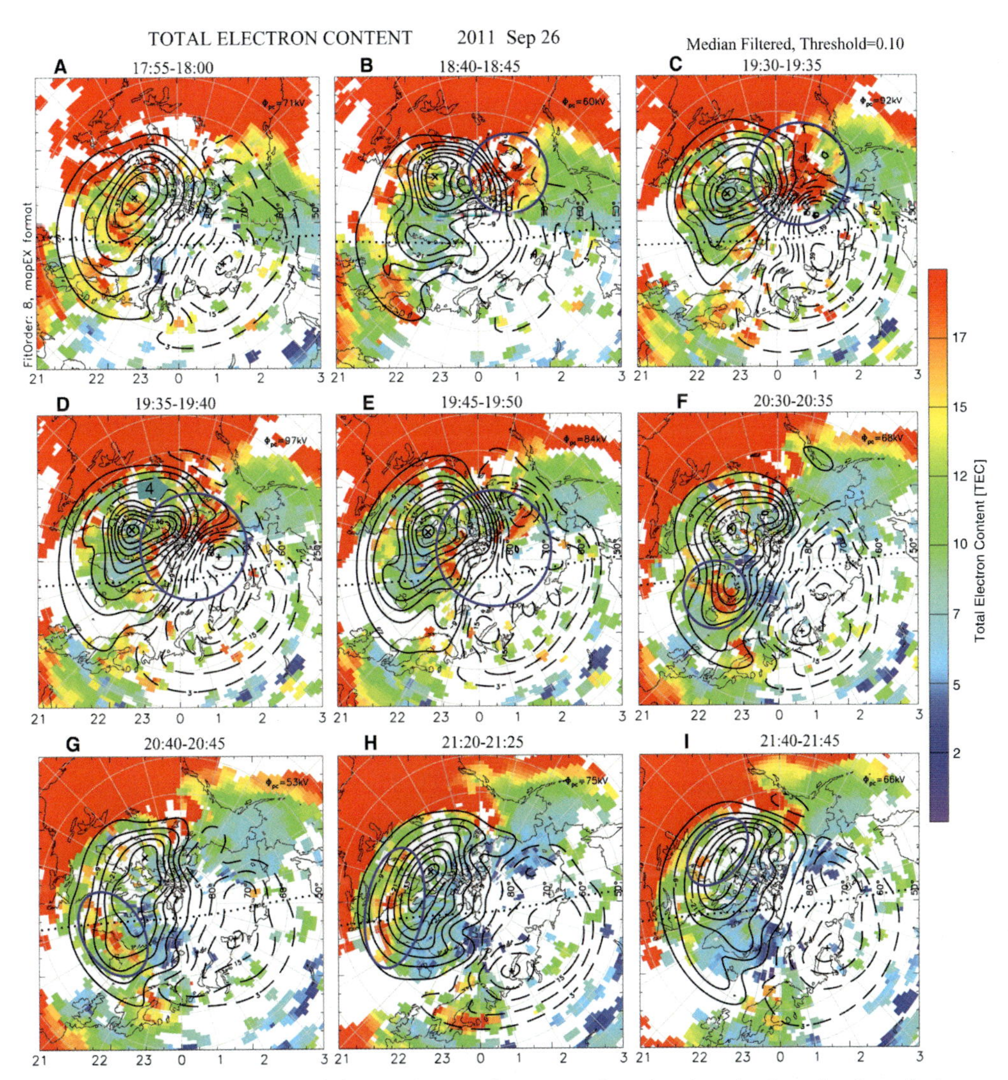

Fig. 3.2 TEC measurements of the northern polar cap region showing the introduction of plasma density enhancements into the region and their transpolar circulation (Zhang et al., 2013). The enhancements *(circled)* are transported through the region via the high-latitude plasma convection streams. The TEC maps are presented in a geomagnetic polar format in which the lines of longitude are magnetic local times (MLT), which remain fixed while the Earth rotates in a counterclockwise direction in the figure.

$$O_2^+ + e \rightarrow O + O^* \tag{3.8}$$

where O_2^+ is a product of the depletion of O^+, described earlier in Eq. (3.4).

A plasma density enhancement will give way to more recombination and thus, more intense optical emissions than the surrounding region, which can be used to detect and track plasma density enhancements. Seminal work in this technique includes (Hosokawa et al., 2006; Lorentzen et al., 2004; Weber et al., 1984).

There are two caveats to this technique that should be discussed. First, the reaction described by Eq. (3.4) is determined by the concentration of O_2; thus, the 630 nm optical emission is altitude dependent. Below 200 km, O^* becomes "quenched" by collisional deactivation with the ambient atmosphere. At higher altitudes, O_2 becomes too tenuous to produce O_2^+ for the emission. Therefore, a bias exists in the technique of studying plasma density enhancements using the 630 nm emission; only enhancements in an altitude range of the order of 200–400 km are likely to be observable.

Fig. 3.3, reproduced from (Perry and St.-Maurice, 2018) provides a demonstration of the 630 nm emission and its altitude dependence. In the top panel are data from an Optical Mesosphere Thermosphere Imager (OMTI) (Shiokawa et al., 2000) located at Resolute Bay. 630-nm emissions presented in a "keogram" format in which a longitudinal slice of the imager's field of view is plotted as a function of time. Starting at 05:00 UT, a steady stream of emission enhancements, corresponding to plasma density enhancements, is seen to move in a north-to-south direction.

The bottom panel of Fig. 3.3 presents plasma density measurements taken by RISR-N, which samples a common volume with the imager. There is a clear correlation between the presence of plasma density enhancements measured by RISR-N and the 630-nm emission enhancements detected by OMTI. There is also a very noticeable connection between the increased altitude of the enhancements, measured by RISR-N, and the optical intensity, measured by OMTI. This is especially apparent just before and after 08:00 UT. The enhancements at higher altitudes have lower optical intensities because the concentration of molecular oxygen decreases with increasing altitude.

The second caveat related to this technique has to do with the nonnegligible vertical component of the F-region plasma's convection velocity in the polar cap ionosphere, a result of the fact that the magnetic field at high latitudes is not entirely vertical. The vertical plasma drift causes plasma density enhancements to also move vertically. This can lead to an increase (downward motion) or decrease (upward motion) of the enhancement's 630-nm emissions. This is due to the fact that the enhancement is driven into a denser neutral atmosphere, increasing the chemistry related to its emissions (Perry et al., 2013). Therefore, when interpreting the 630-nm emissions of plasma density enhancements in the polar cap, one has to consider the fact that the emissions depend on both the altitude and the vertical motion of the enhancements.

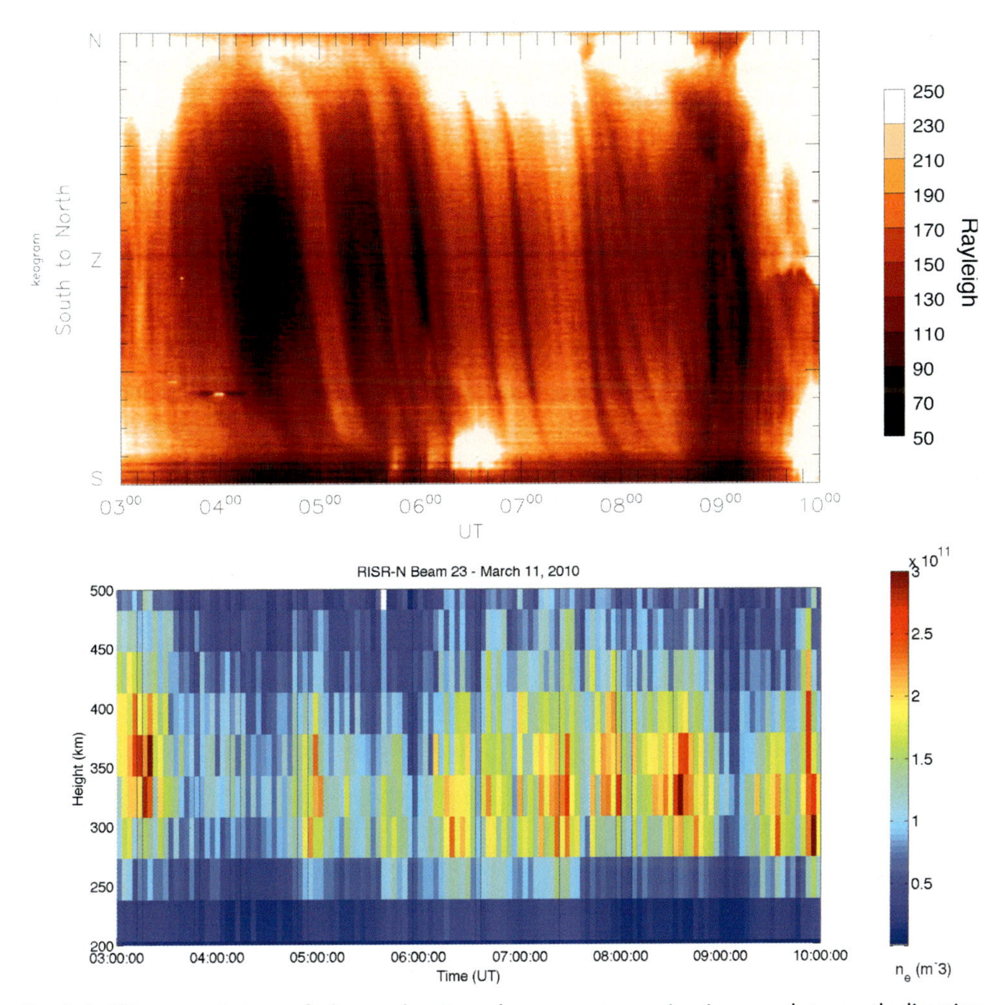

Fig. 3.3 630-nm emissions of plasma density enhancements moving in a north-to-south direction, measured at Resolute Bay (top panel), with simultaneous plasma density measurements from RISR-N (bottom panel). Just before and after 08:00 UT, the optical emissions decrease, corresponding to the measured increase in altitude of the plasma density enhancements (Perry and St.-Maurice, 2018).

3.1.2.2 Sources of plasma density enhancements

Now that some of the properties and detection methods of plasma density irregularities at $\lambda > 10$km have been discussed, it is logical to discuss their source and generation mechanisms.

Impact ionization by particle precipitation is one way of generating plasma in the M-I-T system. Plasma density irregularity production due to particle precipitation transpires in the cusp region (Goodwin et al., 2015; Kelley et al., 1982; MacDougall and

Jayachandran, 2007; Moen et al., 2012; Oksavik et al., 2006; Walker et al., 1999; Weber et al., 1984) and the deep, nighttime polar cap (Perry and St.-Maurice, 2018). However, Fig. 3.1 shows that interspersed on the nightside are plasma density enhancements that are of the order of the dayside plasma density observations measured earlier in the day (<3 UT). That is, the measured density of the irregularities is higher than what is typically observed for impact ionization, and more consistent with photoproduction.

The dayside ionosphere was recognized and shown to be a source of irregularities almost as soon as the appropriate sensors and measurements were available (Hill, 1963). Since then, more contemporary measurements, i.e., Fig. 3.1, showing that the irregularities are moving away from the dayside ionosphere, into the polar cap, and Fig. 3.2, which undeniably shows irregularities emerging from the sunlit ionosphere, have upheld the notion of the dayside ionosphere as a "resevoir" (Carlson, 2012) for plasma density irregularities in the polar cap ionosphere.

The transport of dayside plasma to the nightside ionosphere is often referred to as the tongue of ionization (TOI), owing to its resemblance of a tongue of high-density plasma protruding into the nightside ionosphere from the dayside along plasma convection streams (Knudsen, 1974; Sato, 1959). However, this terminology is misleading in that it presents the image of a steady and uniform stream of plasma moving from the dayside into the nightside polar cap region. TEC maps may give this impression, however, their resolution is fairly coarse. In reality, more finely resolved spatial and temporal measurements of such events reveal that a TOI is a highly structured stream of irregularities (Hosokawa et al., 2009b).

Nevertheless, several observations have shown that, indeed, dense sunlit plasma often moves to higher geomagnetic latitudes in a narrow longitudinal channel, under the influence of high-latitude plasma convection streams. A striking example of this is provided in Fig. 3.4, which shows a low-latitude plume of enhanced plasma during a geomagnetic storm, a storm-enhanced density (SED) plume, moving along high-latitude plasma convection streams and crossing the polar cap ionosphere (Foster et al., 2005). The question remains, however: what processes are responsible for transporting high-density plasma into the polar cap, and thus initiating plasma irregularities—localized plasma density enhancements—which are a seemingly ubiquitous feature of the polar cap ionosphere, including within TOIs?

Several mechanisms have been put forward to explain the generation and introduction of plasma density irregularities in the polar cap ionosphere, with some observational evidence supporting each. The question that still remains unanswered is "which mechanism is most important?" (Carlson, 2012). It is beyond the scope of this work to thoroughly detail each hypothesized generation mechanism and their relative importance. We will, however, discuss a mechanism which appears to be a strong candidate for generating the bulk of the plasma density enhancement irregularities. The hypothesis is that the bulk of the observed irregularities are the ionospheric byproducts of magnetic

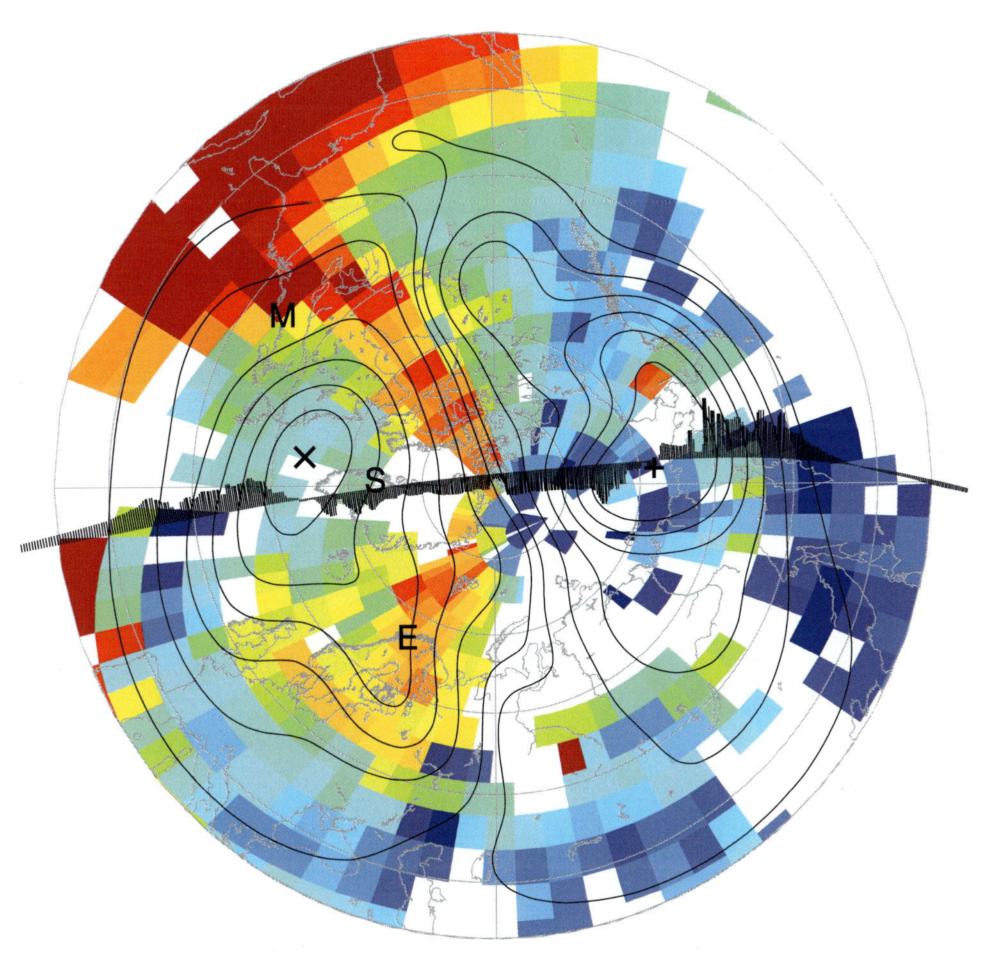

Fig. 3.4 Emerging from an SED plume at lower latitudes during a geomagnetic storm, a TOI is formed as high-density plasma moves into and across the polar cap along plasma convection streams (Foster et al., 2005). The format of this plot is identical to that presented earlier in Fig. 3.2.

reconnection dynamics occurring between the IMF and the terrestrial magnetic field (Lockwood and Carlson, 1992), a process that is also referred to as a flux transfer event (FTE) (Russell and Elphic, 1979) (see also Section 1.2). The ionospheric signature of two consecutive FTEs is depicted in Fig. 3.5, displayed in the same polar format as Figs. 3.2 and 3.4.

Fig. 3.5 illustrates the effect of an FTE that occurs under a purely southward IMF, that is, one in which the z-component is negative and is the only nonzero component of the IMF's vector, in geocentric solar magnetospheric (GSM) coordinates. The demarcation between the lower latitude, dense, photoionized plasma, and the more tenuous nighttime plasma is depicted as a horizontal line in Fig. 3.5, 1. The electrodynamics of an FTE map

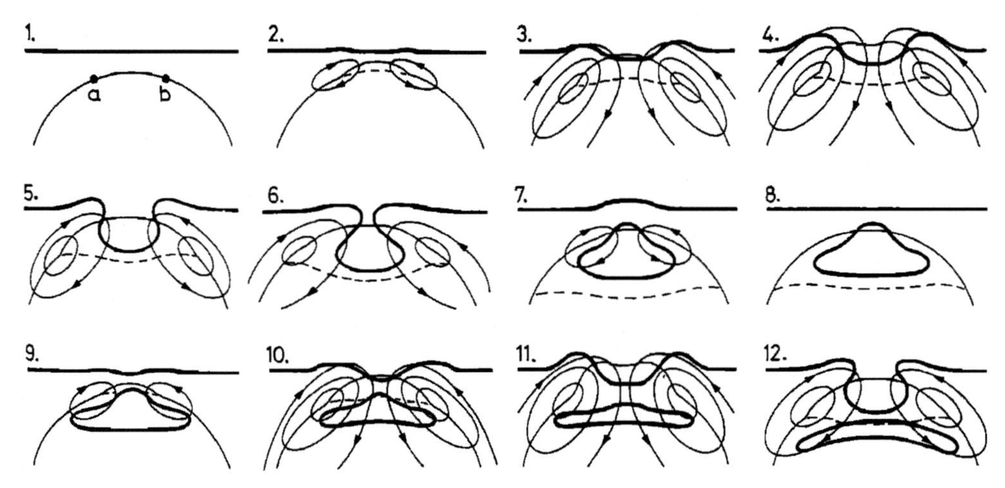

Fig. 3.5 The suggested mechanism for the generation of plasma density enhancements as a results of FTEs (Lockwood and Carlson, 1992). A demarcation between low and high-density plasma is illustrated with a *bold horizontal line.*

from the magnetopause to the high-latitude ionosphere along magnetic field lines, resulting in an electric potential between points "*a*" and "*b*" on the open-closed magnetic field line boundary, is often referred to as the "merging gap." In the reference frame of the boundary, an electric field, directed along the arc $b \rightarrow a$ is established, moving the plasma in an antisunward direction, through the gap, via the $\vec{E} \times \vec{B}$ drift ($\vec{B}$ into the page). In the Earth's reference frame, the FTE will result in either a fixed merging gap with an antisunward directed plasma flowing through it, a stationary plasma with an equatorward expanding gap, or a combination of both (Cowley and Lockwood, 1992; Lockwood and Carlson, 1992).

The electric field directed along the arc joining $b \rightarrow a$ produces a convection drift perpendicular to that arc, resulting in the circular front of protruding plasma shown in Fig. 3.5, 3–5. As the FTE event subsides, the electric potential from $b \rightarrow a$ decreases. As a response, the convection flow decreases, or equivalently, the open-closed field line boundary contracts poleward, resulting in a teardrop shaped volume of enhanced plasma moving into the polar cap region, i.e., Fig. 3.5, 6–7. Consecutive FTE events create a stream of plasma density irregularities moving antisunward. Indeed, this is commonly observed in the polar cap ionosphere when the IMF has a southward component, hence the TOI. FTEs are often sporadic, nonetheless, the frequency of such irregularities closely matches that of FTEs. This mechanism involves a specific ordering of geophysical events and signatures which have not been detailed here, but all of them have been experimentally confirmed with various experiments since the hypothesis was put forward (Carlson, 2012).

3.1.3 Plasma density depletions

Plasma density depletions are the counterpart to plasma density irregularities that are enhancements. Like the enhancements, which we have focused on up to this point, depletions are also described by Eq. (3.7). Accordingly, in the absence of plasma production, e.g., in the nighttime polar ionosphere, a plasma density depletion at scale of $\lambda >$ 10 km can also have a lifetime of several hours and be transported throughout the high-latitude region via the convection streams. This is strongly dependent on many circumstances including the time of day, geomagnetic conditions, and the convection path the irregularity takes through the high-latitude region (Sojka et al., 1981a, b). The properties of depletions can be observed with the same instruments as plasma density enhancements, although it is more difficult given that their densities can be near or below the sensitivity of the instruments. This can be seen in Fig. 3.1, where there are more data gaps at night than during the dayside due to the challenges of measuring low plasma densities with an ISR.

3.1.3.1 Sources of plasma density depletions

Enhanced Joule heating

A mechanism by which F-region plasma density depletion structures are formed is through the enhanced recombination of O^+ via the increased reaction rates $k_1(T)$ and $k_2(T)$ in Eqs. (3.4) and (3.5), relative to the surrounding plasma. These rates are sensitive to increases in the ion plasma temperature. Joule heating will increase the ion plasma temperature in the high-latitude F-region ionosphere. This process involves the sum of the heat exchange between ion, electron, and neutral gases and frictional heating. The latter is proportional to the square of the differential velocity between the ion and neutral gases (St.-Maurice and Hanson, 1982).

Narrow channels of plasma flow enhancements colocated with ion temperature enhancements and plasma density depletions (and frequently particle precipitation) are often observed in the high-latitude and polar cap region. For example, in one observation, a plasma density trough was detected on the edge of auroral arc (Opgenoorth et al., 1990) and polar sun-aligned arc (Perry et al., 2015) in another observation (see also Section 2.3). In the case of the latter, the trough was seen to move through the region in conjunction with the arc showing that not only could plasma density depletions be created deep in the polar cap, but that they could also be transported across significant distances.

Plasma evacuation

The aforementioned sun-aligned arc observation also invoked another depletion mechanism associated with intense field-aligned-currents (FACs) and aurora (see also Section 2.1): plasma evacuation (Perry et al., 2015). This mechanism was first broached

(Doe et al., 1993; Zettergren and Semeter, 2012) to help explain "cavities" frequently observed in the high-latitude ionosphere. An auroral arc can be represented by a FAC pair, where the upward FAC is carried by downward moving electrons, producing the arc's optical emissions. When the E-region conductivity is low, the horizontal currents connected to the downward FAC component of the arc are completed in the lower F region. Plasma will be evacuated from the downward FAC region: the ions move horizontally away from the downward FAC, while the electrons will move vertically upward along the field line. This seeds a plasma depletion cavity, which leads to an enhanced electric field (to maintain the current flow), strengthened plasma convection drifts, frictional heating, and increased reaction rates for the chemical recombination of the ions. This feedback effect accelerates, establishing a plasma density depletion. Although it does not appear that the evacuation mechanism can create a F region depletion on its own, indications are that it works well in conjunction with the chemical recombination mechanism, where the evacuation mechanism contributes strongest at lower F-region and E-region altitudes, and the chemical recombination dominates higher-up (Perry et al., 2015; Zettergren and Semeter, 2012).

Plasma convection

If the high-latitude convection velocity associated with a parcel of plasma is slow, the parcel can remain in the dark polar cap for an extended period of time. This would provide the plasma with an extended period of time to become depleted, assuming no significant plasma production mechanisms are present. This plasma stagnation can result in a plasma depletion structure in the high-latitude region: the "polar hole" (Brinton et al., 1978; Sojka et al., 1981b). The stagnation related hole generally occurs during periods of quiet geomagnetic activity.

A polar hole may also form during active geomagnetic conditions. This feature is not due to stagnation but rather enhanced plasma convection velocities. At polar latitudes, the geomagnetic field has a small, nonnegligible, horizontal component, producing a vertical $\vec{E} \times \vec{B}$ drift. When plasma convection becomes enhanced, the downward component of the plasma convection will also become enhanced, driving the plasma vertically downwards into the dense neutral atmosphere, increasing the chemical recombination of plasma, producing a region of depleted plasma—a hole (Sojka et al., 1981a).

Another way by which plasma depletions can be introduced into the polar ionosphere is by periodically changing the magnetic longitude of the magnetic cusp, the entry point of plasma into the polar region (Milan et al., 2002). In general, plasma entering the polar ionosphere through a cusp located in the afternoon MLT sector convects from the nightside ionosphere along the sunlit evening flank, in a similar fashion to the irregularities observed in Fig. 3.2E–I. On the other hand, plasma entering through a morning MLT cusp will have convected from the nightside along the morning sector; the difference being that this route generally lacks a source of significant plasma production—photoionization. Thus, a mixture of both low and high-density plasma flows into the polar cap if the position of the cusp oscillates between an afternoon and morning source region due to a varying y-component of the IMF. This effect has both been observed by Milan et al. (2002) and Sakai et al. (2013) and modeled by Sojka et al. (1993, 1994). It is also important to point out that this mechanism, examined from another point of view, may be considered a density enhancement generation mechanism as well. If the source of the plasma streaming into the polar cap is the morning MLT, with sporadic excursions to the afternoon MLT sector, the mostly tenuous stream of plasma would be populated by infrequent regions of enhanced plasma density, compared to the ambient ionosphere.

3.1.4 Plasma density irregularities as an agent of space weather

"Space weather" is a term that describes plasma processes in the near-Earth geospace environment that can affect society. These processes can be initiated externally, for example, by solar coronal mass ejections, or internally by M-I-T coupling processes. Plasma density irregularities (enhancements and depletions) are a major focus of M-I-T coupling and space weather research for two prime reasons. First, they are the byproduct of multiscale M-I-T coupling processes, such as FTEs or enhanced Joule heating events. Second, they can pose a significant hazard to radio wave communications in the high-latitude and polar regions. Radio communications in these regions are becoming increasingly important as civilian and economic entities, such as commercial shipping and airlines, escalate their activities there. By studying the characteristics of irregularities, understanding their sources and sinks, and specifying their effects on radio wave communications, we can mitigate their negative effects on society.

In this section, we will focus on the effects of the irregularities on radio communications in the high-frequency (HF; 3–30 MHz) portion of the spectrum. The effects on radio transmissions on other regions of the radio spectrum will be covered elsewhere in this compendium. HF communications are commonly used in remote geographical regions, such as the polar cap, because of their ability to propagate to points "over the horizon." Specifically, the signals propagate from the transmitter to the ionosphere and are reflected back toward the ground.

HF transmissions whose wave vectors are at an acute angle with respect to the vertical can reach receivers that are well beyond the line-of-sight—over the horizon—using the ionosphere as a reflecting medium. This mode of radio propagation can be described by refraction and total internal reflection. From a simplified version of the Appleton-Hartree equation, the index of refraction, n, of a radio wave is

$$n = \sqrt{1 - \left(\frac{f_p}{f_0}\right)^2} \tag{3.9}$$

where f_o is the frequency of the radio wave, and f_p is the plasma frequency. A quick calculation will show that ionospheric plasma frequencies and HF frequencies are of the same order; therefore, $n \neq 1$ for HF communications is common (Gillies et al., 2012). In order to achieve high fidelity in radio communications, a smooth and uniform ionosphere, free of irregularities, is preferred for an HF communications system that is transmitting over-the-horizon. Otherwise, plasma density irregularities would present as irregularities in the value of the index of refraction; the transmissions would be refracted and reflected in an irregular way, degrading the communications link.

3.1.4.1 Off great-circle path deviations

Some of the most dramatic and compelling evidence of the ramifications of plasma density irregularities on HF radio links in the polar cap region is provided in Fig. 3.6 (Warrington et al., 1997), which shows the bearing (angle of arrival) of radio transmissions received at Alert, Canada, originating from Thule, Greenland, as a function of universal time (UT). Transmissions arriving along the great-circle path between the two points have a bearing of approximately 200 degrees. As Fig. 3.6 shows, the bearing of the received transmissions varies quite dramatically, especially between 8 and 13 UT, where the angle varies by approximately 100 degrees.

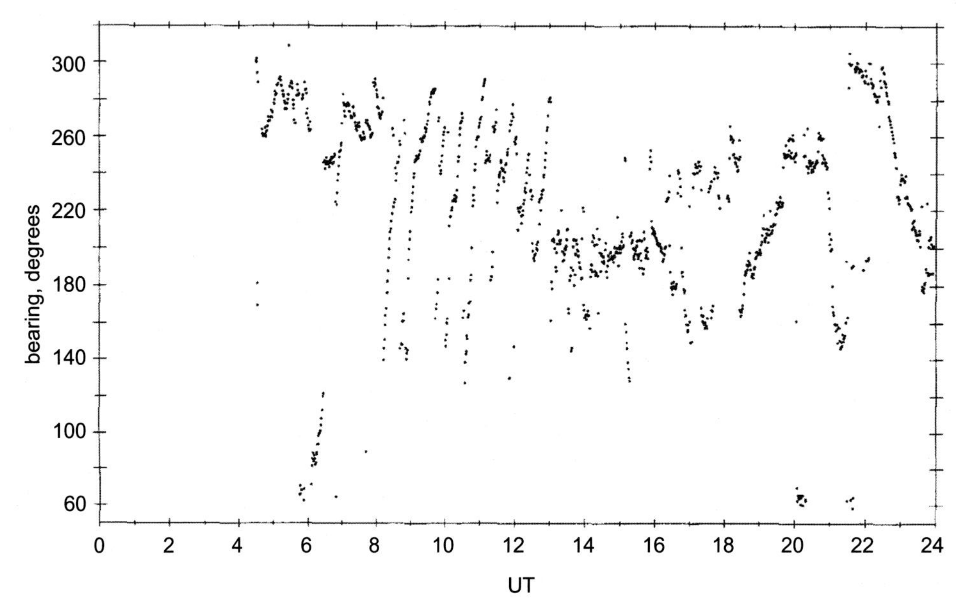

Fig. 3.6 Data from an HF radio link between Alert, Canada and Thule, Greenland showing evidence of significant deviations in the trajectory of the transmitted/received signal. The nominal bearing for the radio link is approximately 200 degrees (Warrington et al., 1997).

The deviations in Fig. 3.6 were attributed to the presence of plasma density irregularities—enhancements—along the ray path between the transmitter and receiver stations. Irregularities introduce significant gradients in plasma density into the ray propagation path which, in accordance with the principles of refraction and internal reflection, will cause the transmissions to deviate both vertically and laterally from their great-circle paths. As one can imagine, these substantial deviations pose a hazard to HF systems which rely on great-circle path propagation to operate effectively.

3.1.4.2 A source of small-scale irregularities and radar backscatter

Plasma density irregularities at scale of $\lambda > 10$km are not only hazardous to radio communications in the high-latitude region because they can deviate the general trajectory of transmissions, but also because they are a large backscatter target of HF radio waves. HF coherent backscatter was recognized very early on in the advent of radar and has been studied extensively in the context of solar-terrestrial physics for decades.

When a radio wave interacts with a plasma population, the incident electric field of the radio wave accelerates the charged particles, which then scatters radiation at a Doppler shifted frequency (Jackson, 2007). At radio frequencies comparable to the plasma frequency (i.e., at HF), a radio wave incident on a volume of plasma will interact with the plasma, which will, in response, emit scattered radiation. The foremost hypothesis regarding radar backscatter originating from ionospheric altitudes is that the scattering processes are analogous to Bragg scattering when $\vec{k} = \pm 2\vec{k}_p$, where $\vec{k}$ is the wave vector of plasma density irregularities and $\vec{k}_p$ is the wave vector of the incident radio wave (Fejer and Kelley, 1980; Milan et al., 1997). The Bragg condition reveals, however, that the irregularities responsible for the scattering have a decameter-scale size (e.g., the 3–30 MHz HF band corresponds to a Bragg scale of 50–5 m)—several orders of magnitude below the scale which we are interested in. The connection between HF radar backscatter from these decameter irregularities and plasma density irregularities at scale of $\lambda > 10$ km is that the latter seed plasma instability processes, which generate the decameter irregularities responsible for coherent backscatter.

The gradient-drift instability (GDI) is generally regarded as a primary source of decameter-scale irregularities in the high-latitude ionosphere (Cerisier et al., 1985; Simon, 1963; Tsunoda, 1988). The onset of the GDI occurs on a plasma density gradient, which are associated irregularities. If an electric field is applied, such as the convection electric field driving $\vec{E} \times \vec{B}$ plasma drifts in the high-latitude regions, the edges whose gradient is parallel to the convection velocity will be unstable to the GDI, while those that are antiparallel will be stable (Keskinen and Ossakow, 1982). Accordingly, in the situation of a plasma density enhancement in the polar cap, the "trailing" of the enhancement moving through the polar cap is unstable to GDI, while the "leading" is stable. As a result, decameter-scale irregularity growth is strongest on the trailing. We should, therefore, expect that coherent backscatter should also be strongest on the trailing. As it turns out, this is not necessarily the case.

Fig. 3.7 (Dahlgren et al., 2012) shows a plasma density enhancement captured by three separate instruments: the RISR-N ISR, the OMTI imager, and the HF Super Dual Auroral Radar Network (SuperDARN) (Chisham et al., 2007; Greenwald et al., 1995) system located at Rankin Inlet, Nunavut. The strength of coherent backscatter received by SuperDARN is indicated in the grayscale tiling; the colored contours indicate plasma density measured by RISR-N; and, the dashed line indicates the outline of the irregularity's optical signature measured by OMTI. Two things are apparent in this figure. First, enhancements are not the only source of coherent backscatter in the high-latitude region. It is believed that they are a major source; however, there is insufficient evidence as of yet to support this. Second, the agreement on the position of an enhancement by three separate instruments using three different techniques is clear, validating the efficacy of the techniques discussed earlier, in Section 3.1.2.1.

The inset in Fig. 3.7 shows that backscatter is present throughout the enhancement. There is a region of elevated backscatter on the northeast portion of the enhancement, which was a leading edge according to the drift of the irregularity in this event. This is contradictory to the widely held belief that radar echoes should be strongest on the trailing edges of such enhancements.

It is important to note that we are not taking into account the fact that the geolocation of the echoes may be inaccurate due to the off great-circle path deviations discussed earlier. This is not taken into account in SuperDARN's geolocation algorithms. Another factor that must be considered with HF coherent backscatter in general is that the decameter-scale irregularities are field aligned. The Bragg condition can only be satisfied when the incident radar wave vector is close to perpendicularity with the local magnetic field direction. It is in this direction that plasma diffusion is slowest (relative to diffusion along the magnetic field line), allowing for a longer irregularity lifetime (Tsunoda, 1988). Nevertheless, observations (Hosokawa et al., 2009a) and simulations (Gondarenko and Guzdar, 2004b) have consistently shown that enhancements such as the one shown in Fig. 3.7 are seemingly saturated with irregularities.

As a strong scattering target, irregularities such as the one shown in Fig. 3.7 are a space weather hazard to HF radio communications and other systems as well, including GNSS. Details of the latter will be provided elsewhere in this compendium, but, in short, the decameter-scale irregularities seeded by irregularities at scale of $\lambda > 10$ km are believed to be the source of both amplitude and phase scintillation of signals into the very- and ultrahigh frequency, and L radio spectrum bands.

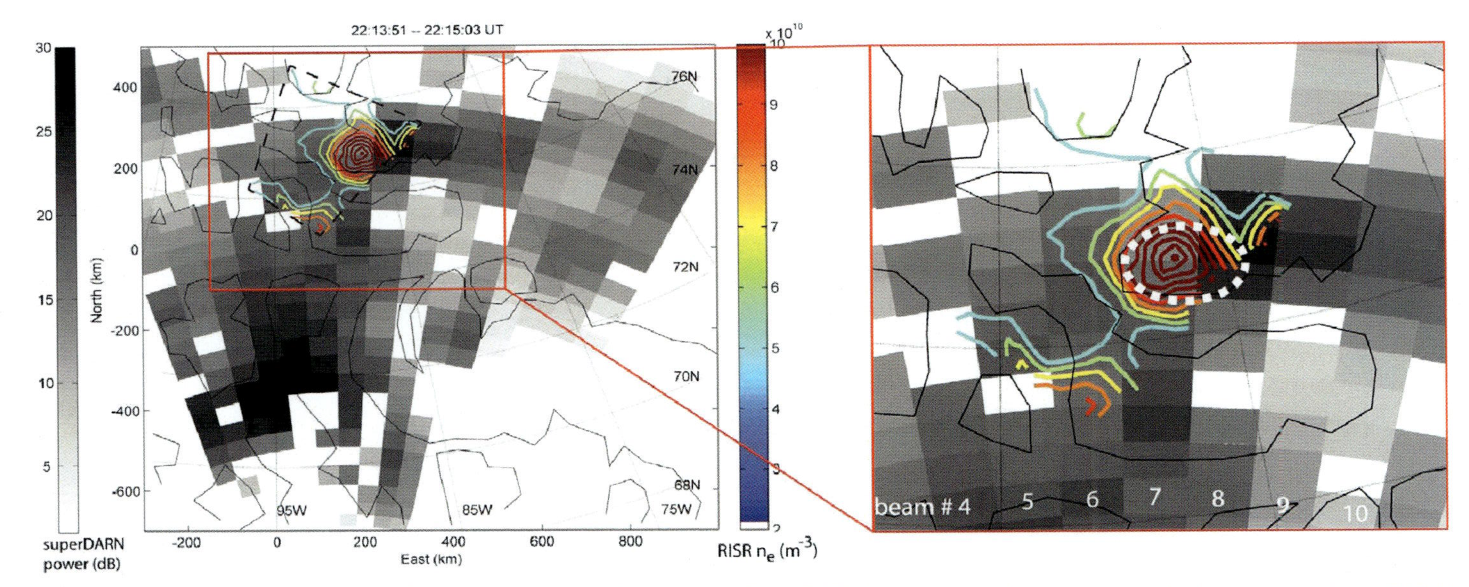

Fig. 3.7 Coincident measurements of a plasma density enhancement. SuperDARN backscatter echoes are presented in *grayscale tiling*, RISR-N plasma density observations are given in *colored contours*, and the *white dashed line* indicates the outline of the irregularity's optical signature measured by an OMTI imager (Dahlgren et al., 2012).

3.1.5 Plasma density irregularities in the M-I-T system

The ionosphere is not a closed system, and it is strongly coupled to the magnetosphere and the thermosphere—the M-I-T system. For example, recall the feedback effect discussed earlier in Section 3.1.3.1 related to the generation of plasma density depletions via enhanced Joule heating, and its dependence on the concentration of thermospheric constituents. Up to this point, the irregularities have been contextualized as products of dynamic M-I-T coupling processes, for example, as the consequence of FTEs or SED plumes. One may get the impression that they are passive features of the larger M-I-T system; however, there is sufficient evidence showing that this is not necessarily the case. That is, the irregularities actively participate in—even incite—M-I-T coupling processes as they move throughout the polar cap.

Regardless of their genesis, plasma density irregularities—enhancements and depletions—constitute conductivity gradients in the ionosphere. Accordingly, the irregularities are often colocated with dynamic plasma phenomena such as elevated plasma flows (Kivanç and Heelis, 1997). The interconnections between these phenomena can be expressed mathematically in the F region (Perry et al., 2015; Sofko et al., 1995):

$$\vec{J}_{\parallel} = -\Sigma_P \nabla \cdot \vec{E}_{\perp} - \vec{E}_{\perp} \cdot \nabla \Sigma_P - \nabla \Sigma_H \cdot \hat{b} \times \vec{E}_{\perp}, \tag{3.10}$$

in which $\vec{J}_{\parallel}$ is the parallel field aligned current density, $\hat{b}$ is the unit vector along the terrestrial magnetic field ($\vec{J}_{\parallel} \cdot \hat{b} > 0$; a positive current density directed parallel to the magnetic field), $\vec{E}_{\perp}$ is the electric field driving plasma convection, and Σ_P and Σ_H are the height integrated Pedersen and Hall conductivity, respectively.

The first term on the right-hand side of Eq. (3.10) has been referred to as the "magnetospheric component" while the remaining two terms have been referred to as the "ionospheric component" (Sofko et al., 1995). Above 200 km, $\Sigma_P > \Sigma_H$. The introduction of a plasma density irregularity, expressed by a nonnegligible value of $\nabla \Sigma_P$, into an otherwise uniform plasma field would need to be supported by changes in the magnitude of $\vec{E}_{\perp}$ and/or $\vec{J}_{\parallel}$ to maintain current continuity, $\nabla \cdot \vec{J} = 0$ (where $\vec{J} = \vec{J}_{\perp} + \vec{J}_{\parallel}$), demonstrating the M-I-T connection. Note that the role of the thermosphere is embedded in the value of Σ_P.

Other work has shown that some enhancements have high-electron temperatures (Ma et al., 2018; Zhang et al., 2017a), compared to others, indicating that they are subject to low-energy particle precipitation as they move through the polar region. Thus, the notion of irregularities being a remnant or a "fossil" of a bygone M-I-T coupling process is inaccurate. One of the more compelling lines of research concerning irregularities and M-I-T coupling has to do with their connection to magnetospheric phenomena, such as ion outflow and merging. In a rather serendipitous observation, an ISR captured a plasma density enhancement moving into a region of ion upflow, demonstrating a "connection between horizontal plasma transport … and vertical transport along the nightside polar cap boundary" (Semeter et al., 2003). This process has also been measured in situ and linked to the convection velocity of the enhancements (Zhang et al., 2016).

Modeling efforts have indicated that enhancements may be a source of enhanced polar wind streams (Schunk et al., 2005), and that polar cap flow channels, which are often observed in conjunction with enhancements, are an ionospheric manifestation of structures in the magnetosphere lobe, related to magnetic merging in the magnetotail (Nishimura and Lyons, 2016). The latter is consistent with spacecraft observations, taken deep in magnetosphere (Goodwin et al., 2019), which have shown that enhancements and their associated flow channels carry a signature in the magnetosphere and, therefore, are heavily coupled with magnetospheric dynamics. This remains a largely unresolved topic at the moment, particularly the notion of a link between geomagnetic storms and substorms, their impact on dayside plasma flows, and interconnection with plasma density irregularities. Substorms have been linked to significant changes in high-latitude convection (Bristow and Jensen, 2007; Clausen et al., 2013) as well as the production of TOIs and enhancements in the polar region (Goodwin et al., 2020), although the causal link between these dynamics remains elusive.

This area of research enjoyed an increased amount of activity in recent years—the connection between plasma density irregularities in the polar region and magnetospheric dynamics. This activity is driven, in part, by the increased access to insightful ground- and space-based measurements of the coupled M-I-T system. The results, mentioned previously, have provided an indication that there is a strong link between processes in the M-I-T system, transpiring across multiple spatiotemporal scales, which are strongly coupled and, therefore, should be considered together when describing the coupled M-I-T system.

3.2 Modeling high-latitude F-region ionospheric fluid instabilities: Linear and nonlinear evolution and observational signatures

K.B. Deshpande[a,], M.D. Zettergren[a,]*, A. Spicher[b], L. Lamarche[c], M. Hirsch[d], and M. Redden[a]*

[a]Department of Physical Sciences and Center for Space and Atmospheric Research (CSAR), Embry-Riddle Aeronautical University (ERAU), Daytona Beach, FL, United States

[b]Department of Physics and Technology, UiT the Arctic University of Norway, Tromsø, Norway

[c]SRI International, Menlo Park, CA, United States

[d]Boston University, Boston, MA, United States

* Equal contribution

3.2.1 Introduction and background

The high-latitude ionospheric plasma exhibits a high degree of spatial structure, which is known to evolve dynamically under the influence of magnetospheric forcing in addition to internal ionospheric processes. Plasma structuring at scale of $\lambda > 10$km, as discussed in Section 3.1.1, is thought to cascade into small scales through a variety of instability mechanisms and structuring processes, but the details of how plasma structuring evolves over time, especially nonlinearly, are poorly constrained. Intermediate-scale sized plasma structures (electron density irregularities) modify the phase and amplitude of transionospheric radio signals, a process referred to herein as scintillation. Rapid variations in the amplitude and phase of the radio signals such as the GNSS signals resulting from these irregularities in the ionosphere can vary with the latitude, season, and geomagnetic conditions. These effects can be detrimental to communication and navigation systems, but are also used as helpful remote sensing diagnostics for the characteristics of fundamental plasma processes related to scintillation.

Ionospheric fluid instabilities that are driven by collisional or inertial processes have received a lot of attention recently since they appear to be able to at least partially explain the generation of intermediate-scale plasma density irregularities (e.g., Kintner and Seyler, 1985; Tsunoda, 1988). Of these types of instabilities, the most commonly invoked (for reasons to be discussed below) at high latitudes are the GDI and *Kelvin-Helmholtz instability* (hereafter "KHI"). GDI occurs in situations of inhomogeneity in ionospheric plasma density, specifically in regions where the plasma drift has a component *along the direction of the background density gradient.* In the presence of seed structures, this configuration leads to Pedersen currents that cause charge configurations responsible for the growth of the seed structures. In the nonlinear stage, GDI appears as alternating "fingers" (i.e., elongated structures along the drift direction) of high and low-density and extremely steep-density gradients in these finger-like structures. Whereas GDI depends on Pedersen currents (collisions), KHI is a fundamentally inertial instability that owes its existence, in the context of the ionospheric plasma, to polarization currents. Configurations having a background shear in the plasma drift (e.g., KHI) can be unstable to perturbations in drift which cause polarization currents and charge accumulation that result in the formation of perturbation electric fields and drift which reinforce the perturbations and cause them to grow. In the context of nonlinear growth of fluid instabilities, there is a tendency for structures produced by a primary instability to become susceptible to secondary instabilities, resulting in a *cascade* to different spatial scales. Cascading, the exchange of energy between different unstable modes, and stabilizing processes at small scales (viscous in nature), generally lead to *turbulence*—a term we use in this review to describe a state characterized by a relatively well-defined spectrum of irregularities. More details of these instabilities are further discussed and reviewed in Section 3.2.2.

Plasma instabilities such as GDI and KHI create intermediate-scale sized structures which in turn cause scintillation of transionospheric radio signals. Ionospheric scintillations are more pronounced in the equatorial and high-latitude regions with irregularities responsible for producing scintillations being predominantly in F layer at altitude ranging from 200 to 1000 km and E-layer altitudes from 90 to 120 km regions (Aarons, 1982). In the auroral and polar regions, the dominant factors in the production of the ionospheric irregularities are current systems (causing GDI and KHI) and energetic particle precipitation.

Complimentary to existing reviews of ionospheric scintillation of radio signals (Yeh and Liu, 1982; Aarons, 1982; Bhattacharyya et al., 1992), the purpose of this chapter is to discuss recent progress specifically in the nonlinear simulation of irregularities and scintillation and application of these approaches to the interpretation of radio data. In this section, we first provide a geophysical context to the different types of instabilities followed by a brief survey of relevant review articles on ionospheric irregularities. We then provide a summary of contextual observations used in prior work to establish connections of irregularities and scintillation to larger structures. This is followed by a review of the linear theory of F-region plasma instabilities thought to be the culprit for a large fraction of scintillation at high latitudes (Section 3.2.2). The bulk of the section is devoted to summarizing past and current approaches to nonlinear simulation of plasma instabilities (Section 3.2.3) and review of theory and simulation of radiowave propagation through turbulent and evolving media (Section 3.2.4). We also provide some example simulation codes to help understand basic aspects of the linear and nonlinear theory for GDI and KHI. Finally, we conclude with a summary and specific discussion of future steps required to further nonlinear simulations efforts and their application to extant and future observations.

3.2.1.1 Geophysical context in which instabilities are thought to play a role

Different high-latitude regions and the scintillation-producing instabilities that are typically found in those regions are summarized in Fig. 3.8; this section focuses on scintillation in the cusp, polar cap, and auroral regions labeled on this diagram.

A recent monograph by Zhang and Paxton (2015) presents a comprehensive review of the auroral dynamics and its space weather effects. While GDI is believed to be dominant in the polar cap, it also exists inside structured plasma density enhancements in the auroral oval known as "blobs" (Basu et al., 1990). KHI that develops in strong velocity shear regions can be a prevalent source of the auroral irregularities (Basu et al., 1988). "Hard precipitation" (1–100 keV electrons) results mainly in discrete or diffuse auroras in the E region, whereas low-energy electrons (<1 keV) can generate "soft precipitation" in the F region. Auroras form due to solar wind-magnetosphere and ionosphere interaction during which the precipitating energetic auroral particles collide with the neutrals in the Earth's upper atmosphere (Störmer, 1955). Fig. 3.9A shows an example of an

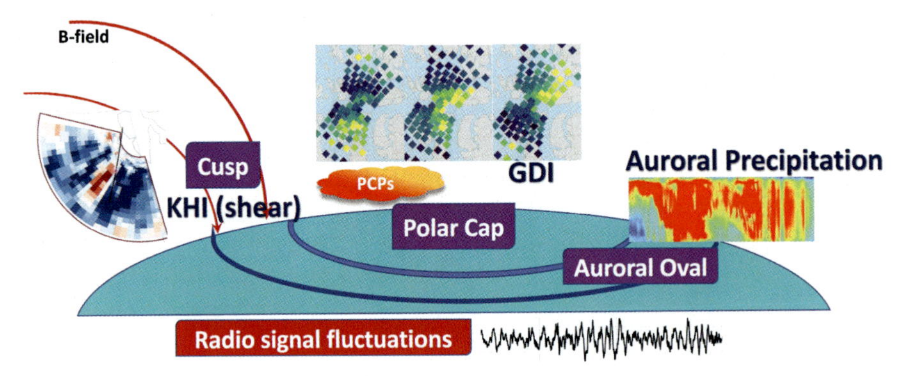

Fig. 3.8 Illustration showing different types of instabilities found in different regions (polar cap, cusp, auroral oval) that can cause scintillation-producing irregularities. The Sun is on the left side of the figure, while midnight is on the right. *Thanks to: Don Hampton (for auroral Image) and Spicher, A., Ilyasov, A.A., Miloch, W.J., Chernyshov, A.A., Clausen, L.B.N., Moen, J.I., Abe, T., Saito, Y., 2016. Reverse flow events and small-scale effects in the cusp ionosphere. J. Geophys. Res. Space Phys. 121 (10), 10466–10480. https://doi.org/10.1002/2016JA022999 (for cusp image).*

auroral arc. The typical boundary of the auroral oval can extend both equatorward and poleward during heightened geomagnetic activity (Aarons, 1997), such as during an auroral substorm (Akasofu, 1968) and geomagnetic storms. The auroral structures due to particle precipitation or formation of different instabilities can result in magnetic field aligned irregularities at different altitudes and can cause scattering of radio waves in HF to UHF range. The reader is referred to a comprehensive review of auroral arcs by Karlsson et al. (2020). In terms of magnetospheric connection, the auroral zone connects to the plasmasheet in the Earth's magnetosphere while cusp connects to the solar wind (Kelley, 2009). Thus, the particles precipitating in the auroral zone are mainly plasmasheet electrons. Although less common, precipitation in the polar cap can also cause sun-aligned arcs or theta aurora (Zhu et al., 1997). An example of such aurora is shown in Fig. 3.9B (adopted from Mailyan et al., 2015). Strong mesoscale flow shears are thought to exist within and around these arcs, which when embedded in a larger polar cap convection pattern could be an important area for instability development (Koustov et al., 2008; Lyons et al., 2016; Robinson et al., 1987; Valladares and Carlson, 1991; Weiss et al., 1993).

Inhomogeneous flows are also common in the cusp region (e.g., Heppner et al., 1993; Moen et al., 2008; Oksavik et al., 2004, 2005; Rinne et al., 2007). One particular cusp phenomenon containing cross-field sheared plasma where electron density irregularities with scales reaching down to tens of meters have been observed is the reversed flow event (RFE) (Moen et al., 2008; Oksavik et al., 2011; Rinne et al., 2007; Spicher et al., 2016). An example of an RFE is shown in Fig. 3.9C. RFEs are longitudinally extended (>400–600 km) but latitudinally narrow (~50–250 km) flow channels with flow direction opposite to that of the large-scale convection background (Rinne et al., 2007). RFEs

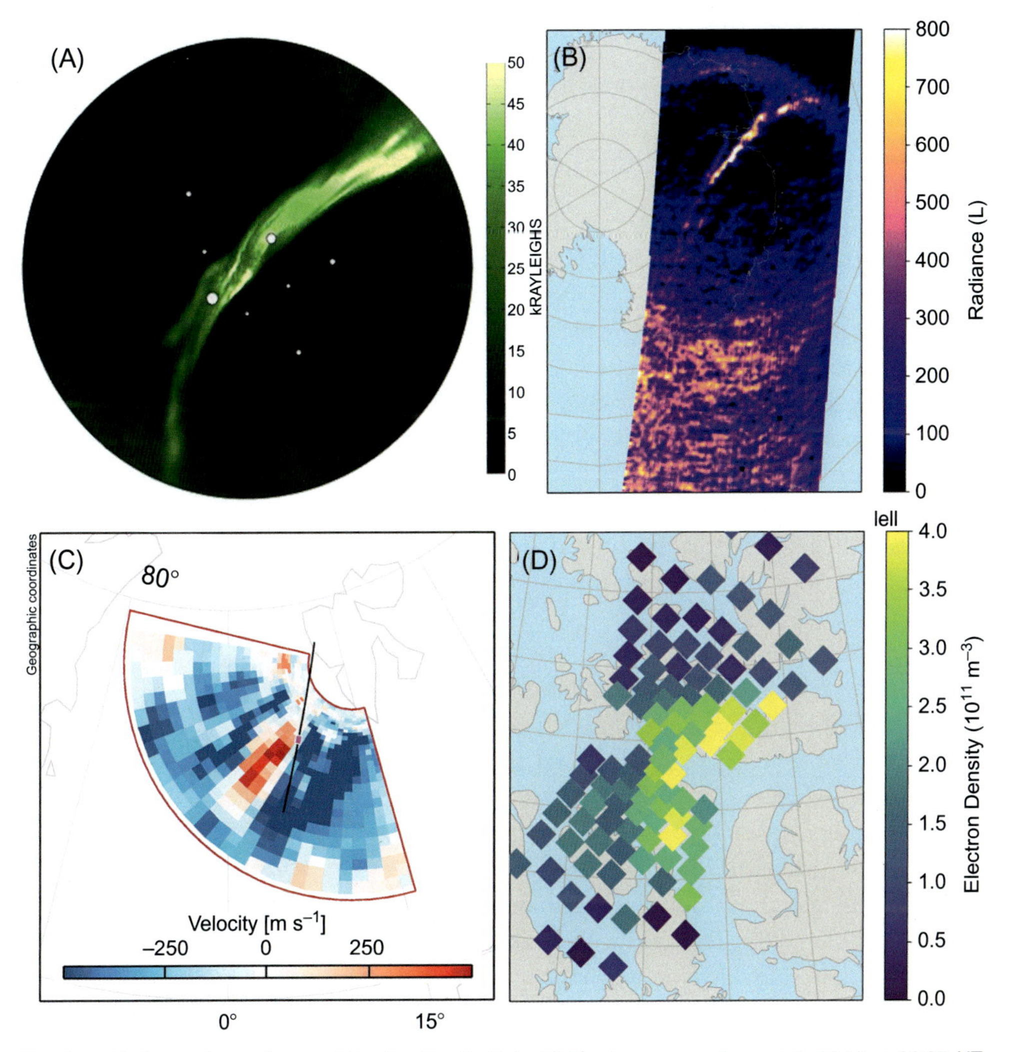

Fig. 3.9 (A) Auroral arc observed by the South Pole All-Sky Imager on August 9, 2010 at 20:37 UT. (B) Sun-aligned arc in the Southern Hemisphere polar cap observed by the Global Ultraviolet Imager (GUVI) on September 27, 2004. (C) Reverse flow event observed by the EISCAT Svalbard Radar on December 3, 2011 at 7:27 UT. (D) Polar cap patch observed by the Resolute Bay Incoherent Scatter Radars North and Canada (RISR-N and RISR-C) on November 21, 2017 at 18:46 UT. *Panel (A): Adapted from Kinrade, J., Mitchell, C.N., Smith, N.D., Ebihara, Y., Weatherwax, A.T., Bust, G.S., 2013. GPS phase scintillation associated with optical auroral emissions: first statistical results from the geographic south pole. J. Geophys. Res. Space Phys. 118 (5), 2490–2502. https://doi.org/10.1002/jgra.50214; Panel B: Adopted from Mailyan, B., Shi, Q.Q., Kullen, A., Maggiolo, R., Zhang, Y., Fear, R.C., Zong, Q.-G., Fu, S.Y., Gou, X.C., Cao, X., Yao, Z.H., Sun, W.J., Wei, Y., Pu, Z.Y., 2015. Transpolar arc observation after solar wind entry into the highlatitude magnetosphere. J. Geophys. Res. Space Phys. 120 (5), 3525–3534. https://doi.org/10.1002/2014JA020912; Panel C: Adapted from Fig. 3 of Spicher, A., Ilyasov, A.A., Miloch, W.J., Chernyshov, A.A., Clausen, L.B.N., Moen, J.I., Abe, T., Saito, Y., 2016. Reverse flow events and small-scale effects in the cusp ionosphere. J. Geophys. Res. Space Phys. 121 (10), 10466–10480. https://doi.org/10.1002/2016JA022999.*

are usually observed around magnetic noon, occurring more often during a dominant B_y IMF, i.e., $|B_y| > |B_z|$, and are believed to be signatures of a flux transfer event (FTE) (Moen et al., 2008; Oksavik et al., 2011; Rinne et al., 2007). Due to RFE, the shears in the velocity perpendicular to the magnetic field can initiate KHI (e.g., Hasegawa, 1975; Keskinen et al., 1988), making them a candidate for the creation of irregularities in the cusp (Basu et al., 1988, 1994; Carlson et al., 2007; Kersley et al., 1988; Keskinen et al., 1988; Moen et al., 2013; Oksavik et al., 2011).

Polar cap patches (Fig. 3.9D) are plasma density enhancements that convect across the polar cap (Crowley, 1996; Weber et al., 1984). Patches form either from dense dayside plasma transported into the polar cap by the background convection pattern or through a localized region of precipitation (Lockwood and Carlson, 1992; Rodger et al., 1994). In addition, the polar cap can also contain plasma depletions known as polar holes (Brinton et al., 1978; Makarevich et al., 2015). In both cases, ionospheric structures create density gradients in the polar cap plasma, on which a variety of instability mechanisms can act and create plasma irregularities. See Section 3.1.1 for more on polar cap patches and depletions.

3.2.1.2 Prior review articles concerning ionospheric irregularities

There are a large number of extant reviews covering various aspects of measurements and theories of ionospheric irregularities and radio scintillation caused by turbulent media. We first discuss reviews on radio propagation and then move to those concerning formation of irregularities. Collectively these reviews on scintillation phenomena and their connection to irregularities suggest use of physics-based scintillation models to understand the connection to cascading mechanisms.

Yeh and Liu (1982) described what has become a standard framework for electromagnetic wave scattering due to ionospheric irregularities. They presented the different spatial scales of density structures associated with phase and amplitude scintillations that fall in the refractive or diffractive regimes. The major contribution to the amplitude fluctuations on the ground comes from the phase front deviations caused by irregularities of the Fresnel scale size $\sqrt{(\lambda H_{iono})}$, where λ is the wavelength of the radio wave and H_{iono} is the height of the ionospheric structures, mainly due to diffractive effects. Irregularities larger than Fresnel scales produce refractive effects and dominate the phase fluctuations. The irregularity spectrum follows a power law form with respect to the wavenumber that suggests the coexistence of intermediate-scale sized irregularities that produce only refractive effects and small irregularities that produce only diffractive effects. In case of weak scattering assumption, a phase screen model can be used to simulate these effects on the radio signals. Such a model treats the irregularity layer at a given height as a slab of limited thickness through which the radio wave propagation may be numerically computed. The problem of radio wave propagation through random media can be solved using the parabolic equation method (PEM) under certain assumptions, such as the temporal variations of the irregularities are much slower than the wave period, and the characteristic size of the irregularities is much greater than the wavelength and forward scattering assumption. The ionospheric density irregularity can be represented as an electron density distribution obtained from a full physics-based plasma model, a spectral model, or an empirical model. Again, when a radio signal propagates through this irregularity, it experiences signal fluctuations.

Fejer and Kelley (1980) presented both ground-based and in situ measurement techniques, as well as a theoretical framework, developed to study and explain ionospheric plasma irregularities. They discussed E- and F-region irregularities at high and low latitudes, with more detail on equatorial regions where more progresses had been made. For high-latitude F-region irregularities, they grouped the possible sources in three categories: particle precipitation, electrostatic turbulence (spatially varying electric field), and plasma instabilities.

Tsunoda (1988) provided a review of high-latitude F-region ionospheric irregularities with emphasis on fluid interchange instabilities. After introducing some of the density irregularity characteristics such as spectral shapes, scale sizes, and distributions, the author reviewed larger-scale (>10 km) high-latitude plasma structures comprised of polar cap patches, blobs, Sun-aligned arcs, as well as density enhancements and depletions caused by soft particle precipitation and strong electric fields. The author then provided a descriptive model where <10 km structures are created by fluid interchange instabilities acting on the larger-scale density structures mentioned above. The basic properties of interchange instabilities and their linear growth rates are reviewed in detail, especially that of the $\mathbf{E} \times \mathbf{B}$ instability, i.e., GDI, and of the current-convective instability (CCI). Based on evidence from observations, the production of small-scale irregularities seems consistent with the model where GDI plays a dominant role.

A notable past review by Kintner and Seyler (1985) presented a rigorous theory of irregularity formation based on a fluid model of the ionospheric plasma. In this work, they first discussed electrically neutral fluid turbulence within the framework of the Navier-Stokes (N-S) equations and discussed how a classical Kolmogorov (energy-based) approach leads to the identification of a spectrum of turbulence with a $k^{-5/3}$ wavenumber dependence. Ionospheric measurements implying fluid-like behavior were presented and used to motivate their presentation of a fluid model of the ionosphere that includes inertial effects—thus rendering the ionospheric equations in a form similar to the N-S equations used to study neutral turbulence. Kintner and Seyler (1985) included ion inertia, to leading order, via the polarization current, which figures into the current continuity equation (cf. also discussion by Lotko, 2004; Mitchell et al., 1985), while still retaining resistive instabilities (e.g., discussion by Keskinen and Ossakow, 1983). Their formulation naturally describes both gradient and shear-driven instabilities in a single framework and has become the basis for a number of numerical models since developed (e.g., Gondarenko and Guzdar, 2006; Huba et al., 1988; Keskinen et al., 1988; Zettergren et al., 2015). While Kintner and Seyler (1985) discussed the spectra of plasma fluid turbulence in the context of ionospheric irregularities, they noted that further numerical work is required to fully understand the ionospheric situation.

Wernik et al. (2003) reviewed the connection between ionospheric irregularities and scintillation parameters. For example, the phase scintillation spectrum is related to the shape of the irregularity spectrum. Phase and amplitude scintillations are found to be proportional to the fluctuations in plasma density ΔN_e, which may not follow the distribution of the irregularity amplitude $\Delta N_e/N$ (Kivanç and Heelis, 1998). This causes seasonal variations, for example, nighttime scintillation over auroral oval during the summer is higher than in winter, while polar cap scintillation minimizes in summer due to the rapid decay of convecting irregularities. In cusp and nighttime auroral oval, structures at scales greater than 50 km are created by structured fluxes of precipitating electrons, which may become unstable under certain conditions and cause smaller irregularities due to wave-wave interaction and cascading.

3.2.1.3 Detailed motivating observations: In situ

A significant number of in situ studies have investigated high-latitude F-region irregularities. Density structures ranging from hundreds of kilometers to a few meters have been observed, with power spectral density for scintillation-producing scales (<10 km) commonly obeying power laws (in log-log) with a single spectral slope p or double slopes (e.g., Basu et al., 1988; Dyson et al., 1974; Fremouw et al., 1985; Ivarsen et al., 2019; Kintner and Seyler, 1985; Lagoutte et al., 1992; Mounir et al., 1991; Phelps and Sagalyn, 1976; Singh et al., 1985; Spicher et al., 2014; Tsunoda, 1988; Villain et al., 1986). Irregularity production and decay can affect p, for example, Atmosphere Explorer D satellite's in situ data analysis by Basu et al. (1984) revealed a systematic variation of p with auroral parameters and Ivarsen et al. (2019) showed a correlation between p and solar illumination. The spectral shape of irregularities can thus help in characterizing and classifying plasma structuring processes (e.g., Kintner and Seyler, 1985). It is worth mentioning here that the in situ one-dimensional spectral slopes p may also be of interest with respect to scintillation data, as spectral indices derived from scintillation phase measurements are expected to be steeper by unity compared with p (Tsunoda, 1988, and references herein). For example, a comparison between the slopes of power spectra derived from in situ (sounding rocket) electron density data and ground-based GPS carrier measurements within an RFE showed reasonable agreement with this prediction, suggesting that high-resolution in situ data could in principle be used for modeling GPS scintillations (Jin et al., 2019). In general, the power laws commonly observed seem consistent with fully developed turbulence generated through an instability mechanism responsible for cascading evolution of large-scale irregularities to progressively small-scale irregularities (e.g., Dyson et al., 1974; Kintner and Seyler, 1985; Tsunoda, 1988). The two main classes of macroscale instability mechanisms that have gained significant attention to explain high-latitude F-region irregularities are GDI and flow shear-driven processes such as KHI.

As discussed in Section 3.2.2.2, the growth rate of GDI has a preferred direction. Consequently, a distinctive feature of GDI is that the resulting structures are expected to develop anisotropically, e.g., with a stable leading edge and a structured trailing edge. Observations of such asymmetries were presented by Cerisier et al. (1985), which commonly (but not consistently) observed different characteristics of density irregularities on both sides of density enhancements, i.e., a stable and an unstable side. The authors also suggested that their observations were consistent with GDI being active as a source mechanism. Asymmetries in the spectral slopes of 1D cuts through patches parallel and perpendicular to the background plasma convection were also discovered by Basu et al. (1990), where the GDI was believed to be operating. Furthermore, Spicher et al. (2015a) and Goodwin et al. (2015) showed in situ measurements of polar cap patches both close to their origin on the dayside, and as they had traveled deeper into the polar cap toward nightside. The latter patches exhibited shorter leading edges consistent with the GDI structuring in the patches as they traveled across the polar cap with significantly more structured and elongated trailing edges. Additional in situ studies based on satellite data (e.g., Lagoutte et al., 1992; Mounir et al., 1991; Singh et al., 1985; Villain et al., 1986) and sounding rocket (Moen et al., 2012; Oksavik et al., 2012; Spicher et al., 2015b) provided further support to the view that high-latitude F-region irregularity creation could be at least partially due to GDI and cascading.

Early sounding rocket (Kelley et al., 1980) and satellite (Mounir et al., 1991) observations suggested that the evolution of irregularities could be significantly different in the polar cap and in the auroral zone. In the polar cap region, the turbulences are related to GDI, while in the auroral region, the major sources of ionospheric irregularity are shears, precipitation, and FACs (Mounir et al., 1991). The importance of velocity shears has also been emphasized for the cusp (e.g., Chernyshov et al., 2018; Heppner et al., 1993; Spicher et al., 2016), and for the auroral oval by, e.g., Basu et al. (1984) and Basu et al. (1988). In particular, Basu et al. (1988) studied the power spectra of electric field and electron density fluctuations coinciding with strong and moderate shears. For moderate shears, the spectra obtained were in agreement with predictions for KHI (Basu et al., 1988; Keskinen et al., 1988), while for the strongly sheared regions where large FACs were also observed, Basu et al. (1988) suggested that other mechanisms such as current-driven ion-cyclotron waves, current-convective or thermal instabilities could contribute (Basu et al., 1988, and references herein). Kelley and Carlson (1977) also presented sounding rocket observation of strong F-region flow shears on the edge of a substorm arc collocated with intense relative plasma perturbations. The authors discussed the observations with respect to flow shear instabilities, including long-wavelength regime KHI.

Despite the significant number of in situ studies providing observations of irregularity structures spanning a wide range of scales down to scintillation-producing scales, the physical sources for their creation are still poorly understood. While generally attributed to GDI and shear-driven KHI, the detailed time-dependent evolution, signatures, and their relative contributions in scintillation are still unresolved and not presently predictable, thus motivating further modeling work of these two macroscale instability mechanisms.

3.2.1.4 Detailed motivating observations: Scintillation

Fluctuations in radio signals, such as those from polar-orbiting satellites, beacons, and GPS and GNSS satellites have been observed at high latitudes for at least five decades. Before that, scientists looked at the sporadic scintillation of radio sources to study the ionosphere (e.g., Briggs, 1964). Although the global scintillation occurrence was well known, the early scintillation observations were mostly from the lower latitudes due to the dearth of high-latitude data (e.g., Aarons, 1982, 1993; Basu et al., 2001; Wernik et al., 2003). With the development of infrastructure in the high latitudes, more scintillation studies have been possible in these regions in recent decades. Multifrequency and multireceiver scintillation observations were used to derive turbulence parameters based on the weak scattering assumptions (Bhattacharyya et al., 1992). Pioneering work on high-latitude scintillation studies on understanding scintillation morphology, plasma structuring during different instability processes over polar cap and auroral regions derived from in situ as well as scintillation measurements is described in Basu et al. (1985, 1988, 1990, 1994, 1998). More recently, presence of ionospheric irregularities responsible for GPS (L-band) scintillation was reported to be associated with E-region particle precipitation (Kinrade et al., 2012) with scintillations occurring close to or on the trailing edge of arcs (Datta-Barua et al., 2015; Mrak et al., 2018; Semeter et al., 2017).

In addition to L-band GPS signals, scintillation has also been observed in the transionospheric UHF and VHF signals (Fremouw et al., 1978; Lamarche et al., 2020). These are typically observed from networks of ground-based receivers that detect signals from beacons transmitting on a variety of LEO satellites (Bernhardt and Siefring, 2006; Siefring et al., 2015). Although these networks are less extensive than the GNSS network, they provide important information about frequency dependence in scintillation.

A significant number of studies examining high-latitude ground-based GNSS data have been performed, comprising statistical analyses (e.g., Alfonsi et al., 2011; Jin et al., 2015, 2018; Kinrade et al., 2013; Meziane et al., 2020; Prikryl et al., 2010, 2011, 2015; Spogli et al., 2009; Spogli et al., 2013) and studies investigating scintillations in different geophysical contexts, including cusp aurora (e.g., Jin et al., 2015; Oksavik et al., 2015), auroral arcs (e.g., Forte et al., 2017; Mrak et al., 2018; Semeter et al., 2017; van der Meeren et al., 2015), polar cap patches (e.g., Jayachandran et al., 2017; Jin et al., 2014, 2016, 2017; Mitchell et al., 2005; van der Meeren et al., 2015; Zhang et al., 2017a), auroral blobs (e.g., Jin et al., 2014, 2016), the storm-enhanced density and tongue of ionization (e.g., van der Meeren et al., 2014; Wang et al., 2016, 2018), reversed flow events (Jin et al., 2019; Spicher et al., 2020), etc.

At high latitudes, enhanced phase scintillations generally peak in the cusp and in the auroral oval (Aarons, 1997; Jin et al., 2015; Prikryl et al., 2010, 2011, 2015; Spogli et al., 2009), and are associated with polar cap patches/tongue of ionization in the polar caps (e.g., Prikryl et al., 2010). In particular Jin et al. (2017) presented evidence of the combination of polar cap patches and cusp auroral dynamics causing an increased level of phase scintillation. The strongest phase scintillations occur when polar cap patches have entered the nightside auroral oval (thus termed auroral blobs) implying that auroral dynamics including shear and precipitation play an important role in enhancing plasma structuring at phase scintillation scales (Clausen et al., 2016; Jin et al., 2014, 2016). Kersley et al. (1988) also observed strong scintillations and flow shears. While it has been suggested that sun-aligned arcs and associated shears may also cause space weather issues for northward IMF (e.g., Carlson, 2012; Moen et al., 2013), a very limited number of studies exist on the relationship between scintillations and transpolar arcs. For example, van der Meeren et al. (2016) showed weak scintillation and irregularities observed on the nightside of a long-lived transpolar arc.

3.2.2 F-region fluid instabilities: Linear theory

This section summarizes key results from the linear theory of GDI and KHI—instabilities that are believed to play a critical role in high-latitude irregularity formation and scintillation. We do not attempt to provide detailed derivations of how these are obtained (viz containing all algebraic steps), but instead list governing equations, discuss assumptions, and present some key results to be compared against nonlinear simulations in later sections.

Fig. 3.10 shows an illustration of fundamental physical processes responsible for the initiation of GDI. In this diagram, a density perturbation is applied to a background

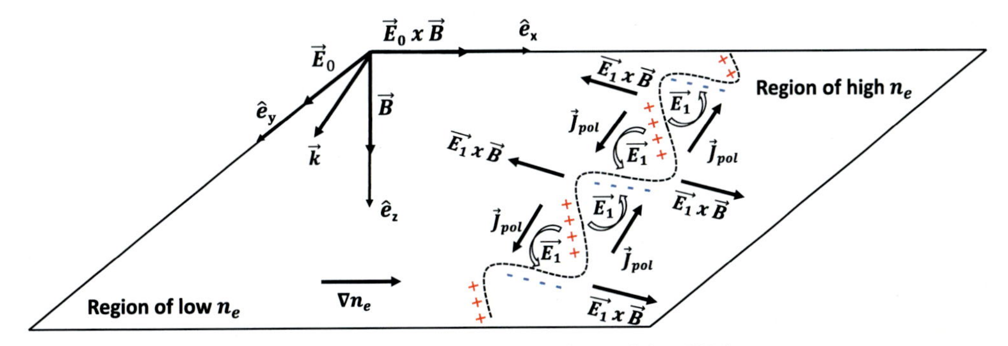

Fig. 3.10 Charge accumulation resulting in gradient-drift instability (GDI).

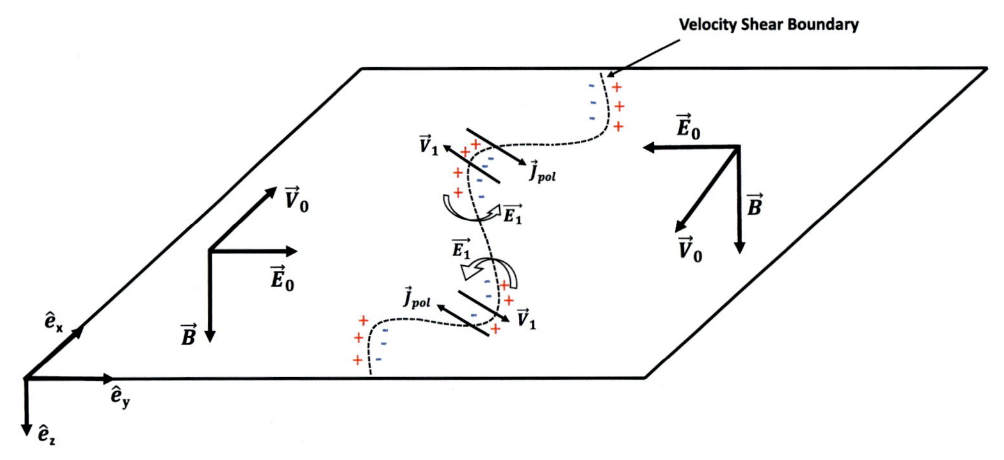

Fig. 3.11 Kelvin-Helmholtz instability (KHI) resulting from a flow shear.

density gradient and subjected to a background electric field. This background field, in the presence of variable ionospheric conductance [density], leads to larger [smaller] Pedersen currents—not shown to reduce visual clutter in the figure—in the high [low]-density region. The divergence [convergence] of this current then results in negative [positive] charge accumulation shown. Perturbation electric fields, $\mathbf{E}_1$ in this diagram, result in $\mathbf{E}_1 \times \mathbf{B}$ drifts that reinforce the seed density perturbations, resulting in unstable growth. This growth can be suppressed by the polarization current (see Section 3.2.2.2) which is proportional to $\partial \mathbf{E}_1/\partial t$ and leads to a partial shorting out of the accumulated charge and a reduction in the response drift and instability growth—note the oppositely directed Pedersen currents (omitted to reduce figure clutter but flowing from the negative to positive charges on the diagram, i.e., they cause these charges to accumulate) and polarization currents in Fig. 3.10.

Ion inertial effects are responsible for destabilizing KHI (Keskinen et al., 1988), which can alternatively be stabilized by conduction currents (collisions). The key ingredient for KHI in the ionosphere is a background plasma velocity shear across a relatively narrow boundary layer. If the component of velocity normal to the boundary layer formed by the shear region $\mathbf{v}_1$ is perturbed, it results in a nonuniform polarization current that causes charge accumulation (Fig. 3.11). The electric field from this charge distribution reinforces the original velocity perturbation and results in the growth of the instability. Pedersen currents can partially short out the charge accumulation from the polarization current and can effectively cause damping of the instability—note, again, the oppositely directed polarization vs Pedersen currents in Fig. 3.11 (not shown but flowing from the positive to negative charges, i.e., they are caused by response fields produced by these charges).

3.2.2.1 Assumptions and governing equations

Instabilities discussed herein are all fundamentally (quasi-)electrostatic in nature—they all involve, at some level, the assumption $\mathbf{E} = -\nabla\Phi$ and the response fields that drive the instabilities that are related to charge accumulation. In this subsection, we will outline linear analysis of these instabilities for conditions representative of the ionospheric F region at high latitudes. First, the plasma will be assumed to drift at the $\mathbf{E} \times \mathbf{B}$ velocity (F region conditions). We also neglect Hall currents (they are very small in the F region) so that the conduction current is given by $\mathbf{J}_c = \sigma_P \mathbf{E}$, where σ_P is the Pedersen conductivity. Depending on the instability and conditions of interest, the ionospheric current density $\mathbf{J}$ will include conduction currents (viz those proportional to conductivity), and (for some analyses) polarization currents (related to ion inertia).

For linear analysis of GDI and KHI, an equipotential field line (EFL) approximation is used; this assumption requires that the electric potential Φ *does not vary along the geomagnetic field lines*. Justification for this is based on the analysis in Farley (1959) and is valid so long as the length scales perpendicular to the field line are not too small ("small" as used here is qualified further in Section 3.2.3.5). The EFL assumption allows equations to be recast in terms of field-integrated plasma quantities like plasma column density ($N(x,y) \equiv \int n(x,y,z)\, dz$, where z is the field line coordinate).

Assumptions outlined above may be combined with the plasma continuity and current continuity equations to produce a system describing the *local* F-region plasma:

$$\frac{\partial n}{\partial t} + \nabla \cdot (n\mathbf{v}) = 0 \tag{3.11}$$

$$\nabla \cdot \mathbf{J} = 0 \tag{3.12}$$

$$\mathbf{v} \approx \frac{\mathbf{E} \times \mathbf{B}}{B^2} \tag{3.13}$$

$$\mathbf{E} = -\nabla\Phi. \tag{3.14}$$

The role of ionospheric conduction (viz collisions) and ion inertia is captured by the particular choice of terms included when computing the current density $\mathbf{J}$—further in Sections 3.2.2.2 and 3.2.2.3.

Eqs. (3.11)–(3.14) can be integrated along the field line (using the EFL assumption) resulting in an analogous set of equations involving column density/conductance; usually we also reduce the system to a set of equations for the N, Φ only by substitution. These equations can then be linearized in the usual fashion by decomposing solutions into the background (N_0, Φ_0) and perturbations (N_1, Φ_1), viz $N = N_0 + N_1$, $\Phi = \Phi_0 + \Phi_1$, and assuming that the perturbations are small so that higher-order nonlinear terms may be neglected.

3.2.2.2 GDI linear theory

Classical linear growth rates for GDI may be derived by considering only the conduction current, viz $\mathbf{J} = \mathbf{J}_c = \sigma_P \mathbf{E}$. The density has a background gradient, here assumed to be in the x-direction $\nabla N_0 = \partial N_0/\partial x\, \hat{\mathbf{e}}_x$ while the background electric field is in the y--direction, $\nabla \Phi_0 = -E_0 \hat{\mathbf{e}}_y$, as shown in Fig. 3.10. Perturbations in column density and potential (N_1, Φ_1) are taken to be proportional to $e^{iky-i\omega t}$,

$$i\omega \tilde{N}_1 + \frac{ik\frac{\partial N_0}{\partial x}}{B}\tilde{\Phi}_1 = 0 \tag{3.15}$$

$$N_0 k^2 \tilde{\Phi}_1 + ik\tilde{N}_1 E_0 = 0 \tag{3.16}$$

These may be solved to yield a solution for the frequency:

$$\omega = i\frac{E_0}{\ell B}, \quad \ell \equiv \left(\frac{1}{N_0}\frac{\partial N_0}{\partial x}\right)^{-1}. \tag{3.17}$$

The parameter ℓ here is referred to as the "scale length" of the background gradient. As the frequency is purely imaginary and positive the perturbation grows exponentially in time at a characteristic *growth rate* of

$$\gamma \equiv \Im\{\omega\} = E_0/(\ell B) \tag{3.18}$$

Growth rates for GDI in the presence of ion inertia can be derived by including the polarization current (e.g., Lotko, 2004) in the linear analysis of the instability (Ossakow et al., 1978), in this case we take:

$$\mathbf{J} = \sigma_P \mathbf{E} + c_m\left(\frac{\partial}{\partial t} + \mathbf{v}\cdot\nabla\right)\mathbf{E}, \tag{3.19}$$

where $\mathbf{v} = \mathbf{E} \times \mathbf{B}/B^2$ and c_m is the inertial capacitance defined in (Mitchell et al., 1985):

$$c_m \equiv \frac{\sum_s n_s m_s}{B^2}; \quad C_m \equiv \int c_m(z)dz. \tag{3.20}$$

The z-coordinate here is again used to represent the distance along the geomagnetic field line. Repeating the linearization and Fourier analysis procedure with these assumptions (e.g., Ossakow et al., 1978), and neglecting background drift normal to the density gradient, gives for continuity and current continuity equations:

$$i\omega \tilde{N}_1 + \frac{ik\frac{\partial N_0}{\partial x}}{B}\tilde{\Phi}_1 = 0 \tag{3.21}$$

$$-ikE_0\,\tilde{\nu}\,\tilde{N}_1 - \tilde{\nu}\,k^2 N_0\tilde{\Phi}_1 + i\omega k^2 N_0\tilde{\Phi}_1 = 0 \tag{3.22}$$

Solving this system of equations gives frequencies of

$$\omega = -\frac{1}{2}i\tilde{\nu} \pm \frac{1}{2}i\sqrt{\tilde{\nu}^2 + 4\tilde{\nu}\frac{E_0}{\ell B}}, \tag{3.23}$$

where $\tilde{\nu} \equiv \Sigma_P / C_m$ is the ratio of Pedersen conductance to integrated inertial capacitance. The largest growth rate corresponds to the positive root in this equation, and represents the growth rate for GDI in the presence of ion inertial effects.

Two limiting cases of interest for GDI with inertial growth are the small inertia limit ($\tilde{\nu} \gg 1$), in which the growth rate implied by Eq. (3.23) reduces to that already given in Eq. (3.17). Alternatively, in the inertia-dominated case $\tilde{\nu} \ll 1$, we have a frequency that is, to the leading order:

$$\omega = i\sqrt{\tilde{\nu}\frac{E_0}{\ell B}} \approx i\sqrt{\nu_s\frac{E_0}{\ell B}}, \tag{3.24}$$

where ν_s is the ion-neutral collision frequency. From Eq. (3.24) dependence of the growth rate on ion inertia (characterized by the inertial capacitance parameter) can be written as

$$\mathcal{I}\{\omega\} = \sqrt{\frac{\Sigma_P}{C_m}\frac{E_0}{\ell B}} \propto C_m^{-1/2} \tag{3.25}$$

This expression clearly shows that increasing the value of the integrated capacitance (i.e., representing an increased importance for ion inertia) works to reduce the linear growth of GDI.

The simplified treatment of linear GDI shown above assumes a specific altitude regime and geometry of the background gradient and drift velocity vectors—namely that they are parallel. Although this is useful for a theoretical understanding, contemporary work has made significant progress in defining dispersion relations that are valid in a wide range of altitudes for arbitrary vector orientations and can describe the Farley-Buneman and current-convective plasma instabilities along with GDI in a single theoretical formulation (Makarevich, 2014, 2016a, b, 2019). For the specific case of long-wavelength GDI in the F region, the following growth rate expression can be extracted from the generalized treatment of Makarevich (2019)

$$\gamma = -\frac{1}{k_\perp^2}\left(\frac{\nabla n}{n}\cdot \mathbf{k}\times\hat{\mathbf{e}}_B\right)\left(\frac{\nu_i}{\Omega_i}\mathbf{v} - \frac{\mathbf{E}_\perp}{B}\right)\cdot\mathbf{k} \tag{3.26}$$

Because no particular relative orientation of the density gradient, electric field, and wave propagation direction is assumed, all these quantities must be expressed as vectors ($\nabla n/n$, $\mathbf{E}$, and $\mathbf{k}$, respectively). The unit vector $\hat{\mathbf{e}}_B$ is in the direction of the geomagnetic field, viz the $\hat{\mathbf{e}}_z$-direction in our diagrams of Fig. 3.10, and ν_s and Ω_s are the ion collision and gyrofrequencies, respectively. Under the assumption $\mathbf{k} \perp \mathbf{B} \perp \mathbf{v} \parallel \nabla n$, this expression simplifies to Eq. (3.18). The generalized GDI growth rates depend not only on the relative directions and magnitudes of the density gradient and electric field but also the propagation direction of the wave (Keskinen and Ossakow, 1981; Makarevich, 2014). The $1/k_\perp^2$ factor effectively normalizes the perpendicular components of both **k** vectors in the expression such that the growth rate depends on the perpendicular direction of the wavevector but not its magnitude, indicating that GDI causes structures to grow preferentially in a particular direction, but uniformly across all scales (Makarevich, 2014).

The plasma structures resulting from GDI are then anisotropic, evidenced by the characteristic "fingers" that grow perpendicular to a density enhancement's trailing edge (Deshpande and Zettergren, 2019; Gondarenko and Guzdar, 2004b; Hosokawa et al., 2016). This is not the same effect as the leading/trailing edge asymmetry that is often discussed with GDI acting on polar cap patches (Lamarche and Makarevich, 2017). If the gradient and plasma drift velocity are antiparallel (such as the leading edge of a patch), there are almost no wavevector directions that result in a positive growth rate, while if they are parallel (trailing edge of a patch) the growth rate is almost always positive. Hence, GDI tends to be active on the trailing edge and suppressed on the leading edge, which explains many observations of small-scale structuring primarily on the trailing edge of patches (asymmetry) (Hosokawa et al., 2013, 2016; Milan et al., 2002; Moen et al., 2012; Weber et al., 1984). This has also been discussed in Section 3.1.1. The fact that the GDI growth rate is wavevector-dependent further predicts that the trailing edge structuring will form preferentially in a particular direction (anisotropy).

3.2.2.3 Linear KHI in the ionosphere

KHI occurs in the presence of shear plasma flows; typically analysis will assume either a sharp (infinitesimal) boundary, in which case the problem may be treated analytically (cf. Keskinen et al., 1988, Appendix A) or a far more realistic smooth drift profile, which then requires a numerical solution of the resulting linear system of differential equations (e.g. Berlok and Pfrommer, 2019) as discussed further below. The instability growth is a function of the ratio of Pedersen conductance to inertial capacitance $\tilde{\nu} \equiv \Sigma_P / C_m$ and also of the scale length of the shear layer ℓ, often parameterized through a hyperbolic tangent profile, in this article we take:

$$v_x(y) = v_0 \tanh\left(\frac{y}{\ell}\right) - v_n \tag{3.27}$$

A starting, equilibrium state for KHI (needed for the linear analysis) requires additional assumptions. Particularly the steady-state current continuity equation $\nabla \cdot \mathbf{J} = 0$ must hold for a straightforward analysis, which means that the plasma velocity profile (i.e., variation with the y-direction) should be asymmetric in the neutral atmosphere frame of reference (or equivalently there must be a nonzero neutral wind in the frame where the plasma velocity profile is symmetric). This is required to balance the divergent Pedersen current implied by a symmetric electric field configuration. Because of this the neutral wind required (in the frame where the plasma velocity is symmetric) is uniquely determined by the choice of plasma velocity amplitude v_0 and density jump across the shear layer (Keskinen et al., 1988) given by

$$\frac{\lim_{y\to\infty} n(y)}{\lim_{y\to-\infty} n(y)}. \tag{3.28}$$

For linear KHI analysis, the full current density given by Eq. (3.19) is required. The geometry for ionospheric KHI is fundamentally different from the GDI cases considered to this point; here we suppose a background state where (see also Fig. 3.11):

$$\nabla_\perp N_0 = \frac{\partial N_0}{\partial y}\hat{\mathbf{e}}_y; \quad \nabla_\perp \Phi_0 = \frac{\partial \Phi_0}{\partial y}\hat{\mathbf{e}}_y = -E_0\hat{\mathbf{e}}_y, \tag{3.29}$$

and for our linear perturbation analysis, we must allow the perturbed quantities to vary continuously in the y-direction and harmonically in the x-direction:

$$\Phi_1(x,y,t) = \Re\left\{\tilde{\Phi}_1(y)e^{ikx-i\omega t}\right\}; \quad N_1(x,y,t) = \Re\left\{\tilde{N}_1(y)e^{ikx-i\omega t}\right\}, \tag{3.30}$$

The algebra is substantial but linearizing the plasma continuity and current continuity equations and plugging in these perturbations yields a system of ordinary differential equations (viz an eigenvalue problem) to be solved.

$$(\omega - kv_0)\tilde{N}_1 - k\frac{1}{B}\frac{\partial N_0}{\partial y}\tilde{\Phi}_1 = 0 \tag{3.31}$$

$$\left[\frac{\partial}{\partial y}\left(\frac{N_0\partial\tilde{\Phi}_1}{\partial y}\right) - k^2N_0\tilde{\Phi}_1\right](\omega - kv_0 + i\tilde{\nu}) + i\tilde{\nu}\,(v_n - v_0)BN_0\frac{\partial}{\partial y}\left(\frac{\tilde{N}_1}{N_0}\right) +$$

$$k\tilde{\Phi}_1\frac{\partial}{\partial y}\left(N_0\frac{\partial v_0}{\partial y}\right) = 0. \tag{3.32}$$

where, as in prior equations, $v_0 \equiv E/B$, and v_n is the neutral wind velocity. This eigenvalue problem must be solved numerically for the growth rates—e.g., the values of ω for which these equations are valid in the nontrivial sense $\tilde{N}_1, \tilde{\Phi}_1 \neq 0$ and the corresponding eigenfunctions (solutions $\tilde{N}_1$, $\tilde{\Phi}_1$ corresponding to "allowable" values of frequency).

An example of growth rate solutions by this method from Keskinen et al. (1988) is shown in Fig. 3.12. Note that this figure is a reproduction and differs in the notation for the rest of this article; these may be translated as $L \equiv \ell$ and $\nu \equiv \tilde{\nu}$. As shown in this result,

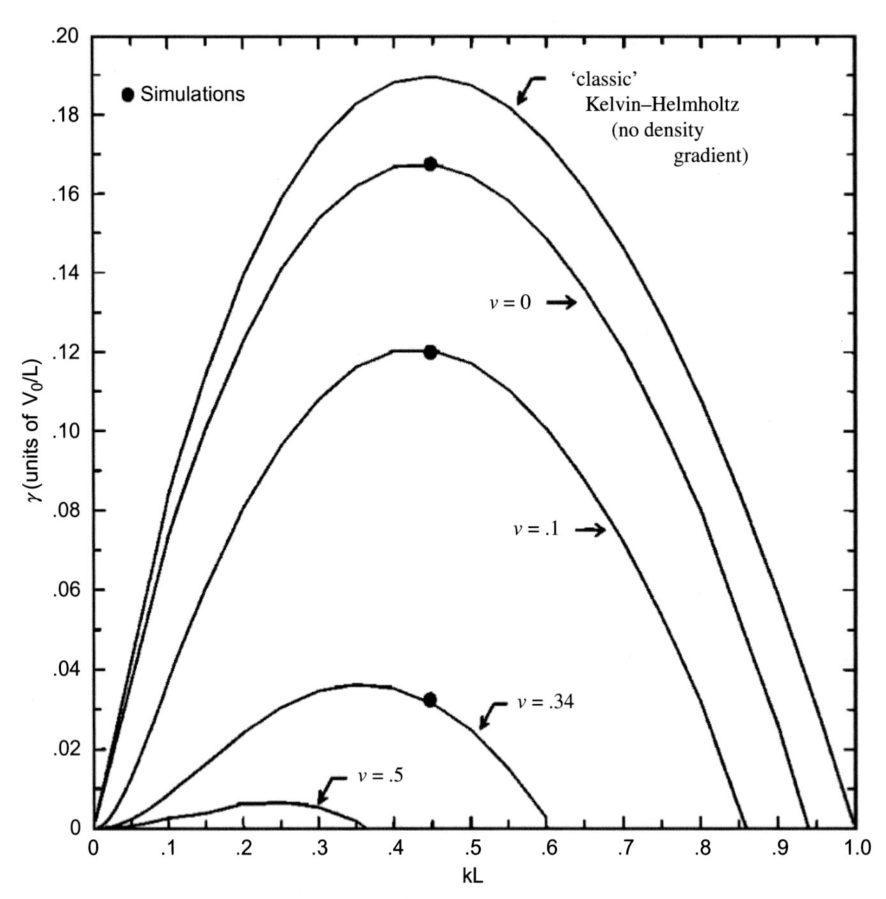

Fig. 3.12 KHI linear growth rate solutions. For purposes of comparing to the rest of this article note that $L \equiv \ell$ and $\nu \equiv \tilde{\nu}$. *Reprinted from Keskinen, M.J., Mitchell, H.G., Fedder, J.A., Satyanarayana, P., Zalesak, S. T., Huba, J.D., 1988. Nonlinear evolution of the Kelvin-Helmholtz instability in the high-latitude ionosphere. J. Geophys. Res. Space Phys. 93 (A1), 137–152.*

the maximum growth rates occur for modes having $k\ell \approx 0.4 - 0.5$, such that the *fastest growing* mode has wavelength $\lambda \approx 2\pi\ell/0.44$. Fig. 3.12 also shows that the particular value for the growth rate at this maximum is strongly dependent on the parameter $\tilde{\nu}$ (the ratio of conductance to capacitance)—generally larger values of this parameter result in more damping via Pedersen currents, which "short out" KHI, and decrease overall growth rate.

3.2.2.4 Applications of linear theory

As outlined in Sections 3.2.2.2 and 3.2.2.3, relatively simple expressions can be obtained for the growth rates of GDI and KHI using linear theory, subject to certain assumptions.

These growth rates depend only on a few parameters, which can be measured both from the ground or in situ. Thus, several studies have applied results from linear theory to observations to evaluate the role of different instability mechanisms in producing irregularities.

To explain the presence of strong HF backscatter echoes in the cusp due to decameter-scale irregularities, Moen et al. (2002) combined tomographic images based on TEC and line-of-sight velocity measurements from the CUTLASS Finland radar and estimated the linear growth rate of GDI (using Eq. 3.17). They found that irregularities would develop within 12 min due to GDI acting on the large-scale gradient. They suggested that particle precipitation could act as the primary mechanism, or that GDI could occur on intermediate-scale-sized structures. Later, Carlson et al. (2007, 2008) proposed a framework where the KHI would initialize plasma structuring of polar cap patches as they transit through the cusp entry region, providing seed irregularities enabling GDI to develop faster as a secondary mechanism. Using the fastest growing mode (based on Keskinen et al., 1988), they estimated KHI to have a growth time of about 50 s. Consistent values of KHI growth times were then calculated by Oksavik et al. (2011) using SuperDARN HF data observations of RFEs. Later, Moen et al. (2012) used high-resolution electron density data from the Investigation of Cusp Irregularities-2 (ICI-2) sounding rocket to estimate the growth time for GDI on the trailing edge of a density enhancement in the cusp. Using Eq. (3.17), they found a value of 47.6 s, and suggested that the growth time could be as fast as about 10 s, which is 50–70 times faster than previously estimated (Moen et al., 2002). Similar growth times were then reported using data from the Swarm satellites (Spicher et al., 2015a). Additionally, also using ICI-2 data, Oksavik et al. (2012) investigated in detail the minimum linear growth times of GDI and KHI to assess which mechanism would grow faster and at what scales. The authors found shorter growth times for GDI than for KHI for most of the flights, suggesting that GDI would be dominating over KHI during that event.

Also based on in situ measurements, Burston et al. (2016) used DE-2 satellite data to compare the linear growth rates of GDI, CCI, KHI, and "stirring" (defined in Section 3.2.3.7), in both the inertial and collisional regimes. Note that the authors used the word turbulence instead of stirring (Burston et al., 2016; Kelley, 2009), but we here opt for the latter name to avoid confusion with turbulence generated from primary ionospheric instabilities. Based on the linear growth rates, they found inertial stirring to be dominant, followed by inertial GDI. The authors also showed that these two processes and the collisional CCI can regularly grow within timescales shorter than 60 s in the linear regime, while the KHI did not meet the instability condition in any of their cases. It is worth mentioning that Burston et al. (2016) used a different KHI growth rate compared to Oksavik et al. (2012).

Burston et al. (2009) also made use of linear theory and estimated the amplitude of GDI waves with the help of ionospheric imaging reconstruction (MIDAS 2.0) for

13 days during magnetic storms. The authors used Eq. (3.24) and assumed that the amplitude of GDI waves were proportional to $\exp \gamma t$, where t is the time. Burston et al. (2009) were able to estimate GDI growth rates γ for the entire part of the reconstructed ionosphere at each time step, to calculate how the GDI wave amplitude varied with time, and to correlate their calculations with scintillation indices recorded at several stations. They found weak but statistically significant correlation between the scintillation indices and the GDI mechanism for causing irregularities during storm time. Burston et al. (2010) then used the same method to correlate scintillation indices with the amplitude of waves caused by GDI and by stirring, and found in general a better correlation for GDI.

Several studies have also evaluated the GDI growth rates based on remote sensing imaging (Incoherent Scatter Radar (ISR) and All-Sky Imager (ASI)) techniques. Hosokawa et al. (2013) estimated growth rates on the order of $10^{-1}\ \mathrm{s}^{-1}$–$10^{-2}\ \mathrm{s}^{-1}$ while Lamarche and Makarevich (2017) and Lamarche et al. (2020) found rates on the order of $10^{-2}\ \mathrm{s}^{-1}$–$10^{-4}\ \mathrm{s}^{-1}$ using measured gradient and velocity magnitude and directions. In general, these studies found that qualitatively linear growth rates often agree with measurements where plasma irregularity observations occur, but as for the studies based on in situ observations, it is challenging to use them to fully explain the irregularity evolution, especially when the instability is in the nonlinear stage or likely to be part of a complex structuring process.

One important factor to take into account when interpreting the results from different observational studies is the inherent limitations of different observational techniques. In general, in situ instruments on sounding rockets will have the highest spatial and temporal resolution, followed by those on satellites (although both suffer from space-time ambiguity issues), with ground-based remote sensing instruments like ISRs or ASIs having the lowest resolution, but largest region of coverage. The spatial resolution of a particular measurement technique controls the level of detail at which the structuring can be measured, which in turn impacts the gradients and shears measured and the resulting linear growth rates. This may help explain why the growth rates derived from the ICI-2 sounding rocket Langmuir probe (Moen et al., 2012) are an order of magnitude greater than those estimated from ISR densities (Lamarche and Makarevich, 2017). It is also important to remember that satellites and sounding rockets typically measure a 1D slice through the density structure, and unless this slice is perpendicular to the edge of the structure it will underestimate the steepness of the edge gradient. Finally, we note that the impact of the neutral dynamics has generally been omitted for observation-based linear growth rate calculations.

These issues highlight the fact that the edges of large-scale irregularities are likely to be quite complex, with smaller density perturbations overlying larger background gradients. It is not clear whether it is appropriate to apply linear growth rates to the steeper perturbation gradients, the large-scale edge gradient, or both. Nonlinear modeling can help address how structuring develops on an arbitrary superposition of structures at a variety of scales, but only if the model is initialized with a realistic spectrum of structures. Measuring details of the spectrum of ionospheric seed structures on which instability mechanisms act as key to advancing our understanding of how instabilities evolve.

3.2.2.5 Diffusive stabilization at small scales

At the smallest scales we consider (likely below ~100 m), pressure effects like diamagnetic drift and cross-field diffusion may become significant. Diamagnetic drifts lead to additional flows to be considered, while at F region and lower altitudes, where collisions are frequent, charged particles can also diffuse across geomagnetic field lines (e.g., Kelley, 2009). Ions are less magnetized than electrons and so are more able to diffuse across geomagnetic field lines. This can lead to charge separation and a *perpendicular* ambipolar electric field that will generally act to suppress further diffusion. As a consequence of these processes, encapsulating small-scales requires incorporating pressure terms self-consistently into both drift equations and electromagnetic solutions. Most models of GDI and KHI, to date, have not fully incorporated pressure effects (Zettergren et al., 2015) or have been run at resolutions where they are not the dominant contributor to irregularity evolution (e.g., Gondarenko and Guzdar, 2006; Huba et al., 1988, and related studies).

One may approximate the scale at which plasma pressure effects become substantial via examination of the diamagnetic drift expression (e.g., Schunk and Nagy, 2009). Typical plasma drifts in GDI or KHI unstable situations are likely to be in the 250–1000 m/s range, roughly corresponding to 12.5–50 mV/m electric fields. Using representative plasma parameters, one needs approximately a ~10 m scale length to generate the diamagnetic drift equally as strong as the drift generated by a 12.5 mV/m electric field. This suggests that diamagnetic drift effects can be safely ignored for all but the smallest irregularities considered in the ionosphere.

In addition to diamagnetic effects, one must also consider cross-field diffusion. A generic diffusion equation (derived under assumptions of constant diffusion rate, for the sake of simplicity and illustration) describing F-region plasma transport may be written in the following form.

$$\frac{\partial n}{\partial t} + D\frac{\partial^2 n}{\partial x^2} = 0; \quad n(x,t) = \int_{-\infty}^{\infty} A(k)e^{-k^2 Dt}e^{ikx}\,dk \tag{3.33}$$

where, D is the *diffusion coefficient*, and the solution for density n is written for the case of an unbounded domain with density decaying to zero at infinity. This shows that a single spatial mode with wavenumber k decays exponentially at a rate given by

$$k^2 D = \frac{4\pi^2 D}{\lambda^2} \tag{3.34}$$

showing the well-known fact that small-scale modes decay faster. If the instability growth rate, γ, is known (for either GDI, KHI, or some other mechanism), the net irregularity growth/decay rate in the linear regime is $\gamma - k^2 D$ (Keskinen and Ossakow, 1981). By setting this net growth/decay rate to zero and solving for λ, we can identify a critical scale, above which the growth rate is larger than the decay rate (and irregularities will grow) and below which the decay rate is larger than the growth rate (and irregularities will dissipate) (Lamarche et al., 2020). For a scale-independent growth rate (such as GDI without ion inertial effects, Section 3.2.2.2), this calculation is straightforward.

$$\frac{E}{\ell B} - \frac{4\pi^2 D}{\lambda^2} = 0 \Longrightarrow \lambda = 2\pi\sqrt{\frac{D\ell B}{E}} \tag{3.35}$$

The critical scale, λ, can be interpreted as the minimum scale at which GDI growth will occur under given conditions. As the growth rate ($E/\ell B$) increases, small-scale structures may move from the diffusion regime into the growth regime (cf. Lamarche et al., 2020, Fig. 12).

The appropriate form of the diffusion coefficient, D, depends on the background conditions of interest, particularly the particle collision frequencies. In the F region, the ion diffusion coefficient for motion across the geomagnetic field is given by (Kelley, 2009; Perkins et al., 1973):

$$D = \frac{v_{ths}^2 \nu_s}{\nu_s^2 + \Omega_s^2} \tag{3.36}$$

where $v_{ths} = \sqrt{k_B T_s / m_s}$ is the ion thermal speed and ν_s and Ω_s are the ion-neutral collision and ion gyrofrequencies, respectively. The diffusion coefficient is dependent on which modalities of cross-field transport are considered to be important, so variations on the above expression can be found in the literature. For example, a robust E-region will have a significant impact on diffusion through E-region shorting (e.g., Vickrey and Kelley, 1982). This is a nonlocal effect (discussed in Section 3.2.3.5) and is challenging to quantify in linear interpretations because it requires knowledge of the plasma conductance along the entire field line (viz it is inherently nonlocal).

3.2.3 Simulating ionospheric F-region instability and turbulence: A nonlinear modeling tutorial

Linear theory presented in Section 3.2.2, while useful for a wide range of studies discussed in Section 3.2.2.4, does not address mid- or late-stage development of irregularities—an essential aspect relevant to the resulting scintillation of radio signals as they pass through those irregularities. This section presents nonlinear aspects of ionospheric instability behavior within a framework of numerical simulation studies both past and current. For the recent modeling results presented, we include links to source code for demonstration simulations so the readers may reproduce, modify, and study these demos or use them for their own instability-related research.

3.2.3.1 Summary of nonlinear modeling approaches

Various nonlinear ionospheric models of GDI and KHI, which incorporate ion inertia effects, have been developed and applied in prior work to study mid-to-late-stage evolution of the instabilities. While ultimately limited in resolution, these models have served to illustrate important aspects of irregularities not apparent from the linear theory. Here we focus specifically on more recent models that have incorporated ion inertia, e.g., in the manner introduced by Kintner and Seyler (1985).

Mitchell et al. (1985) simulated high-latitude F-region GDI with a two dimensional model (the plane perpendicular to the geomagnetic field) that included both polarization and Pedersen currents. This model included ionospheric and magnetospheric slabs and was used to show the relative effects of polarization drifts on the initiation and evolution of GDI, generally showing that the polarization drifts slow progression of the instability and lead to strong large-wavelength features in the nonlinear stage. Keskinen et al. (1988) applied the same model to study linear and nonlinear evolution of KHI under conditions of strong background plasma flow shears. Their analysis of the linear stage of KHI was used to derive growth rates for the instability under various shear conditions and was critically compared against nonlinear simulations for purposes of validation. Work by Huba et al. (1988) synthesized these nonlinear studies to demonstrate roles of polarization current as a stabilizing mechanism for GDI and Pedersen currents as a stabilizing mechanism for KHI while suggesting likely appropriateness of slab models of these instabilities for km-scale structures, i.e., the EFL assumption is appropriate for such structures. Ganguli et al. (1994) examined the interplay between fluid-scale and microscale processes in shear instabilities and parameterized microinstability effects into their fluid model of KHI via an anomalous viscosity in the model.

Three-dimensional, nonlinear studies of the role of GDI in structuring plasma patches have sought to explain various observed aspects of the progression rate of the instability. Guzdar et al. (1998) focused on effects of including field line resolved dynamics on the growth of the instability compared to prior 2D, slab modeling efforts. These studies were later extended by Gondarenko and Guzdar (1999) to include effects of ion inertia, which was shown to slow the growth of irregularities, similar to prior studies, and led to the suggestion that the persistence of polar cap patches could be a consequence of slowed growth due to parallel to B (shorting) and inertial effects. Results from this model were compared against data from the DE satellites (Gondarenko and Guzdar, 2004a, b)—reasonable agreement further indicated that GDI was a likely source of the polar cap irregularities, and that ion inertia played a key role in secondary instabilities (e.g., KHI) and nonlinear evolution of GDI. Effects of variable magnetospheric electric field (including field reversals) (Gondarenko et al., 2003) on the development of irregularities have illustrated the important role of time-dependent forcing (elsewhere referred to as "stirring") on the character of density structures. Further work using the same models explored mixed-mode instability development due to the combined presence of flow shears on the leading edge of a polar cap patch and unstable density gradients on the trailing edge (Gondarenko and Guzdar, 2006)—in particular, these studies have suggested a mode through which the density irregularities can be isotropized as compared to GDI acting alone (which overwhelmingly favors anisotropic development of plasma structures).

Most recent modeling efforts have examined the behavior of GDI and KHI under a range of data-inspired conditions and have self-consistently included thermodynamic and chemical effects on the development of these high-latitude instabilities. Simulations by Zettergren et al. (2015) have investigated plasma depletion processes, resulting from strong frictional heating of the ionosphere and chemical alterations (e.g., Zettergren and Semeter, 2012, and references therein), as a potential source for unstable density gradients. They show that the cavity formation process critically affects the character of subsequent instability, particularly during nonlinear development. Deshpande and Zettergren (2019) coupled the Zettergren et al. (2015) model of plasma instabilities (subsequently named GEMINI: Geospace Environment Model of Ion-Neutral Interactions) to the SIGMA (Satellite-beacon Ionospheric-scintillation Global Model of the upper Atmosphere) model of radio propagation (Deshpande et al., 2014) to demonstrate that GDI unstable polar cap patches cause scintillation of a character similar to observed GNSS data from Resolute Bay, Canada. They have further suggested the possible use of their full physics-based toolchain for studying specific plasma processes resulting in L-band scintillation. This GEMINI-SIGMA toolchain has also been used to study KHI generated by polar cusp flow shears by Spicher et al. (2020). They showed that shear-flow conditions observed from the EISCAT Svalbard Radar (ESR) and DMSP were capable of producing ionospheric KHI and that scintillation modeled from KHI irregularities was reminiscent of observations from the Svalbard GPS network.

3.2.3.2 GDI simulations: Comparisons against linear theory and basic nonlinear features

Before discussing nonlinear features of the model results, we first compare GDI simulations against the linear growth rates presented in Sections 3.2.2.2 and 3.2.2.3 for three different cases (a) classical GDI (no inertia), (b) GDI with ionospheric inertial effects, and (c) GDI with a magnetospheric capacitance added (conceptually this can be understood as including a magnetospheric slab in calculations of capacitance along with the ionospheric contribution). These comparisons are summarized in Fig. 3.13 and can be compared against growth rates from Section 3.2.2.2. In order to compare the simulations against the linear theory, we extract plasma density along a line in the vertical (y) direction at the center of the region containing the unstable gradient. Because this region moves

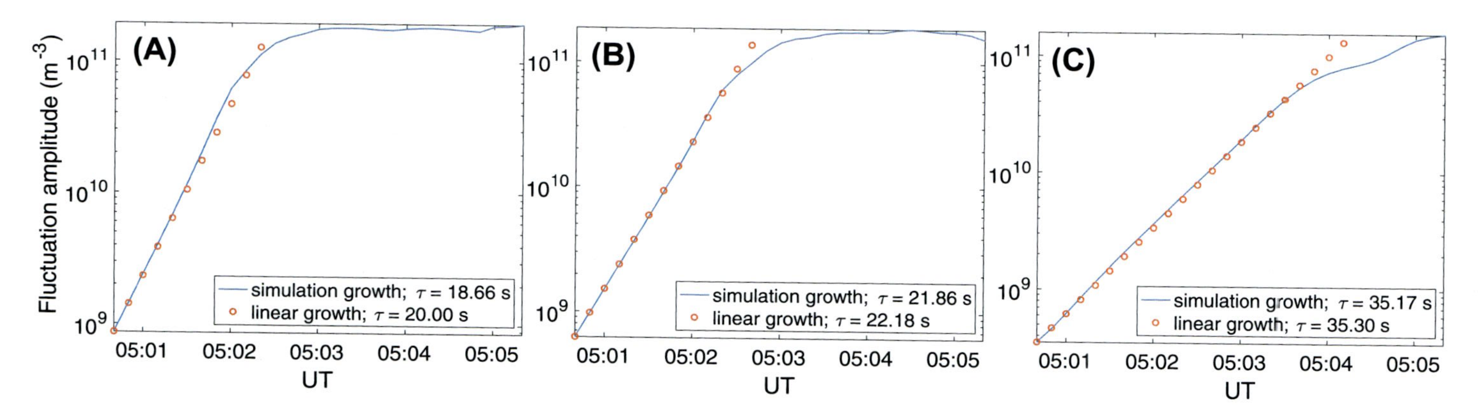

Fig. 3.13 A comparison of linear growth times $\tau \equiv 1/\gamma$ vs nonlinear simulation for gradient-drift instability. (A) Without ion inertia, (B) with ionospheric inertia, and (C) with an added magnetospheric capacitance.

with time, the location where the line is taken must be different at each time step in the simulation. Fluctuations due to the instability are separated from the background by subtracting off the mean density along the extracted line and then a standard deviation of the background-subtracted fluctuations is used to define the amplitude of the fluctuations along the line as a function of time. Perturbation amplitudes are plotted in Fig. 3.13 alongside a pure exponential function computed from the linear growth rate and density following initial settling in the simulation (settling appears as a brief, initial decay in noise due to the arbitrary perturbations imposed at the beginning).

All three simulations match the linear theory quite well during the initial stage of growth. Comparisons show clear deviations near saturation when the amplitude of the fluctuations is no longer growing strongly—following 5:02 UT in panel (A); following 5:02:30 UT in panel (B); and following 5:03:30 UT in panel (C).

Fig. 3.14 shows different example simulations including (A) the initial state for each example simulation, (B) classical GDI (no inertia), (C) GDI with ionospheric inertial effects, and (E) GDI with a magnetospheric capacitance added. Additionally, a simulation containing a more robust E region was included to illustrate the effects of a higher conductivity on the development of the instability in (D).

From Fig. 3.14 it is clear that the physical processes included in the simulation, viz inertia, MI-coupling, and E-region conductance, greatly impact the character of the resulting density structures. Comparison of panels (B) and (C) show that the inclusion of ionospheric inertia has the effect of slightly slowing the growth of the instability (also evident in Fig. 3.13) and that the plasma density structures have a slightly larger (y) extent and tend to be somewhat less narrow than those computed without ionospheric inertia (cf. Huba et al., 1988). Inclusion of stronger E-region (generated from an energy flux of 0.05 mW/m^2 electron precipitation) creates a shorting pathway for the charge and greatly suppresses the growth of structures as shown in panel (D). Finally, panel (E) demonstrates that a large magnetospheric capacitance (30 F in this case) can also greatly slow the growth of the instability.

3.2.3.3 KHI simulations: Comparisons against linear theory and basic nonlinear features

KHI growth is wavenumber dependent (see Section 3.2.2.3) so if the instability is seeded from noise one may, in principle, expect a range of wavelengths to visibly appear as the instability passes through the linear stage. In practice, however, we find that the fastest growing mode fairly clearly (at least visually) dominates the linear and early nonlinear evolution of the instability such that one does not need to be too careful about the seeding specific wavenumbers of interest.

For our cursory comparisons against linear theory, we initialize our simulations with white noise perturbations in density. Because these are not accompanied by potential perturbations defined through Eq. (3.31) some settling in the simulation will occur while the

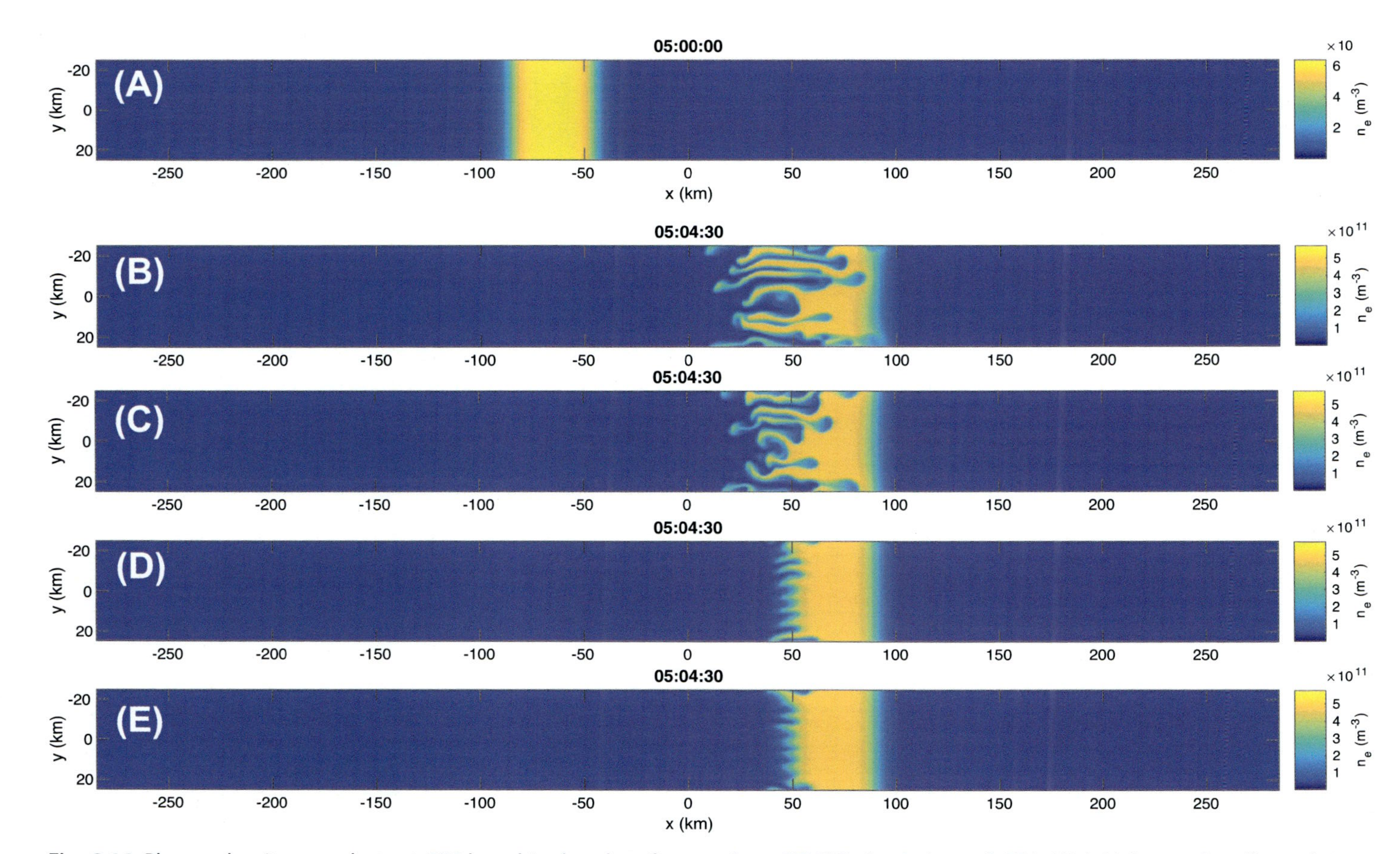

Fig. 3.14 Plasma density snapshots at 300 km altitude taken from various GEMINI simulations of GDI. (A) Initial state for all simulations, (B) simulation without ion inertia, (C) simulation with ionospheric inertia included, (D) simulation with modest E-region conductivity, and (E) simulation with a magnetospheric capacitance added.

simulation established a density/potential configuration consistent with this equation, which will then be destabilize. This process of seeding the simulation is obviously sub-optimal as comparisons with the linear theory must occur after the initial settling. The relatively large capacitance used in KHI simulations, i.e., the time constant for the simulation to settle into a consistent potential perturbation, is long enough for this to require more than a minute in our simulations. One proper solution to the issue of providing consistent seed to the nonlinear simulation would be to solve the eigenvalue problem encoded in Eqs. (3.31) and (3.32) for the eigenvalues and eigenfunctions and use these to define inputs for both density and potential perturbations. While more elegant, this approach requires a substantial set of additional software in order to be used, so we instead use the noise-seeded approach with the caveat that we can only sensibly compare to the linear theory after some settling has occurred.

When doing KHI simulations, in particular, one must be careful to properly observe background equilibrium conditions and grid size. One must be careful that the fastest growing mode that is a harmonic of the grid size in the periodic direction; otherwise, spatial aliasing of the instability growth could occur. Finally, one must take care, when comparing with linear theory, to initialize the simulation background state in a manner that corresponds to the balanced state consistent with that defined by Keskinen et al. (1988). To do otherwise will cause the simulation to again experience a settling process in which an equilibrium is first reached before instability growth occurs.

Example simulations included in this article use a background state corresponding to boundary layer width $\boldsymbol{\ell} = 3.15$ km; this results in the fastest growth mode of ~45 km according to the linear theory shown in Fig. 3.12 (Keskinen et al., 1988). The velocity shear for our simulations is set by the parameter $v_0 = -500$ m/s, corresponding to a 3 to 1 density jump across the shear boundary layer (Keskinen et al., 1988). The example simulation is conducted in the neutral wind frame of reference (required for the initial balanced state Keskinen et al. (1988)) per Eq. (3.27). Fig. 3.15A shows the linear growth rate compared with the nonlinear simulations for the small $\widetilde{\nu}$ limit demonstrating a generally reasonable match particularly during the early stages of the instability. Inherent complexities in seeding the simulations and nonlinear evolution make comparing the linear theory to the simulations challenging and we do not pursue those further.

Fig. 3.15B–E shows snapshots of the nonlinear stage density resulting from primary KHI at various times during our example simulation. Panel (B) shows initial appearance of the characteristic vortices forming near the beginning of the nonlinear stage of evolution—note consistency of the wavelength of these perturbations with the theoretical fastest growth mode (~45 km wavelength). In panel (C) both the effects of the forward and inverse cascade can be seen as what were initially two well-formed vortices have merged into a single large vortex with small-scale structures evident inside. Panels (D)–(E) show late-stage evolution of the instability which is characterized by turbulence

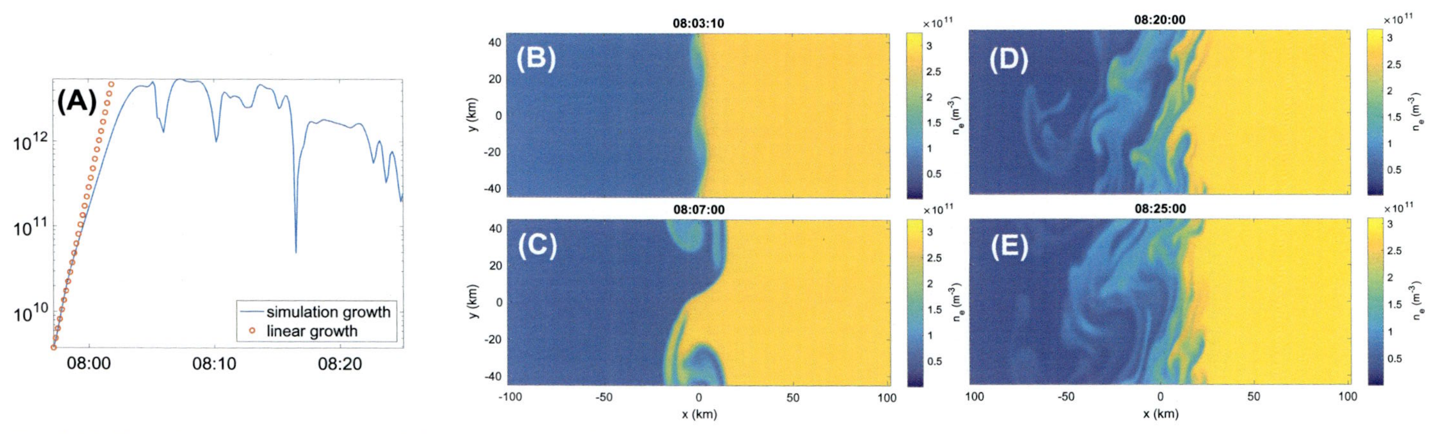

Fig. 3.15 Time-dependent evolution of KHI in our example simulations.

(wide range of spatial scales apparent) and showing intrusion of density irregularities into regions well outside the initially narrow boundary layer.

3.2.3.4 Secondary instabilities, time-dependent cascade

Anecdotal evidence of secondary instability exists in prior simulation studies that have simulated GDI and KHI. Mitchell et al. (1985) and Huba et al. (1988) discussed the effects of capacitance on the nonlinear development of GDI in terms of the effects of the size of the late-stage bubble structures—there is a tendency toward larger structures when polarization currents are included—similar to what we see in the example simulations in Fig. 3.14. Gondarenko and Guzdar (1999) showed that inertial effects also lead to pronounced tilting and further structuring of the elongated "fingers" that develop due to primary GDI. These studies have generally made it clear that the presence of electrodynamic effects (encapsulated in polarization currents) impacts the spectrum of irregularities present, particularly during the late-stage evolution of the instabilities. KHI modeling by Keskinen et al. (1988) shows that the cascade of density structures to small scales in the vortexes that form as secondary KHI waves are generated on the edge of the primary waves as they steepen and begin to "break." Since they initialize their simulations with a harmonic mode of the full grid, inverse cascade (i.e., formation of larger vortices) is not present, though it is readily apparent in our examples in Fig. 3.16.

Recent high-resolution modeling studies of GDI in the polar cap and KHI in the cusp show the evidence of secondary instability in the form of highly turbulent density fields suggestive of a combination of gradient and shear-driven effects. Fig. 3.16 (left), following Deshpande and Zettergren (2019), shows a simulation of late-stage GDI in which

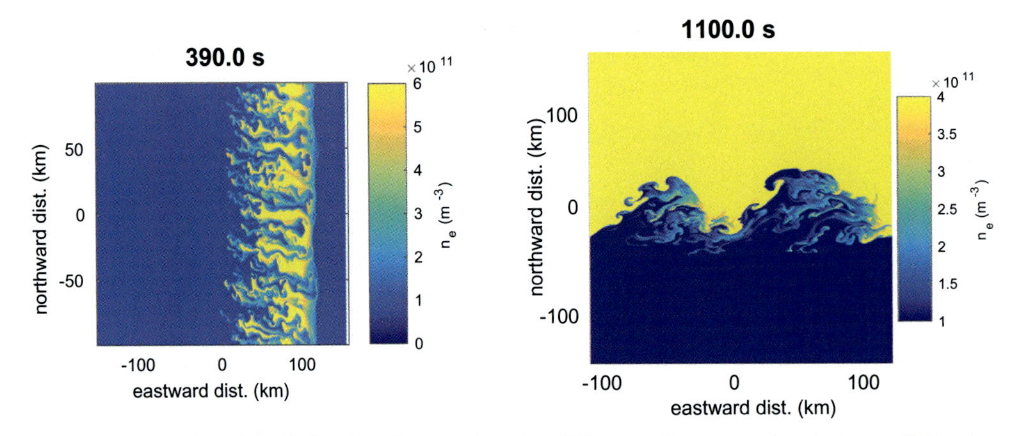

Fig. 3.16 Examples of well-developed secondary instability tending toward turbulence (following Deshpande and Zettergren (2019) and Spicher et al. (2020)). Left: Density structures at 300 km altitude during the nonlinear stage of GDI. Right: Density structures at 300 km altitude during the nonlinear stage of KHI.

wave-like structures can be seen to have a grown on the edges of the density "fingers." Similarly, Fig. 3.16 (right) shows KHI modeling from Spicher et al. (2020) shows both the formation of vortexes-within-vortexes and regions of bubble-like structures reminiscent of secondary GDI.

It can be seen from the simple example simulations in Figs. 3.14 and 3.15 that the primary GDI naturally results in situations favorable for secondary KHI (and vice versa). Fig. 3.17 shows both density and flow structures during the nonlinear stage of each of these instabilities. Fig. 3.17A and B shows density and x-component of plasma drift for the GDI simulation with ionospheric inertia. Fig. 3.17C and D shows density and y-component of drift for the example KHI simulation. Comparison of Fig. 3.17A and B shows clearly that sheared flows develop in the GDI "fingers," with slower flows in the high-density regions and faster flows where density is lower. Fig. 3.17C and D shows collocated drifts along density gradients formed from primary KHI.

Further complications involving excitation of secondary instabilities prevent a simple and fully accurate description of such dynamics. M-I coupling, encapsulated by the inertial capacitance parameters in our analysis, plays the dual role of destabilizing KHI while stabilizing GDI. As such secondary instabilities (KHI leading into GDI and vice versa) will be very sensitive to the ratio of conductance to capacitance $\widetilde{\nu}$. The particular, quantitative effects of this parameter on secondary instabilities have not been studied carefully as far as we know. The presence of plasma drifts at an angle other than normal or tangent to background density gradients has the potential to create a wider range of density irregularities since both primary GDI and KHI (if shears are also present) may be excited in a way that interacts strongly, e.g., with each instability providing robust seed structures for the other. Such "mechanics" have been suggested in prior studies as possibly contributing to fast structuring of polar cap patches leaving the cusp region (Carlson et al., 2007), but to our knowledge have not been fully studied with a robust modeling approach and an eye toward self-consistent comparisons with available data.

3.2.3.5 Mapping of electric fields: E-region shorting

Instabilities discussed herein are quasi-static in nature (viz involving charge accumulation) and are affected by the presence of any type of conducting pathway that allows neutralization of this charge. The E region can be strongly conducting by comparison to the F region where these instabilities initiate, even for a relatively modest plasma density and may be understood to generally suppress irregularity growth. A complicating factor to this is that for small-scale sizes perpendicular to the geomagnetic field the electric fields will no longer map large distances (Farley, 1959; Kintner and Seyler, 1985) and the unstable modes will tend to localize along the field line, to a degree. There are several different limiting descriptions, then, of the behavior of the instabilities (a) purely local—wherein only the F region is considered, (b) perfect mapping of fields between the F and E region, which can describe the suppression of instabilities, and (c) full 3D treatments, which will

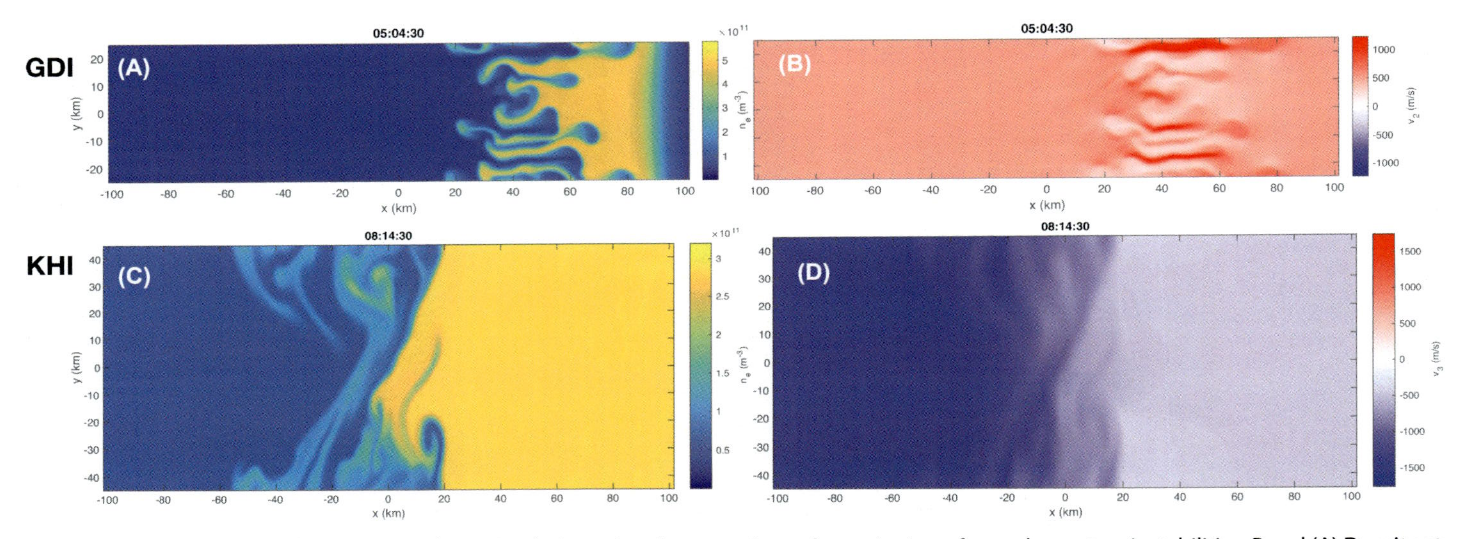

Fig. 3.17 Density and drift for the GDI and KHI simulations showing a tendency for excitation of complementary instabilities. Panel (A) Density at 300 km altitude from the GDI simulation with ionospheric inertia. Panel (B) *x*-component of the plasma drift velocity exhibiting small-scale shearing at the interface of the density "fingers." Panel (C) density structuring during the nonlinear stage of the KHI simulation. Panel (D) *y*-component of the plasma drift illustrating a substantial component along the density gradient in some locations.

describe both the shorting process and the localization of modes as they cascade to the smallest scales. We consider, in simulations presented in this work, scenario (b) which is valid in the limit of perfect mapping of the fields.

Following Farley (1959), Kintner and Seyler (1985), and Huba et al. (1988), we may roughly estimate the field line mapping distance of potential for a structure by considering the electrostatic form of the current continuity equation under assumptions of constant conductivity (for simplicity and sake of illustration).

$$\sigma_{\|}\frac{\partial^2\Phi}{\partial z^2}+\sigma_P\frac{\partial^2\Phi}{\partial x^2}=0 \tag{3.37}$$

Applying a trial solution of $\Phi(x,z)=\Re\{\tilde{\Phi}\, e^{ik_{\|}z+ik_{\perp}x}\}$ to this equation yields:

$$k_{\|}^2\sigma_{\|}\,\tilde{\Phi}+k_{\perp}^2\sigma_P\,\tilde{\Phi}=0;\quad k_{\|}=ik_{\perp}\sqrt{\frac{\sigma_P}{\sigma_{\|}}} \tag{3.38}$$

Thus, for a perpendicular oscillation having real wavenumber $k_{\perp}$ the parallel variation of potential is imaginary and positive such that it *decays exponentially* along the geomagnetic field line with a characteristic length scale of

$$\lambda_{\|}=\lambda_{\perp}\sqrt{\frac{\sigma_{\|}}{\sigma_P}} \tag{3.39}$$

In the F region, the conductance ratio is quite large, indicating that electric fields generated here due to charge accumulation (i.e., fields from charge accumulation) will map fairly effectively until the scales become quite small. For an F-region Pedersen conductivity of 5×10^{-6} S/m and a parallel conductivity of 25 S/m (conditions somewhat similar to many of our simulations), a $\lambda_{\perp}=1$ km structure will map about 2200-km along the field line (i.e., well into the magnetosphere).

Simple simulations presented in Fig. 3.14 have suggested, as discussed, that the presence of any type of E-region, however, modest, slows the growth. This particular example shows that even 0.05 mW/m^2 of precipitation particles will damp growth of GDI quite substantially such that it appears to grow much slower than the linear and local analysis suggests and could potentially play a role in stabilizing drifting patches.

While finite parallel conductivity will tend to suppress mapping of fields between the E and F regions, the coupling with the magnetosphere is largely dictated by ion inertial effects. The inclusion of these effects through a simplification of the generalized Ohm's law (e.g., Kintner and Seyler, 1985) removes the fundamental physics of wave propagation (Alfvén waves) from the system. Such assumptions effectively model the ionosphere as a distributed capacitance rather than a system that exchanges information with the magnetosphere via transmission and reflection of waves. For waves with a sufficiently long wavelength (low frequency), the adopted model is probably reasonable, but it is clear that applicability will not extend to higher frequency waves. The typical Alfvén speed in the topside ionosphere is in the ballpark of ~100 km/s (obviously this will be sensitive to geomagnetic conditions). For a distance of 1000 km, this corresponds to a 10 s time scale; in some cases, this may be comparable to linear growth rates estimated above. Thus, our "local" (viz nonwavelike) model of ion inertia can only be considered an approximation and there are additional dynamics concerning Alfvén waves that may, in principle, need to be resolved in future modeling efforts.

3.2.3.6 Nonlinear steepening

The nonlinear character of GDI naturally results in steepening of initial density irregularities as the instability proceeds. This can be understood as a consequence of the finite geometry (y-direction) of the irregular "finger-like" density structures as they form. The accumulation of charge on the lateral edges (y extents) of these structures results in an electrostatic field and response $\mathbf{E} \times \mathbf{B}$ drift that slows the high-density region such that it moves more slowly that the surrounding region. At the same time, the velocity field tends to be convergent on the trailing edge of the structure so that the density gradient steepens. This steepening dynamic has been discussed in the context of finite plasma clouds in studies by Perkins et al. (1973) and Zabusky et al. (1973). Similarly the divergent nature of the flows on the leading edge of a finite high-density structure lends a more diffuse character to its evolution (this can be seen in simulations where the bubbles "break through" the front edge of the patch).

Steepening and cascade to small scales have important consequences for how the instability can be described during its various phases. During the initial stages of the instability, when the irregularities have not yet steepened fully, the perturbation potential structures can be expected to map efficiently along geomagnetic field lines. As steepening and cascade occurs, smaller structures naturally develop such that at some point it is likely that the EFL approximation will be violated. Coupled with this, at the smallest scales we consider (10s of meters), cross-field diffusion, and associated charge accumulation will generally act to stabilize the progression of GDI (Kelley, 2009). Consequently, to fully describe evolution at these scales one needs to account for finite perpendicular pressure and fully three-dimensional potential variations; this is extremely challenging from a numerical point of view as it places rather extreme requirements on the number of grid points needed for such simulations. To date models have not been able to resolve these scales and physical processes fully.

3.2.3.7 Stirring of ionospheric plasma via external forcing

Another mechanism that has gained significant attention to explain ionospheric irregularities is referred to as *stirring* (or sometimes "turbulent mixing," though we avoid this term) (Kelley, 2009). The distinction between stirring and the ionospheric instabilities (and resulting turbulence) discussed to this point is that processes labeled as stirring are *externally imposed* on the ionospheric plasma. A spatially inhomogeneous background electric field, which could be generated by magnetospheric structuring will, e.g., result in the strong growth of GDI in specific directions (e.g., Burston et al., 2016; Kelley, 2009) leading to a potentially complicated, inhomogeneous field of irregularities. Additionally, turbulence in the neutral atmosphere can likewise stir the plasma and create irregularities through ion-neutral collisional momentum transfer (e.g., Kintner and Seyler, 1985). In both of these cases, the irregularities partially result from influences outside the ionospheric plasma rather than purely due to internal plasma dynamics in the ionosphere.

Various studies have considered the impacts of stirring on irregularity generation. For instance, using sounding rocket data, Earle and Kelley (1993) concluded that stirring would be as probable as linear instability theory to explain their observations. Additionally, Kivanc and Heelis (1997) studied the spectral characteristics of density and velocity structures associated with 18 polar cap patches and found the presence of structures on most of the patches identified, regardless of their location. The authors suggested that while the GDI could contribute to the development of the irregularities, their creation could be more easily explained by a stirring mechanism. This was further supported by Kivanç and Heelis (1998) who based on a statistical analysis, suggested that GDI was not the dominant mechanism for creating the density and velocity irregularities observed on scale-size ranging from 30 km to 300 m and that the majority of 1-km scale size density irregularities was due to inhomogeneous velocity structures of magnetospheric origin; these could stir density gradients or initiate KHI (Kivanç and Heelis, 1998). Finally, as already discussed in Section 3.2.2.4, Burston et al. (2016) compared linear growth rates of different mechanisms including GDI, KHI, CCI, and found inertial turbulent mixing to be dominant, followed by inertial GDI.

3.2.4 Simulating observable effects of ionospheric instability: Remote sensing tutorial

Although plasma irregularity cascading and turbulence are known to be important factors in ionospheric physics, the fundamental physics behind these processes is still relatively poorly understood. In situ measurements are expensive and are limited to confined spatial regions along the spacecraft trajectory. Thus, remote sensing of the ionosphere using a veritable "ocean of information" from the navigation and communication satellites presents a unique tool to investigate the ionospheric instabilities and helps us understand their effects. However, scintillation observations alone are insufficient to understand responsible plasma physics. Thus, in order to solve this problem, we need other supplementary approaches such as modeling, including full physics-based simulation tool that can capture the observable effects of ionospheric instabilities and can assist in scintillation prediction. In this section, we will review the theory of propagation of signals in random (viz turbulent) media followed by a discussion about the current propagation modeling efforts.

Basu et al. (1981, 1988) used in situ satellite data from Atmospheric Explorer D (AE-D) to model scintillations at high latitudes; note, however, that these measurements are not sufficient to represent geomagnetic and altitudinal variations of density fluctuations. Fremouw and Rino (1973) presented the first empirical model that could estimate the S4 scintillation index (normalized standard deviation of amplitude fluctuations) for VHF/UHF under weak scattering conditions, which often underestimate scintillations at auroral latitudes. Two climatological models, Global Ionospheric Scintillation Model (GISM) and WBMOD (for WideBand MODel) ionospheric scintillation model have had success in predicting the overall trend of scintillations. However, they lack the capability of resolving the scintillation effects of irregularities in localized regions. WBMOD also fails to predict the day-to-day and short-term (during a magnetic storm) variability of scintillations (Wernik et al., 2003) (see also Priyadarshi (2015) for a review of these scintillation models). A more recently developed global full 3D forward propagation model Satellite-beacon Ionospheric-scintillation Global Model of the upper Atmosphere (SIGMA) attempts to address many of these issues (Chartier et al., 2016; Deshpande and Zettergren, 2019; Deshpande et al., 2014; Deshpande et al., 2016). SIGMA takes electron density distributions from an empirical, spectral, or physics-based plasma model. SIGMA is based on multiple-phase screen (MPS) method and that solves forward propagation equation according to the theoretical presentation by Rino (2010). The use of this model is discussed further in detail in Sections 3.2.4.2 and 3.2.4.4. In this section, we will review the theory of propagation through a turbulent ionosphere in Section 3.2.4.1, followed by a discussion of some of the state-of-the-art propagation models in Section 3.2.4.2. We will then compare the spectral and physics-based models of ionospheric irregularities in Section 3.2.4.3. Some specific signatures of scintillations observed on the ground when the signal passes through GDI or KHI are discussed in Section 3.2.4.4. Signal received on the ground can be used to derive physical properties of the ionospheric irregularities the signal passes on its way using an inverse method. This is discussed in Section 3.2.4.5. Different radio frequencies are sensitive to different scales of ionospheric structures as they propagate through those as discussed in Section 3.2.4.6. Finally, to complete the discussion of effect of the irregularities on radio waves, we briefly touch the topic of high-frequency (HF) reflections in Section 3.2.4.7 as a means to probe some of the ionospheric irregularities.

3.2.4.1 Theory of propagation through a turbulent ionosphere

Radio signals propagating through the turbulent ionosphere is a problem of propagation of electromagnetic waves through *random media*; the signal undergoes reflection, refraction, diffraction, and interference. Standard scintillation theory simplifies this complicated problem by assuming that there is no backscatter (only forward scatter) and it is mainly weak scattering (Rino, 2010) which mostly holds except for some strong auroral scintillation instances. In this section, we present a brief summary of the theory of propagation of transionospheric signals. The reader is referred to the classic reviews by Yeh and Liu (1982) and Bhattacharyya et al. (1992) for more detailed descriptions of many scintillation-related concepts and propagation of transionospheric signals.

Propagation of radio waves through the constantly changing ionosphere can be solved by using a multiple phase screen approach. Simply speaking, a wavefront from radio source (satellite or any beacon) gets distorted while passing through the ionosphere and emerges having a random phase change, because of the turbulent variations of the refractive index of the media (Briggs, 1975). There are a few basic assumptions that need to be stated before describing the solution to the propagation problem. Scintillations observed by a receiver are assumed to be related to ionospheric irregularities that are "frozen-in" with respect to the magnetic field, which is due to electron density structures that are convecting at a uniform velocity along the field lines. Thus, the receiver would record the motion of the diffraction pattern on the ground due to either the motion of the irregularities or the satellite motion or a combination of both. The "non-frozen-in" effects comprise of random flow velocities, however, they are believed to make up only a small contribution toward the high-frequency part of the scintillation spectrum. It should be noted here that a full physics-based plasma model can simulate these effects well. Another assumption is that the scale size of the irregularities that cause scintillations (Fresnel scale and above) is much greater than the wavelength of the radio signal and is smaller than the distance traveled by the signal from the irregularity to the ground. Thus, the radio wave gets scattered only in the forward direction. Furthermore, the last assumption for weak scattering is that the diffraction effects by Fresnel scale-size structures are dominant. This allows us to solve the propagation problem by multiple phase screen approach or Rytov approximation.

The electron density in an ionosphere fluctuates causing the refractive index to fluctuate as well. The deviation of density from its background value ΔN_e is a function of space (x, y, z) and time. The variations Δn_i in the refractive index are approximately linearly proportional to ΔN_e at VHF/UHF under the high-frequency or short wavelength assumption. As a radio wave propagates through the density fluctuations, only the phase is affected by the random fluctuations in the refractive index (to the first order). Fluctuations in optical path are obtained by integrating Δn_i along the vertical direction and are related to the phase deviation by the free-space wavenumber k.

It must be noted that this section uses engineering notation ($j \equiv \sqrt{-1}$ and an $e^{j\omega t}$ time-harmonic convention), in contrast to Section 3.2.2 which uses ($i \equiv \sqrt{-1}$ and an $e^{-i\omega t}$ time-harmonic convention).

The phase fluctuations on a radio signal of wavelength λ as it passes through the irregularity are given by

$$\varphi(\vec{\rho}, t) = \frac{2\pi}{\lambda} \int \Delta n_i(\vec{\rho}, z, t)\, dz, \tag{3.40}$$

where $\vec{\rho} = (x, y)$ is the transverse coordinate and z is the vertical distance. The phase deviation can be written in terms of the TEC ΔN_T along the signal's path as

$$\varphi(\vec{\rho}, t) = -\lambda r_e \Delta N_T(\vec{\rho}, t), \tag{3.41}$$

where

$$\Delta N_T(\vec{\rho}, t) = \int \Delta N_e(\vec{\rho}, z, t)\, dz, \tag{3.42}$$

and $r_e (= e^2/4\pi m \epsilon_0 c^2)$ is the classic electron radius. A plane wave coming from a source that is a lot farther from the ionosphere than the receiver on the ground is from the ionosphere, is represented by its electric field with constant amplitude A_0. Upon emerging from the ionosphere, it takes the following form:

$$\vec{u_0}\,(\vec{\rho}, t) = A_0 e^{-j\varphi(\vec{\rho}, t)} \tag{3.43}$$

For the phase screen approach where the random media with density fluctuations ΔN_e is represented as an infinitesimally thin screen that imposes only phase variations (from Eq. 3.41) on the incoming wave. The diffraction pattern of the plane wave beyond the phase screen can be obtained using Kirchhoff's diffraction formula, under the forward scattering assumption. This is described in detail by Ratcliffe (1956).

$$\vec{u}\,(\vec{\rho}, z, t) = \frac{jkA_0}{2\pi z} \iint e^{-j[\varphi(\vec{\rho}', t) + (k/2z)|\vec{\rho} - \vec{\rho}'|^2]}\, d^2\rho', \tag{3.44}$$

where $k = 2\pi/\lambda$ is the wavenumber. $\vec{u}\,(\vec{\rho}, z, t)$ is the field at a distance z from the screen and is obtained by integrating the wave field $\vec{u}\,(\mathbf{r}, t)$ (where $\vec{r} = (x, y, z)$ is a 3D spatial coordinate) and its spatial derivative with the simplification from the forward scattering assumption (see Ratcliffe (1956) for detailed derivation). This signal is then propagated through the free space between the phase screens and then from the bottom of the phase screen to the ground, where both the phase and amplitude of the signal vary. The MPS technique (Knepp, 1983) is combined with a split-step solution (a method that treats a problem in smaller steps to separate out linear and nonlinear parts in Fourier and spatial domains) to the forward propagation equation (FPE) that represents all the above steps to solve the scintillation problem in the most general way (Rino, 2010). In this method, the signal experiences phase fluctuations as it passes through the phase screens and in between the phase screens and from the bottom of the irregularity to the ground, it experiences free-space propagation. FPE encompasses refraction, diffraction, scattering, and interference effects on the radio wave propagating through the random ionosphere. The detailed mathematics for FPE is beyond the scope of this review. Thus, the reader is referred to the scintillation theory reviews by Yeh and Liu (1982), Bhattacharyya et al. (1992), and Rino (2010). Yeh and Liu (1982) also gives details about another theory for weak scintillations called Rytov Solution, which considers the effects of scattering on the amplitude of the wave inside the irregularity. This is valid for the case when multiple scattering effects can be neglected and amplitude scintillations are caused by the irregularities with scales size of the order of the first Fresnel zone. The following section presents some of the current propagation modeling approaches.

3.2.4.2 State-of-the-art approaches

Global climatological models such as WBMOD or GISM provide a global distribution of scintillations and allow us to understand the overall behavior of the responsible irregularities. Recent analytical models, such as a three-dimensional model of plasma plumes that are caused by interchange instabilities developed by Retterer (2010), can be used to derive the strength of scintillations in terms of S4. There are very few recent propagation and scintillation models that have been developed in recent years to study irregularities that cause radio scintillations, especially in high-latitude regions. Deshpande et al. (2014) developed a full 3D forward propagation model SIGMA following Rino (2010) and Rino and Carrano (2011). The model SIGMA solves the FPE to encapsulate the scattering, interference, refraction, and diffraction effects on a radio signal caused by ionospheric irregularities. It can work anywhere on the globe and can accept the electron densities from any form of irregularity model, such as a spectral model (Chartier et al., 2016; Deshpande et al., 2016), an empirical model or a physics-based plasma model, e.g., GEMINI (Deshpande and Zettergren, 2019). Other models based on phase screen theory include a model by Rino et al. (2018a) where a compact parameter set was utilized to simplify the model at equatorial regions and was used for GPS performance analysis at three different frequencies (L1, L2, and L5). A new stochastic structure model called configuration space model was developed to generate realizations of number density that simulates extended highly anisotropic media (Rino et al., 2018b, 2019). This model can be used in propagation simulations applicable at equatorial as well as high latitudes.

3.2.4.3 Physics-based vs spectral approaches to irregularities

Ionospheric irregularities can be modeled using a physics-based model or realization of spatial electron density distributions derived from an irregularity spectrum. The latter is based on an assumption that irregularities are random media with certain statistical properties, e.g., turbulence typically follows a power law spectrum. The irregularity spectrum can then be presented in terms of a single-slope or double-slope power law spectrum. Some of the notable spectral models of irregularities encompassing the constraints on the ionosphere, such as high latitude, weak scattering etc., are those presented by Shkarofsky (1968), Costa and Kelley (1977), Rino (1979a, b), and Wernik et al. (1990). Analytical models by Fremouw and Rino (1973) can estimate the root-mean-square fluctuations in the scintillation index of the received radio signal under averaged scintillation activity. There are other spectral models that have been derived from satellite data (Basu et al., 1988). Some of the early in situ satellite measurements have pointed to a one-dimensional spectrum following a power law. The 3D spectra of electron density of ionospheric irregularities can be derived easily from the 1D spectra if the irregularities are assumed to be isotropic. The spectra can then be used to obtain a realization of electron number density distribution in the irregularities. However, the spectral models do not encompass the different stages of instability development. Thus, spectral models are only valid for fully developed turbulence which is the reason why we need to consider physics-based models to represent the *evolution* of ionospheric irregularities.

There are very few full physics-based 3D models that can generate plasma structures locally. For example, SAMI3 (Huba et al., 2000, 2008) works at low-to-midlatitudes, calculates the plasma and chemical evolution, and drifts of several ion species given neutral composition, wind, and temperature as input. However, SAMI3 has limited spatial resolution which to date has not been used to resolve scintillation scales. Another similar, high-resolution model for low-latitude phenomena like equatorial plasma bubbles has been presented by Yokoyama (2017).

The ionospheric model GEMINI (Zettergren et al., 2015) includes both aeronomical and electrodynamic processes relevant to the formation of ionospheric fluid instabilities (viz. GDI and KHI) at high latitudes (Zettergren et al., 2015). It has been recently interfaced with the propagation model SIGMA and used with resolutions high enough to resolve structures close to Fresnel scale size (Deshpande and Zettergren, 2019). The models together were also used to study how the difference in the development of KHI (e.g., through changing inertial capacitance) can affect the scintillation onset times as well as the strength of GPS scintillations in cusp regions (Spicher et al., 2020).

3.2.4.4 Scintillation signatures of ionospheric instability

Deshpande and Zettergren (2019) demonstrated that a combination of a physics-based plasma model and a propagation model can be used to study radio signal fluctuations on the ground as the signals pass through an instability, such as GDI, and that with reasonable initial conditions, the instability generated from GEMINI can result in GPS signatures on the ground comparable to observations. Fig. 3.18 shows an example of scintillations produced by GDI. Panel A of Fig. 3.18 shows an illustration of SIGMA where a signal from a satellite scatters between the layers (or phase screens) and propagates to the ground experiencing phase and amplitude fluctuations. Fig. 3.18B shows three different snapshots of 2D electron density at the height of maximum F region; Fig. 3.18C shows the snapshots of the corresponding propagated 2D phase on the ground, and Fig. 3.18D shows the phase time series which are basically obtained with the value at receiver shown by the red cross in each snapshot of panel (B). Phase scintillations seem to appear when the signal passes through the trailing edges of GDI or the patch here, especially when the large-scale structures in the GDI break into small-scale structures. During the earlier time of GDI development when the patch is starting

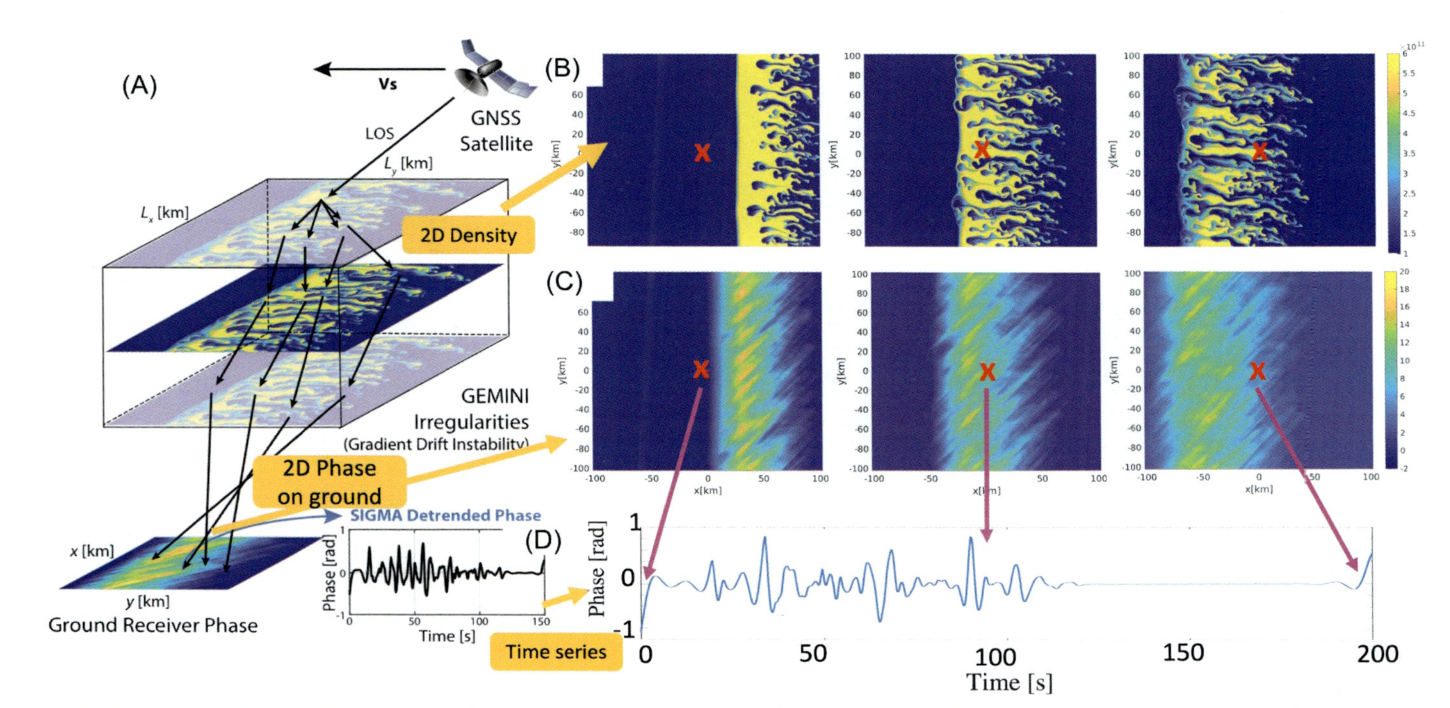

Fig. 3.18 (A) Illustration of SIGMA showing the density cube where scattering of radio signal occurs and the fluctuations are observed on the ground in the form of time series as well as simulated propagated 2D phase pattern on the ground. At different stages of development of the density structures represented in (B) as 2D snapshots of density, (C) the 2D phase pattern on the ground and thus (D) the fluctuations in the time series on the ground, all change.

to break into structures that are still too large compared with scintillation scales, the radio signal passing through those structures does not undergo any scintillation (not shown in this figure). Predominant spectral power appears to be concentrated in the spatial scales of 1–4 km for the case shown in Fig. 3.18.

Very recently, using observations and modeling, Spicher et al. (2020) presented a nonlinear, quantitative analysis of the KHI as the possible source of irregularities causing GPS phase scintillations detected in a cusp flow channel, and how it causes GPS phase variations. Using GEMINI-SIGMA they observed that, compared to GDI, KHI displays a specific pattern in scintillation occurrence, that is, with respect to the density boundaries, vortices, and small-scale structures developing around those vortices. Under reasonable assumptions consistent with observations, their modeling results showed that KHI could be responsible for the creation of density irregularities within minutes, assuming a shear scale length of $\ell \sim 1$ km. They studied how simulated KHI or shear (of varying inertial capacitances of ~10 F and ~30 F and maximum Pederson conductance ~1.2 S) affects scintillations at different stages of KHI development. For the numerical simulations performed, their findings include, for example, with larger capacitance KHI generates vortices and structures of scintillation scale earlier than that with the lower capacitance.

3.2.4.5 Inverse analysis

The technique of data inversion allows one to understand the physics behind remotely sensed observations. For GNSS scintillation data, there have been studies that used an inverse technique to infer ionospheric irregularity parameters such as turbulence strength, the height of irregularity, density in the background, etc. at low and high latitudes (Carrano and Rino, 2019; Carrano et al., 2012; Deshpande et al., 2016; Keskinen, 2006). All of these studies are essentially trying to fit N data points to a scintillation or propagation model that has M adjustable parameters and which predicts a functional relationship between the measured independent and dependent variables. The typically used method for the fitting is the least-squares fit (the maximum likelihood estimator (Press et al., 1992) for assumed Gaussian statistics). An inverse technique, however, calls for running the full propagation models over a multidimensional parametric space, which is possible only if the computation time for each run is not overly burdensome. Having said that, coupled models such as GEMINI-SIGMA may take a few days for a single run, preventing their use for an inverse analysis. For scintillation models with more than half-a-dozen parameters to fit, it is prudent to use data from other instruments such as radars, satellite data, ASI, etc. to provide a priori specification of some parameters. Another tool that can be extremely useful for inverse analysis for radio scintillations are groups of radio receivers which permit using a spaced-receiver analysis technique to estimate ionospheric drifts for irregularities "frozen-in" the magnetic field lines (Costa et al., 1988; Datta-Barua et al., 2015; Rino and Livingston, 1982). Furthermore, if signal fluctuations can be fit using a Rytov spectrum, i.e., assuming that weak scattering is occurring, it is possible to use an inverse analysis to estimate the height of the scattering layer, and its thickness (Yeh and Liu, 1982).

3.2.4.6 Scale-dependent scintillation

The majority of observational techniques presently used can either achieve good coverage across large areas of space and time (i.e., ground-based radars and imagers) or high resolution (i.e., instruments on satellites and sounding rockets), but generally not both. This makes it challenging to investigate the full range of dynamics and coupling at all scales using observations alone, but they can play an important role in constraining model parameters. Using ISRs, the approximate size, shape, and basic characteristics of patches can be determined and used to initialize realistic model runs. Furthermore, the output of scintillation models can be compared with real scintillation measurements to determine how closely the models captured the physical processes that occurred in the ionosphere. This is particularly important for tuning model input that cannot be measured directly.

Radio waves are most sensitive to scintillation by ionospheric structures at the Fresnel scale (Yeh and Liu, 1982). The Fresnel scale depends on both the distance between the irregularity layer and the observing receiver and the frequency of the incident radio wave. This means that radio waves of different frequencies will be sensitive to structuring at different scales. GPS and GNSS satellites often transmit multiple frequencies, but they are usually all fairly close together in the L-band (1–2 GHz). To probe different scales in the ionosphere, it is useful to also consider the scintillation of ultra high-frequency (UHF) and very high-frequency (VHF) signals, where available. Some low Earth orbit satellites carry Coherent Electromagnetic Radio Tomography (CERTO) beacon transmitters, which continuously transmit VHF (150.012 MHz), UHF (400.032 MHz), and L-band (1066.752 MHz) radio waves (Bernhardt and Siefring, 2006). These signals can be detected by ground receivers and analyzed for scintillation similar to those observed on the GPS signals. Scintillation observations at multiple frequencies allow one to consider if ionospheric irregularities are uniformly distributed over a range of spatial scales or if a particular size is favored under some conditions.

3.2.4.7 HF effects

Propagation of high-frequency (HF) signals, generally identified as 3–30 MHz, can also be influenced by plasma irregularities. HF signals experience significantly more refraction and diffraction in dense ionospheric plasma than the higher frequency signals discussed previously in this chapter. This extreme refraction allows oblique HF radio signals to be "reflected" off the bottom side of the ionosphere and permits over-the-horizon communication. A substantial amount of work has been done to create raytracing algorithms that determine the propagation path of a HF signal through an inhomogeneous ionosphere

(Frissell et al., 2018; Michael et al., 2020; Ravindran Varrier, 2010; Theurer, 2012). Additionally, Smith et al. (2020) applied the finite-difference time-domain method to solve for electromagnetic wave interactions with subkilometer scale plasma irregularities and determine the transmission through a perturbed ionosphere.

3.2.5 Synthesis, outstanding issues and future needs

Evaluating contributions of precipitation to scintillation Particle precipitation is believed to contribute to plasma structuring in the cusp region at F-region altitudes, where low-energy ($\leq$1000 eV) electrons deposit most of their energy (e.g., Dyson et al., 1974; Kelley et al., 1982; Millward et al., 1999). While Moen et al. (2002) hypothesized that the source of the decameter-scale irregularities associated with HF backscatter in the cusp might be due to the structures within particle precipitation, evidence suggests that precipitation acts as a direct source of density fluctuations with scale sizes about $\lambda \geq$ 7 km (Dyson et al., 1974; Kelley et al., 1982; Labelle et al., 1989), or as seed irregularities on which the GDI could operate causing small-scale irregularities (Moen et al., 2012; Oksavik et al., 2012).

Recent observations also support the view that structured precipitation and associated FACs are an essential component for the creation of severe scintillations around the dayside auroral region (Fæhn Follestad et al., 2020). Furthermore, on the nightside, Jin et al. (2014, 2016) and Clausen et al. (2016) showed that the strongest phase scintillations occur when polar cap patches have entered the nightside auroral oval, implying that auroral dynamics including precipitation and/or field-aligned currents and flow shears play important roles in enhancing plasma structuring at phase scintillation scales. Additionally, Kinrade et al. (2013) presented statistical evidence of a correlation between enhanced σ_φ (standard deviation of phase fluctuations) and auroral intensity, and Semeter et al. (2017) and Mrak et al. (2018), using precise multiinstrument observations, showed that strong phase fluctuations and signal loss of locks were collocated with trailing edges of nightside auroral arcs in the E region. Chartier et al. (2016) presented observations of an auroral E-region ionization enhancement occurred with associated phase scintillations, modeling of which showed that an enhanced E-region density on top of the mean EISCAT density was needed to explain the observed phase scintillations.

Evaluating the effects of auroral precipitation and/or FACs on irregularity creation at different altitudes, and comparing it with respect to instabilities such as KHI or GDI is thus essential to advance our understanding and for differentiating the importance and contributions from these different sources of density irregularities. There are currently no detailed modeling studies that have evaluated how precipitation may directly induce irregularities or serve to generate seed structures; however, contemporary instability models now contain physics-based descriptions of energetic electron transport (e.g.,

Deshpande and Zettergren, 2019; Spicher et al., 2020) so these studies appear to now be feasible.

Small-scale density irregularity evolution Despite the importance of density irregularities, a complete physical description of how structures cascade from ~10–100 s of km scales to scintillation-producing scales (~100 s of m) is still lacking. Effects of secondary physical processes and instabilities (which will affect the smallest scales), along with diffusive stabilization, E-region, and polarization shorting are poorly constrained. Furthermore, evaluation of instability behavior in realistic situations (nontrivial initial and boundary conditions) has not been conducted. Of particular importance to resolve, is the affect of diffusive processes perpendicular to the geomagnetic fields and the behavior of the ionospheric potential in three dimensions at high resolution (i.e., there is a need to remove the EFL assumption discussed herein).

Encapsulating magnetospheric coupling Simulation examples presented and referenced herein demonstrate a strong influence of MI coupling, encapsulated in the magnetospheric capacitance parameter in our models, on the development and details of irregularities during the linear and nonlinear stages of development. Nevertheless, these rely on assumptions about how the currents close either in or outside the model (e.g., via conductance or polarization currents in the ionosphere or through magnetospheric polarization currents, which are assumed to exist but not self-consistently modeled). It is important to note that the models discussed in this review are not proper magnetospheric models in the sense that they do not resolve the propagation of Alfvén waves. The characterization of the magnetosphere in these models is therefore overly simple and furthermore relies on ad hoc assumptions about conditions in the overlying magnetosphere, encoded in the model capacitance parameter C_m.

To our knowledge, there have not been concerted modeling efforts aimed at developing a quantitative description of the ionospheric stirring process (e.g., forcing via space and time-dependent magnetospheric activity), although some studies have examined basic effects on convection reversals (global scale) on the evolution of GDI (Gondarenko and Guzdar, 2004b). To properly account for stirring will require a good observational characterization of the spectrum of the electric field fluctuations in the ionosphere (both in wavenumber and time) to serve as constraints for event-based studies, which would seem like a very promising future avenue for future modeling/theoretical work. In a similar vein, we are unaware of any study that has definitively evaluated the effects of neutral atmospheric turbulence or structured winds on ionospheric irregularities (we also consider this to be a stirring process).

On amplitude scintillation and diffractive effects As mentioned in Section 3.2.1.2, amplitude scintillations are attributed to diffraction due to Fresnel scale or smaller ionospheric structures, while phase fluctuations are due to structures larger than Fresnel scale size. The Fresnel scale depends on the wavelength of radio signals and the height of the structures in the ionosphere, resulting in small-scale sizes for higher frequency. The current state-of-the-art models for simulating the plasma instabilities or the ionospheric structures in the irregularities are limited in resolution. For example, the plasma physics model GEMINI can model the structures in a 3D space of few hundreds of kilometers in each direction with the highest resolution of ~200 m. This could still limit resolving sub-Fresnel scale structures responsible for some of the diffractive scintillations associated with amplitude scintillations at GPS frequencies but it could be sufficient to model amplitude scintillations at lower UHF (400 MHz) and VHF (150 MHz) frequencies. It is computationally expensive to increase the resolution of the modeled plasma structures for both plasma modeling as well as propagation modeling. However, in order to study amplitude scintillations in detail, both low-frequency observations and improvement the spatial resolutions in these physics-based models are necessary.

Toward a more quantitative approach to interpreting scintillation data Future studies of ionospheric instabilities and scintillation would benefit greatly from more quantitative methods for analyzing scintillation data. Recently developed physics-based simulations and data processing tools would seem to be a promising avenue for predicting scintillation time series resulting from different types of instabilities as a function of progression through the linear and nonlinear stages and into turbulence. This opens up the possibility of being able to identify unique observational signatures of different instabilities as a function of stage of progress, angle of observation, background conditions, etc. Such a possibility represents a potentially powerful, untapped tool for studying instabilities in different geophysical situations, especially since scintillation observations are numerous and can leverage existing, dense GPS receiver networks.

3.2.5.1 Future needs

Observations and analysis needed to advance the field Many of the current uncertainties in how to correctly initialize and drive plasma irregularity and scintillation models will have to be resolved with improved observations of the ionospheric state. Currently, 3D mappings of the initial state are only achievable with incoherent scatter radars, which tend to have a minimum resolution in the F region on the order of 10s of kilometers. This is comparable to high-resolution global models (25–200 km), but inadequate for local models that can resolve features down to 200 m. There are other observational techniques capable of much higher spatial and temporal resolutions (e.g., in situ rocket and satellite measurements), but these tend to sample an extremely small area in the ionosphere. Single spacecraft observations of plasma turbulence/irregularities are also generally subject to space-time ambiguities, and require assumptions for analysis (e.g., Taylor's "frozen in" hypothesis) (Narita, 2012; Paschmann and Daly, 1998, and references therein). Fundamental properties of fluctuating fields such as relationships between frequency and wave numbers, wave vectors and wave propagation, energy transfer across scales, growth or damping, etc. may be extracted from multipoint measurements (e.g., de Wit et al., 1999; Narita et al., 2010), making (future) ionospheric multispacecraft and multipoint analysis techniques essential to advance our knowledge of plasma structuring in the ionosphere. Future radar experiments such as EISCAT_3D (McCrea et al., 2015) and advancements like interferometry full-profile ISR fitting may allow subbeam and subrange gate resolution and significantly improve the description of the initial state for irregularity models (Holt et al., 1992; Hysell et al., 2008, 2015). In addition, more work needs to be done in developing methods to merge disparate data sets at different scales (such as All-Sky Imager and ISR data with rocket and satellite data) so that information about both large-scale evolution and dynamics and small-scale initial structure can be made available. Scintillation modeling would also be improved with a more thorough parameterization of spectral models under different ionospheric conditions.

Observationally, it is also essential, moving forward, to obtain a better characterization of physical conditions accompanying inhomogeneous plasma flows, and to be able to quantify shear strengths, shear scale lengths, Pedersen conductance to inertial capacitance ratios, effect of the neutrals, etc. These quantities strongly impact simulation results and would help constrain models to better understand the relationship between different instability mechanisms and irregularities causing scintillations (e.g., Spicher et al., 2020).

Modeling While recently developed nonlinear ionospheric model tools have begun to capture some aspects of ionospheric instability at high latitudes, there remains much work that can, in principle, be done with these tools. Very little of the parameter space of these models has been explored in published works (mostly due to the computational cost of running simulations with large numbers of grid points) so we currently have only very rudimentary ideas of how precipitation, background conditions, seeding, and nonideal effects (i.e., those not encapsulated in the standard linear theory) may affect the onset and progression of instability.

Ultimately, a distributed and dynamic model of the magnetosphere-ionosphere system (including finite propagation times) would be desirable, but due to the fast Alfvén speed in the ionosphere, such an approach would come with a rather extreme computational cost. This is further compounded by the computational expense of running the existing simulations, which must include three spatial dimensions and extremely high resolution to capture Fresnel scales for signals of interest. Indeed existing simulations are already CPU bound and it has proven very troublesome to do something even as conceptually straightforward as relax/replace the EFL assumption with a full 3D potential solution. Thus, it may be quite a while before a full 3D potential solver or a full description of magnetospheric dynamics can be applied to study ionospheric instabilities. Notwithstanding such issues, there are still more incremental improvements that can be made to better understand ionospheric turbulence, e.g., examination of stirring via time-dependent boundary conditions, etc.

Coupling with local MHD codes with plasma models may also be a future pathway to including a more refined description of magnetospheric aspects of coupling. The ionospheric model itself could be represented via a lumped parameter conductance and capacitance, which could then be connected to a fully electromagnetic model of the overlying magnetosphere, and vice versa. In essence the two codes could, in principle, provide improved boundary conditions to each other in a way that should improve characterizations of magnetospheric effects on the development of ionospheric turbulence.

It is important to note that no simulation studies to date have resolved the full spectrum of turbulence that may be expected to result from the nonlinear coupling of GDI and KHI (i.e., these models are not direct numerical simulations, DNSs). Recent works (Deshpande and Zettergren, 2019; Spicher et al., 2020) have resolved 200 m scales in their simulations; yet higher resolution is needed to study the full instability progression. The likely stabilizing processes leading to a saturated state for the instabilities we study are cross-field diffusion which will tend to become significant at roughly decameter scales (Section 3.2.2.5). A full accounting for turbulence and scintillation that it creates will be needed to resolve these scales as these scales will produce diffractive effect contributing greatly to the scintillation. Additionally, these scales are likely too small to apply the EFL assumption so this will also probably need to be relaxed for DNS simulation of ionospheric instabilities.

We strongly emphasize the importance of improved modeling of precipitating electron population, which can affect conductivity, heating, and other aspects of plasma behavior. Presently GEMINI includes many of these processes but they are based on semiempirical methods which should be compared against physics-based approaches and used to examine the myriad ways in which precipitation can contribute to the formation of density irregularities.

Additionally, there is a need to explore how the spectral slopes of ionospheric density structures change for various stages of different instabilities, how they are related to the in situ measurements as discussed in Section 3.2.1.3, and if there is a correlation of these slopes to the slopes in scintillation data on the ground. This investigation can only be done fully using physics-based models instead of spectral models in the ionospheric irregularity study.

Finally, working with coupled models, such as GEMINI-SIGMA at higher resolution is challenging because of computational constraints, such as memory required for density blocks, time of computation, Fourier analysis on huge matrices, etc. Furthermore, full inverse analysis is impossible for expensive simulations such as those with GEMINI-SIGMA. For future studies, these models need to be updated to work more efficiently with larger matrices so that high-resolution plasma densities can be enabled. Current propagation models also work with just E or F region. There is a need to meticulously include both E- and F-region effects in propagation modeling, for example, by splitting the propagation into two parts of the ionosphere.

3.3 Ionospheric electron density large gradients at midlatitudes

Shun-Rong Zhang and Ercha Aa

MIT Haystack Observatory, Westford, MA, United States

3.3.1 Introduction

Ionospheric plasma is produced primarily by solar photoionization that varies smoothly on either global or regional scales for any temporal scale. At high latitudes, additional ionization is available due to energetic particle precipitation. Once the ionization is created, many physical and chemical processes can either smooth out the plasma spatial variation or build up large-scale structures. During geospace storms or substorms, some of these processes could be intensified and exhibit significant variability in space and time. These structures often exhibit substantial spatial gradients and thus can impose detrimental effects on modern navigation and communication systems, forming potential space weather hazards. Radio scintillation at L-band frequencies induced by significant density gradients, for example, can cause signal phase disruption and lead to telecommunication difficulties and satellite navigation availability reduction.

In this chapter, we will characterize several typical large-scale plasma density gradient structures, including SED, midlatitude ionospheric trough, and polar cap patch and tongue of ionization (TOI). Fig. 3.19 shows some of these simultaneous structures in TEC and their spatial context during an intense geomagnetic storm. We will discuss some of the fundamental physical and chemical processes. It should be noted that our understanding of these processes remains limited and will continue to evolve as observation, modeling, and analysis efforts make new progresses.

3.3.2 Storm-enhanced density

SED, first coined by Foster (1993), is a highly structured ionospheric disturbance that manifests as a large TEC/*Ne* enhancement channel elongated primarily sunward (azimuthally) and sometimes partially poleward in the local afternoon at mid- and subauroral latitudes. This enhancement is coincident spatially and temporally with storm-time ionospheric characteristics historically named as the “dusk effect” (Buonsanto, 1999;

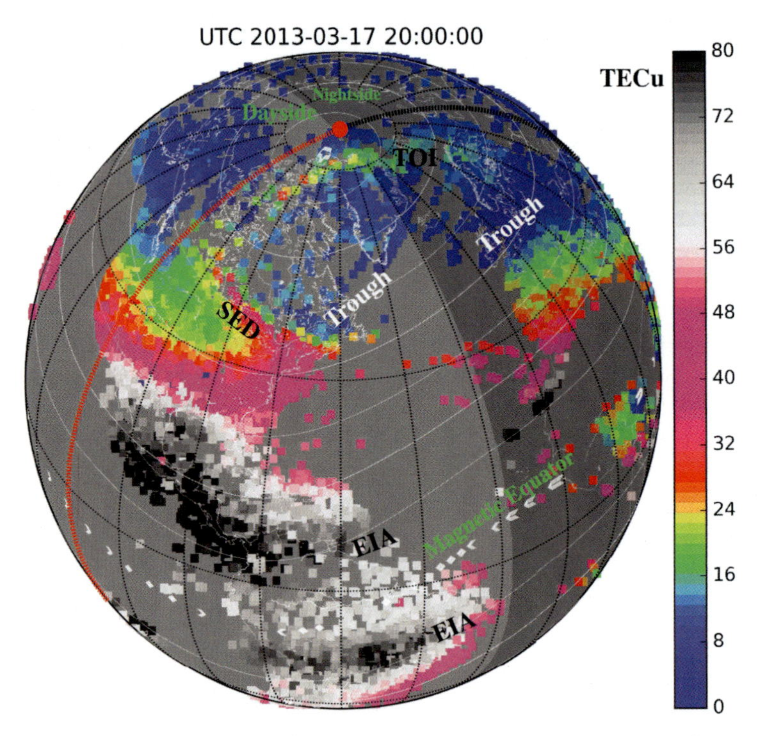

Fig. 3.19 A global view of ground-based GNSS TEC observation during the March 17, 2013 storm. Several plasma density structures are very significant, including storm-enhanced density, the midlatitude ionospheric trough, the polar tongue of ionization, as well as the equatorial ionization anomaly crests.

Mendillo, 2006). This regional ionospheric positive storm phase at dusk was the topic of intense debate in the 1970s (Prölss, 1995). High-latitude convection penetration and disturbance meridional neutral winds were considered as the main processes responsible for the dusk effect.

With the wide availability of GNSS data and coincident satellite in situ measurements in the plasmasphere and magnetosphere, our SED plume understanding evolved to become part of a larger set of geospace storm processes, associated particularly with the erosion of the outer plasmasphere by strong subauroral polarization stream (SAPS) electric fields (Foster and Vo, 2002). SED is associated with the important supply of cold, dense O^+ rich plasma of the ionospheric origin to the inner magnetosphere and eventually to the magnetopause (Walsh et al., 2014). The SED plasma also convects through the cusp into the polar cap, providing significant ion mass flux with high number densities to form the polar Tongue of Ionization (TOI). In general, previous efforts have revealed significant SED longitudinal and UT dependence (Coster et al., 2007; Yizengaw et al., 2006) and magnetic conjugacy (Foster and Rideout, 2007) among other characteristics.

3.3.2.1 Dusk effect

To start with the SED discussion, let us analyze what has been extensively explored under the name of "dusk effect" in the pre-SED era which characterized the SED-type plasma density enhancement as ionospheric positive phase storm. The positive phase ionospheric deviation is typical during the storm's early stage (e.g., the storm main phase). The dusk effect is an electron density enhancement that often occurs at subauroral latitudes, frequently reported in the American sector, in the afternoon-evening sector. The enhancement in both peak density NmF2 and TEC can be substantial, sometimes 50%, lasting for 1–2 h or longer. Fig. 3.20 shows an example of the storm-time F-region electron density observations over Millstone Hill (42.6°N, 288.5°E) and its vicinity. The dusk effect with elevated electron density occurred to the south, east, and west but not quite much to the north of Millstone Hill. To the north is the midlatitude ionospheric trough with low-electron density.

To explain this regional feature of density enhancements, many mechanisms have been suggested including both neutral and electromagnetic effects. Storm-time penetration or expansion of high-latitude convection electric fields with their magnetospheric

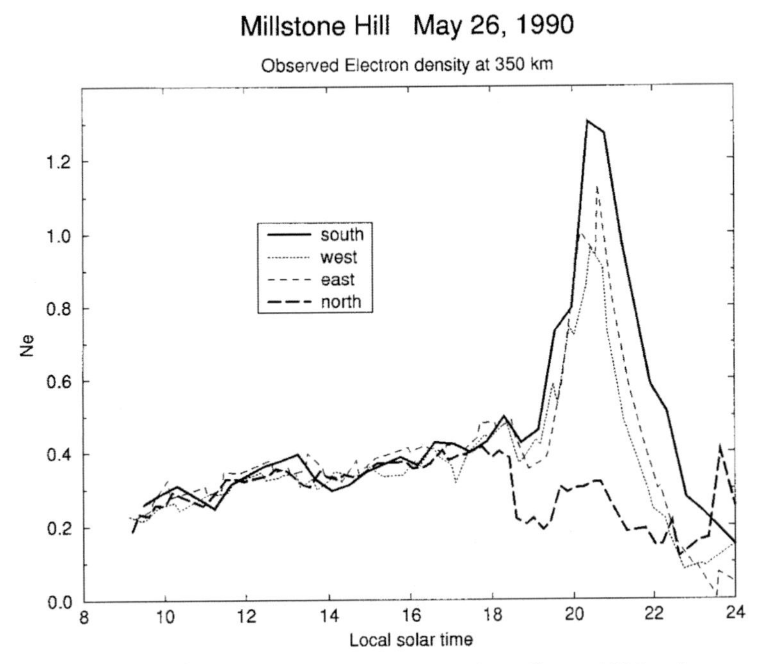

Fig. 3.20 Electron density at 350 km altitude observed by the Millstone Hill incoherent scatter radar steerable antenna pointing at 45 degrees elevation angle to the south, west, east, and north. The dusk effect appears as an exceptionally large north-south density gradient during evening hours. *After Buonsanto, M.J., 1995. A case study of the ionospheric storm dusk effect. J. Geophys. Res. Space Phys. 100 (A12), 23857–23869.*

origin (Kelley et al., 1979) can easily reach Millstone Hill at subauroral latitudes. This dayside electric field has both eastward and polar components, driving poleward and westward plasma drifts, which are perpendicular to the geomagnetic field **B**. The poleward drift perpendicular to **B**, $\mathbf{V}_p^N$, has an upward component which can push ions to high altitudes where they can survive for a longer time due to slower chemical loss rates than they would at low altitudes (Fig. 3.21), thus effectively the upward drift raises the column electron density after a certain time. Also, because of the parabola-type height profile of the F2 region ion density, when the ions are moved upward, the electron density in the topside increases and the bottomside decreases. Indeed, there are both observational and theoretical supports for the penetration or expansion electric field during the period of IMF Bz negative (Fejer et al., 1990; Huang et al., 2006; Jaggi and Wolf, 1973; Nishida, 1968; Spiro et al., 1988). As discussed further in Section 3.3.3.2, a specific pattern of the electric field spatial distribution is needed to account for specific regional patterns of the dusk effects.

An enhanced equatorward neutral wind can induce an ion upward motion along the magnetic field lines to higher altitudes where ions have slower chemical loss rate and longer lifetime, thus enhancing the column electron density. This is very much similar to the effect of the upward component of $\mathbf{V}_p^N$ (Fig. 3.21). Also similarly, enhanced equatorward

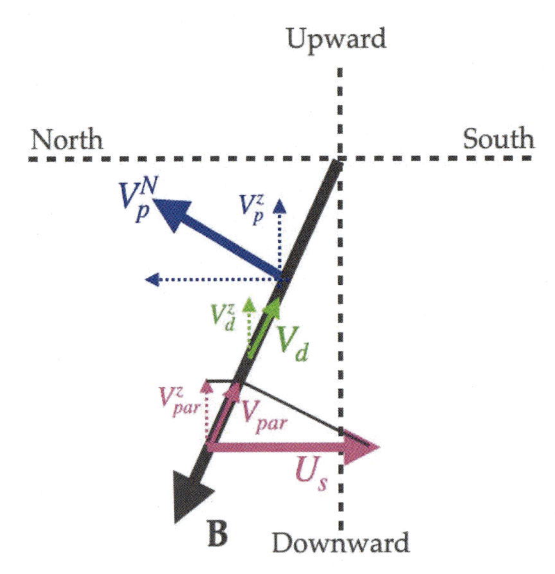

Fig. 3.21 The F region ion motions in the magnetic meridional plane in the northern hemisphere induced by a zonal electric field, a meridional neutral wind, and ambipolar diffusion. $\mathbf{V}_p^N$ is the northward ion drift perpendicular to magnetic field **B**, induced by the eastward electric field **E** through **E** $\times$**B**; $\mathbf{V}_p^N$ has a vertical component $\mathbf{V}_p^z$. $\mathbf{V}_d$ is the ambipolar ion velocity (parallel to **B**) which has a vertical component $\mathbf{V}_d^z$. $\mathbf{V}_{par}$ is the ion velocity induced by the meridional wind $\mathbf{U}_s$; $\mathbf{V}_{par}$ has a vertical component $\mathbf{V}_{par}^z$.

winds can directly enhance the ionospheric density in the topside by the upward transport. The needed equatorward neutral wind enhancement (sometimes a "surge") is a well-established storm-time neutral dynamical phenomenon (Fuller-Rowell et al., 1994; Lu et al., 2020). With the storm onset, Joule and particle precipitation heating within the aurora set up upward and meridional pressure gradients in the upper neutral atmosphere and cause equatorward wind enhancements at midlatitudes. These aurora-driving neutral wind disturbances may be up to a few hundred m/s. They can sometimes reach subauroral latitudes within 1 h, which is faster than the propagation of auroral induced gravity waves (Fuller-Rowell et al., 2002), depending on the exact distance from the auroral zone that is expanding equatorward.

3.3.2.2 SED phenomenology

SED characterization remains an active research frontier involving analysis of available observations primarily from GNSS TEC and incoherent scatter radars at mid- and high latitudes. The essential features can be summarized in the following prospects:

(1) Occurrence. SED is a spatially continuous large-scale feature spanning predominately mid- and subauroral latitudes in the afternoon-evening sectors. It occurs most frequently in the American sector, but evidence in the European sectors and in the Antarctic area exists as well. Fig. 3.22 shows typical SED plumes from TEC observations in North America. The TEC enhancement with a narrow channel appears to have a base in the Northeast US and extend westward in the central US, and northwestward near the northwestern US bordering Canada. It eventually reaches the noon sector near the cusp region, and convects further into the polar cap,

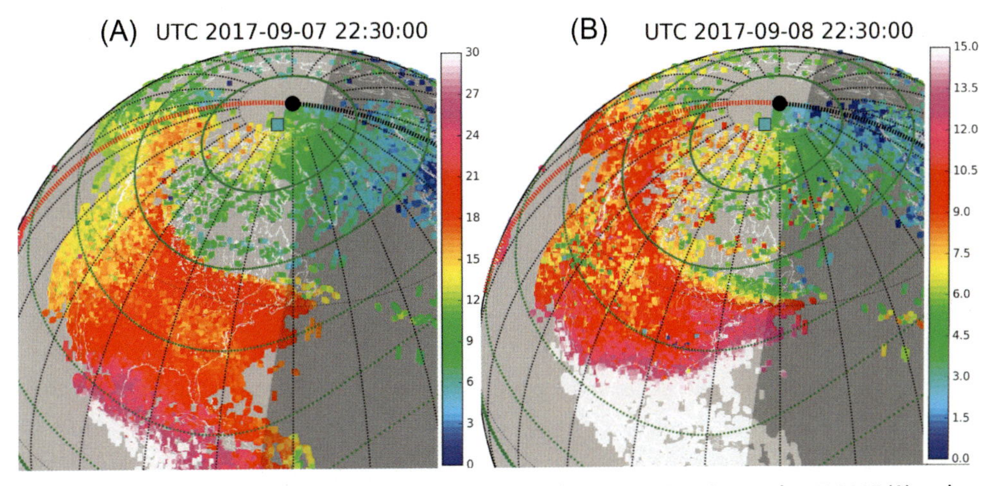

Fig. 3.22 GNSS TEC observations of SED plumes observed at 22:30 UT on September 7, 2017 (A) and on the following day September 8, 2017 (B). Notice the color scale change over the two panels.

contributing to the polar plasma density structures such as Tongue of Ionization (TOI) and patches.

(2) The intensity depends strongly on the level of geomagnetic disturbances. For superstorms (with Dst index less than ~ -250 nT), the SED plume in TEC can elevate above the background by 100 TECu (1TECu $= 10^{16}$ el/m^2). However, the intensity is also expected to be dependent on background conditions of the ionosphere which provides the source plasma. Fig. 3.22 displays SED plume examples over the same region at the same UT but on two consecutive days (thus influences on season, local time, and UT are minimized). Note that the color scales change by a factor of two. By 22:30 UT on September 2017 when the plume was observed, the IMF Bz drop which had started an hour before reached ~ -10 nT causing ~ -30 nT Dst index. The continuous Bz drop and solar wind speed elevation followed and triggered substantial ionospheric and thermospheric disturbances, including negative phase ionospheric variations with electron density depletion on the following day. 24 h later, the ionosphere was recovering from the impact of another CME causing strong magnetic activities 10 h ago, but Dst remained at ~ -100 nT. The SED plumes appear to be a very repeatable phenomenon, but the second SED was weak with low-electron density elevation when the background electron density in the ionospheric recovery phase was low too, and its northeastern base was more equatorward than it was on the previous day.

3.3.3 Subauroral dynamics and SEDs

Fundamental dynamical processes at subauroral latitudes include intense ionospheric electric fields of various origins and associated plasma drifts, ionosphere-plasmasphere diffusion, and thermospheric neutral wind and composition changes. All these processes result from strong M-I-T coupling.

3.3.3.1 SAPS and ion-neutral coupling

Ionospheric observations at Millstone Hill often indicate a close correlation between the SED occurrence and the fast westward (sunward) convection of ions in the dusk sector. A SED plume is found typically at the equatorward wall of the midlatitude ionospheric trough, whereas the ion fast convection takes place in the SED vicinity but more specifically on the poleward edge of it where the electron density is low. Fig. 3.23 shows this connection using ground-based GNSS TEC and in situ cross-track ion velocity observations from DMSP satellites (Zhang et al., 2017). The ion velocity in this figure is characterized by two sunward enhancements that are located near the trough's equatorward wall and poleward of the wall, respectively, as well as an antisunward enhancement located further poleward. These latter sunward and antisunward ion velocity channels away from the equatorward wall of the main trough are high-latitude convection which

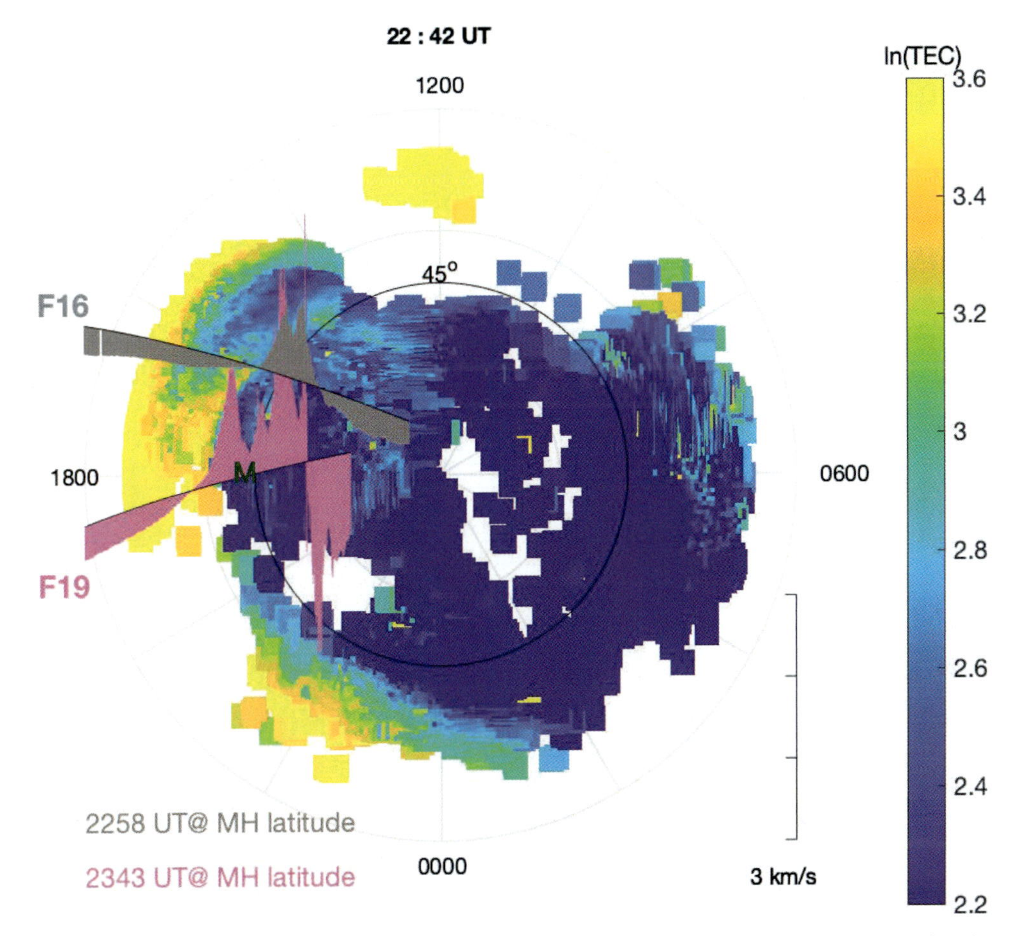

Fig. 3.23 Observations of a SED plume in TEC and SAPS from DMSP F16/F19 cross-track ion drifts. The SAPS channel of sunward ion flow was found near the equatorial wall of midlatitude ionospheric trough region. *After Zhang, S.-R., Erickson, P.J., Zhang, Y., Wang, W., Huang, C., Coster, A.J., Holt, J.M., Foster, J.F., Sulzer, M., Kerr, R., 2017. Observations of ionneutral coupling associated with strong electrodynamic disturbances during the 2015 St. Patrick's Day storm. J. Geophys. Res. Space Phys. 122 (1), 1314–1337. https://doi.org/10.1002/2016JA023307.*

can often expand into subauroral latitudes during storms. The large sunward ion drift near the equatorward wall of the trough is a storm-time feature termed as SAPS.

Storm-time electric fields in the subauroral region, equatorward of auroral electron precipitation, occur frequently and exhibit highly dynamical features as a result of the enhanced coupling of the interplanetary medium to the terrestrial magnetosphere-ionosphere system. SAPS is coined to encompass these subauroral electric fields, including those intense electric field structures in narrow channels shown as polarization jet (PJ)

(Galperin and Zosimova, 1974) and subauroral ion drift (SAID) (Spiro et al., 1979), as well as the broader subauroral regions of enhanced poleward electric fields (Yeh et al., 1991).

SAPS represents a particular signature of magnetosphere-ionosphere coupling of electric fields and currents, in tight feedback loops with the ionosphere in the dusk sector. It overlaps with Region-2 (R2) field-aligned currents (FACs) (Fig. 3.24) which are

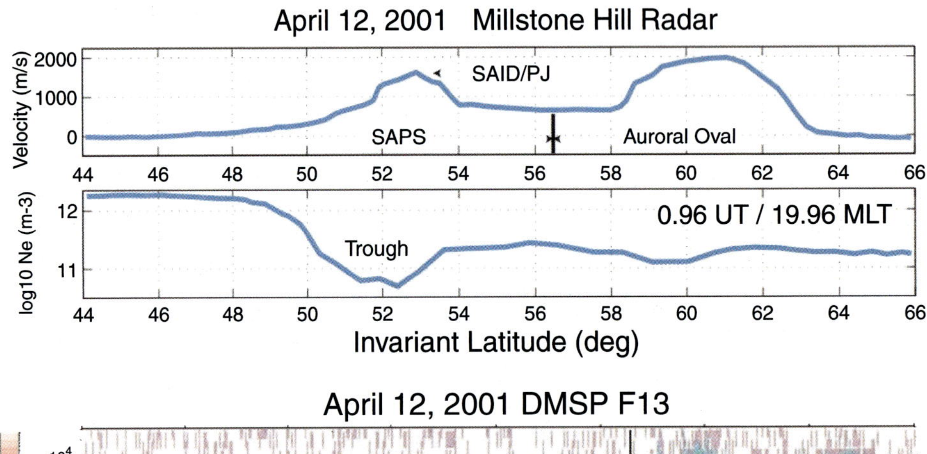

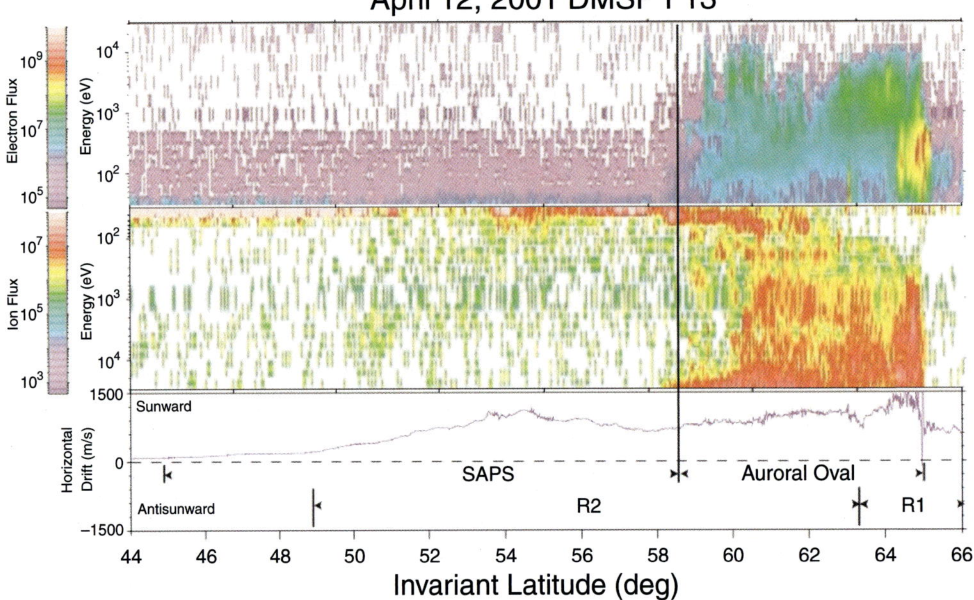

Fig. 3.24 Simultaneous radar and in situ measurements of auroral and SAPS plasma characteristics during a dusk overflight of the Millstone Hill incoherent scatter radar by the DMSP F13 satellite near 01:00 UT (20:00 MLT) on April 12, 2001. The SAPS is seen as a broad region of strong sunward plasma convection centered at 53°N apex latitude, equatorward of the auroral two-cell convection and coincident with a deep ionospheric trough. Region 1 (R1) and region 2 (R2) field-aligned currents have been determined using the DMSP magnetometer data. *After Foster, J.C., Vo, H.B., 2002. Average characteristics and activity dependence of the subauroral polarization stream. J. Geophys. Res. Space Phys. 107 (A12), 1475. https://doi.org/10.1029/2002JA009409.*

believed to be driven by plasma pressure gradients in the inner magnetosphere. A fraction of the ring current and FAC flows into the ionosphere in the main trough region (well equatorward from the electron precipitation auroral region, see Section 3.3.5) of very low conductance. Such a magnetospheric current generator causes a poleward polarization electric field, which is set up to maintain the electric current continuity. An alternative magnetospheric voltage generator, arising from the misalignment of the ion and electron inner boundaries of the plasma sheet, can possibly generate a radially outward polarization electric field in the inner magnetosphere which is poleward when mapping to the ionosphere along the equipotential field lines.

SAPS induced F region ion drifts have been analyzed by using incoherent scatter radars (at Millstone Hill and Poker Flat), SuperDARN systems at midlatitudes, and satellite in situ observations (Erickson et al., 2011; Foster and Vo, 2002; He et al., 2014; Kunduri et al., 2017; Wang et al., 2008). SAPS spans the afternoon through dusk to the early morning sector for all Kp greater than 4, with narrow channels of a few degrees of latitudes. The SAPS ion velocity varies in a large range from a few hundred m/s up to 2 km/s. The magnetic latitude peak location of the SAPS flow channels decreases with both Dst and MLT. The SAPS carries westward ion flow with a flux magnitude between 3×10^{13} and 3×10^{14} $m^{-2}s^{-1}$ in the F region in a manner nearly invariant to the geomagneic activity level (Erickson et al., 2011).

Regardless of how it is generated through either the voltage or the current generator, this polarization electric field is intense (~50 mV/m) because of low conductivity (Wang et al., 2008) in the subauroral ionosphere, and it drives large sunward ion drifts, with speeds much higher than the neutral wind speeds. Although the ion density is low, substantial frictional heating can develop and considerably raise ion and neutral temperatures, leading to excessive plasma loss due to enhanced rates of charge exchange and recombination reactions, which are temperature-dependent (Schunk et al., 1976). This further reduces conductance over the polarization electric region, provides positive feedback to SAPS electric field, and causes the intensification of the associated ion drifts.

Under the SAPS influence (or a similar influence by the expanded auroral convection as will be further discussed in the following Section 3.3.3.2), the ionospheric plasma convect horizontally sunward with a large flux. This sunward flow moves plasma from the nightside into the dayside. In contrast, the nightside ionosphere does not produce a significant amount of fresh plasma by solar irradiation nor transport sufficient plasma into the region to compensate for this SAPS/convection transportation loss, and therefore the SAPS/convection dynamics likely deeps the main trough. On the dayside, the ionospheric plasma may or may not be accumulated over time by the influx, depending on the sunward gradient

$$\partial(\text{flux})/\partial x = V_{\text{SAPS}}\partial N/\partial x + N\partial V_{\text{SAPS}}/\partial x \tag{3.45}$$

where x is in the sunward direction, N is plasma density, and V_{SAPS} is the SAPS velocity. It is obvious that in the interface region where SAPS is terminated, $|\partial V_{SAPS}/\partial x|$ is maximized, and SAPS most likely contributes to the SED development by providing a horizontal influx gradient. Theoretically, it is possible to infer the approximate SAPS/convection influence region on the SED formation by examining the SED sunward gradient.

SAPS also provides strong forcing on the neutrals via an ion drag effect. The intense sunward ion flows, sometimes up to 1–2 km/s, are much higher than the neutral wind speeds. At midlatitudes, the neutral wind speeds are normally 100–200 m/s or somewhat higher than these but well less than 500 m/s even during periods of neutral wind equatorward surge. Thus SAPS drives neutrals which then continue to spin like a flying wheel. Zonal winds will be strongly westward (Wang et al., 2011; Zhang et al., 2015b). These enhanced westward winds at midlatitude may cause a poleward wind disturbance in certain lag time (e.g., 30–60 min) due to Coriolis forcing (Zhang et al., 2015b). This Coriolis effect contributes to the observed poleward winds following a SAPS event (Fig. 3.25). The frictional heating by SAPS (Zhang et al., 2017) can modify thermospheric composition via local upwelling, in a way very similar to heating in the auroral zone (Wang et al., 2012). Thus local increases in O/N_2 can contribute to SED by enhancing plasma density, and O/N_2 depletion in nearby areas can deplete ionospheric density surrounded the SED and reinforce SED.

3.3.3.2 Influences of other dynamical processes

The subauroral ionospheric electrodynamics is easily affected by the expansion or penetration of high-latitude electric fields. Their zonal and meridional components produce the plasma flow with upward and sunward components at ionospheric heights in the afternoon sector and can make a substantial contribution to the SED plume formation and evolution (Deng and Ridley, 2006; Heelis et al., 2009; Huba et al., 2017; Liu et al., 2016; Lu et al., 2012; Zhang et al., 2017; Zou et al., 2014).

Observations indicate that an eastward electric field may be established accompanying either SAPS or the expansion of auroral convection electric field (Fig. 3.25). At mid- and subauroral latitudes where the magnetic field is inclined, the eastward electric field yields perpendicular ion drift pointing poleward (V_p^N in Fig. 3.21) with a large vertical component; at low and equatorial latitudes, this penetration electric field yields a largely upward ion drift; at high latitudes, this electric field yields poleward drift predominantly in the horizontal direction but with a small vertical component (Deng and Ridley, 2006). This eastward electric field induces a vertical ion drift that pushes plasma upward from low altitudes where charge exchange and recombination rates are fast to high altitudes where these rates are reduced substantially due to their dramatic variation in height. Thus, the eastward electric field effectively enhances the column density as well as electron density in the topside, and reduces the density in the bottomside where refresh solar-

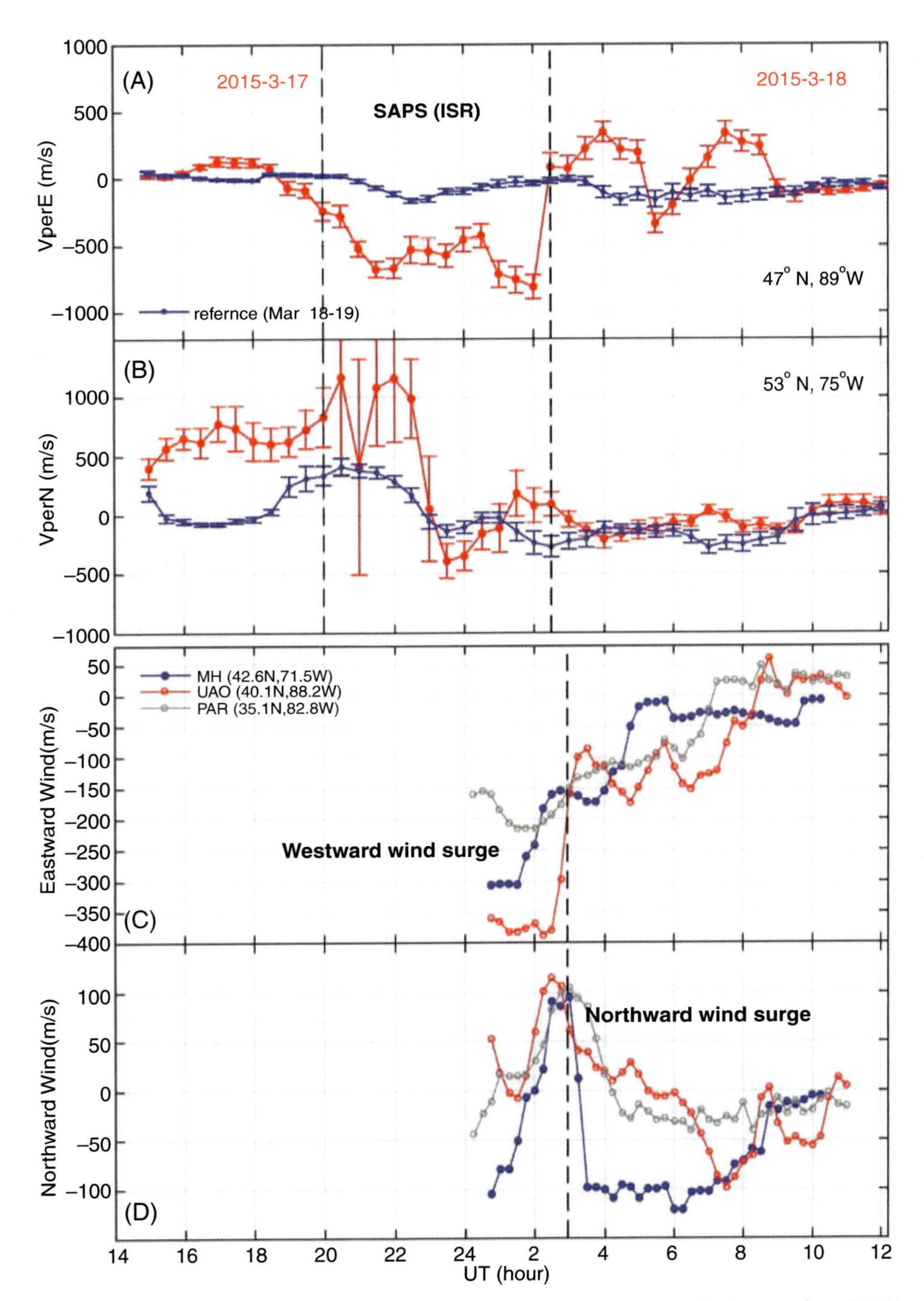

Fig. 3.25 Millstone Hill ISR measurements of plasma drifts (A) **V**perE (perpendicular east) for ~(89°W, 47°N), (B) **V**perN ($\mathbf{V}_p^N$) (perpendicular north) for ~(75°W, 53°N), Fabry-Perot Interferometer measurements of (C) eastward neutral winds and (D) meridional neutral winds at Millstone Hill and other nearby midlatitude sites. *Adapted from Figs. 2 and 3 in Zhang, S.-R., Erickson, P.J., Foster, J.C., Holt, J.M., Coster, A.J., Makela, J.J., Noto, J., Meriwether, J.W., Harding, B.J., Riccobono, J., Kerr, R.B., 2015b. Thermospheric poleward wind surge at midlatitudes during great storm intervals. Geophys. Res. Lett. 42 (13), 5132–5140.*

produced plasma could compensate partially for this loss. In short, the vertical component (V_p^z) of V_p^N induced by this electric field can also contribute to the SED development by providing an important plasma source and lifting the F region (Liu et al., 2016). This vertical transport is efficient because, similar to Eq. (3.45), the influx in the ion continuity equation $-\partial(\text{flux})/\partial z = -V_\text{p}^z \partial N/\partial z - N\partial V_\text{p}^z/\partial z$ will be always positive in the topside with $-\partial N/\partial z > 0$ and $\partial V_\text{p}^z/\partial z \sim 0$. The question is, however, the specific shape and magnitude in the gradient of latitudinal profiles of V_P^z variation which is responsible for the characteristic narrow channel of SED. Unless the penetration or expansion of the auroral convection electric field is imposed with a clear boundary, this dynamic forcing alone appears insufficient to account for the key SED plume morphology.

Other factors may affect, contribute to, and perhaps shape the SED formation and evolution. Influences from the storm-time expansion of equatorial ionization anomaly (EIA) to midlatitudes, an equatorward surge of meridional winds (see Section 3.3.2.1), and possibly other processes are among these important factors (Anderson, 1976; Gardner et al., 2018; Horvath and Lovell, 2011; Kelley et al., 2004; Lu et al., 2020; Moldwin et al., 2016). The diversity of these studies clearly demonstrates that the physical mechanism for SED formation remains a very active research frontier.

3.3.4 Polar cap structures and dynamics

The sunward convection entrains the ionospheric plasma in the afternoon sector at subauroral latitudes into the cusp near noon. This region overlaps the open-close field line boundary (OCB) in the magnetosphere. During the period of southward IMF Bz, magnetopause reconnection forcing takes the frozen-in plasma to streamline inward/poleward into the open field line region in the polar cap. At the ionospheric footprint, the poleward (perpendicular to **B**) motion has a vertically upward component that provides uplifting of the plasma to high altitudes where they survive for more time from the chemical loss processes than they would at lower altitudes. If stable IMF Bz southward conditions sustain, the stable antisunward convection in the polar cap carries the plasma originated from the midlatitude plasma such as SED and from particle precipitating effect to travel through the polar cap, forming a tongue of ionization (TOI) structure with elevated ionospheric density elongated along the day-night alignment (Dang et al., 2019; Foster et al., 2005; Hosokawa et al., 2010; Knudsen, 1974; Thomas et al., 2013). The TOI further runs across the nightside OCB where the reconnection forcing at the magnetotail drives the plasma exit from the polar cap and return into auroral latitudes on both dusk and dawn sectors. They continue to flow sunward and complete the large-scale Dungey circulation cycle (Dungey, 1961, 1963; Zhang et al., 2015a) of horizontal plasma transportation at high latitudes (Fig. 3.26).

The TOI occurrence is correlated well with the presence of SED plume; clearly, the SED plume contributes the source plasma to TOI in the polar cap. Of course, even

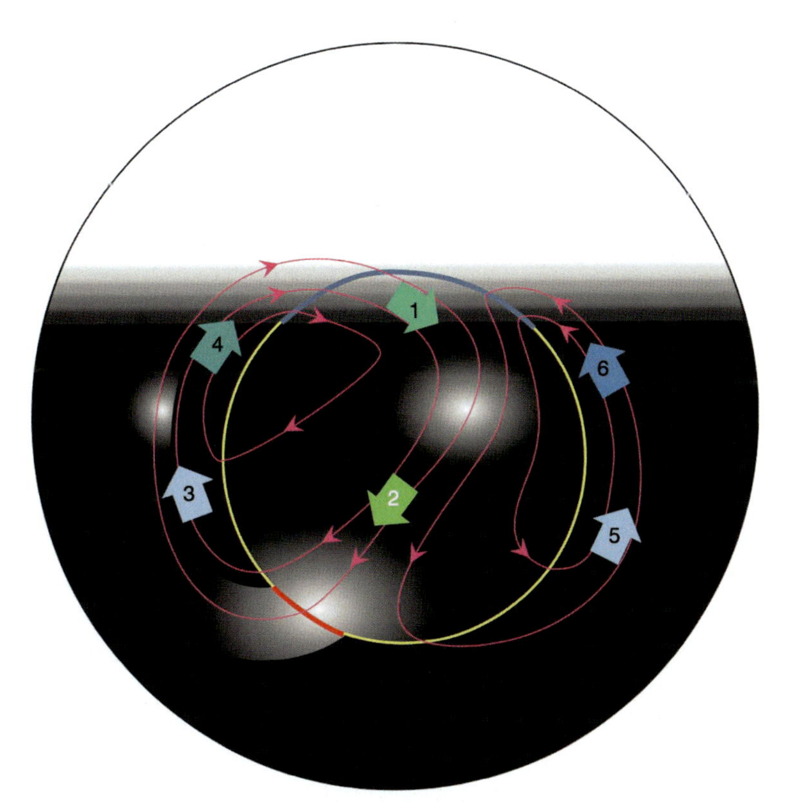

Fig. 3.26 Schematic of the northern polar ionosphere during a substorm growth phase with southward IMF and By > 0. *Blue* and *yellow arc* portions mark the open-close field line boundary (OCB) where plasma entry or exit the polar cap as a result of reconnection forcing at the magnetopause and magnetotail. The dayside *(white)* is solar-produced high-electron area and nightside *(black)* has low-electron density due to a lack of direct solar irradiation. The white spot inside the polar cap is the polar patch, and in the auroral zone is a "blob". *After Zhang, Q.-H., Zhang, B.-C., Lockwood, M., Hu, H.-Q., Moen, J., Ruohoniemi, J.M., Thomas, E.G., Zhang, S.-R., Yang, H.-G., Liu, R.-Y., McWilliams, K.A., Baker, J.B.H., 2013. Direct observations of the evolution of polar cap ionization patches. Science 339 (6127), 1597–1600. https://doi.org/10.1126/science.1231487.*

without SED, TOI could still develop through entraining midlatitude or cusp plasma by the enhanced convection. The IMF conditions play an important role in the polar cap plasma dynamics and structures, affecting plasma entry cross the dayside OCB, transpolar antisunward convection, and exit from the polar cap cross the nightside OCB. When Bz is southward turning or subject to substantial variations, the continuous plasma flow is possibly segmented into patches of small-scale-enhanced plasma density clouds in the polar cap.

Several segmentation processes resulting from IMF variations are plausible. The convection pattern expansion, distortion, speed change, and other variations associated with transient dayside magnetic reconnection due to IMF changes in Bz and sometimes in By can create inhomogeneity in forcing on the ionospheric plasma (Lockwood and Carlson, 1992; Rodger et al., 1994). The development of plasma flow channels leading to enhanced ion-neutral collision and heating can enhance the chemical loss and therefore "meltdown" plasma within the channel and eventually form separated density structures (Ren et al., 2020). When formed, these patches are often mesoscale usually spanning 200–1000 km in longitude and 2–3 degrees wide in latitude (Rodger et al., 1994; Weber et al., 1984), and can be 200% above the nearby background ionospheric density (Crowley, 1996).

Similar to a TOI, they undergo transpolar antisunward flow, exit from the polar cap and return to the auroral zone. Insider the aurora latitude, they are named as "blobs" (Rodger et al., 1986) (Fig. 3.26).

3.3.5 Main ionospheric trough

The main ionospheric trough, also termed midlatitude ionospheric trough, is a plasma depletion structure in the nighttime F region and topside ionosphere at subauroral latitudes. The main trough occurs for all levels of geomagnetic activity, and its electron density around 300 km can be as low as 10^3 cm^{-3} during quiet geomagnetic conditions, which is significantly lower compared with its adjacent poleward and equatorward regions (Brinton et al., 1978). As is shown in Fig. 3.27, the main trough normally situates between the footprints of the plasmaspheric boundary layer and the equatorward boundary of the auroral oval, spanning a narrow latitudinal extent for only few degrees, but covering an extended longitudinally range for several hours of local time from dusk to dawn (Carpenter and Lemaire, 2004; Moffett and Quegan, 1983; Pierrard and Voiculescu, 2011; Rodger et al., 1992).

There are three major components of a main trough: (1) a sharp poleward edge, which is mainly formed by additional local ionization at both E and F region due to auroral particle precipitation and/or partially by the high-density plasma transportation across the polar cap from the dayside, i.e., boundary blobs (Rodger et al., 1986); (2) a less-steep equatorward edge, which is usually formed by the replenishment of plasma from the nightside plasmasphere (Yizengaw and Moldwin, 2005) and/or built by the sunlit midlatitude plasma that gradually decays as it corotates into the darkness (Voiculescu et al., 2010); (3) the trough minimum region between the two edges mentioned above, where the plasma density depletion can exceed an order of magnitude in the ionospheric F region.

Since first identified by a topside sounder (Muldrew, 1965), the main trough has been extensively studied for several decades via different observational and modeling techniques (see the review by Rodger (2008) and references therein). The strong density gradient in the trough area can significantly affect the propagation of

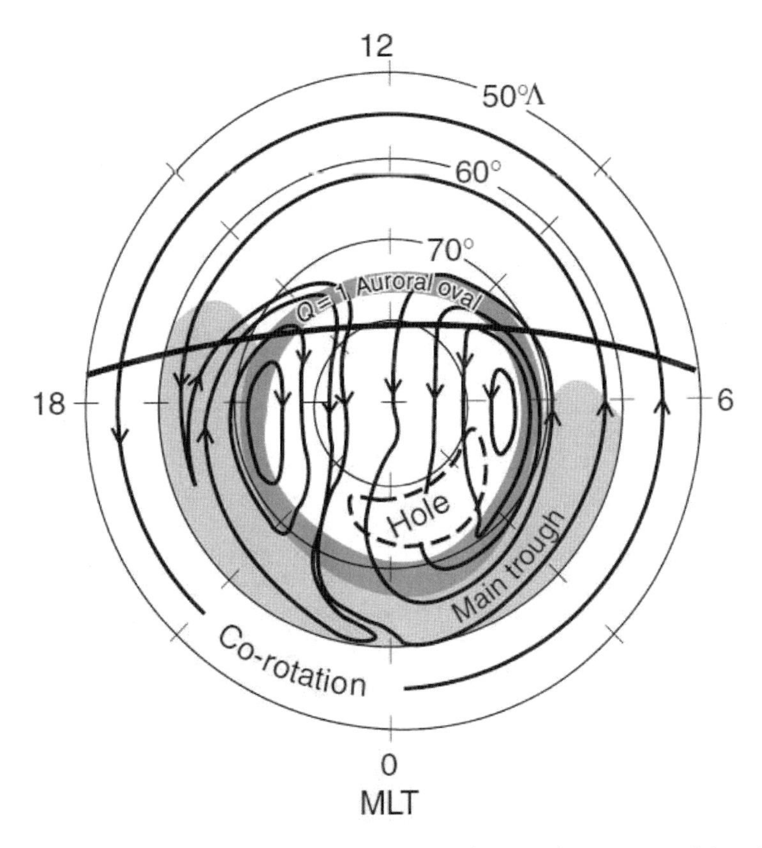

Fig. 3.27 A schematic diagram of the Earth's polar region showing the main trough location in a MLT and invariant latitude. Also shown are the two-cell plasma convection trajectories at 300 km *(solid lines)*, the corotation flow, the quiet-time auroral oval, and the high-latitude ionization hole. *After Brinton, H.C., Grebowsky, J.M., Brace, L.H., 1978. The highlatitude winter F region at 300 km: thermal plasma observations from AE-C. J. Geophys. Res. Space Phys. 83 (A10), 4767–4776. https://doi.org/10.1029/JA083iA10p04767.*

transionospheric radio signals. In addition, the trough edges are often regions of high plasma irregularity occurrence that can impose detrimental effects on modern navigation and communication systems, such as GNSS and Wide Area Augmentation System (WAAS). Thus, the fine structures and dynamic variation of the main trough are of great space weather importance. The content of this section will focus on three objectives: (1) to explain the fundamental formation mechanisms of the main trough, (2) to summarize the major temporal and spatial characteristics of the main trough, and (3) to highlight some of the recent scientific progresses associated with the main trough.

3.3.5.1 Trough formation mechanisms

Stagnation mechanism

The upper atmosphere is partially ionized, and the ion-neutral collision is frequent and cannot be neglected. Thus, in the low and midlatitude region, the neutral thermosphere particles that corotate with Earth also bring the ionospheric plasma into corotation via ion-neutral collisions, which is equivalent to a corotation electric field that directs radially inward in an inertial reference (Baumjohann and Treumann, 1996). On the other hand, the high-latitude ionosphere is dominated by the solar wind-induced convection electric field with a strong duskward component. Fig. 3.28 shows a schematic illustration of the total electric potential given by the sum of convection and corotation electric field in the equatorial plane. A stagnation point (the equipotential contour crosses itself) can be found in the duskside plasmapause where the convection and corotation electric fields are approximately oppositely directed with the same value.

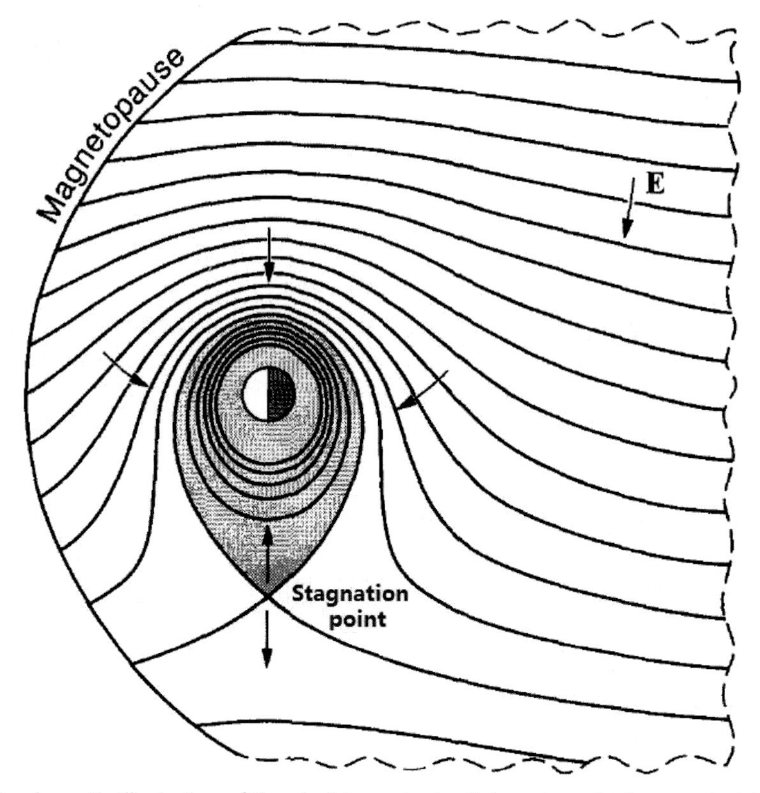

Fig. 3.28 A schematic illustration of the electric equipotential contours in the equatorial plane. The shaded area represents the plasmasphere. *After Berlok, T., Pfrommer, C., 2019. On the Kelvin-Helmholtz instability with smooth initial conditions-linear theory and simulations. Mon. Not. R. Astron. Soc. 485 (1), 908–923.*

The generation of the main trough can be traced to the footprints of the plasmapause around the stagnation point, i.e., the low-speed subauroral ionosphere in the dusk sector to the east of sunset terminator. In this plasma stagnation area, the westward convection returning flow counteracts with the eastward corotation flow. As a result, the plasma in this region has a longer residence time in darkness than at lower latitudes, and the prolonged recombination processes therein cause the plasma density decay to very low levels to form the main trough (Nilsson et al., 2005; Spiro et al., 1978). Eventually, the low-density plasma could slowly drift out of this quasi-stagnation area and extend to the whole night sector that corotates and/or convects toward dawn. Moreover, the convection flow could transport a low influx into the dayside sector, sometimes forming a dayside trough (Pryse et al., 1998; Rodger et al., 1992; Whalen, 1989).

Subauroral ion heating mechanism

During geomagnetically active periods, the morphology and dynamics of the main trough can be greatly complicated by the spatiotemporal variation of electric fields in the vicinity of subauroral ionosphere, and the existence of SAPS, which is extensively discussion in Section 3.3.3.1.

As pointed out earlier, SAPS-driven ion velocities are significantly larger than neutral wind velocities. Thus, the ion temperature in the vicinity of SAPS will be increased considerably due to enhanced ion-neutral frictional heating (Zhang et al., 2017), which in turn accelerates the ion loss process via the following charge exchange reaction (Schunk et al., 1976):

$$N_2 + O^+ = NO^+ + N, \tag{3.46}$$

where the resulted molecular ion, NO^+, has much faster recombination rate with electrons than that of O^+ (Schunk and Nagy, 2000). This leads to rapid depletion of ionization and thus a deep trough can be quickly formed. It should be noted that this plasma loss process can be highly nonlinear. As can be seen from Fig. 3.29, when the relative ion-neutral velocity increases from 1 to 2 km/s, the rate coefficients of the loss process from O^+ to N_2 as mentioned above will rise by one order of magnitude (Rodger, 2008; Schunk et al., 1976). Moreover, the enhanced frictional heating and associated plasma thermal expansion will cause field-aligned plasma upflow, which also contributes to the formation of the main trough (Anderson et al., 1991; Voiculescu and Roth, 2008).

3.3.5.2 Morphology and dynamics of the main trough

Diurnal variation: Although the main trough can drift into the dayside ionosphere, it is basically a nocturnal phenomenon that longitudinally extends from dusk to dawn and is limited mainly between 55 and 65 degrees of geomagnetic latitude (MLAT) to the equatorward side of the auroral oval. The trough minimum position normally occurs at higher latitudes in the afternoon sector and migrates to lower latitudes with a later

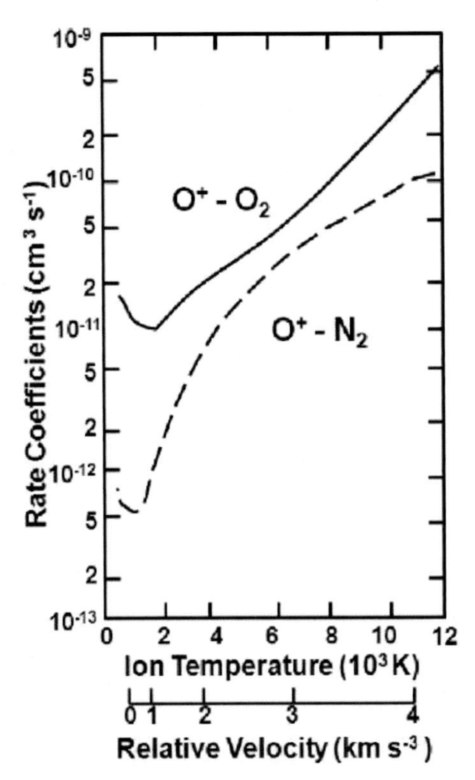

Fig. 3.29 Rate coefficients for the two major loss reactions at F layer as a function of ion temperature and velocity in the rest frame of the neutral particles. *After Rodger, A.S., 2008. The Mid-Latitude Trough—Revisited, Geophysical Monograph Series, vol. 181. American Geophysical Union, Washington, DC, pp. 25–33. https://doi.org/10.1029/181GM04.*

MLT. It usually reaches its equatorward-most position in the early morning hours (02–04 MLT) and then quickly retreats back to higher latitudes (Karpachev, 2003; Lee et al., 2011; Mallis and Essex, 1993; Werner and Prölss, 1997).

Seasonal and hemispheric variation: The main trough is primarily observed in winter and equinoxes. Under the polar night and quietest geomagnetic conditions, the trough can be seen at all local times around midwinter. In summer, however, the trough is less common, exhibiting a very limited occurrence centered around midnight (Horvath and Essex, 2003; Rodger, 2008). A trough tends to occur at higher latitudes with broader width in winter than those in summer (Karpachev, 2003; Voiculescu et al., 2006). In addition, there is also a hemispheric asymmetry, in that the Northern Hemisphere has a higher trough occurrence rate than the Southern Hemisphere during winter and equinoctial periods. This is due to the smaller offset between the geomagnetic and geographic poles in the Northern Hemisphere, which makes the trough zone taken as a whole closer to the darkness in the winter/equinoctial time (Aa et al., 2020).

Longitudinal variation: The location of the main trough exhibits strong longitudinal dependence. Typically, the deepest trough is located to the west of the geomagnetic poles, which is around 130°W in the Northern Hemisphere and around 60°E in the Southern Hemisphere. The trough usually occurs at a higher latitude in the longitudinal sector that contains the magnetic pole (He et al., 2011; Karpachev et al., 2019). Moreover, the trough distribution exhibits clear east-west hemispherical preferences. Fig. 3.30 shows that the trough has a higher occurrence rate in eastern longitudes during the December solstice and in western longitudes during the June solstice. These longitudinal variations are caused by the illumination difference along a fixed MLAT line within the trough due to Earth's rotation and the offsets between the geographic and geomagnetic poles. For instance, in the December solstice, the eastern longitudes along 62 degrees MLAT are closer to the polar night in Northern Hemisphere and far from the polar day in Southern Hemisphere than their western counterparts, thus the trough therein is likely to have a higher occurrence rate.

Geomagnetic and solar cycle dependence: The trough shifts progressively toward lower latitudes with increasing geomagnetic activity due to the expansion of auroral oval (Prölss, 2007; Werner and Prölss, 1997; Zou et al., 2011). The trough tends to have a higher occurrence rate and becomes deeper during geomagnetically active times due to the enhancement of subauroral electric field and associated plasma heating as mentioned above. Conditions for the trough occurrence are more favored in low solar activity periods, and the trough depth were found to increase with the 10.7-cm solar radiation flux F10.7 (Ishida et al., 2014; Karpachev et al., 2019; Yang et al., 2015).

3.4 Conclusion

3.4.1 Conclusion of Section 3.1

F-region plasma density irregularities have been the focus of M-I-T coupling research for over. Every year, new and exciting discoveries are made regarding their generation mechanisms, characteristics and morphology, effects on technology, and overall role in the coupled M-I-T system. By and large, this scientific progress is driven by advances in geospace sensor technology, including in situ and remote techniques, which were discussed earlier, and the proliferation of these sensors throughout the high-latitude region.

There is no sign that interest in the high-latitude regions and its plasma density irregularities is subsiding. In fact, it appears that, at least anecdotally, the opposite is occurring. Without question, there is a clear societal need for future scientific investigations. Commercial airlines that are seeking to shorten long-haul flights using transpolar routes are becoming more reliant on radio communications that are vulnerable to irregularities and associated phenomena. Furthermore, as the high-latitude waterways open up due to the decline in ice coverage, long-range maritime monitoring and navigation will

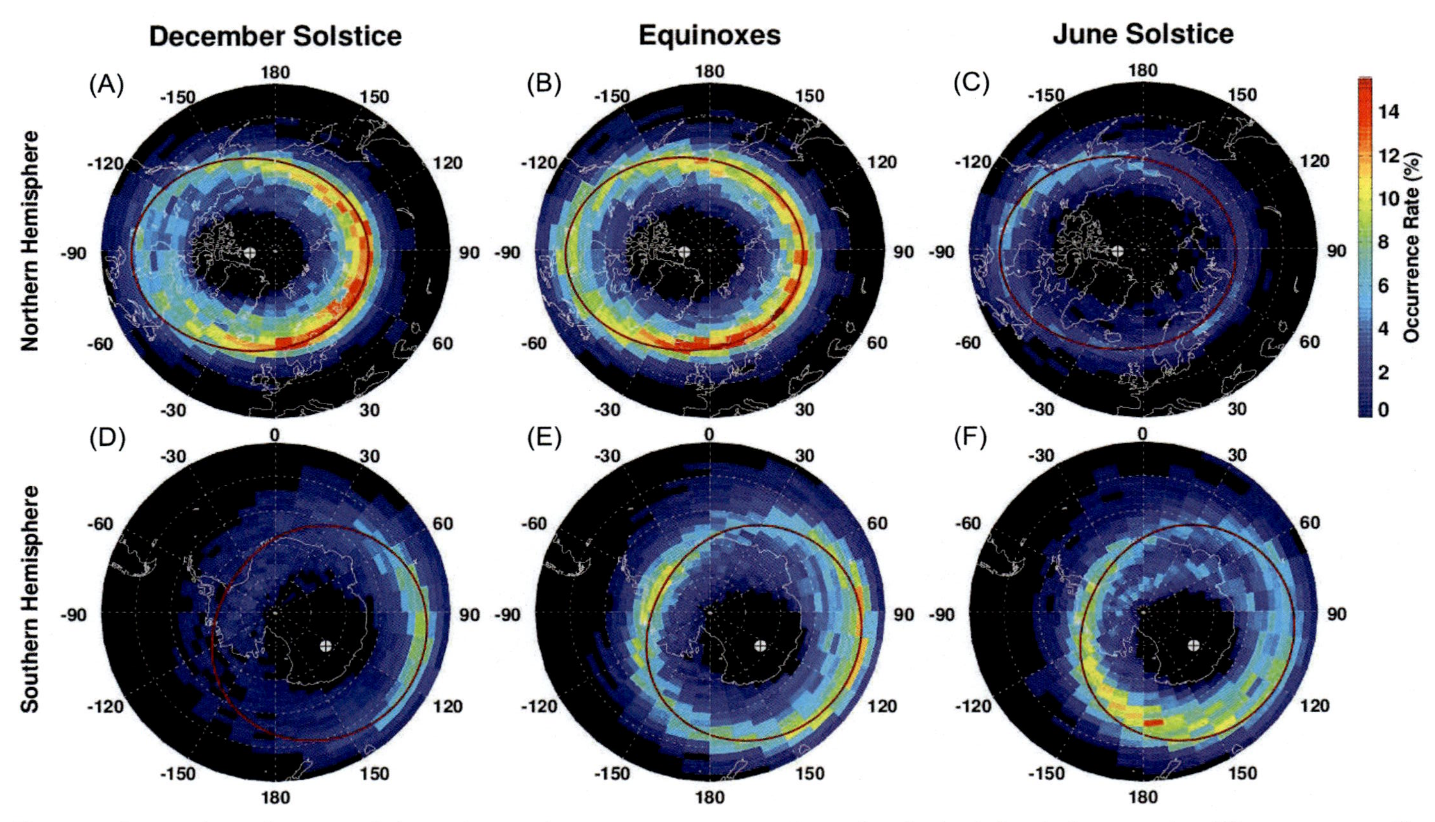

Fig. 3.30 Geographic polar view of the main trough occurrence rate at 400–500 km for both hemispheres under different seasons. The concentric circles are plotted in 10 degrees interval with outermost one representing latitude of 40 degrees. The geomagnetic poles are marked with *crosses*. The ± 62 degrees MLAT lines are also plotted in *solid red lines*. *After Aa, E., Zou, S., Erickson, P.J., Zhang, S.-R., Liu, S., 2020. Statistical analysis of the main ionospheric trough using Swarm in situ measurements. J. Geophys. Res. Space Phys. 125 (3), e2019JA027583.*

become more common, requiring a more detailed and reliable understanding of the signal propagation characteristics in the region.

From our perspective, the amount of valuable scientific data that has been generated by the abundance and variety of sensors in the high-latitude regions has, by far, outpaced their assimilation and transformation into new knowledge by inquiring scientists. Undoubtedly, the necessary measurements for resolving the question of the dominant generation mechanism for plasma density enhancements have already been collected—they only have to be pieced together. Other outstanding questions, like those related to the instability mechanisms associated with plasma irregularities, may be out of reach of experiment until more highly resolved and insightful measurement techniques become available.

It is without question that we, as solar-terrestrial physicists, have only scratched the surface on understanding the interconnections within the M-I-T system. Indeed, two major conclusions of our collective efforts and progress in resolving finer spatial scale and faster cadences are that, first, influential M-I-T coupling phenomena are transpiring across multiple spatiotemporal scales. Second, the phenomena have major impacts on the M-I-T from bottomside F-region ionosphere all the way out to the distant reaches of the terrestrial magnetotail.

3.4.2 Concluding remarks and future work of Section 3.2

This review has focused on theories and simulations of high-latitude ionospheric density irregularities and the role these play in altering the terrestrial radio propagation environment, specifically through refractive and diffractive effects collectively referred to herein as "scintillation" resulting in fluctuations in radio signals. Prior reviews concerning observations and theories of ionospheric structures have been discussed and summarized. These are mostly conducted under the framework of linear theory, whereas the present review article focuses on the nonlinear evolution of ionospheric instabilities. Because scintillation occurs from structures at ~100 m scales and above, the features of such density structures and how they affect propagation are inherently the product of nonlinear cascading/steepening processes.

To motivate discussions of GDI and KHI we first summarize salient work on the linear theory of these instabilities. As shown in this section, linear theory is valuable for quick interpretations of data and is also an important validation point for more complicated nonlinear simulations. We review recent nonlinear simulation work and then provide a set of reproducible simulation examples, using the GEMINI model, for the reader to access and run those optionally. These examples illustrate some important properties of the instability nonlinear behavior, namely the sensitivity to parameters like background conductance (precipitation) and inertial capacitance (amount of MI coupling).

Ionospheric instabilities, such as KHI and GDI can cause scintillation-scale structures which cause fluctuations in the radio signals observed on the ground. We review the theory of how the propagation through the turbulent ionosphere can be simulated and mapped to signal fluctuations on the ground. We also emphasize how the physics-based models, such as GEMINI-SIGMA, can simulate the scintillations during different growth stages of instabilities. We briefly discuss the scintillation signatures related to different instabilities and review inverse analysis to retrieve physics of ionosphere from the observed scintillations on the ground. We also touch upon scale-dependent scintillations and HF effects. Finally, we present some of the challenges and unresolved issues in instability and scintillation studies, and make some suggestions for fruitful future work to advance this field.

3.4.3 Concluding remarks of Section 3.3

We have provided an overview on the substantial electron density gradients in the ionosphere with focus on their fundamental morphology. Many of these have been known for several decades, however, it is only until recent years the observational coverage in subauroral and high altitudes has dramatically improved, especially due to the use of GNSS TEC, SuperDARN, ISRs, All-sky Imagers, and other radio and optical technologies. Coupling models with magnetosphere, ionosphere, and thermosphere components are more and more capable due to improved resolution and specification of essential drivers. It is fair to state, however, that there are still very significant knowledge gaps. The SED morphology, its dependency on solar-terrestrial drivers, and its evolution in space and time are not well established. Furthermore, the known physics processes, which could be subject to variability from storm to storm, remain highly debatable. There are also some renewed interests in the main midlatitude trough, particularly, its relationship with the SAPS presence and SED location. We believe research toward improved understanding of the ionospheric large gradients will be staying very active as one of the frontiers of geospace coupling and systems science.

Acknowledgments

Section 3.1

This work is dedicated to Dr. Jean-Pierre St. Maurice, our mentor. We stand on his shoulders. LVG is supported by the NASA Living With a Star Jack Eddy Postdoctoral Fellowship Program, administered by UCAR's Cooperative Programs for the Advancement of Earth System Science (CPAESS) under award #NNX16AK22G.

This material is based upon work supported by the Resolute Bay Observatory, which is a major facility funded by the National Science Foundation through cooperative agreement AGS-1840962 to SRI International. RISR-C is funded by the Canada Foundation for Innovation and led by the University of Calgary's Auroral Imaging Group, in partnership with UofC Geomatic Engineering, University of Saskatchewan, Athabasca University and SRI International (http://aurora.phys.ucalgary.ca/resu/). RISR-C data are available from http://data.phys.ucalgary.ca/ and https://madrigal.phys.ucalgary.ca/.

Section 3.2

KD was supported by NSF grants 1651410 (for work on propagation modeling theory and inverse technique) and NSF CAREER grant AGS-1848207 (for work on scintillation signatures through different types on instability in the cusp and polar cap).

Development of the GEMINI model was supported by NSF CAREER grant AGS-1255181 (supporting MZ) and NASA HDEE grant 80NSSC20K0176 (supporting MH).

AS acknowledges funding support from the Research Council of Norway, grant number 275653.

The material on current and future ISR observations, contributed by LL, is based on work supported by the Resolute Bay Observatory which is a major facility funded by the National Science Foundation through cooperative agreement AGS-1840962 to SRI International. EISCAT is an international association supported by research organisations in China (CRIRP), Finland (SA), Japan (NIPR and ISEE), Norway (NFR), Sweden (VR), and the United Kingdom (UKRI).

We also acknowledge NSF collaborative grants AGS-2027308 (LL) and AGS-2027300 (MZ,KD, and MR) which contributed to development of many of the ideas presented in this article.

Finally, MZ and KD gratefully acknowledge use of the ERAU VEGA supercomputing system for some of the simulations presented in this article.

Section 3.3

This work was supported by AFOSR MURI project FA9559-16-1-0364, ONR grant N00014-17-1-2186, and NSF awards AGS-1952737 and AGS-2033787.

Appendix: Simulation software used for examples in Section 3.2

This section presents results from basic, reproducible simulations that illustrate some important features of the linear and nonlinear evolution of GDI and KHI. The model used to generate the results, GEMINI (Zettergren and Semeter, 2012; Zettergren and Snively, 2015; Zettergren et al., 2015), is freely available to the readers, who can download the source code, compile it, and generate results similar to those presented in this paper (or extend these examples for their own work). These examples can be run by the reader by installing the GEMINI ionospheric model repositories, which can be downloaded from https://github.com/gemini3d/. The examples used for this article (Zettergren and Hirsch, 2020) have additional documentation explaining their use under these directories:

- GDI: INIT/GDI_PERIODIC_LOWRES?
- KHI: INIT/KHI_PERIODIC_LOWRES?

The example scripts for basic simulations will run on a typical laptop or desktop computer running Linux, MacOS or Windows. An HPC will reduce the processing time according to the number of CPUs available; the code scales efficiently up to at least hundreds of cores. Software requirements are

- FORTRAN 2008 COMPILER (GCC OR INTEL FORTRAN)?
- PYTHON OR MATLAB®?

Our testing indicates that an Intel i7 quad-core CPU runs the example GDI simulation in less than 2 h. The example KHI simulation takes about 2.5 h on a 16-core Intel Skylake HPC CPU.

The numerical core of GEMINI is written in object-oriented Fortran 2018. A high-level view of the software library interfaces is shown in Fig. 3.31.

GEMINI uses sparse and linear algebra libraries MUMPS (Amestoy et al., 2019), ScaLAPACK (Blackford et al., 1997) and LAPACK (Anderson et al., 1999), and benefits from using an MPI-2 (Gropp et al., 1999)

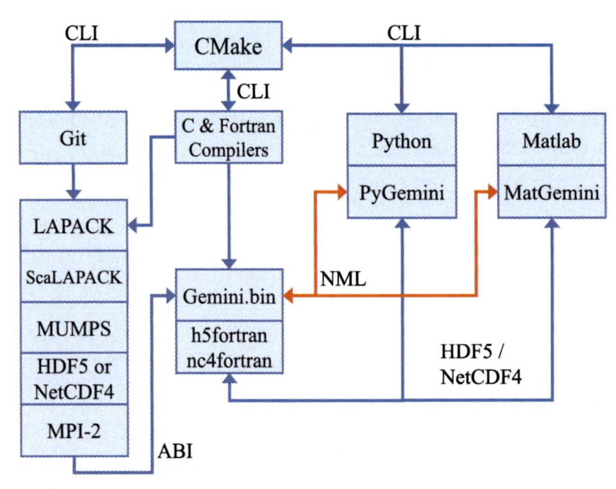

Fig. 3.31 GEMINI3D software suite interconnections.

message passing interface such as OpenMPI, MPICH, Intel MPI, MS-MPI, etc. to enable domain parallelization. GEMINI can also run without an MPI library using a single CPU core. Data input and output files use the HDF5 (Folk et al., 2011) format. Configuration parameters and metadata are communicated via plain-text, Fortran standard Namelist (NML) files.

Care has been taken to remain agnostic of the compiler, operating system, and CPU architecture throughout the GEMINI software stack. This also applies to the Python interface to GEMINI (PyGemini (Hirsch, 2020)) and the MATLAB interface (MatGemini (Hirsch and Zettergren, 2020)). PyGemini and MatGemini each automatically determine the optimal grid—worker partitioning to maximize CPU core usage. Over 100 unit tests and project registration tests are built-in and run automatically upon first install. This helps identify problems with the user's software stack. Gemini3D will attempt to download and build missing or incompatible libraries. Plotting and utility scripts exist independently in Python and MATLAB (Hirsch, 2020; Hirsch and Zettergren, 2020).

The main scripting frontend program "gemini3d.run" generates a Namelist NML file and HDF5 files containing the simulation initialization data. "gemini3d.run" can auto-generate an HPC job file or run the simulation on the local computer. A second program "gemini3d.plot" generates numerous plots per simulation time step.

The script for generating Fig. 3.13 (Zettergren and Hirsch, 2020) is run from MATLAB function:

```
gemscr.postprocess.GDIgrowth?
```

References

Aa, E., Zou, S., Erickson, P.J., Zhang, S.-R., Liu, S., 2020. Statistical analysis of the main ionospheric trough using Swarm in situ measurements. J. Geophys. Res. Space Phys. 125 (3). e2019JA027583.

Aarons, J., 1982. Global morphology of ionospheric scintillations. IEEE Proc. 70, 360–378.

Aarons, J., 1993. The longitudinal morphology of equatorial F-layer irregularities relevant to their occurrence. Space Sci. Rev. 63 (3–4), 209–243. https://doi.org/10.1007/BF00750769.

Aarons, J., 1997. Global positioning system phase fluctuations at auroral latitudes. J. Geophys. Res. Space Phys. 102 (A8), 17219–17231. https://doi.org/10.1029/97JA01118.

Akasofu, S.I., 1968. Polar and Magnetosphere Substorms. vol. 11 Astrophysics and Space Science Library, Springer, https://doi.org/10.1007/978-94-010-3461-6.

Alfonsi, L., Spogli, L., De Franceschi, G., Romano, V., Aquino, M., Dodson, A., Mitchell, C.N., 2011. Bipolar climatology of GPS ionospheric scintillation at solar minimum. Radio Sci. 46 (3). https://doi.org/10.1029/2010RS004571.

Amestoy, P.R., Buttari, A., L'Excellent, J.-Y., Mary, T., 2019. Performance and scalability of the block low-rank multifrontal factorization on multicore architectures. ACM Trans. Math. Softw. 45, 2:1–2:26.

Anderson, D.N., 1976. Modeling the midlatitude F-region ionospheric storm using east-west drift and a meridional wind. Planet. Space Sci. 24 (1), 69–77. https://doi.org/10.1016/0032-0633(76)90063-5.

Anderson, P.C., Heelis, R.A., Hanson, W.B., 1991. The ionospheric signatures of rapid subauroral ion drifts. J. Geophys. Res. Space Phys. 96 (A4), 5785–5792. https://doi.org/10.1029/90JA02651.

Anderson, E., Bai, Z., Bischof, C., Blackford, S., Demmel, J., Dongarra, J., Du Croz, J., Greenbaum, A., Hammarling, S., McKenney, A., Sorensen, D., 1999. LAPACK Users' Guide, third ed. Society for Industrial and Applied Mathematics, Philadelphia, PA.

Bahcivan, H., Tsunoda, R., Nicolls, M., Heinselman, C., 2010. Initial ionospheric observations made by the new resolute incoherent scatter radar and comparison to solar wind IMF. Geophys. Res. Lett. 37 (15), L15103. https://doi.org/10.1029/2010GL043632.

Basu, S., Basu, S., Hanson, W.B., 1981. The role of in-situ measurements in scintillation modelling. In: Symposium on the Eect of the Ionosphere on Radiowave Systems.

Basu, S., Basu, S., MacKenzie, E., Coley, W.R., Hanson, W.B., Lin, C.S., 1984. F region electron density irregularity spectra near auroral acceleration and shear regions. J. Geophys. Res. Space Phys. 89 (A7), 5554–5564. https://doi.org/10.1029/JA089iA07p05554.

Basu, S., Basu, S., MacKenzie, E., Whitney, H.E., 1985. Morphology of phase and intensity scintillations in the auroral oval and polar cap. Radio Sci. 20 (3), 347–356. https://doi.org/10.1029/RS020i003p00347.

Basu, S., Basu, S., MacKenzie, E., Fougere, P.F., Coley, W.R., Maynard, N.C., Winningham, J.D., Sugiura, M., Hanson, W.B., Hoegy, W.R., 1988. Simultaneous density and electric field fluctuation spectra associated with velocity shears in the auroral oval. J. Geophys. Res. Space Phys. 93 (A1), 115–136. https://doi.org/10.1029/JA093iA01p00115.

Basu, S., Basu, S., Weber, E.J., Coley, W.R., 1988. Case study of polar cap scintillation modeling using DE 2 irregularity measurements at 800 km. Radio Sci. 23 (4), 545–553. https://doi.org/10.1029/RS023i004p00545.

Basu, S., Basu, S., MacKenzie, E., Coley, W.R., Sharber, J.R., Hoegy, W.R., 1990. Plasma structuring by the gradient drift instability at high latitudes and comparison with velocity shear driven processes. J. Geophys. Res. Space Phys. 95 (A6), 7799–7818. https://doi.org/10.1029/JA095iA06p07799.

Basu, S., Basu, S., Chaturvedi, P.K., Bryant Jr, C.M., 1994. Irregularity structures in the cusp/cleft and polar cap regions. Radio Sci. 29 (1), 195–207. https://doi.org/10.1029/93RS01515.

Basu, S., Weber, E.J., Bullett, T.W., Keskinen, M.J., MacKenzie, E., Doherty, P., Sheehan, R., Kuenzler, H., Ning, P., Bongiolatti, J., 1998. Characteristics of plasma structuring in the cusp/cleft region at Svalbard. Radio Sci. 33 (6), 1885–1899. https://doi.org/10.1029/98RS01597.

Basu, S., Basu, S., Valladares, C.E., Yeh, H.-C., Su, S.-Y., MacKenzie, E., Sultan, P.J., Aarons, J., Rich, F.J., Doherty, P., Groves, K.M., Bullett, T.W., 2001. Ionospheric effects of major magnetic storms during the international space weather period of September and October 1999: GPS observations, VHF/UHF scintillations, and in situ density structures at middle and equatorial latitudes. J. Geophys. Res. Space Phys. 106 (A12), 30389–30413. https://doi.org/10.1029/2001JA001116.

Baumjohann, W., Treumann, R.A., 1996. Basic Space Plasma Physics. Imperial College, London.

Berlok, T., Pfrommer, C., 2019. On the Kelvin-Helmholtz instability with smooth initial conditions-linear theory and simulations. Mon. Not. R. Astron. Soc. 485 (1), 908–923.

Bernhardt, P.A., Siefring, C.L., 2006. New satellite-based systems for ionospheric tomography and scintillation region imaging. Radio Sci. 41 (5), RS5S23. https://doi.org/10.1029/2005RS003360.

Bhattacharyya, A., Yeh, K.C., Franke, S.J., 1992. Deducing turbulence parameters from transionospheric scintillation measurements. Space Sci. Rev. 61 (3-4), 335–386. https://doi.org/10.1007/BF00222311.

Blackford, L.S., Choi, J., Cleary, A., D'Azevedo, E., Demmel, J., Dhillon, I., Dongarra, J., Hammarling, S., Henry, G., Petitet, A., Stanley, K., Walker, D., Whaley, R.C., 1997. ScaLAPACK Users' Guide. Society for Industrial and Applied Mathematics, Philadelphia, PA.

Briggs, B.H., 1964. Observations of radio star scintillations and spread-F echoes over a solar cycle. J. Atmos. Terr. Phys. 26 (1), 1–23. https://doi.org/10.1016/0021-9169(64)90104-7.

Briggs, B.H., 1975. Ionospheric irregularities and radio scintillations. Contemp. Phys. 16, 469–488. https://doi.org/10.1080/00107517508210825.

Brinton, H.C., Grebowsky, J.M., Brace, L.H., 1978. The high-latitude winter F region at 300 km: thermal plasma observations from AE-C. J. Geophys. Res. Space Phys. 83 (A10), 4767–4776.

Brinton, H.C., Grebowsky, J.M., Brace, L.H., 1978. The high-latitude winter F region at 300 km: thermal plasma observations from AE-C. J. Geophys. Res. Space Phys. 83 (A10), 4767–4776. https://doi.org/10.1029/JA083iA10p04767.

Bristow, W.A., Jensen, P., 2007. A superposed epoch study of SuperDARN convection observations during substorms. J. Geophys. Res. Space Phys. 112 (A6).

Buonsanto, M.J., 1999. Ionospheric storms—a review. Space Sci. Rev. 88, 563–601. https://doi.org/10.1023/A:1005107532631.

Burston, R., Astin, I., Mitchell, C., Alfonsi, L., Pedersen, T., Skone, S., 2009. Correlation between scintillation indices and gradient drift wave amplitudes in the northern polar ionosphere. J. Geophys. Res. Space Phys. 114 (A7). https://doi.org/10.1029/2009JA014151.

Burston, R., Astin, I., Mitchell, C., Alfonsi, L., Pedersen, T., Skone, S., 2010. Turbulent times in the northern polar ionosphere? J. Geophys. Res. Space Phys. 115 (A4). https://doi.org/10.1029/2009JA014813.

Burston, R., Mitchell, C., Astin, I., 2016. Polar cap plasma patch primary linear instability growth rates compared. J. Geophys. Res. Space Phys. 121 (4), 3439–3451. https://doi.org/10.1002/2015JA021895.

Carlson, H.C., 2012. Sharpening our thinking about polar cap ionospheric patch morphology, research, and mitigation techniques. Radio Sci. 47 (4), RS0L21. https://doi.org/10.1029/2011RS004946.

Carlson, H.C., Oksavik, K., Moen, J., Pedersen, T., 2004. Ionospheric patch formation: direct measurements of the origin of a polar cap patch. Geophys. Res. Lett. 31 (8), L08806. https://doi.org/10.1029/2003GL018166.

Carlson, H.C., Pedersen, T., Basu, S., Keskinen, M., Moen, J., 2007. Case for a new process, not mechanism, for cusp irregularity production. J. Geophys. Res. Space Phys. 112 (A11). https://doi.org/10.1029/2007JA012384.

Carlson, H.C., Oksavik, K., Moen, J., 2008. On a new process for cusp irregularity production. Ann. Geophys. 26 (9), 2871–2885. https://doi.org/10.5194/angeo-26-2871-2008.

Carpenter, D., Lemaire, J., 2004. The plasmasphere boundary layer. Ann. Geophys. 22 (12), 4291–4298. https://doi.org/10.5194/angeo-22-4291-2004.

Carrano, C.S., Rino, C.L., 2019. Irregularity parameter estimation for interpretation of scintillation doppler and intensity spectra. In: 2019 United States National Committee of URSI National Radio Science Meeting (USNC-URSI NRSM), pp. 1–2.

Carrano, C.S., Groves, K.M., Caton, R.G., 2012. Simulating the impacts of ionospheric scintillation on L band SAR image formation. Radio Sci. 47. https://doi.org/10.1029/2011RS004956.

Cerisier, J.C., Berthelier, J.J., Beghin, C., 1985. Unstable density gradients in the high-latitude ionosphere. Radio Sci. 20 (4), 755–761. https://doi.org/10.1029/RS020i004p00755.

Chartier, A., Forte, B., Deshpande, K., Bust, G., Mitchell, C., 2016. Three-dimensional modeling of high-latitude scintillation observations. Radio Sci. 51 (7), 1022–1029. https://doi.org/10.1002/2015RS005889.

Chartier, A.T., Mitchell, C.N., Miller, E.S., 2018. Annual occurrence rates of ionospheric polar cap patches observed using Swarm. J. Geophys. Res. Space Phys. 123 (3), 2327–2335. https://doi.org/10.1002/2017JA024811.

Chartier, A.T., Huba, J.D., Mitchell, C.N., 2019. On the annual asymmetry of high-latitude sporadic F. Space Weather 17 (11), 1618–1626. https://doi.org/10.1029/2019SW002305.

Chernyshov, A.A., Spicher, A., Ilyasov, A.A., Miloch, W.J., Clausen, L.B.N., Saito, Y., Jin, Y., Moen, J.I., 2018. Studies of small-scale plasma inhomogeneities in the cusp ionosphere using sounding rocket data. Phys. Plasmas 25 (4), 042902. https://doi.org/10.1063/1.5026281.

Chisham, G., Lester, M., Milan, S.E., Freeman, M.P., Bristow, W.A., Grocott, A., McWilliams, K.A., Ruohoniemi, J.M., Yeoman, T.K., Dyson, P.L., Greenwald, R.A., Kikuchi, T., Pinnock, M., Rash, J.P.S., Sato, N., Sofko, G.J., Villain, J.-P., Walker, A.D.M., 2007. A decade of the Super Dual Auroral Radar Network (SuperDARN): scientific achievements, new techniques and future directions. Surv. Geophys. 28 (1), 33–109. https://doi.org/10.1007/s10712-007-9017-8.

Clausen, L.B.N., Baker, J.B.H., Ruohoniemi, J.M., Milan, S.E., Coxon, J.C., Wing, S., Ohtani, S., Anderson, B.J., 2013. Temporal and spatial dynamics of the regions 1 and 2 Birkeland currents during substorms. J. Geophys. Res. Space Phys. 118 (6), 3007–3016. https://doi.org/10.1002/jgra.50288.

Clausen, L.B.N., Moen, J.I., Hosokawa, K., Holmes, J.M., 2016. GPS scintillations in the high latitudes during periods of dayside and nightside reconnection. J. Geophys. Res. Space Phys. 121 (4), 3293–3309. https://doi.org/10.1002/2015JA022199.

Costa, E., Kelley, M.C., 1977. Ionospheric scintillation calculations based on in situ irregularity spectra. Radio Sci. 12, 797–809. https://doi.org/10.1029/RS012i005p00797.

Costa, E., Fougere, P.F., Basu, S., 1988. Cross-correlation analysis and interpretation of spaced-receiver measurements. Radio Sci. 23 (2), 141–162. https://doi.org/10.1029/RS023i002p00141.

Coster, A.J., Colerico, M.J., Foster, J.C., Rideout, W., Rich, F., 2007. Longitude sector comparisons of storm enhanced density. Geophys. Res. Lett. 34 (18), L18105. https://doi.org/10.1029/2007GL030682.

Cowley, S.W.H., Lockwood, M., 1992. Excitation and decay of solar wind-driven flows in the magnetosphere-ionosphere system. Ann. Geophys., 103–114.

Crowley, G., 1996. Critical review of ionospheric patches and blobs. In: Stone, W.R. (Ed.), The Review of Radio Science 1992–1996. Oxford Univ. Press, New York, pp. 619–648.

Crowley, G., 1996. Critical review of ionospheric patches and blobs. In: Stone, W.R. (Ed.), The Review of Radio Science 1992-1996. Oxford Univ. Press, New York, pp. 619–648. chap. 27.

Dahlgren, H., Perry, G.W., Semeter, J.L., St.-Maurice, J.-P., Hosokawa, K., Nicolls, M.J., Greffen, M., Shiokawa, K., Heinselman, C., 2012. Space-time variability of polar cap patches: direct evidence for internal plasma structuring. J. Geophys. Res. Space Phys. 117 (A9), A09312. https://doi.org/10.1029/2012JA017961.

Dang, T., Lei, J., Wang, W., Wang, B., Zhang, B., Liu, J., Burns, A., Nishimura, Y., 2019. Formation of double tongues of ionization during the 17 March 2013 geomagnetic storm. J. Geophys. Res. Space Phys. 124 (12), 10619–10630. https://doi.org/10.1029/2019JA027268.

Datta-Barua, S., Su, Y., Deshpande, K., Miladinovich, D., Bust, G.S., Hampton, D., Crowley, G., 2015. First light from a kilometer-baseline scintillation auroral GPS array. Geophys. Res. Lett. 42 (10), 3639–3646. https://doi.org/10.1002/2015GL063556.

David, M., Sojka, J.J., Schunk, R.W., Coster, A.J., 2019. Hemispherical shifted symmetry in polar cap patch occurrence: a survey of GPS TEC maps from 2015–2018. Geophys. Res. Lett., 726–734. https://doi.org/10.1029/2019gl083952.

de la Beaujardière, O., Wickwar, V.B., Caudal, G., Holt, J.M., Craven, J.D., Frank, L.A., Brace, L.H., Evans, D.S., Winningham, J.D., Heelis, R.A., 1986. Correction to "Universal time dependence of nighttime F region densities at high latitudes". J. Geophys. Res. Space Phys. 91 (A1), 381. https://doi.org/10.1029/ja091ia01p00381.

de Wit, T.D., Krasnosel'skikh, V.V., Dunlop, M., Lühr, H., 1999. Identifying nonlinear wave interactions in plasmas using two-point measurements: a case study of short large amplitude magnetic structures (SLAMS). J. Geophys. Res. Space Phys. 104 (A8), 17079–17090. https://doi.org/10.1029/1999JA900134.

Deng, Y., Ridley, A.J., 2006. Role of vertical ion convection in the high-latitude ionospheric plasma distribution. J. Geophys. Res. Atmos. 111 (A9), A09314. https://doi.org/10.1029/2006JA011637.

Deshpande, K.B., Zettergren, M.D., 2019. Satellite-beacon ionospheric-scintillation global model of the upper atmosphere (SIGMA) III: scintillation simulation using a physics-based plasma model. Geophys. Res. Lett. 46 (9), 4564–4572.

Deshpande, K.B., Bust, G.S., Clauer, C.R., Rino, C.L., Carrano, C.S., 2014. Satellite-beacon ionospheric-scintillation global model of the upper atmosphere (SIGMA) I: high latitude sensitivity study of the model parameters. J. Geophys. Res. Space Phys. 119, 4026–4043. https://doi.org/10.1002/2013JA019699.

Deshpande, K.B., Bust, G.S., Clauer, C.R., Scales, W.A., Frissell, N.A., Ruohoniemi, J.M., Spogli, L., Mitchell, C., Weatherwax, A.T., 2016. Satellite-beacon ionospheric-scintillation global model of the upper atmosphere (SIGMA) II: inverse modeling with high-latitude observations to deduce irregularity physics. J. Geophys. Res. Space Phys. 121 (9). https://doi.org/10.1002/2016JA022943.
Doe, R.A., Mendillo, M., Vickrey, J.F., Zanetti, L.J., Eastes, R.W., 1993. Observations of nightside auroral cavities. J. Geophys. Res. Space Phys. 98 (A1), 293–310. https://doi.org/10.1029/92JA02004.
Dungey, J.W., 1961. Interplanetary magnetic field and the auroral zones. Phys. Rev. Lett. 6 (2), 47.
Dungey, J.W., 1963. The structure of the exosphere or adventures in velocity space. In: deWitt, C., Hieblot, J., Lebeau, L. (Eds.), Geophysics: the earth's environment. Gordon and Breach, p. 503.
Dyson, P.L., McClure, J.P., Hanson, W.B., 1974. In situ measurements of the spectral characteristics of F region ionospheric irregularities. J. Geophys. Res. Space Phys. 79 (10), 1497–1502. https://doi.org/10.1029/JA079i010p01497.
Earle, G.D., Kelley, M.C., 1993. Spectral evidence for stirring scales and two-dimensional turbulence in the auroral ionosphere. J. Geophys. Res. Space Phys. 98 (A7), 11543–11548. https://doi.org/10.1029/93JA00632.
Erickson, P.J., Beroz, F., Miskin, M.Z., 2011. Statistical characterization of the American sector subauroral polarization stream using incoherent scatter radar. J. Geophys. Res. Space Phys. 116, A00J21. https://doi.org/10.1029/2010JA015738.
Fæhn Follestad, A., Herlingshaw, K., Ghadjari, H., Knudsen, D.J., McWilliams, K.A., Moen, J.I., Spicher, A., Wu, J., Oksavik, K., 2020. Dayside field-aligned current impacts on ionospheric irregularities. Geophys. Res. Lett. 47 (11). https://doi.org/10.1029/2019GL086722. e2019GL086722.
Farley Jr, D.T., 1959. A theory of electrostatic fields in a horizontally stratified ionosphere subject to a vertical magnetic field. J. Geophys. Res. Space Phys. 64 (9), 1225–1233. https://doi.org/10.1029/JZ064i009p01225.
Fejer, B.G., Kelley, M.C., 1980. Ionospheric irregularities. Rev. Geophys. 18 (2), 401–454. https://doi.org/10.1029/RG018i002p00401.
Fejer, B.G., Spiro, R.W., Wolf, R.A., Foster, J.C., 1990. Latitudinal variation of perturbation electric fields during magnetically disturbed periods-1986 sundial observations and model results. Ann. Geophys. 8, 441–454.
Folk, M., Heber, G., Koziol, Q., Pourmal, E., Robinson, D., 2011. An overview of the HDF5 technology suite and its applications. In: AD '11. Proceedings of the EDBT/ICDT 2011 Workshop on Array Databases, Association for Computing Machinery, New York, NY, USA, pp. 36–47, https://doi.org/10.1145/1966895.1966900.
Forte, B., Coleman, C., Skone, S., Häggström, I., Mitchell, C., Da Dalt, F., Panicciari, T., Kinrade, J., Bust, G., 2017. Identification of scintillation signatures on GPS signals originating from plasma structures detected with EISCAT incoherent scatter radar along the same line of sight. J. Geophys. Res. Space Phys. 122 (1), 916–931. https://doi.org/10.1002/2016JA023271.
Foster, J.C., 1993. Storm time plasma transport at middle and high latitudes. J. Geophys. Res. Space Phys. 98 (A2), 1675–1690. https://doi.org/10.1029/92JA02032.
Foster, J.C., Rideout, W., 2007. Storm enhanced density: magnetic conjugacy effects. Ann. Geophys. 25 (8), 1791–1799. https://doi.org/10.5194/angeo-25-1791-2007.
Foster, J.C., Vo, H.B., 2002. Average characteristics and activity dependence of the subauroral polarization stream. J. Geophys. Res. Space Phys. 107 (A12), 1475. https://doi.org/10.1029/2002JA009409.
Foster, J.C., Coster, A.J., Erickson, P.J., Holt, J.M., Lind, F.D., Rideout, W., McCready, M., van Eyken, A., Barnes, R.J., Greenwald, R.A., Rich, F.J., 2005. Multiradar observations of the polar tongue of ionization. J. Geophys. Res. Space Phys. 110 (A9), A09S31. https://doi.org/10.1029/2004JA010928.
Fremouw, E.J., Rino, C.L., 1973. An empirical model for average F-layer scintillation at VHF/UHF. Radio Sci. 8 (3), 213–222. https://doi.org/10.1029/RS008i003p00213.
Fremouw, E.J., Leadabrand, R.L., Livingston, R.C., Cousins, M.D., Rino, C.L., Fair, B.C., Long, R.A., 1978. Early results from the DNA wideband satellite experiment–complex-signal scintillation. Radio Sci. 13 (1), 167–187. https://doi.org/10.1029/RS013i001p00167.
Fremouw, E.J., Secan, J.A., Lansinger, J.M., 1985. Spectral behavior of phase scintillation in the nighttime auroral region. Radio Sci. 20 (4), 923–933. https://doi.org/10.1029/RS020i004p00923.

Frissell, N.A., Katz, J.D., Gunning, S.W., Vega, J.S., Gerrard, A.J., Earle, G.D., Moses, M.L., West, M.L., Huba, J.D., Erickson, P.J., Miller, E.S., Gerzoff, R.B., Liles, W., Silver, H.W., 2018. Modeling amateur radio soundings of the ionospheric response to the 2017 great American eclipse. Geophys. Res. Lett. 45 (10), 4665–4674. https://doi.org/10.1029/2018GL077324.

Fuller-Rowell, T.J., Codrescu, M.V., Moffett, R.J., Quegan, S., 1994. Response of the thermosphere and ionosphere to geomagnetic storms. J. Geophys. Res. Space Phys. 99 (A3), 3893–3914.

Fuller-Rowell, T.J., Millward, G.H., Richmond, A.D., Codrescu, M.V., 2002. Storm-time changes in the upper atmosphere at low latitudes. J. Atmos. Sol. Terr. Phys. 64 (12–14), 1383–1391.

Galperin, Y.I., Zosimova, A.G., 1974. Plasma convection in polar ionosphere. Ann. Geophys. 30 (1), 1–7.

Ganguli, G., Keskinen, M.J., Romero, H., Heelis, R., Moore, T., Pollock, C., 1994. Coupling of microprocesses and macroprocesses due to velocity shear: an application to the low-altitude ionosphere. J. Geophys. Res. Space Phys. 99 (A5), 8873–8889.

Gardner, L.C., Schunk, R.W., Scherliess, L., Eccles, V., Basu, S., Valladeres, C., 2018. Modeling the mid-latitude ionosphere storm-enhanced density distribution with a data assimilation model. Space Weather 16 (10), 1539–1548. https://doi.org/10.1029/2018SW001882.

Gillies, R.G., Hussey, G.C., Sofko, G.J., McWilliams, K.A., 2012. A statistical analysis of SuperDARN scattering volume electron densities and velocity corrections using a radar frequency shifting technique. J. Geophys. Res. Space Phys. 117 (A8), A08320. https://doi.org/10.1029/2012JA017866.

Gillies, R.G., van Eyken, A., Spanswick, E., Nicolls, M., Kelly, J., Greffen, M., Knudsen, D., Connors, M., Schutzer, M., Valentic, T., Malone, M., Buonocore, J., St.-Maurice, J.P., Donovan, E., 2016. First observations from the RISR-C incoherent scatter radar. Radio Sci. 51 (10), 1645–1659. https://doi.org/10.1002/2016RS006062.

Gondarenko, N.A., Guzdar, P.N., 1999. Gradient drift instability in high latitude plasma patches: ion inertial effects. Geophys. Res. Lett. 26 (22), 3345–3348.

Gondarenko, N.A., Guzdar, P.N., 2004a. Density and electric field fluctuations associated with the gradient drift instability in the high-latitude ionosphere. Geophys. Res. Lett. 31 (11), L11802.

Gondarenko, N.A., Guzdar, P.N., 2004b. Plasma patch structuring by the nonlinear evolution of the gradient drift instability in the high-latitude ionosphere. J. Geophys. Res. Space Phys. 109 (A9), A09301.

Gondarenko, N.A., Guzdar, P.N., 2006. Nonlinear three-dimensional simulations of mesoscale structuring by multiple drives in high-latitude plasma patches. J. Geophys. Res. Space Phys. 111 (A8), A08302.

Gondarenko, N.A., Guzdar, P.N., Sojka, J.J., David, M., 2003. Structuring of high latitude plasma patches with variable drive. Geophys. Res. Lett. 30 (4).

Goodwin, L., St.-Maurice, J.P., Richards, P., Nicolls, M., Hairston, M., 2014. F region dusk ion temperature spikes at the equatorward edge of the high-latitude convection pattern. Geophys. Res. Lett. 41 (2), 300–307. https://doi.org/10.1002/2013GL058442.

Goodwin, L.V., Iserhienrhien, B., Miles, D.M., Patra, S., van der Meeren, C., Buchert, S.C., Burchill, J., Clausen, L.B.N., Knudsen, D.J., McWilliams, K.A., Moen, J., 2015. Swarm in situ observations of F region polar cap patches created by cusp precipitation. Geophys. Res. Lett. 42 (4), 996–1003. https://doi.org/10.1002/2014GL062610.

Goodwin, L.V., Nishimura, Y., Zou, Y., Shiokawa, K., Jayachandran, P.T., 2019. Mesoscale convection structures associated with airglow patches characterized using cluster-imager conjunctions. J. Geophys. Res. Space Phys. 124 (9), 7513–7532. https://doi.org/10.1029/2019ja026611.

Goodwin, L.V., Nishimura, Y., Coster, A.J., Zhang, S., Nishitani, N., Ruohoniemi, J.M., Anderson, B.J., Zhang, Q.-H., 2020. Dayside polar cap density enhancements formed during substorms. J. Geophys. Res. Space Phys. 125. https://doi.org/10.1029/2020JA028101. e2020JA028101.

Greenwald, R.A., Baker, K.B., Dudeney, J.R., Pinnock, M., Jones, T.B., Thomas, E.C., Villain, J.-P., Cerisier, J.-C., Senior, C., Hanuise, C., et al., 1995. DARN/SuperDARN. Space Sci. Rev. 71 (1–4), 761–796. https://doi.org/10.1007/BF00751350.

Gropp, W., Thakur, R., Lusk, E., 1999. Using MPI-2: Advanced Features of the Message Passing Interface, second ed. MIT Press, Cambridge, MA.

Guzdar, P.N., Gondarenko, N.A., Chaturvedi, P.K., Basu, S., 1998. Three-dimensional nonlinear simulations of the gradient drift instability in the high-latitude ionosphere. Radio Sci. 33 (6), 1901–1913.

Hasegawa, A., 1975. Macroinstabilities—Instabilities Due to Coordinate Space Nonequilibrium. Springer Berlin Heidelberg, Berlin, Heidelberg, pp. 110–144, https://doi.org/10.1007/978-3-642-65980-5_3.
He, M., Liu, L., Wan, W., Zhao, B., 2011. A study on the nighttime midlatitude ionospheric trough. J. Geophys. Res. Space Phys. 116 (A5), A05315. https://doi.org/10.1029/2010JA016252.
He, F., Zhang, X., Chen, B., 2014. Solar cycle, seasonal, and diurnal variations of subauroral ion drifts: statistical results. J. Geophys. Res. Space Phys. 119 (6), 5076–5086. https://doi.org/10.1002/2014JA019807.
Heelis, R.A., Sojka, J.J., David, M., Schunk, R.W., 2009. Storm time density enhancements in the middle-latitude dayside ionosphere. J. Geophys. Res. Space Phys. 114 (A3), A03315. https://doi.org/10.1029/2008JA013690.
Heppner, J.P., Liebrecht, M.C., Maynard, N.C., Pfaff, R.F., 1993. High-latitude distributions of plasma waves and spatial irregularities from DE 2 alternating current electric field observations. J. Geophys. Res. Space Phys. 98 (A2), 1629–1652. https://doi.org/10.1029/92JA01836.
Hill, G.E., 1963. Sudden enhancements of F-layer ionization in polar regions. J. Atmos. Sci. 20 (6), 492–497. https://doi.org/10.1175/1520-0469(1963)020¡0492:SEOLII¿2.0.CO;2.
Hirsch, M., 2020. gemini3d/pygemini., https://doi.org/10.5281/zenodo.3910039.
Hirsch, M., Zettergren, M.D., 2020. gemini3d/mat_gemini., https://doi.org/10.5281/zenodo.3987705.
Holt, J.M., Rhoda, D.A., Tetenbaum, D., van Eyken, A.P., 1992. Optimal analysis of incoherent scatter radar data. Radio Sci. 27 (3), 435–447. https://doi.org/10.1029/91RS02922.
Horvath, I., Essex, E.A., 2003. The southern-hemisphere mid-latitude day-time and night-time trough at low-sunspot numbers. J. Atmos. Sol. Terr. Phys. 65 (8), 917–940. https://doi.org/10.1016/S1364-6826(03)00113-5.
Horvath, I., Lovell, B.C., 2011. Storm-enhanced plasma density (SED) features, auroral and polar plasma enhancements, and rising topside bubbles of the 31 March 2001 superstorm. J. Geophys. Res. Space Phys. 116 (A4), A04307. https://doi.org/10.1029/2010JA015514.
Hosokawa, K., Shiokawa, K., Otsuka, Y., Nakajima, A., Ogawa, T., Kelly, J.D., 2006. Estimating drift velocity of polar cap patches with All-Sky Airglow Imager at Resolute Bay, Canada. Geophys. Res. Lett. 33 (15), L15111. https://doi.org/10.1029/2006GL026916.
Hosokawa, K., Shiokawa, K., Otsuka, Y., Ogawa, T., St.-Maurice, J.-P., Sofko, G.J., Andre, D.A., 2009a. Relationship between polar cap patches and field-aligned irregularities as observed with an All-Sky Airglow Imager at Resolute Bay and the PolarDARN radar at Rankin Inlet. J. Geophys. Res. Space Phys. 114 (A3), A03306. https://doi.org/10.1029/2008JA013707.
Hosokawa, K., Tsugawa, T., Shiokawa, K., Otsuka, Y., Ogawa, T., Hairston, M.R., 2009b. Unusually elongated, bright airglow plume in the polar cap F region: is it a tongue of ionization? Geophys. Res. Lett. 36 (7), L07103. https://doi.org/10.1029/2009GL037512.
Hosokawa, K., Tsugawa, T., Shiokawa, K., Otsuka, Y., Nishitani, N., Ogawa, T., Hairston, M.R., 2010. Dynamic temporal evolution of polar cap tongue of ionization during magnetic storm. J. Geophys. Res. Space Phys. 115 (A12), A12333.
Hosokawa, K., Taguchi, S., Ogawa, Y., Sakai, J., 2013. Two-dimensional direct imaging of structuring of polar cap patches. J. Geophys. Res. Space Phys. 118, 6536–6543. https://doi.org/10.1002/jgra.50577.
Hosokawa, K., Taguchi, S., Ogawa, Y., 2016. Edge of polar cap patches. J. Geophys. Res. Space Phys. 121, 3410–3420. https://doi.org/10.1002/2015JA021960.
Huang, C., Sazykin, S., Spiro, R., Goldstein, J., Crowley, G., Ruohoniemi, J.M., 2006. Storm-time penetration electric fields and their effects. Eos Trans. AGU 87 (13), 131. 131.
Huba, J.D., Mitchell, H.G., Keskinen, M.J., Fedder, J.A., Satyanarayana, P., Zalesak, S.T., 1988. Simulations of plasma structure evolution in the high-latitude ionosphere. Radio Sci. 23 (4), 503–512.
Huba, J.D., Joyce, G., Fedder, J.A., 2000. Sami2 is another model of the ionosphere (SAMI2): a new low-latitude ionosphere model. J. Geophys. Res. Space Phys. 105 (A10), 23035–23053. https://doi.org/10.1029/2000JA000035.
Huba, J.D., Joyce, G., Krall, J., 2008. Three-dimensional equatorial spread F modeling. Geophys. Res. Lett. 35 (10). https://doi.org/10.1029/2008GL033509.
Huba, J.D., Sazykin, S., Coster, A., 2017. SAMI3-RCM simulation of the 17 March 2015 geomagnetic storm. J. Geophys. Res. Space Phys. 122 (1), 1246–1257. https://doi.org/10.1002/2016JA023341.

Hysell, D.L., Rodrigues, F.S., Chau, J.L., Huba, J.D., 2008. Full profile incoherent scatter analysis at Jicamarca. Ann. Geophys. 26, 59–75. https://doi.org/10.5194/angeo-26-59-2008.
Hysell, D.L., Milla, M.A., Rodrigues, F.S., Varney, R.H., Huba, J.D., 2015. Topside equatorial ionospheric density, temperature, and composition under equinox, low solar flux conditions. J. Geophys. Res. Space Phys. 120, 3899–3912. https://doi.org/10.1002/2015JA021168.
Ishida, T., Ogawa, Y., Kadokura, A., Hiraki, Y., Häggström, I., 2014. Seasonal variation and solar activity dependence of the quiet-time ionospheric trough. J. Geophys. Res. Space Phys. 119, 6774–6783. https://doi.org/10.1002/2014JA019996.
Ivarsen, M.F., Jin, Y., Spicher, A., Clausen, L.B.N., 2019. Direct evidence for the dissipation of small-scale ionospheric plasma structures by a conductive E region. J. Geophys. Res. Space Phys. 124 (4), 2935–2942. https://doi.org/10.1029/2019JA026500.
Jackson, J.D., 2007. Classical Electrodynamics. John Wiley & Sons.
Jaggi, R.K., Wolf, R.A., 1973. Self-consistent calculation of the motion of a sheet of ions in the magnetosphere. J. Geophys. Res. Space Phys. 78 (16), 2852–2866.
Jayachandran, P.T., Hamza, A.M., Hosokawa, K., Mezaoui, H., Shiokawa, K., 2017. GPS amplitude and phase scintillation associated with polar cap auroral forms. J. Atmos. Sol. Terr. Phys. 164, 185–191. https://doi.org/10.1016/j.jastp.2017.08.030.
Jin, Y., Moen, J.I., Miloch, W.J., 2014. GPS scintillation effects associated with polar cap patches and substorm auroral activity: direct comparison. J. Space Weather Space Clim. 4, A23. https://doi.org/10.1051/swsc/2014019.
Jin, Y., Moen, J.I., Miloch, W.J., 2015. On the collocation of the cusp aurora and the GPS phase scintillation: a statistical study. J. Geophys. Res. Space Phys. 120 (10), 9176–9191. https://doi.org/10.1002/2015JA021449.
Jin, Y., Moen, J.I., Miloch, W.J., Clausen, L.B.N., Oksavik, K., 2016. Statistical study of the GNSS phase scintillation associated with two types of auroral blobs. J. Geophys. Res. Space Phys. 121 (5), 4679–4697. https://doi.org/10.1002/2016JA022613.
Jin, Y., Moen, J.I., Oksavik, K., Spicher, A., Clausen, L.B.N., Miloch, W.J., 2017. GPS scintillations associated with cusp dynamics and polar cap patches. J. Space Weather Space Clim. 7, A23. https://doi.org/10.1051/swsc/2017022.
Jin, Y., Miloch, W.J., Moen, J.I., Clausen, L.B.N., 2018. Solar cycle and seasonal variations of the GPS phase scintillation at high latitudes. J. Space Weather Space Clim. 8, A48. https://doi.org/10.1051/swsc/2018034.
Jin, Y., Moen, J.I., Spicher, A., Oksavik, K., Miloch, W.J., Clausen, L.B.N., PoŻoga, M., Saito, Y., 2019. Simultaneous rocket and scintillation observations of plasma irregularities associated with a reversed flow event in the cusp ionosphere. J. Geophys. Res. Space Phys. 124 (8), 7098–7111. https://doi.org/10.1029/2019JA026942.
Karlsson, T., Andersson, L., Gillies, D.M., Lynch, K., Marghitu, O., Partamies, N., Sivadas, N., Wu, J., 2020. Quiet, discrete auroral arcs–observations. Space Sci. Rev. 216 (1), 16. https://doi.org/10.1007/s11214-020-0641-7.
Karpachev, A.T., 2003. The dependence of the main ionospheric trough shape on longitude, altitude, season, local time, and solar and magnetic activity. Geomagn. Aeron. 43 (2), 239–251.
Karpachev, A.T., Klimenko, M.V., Klimenko, V.V., 2019. Longitudinal variations of the ionospheric trough position. Adv. Space Res. 63 (2), 950–966. https://doi.org/10.1016/j.asr.2018.09.038.
Kelley, M.C., 2009. The Earth's Ionosphere: Plasma Physics and Electrodynamics. Elsevier Science.
Kelley, M.C., Carlson, C.W., 1977. Observations of intense velocity shear and associated electrostatic waves near an auroral arc. J. Geophys. Res. Space Phys. 82 (16), 2343–2348. https://doi.org/10.1029/JA082i016p02343.
Kelley, M.C., Fejer, B.G., Gonzales, C.A., 1979. An explanation for anomalous equatorial ionospheric electric fields associated with a northward turning of the interplanetary magnetic field. Geophys. Res. Lett. 6 (4), 301–304.
Kelley, M.C., Baker, K.D., Ulwick, J.C., Rino, C.L., Baron, M.J., 1980. Simultaneous rocket probe, scintillation, and incoherent scatter radar observations of irregularities in the auroral zone ionosphere. Radio Sci. 15 (3), 491–505. https://doi.org/10.1029/RS015i003p00491.

Kelley, M.C., Vickrey, J.F., Carlson, C., Torbert, R., 1982. On the origin and spatial extent of high-latitude F region irregularities. J. Geophys. Res. Space Phys. 87 (A6), 4469–4475. https://doi.org/10.1029/JA087iA06p04469.

Kelley, M.C., Vlasov, M.N., Foster, J.C., Coster, A.J., 2004. A quantitative explanation for the phenomenon known as storm-enhanced density. Geophys. Res. Lett. 31 (19), L19809. https://doi.org/10.1029/2004GL020875.

Kersley, L., Pryse, S.E., Wheadon, N.S., 1988. Small scale ionospheric irregularities near regions of soft particle precipitation: scintillation and EISCAT observations. J. Atmos. Terr. Phys. 50 (12), 1047–1055. https://doi.org/10.1016/0021-9169(88)90094-3.

Keskinen, M.J., 2006. GPS scintillation channel model for the disturbed low-latitude ionosphere. Radio Sci. 41 (04), 1–7.

Keskinen, M.J., Ossakow, S.L., 1981. On the spatial power spectrum of the $\mathbf{E} \times \mathbf{B}$ gradient drift instability in ionospheric plasma clouds. J. Geophys. Res. Space Phys. 86, 6947–6950. https://doi.org/10.1029/JA086iA08p06947.

Keskinen, M.J., Ossakow, S.L., 1982. Nonlinear evolution of plasma enhancements in the auroral ionosphere, 1, Long wavelength irregularities. J. Geophys. Res. Space Phys. 87 (A1), 144–150. https://doi.org/10.1029/JA087iA01p00144.

Keskinen, M.J., Ossakow, S.L., 1983. Theories of high-latitude ionospheric irregularities: a review. Radio Sci. 18 (06), 1077–1091.

Keskinen, M.J., Mitchell, H.G., Fedder, J.A., Satyanarayana, P., Zalesak, S.T., Huba, J.D., 1988. Nonlinear evolution of the Kelvin-Helmholtz instability in the high-latitude ionosphere. J. Geophys. Res. Space Phys. 93 (A1), 137–152.

Kinrade, J., Mitchell, C.N., Yin, P., Smith, N., Jarvis, M.J., Maxfield, D.J., Rose, M.C., Bust, G.S., Weatherwax, A.T., 2012. Ionospheric scintillation over Antarctica during the storm of 5–6 April 2010. J. Geophys. Res. Space Phys. 117 (A5). https://doi.org/10.1029/2011JA017073.

Kinrade, J., Mitchell, C.N., Smith, N.D., Ebihara, Y., Weatherwax, A.T., Bust, G.S., 2013. GPS phase scintillation associated with optical auroral emissions: first statistical results from the geographic south pole. J. Geophys. Res. Space Phys. 118 (5), 2490–2502. https://doi.org/10.1002/jgra.50214.

Kintner, P.M., Seyler, C.E., 1985. The status of observations and theory of high latitude ionospheric and magnetospheric plasma turbulence. Space Sci. Rev. 41 (1-2), 91–129.

Kivanç, Ö., Heelis, R.A., 1997. Structures in ionospheric number density and velocity associated with polar cap ionization patches. J. Geophys. Res. Space Phys. 102 (A1), 307–318. https://doi.org/10.1029/96JA03141.

Kivanc, O., Heelis, R.A., 1997. Structures in ionospheric number density and velocity associated with polar cap ionization patches. J. Geophys. Res. Space Phys. 102 (A1), 307–318. https://doi.org/10.1029/96JA03141.

Kivanç, Ö., Heelis, R.A., 1998. Spatial distribution of ionospheric plasma and field structures in the high-latitude F region. J. Geophys. Res. Space Phys. 103 (A4), 6955–6968. https://doi.org/10.1029/97JA03237.

Knepp, D.L., 1983. Multiple phase-screen calculation of the temporal behavior of stochastic waves. IEEE Proc. 71, 722–737.

Knudsen, W.C., 1974. Magnetospheric convection and the high-latitude F_2 ionosphere. J. Geophys. Res. Space Phys. 79 (7), 1046–1055. https://doi.org/10.1029/JA079i007p01046.

Koustov, A., Hosokawa, K., Nishitani, N., Ogawa, T., Shiokawa, K., 2008. Rankin Inlet PolarDARN radar observations of duskward moving sun-aligned optical forms. Ann. Geophys. 26 (9), 2711–2723. https://doi.org/10.5194/angeo-26-2711-2008.

Kunduri, B.S.R., Baker, J.B.H., Ruohoniemi, J.M., Thomas, E.G., Shepherd, S.G., Sterne, K.T., 2017. Statistical characterization of the large-scale structure of the subauroral polarization stream. J. Geophys. Res. Space Phys. 122 (6), 6035–6048. https://doi.org/10.1002/2017JA024131.

Labelle, J., Sica, R.J., Kletzing, C., Earle, G.D., Kelley, M.C., Lummerzheim, D., Torbert, R.B., Baker, K. D., Berg, G., 1989. Ionization from soft electron precipitation in the auroral F region. J. Geophys. Res. Space Phys. 94 (A4), 3791–3798. https://doi.org/10.1029/JA094iA04p03791.

Lagoutte, D., Cerisier, J.C., Plagnaud, J.L., Villain, J.P., Forget, B., 1992. High-latitude ionospheric electrostatic turbulence studied by means of the wavelet transform. J. Atmos. Terr. Phys. 54 (10), 1283–1293. https://doi.org/10.1016/0021-9169(92)90037-L.

Lamarche, L.J., Makarevich, R.A., 2017. Radar observations of density gradients, electric fields, and plasma irregularities near polar cap patches in the context of the gradient-drift instability. J. Geophys. Res. Space Phys. 122, 3721–3736. https://doi.org/10.1002/2016JA023702.

Lamarche, L., Deshpande, Bharat, K., Zettergren, M.D., Varney, R., 2020. Satellite-beacon ionospheric-scintillation global model of the upper atmosphere (SIGMA) III: Scintillation simulation using a physics-based plasma model. J. Geophys. Res. 46 (9), 4564–4572.

Lee, I.T., Wang, W., Liu, J.Y., Chen, C.Y., Lin, C.H., 2011. The ionospheric midlatitude trough observed by FORMOSAT-3/COSMIC during solar minimum. J. Geophys. Res. Space Phys. 116 (A6), A06311. https://doi.org/10.1029/2010JA015544.

Liang, J., Donovan, E., Reimer, A., Hampton, D., Zou, S., Varney, R., 2018. Ionospheric electron heating associated with pulsating auroras: joint optical and PFISR observations. J. Geophys. Res. Space Phys. 123 (5), 4430–4456.

Liu, J., Wang, W., Burns, A., Solomon, S.C., Zhang, S., Zhang, Y., Huang, C., 2016. Relative importance of horizontal and vertical transports to the formation of ionospheric storm-enhanced density and polar tongue of ionization. J. Geophys. Res. Space Phys. 121 (8), 8121–8133. https://doi.org/10.1002/2016JA022882.

Lockwood, M., Carlson, H.C., 1992. Production of polar cap electron density patches by transient magnetopause reconnection. Geophys. Res. Lett. 19 (17), 1731–1734.

Lockwood, M., Carlson Jr, H.C., 1992. Production of polar cap electron density patches by transient magnetopause reconnection. Geophys. Res. Lett. 19 (17), 1731–1734. https://doi.org/10.1029/92GL01993.

Lorentzen, D.A., Shumilov, N., Moen, J., 2004. Drifting airglow patches in relation to tail reconnection. Geophys. Res. Lett. 31 (2), L02806. https://doi.org/10.1029/2003GL017785.

Lotko, W., 2004. Inductive magnetosphere-ionosphere coupling. J. Atmos. Sol. Terr. Phys. 66 (15–16), 1443–1456.

Lu, G., Goncharenko, L., Nicolls, M.J., Maute, A., Coster, A., Paxton, L.J., 2012. Ionospheric and thermospheric variations associated with prompt penetration electric fields. J. Geophys. Res. Space Phys. 117 (A8), A08312.

Lu, G., Zakharenkova, I., Cherniak, I., Dang, T., 2020. Large-scale ionospheric disturbances during the 17 March 2015 storm: a model-data comparative study. J. Geophys. Res. Space Phys., e2019JA027726.

Lyons, L.R., Nishimura, Y., Zou, Y., 2016. Unsolved problems: mesoscale polar cap flow channels' structure, propagation, and effects on space weather disturbances. J. Geophys. Res. Space Phys. 121 (4), 3347–3352. https://doi.org/10.1002/2016JA022437.

Ma, Y.Z., Zhang, Q.H., Xing, Z.Y., Heelis, R.A., Oksavik, K., Wang, Y., 2018. The ion/electron temperature characteristics of polar cap classical and hot patches and their influence on ion upflow. Geophys. Res. Lett. 45 (16), 8072–8080. https://doi.org/10.1029/2018GL079099.

MacDougall, J.W., Jayachandran, P.T., 2007. Polar patches: auroral zone precipitation effects. J. Geophys. Res. Space Phys. 112 (A5), 1–16. https://doi.org/10.1029/2006JA011930.

Makarevich, R.A., 2014. Symmetry considerations in the two-fluid theory of the gradient-drift instability in the lower ionosphere. J. Geophys. Res. Space Phys. 119. https://doi.org/10.1002/2014JA020292.

Makarevich, R.A., 2016a. Towards an integrated view of ionospheric plasma instabilities: 2. Three inertial modes of a cubic dispersion relation. J. Geophys. Res. Space Phys. 121, 6855–6869. https://doi.org/10.1002/2016JA022864.

Makarevich, R.A., 2016b. Towards an integrated view of ionospheric plasma instabilities: altitudinal transitions and strong gradient case. J. Geophys. Res. Space Phys. 121, 3634–3647. https://doi.org/10.1002/2016JA022515.

Makarevich, R.A., 2019. Toward an integrated view of ionospheric plasma instabilities: 3. Explicit growth rate and oscillation frequency for arbitrary altitude. J. Geophys. Res. Space Phys. 124 (7), 6138–6155. https://doi.org/10.1029/2019JA026584.

Makarevich, R.A., Lamarche, L.J., Nicolls, M.J., 2015. Resolute Bay incoherent scatter radar observations of plasma structures in the vicinity of polar holes. J. Geophys. Res. Space Phys. 120 (9), 7970–7986. https://doi.org/10.1002/2015JA021443.
Mallis, M., Essex, E.A., 1993. Diurnal and seasonal variability of the southern-hemisphere main ionospheric trough from differential-phase measurements. J. Atmos. Terr. Phys. 55 (7), 1021–1037. https://doi.org/10.1016/0021-9169(93)90095-G.
Mannucci, A.J., Wilson, B.D., Yuan, D.N., Ho, C.H., Lindqwister, U.J., Runge, T.F., 1998. A global mapping technique for GPS-derived ionospheric total electron content measurements. Radio Sci. 33 (3), 565–582. https://doi.org/10.1029/97RS02707.
McCrea, I., Aikio, A., Alfonsi, L., Belova, E., Buchert, S., Clilverd, M., Engler, N., Gustavsson, B., Heinselman, C., Kero, J., Kosch, M., Lamy, H., Leyser, T., Ogawa, Y., Oksavik, K., Pellinen-Wannberg, A., Pitout, F., Rapp, M., Stanislawska, I., Vierinen, J., 2015. The science case for the EISCAT_3D radar. Prog. Earth Planet. Sci. 2 (1), 21. https://doi.org/10.1186/s40645-015-0051-8.
Meek, J.H., 1949. Sporadic ionization at high latitudes. J. Geophys. Res. Space Phys. 54 (4), 339–345. https://doi.org/10.1029/JZ054i004p00339.
Mendillo, M., 2006. Storms in the ionosphere: patterns and processes for total electron content. Rev. Geophys. 44 (4), RG4001. https://doi.org/10.1029/2005RG000193.
Meziane, K., Kashcheyev, A., Patra, S., Jayachandran, P.T., Hamza, A.M., 2020. Solar cycle variations of GPS amplitude scintillation for the polar region. Space Weather 18 (8). https://doi.org/10.1029/2019SW002434. e2019SW002434.
Michael, C.M., Yeoman, T.K., Wright, D.M., Milan, S.E., James, M.K., 2020. A ray tracing simulation of HF ionospheric radar performance at African equatorial latitudes. Radio Sci. 55 (2). https://doi.org/10.1029/2019RS006936. e2019RS006936.
Milan, S.E., Yeoman, T.K., Lester, M., Thomas, E.C., Jones, T.B., 1997. Initial backscatter occurrence statistics from the CUTLASS HF radars. Ann. Geophys. 15 (6), 703–718. https://doi.org/10.1007/s00585-997-0703-0.
Milan, S.E., Lester, M., Yeoman, T.K., 2002. HF radar polar patch formation revisited: summer and winter variations in dayside plasma structuring. Ann. Geophys. 20 (4), 487–499. https://doi.org/10.5194/angeo-20-487-2002.
Millward, G.H., Moffett, R.J., Balmforth, H.F., Rodger, A.S., 1999. Modeling the ionospheric effects of ion and electron precipitation in the cusp. J. Geophys. Res. Space Phys. 104 (A11), 24603–24612. https://doi.org/10.1029/1999JA900249.
Mitchell, C.N., Alfonsi, L., De Franceschi, G., Lester, M., Romano, V., Wernik, A.W., 2005. GPS TEC and scintillation measurements from the polar ionosphere during the October 2003 storm. Geophys. Res. Lett. 32 (12). https://doi.org/10.1029/2004GL021644.
Mitchell Jr, H.G., Fedder, J.A., Keskinen, M.J., Zalesak, S.T., 1985. A simulation of high latitude F-layer instabilities in the presence of magnetosphere-ionosphere coupling. Geophys. Res. Lett. 12 (5), 283–286.
Moen, J., Walker, I.K., Kersley, L., Milan, S.E., 2002. On the generation of cusp HF backscatter irregularities. J. Geophys. Res. Space Phys. 107 (A4). https://doi.org/10.1029/2001JA000111. SIA 3-1–SIA 3-5.
Moen, J., Rinne, Y., Carlson, H.C., Oksavik, K., Fujii, R., Opgenoorth, H., 2008. On the relationship between thin Birkeland current arcs and reversed flow channels in the winter cusp/cleft ionosphere. J. Geophys. Res. Space Phys. 113 (A9), A09220. https://doi.org/10.1029/2008JA013061.
Moen, J., Oksavik, K., Abe, T., Lester, M., Saito, Y., Bekkeng, T.A., Jacobsen, K.S., 2012. First in-situ measurements of HF radar echoing targets. Geophys. Res. Lett. 39 (7), L07104. https://doi.org/10.1029/2012GL051407.
Moen, J., Oksavik, K., Alfonsi, L., Daabakk, Y., Romano, V., Spogli, L., 2013. Space weather challenges of the polar cap ionosphere. J. Space Weather Space Clim. 3, A02. https://doi.org/10.1051/swsc/2013025.
Moffett, R.J., Quegan, S., 1983. The mid-latitude trough in the electron concentration of the ionospheric F-layer—a review of observations and modelling. J. Atmos. Terr. Phys. 45, 315–343. https://doi.org/10.1016/S0021-9169(83)80038-5.
Moldwin, M.B., Zou, S., Heine, T., 2016. The story of plumes: the development of a new conceptual framework for understanding magnetosphere and ionosphere coupling. Ann. Geophys. 34 (12), 1243–1253. https://doi.org/10.5194/angeo-34-1243-2016.

Mounir, H., Berthelier, A., Cerisier, J.C., Lagoutte, D., Beghin, C., 1991. The small-scale turbulent structure of the high latitude ionosphere—Arcad-Aureol-3 observations. Ann. Geophys. 9, 725–737.
Mrak, S., Semeter, J., Hirsch, M., Starr, G., Hampton, D., Varney, R.H., Reimer, A.S., Swoboda, J., Erickson, P.J., Lind, F., Coster, A.J., Pankratius, V., 2018. Field-aligned GPS scintillation: multisensor data fusion. J. Geophys. Res. Space Phys. 123 (1), 974–992. https://doi.org/10.1002/2017JA024557.
Muldrew, D.B., 1965. F-layer ionization troughs deduced from Alouette data. J. Geophys. Res. Space Phys. 70 (11), 2635–2650. https://doi.org/10.1029/JZ070i011p02635.
Narita, Y., 2012. Multi-spacecraft Measurements. Springer Berlin Heidelberg, Berlin, Heidelberg, pp. 39–65, https://doi.org/10.1007/978-3-642-25667-7_3.
Narita, Y., Glassmeier, K.-H., Motschmann, U., 2010. Wave vector analysis methods using multi-point measurements. Nonlinear Process. Geophys. 17 (5), 383–394. https://doi.org/10.5194/npg-17-383-2010.
Nicolls, M.J., Heinselman, C.J., 2007. Three-dimensional measurements of traveling ionospheric disturbances with the poker flat incoherent scatter radar. Geophys. Res. Lett. 34 (21), L21104.
Nilsson, H., Sergienko, T.I., Ebihara, Y., Yamauchi, M., 2005. Quiet-time mid-latitude trough: influence of convection, field-aligned currents and proton precipitation. Ann. Geophys. 23 (10), 3277–3288. https://doi.org/10.5194/angeo-23-3277-2005.
Nishida, A., 1968. Coherence of geomagnetic DP2 fluctuations with interplanetary magnetic variations. J. Geophys. Res. Space Phys. 73 (5), 1795–1803.
Nishimura, Y., Lyons, L.R., 2016. Localized reconnection in the magnetotail driven by lobe flow channels: global MHD simulation. J. Geophys. Res. Space Phys. 121 (2), 1327–1338. https://doi.org/10.1002/2015JA022128.
Noja, M., Stolle, C., Park, J., Lühr, H., 2013. Long-term analysis of ionospheric polar patches based on CHAMP TEC data. Radio Sci. 48 (3), 289–301. https://doi.org/10.1002/rds.20033.
Oksavik, K., Moen, J., Carlson, H.C., 2004. High-resolution observations of the small-scale flow pattern associated with a poleward moving auroral form in the cusp. Geophys. Res. Lett. 31 (11). https://doi.org/10.1029/2004GL019838.
Oksavik, K., Moen, J., Carlson, H.C., Greenwald, R.A., Milan, S.E., Lester, M., Denig, W.F., Barnes, R.J., 2005. Multi-instrument mapping of the small-scale flow dynamics related to a cusp auroral transient. Ann. Geophys. 23 (7), 2657–2670. https://doi.org/10.5194/angeo-23-2657-2005.
Oksavik, K., Ruohoniemi, J.M., Greenwald, R.A., Baker, J.B.H., Moen, J., Carlson, H.C., Yeoman, T.K., Lester, M., 2006. Observations of isolated polar cap patches by the European Incoherent Scatter (EISCAT) Svalbard and Super Dual Auroral Radar Network (SuperDARN) Finland radars. J. Geophys. Res. Space Phys. 111 (A5). https://doi.org/10.1029/2005JA011400.
Oksavik, K., Moen, J.I., Rekaa, E.H., Carlson, H.C., Lester, M., 2011. Reversed flow events in the cusp ionosphere detected by SuperDARN HF radars. J. Geophys. Res. Space Phys. 116 (A12). https://doi.org/10.1029/2011JA016788.
Oksavik, K., Moen, J., Lester, M., Bekkeng, T.A., Bekkeng, J.K., 2012. In situ measurements of plasma irregularity growth in the cusp ionosphere. J. Geophys. Res. Space Phys. 117 (A11). https://doi.org/10.1029/2012JA017835.
Oksavik, K., van der Meeren, C., Lorentzen, D.A., Baddeley, L.J., Moen, J., 2015. Scintillation and loss of signal lock from poleward moving auroral forms in the cusp ionosphere. J. Geophys. Res. Space Phys. 120 (10), 9161–9175. https://doi.org/10.1002/2015JA021528.
Opgenoorth, H.J., Hägström, I., Williams, P.J.S., Jones, G.O.L., 1990. Regions of strongly enhanced perpendicular electric fields adjacent to auroral arcs. J. Atmos. Terr. Phys. 52 (6-8), 449–458. https://doi.org/10.1016/0021-9169(90)90044-N.
Ossakow, S.L., Chaturvedi, P.K., Workman, J.B., 1978. High-altitude limit of the gradient drift instability. J. Geophys. Res. Space Phys. 83 (A6), 2691–2693.
Paschmann, G., Daly, P.W., 1998. Analysis Methods for Multi-Spacecraft Data. ISSI Scientific Reports Series SR-001, ESA/ISSI. vol. 1. ISSI Scientific Reports Series.
Perkins, F.W., Zabusky, N.J., Doles, J.H., 1973. Deformation and striation of plasma clouds in the ionosphere, I. J. Geophys. Res. Space Phys. 78 (4), 697–709.

Perkins, F.W., Zabusky, N.J., Doles III, J.H., 1973. Deformation and striation of plasma clouds in the ionosphere: 1. J. Geophys. Res. Space Phys. 78 (4), 697–709.
Perry, G.W., St.-Maurice, J.-P., 2018. A polar-cap patch detection algorithm for the advanced modular incoherent scatter radar system. Radio Sci. 53 (10), 1225–1244. https://doi.org/10.1029/2018RS006600.
Perry, G.W., St.-Maurice, J.-P., Hosokawa, K., 2013. The interconnection between cross-polar cap convection and the luminosity of polar cap patches. J. Geophys. Res. Space Phys. 118 (11), 7306–7315. https://doi.org/10.1002/2013JA019196.
Perry, G.W., Dahlgren, H., Nicolls, M.J., Zettergren, M., St.-Maurice, J.-P., Semeter, J.L., Sundberg, T., Hosokawa, K., Shiokawa, K., Chen, S., 2015. Spatiotemporally resolved electrodynamic properties of a sun-aligned arc over Resolute Bay. J. Geophys. Res. Space Phys., 9977–9987. https://doi.org/10.1002/2015JA021790.
Phelps, A.D.R., Sagalyn, R.C., 1976. Plasma density irregularities in the high-latitude top side ionosphere. J. Geophys. Res. Space Phys. 81 (4), 515–523. https://doi.org/10.1029/JA081i004p00515.
Pierrard, V., Voiculescu, M., 2011. The 3D model of the plasmasphere coupled to the ionosphere. Geophys. Res. Lett. 38 (12), L12104. https://doi.org/10.1029/2011GL047767.
Press, W., Flannery, B., Teukolsky, S., Vetterling, W., 1992. Numerical Recipes in C: The Art of Scientific Computing. Cambridge University Press.
Prikryl, P., Jayachandran, P.T., Mushini, S.C., Pokhotelov, D., MacDougall, J.W., Donovan, E., Spanswick, E., St.-Maurice, J.-P., 2010. GPS TEC, scintillation and cycle slips observed at high latitudes during solar minimum. Ann. Geophys. 28 (6), 1307–1316. https://doi.org/10.5194/angeo-28-1307-2010.
Prikryl, P., Jayachandran, P.T., Mushini, S.C., Chadwick, R., 2011. Climatology of GPS phase scintillation and HF radar backscatter for the high-latitude ionosphere under solar minimum conditions. Ann. Geophys. 29 (2), 377–392. https://doi.org/10.5194/angeo-29-377-2011.
Prikryl, P., Jayachandran, P.T., Chadwick, R., Kelly, T.D., 2015. Climatology of GPS phase scintillation at northern high latitudes for the period from 2008 to 2013. Ann. Geophys. 33 (5), 531–545. https://doi.org/10.5194/angeo-33-531-2015.
Priyadarshi, S., 2015. A review of ionospheric scintillation models. Surv. Geophys. 36 (2), 295–324. https://doi.org/10.1007/s10712-015-9319-1.
Prölss, G.W., 2007. The equatorward wall of the subauroral trough in the afternoon/evening sector. Ann. Geophys. 25 (3), 645–659. https://doi.org/10.5194/angeo-25-645-2007.
Prölss, G.W., 1995. Ionospheric F-region storms. In: Handbook of Atmospheric Electrodynamics, vol. 2. CRC Press, Boca Raton, FL.
Pryse, S.E., Kersley, L., Williams, M.J., Walker, I.K., 1998. The spatial structure of the dayside ionospheric trough. Ann. Geophys. 16 (10), 1169–1179. https://doi.org/10.1007/s00585-998-1169-4.
Ratcliffe, J.A., 1956. Some aspects of diffraction theory and their application to the ionosphere. Rep. Prog. Phys. 19 (1), 188–267. https://doi.org/10.1088/0034-4885/19/1/306.
Ravindran Varrier, N., 2010. Ray Tracing Analysis for the Mid-latitude SuperDARN HF Radar at Blackstone Incorporating the IRI-2007 Model (Master's thesis). Virginia Polytechnic Institute and State University.
Ren, J., Zou, S., Gillies, R.G., Donovan, E., Varney, R.H., 2018. Statistical characteristics of polar cap patches observed by RISR-C. J. Geophys. Res. Space Phys. 123, 6981–6995. https://doi.org/10.1029/2018JA025621.
Ren, J., Zou, S., Kendall, E., Coster, A., Sterne, K., Ruohoniemi, M., 2020. Direct observations of a polar cap patch formation associated with dayside reconnection driven fast flow. J. Geophys. Res. Space Phys. 125 (4). e2019JA027745.
Retterer, J.M., 2010. Forecasting low-latitude radio scintillation with 3-D ionospheric plume models: 2. Scintillation calculation. J. Geophys. Res. Space Phys. 115 (A3). https://doi.org/10.1029/2008JA013840.
Rinne, Y., Moen, J., Oksavik, K., Carlson, H.C., 2007. Reversed flow events in the winter cusp ionosphere observed by the European incoherent scatter (EISCAT) Svalbard radar. J. Geophys. Res. Space Phys. 112 (A10). https://doi.org/10.1029/2007JA012366.

Rino, C.L., 1979a. A power law phase screen model for ionospheric scintillation. I—Weak scatter. Radio Sci. 14, 1135–1145. https://doi.org/10.1029/RS014i006p01135.

Rino, C.L., 1979b. A power law phase screen model for ionospheric scintillation. II—Strong scatter. Radio Sci. 14, 1147–1155. https://doi.org/10.1029/RS014i006p01135.

Rino, C.L., 2010. The Theory of Scintillation with Applications in Remote Sensing. John Wiley & Sons.

Rino, C.L., Carrano, C.S., 2011. The application of numerical simulations in Beacon scintillation analysis and modeling. Radio Sci. 46 (3), RS0D02. https://doi.org/10.1029/2010RS004563.

Rino, C.L., Livingston, R.C., 1982. On the analysis and interpretation of spaced-receiver measurements of transionospheric radio waves. Radio Sci. 17 (4), 845–854. https://doi.org/10.1029/RS017i004p00845.

Rino, C., Breitsch, B., Morton, Y., Jiao, Y., Xu, D., Carrano, C., 2018a. A compact multi-frequency GNSS scintillation model. Navigation 65 (4), 563–569. https://doi.org/10.1002/navi.263.

Rino, C., Carrano, C., Groves, K., Yokoyama, T., 2018b. A configuration space model for intermediate-scale ionospheric structure. Radio Sci. 53 (12), 1472–1480. https://doi.org/10.1029/2018RS006678.

Rino, C., Carrano, C., Groves, K., 2019. Wave field propagation in extended highly anisotropic media. Radio Sci. 54 (7), 646–659. https://doi.org/10.1029/2019RS006793.

Rishbeth, H., Hanson, W.B., 1974. A comment on plasma 'pile-up' in the F-region. J. Atmos. Terr. Phys. 36 (4), 703–706. https://doi.org/10.1016/0021-9169(74)90094-4.

Robinson, R.M., Vondrak, R.R., Friis-Christensen, E., 1987. Ionospheric currents associated with a sun-aligned arc connected to the auroral oval. Geophys. Res. Lett. 14 (6), 656–659. https://doi.org/10.1029/GL014i006p00656.

Rodger, A.S., 2008. The Mid-Latitude Trough—Revisited, Geophysical Monograph Series, vol. 181. In: American Geophysical Union. Washington, DC, pp. 25–33.

Rodger, A.S., Brace, L.H., Hoegy, W.R., Winningham, J.D., 1986. The poleward edge of the mid-latitude trough—its formation, orientation and dynamics. J. Atmos. Terr. Phys. 48, 715–728. https://doi.org/10.1016/0021-9169(86)90021-8.

Rodger, A.S., Moffett, R.J., Quegan, S., 1992. The role of ion drift in the formation of ionisation troughs in the mid- and high-latitude ionosphere—a review. J. Atmos. Terr. Phys. 54, 1–30. https://doi.org/10.1016/0021-9169(92)90082-V.

Rodger, A.S., Pinnock, M., Dudeney, J.R., Baker, K.B., Greenwald, R.A., 1994. A new mechanism for polar patch formation. J. Geophys. Res. Space Phys. 99 (A4), 6425–6436. https://doi.org/10.1029/93JA01501.

Ruohoniemi, J.M., Baker, K.B., 1998. Large-scale imaging of high-latitude convection with super dual auroral radar network HF radar observations. J. Geophys. Res. Space Phys. 103 (A9), 20797–20811. https://doi.org/10.1029/98ja01288.

Ruohoniemi, J.M., Greenwald, R.A., 2005. Dependencies of high-latitude plasma convection: consideration of interplanetary magnetic field, seasonal, and universal time factors in statistical patterns. J. Geophys. Res. Space Phys. 110 (A9), A09204. https://doi.org/10.1029/2004JA010815.

Russell, C.T., Elphic, R.C., 1979. ISEE observations of flux transfer events at the dayside magnetopause. Geophys. Res. Lett. 6 (1), 33–36. https://doi.org/10.1029/GL006i001p00033.

Sakai, J., Taguchi, S., Hosokawa, K., Ogawa, Y., 2013. Steep plasma depletion in dayside polar cap during a CME-driven magnetic storm. J. Geophys. Res. Space Phys. 118 (1), 462–471.

Sato, T., 1959. Morphology of ionospheric F2 disturbances in the polar regions. Rep. Ionos. Space Res. Jpn 13, 91–95.

Schunk, R.W., Nagy, A., 2000. Ionospheres: Physics, Plasma Physics, and Chemistry. Cambridge Atmospheric and Space Science Series, Cambridge University Press.

Schunk, R., Nagy, A., 2009. Ionospheres: Physics, Plasma Physics, and Chemistry. Cambridge University Press.

Schunk, R.W., Banks, P.M., Raitt, W.J., 1976. Effects of electric fields and other processes upon the night-time high-latitude F layer. J. Geophys. Res. Space Phys. 81 (19), 3271. https://doi.org/10.1029/JA081i019p03271.

Schunk, R.W., Demars, H.G., Sojka, J.J., 2005. Propagating polar wind jets. J. Atmos. Sol. Terr. Phys. 67 (4), 357–364. https://doi.org/10.1016/j.jastp.2004.09.005.

Semeter, J., Heinselman, C.J., Thayer, J.P., Doe, R.A., Frey, H.U., 2003. Ion upflow enhanced by drifting F-region plasma structure along the nightside polar cap boundary. Geophys. Res. Lett. 30 (22), 2139. https://doi.org/10.1029/2003GL017747.

Semeter, J., Mrak, S., Hirsch, M., Swoboda, J., Akbari, H., Starr, G., Hampton, D., Erickson, P., Lind, F., Coster, A., Pankratius, V., 2017. GPS signal corruption by the discrete aurora: precise measurements from the Mahali experiment. Geophys. Res. Lett. 44 (19), 9539–9546. https://doi.org/10.1002/2017GL073570.

Shiokawa, K., Katoh, Y., Satoh, M., Ejiri, M.K., Ogawa, T., 2000. Integrating-sphere calibration of all-sky cameras for nightglow measurements. Adv. Space Res. 26 (6), 1025–1028. https://doi.org/10.1016/S0273-1177(00)00052-1.

Shkarofsky, I.P., 1968. Generalized turbulence space-correlation and wave-number spectrum-function pairs. Can. J. Phys. 46, 2133. https://doi.org/10.1139/p68-562.

Siefring, C.L., Bernhardt, P.A., James, H.G., Parris, R.T., 2015. The CERTO beacon on CASSIOPE/e-POP and experiments using high-power HF ionospheric heaters. Space Sci. Rev. 189 (1–4), 107–122. https://doi.org/10.1007/s11214-014-0110-2.

Simon, A., 1963. Instability of a partially ionized plasma in crossed electric and magnetic fields. Phys. Fluids 6 (3), 382. https://doi.org/10.1063/1.1706743.

Singh, M., Rodriguez, P., Szuszczewicz, E.P., 1985. Spectral classification of medium-scale high-latitude F region plasma density irregularities. J. Geophys. Res. Space Phys. 90 (A7), 6525–6532. https://doi.org/10.1029/JA090iA07p06525.

Smith, D.R., Huang, C.Y., Dao, E., Pokhrel, S., Simpson, J.J., 2020. FDTD modeling of high-frequency waves through ionospheric plasma irregularities. J. Geophys. Res. Space Phys. 125 (3). https://doi.org/10.1029/2019JA027499. e2019JA027499.

Sofko, G.J., Greenwald, R., Bristow, W., 1995. Direct determination of large-scale magnetospheric field-aligned currents with SuperDARN. Geophys. Res. Lett. 22 (15), 2041–2044. https://doi.org/10.1029/95GL01317.

Sojka, J.J., Raitt, W.J., Schunk, R.W., 1981a. Plasma density features associated with strong convection in the winter high-latitude F region. J. Geophys. Res. Space Phys. 86 (A8), 6908–6916.

Sojka, J.J., Raitt, W.J., Schunk, R.W., 1981b. A theoretical study of the high-latitude winter F region at solar minimum for low magnetic activity. J. Geophys. Res. Space Phys. 86 (A2), 609–621.

Sojka, J.J., Bowline, M.D., Schunk, R.W., Decker, D.T., Valladares, C.E., Sheehan, R., Anderson, D.N., Heelis, R.A., 1993. Modeling polar cap F-region patches using time varying convection. Geophys. Res. Lett. 20 (17), 1783–1786. https://doi.org/10.1029/93GL01347.

Sojka, J.J., Bowline, M.D., Schunk, R.W., 1994. Patches in the polar ionosphere: UT and seasonal dependence. J. Geophys. Res. Space Phys. 99 (A8), 14959–14970.

Spicher, A., Miloch, W.J., Moen, J.I., 2014. Direct evidence of double-slope power spectra in the high-latitude ionospheric plasma. Geophys. Res. Lett. 41 (5), 1406–1412. https://doi.org/10.1002/2014GL059214.

Spicher, A., Cameron, T., Grono, E.M., Yakymenko, K.N., Buchert, S.C., Clausen, L.B.N., Knudsen, D. J., McWilliams, K.A., Moen, J.I., 2015a. Observation of polar cap patches and calculation of gradient drift instability growth times: a Swarm case study. Geophys. Res. Lett. 42 (2), 201–206. https://doi.org/10.1002/2014GL062590.

Spicher, A., Miloch, W.J., Clausen, L.B.N., Moen, J.I., 2015b. Plasma turbulence and coherent structures in the polar cap observed by the ICI-2 sounding rocket. J. Geophys. Res. Space Phys. 120 (12). https://doi.org/10.1002/2015JA021634. 10,959–10,978.

Spicher, A., Ilyasov, A.A., Miloch, W.J., Chernyshov, A.A., Clausen, L.B.N., Moen, J.I., Abe, T., Saito, Y., 2016. Reverse flow events and small-scale effects in the cusp ionosphere. J. Geophys. Res. Space Phys. 121 (10), 10466–10480. https://doi.org/10.1002/2016JA022999.

Spicher, A., Deshpande, K., Jin, Y., Oksavik, K., Zettergren, M.D., Clausen, L.B.N., Moen, J.I., Hairston, M.R., Baddeley, L., 2020. On the production of ionospheric irregularities via Kelvin-Helmholtz instability associated with cusp flow channels. J. Geophys. Res. Space Phys. 125 (6). https://doi.org/10.1029/2019JA027734. e2019JA027734.

Spiro, R.W., Heelis, R.A., Hanson, W.B., 1978. Ion convection and the formation of the mid-latitude F region ionization trough. J. Geophys. Res. Space Phys. 83 (A9), 4255–4264. https://doi.org/10.1029/JA083iA09p04255.
Spiro, R.W., Heelis, R.A., Hanson, W.B., 1979. Rapid subauroral ion drifts observed by atmosphere explorer C. Geophys. Res. Lett. 6 (8), 657–660. https://doi.org/10.1029/GL006i008p00657.
Spiro, R.W., Wolf, R.A., Fejer, B.G., 1988. Penetrating of high-latitude-electric-field effects to low latitudes during SUNDIAL 1984. Ann. Geophys. 6, 39–49.
Spogli, L., Alfonsi, L., de Franceschi, G., Romano, V., Aquino, M.H.O., Dodson, A., 2009. Climatology of GPS ionospheric scintillations over high and mid-latitude European regions. Ann. Geophys. 27, 3429–3437. https://doi.org/10.5194/angeo-27-3429-2009.
Spogli, L., Alfonsi, L., Cilliers, P.J., Correia, E., De Franceschi, G., Mitchell, C.N., Romano, V., Kinrade, J., Cabrera, M.A., 2013. GPS scintillations and total electron content climatology in the southern low, middle and high latitude regions. Ann. Geophys. 56 (2), R0220.
St.-Maurice, J.-P., Hanson, W.B., 1982. Ion frictional heating at high latitudes and its possible use for an in situ determination of neutral thermospheric winds and temperatures. J. Geophys. Res. Space Phys. 87 (A9), 7580–7602. https://doi.org/10.1029/JA087iA09p07580.
St.-Maurice, J.-P.-P., Torr, D.G.G., 1978. Nonthermal rate coefficients in the ionosphere: the reactions of O+ With N2, O2, and NO. J. Geophys. Res. Space Phys. 83 (7), 969–977. https://doi.org/10.1029/JA083iA03p00969.
Störmer, C., 1955. The Polar Aurora. Oxford University Press.
Theurer, T.E., 2012. Ray Tracing Applications for High-Frequency Radar: Characterizing Artificial Layers and Background Density Perturbations in the Ionosphere (Master's thesis). University of Alaska Fairbanks.
Thomas, E.G., Baker, J.B.H., Ruohoniemi, J.M., Clausen, L.B.N., Coster, A.J., Foster, J.C., Erickson, P.J., 2013. Direct observations of the role of convection electric field in the formation of a polar tongue of ionization from storm enhanced density. J. Geophys. Res. Space Phys. 118 (3), 1180–1189.
Tsunoda, R.T., 1988. High-latitude F region irregularities: a review and synthesis. Rev. Geophys. 26 (4), 719–760. https://doi.org/10.1029/RG026i004p00719.
Valladares, C.E., Carlson Jr, H.C., 1991. The electrodynamic, thermal, and energetic character of intense Sun-aligned arcs in the polar cap. J. Geophys. Res. Space Phys. 96 (A2), 1379–1400. https://doi.org/10.1029/90JA01765.
van der Meeren, C., Oksavik, K., Lorentzen, D., Moen, J.I., Romano, V., 2014. GPS scintillation and irregularities at the front of an ionization tongue in the nightside polar ionosphere. J. Geophys. Res. Space Phys. 119 (10), 8624–8636. https://doi.org/10.1002/2014JA020114.
van der Meeren, C., Oksavik, K., Lorentzen, D.A., Rietveld, M.T., Clausen, L.B.N., 2015. Severe and localized GNSS scintillation at the poleward edge of the nightside auroral oval during intense substorm aurora. J. Geophys. Res. Space Phys. 120 (12), 10607–10621. https://doi.org/10.1002/2015JA021819.
van der Meeren, C., Oksavik, K., Lorentzen, D.A., Paxton, L.J., Clausen, L.B.N., 2016. Scintillation and irregularities from the nightside part of a Sun-aligned polar cap arc. J. Geophys. Res. Space Phys. 121 (6), 5723–5736. https://doi.org/10.1002/2016JA022708.
Vickrey, J.F., Kelley, M.C., 1982. The effects of a conducting E layer on classical F region cross-field plasma diffusion. J. Geophys. Res. Space Phys. 87 (A6), 4461–4468.
Vickrey, J.F., Rino, C.L., Potemra, T.A., 1980. Chatanika/Triad observations of unstable ionization enhancements in the auroral F-region. Geophys. Res. Lett. 7 (10), 789–792. https://doi.org/10.1029/GL007i010p00789.
Villain, J.P., Hanuise, C., Beghin, C., 1986. ARCAD3-SAFARI coordinated study of auroral and polar F-region ionospheric irregularities. Ann. Geophys. 4, 61–68.
Voiculescu, M., Roth, M., 2008. Eastward sub-auroral ion drifts or ASAID. Ann. Geophys. 26 (7), 1955–1963. https://doi.org/10.5194/angeo-26-1955-2008.
Voiculescu, M., Virtanen, I., Nygrén, T., 2006. The F-region trough: seasonal morphology and relation to interplanetary magnetic field. Ann. Geophys. 24 (1), 173–185. https://doi.org/10.5194/angeo-24-173-2006.

Voiculescu, M., Nygrén, T., Aikio, A., Kuula, R., 2010. An olden but golden EISCAT observation of a quiet-time ionospheric trough. J. Geophys. Res. Space Phys. 115 (A10), A10315. https://doi.org/10.1029/2010JA015557.
Copernicus GmbH Walker, I., Moen, J., Kersley, L., Lorentzen, D., 1999. On the possible role of cusp/cleft precipitation in the formation of polar-cap patches. Ann. Geophys. 17 (10), 1298–1305. https://doi.org/10.1007/s00585-999-1298-4.
Walsh, B.M., Foster, J.C., Erickson, P.J., Sibeck, D.G., 2014. Simultaneous ground- and space-based observations of the plasmaspheric plume and reconnection. Science 343 (6175), 1122–1125. https://doi.org/10.1126/science.1247212.
Wang, H., Ridley, A.J., Lühr, H., Liemohn, M.W., Ma, S.Y., 2008. Statistical study of the subauroral polarization stream: its dependence on the cross-polar cap potential and subauroral conductance. J. Geophys. Res. Space Phys. 113 (A12), A12311.
Wang, H., Lühr, H., Häusler, K., Ritter, P., 2011. Effect of subauroral polarization streams on the thermosphere: a statistical study. J. Geophys. Res. Space Phys. 116 (A3), 5785.
Wang, W., Talaat, E.R., Burns, A.G., Emery, B., Hsieh, S.Y., Lei, J., Xu, J., 2012. Thermosphere and ionosphere response to subauroral polarization streams (SAPS): model simulations. J. Geophys. Res. Space Phys. 117 (A7), A07301.
Wang, Y., Zhang, Q.-H., Jayachandran, P.T., Lockwood, M., Zhang, S.-R., Moen, J., Xing, Z.-Y., Ma, Y.-Z., Lester, M., 2016. A comparison between large-scale irregularities and scintillations in the polar ionosphere. Geophys. Res. Lett. 43 (10), 4790–4798. https://doi.org/10.1002/2016GL069230.
Wang, Y., Zhang, Q.-H., Jayachandran, P.T., Moen, J., Xing, Z.-Y., Chadwick, R., Ma, Y.-Z., Ruohoniemi, J.M., Lester, M., 2018. Experimental evidence on the dependence of the standard GPS phase scintillation index on the ionospheric plasma drift around noon sector of the polar ionosphere. J. Geophys. Res. Space Phys. 123 (3), 2370–2378. https://doi.org/10.1002/2017JA024805.
Warrington, E.M., Rogers, N.C., Jones, T.B., 1997. Large HF bearing errors for propagation paths contained within the polar cap. IEE Proc. Microwaves Antenn. Propag. 144 (4), 241–249. https://doi.org/10.1049/ip-map:19971187.
Watson, C., Jayachandran, P.T., Macdougall, J.W., 2016. Characteristics of GPS TEC variations in the polar cap ionosphere. J. Geophys. Res. A Space Phys. 121, 4748–4768. https://doi.org/10.1002/2015JA022275.Received.
Weber, E.J., Buchau, J., Moore, J., Sharber, J., Livingston, R., Winningham, J.D., Reinisch, B., 1984. F layer ionization patches in the polar cap. J. Geophys. Res. Space Phys. 89 (A3), 1683–1694. https://doi.org/10.1029/JA089iA03p01683.
Weber, E.J., Buchau, J., Moore, J.G., Sharber, J.R., Livingston, R.C., Winningham, J.D., Reinisch, B.W., 1984. F layer ionization patches in the polar cap. J. Geophys. Res. Space Phys. 89 (A3), 1683–1694.
Weiss, L.A., Weber, E.J., Reiff, P.H., Sharber, J.R., Winningham, J.D., Primdahl, F., Mikkelsen, I.S., Seifring, C., Wescott, E.M., 1993. Convection and electrodynamic signatures in the vicinity of a sun-aligned arc: results from the polar acceleration regions and convection study (Polar ARCS). In: Lysak, R.L. (Ed.), Aurora1 Plasma Dynamics. Geophys. Monog., vol. 80. American Geophysical Union (AGU), p. 69.
Werner, S., Prölss, G.W., 1997. The position of the ionospheric trough as a function of local time and magnetic activity. Adv. Space Res. 20 (9), 1717–1722. https://doi.org/10.1016/S0273-1177(97)00578-4.
Wernik, A.W., Gola, M., Liu, C.H., Franke, S.J., 1990. High-latitude irregularity spectra deduced from scintillation measurements. Radio Sci. 25, 883–895. https://doi.org/10.1029/RS025i005p00883.
Wernik, A.W., Secan, J.A., Fremouw, E.J., 2003. Ionospheric irregularities and scintillation. Adv. Space Res. 31, 971–981. https://doi.org/10.1016/S0273-1177(02)00795-0.
Whalen, J.A., 1989. The daytime F layer trough and its relation to ionospheric-magnetospheric convection. J. Geophys. Res. Space Phys. 94 (A12), 17169–17184. https://doi.org/10.1029/JA094iA12p17169.
Wickwar, V.B., Cogger, L.L., Carlson, H.C., 1974. The 6300 ÅO1D airglow and dissociative recombination. Planet. Space Sci. 22 (5), 709–724. https://doi.org/10.1016/0032-0633(74)90141-X.
Yang, N., Le, H., Liu, L., 2015. Statistical analysis of ionospheric mid-latitude trough over the Northern Hemisphere derived from GPS total electron content data. Earth Planets Space 67, 196. https://doi.org/10.1186/s40623-015-0365-1.
Yeh, K.C., Liu, C.-H., 1982. Radio wave scintillations in the ionosphere. IEEE Proc. 70, 324–360.
Yeh, H.-C., Foster, J.C., Rich, F.J., Swider, W., 1991. Storm time electric field penetration observed at mid-latitude. J. Geophys. Res. Space Phys. 96 (A4), 5707–5721.

Yizengaw, E., Moldwin, M., 2005. The altitude extension of the mid-latitude trough and its correlation with plasmapause position. Geophys. Res. Lett. 32 (9), L09105. https://doi.org/10.1029/2005GL022854.

Yizengaw, E., Moldwin, M.B., Galvan, D.A., 2006. Ionospheric signatures of a plasmaspheric plume over Europe. Geophys. Res. Lett. 33 (17), L17103. https://doi.org/10.1029/2006GL026597.

Yokoyama, T., 2017. A review on the numerical simulation of equatorial plasma bubbles toward scintillation evaluation and forecasting. Prog. Earth Planet. Sci. 4 (1), 37. https://doi.org/10.1186/s40645-017-0153-6.

Zabusky, N.J., Doles III, J.H., Perkins, F.W., 1973. Deformation and striation of plasma clouds in the ionosphere: 2. Numerical simulation of a nonlinear two-dimensional model. J. Geophys. Res. Space Phys. 78 (4), 711–724.

Zettergren, M.D., Hirsch, M., 2020. gemini3d/gemini-examples., https://doi.org/10.5281/zenodo.3992594.

Zettergren, M., Semeter, J., 2012. Ionospheric plasma transport and loss in auroral downward current regions. J. Geophys. Res. Space Phys. 117 (A6), A06306. https://doi.org/10.1029/2012JA017637.

Zettergren, M.D., Snively, J.B., 2015. Ionospheric response to infrasonic-acoustic waves generated by natural hazard events. J. Geophys. Res. Space Phys. 120 (9), 8002–8024.

Zettergren, M.D., Semeter, J.L., Dahlgren, H., 2015. Dynamics of density cavities generated by frictional heating: formation, distortion, and instability. Geophys. Res. Lett. 42 (23), 10–120.

Zhang, Y., Paxton, L.J., 2015. Auroral Dynamics and Space Weather, Geophysical Monograph Series, vol. 215. In: American Geophysical Union (AGU). Washington, DC. https://doi.org/10.1002/9781118978719.

Zhang, Q.-H., Zhang, B.-C., Lockwood, M., Hu, H.-Q., Moen, J., Ruohoniemi, J.M., Thomas, E.G., Zhang, S.-R., Yang, H.-G., Liu, R.-Y., McWilliams, K.A., Baker, J.B.H., 2013. Direct observations of the evolution of polar cap ionization patches. Science 339 (6127), 1597–1600. https://doi.org/10.1126/science.1231487.

Zhang, Q.H., Lockwood, M., Foster, J.C., Zhang, S.-R., Zhang, B.C., McCrea, I.W., Moen, J., Lester, M., Ruohoniemi, J.M., 2015a. Direct observations of the full Dungey convection cycle in the polar ionosphere for southward interplanetary magnetic field conditions. J. Geophys. Res. Space Phys. 120 (6), 4519–4530.

Zhang, Q.-H., Zong, Q.-G., Lockwood, M., Heelis, R.A., Hairston, M., Liang, J., McCrea, I., Zhang, B.-C., Moen, J., Zhang, S.-R., Zhang, Y.-L., Ruohoniemi, J.M., Lester, M., Thomas, E.G., Liu, R.-Y., Dunlop, M.W., Liu, Y.C.-M., Ma, Y.-Z., 2016. Earth's ion upflow associated with polar cap patches: global and in situ observations. Geophys. Res. Lett. 43 (5), 1845–1853. https://doi.org/10.1002/2016GL067897.

Zhang, Q.-H., Ma, Y.-Z., Jayachandran, P.T., Moen, J., Lockwood, M., Zhang, Y.-L., Foster, J.C., Zhang, S.-R., Wang, Y., Themens, D.R., et al., 2017a. Polar cap hot patches: Enhanced density structures different from the classical patches in the ionosphere. Geophys. Res. Lett. 44 (16), 8159–8167. https://doi.org/10.1002/2017GL073439.

Zhang, S.-R., Erickson, P.J., Foster, J.C., Holt, J.M., Coster, A.J., Makela, J.J., Noto, J., Meriwether, J.W., Harding, B.J., Riccobono, J., Kerr, R.B., 2015b. Thermospheric poleward wind surge at midlatitudes during great storm intervals. Geophys. Res. Lett. 42 (13), 5132–5140.

Zhang, S.-R., Erickson, P.J., Zhang, Y., Wang, W., Huang, C., Coster, A.J., Holt, J.M., Foster, J.F., Sulzer, M., Kerr, R., 2017. Observations of ion-neutral coupling associated with strong electrodynamic disturbances during the 2015 St. Patrick's Day storm. J. Geophys. Res. Space Phys. 122 (1), 1314–1337. https://doi.org/10.1002/2016JA023307.

Zhu, L., Shunk, R.W., Sojka, J.J., 1997. Polar cap arcs: a review. J. Atmos. Sol. Terr. Phys. 59 (10), 1087–1126.

Zou, S., Moldwin, M., Coster, A., Lyons, L., Nicolls, M., 2011. GPS TEC observations of dynamics of the mid-latitude trough during substorms. Geophys. Res. Lett. 38 (14), L14109. https://doi.org/10.1029/2011GL048178.

Zou, S., Moldwin, M.B., Ridley, A.J., Nicolls, M.J., Coster, A.J., Thomas, E.G., Ruohoniemi, J.M., 2014. On the generation/decay of the storm-enhanced density plumes: role of the convection flow and field-aligned ion flow. J. Geophys. Res. Space Phys. 119 (10), 8543–8559. https://doi.org/10.1002/2014JA020408.

Further reading

Mailyan, B., Shi, Q.Q., Kullen, A., Maggiolo, R., Zhang, Y., Fear, R.C., Zong, Q.-G., Fu, S.Y., Gou, X. C., Cao, X., Yao, Z.H., Sun, W.J., Wei, Y., Pu, Z.Y., 2015. Transpolar arc observation after solar wind entry into the high-latitude magnetosphere. J. Geophys. Res. Space Phys. 120 (5), 3525–3534. https://doi.org/10.1002/2014JA020912.

CHAPTER 4

Energetic particle dynamics, precipitation, and conductivity

Christine Gabrielse[a], Stephen R. Kaeppler[b], Gang Lu[c], Chih-Ping Wang[d], and Yiqun Yu[e]
[a]The Aerospace Corporation, El Segundo, CA, United States
[b]Department of Physics and Astronomy, Clemson University, Clemson, SC, United States
[c]High Altitude Observatory, National Center for Atmospheric Research, Boulder, CO, United States
[d]UCLA, Los Angeles, CA, United States
[e]School of Space and Environment, Beihang University, Beijing, China

Chapter outline

Cross-Scale Coupling and Energy Transfer in the Magnetosphere-Ionosphere-Thermosphere System
https://doi.org/10.1016/B978-0-12-821366-7.00002-0

Key points

- Spatial distributions of plasma sheet properties that are the result of the source, transport, energization, and loss processes, at different scales.
- Plasma sheet processes that affect the dynamics of FAC and precipitation.
- Basic concepts of pitch angle scattering processes that lead to particle precipitation.
- Recent efforts of modeling the electron and ion precipitation.
- Observation techniques to infer convection, precipitation, and ionospheric conductance.
- How convection, precipitation, and conductance interplay across scale sizes due to magnetosphere-ionosphere coupling.

4.1 Energetic particles in the Earth's plasma sheet and contribution to the magnetosphere-ionosphere coupling

4.1.1 Introduction to the plasma sheet properties

The plasma sheet lies in the magnetosphere within the closed magnetic field-line region (a closed field line means the field line that has both ends connecting to the Earth, see also Section 1.1). It is a sheet-like structure located around the equatorial plane (or the neutral plane), where magnetic fields reverse directions from pointing away from the Earth below the equatorial plane to pointing toward the Earth above the equatorial plane. The term "plasma sheet" was first used in a study by Bame et al. (1967) to describe "the region of enhanced plasma associated with neutral sheet." The thickness of the plasma sheet is on the order of several Earth radii (R_E). It is sandwiched by the cusps on the dayside (in the opposite hemispheres) and the lobes on the nightside. The plasma sheet is illustrated in Fig. 4.1A and B in the *X*-Z and *Y*-*Z* planes (the positive *X* is pointed toward the Sun, the positive *Y* is pointed toward dusk, and the positive *Z* is pointed toward the north), respectively, together with the solar wind, bow shock, magnetosheath, magnetopause, cusp, and lobe/mantle. The earthward edge (or the inner edge) of the plasma sheet connects with the inner magnetosphere (at radial distance $r < \sim 6\ R_E$) where the plasmasphere, ring current, and radiation belts are colocated. Its outer edge is the magnetopause on the dayside and the flanks, and its tailward edge extends to several tens of R_E to $>100\ R_E$.

Fig. 4.1C and D compare the densities and temperatures observed in the different plasma regimes of the solar wind and magnetosphere shown in Fig. 4.1A and B (except for the cusp). The data in the solar wind were measured by WIND, and the data in the magnetosheath and magnetosphere were measured by THEMIS/ARTEMIS satellites (Angelopoulos, 2008, 2011). Description of the dataset used in Figs. 4.1–4.3 is given

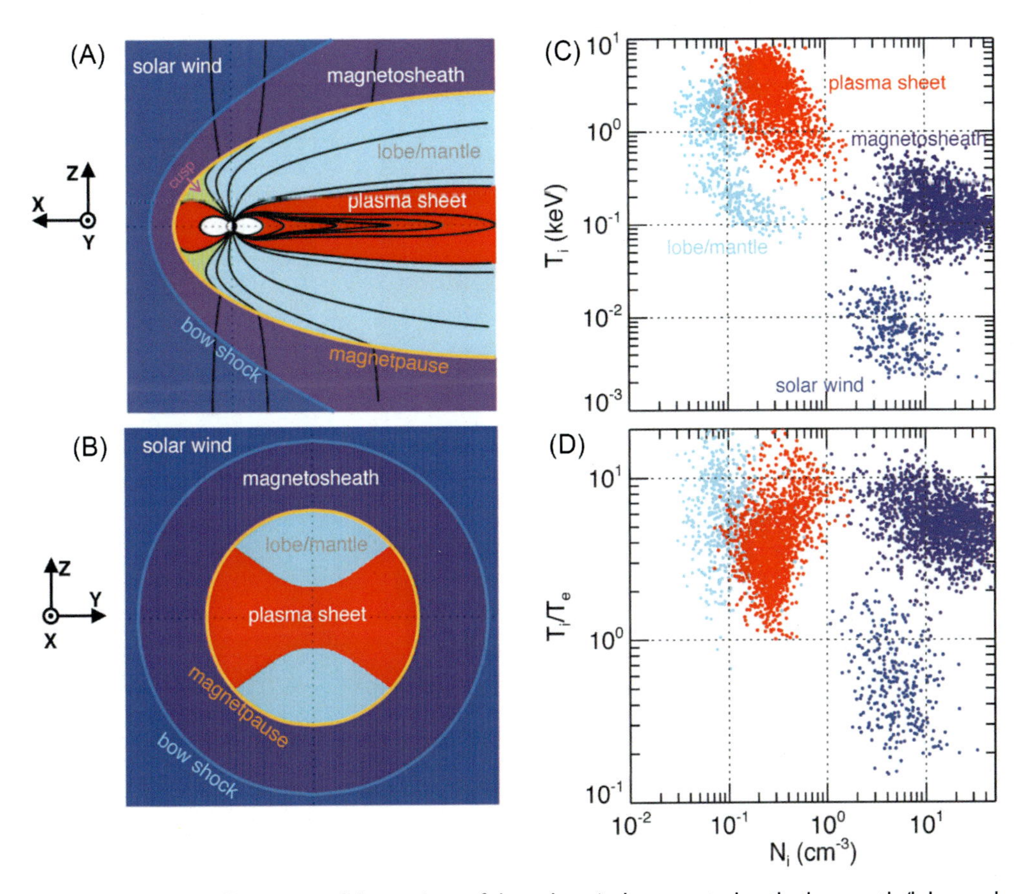

Fig. 4.1 Schematic illustration of the regions of the solar wind, magnetosheath, the mantle/lobe, and plasma sheet in (A) *X-Z* plane and (B) *Y-Z* plane. Observed (C) ion temperatures and (D) ion to electron temperature ratios as a function of ion densities in different plasma regimes.

in a study by Wang et al. (2012a, b). The plasma sheet consists of energetic ions (a few keV) and electrons (a few hundreds of eV to a few keV) with densities of $\sim 0.1\text{–}1\ \mathrm{cm}^{-3}$. The plasma sheet ion densities are similar to the electron densities. The spatial and temporal variations of the plasma sheet, plasma properties, and magnetic fields are presented in Section 4.1.2. The plasma sheet densities are lower than those of the solar wind and the magnetosheath plasma because only a small portion of the solar wind plasma can eventually enter the plasma sheet. On the other hand, higher plasma sheet temperature as compared to the solar wind and magnetosheath plasma is a result of energization. The plasma entry, transport, and energization processes are presented in Section 4.1.2.

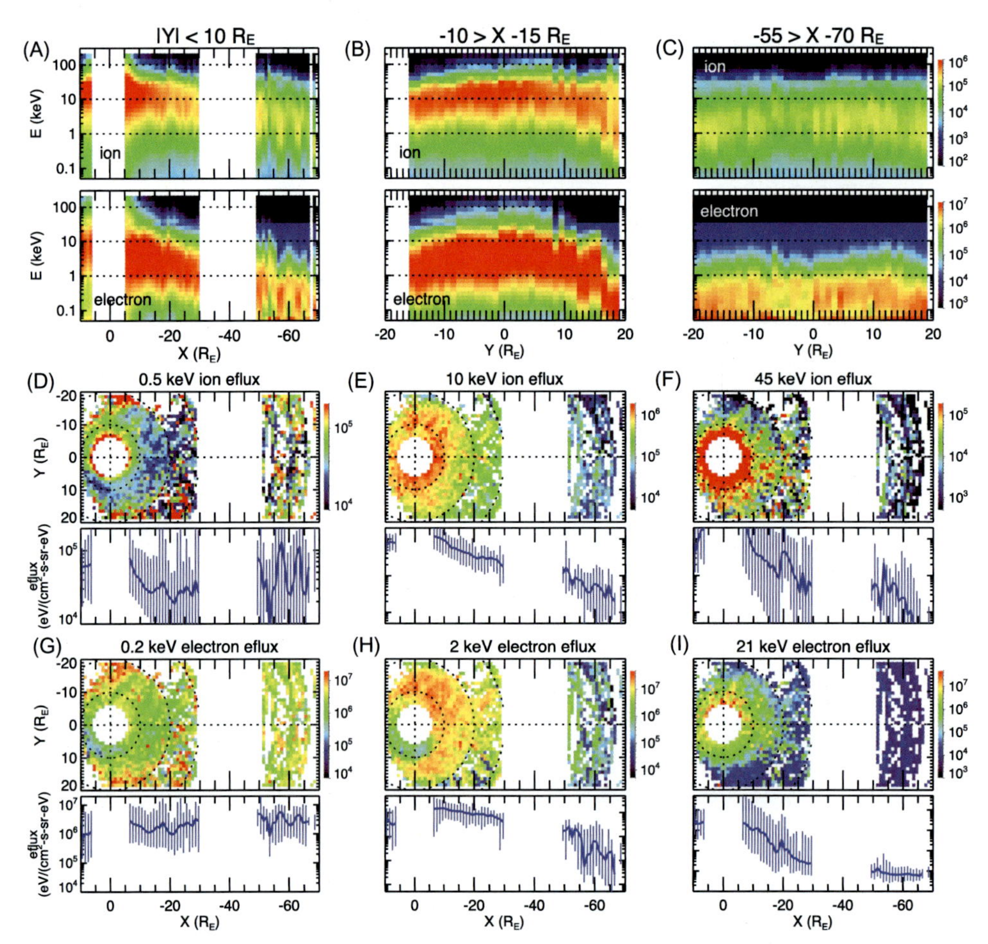

Fig. 4.2 (A) X distributions at $|Y| < 10\ R_E$, (B) Y distributions at $X = -10$ to $-15\ R_E$, and (C) Y distributions at $X = -55$ to $-70\ R_E$ of the medium values of energy fluxes (eV/(s-sr-cm^2-eV)) for ions (top) and electrons (bottom). The X-Y distributions (top) and X-profiles (bottom) of energy fluxes for (D) 0.5 keV ions, (E) 10 keV ions, (F) 45 keV ions, (G) 0.2 keV electrons, (H), 2 keV electrons, and (I) 21 keV electrons. The *blue curves* in the X-profiles are the medium values and the *blue vertical lines* indicate 10% and 90% percentiles.

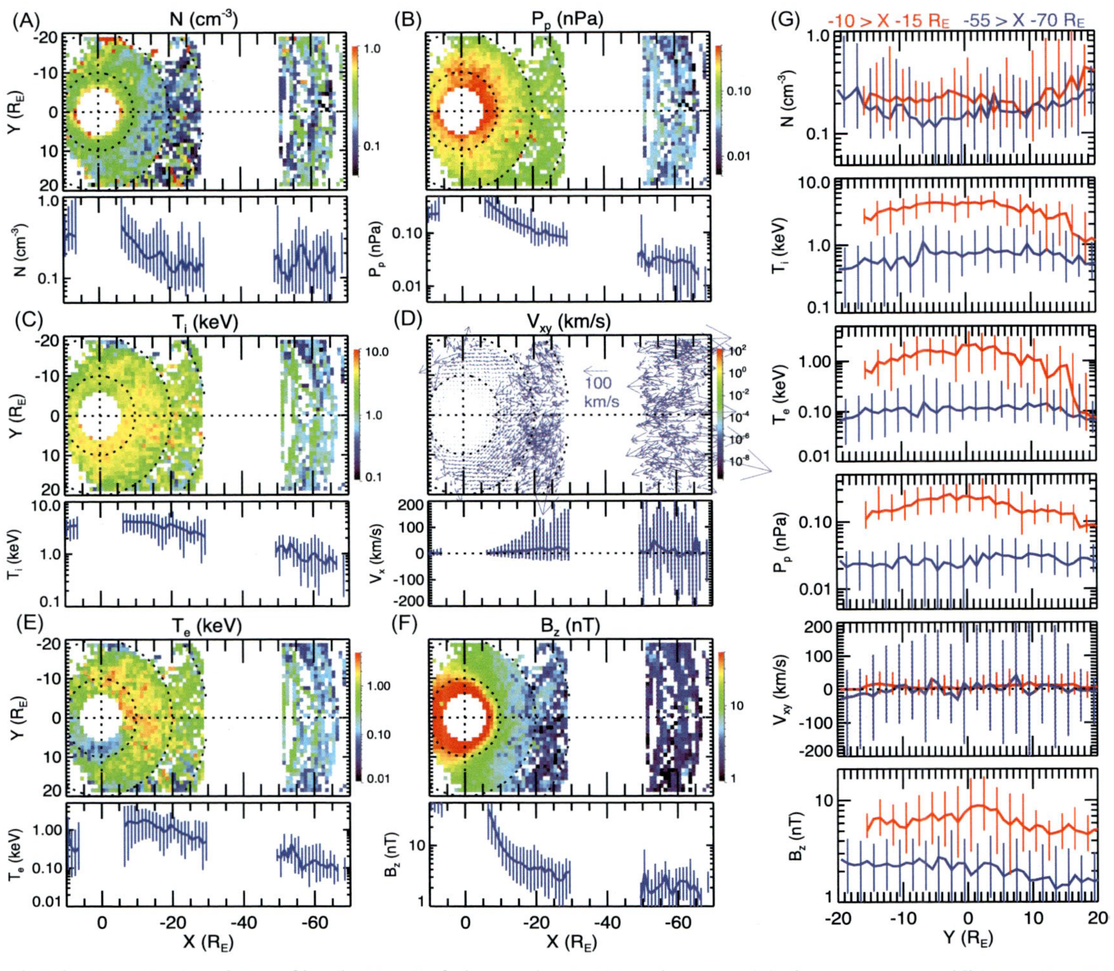

Fig. 4.3 The *X-Y* distributions (top) and *X*-profiles (bottom) of plasma sheet (A) ion densities, (B) plasma pressure, (C) ion temperatures, (D) $Vx - Vy$ vectors, (E) electron temperatures, and (F) B_Z. (G) *Y*-profiles for the regions of $X = -10$ to -15 R_E *(red)* and $X = -55$ to -70 R_E *(blue)*. The colors in the *X-Y* plots of (A)–(F) show the medium values. The curves in the *X*-profiles and *Y*-profiles are the medium values and vertical lines indicate 10% and 90% percentiles.

Thermal pressure (P_p) in the plasma sheet is high and magnetic field strength is low so that plasma beta ($\beta = P_p/P_B$, where P_B is magnetic pressure) is high ($>\sim 1$). In this high plasma beta regime, magnetic field lines in the plasma sheet are stretched to generate currents (**j**) and to provide a magnetic force ($\mathbf{j} \times \mathbf{B}$) to balance the plasma pressure gradient force (∇P_p). The currents flow in the directions perpendicular and parallel ($j_{||}$) to the magnetic field and satisfy current continuity, $\nabla \cdot \mathbf{j} = 0$. Since the plasma sheet is within the closed field-line region, the field-aligned currents (FACs, that is, $j_{||}$) connect the plasma sheet to the ionosphere. The plasma sheet plasma pressure is dominated by the ion plasma pressure because ion temperatures are higher than electron temperatures (Fig. 4.1D). Thus, ion transport and energization within the plasma sheet play an important role in the dynamics of FACs.

Three basic periodic motions exist for particles around the Earth: gyromotion around the magnetic field line, bounce motion along magnetic field lines between two magnetic mirrors, and azimuthal drift motion around the Earth. According to classic mechanics, each of these periodic motions corresponds to three constants of motion, or adiabatic invariants, provided the forces directing the motions varies infinitely slowly as compared to the characteristic frequency of the particle motion or the length scale of the forces that abruptly changes are significantly larger than the characteristic radius of the particle motion. Violation of the first adiabatic invariant associated with gyration or the second adiabatic invariant associated with bounce changes a particle's pitch angle. Particles with pitch angles smaller than the loss cone precipitate to the ionosphere (Sections 4.3.1.1 and 4.4.1). Precipitation of the plasma sheet particles generates aurora (see Chapter 2) and conductance in the ionosphere. The plasma sheet maps along magnetic field lines into the ionosphere at auroral latitudes. The equatorward edge of the diffuse aurora approximately corresponds to the inner edge of the plasma sheet and the poleward edge of the diffuse aurora approximately separates between the plasma sheet and lobe. The plasma sheet electron precipitation energy fluxes are larger than those of the ions (Hardy et al., 1989) so that they produce larger auroral conductance. Thus, electron dynamics in the plasma sheet is important to precipitation.

FACs and precipitation are the two main inputs from the plasma sheet into the ionosphere and thermosphere, and they lead to two major feedbacks from the ionosphere/thermosphere: one is redistribution of the convection electric field (Sections 4.2.1 and 4.2.2) and the other is ionospheric outflow. The redistributed convection electric field subsequently changes the particle transport and energization in the plasma sheet. The ionospheric outflow supplies new particles, in particular O^+ ions, into the lobe and plasma sheet.

The spatial scales of the plasma sheet plasma are crucial to the redistribution of convection electric field in the ionosphere as described by $\nabla_i \cdot [\Sigma \cdot \mathbf{E}_i] = -j_{||,i} \sin(I)$ [Vasyliunas, 1970, see also equation (1.1)], where the subscript i refers to quantities in the ionosphere, I is dip angle (or local magnetic field inclination), and Σ is the height-integrated conductance. Spatial distributions of the plasma sheet ion and electron precipitation determine the gradients of Σ, and $j_{||}$ in the plasma sheet is determined by the spatial gradients of plasma pressure (Vasyliunas, 1970) and plasma flows (Sonnerup, 1980). The plasma sheet spatial distribution is the result of the source, transport, energization, and loss processes, and each process involves many different mechanisms with different scales. Competition between these mechanisms determines the scales of $j_{||}$ and precipitation. In Sections 4.1.2 and 4.1.2, we focus our descriptions of the plasma sheet variations and processes on large scales (over a spatial scale of ~10 to several tens of R_E or a timescale of hours, see also Section 1.1) and meso-scale (over a spatial scale of a few R_E or a time-scale of ~1 min to 1 h, see also Section 1.2).

4.1.2 Properties of plasma sheet particles, plasma moments, and magnetic fields

4.1.2.1 Particle species

Plasma sheet particles consist of ions and electrons. The main ion species are H^+ and O^+, and minor ion species include He+, He^{++}, and O^{++} (Seki et al., 1996). H^+ ions come from either the solar wind or ionosphere, O^+ ions exclusively come from the ionosphere, and He^{++} ions have exclusively come from the solar wind origin. The O^+/H^+ density ratio varies from ~0.01 during quiet times and low F10.7 levels to ~0.1 during disturbed times and high F10.7 levels (Mouikis et al., 2010). In this section, we describe the variations of ions and electrons observed by THEMIS/ARTEMIS.

4.1.2.2 Particle distributions

In the plasma sheet, the ion and electron particle distributions are commonly observed to be either a Maxwellian or kappa distribution (Christon et al., 1988; Wing and Newell, 1998; Wang et al., 2006). A Maxwellian distribution indicates that the population is in the thermal equilibrium. The kappa distribution has more high-energy particles than does a Maxwellian distribution. The distribution can either be a single component or consists of multiple components (Wing and Newell, 1998, Wang et al., 2007). A multiple component distribution often indicates a mixture of different populations of different origins, for example, one hot population coming from the tail mixes with one cold population coming from the flank (Wang et al., 2012a). Pitch angle distributions can vary from highly isotropic to highly anisotropic, and the anisotropy can vary significantly with species and energies (e.g., Wang et al., 2012b; Walsh et al., 2013). Pitch angle distributions are important to precipitation, and the different mechanisms for changing pitch angle are described in more detail in Section 4.3.1.

4.1.1.1 Statistical spatial distributions of particle fluxes, plasma moments, and magnetic field

The spatial structures of the plasma sheet plasma moments have been established with statistical studies (Lui and Hamilton, 1992; Angelopoulos et al., 1993; Wing and Newell, 1998; Hori et al., 2000; Tsyganenko and Mukai, 2003; Kaufmann et al., 2001, 2004; Wang et al., 2011). Figs. 4.2 and 4.3 show statistical spatial distributions of the plasma sheet obtained from THEMIS/ARTEMIS observations [see Wang et al. (2012a) for the dataset] for 10%, 50% (the medium value), and 90% percentiles. The geocentric solar magnetospheric (GSM) coordinates are used.

Fig. 4.2 shows statistical spatial distributions of the energy fluxes (averaged over all pitch angles) of the plasma sheet ions and electrons. Fig. 4.2A shows the medium energy fluxes as a function of X and energy for ions (top) and electrons around midnight ($|Y| < 10\ R_E$). The thermal energy, as indicated by the peak of energy fluxes, is $\sim$2 keV for ions and 0.2 keV for electrons in the midtail ($X \sim -40$ to $-70\ R_E$). Both the ion and electron thermal energies increase by an order of magnitude with decreasing radial distances to the Earth (r) from $X = -60$ to $-10\ R_E$. Fig. 4.2B and C shows the distributions across the tail in the near-Earth region and midtail, respectively. In the near-Earth region, the thermal energies are relatively higher at midnight than the flanks. Such midnight-flank differences are much weaker in the midtail.

The top panels of Fig. 4.2D–F show the spatial distributions of the medium energy fluxes in the X-Y plane for ions of 0.5 keV (cold ions), 10 keV (thermal ions), and 45 keV (hot ions), respectively. Similarly, the top panels of Fig. 4.2G–I shows the distributions for electrons of 0.2 keV (cold electrons), 2 keV (thermal electrons), and 21 keV (hot electrons). Their X profiles (in the direction of the X-axis) around midnight are shown in the bottom panels of Fig. 4.2D–I (blue curves), together with 10% and 90% percentiles (indicated by the vertical lines). For the cold ions and electrons, their fluxes do not vary significantly from the midtail to the near-Earth region. But there are strong midnight-flank asymmetries in the near-Earth region with higher fluxes closer to the two flanks. These cold particles have their entry from the flanks, and the different entry mechanisms are described in Section 4.1.2.2. For hot ions and electrons, their fluxes increase significantly (by almost two orders of magnitudes) from the midtail to the near-Earth region, and there are strong dawn-dusk asymmetries closer to the Earth. For hot ions, fluxes are higher on the duskside than the dawnside, while the opposite asymmetry is seen in the electrons. The asymmetry is mainly due to magnetic drift transport, which is described in Section 4.1.2.3. For thermal particles, their fluxes increase with decreasing r. The thermal ions have no strong spatial asymmetry. However, the fluxes of thermal electrons in the afternoon MLTs are much lower than other MLTs. This is mainly due to precipitation loss, which is described in Section 4.1.2.5 (see also Section 4.3.1). As indicated by 10% and 90% percentiles, particle fluxes at a fixed location can vary significantly by up to two orders of magnitudes.

Fig. 4.3 shows statistical distributions of the plasma sheet plasma moments and B_z. The top panels of Fig. 4.3A–F show the X-Y distributions of the medium values. The X profiles of 10%, 50%, and 90% percentiles around midnight are shown in the bottom panels of Fig. 4.3A–F, and their Y profiles are shown in Fig. 4.3G for the near-Earth region (red) and the midtail (blue). The medium values of number density (N) are $\sim 0.15\,\text{cm}^{-3}$ and do not change significantly from $X = -60$ to -20 R_E. Density starts to increase quickly inside $r \sim 20$ R_E to $\sim 0.5\,\text{cm}^{-3}$ at $r < 10$ R_E. Ion temperature (T_i), electron temperature (T_e), and plasma pressure (P_p) increase gradually with decreasing r, with $T_i \sim 0.8\,\text{keV}$, $T_e \sim 0.1\,\text{keV}$, and $P_p \sim 0.03\,\text{nPa}$ at $X \sim -60$ R_E increasing to $T_i \sim 4\,\text{keV}$, $T_e \sim 1\,\text{keV}$, and $P_p > \sim 0.3$ nPa at $r < 10$ R_E. The ion bulk flow in the midtail can be either earthward or tailward, but becomes dominantly earthward at $r < 30$ R_E. The medium flow speeds decrease with decreasing r. In the midtail, B_z is $\sim 2\,\text{nT}$. When inside $r = 30$ R_E, B_z first increases gradually with decreasing r, and then increases sharply inside $r \sim 10$ R_E. The region of the sharp B_z increase is often referred to as the "transition region," where the magnetic field configuration transitions from being more stretched downtail to being more dipolar closer to Earth. At a fixed location, the variations in the plasma moments and B_z can be up to one order of magnitude.

4.1.1.2 Example events for large-scale and meso-scale variations

The large plasma sheet variations at a fixed location as shown in Figs. 4.2 and 4.3 can be associated with variations of different scales. Here we show two example events in Figs. 4.4 and 4.5 for large-scale and meso-scale variations, respectively. Fig. 4.4 shows the plasma sheet over a large spatial domain (Fig. 4.4A) observed simultaneously by three THEMIS satellites (P1, P2, and P3) on March 8, 2008. The separations between these satellites were > 10 R_E. Fig. 4.4B shows that the plasma sheet at these different locations was relatively tenuous and hot for hours prior to ~09:30 UT. Then from ~09:30 to 09:50 UT, plasma at all these locations changed to relatively cold and dense and remained in this cold-dense state for ~2 h. These observations indicate that the density and temperature changes observed locally at each satellite location are actually associated with variations of large spatial and temporal scales. In addition, there were also variations with smaller temporal and spatial scales embedded within the large-scale hot-tenuous and the cold-dense states.

Fig. 4.5 shows the plasma sheet observed simultaneously by two THEMIS satellites (P2 and P4) on February 12, 2009. Both P2 and P4 were at $X \sim -10$ R_E but separated by ~10 R_E in the Y direction with P2 (P4) in the premidnight (postmidnight) sector (Fig. 4.5A). Fig. 4.5B shows that, before ~10:00 UT, the plasma moments, magnetic field, and particle fluxes at the P2 and P4 locations were similar and there were no clear perturbations (the plasma sheet had remained relatively steady for ~3 h). Corresponding to this steady plasma sheet plasma and magnetic field state, there were no meso-scale perturbations in the FACs and precipitations. This can be seen in Fig. 4.5C that there were

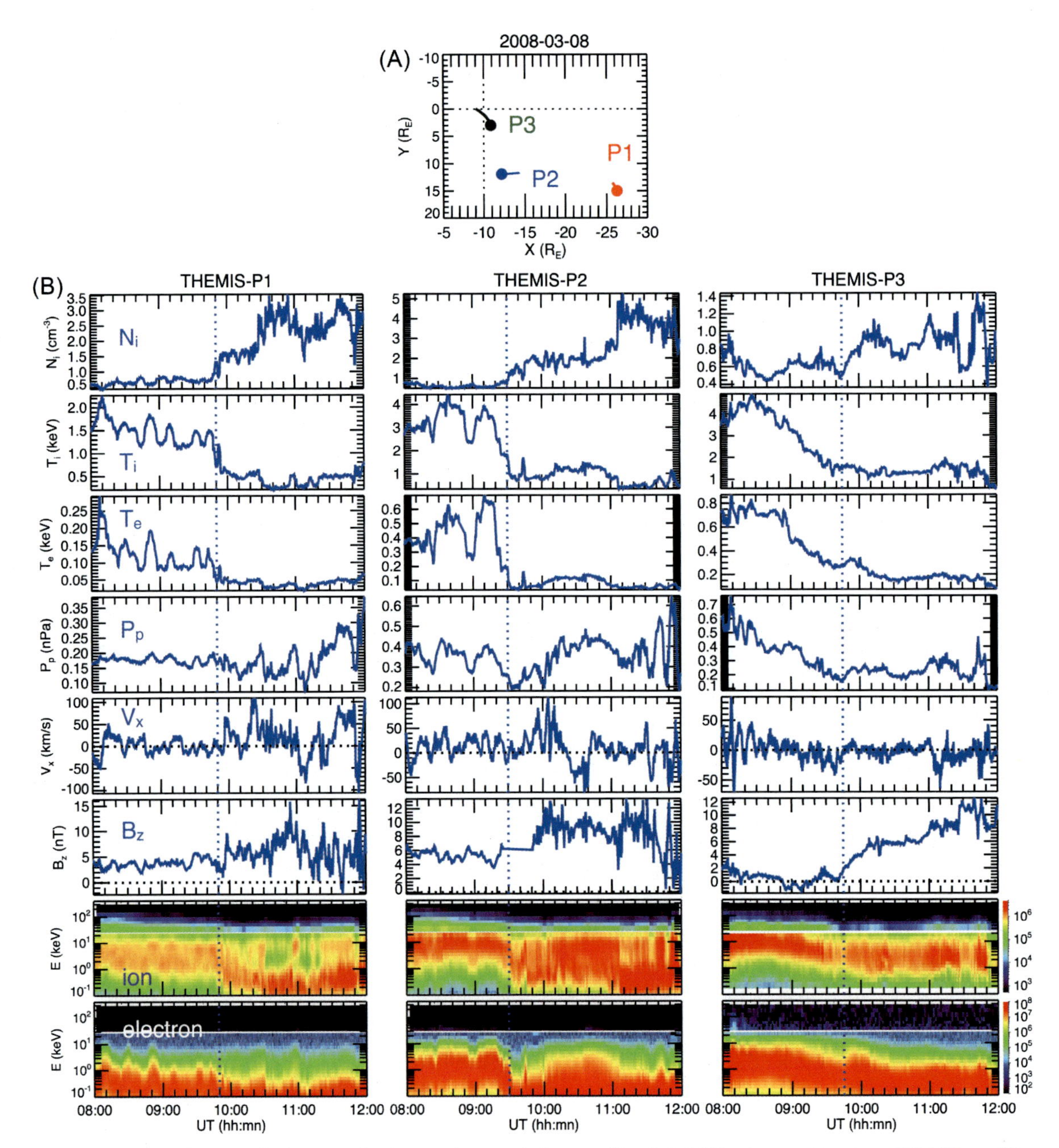

Fig. 4.4 An example event for large-scale plasma sheet variations on March 8, 2008, observed by three THEMIS satellites, P1, P2, and P3. (A) The locations of the satellites in the *X-Y* plane from 08:00 to 12:00 UT. The *solid circles* indicate the locations at 08:00 UT. (B) From top the bottom: Temporal profiles of ion densities, ion temperatures, electron temperatures, plasma pressures, V_x, B_z, and ion and electron energy fluxes [eV/(s-sr-cm^2-eV)] observed by P1 (left panels), P2 (middle panels), and P3 (right panels).

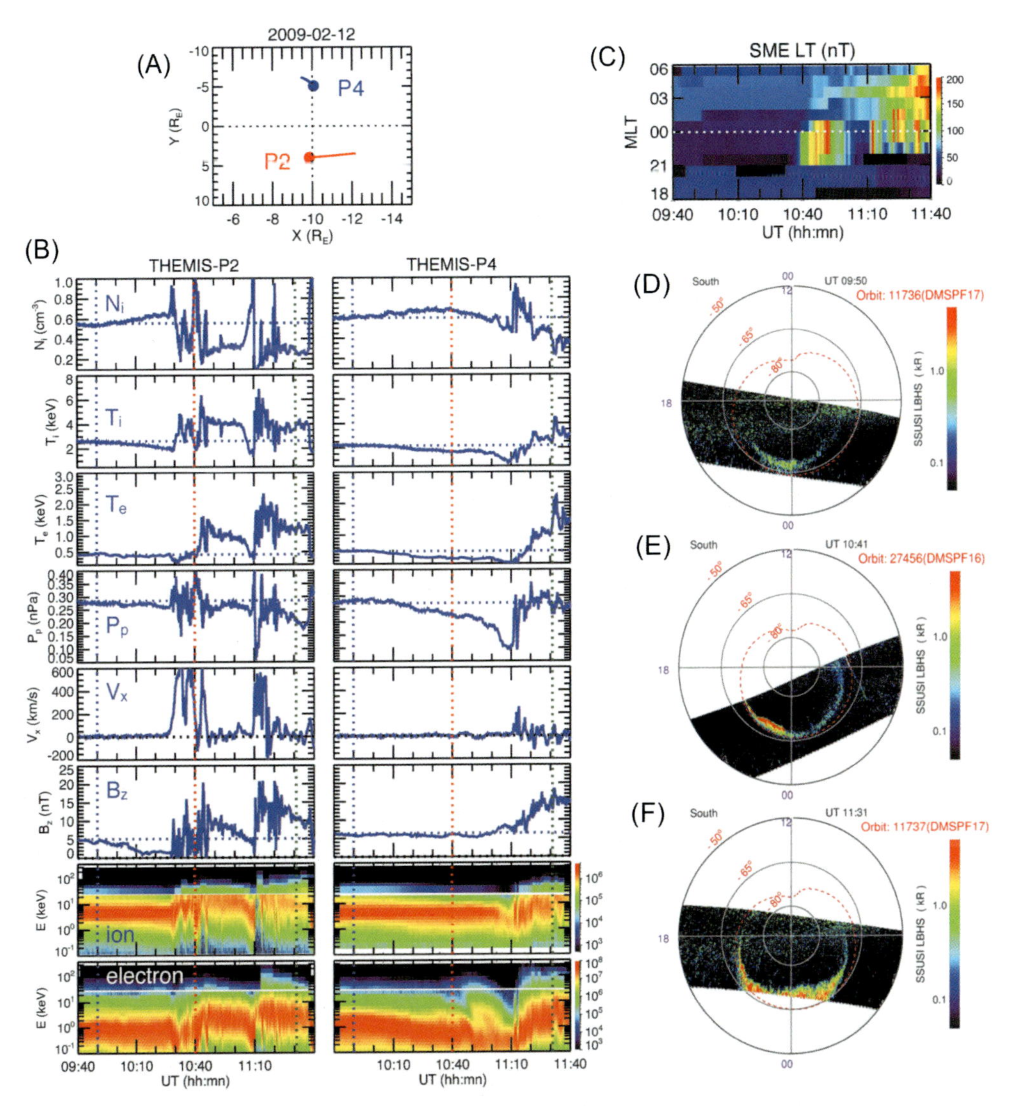

Fig. 4.5 An example event for meso-scale variations on February 12, 2009, observed by two THEMIS satellites, P2 and P4. (A) The locations of P2 and P4 in the *X-Y* plane from 09:40 to 11:40 UT. The *solid circles* indicate the locations at 09:40 UT. (B) From the top to bottom: temporal profiles of ion densities, ion temperatures, electron temperatures, plasma pressures, V_x, B_z, and ion and electron energy fluxes [eV/(s-sr-cm^2-eV)] observed by P2 (left panels) and P4 (right panels). (C) SuperMAG SME-LT index as a function of time and MLT. (D) Aurora image from DMSP SSUSI at (D) 09:50, (E) 10:41, and (F) 11:31 UT [indicated by the *blue*, *red*, and *green vertical dotted lines* in (B), respectively].

no strong magnetic field perturbations on the ground, as indicated by the SME-LT index (Newell and Gjerloev, 2014) from SuperMAG (Gjerloev, 2012), and in Fig. 4.5D there were no localized enhancements in aurora from DMSP SSUSI (Paxton et al., 2018). At ~10:30 UT, P2 observed strong earthward flows and large fluctuations in both plasma and magnetic fields. The perturbations lasted for ~15 min and the plasma sheet became relatively steady again, indicating that this was a meso-scale disturbance. The transient and strong earthward flows are called bursty bulk flows (BBFs) (Angelopoulos et al., 1992, 1994a,b). However, there were no strong perturbations observed at P4, indicating that the disturbances at P2 were spatially limited. The large magnetic field perturbations at P2 created strong FACs and enhanced the magnetic field perturbations on the ground only in the premidnight sector (Fig. 4.5C). Also, the enhanced electron fluxes in the plasma sheet caused auroral enhancement, which is also seen only in the premidnight sector (Fig. 4.5E). This indicates that the M-I coupling during this interval was at meso-scale. At ~10:55 UT, both P2 and P4 started to observe strong perturbations for ~15–30 min, and both the ground magnetic field perturbations and aurora were enhanced in both premidnight and postmidnight sectors (Fig. 4.4C and F). Compared to the first disturbance, the spatial extent of the second meso-scale disturbances was about twice as large.

There are many types of meso-scale perturbations in the plasma sheet, such as magnetic reconnection (Birn and Priest, 2007), BBFs (Angelopoulos et al., 1992), dipolarization fronts (or dipolarizing flux bundles) (Runov et al., 2009; Liu et al., 2013), current sheet flapping (including the kink mode and sausage mode) (Runov et al., 2005; Golovchanskaya and Maltsev, 2005; Gabrielse et al., 2008; Kubyshkina et al., 2014), Kelvin-Helmholtz vortices (Wang et al., 2017b), and ULF waves (McPherron, 2005; Hwang, 2015). They are either driven externally or internally and can be mutually exclusive. Some of them can be associated causally with each other, for example, reconnection is one of the processes that generate BBFs and flapping (Wang et al., 2020).

4.1.1.3 Dependence on solar wind and IMF

The variations of the plasma sheet properties have strong correlations with the external solar wind and interplanetary magnetic field (IMF) conditions. The overall plasma sheet density is higher when the solar wind density is higher, plasma sheet temperature is higher when the solar wind speed is higher, and the plasma sheet plasma pressure is higher when the solar wind dynamic pressure is higher (Borovsky et al., 1998; Tsygenenko and Mukai, 2003; Wang et al., 2007; Nagata et al., 2007). The plasma sheet is colder and denser during northward IMF and hotter and more tenuous during southward IMF (Terasawa et al., 1997; Wing and Newell, 1998; Wang et al., 2006, 2010; Nagata et al., 2007). The plasma sheet properties are also affected by ultralow frequency fluctuations (ULFs) in the solar wind and IMF (Wang et al., 2017a, b). These variations with the solar wind and IMF parameters result from combined processes of sources, entry, transport, energization, and losses. Each process may involve several different mechanisms (not mutually exclusive) and each mechanism may have different spatial and temporal scales.

4.1.2 Processes for determining plasma sheet particles distributions and their scales

The spatial distributions of plasma sheet particles are determined by several processes: source, entry, transport, energization, and losses.

4.1.2.1 The solar wind sources

The particle sources for the plasma sheet are the solar wind and ionosphere outflow (e.g., Welling et al., 2015). As shown in Fig. 4.1C and D, the variations in the solar wind density and temperature are more than one order of magnitude. With increasing solar wind speed, the solar wind density is generally lower, the solar wind ion temperature is higher, and the ULF fluctuations are larger (Wang et al., 2012a, b, 2017a).

Even when there are no variations in the upstream solar wind, large density and temperature variations can be generated in the region of the foreshock (that is, the region in front of quasi-parallel shock). These are the foreshock transients in a timescale of several minutes and spatial scales of $\sim$1 to $>$10 R_E (Zhang and Zong, 2020).

4.1.2.2 Entry

There are two routes for the solar wind particles to enter the plasma sheet: (1) First from the cusp or open-magnetopause to the lobes (becoming mantle plasma), and then to the plasma sheet (Ashour-Abdalla et al., 1996; Wang and Xing, 2021). (2) Direct entry into the plasma sheet (Wing et al., 2014). The plasma entering through route 1 is relatively hotter and tenuous, while the plasma entering through route 2 is colder and denser and appears near the flanks. Route 1 is the main route when the IMF is southward, whereas route 2 is prevalent when the IMF is northward. A mixture of plasma entering through these two routes results in two-component energy spectra (Wing and Newell, 1998; Wang et al., 2012a, b).

For route 1, the mantle plasma is distributed over a large portion of the magnetosphere but can have a dawn-dusk asymmetry depending on the direction of IMF B_y (Maezawa and Hori, 1998; Wang et al., 2014a; Wang and Xing, 2021). The mantle plasma can enter the plasma sheet in the tail through either large-scale steady reconnection (Hietala et al., 2015) or localized reconnection (Wang et al., 2020). The localized tail reconnection has an azimuthal size of $\sim$5–10 R_E and occurs more preferentially in the premidnight than postmidnight sector (Li et al., 2014).

For route 2, there are mainly four mechanisms (Wing et al., 2014): (1) double cusp/lobe reconnection, (2) Kelvin-Helmholtz (K–H) instability, (3) Kinetic Alfvén wave diffusion, and (4) impulsive penetration. The entry through double cusp/lobe reconnection can be slow and extend over the magnetospheric spatial scale, while the entry through the other three mechanisms is relatively localized and transient.

4.1.2.3 Transport

Drift is the main transport mechanism within the plasma sheet. Drift includes electric drift (also called $E \times B$ drift or convection) and magnetic drift. Electric drift does not depend on particle species or energy, while magnetic drift depends on particle species, energy, and pitch angles. Because of magnetic drift, ions and electrons of different energies drift separately and result in the dawn-dusk asymmetries in the spatial distributions of high-energy particle fluxes shown in Fig. 4.2 and the temperatures shown in Fig. 4.3.

Drift is typically slow with speeds <100 km/s (Borovsky, et al., 1997), except, at times, fast flows or BBFs are observed, as shown in Figs. 4.4 and 4.5. The occurrence rate of BBFs is higher at larger r (Wang et al., 2009; Kiehas et al., 2018). The slow drift is large-scale convection driven externally by the solar wind and IMF (Wang et al., 2006) and their ULF fluctuations of IMF (Kim et al., 2009). Fluctuations in the slow drift speed (Borovsky et al., 1997) can lead to slow diffusive transport that brings particles from the flanks inward over a timescale of hours (Antonova, 2006; Wang et al., 2010). On the other hand, BBFs are meso-scale transport with a Y-scale of $\sim 3\ R_E$ in the near-Earth region (Liu et al., 2013) and ~ 5–$10\ R_E$ in the midtail (Li et al., 2014). BBFs are associated with localized electric field variations due to reconnection (Wang et al., 2020) or interchange instability (Chen and Wolf; 1993, 1999; Xing and Wolf, 2007). In addition, meso-scale interchange motion associated with the K–H entry can transport cold particles radially inward from the flanks (Henderson, 2012; Wang et al., 2014b). The ionospheric mapping of these electric field distributions associated with the above drift transport in different scales is further described in Section 4.2.

4.1.2.4 Energization

Plasma sheet particles are energized by adiabatic and nonadiabatic heating, depending on whether the first and second adiabatic invariants are conserved. Adiabatic heating can be associated with drift transport, reconnection, current sheet flapping (the sausage mode), or external magnetospheric compression/decompression. Adiabatic heating associated with the large-scale, slow-flow drift results in the increase in temperatures and plasma pressures from the midtail to the near-Earth as shown in Fig. 4.3. Thus, large-scale adiabatic heating plays an important role in generating large-scale FACs (Yue et al., 2015) and precipitation energy fluxes into the ionosphere.

Wave-particle interaction can result in nonadiabatic heating. Numerous wave modes are capable of heating plasma sheet ions and electrons (Ni et al., 2016). These wave modes may be considered as small scale, but the wave energy sources may come from meso-scale disturbances. For example, BBFs can produce electron cyclotron harmonic and ELF chorus waves.

4.1.2.5 Losses

Plasma sheet particles are on open drift trajectories, so they eventually encounter the magnetopause and are lost. While within the plasma sheet, the main loss is due to precipitation. Ion loss due to charge exchange and electron loss due to Coulomb collisions is only important in the inner magnetosphere (Welling et al., 2015). The precipitation loss is more efficient for electrons than ions. The precipitation loss is due to current sheet scattering and wave-particle scattering. More details of these two mechanisms are described in Section 4.3.

4.2 Observations of multiscale convection, precipitation, and conductivity

Because magnetosphere convection and particle scattering into the ionosphere occur on multiple scales, as shown in examples in Section 4.1.1.2, it follows that ionosphere flows, particle precipitation, and the resulting conductivity variation also occurs on multiple scales. As discussed in Chapter 1, the magnetosphere and ionosphere are coupled via Earth's magnetic field. A magnetic field line's location in the ionosphere is known as its *footprint*. A magnetic field line with a footpoint at high latitudes will extend farther from Earth than a magnetic field line with a footpoint at lower latitudes. Therefore, a phenomenon like a fast plasma flow in Earth's magnetotail that travels Earthward will map to the ionosphere as moving equatorward. The following is a summary of observational techniques and results regarding multiscale flow, precipitation, and conductivity, organized by a scale size. Although this review cannot cover all previous literature on the topics of convection, precipitation, and conductivity, it will be a good starting point for the curious reader. Readers are encouraged to also refer to Sections 1.1–1.3 on basic concepts on the convection, precipitation, and conductance, and Chapter 2 on aurora.

4.2.1 Large-scale convection and related precipitation and conductivity

4.2.1.1 Large-scale convection

As described in Section 1.1, large-scale convection occurs in the magnetosphere and high-latitude ionosphere. The shape and intensity of the two-cell convection pattern differ depending on the solar wind and the IMF. Mapping of the large-scale convection pattern in the high-latitude ionosphere as a function of the IMF *clock angle*—a measure of the degree to which the solar wind magnetic field points in the $\pm B_Z$ vs $\pm B_Y$ direction—has been progressively detailed over the years with increased observations both in space and from the ground [e.g., Heppner (1977) using Ogo 6; Heppner and Maynard (1987) using Dynamics Explorer 2; Rich and Hairston (1994) using DMSP; Weimer (1995) using DE satellite data; Ruohoniemi and Greenwald (1996) using ground-based HF radar data; Papitashvili and Rich (2002) using DMSP; Ruohoniemi and Greenwald (2005) and Cousins and Shepherd (2010) using SuperDARN HF radar data;

Haaland et al. (2007) using Cluster data]. Fig. 4.6, which is replicated from Haaland et al. (2007), illustrates the convection cells. During −IMF B_Y, the convection pattern is symmetric about midnight. During +IMF B_Y, the convection pattern tilts so that the central, faster part of the antisunward convection near the poles lies postmidnight, but the nightside, equatorward portion (in the oval) lies premidnight. When IMF B_Z is negative—resulting in dayside reconnection—large-scale convection is enhanced as compared to +IMF B_Z. During +IMF B_Z, three or even four convection cells may form from

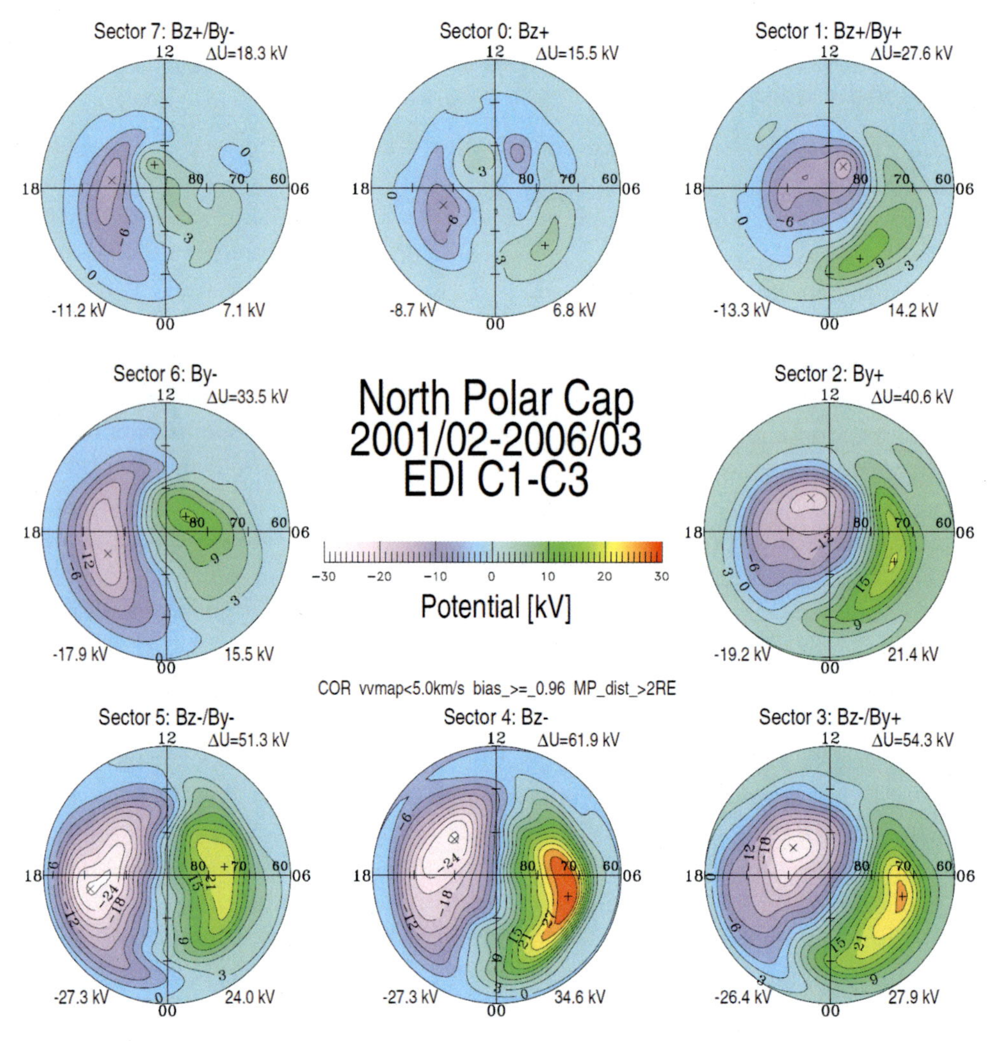

Fig. 4.6 Fig. 7 from Haaland et al. (2007) demonstrating the two-cell (sometimes three- or four-cell) convection orientation as a function of IMF clock angle. Electric potential was obtained statistically from Cluster velocity measurements.

reconnection between the IMF and preexisting open flux in the tail lobe (Dungey, 1963; Reiff and Burch, 1985; Cowley and Lockwood, 1992).

4.2.1.2 Large-scale precipitation

When open-field lines reconnect to each other in Earth's magnetotail, they create stretched, closed field lines that trap plasma between Earth's hemispheres, forming the magnetotail plasma sheet. Particles with some parallel momentum move along the closed field line, bouncing back and forth between Earth's northern and southern hemispheres where the field line's footpoints exist. A particle with only perpendicular momentum (having a 90° pitch angle) will remain at the center of the plasma sheet (where B_X switches sign), whereas particles with increasing parallel momentum (having pitch angles approaching 0° or 180°) will bounce increasingly closer to the ionosphere. The point at which a particle population bounces back away from the ionosphere is called the *mirror point.* A particle with enough parallel momentum will travel so far from the center of the plasma sheet toward Earth that it will interact with the particles comprising the ionosphere prior to reaching its mirror point, and never return. These particles are said to fall into the *loss cone.*

Because the magnetotail is so stretched—especially downtail where reconnection takes place—the field line radius of curvature (R_C) approaches the scale size of the particle gyroradius (ρ). When $R_C/\rho < 8$ (Sergeev et al., 1983), the first adiabatic invariant is violated, causing particles to move chaotically (Lyons and Speiser, 1982) and allowing them to be scattered—or lost—to the ionosphere. The first adiabatic invariant is defined by:

$$\mu = \frac{m v_{\perp}^{2}}{2B} \tag{4.1}$$

where m is a particle mass, $v_{\perp}$ is the particle velocity perpendicular to the magnetic field, B is the magnetic field magnitude. The first adiabatic invariant remains constant (so particle perpendicular energy increases with increasing magnetic field strength) as long as the magnetic field is constant on timescales of the gyrofrequency Eq. (4.2) and spatial scales of the particle gyroradius Eq. (4.3).

$$\omega_g = \frac{|q|B}{m} \tag{4.2}$$

$$\rho = \frac{m v_{\perp}}{|q|B} \tag{4.3}$$

where q is the charge of the particle. A deeper description of field-line curvature scattering can be found in Section 4.3.1.2.

The process of being scattered into the loss cone is a form of *particle precipitation.* Precipitating particles with energies ~1–15 keV interact with the Earth's upper atmosphere, depositing their energy there and creating the diffuse aurora. Particles with much higher energies penetrate even deeper into Earth's atmosphere and, though they cause other phenomena, do not create aurora.

The poleward boundary of the main auroral oval is therefore defined by the open-closed field line boundary. The poleward edge is observationally defined as where the precipitating fluxes drop by a factor of at least 4 over a short distance to values below $3 \cdot 10\ \mathrm{eV/cm^2\ s\ sr}$ (electrons) or $10\,\mathrm{eV/cm^2\ s\ sr}$ (ions) (Newell et al., 1996). This is not to say there is absolutely no precipitation in the polar cap. Weak, uniform electron precipitation called *polar rain* (Winningham and Heikkila, 1974; Gussenhoven et al., 1984) occurs poleward of the dayside auroral oval, gradually decreasing in flux toward the nightside (Newell et al., 1996). There is also a subvisual drizzle of particles on the nightside extending poleward of the main auroral oval, terminating when fluxes go down to background levels. The subvisual drizzle boundary is observationally defined as either the point where polar rain is encountered or the log of electron flux drops below 10.4 and the log of ion flux drops below 9.6 (Newell et al., 1996).

Closer to Earth, where the field lines are not so stretched, the first adiabatic invariant is maintained and particles are no longer pitch angle scattered from crossing stretched field lines. This transition to more dipolar field lines is called the *isotropic boundary* (Sergeev et al., 1983; Sergeev and Bösinger, 1993; Newell et al., 1996, 2009; Newell and Meng, 1998). It is named thusly because poleward (or tailward) of the isotropic boundary, the loss cone population is isotropic since curvature scattering affects particles of all pitch angles. Sergeev et al. (1983) observationally showed the proton isotropic boundary with ESRO-IA, a low-altitude satellite, by locating where the ratio of perpendicular to parallel flux approached unity. Earthward (equatorward) of the isotropic boundary, perpendicular flux dominates and there is a lack of precipitation, as illustrated in Fig. 4.7. Ganushkina et al. (2005) plotted trapped (j_T), or perpendicular, and precipitating (j_P), or parallel, flux from a NOAA 12 satellite that flew poleward over an empty loss cone (j_P) near zero) and over the isotropic boundary (IB), where the parallel fluxes increased to match those of the perpendicular fluxes (the definition of an isotropic population). As the satellite continued poleward, it left the auroral oval and entered the polar cap where fluxes of all pitch angles decreased. As the satellite reentered the auroral oval from the poleward side, isotropic flux increased until it crossed the IB, on the other side of which parallel fluxes dropped off as NOAA 12 continued equatorward.

Because ions have larger gyroradii than electrons, the two species have separate isotropic boundaries. Protons require field lines with a larger radius of curvature to maintain adiabaticity, meaning the proton isotropic boundary is equatorward (or earthward) of the electron isotropic boundary. Furthermore, because energy affects gyroradius, the high-energy protons have an isotropic boundary closer to Earth than the low-energy protons. The high-energy (few to 10s keV) proton isotropic boundary is approximately colocated with the equatorward boundary of the visible diffuse proton aurora and can be considered the closest proxy for the earthward edge of the magnetotail current sheet

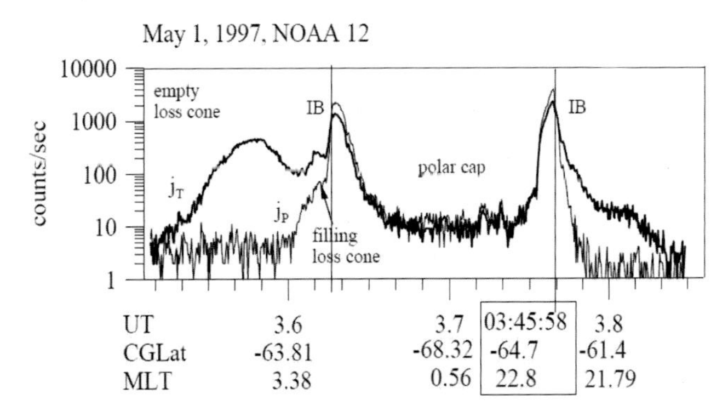

Fig. 4.7 Precipitating (j_P) and trapped (j_T) particles were measured by NOAA 12, as it first traveled poleward across empty loss cones, crossed the isotropic boundary (IB), then saw isotropy in the particle distribution prior to reaching the polar cap. After crossing the polar cap, NOAA 12 traveled equatorward, passing through an isotropic auroral oval until it reached the IB on the other side. *From Ganushkina, N., Yu., T.I., Pulkkinen, M.V., Kubyshkina, V.A., Sergeev, E.A., Lvova, T.A., Yahnina, A.G., Yahnin, and Fritz, T., 2005. Proton isotropy boundaries as measured on mid- and low-altitude satellites, Ann. Geophys. 23, 1839–1847.*

(Newell et al., 1996). Precipitation equatorward (earthward) of the isotropic boundary must have been caused by other mechanisms, such as wave-particle interactions. Newell et al. (1996) defined the ion isotropic boundary as the precipitating ion energy flux peak (integrated over the range 3–30 keV), as observed by a low-altitude satellite such as DMSP. The exact location of the isotropic boundary is dependent on the ion energies, magnetic local time (MLT), and magnetospheric conditions.

In contrast to the proton population, the electron isotropic boundary has no observational definition. Throughout the plasma sheet, electromagnetic waves with sufficiently high frequencies comparable to the trapped particle's gyration frequency can scatter particles into the loss cone, creating the diffuse aurora. At larger distances from Earth ($>\sim 8\ R_E$), electrostatic electron cyclotron harmonic waves are an important source of nightside electron diffuse auroral precipitation. Ions, on the other hand, can be scattered by electromagnetic ion cyclotron waves. Closer to Earth, in the inner magnetosphere, chorus waves cause both the intense electron diffuse aurora on the nightside as well as the weaker dayside electron diffuse aurora. Section 4.3.1.1 discusses wave-particle interactions that cause precipitation in more detail. Ni et al. (2016) presented a detailed review of the types of waves that scatter particles to create diffuse aurora, including where the waves typically occur and how they interact with the particles.

As convection transports particles closer to Earth, the gradient in Earth's magnetic field strength increases and energetic particles begin to gradient-curvature drift around the Earth. More energetic particles are more affected by gradient-curvature drift, meaning increasingly less energetic particles can more closely approach the Earth. The observed precipitation therefore also falls off in energy as a low-altitude satellite approaches the equatorward boundary of the auroral oval. Newell (1996) defined this transition point where gradient-curvature drift starts taking over particle motion as the earthward boundary of the plasma sheet, observed by a low-altitude, polar-orbiting satellite when the average energy starts decreasing with decreasing latitude (or, increasing with increasing latitude).

The gradient-curvature drift of energetic particles means zero-energy particles whose motion is only defined by $\mathbf{E} \times \mathbf{B}$ drift that makes it closest to Earth until the Earth's corotation electric field sweeps them around Earth. Because neither zero-energy electrons nor zero-energy protons are affected by gradient-curvature drift, the zero-energy electron and proton boundaries lie approximately on top of one another. Although in a geomagnetically quiet situation, the zero-energy boundary would be synonymous with the plasmapause, Newell et al. (1996) pointed out that the bursty nature of magnetotail convection means we should keep these definitions separate.

As noted earlier, ions of increasing energy have an isotropic boundary closer to Earth, meaning the average temperature of the precipitating ions increases in the earthward (equatorward) direction (see Fig. 4.8A). However, because less energetic ions can penetrate closer to Earth before gradient-curvature drift alters their trajectories, and if waves are present to scatter them into the loss cone, a low-altitude satellite will observe decreasing average temperatures of the precipitating ions as it travels equatorward of the isotropic boundary (see Fig. 4.8B). This "reverse-dispersion," along with the "normal-dispersion," has been observed and studied with FAST overflights by Donovan et al. (2003) and Liang et al. (2014), who correlated the reverse-dispersion events with EMIC waves in the inner magnetosphere in the postmidnight region.

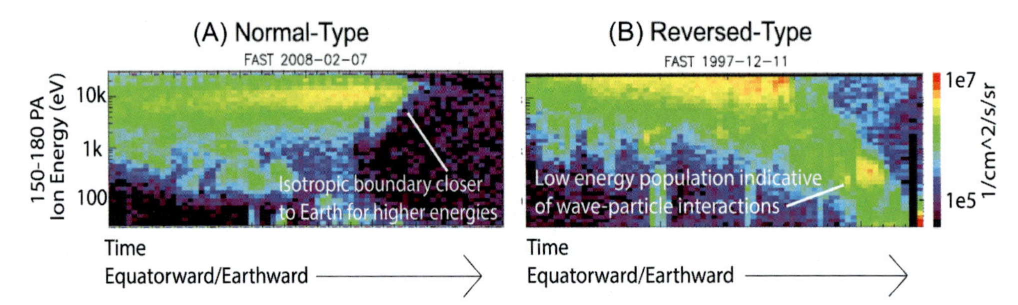

Fig. 4.8 FAST satellite particle observations of (A) typical/normal and (B) reversed-type dispersion. Because higher energies have isotropic boundaries closer to Earth, a satellite traveling equatorward would observe a dispersion in precipitating flux cut-off from low-to-high energies. However, when there are wave-particle interactions, a satellite traveling equatorward would observe a reversed dispersion where the precipitating flux cut-off goes from high-to-low energies. This is because lower energy populations $E \times B$ drift closer to Earth. *Modified from Liang, J., Donovan, E., Ni, B., Yue, C., Jiang, F., Angelopoulos, V., 2014. On an energy-latitude dispersion pattern of ion precipitation potentially associated with magnetospheric EMIC waves. J. Geophys. Res. Space Phys. 119, 8137–8160. https://doi.org/10.1002/2014JA020226.*

This section has attempted to summarize large-scale precipitation by focusing on large-scale precipitation as it relates to large-scale convection. See Newell et al. (1996) for much greater detail including additional auroral boundaries, and Ni et al. (2016) for an in-depth review on diffuse aurora. Fig. 4.9, taken from Newell et al. (1996), presents the particle flux observed by a DMSP satellite as it flew over the auroral oval in the equatorward direction over time. It summarizes the precipitation observations and different boundaries, mapping from downtail at the open-closed field line region toward Earth, until the plasmapause. Note that the ion energy y-axis is flipped compared to Fig. 4.8 so that Fig. 4.9 shows a reverse-dispersion indicative of wave-particle interactions in the inner magnetosphere. Fig. 4.10, taken from Newell et al. (2004), summarizes some of the content discussed above by depicting different precipitation regions in the ionosphere, labeling where they map to in the magnetosphere. Fig. 4.10 is for $-$IMF B_z, though Newell et al. (2004) also included a figure for +IMF B_z. The precipitation regions were drawn from 6 years of DMSP data. The superimposed contours are the equipotentials drawn from SuperDARN statistics taken from similar IMF conditions.

4.2.1.3 Large-scale conductivity

Precipitation affects the ionosphere's conductivity. The ionosphere's conductivity affects how easily currents flow through the ionosphere-thermosphere and, in contrast, how tied the magnetic field lines are to the ionosphere (see Ridley et al., 2004 for a detailed discussion of the ionosphere's control of the magnetosphere via conductance). There are two basic sources of ionization that enhance conductivity: solar extreme ultraviolet (EUV) and particle precipitation. For the latter, as electrons and ions precipitate, they collide with neutral atmospheric constituents that can become excited or ionized. Electron density increases due to this enhanced ionization, which increases ionospheric conductance due to the linear relationship between electron density and conductivity. Starting with Ohm's Law with no magnetic field present, where $\mathbf{J}$ is current density, $\mathbf{E}$ is electric field vector, and σ_0 is conductivity scalar dependent on collision frequencies,

$$\boldsymbol{J} = \sigma_0 \boldsymbol{E} \tag{4.4}$$

This is derived from steady-state momentum equations, where q is the charge of an ion, $m_e$and m_i are the masses of an electron and ion, respectively, and ν_{en} and ν_{in} are the electron and ion collision frequencies, respectively:

$$-q\boldsymbol{E} = m_e \nu_{en} \boldsymbol{u}_e \tag{4.5}$$

$$q\boldsymbol{E} = m_i \nu_{in} \boldsymbol{u}_i \tag{4.6}$$

where

$$\boldsymbol{J} = n_e(\boldsymbol{u_i} - \boldsymbol{u_e}) \tag{4.7}$$

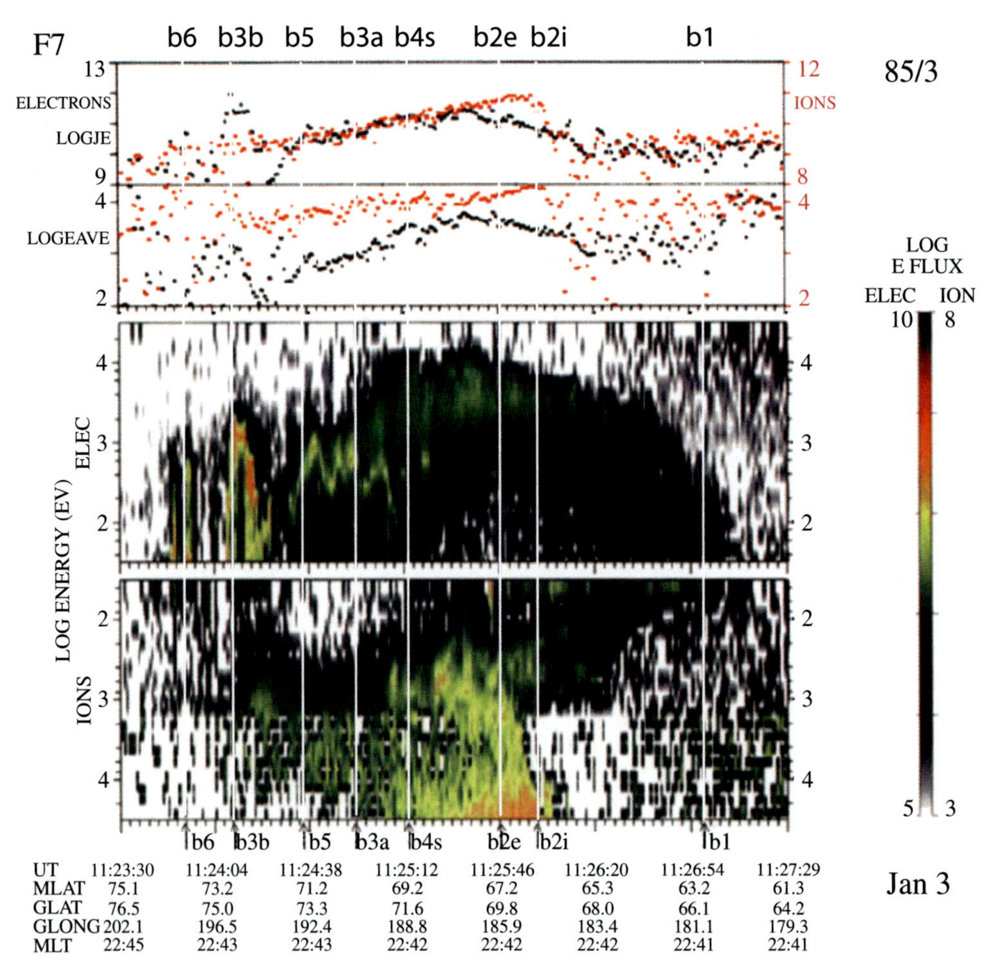

Fig. 4.9 DMSP data as the satellite passed over the nightside auroral oval (see text for details). B1e, b1i: zero-energy particles experiencing zero curvature or gradient drifts, often coinciding with the plasmapause. B2e, b2i: the point where the change in energy over the change in latitude is equal to zero; the start of the main plasma sheet. B3a, b3b: most equatorward and poleward electron spectra showing signs of field-aligned acceleration through a potential drop (monoenergetic peaks), respectively. B4s: electron precipitation near b2 often lacking spatial structure. B5e, b5i: sharp poleward cutoff of the contiguous oval with a drop in fluxes by four times over 0.2° latitudinal range. B6: poleward boundary in quiet conditions, bounds a region of low-energy, highly structured electron, and ion precipitation at low flux levels. This region between b5 and b6 is subvisual drizzle poleward of the oval. *Modified from Newell, P.T., Feldstein, Y.I., Galperin, Y.I., Meng, C.-I., 1996. Morphology of nightside precipitation. J. Geophys. Res. 101, A5, 10,737–10,748. https://doi.org/10.1029/95JA03516.*

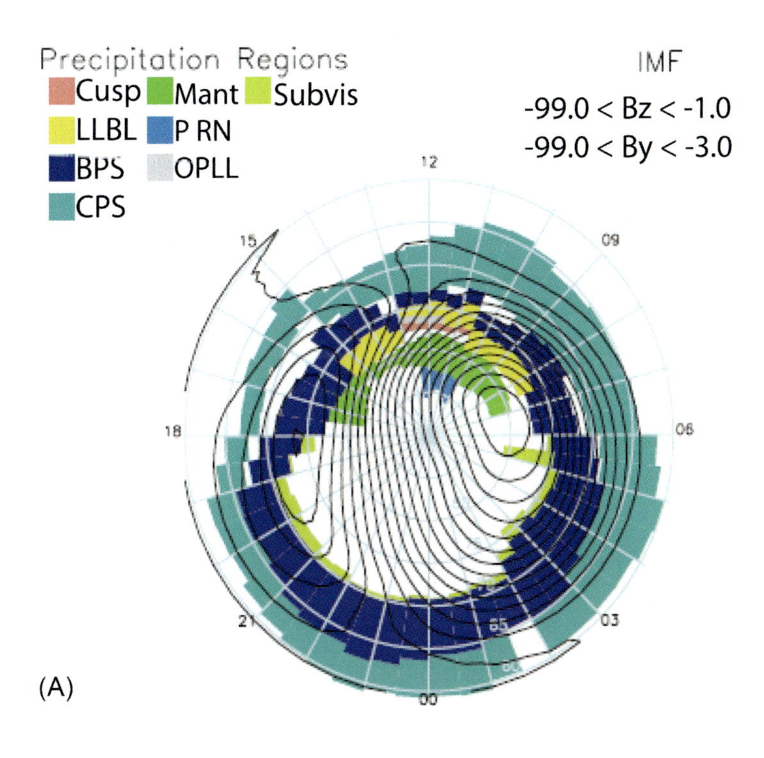

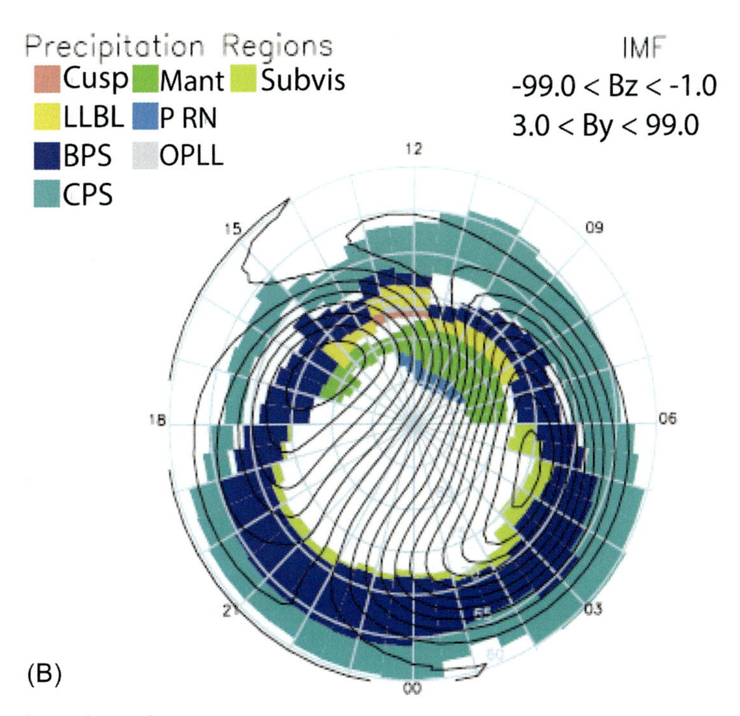

Fig. 4.10 See figure legend on next page

(Continued)

$$\sigma_0 = \left(\frac{1}{m_e \nu_{en}} + \frac{1}{m_i \nu_{in}}\right) n_e e^2 \tag{4.8}$$

Introducing a magnetic field in the z-direction affects the ion and electron motion differently, altering the equation as follows

$$J = \begin{pmatrix} \sigma_P & -\sigma_H & 0 \\ \sigma_H & \sigma_P & 0 \\ 0 & 0 & \sigma_0 \end{pmatrix} \begin{pmatrix} E_X \\ E_Y \\ E_Z \end{pmatrix} \tag{4.9}$$

where

$$\sigma_P = \left(\frac{1}{m_e \nu_{en}} \left(\frac{\nu_{en}{}^2}{\nu_{en}{}^2 + \Omega_e^2}\right) + \frac{1}{m_i \nu_{in}} \left(\frac{\nu_{in}{}^2}{\nu_{in}{}^2 + \Omega_i^2}\right)\right) n_e e^2 \tag{4.10}$$

$$\sigma_H = \left(\frac{1}{m_e \nu_{en}} \left(\frac{\Omega_e \nu_{en}}{\nu_{en}{}^2 + \Omega_e^2}\right) - \frac{1}{m_i \nu_{in}} \left(\frac{\Omega_i \nu_{in}}{\nu_{in}{}^2 + \Omega_i^2}\right)\right) n_e e^2 \tag{4.11}$$

The Pederson conductivity (σ_P) is parallel to the perpendicular electric field, whereas the Hall conductivity (σ_H) is perpendicular to both the perpendicular electric field and the magnetic field. If an electric field exists parallel to the magnetic field (E_Z), its associated conductivity is σ_0, the same as in the no magnetic field case. The parallel electric field is usually assumed to be small in the ionosphere. *Conductance*—the height integration of conductivity—is used in MHD models to simulate the ionospheric response as a resistive load for the magnetosphere (e.g., Boström, 1974).

To date, there are no observational techniques that can directly measure conductivity or conductance, making it a very difficult parameter to specify. Conductivity can be indirectly measured using observations of electron density and temperature, which are used to calculate the ion-neutral collision frequency. Incoherent scattering radar (ISR) provides one way to simultaneously measure these parameters. ISRs have a single-point look

Fig. 4.10, cont'd Depictions of different precipitation regions in the ionosphere were collected from DMSP statistics, with superimposed equipotentials from SuperDARN statistics taken during similar IMF conditions (see text for details). Labels: Cusp (Newell and Meng, 1988: energies and densities approximate that of the frontside magnetosheath), LLBL (low latitude boundary layer: Newell et al., 1991a; Newell and Meng, 1998, can be open or closed field lines), BPS (plasma sheet boundary: here means the region of structured aurora), CPS (central plasma sheet: equatorward boundary defined as b0, the latitude at which significant electron or ion precipitation is observed, often equivalent to b1e, corresponding to plasmapause), Mant (Mantle, Newell et al., 1991b: deaccelerated shocked solar wind), P RN (polar rain), OPLL (open low-latitude boundary layer), Subvis (subvisual). (A) +IMF B_z. (B) − IMF B_z. *Modified from Newell, P.T., Ruohoniemi, J. M., Meng, C.I., 2004. Maps of precipitation by source region, binned by IMF, with inertial convection streamlines. J. Geophys. Res. 109, A10206. https://doi.org/10.1029/2004JA010499.*

direction but can scan across various elevation angles (e.g., Vondrak and Baron, 1976) or, more recently, point in multiple look directions on a pulse-to-pulse basis (e.g., Kelly and Heinselman, 2009). ISR observations are discussed in more detail in Section 4.2.2.3. The gyrofrequencies needed for Eq. (4.8) are determined using a magnetic field model. Electron- and ion-neutral collision frequencies require input from a neutral atmosphere model, which is run for a specified date, time, and geophysical conditions. A key empirical neutral atmosphere model that is used is the NRLMSIS00 model (Picone et al., 2002).

The ion-neutral collision frequencies for nonresonant interactions have the form [eq. (4.88) in Schunk and Nagy, 2009],

$$\upsilon_{in} = 2.21\pi \frac{n_n m_n}{m_i + m_n} \sqrt{\frac{\gamma_n e^2}{\mu_{in}}} \tag{4.12}$$

where n_n is the neutral mass density, m_i and $m_n$are the ion and neutral mass, $\mu_{in}$is the reduced mass, and $\gamma_n$is the neutral polarizability, with values given in table 4.1 in Schunk and Nagy (2009) and not repeated here for brevity. The resonant ion-neutral collision frequencies must also be considered for situations of temperature enhancements, which are described in table 4.5 of Schunk and Nagy (2009) and not repeated here.

As previously noted, the conductivity is linearly related to the electron density. In some cases, the electron density can also be inferred from the number flux associated with particle precipitation. Precipitating energetic particles interact with neutrals in the thermosphere to cause ionization and excitation. The former process leads to enhanced electron density, and the latter produces visible auroras at different wavelengths depending on the energy of the incident particles. It is therefore conceivable that measurements of the energy spectral distributions of precipitating particles and auroral emissions can be used to infer auroral conductances.

How precipitating particle fluxes generate ionization has been extensively investigated (e.g., Rees, 1963; Strickland et al., 1993, 1994; Basu et al., 1993; Fang et al., 2008, 2010) and for brevity, we do not review this topic here. When an appropriate chemistry model for the D- or E-region of the ionosphere is used, the electron density can be determined and then used to calculate the conductivity (e.g., Kaeppler et al., 2020 and references therein). Two defining characteristics of the precipitating electron flux are the energy flux:

$$Q = \int \Phi(E) dE \tag{4.13}$$

and average energy:

$$\langle E \rangle = \frac{\int E\Phi(E) dE}{\int \Phi(E) dE} \tag{4.14}$$

where $\Phi(E)$ is the particle differential number flux. Satellite overpasses from missions such as DMSP measure the energy flux and average energy directly, since the instruments on board the satellites can resolve the differential number flux (e.g., Hardy et al., 1987, Fuller-Rowell and Evans, 1987; Newell et al., 2009; Redmon et al., 2017). Once energy flux and average energy are known, they can be input into ionospheric models such as Boltzmann 3-Constituent (B3C) (Strickland et al., 1993) and GLobal AirglOW (GLOW), which rely on a neutral atmosphere model, to compute conductance (Solomon et al., 1988; Solomon and Abreu, 1989; Bailey et al., 2002; Solomon, 2017).

Early efforts were made to derive auroral conductances from the in situ measurements of precipitating electrons by polar-orbiting satellites (e.g., Wallis and Budzinski, 1981; Spiro et al., 1982; Vondrak and Robinson, 1985). To derive the Pedersen and Hall conductances, Wallis and Budzinski (1981) and Vondrak and Robinson (1985) took a similar approach as Evans et al. (1977) (see Section 4.2.3). Spiro et al. (1982), on the other hand, took a less sophisticated approach by simply applying the empirical results of Vickrey et al. (1981) to infer the conductances directly from the average energies and energy fluxes of precipitating electrons. Vondrak and Robinson (1985) validated the Pedersen and Hall conductances inferred from the satellite measurements with those from the Chatanika radar and found the difference between the satellite and radar-derived conductances to be less than 25%.

Recently, Robinson et al. (2020) investigated the relationship between field-aligned currents from the active magnetosphere and planetary electrodynamics response experiment (AMPERE) (Anderson et al., 2014) and the simultaneous auroral conductances were measured by the PFISR radar during 20 geomagnetically active days. Based on the statistical results, the authors further developed analytical functions to infer the Pedersen and Hall conductances from the intensity and polarity of field-aligned currents. Since the AMPERE project provides global FAC patterns at a cadence of 2–10 min, the newly developed conductance model represents a unique opportunity to specify global auroral conductance patterns in both hemispheres at the same temporal cadence as the AMPERE data can offer. Once this new FACs-driven conductance model is more thoroughly validated, it can greatly improve the specification of ionospheric electrodynamics globally, especially under disturbed conditions.

Another way to indirectly measure conductance from characteristic energies and energy flux was developed by Robinson et al. (1987), who determined empirically based formulas to calculate the Hall and Pederson conductances based on the average energy and energy flux assuming that the precipitation distribution function was Maxwellian—an assumption that adequately describes diffuse aurora (e.g., Rees, 1989; Khazanov et al., 2015).

One of the advantages of using satellite observations, i.e., DMSP observations, to determine precipitating number flux is that they can be used to specify the global high-latitude distribution of conductance statistically. A key disadvantage of DMSP satellite observations is the >90-min revisit time due to its orbit, making it difficult to observe the time evolution of precipitation. A number of investigations have used satellite-based observations to specify the conductance statistically at high latitudes (e.g., Hardy et al., 1987; Fuller-Rowell and Evans, 1987; Robinson et al., 1987; McGranaghan et al., 2015). Recently, McGranaghan et al. (2015) expanded upon these previous investigations and used 6 years of DMSP particle data as input to the GLOW model to quantify conductance variability according to typical diffuse auroral oval structures and quiet time strengthening/weakening of the mean pattern, geomagnetically induced auroral zone expansion, auroral substorm current wedge, and the ionospheric substorm recovery mode.

Global auroral imagers do not have temporal limitation. Craven et al. (1984) and Kamide et al. (1986) were the first to infer global auroral conductances based on global auroral images from the Dynamics Explorer (DE-1) spacecraft. Their approach was rather crude, simply assuming auroral conductance values to be directly proportional to the emission intensity. The estimation of auroral conductance from the DE-1 auroral images was later improved by Lummerzheim et al. (1991). By employing an ion chemistry model (Rees et al., 1988) and the auroral model of Roble and Rees (1977), the authors first retrieved the characteristic energy and energy flux of precipitating electrons based on a pair of auroral emissions in the UV and visible wavelengths, and then calculated the conductances by assuming a Maxwellian energy spectrum.

A similar approach was taken by Aksnes et al. (2002, 2004) to obtain global instantaneous maps of Pedersen and Hall conductances from the UV and X-ray emissions measured by the Ultraviolet Imager (UVI) and the Polar Ionospheric X-ray Imaging Experiment (PIXIE) on board the Polar satellite. Their derivation of energy characteristics was carried out separately for the UVI and PIXIE data. For UVI, the method of Germany et al. (1997) was employed to derive the electron mean energy from the ratio between the two Lyman-Birge-Hopfield (LBH) bands at 140–160 nm (e.g., LBHS) and at 160–180 nm (e.g., LBHL) and the electron energy flux from the intensity of the LBHL emission. For PIXIE, the electron energy spectrum was represented by a four-parameter exponential function based on the X-ray emission intensity together with a lookup table generated from an electron-photon transport code described by Østgaard et al. (2000, 2001). The conductances were calculated from the electron density profiles using the transport code described by Vondrak and Baron (1976) and Vickrey et al. (1981). It was found that while the Pedersen conductance is generally not affected by the additional PIXIE data, the increase in the Hall conductance can exceed 100% locally with the inclusion of the PIXIE data (Aksnes et al., 2004).

Auroral emissions observed by the far ultraviolet (FUV) experiment onboard the Imager for Magnetopause-to-Aurora Global Exploration (IMAGE) satellite were also used to obtain global auroral conductance maps (Coumans et al., 2004). The FUV instrument consisted of 3 imagers: the Wide-band Imaging Camera (WIC) and two Spectrographic Imagers (SI12 and SI13). The SI12 imager observed the brightness of the

Doppler-shifted Lyman-α emission by energetic protons, the SI13 imager measured the OI 135.6 nm emission, and the WIC imager measured emissions from the N_2 LBH-band, atomic NI lines, as well as small contributions from the OI 135.6 nm line. Emissions measured by SI13 and WIC were excited not only by precipitating electrons but also by protons, and the SI12 data was used to subtract the proton contribution from the WIC and SI13 data. The mean energy and energy flux of the incident electrons were then retrieved from the corrected WIC and SI13 data based on the global airglow (GLOW) model (Solomon et al., 1988). Since the electron mean energy derived from the WIC/SI13 ratio appeared to have relatively large uncertainties, the Pedersen and Hall conductances were calculated based on the WIC electron energy flux and the electron mean energy from Hardy et al. (1985) model. In addition to the conductances induced by precipitating electrons, Coumans et al. (2004, 2006) also obtained global conductance maps associated with precipitating protons, which was a unique feature of the IMAGE FUV instrument. Section 4.3 continues the conversation about large-scale conductance by discussing empirical and statistical conductance models based on such datasets. Finally, Section 4.4 explains the important contributions data assimilation provides in determining global auroral conductance patterns.

Complementary to space-borne auroral imagers that have a global view of auroras but at the coarse spatial resolution, ground-based auroral imagers are able to reveal more detailed auroral structures but over a localized region. We discuss these methods to infer meso-scale structures in the following section.

4.2.2 Meso-scale convection and related precipitation and conductivity

Meso-scales have been defined slightly differently across the literature, but are generally considered features ~50–500 km wide (though some definitions may go as low as 10 km and as high as 1000 km). The following discusses meso-scale flows, precipitation, and conductance.

4.2.2.1 Meso-scale convection

Embedded within the large-scale convection pattern are smaller, more intense flows that correspond to meso-scale magnetosphere phenomena (approximately one to a few R_E wide). After impulsive dayside reconnection occurs, flux transfer events (FTEs) carry magnetic flux from the dayside to the nightside (Haerendel et al., 1978; Russell and Elphic, 1978, 1979). These FTEs, with one R_E scale size (Saunders et al., 1984), create antisunward flows in the ionosphere ~100–200 km wide (van Eyken et al., 1984; Southwood, 1985, 1987; Goertz et al., 1985). Meso-scale flows in the sunward direction—opposing convection—also exist. Both the European Incoherent Scatter (EISCAT) radars and the Super Dual Auroral Radar Network (SuperDARN) HF radars have been used to study meso-scale "reversed flow events" in the polar cusp as signatures of dayside reconnection events and are related to distinct auroral arcs (Moen et al., 2008; Rinne et al., 2007; Oksavik et al., 2011).

In the direction of convection, meso-scale flows continue across the polar cap and within the auroral oval. Gabrielse et al. (2018) collected statistics using SuperDARN data that showed that meso-scale flows tend to be embedded on top of the background convection flow in terms of their orientation. Polar cap flow speed increases with F10.7 index. Gabrielse et al. (2019) found the flow speed was greater during Coronal Mass Ejection (CME) storms than high speed stream (HSS) storms, the former occurring more frequently during solar max (and therefore, large F10.7). Storm time polar cap flows of all speeds have a postmidnight preference, which is more apparent during the recovery phase than during the main phase, and during HSSs than during CMEs. Quiet AL appears to be indicative of the postmidnight preference. All this points to a relationship with polar cap arcs, further discussed in Section 4.2.2.2. Gabrielse et al. (2019) also found that meso-scale flows occur less frequently during the main phase than during the recovery phase or during quiet time in the polar cap. This could be due to the fact that flow width increases during the main phase, so more flows may grow beyond the 500 km definition of mesoscale.

Meso-scale flows in the auroral oval are tied to closed magnetic field lines (Fig. 4.11A). After reconnection in the tail, a few R_E wide flows travel earthward, eventually breaking in the near-Earth region when the large magnetic pressure slows and diverts the flows around Earth (e.g., Sergeev et al., 2000, 2012; Panov et al., 2013). These flows map to the ionosphere as meso-scale equatorward flows. Their optical features are called auroral streamers, as discussed in Sections 1.2 and 2.4.

As the plasma sheet flows travel earthward, they form vortices that also map to the ionosphere (Fig. 4.11B). The vortices form poleward flows next to the equatorward flows. Gabrielse et al. (2018) collected statistics on both these equatorward and poleward flows, and found that their speed and occurrence rate increase during substorms. Their speed slows at lower latitudes, which makes sense due to the plasma sheet flows braking closer to Earth. Gallardo-Lacourt et al. (2014) and Gabrielse et al. (2018) found that meso-scale flows in the auroral oval had a premidnight preference during substorms, similar to that of plasma sheet flows during substorms (Nagai & Machida, 1998; Raj et al., 2002; McPherron et al., 2011).

Gabrielse et al. (2018) also found that summer flows are faster than winter flows. This could be caused by the fact that there is more precipitation on the nightside during winter, which increases conductivity and ties magnetic field lines to the ionosphere more strongly (Ohtani et al., 2009). This trend is opposite to that of the background convection, which is slower on the dayside in the summer thanks to the increased amount of sunlight and therefore, increased conductivity (de la Beaujardière et al., 1991).

Gabrielse et al. (2019) specifically studied storm-time meso-scale flows and contrasted their characteristics between quiet time, CME storms, and HSS storms. In addition to faster

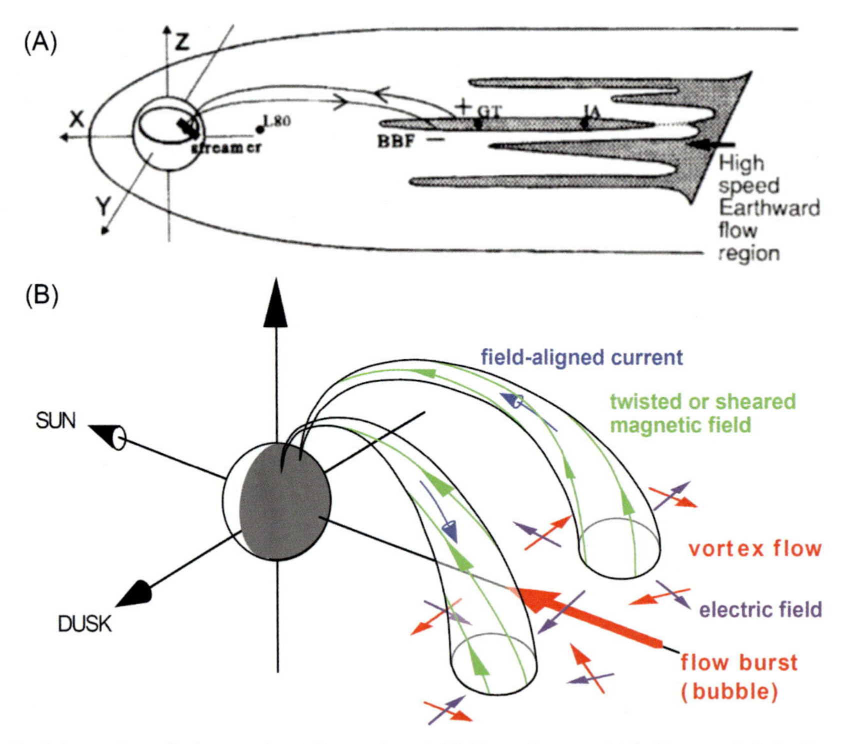

Fig. 4.11 Schematics of plasma sheet flows. *Panel (A) From Sergeev, V.A., Sauvaud, J.-A., Popescu, D., Kovrazhkin, R.A., Liou, K., Newell, P.T., et al., 2000a. Multiple-spacecraft observation of a narrow transient plasma jet in the Earth's plasma sheet. Geophys. Res. Lett. 27(6), 851–854; and Panel (B) from Birn, J., Raeder, J., Want, Y.L., Wolf, R.A., Hesse, M., 2004. On the propagation of bubbles in the geomagnetic tail. Ann. Geophys. 22, 1773–1786.*

flows occurring during CMEs as compared to HSSs, they found that faster equatorward flows are more probable during the storm's main phase. Similar to polar cap meso-scale flows, meso-scale flows occur less frequently during the main phase in the auroral oval as well. Unlike polar cap flows, the preference for midnight flows in the auroral oval during the main phase is strong. Storm time meso-scale flows have a higher occurrence rate in the auroral oval than in the polar cap, similar to meso-scale flows in general.

Meso-scale flows are a source of heating in the ionosphere. Chen and Heelis (2018) used DMSP to study meso-scale flows, finding those flows exceeding 300 m/s must be closed locally with lower magnitude return flows. They suggest that these meso-scale flows are additional sources of frictional heating and momentum transfer to the thermosphere.

4.2.2.2 Meso-scale precipitation

Meso-scale precipitation related to meso-scale flows can occur if waves related to the magnetosphere flows knock particles into the loss cone, or if the flow vortex creates a strong field-aligned current capable of accelerating electrons into the ionosphere. Electrons accelerated through a field-aligned potential drop from the discrete aurora, which is another type of precipitation (see Section 2.1 for more information on discrete aurora formation). As shown in Section 2.4, the streamers are located west of the equatorward flow and east of the poleward flow, where the shear between equatorward and poleward flows is located. Auroral streamers have been studied as the auroral footpoint of plasma sheet flow vortices (Fig. 4.11, Henderson et al., 1998; Sergeev et al., 1999, Sergeev et al., 2000, 2004; Lyons et al., 1999, 2002; Kauristie et al., 2000; Zesta et al., 2000; Zou et al., 2010), consistent with an upward field-aligned current on the duskside of the flow (Amm et al., 1999; Nakamura et al., 2001; Birn et al., 2004; Sergeev et al., 2004). See Chapter 2 for an in-depth discussion on meso-scale aurora in the auroral oval, and Gabrielse et al. (2021) for an analysis of meso-scale precipitation throughout substorms in the auroral oval.

Meso-scale auroral features are not limited to the auroral oval. In contrast to the auroral arcs in the auroral oval that occur more frequently during southward interplanetary magnetic field (IMF), polar cap arcs are more frequent during northward IMF and quiet magnetic conditions (Berkey et al., 1976; Ismail et al., 1977; Lassen and Danielsen, 1978; Valladares et al., 1994; Hosokawa et al., 2011, see also Section 2.3). Gabrielse et al. (2018, 2019) drew an indirect link between polar cap arcs and polar cap meso-scale flows, as both phenomena have a postmidnight/dawnside preference (Hosokawa et al., 2011; Lassen and Danielsen, 1972). Furthermore, both polar cap arcs and polar cap meso-scale flows occur more frequently when AL index is low (quiet time) and when IMF B_z is positive (Gabrielse et al., 2018).

How polar cap arcs are generated has been debated for decades. That discussion included whether or not they form on open (e.g., Hardy et al., 1982) or closed (e.g., Frank et al., 1982) field lines. Studies that find them with polar rain-like electrons with no precipitating ions suggest they form on open field lines. Other studies that find accelerated electron populations, plasma sheet-like electrons and ions, or conjugate arcs occurring in both northern and southern hemispheres suggest they form on closed field lines. Fig. 4.14 is an example of a polar cap arc observed by DMSP GUVI, where a conjugate Cluster flyover observed an inverted V signature suggesting acceleration of the precipitating electrons. Multiple studies have modeled how closed field lines/a plasma sheet source could create auroral arcs in the polar cap, including a highly twisted (Cowley, 1981; Makita et al., 1991) or a bifurcated plasma sheet (Obara et al., 1988) due to a large B_Y component of the IMF, magnetotail reconnection under such conditions (Milan et al., 2005), or ballooning instability in the plasma sheet (Rezhenov, 1995; Golovchanskaya et al., 2006). Zhu et al. (1997) and Hosokawa et al. (2020) present a thorough review of polar cap arcs, their generation mechanism(s), observations, and modeling. See especially Hosokawa et al. (2020) for discussion of potential generation mechanisms.

Polar cap patches (see Section 2.3) are islands of high-density structures that are segmented due to variations in the high-latitude convection pattern. Transient magnetopause reconnection, which has been related to polar cap patches (Carlson et al., 2006; Lockwood and Carlson, 1992), causes localized flow channels and poleward moving auroral forms (PMAFs) which have been related to polar cap patches using imager and radar data (e.g., Nishimura et al., 2014b; Wang et al., 2016). The variations in the convection pattern that can redistribute the high-density islands into smaller patches can be caused by transient reconnection at the magnetopause or by changes in the IMF B_z component (Anderson et al., 1988; Tsunoda, 1988) or IMF By component (Sojka et al., 1993, 1994). These variations can be caused by transient reconnection at the magnetopause (Carlson et al., 2006; Lockwood and Carlson, 1992) or by changes in the IMF B_z component (Anderson et al., 1988; Tsunoda, 1988) or IMF By component (Sojka et al., 1993, 1994; Zhang et al., 2011). IMF By reversals can also cause patches by creating high-speed flow channels that segment the soft particle precipitation near the cusp (Rodger et al., 1994). For more details, Ren et al. (2018) provide a comprehensive review of polar cap patches in their introduction section, including polar cap observations made with satellites (e.g., Coley and Heelis, 1995; Noja et al., 2013), meridian scanning photometers (e.g., McEwen & Harris, 1996; Moen et al., 2007), and all-sky imagers (e.g., Hosokawa et al., 2009).

Polar cap patches (see Section 2.4) generally follow large-scale convection, but they can sometimes be related to flow channels. Zou et al. (2017) studied nonstorm-time patches with FAST data and determined that patches are associated with ionospheric flow channels and localized field-aligned currents. They identified localized precipitation that is enhanced within patches consisting of structured or diffuse soft electron fluxes. They concluded that patches should be regarded as part of a localized magnetosphere-ionosphere coupling system along open magnetic field lines and that their transpolar evolution is a reflection of meso-scale magnetotail lobe processes. Further, Nishimura et al. (2014b) observed polar cap patches propagating all the way from the dayside to the nightside. They observed a PMAF associated with fast flows evolve into a polar cap airglow patch associated with antisunward flows that propagated across the polar cap until it reached the nightside auroral oval poleward boundary. There, a poleward boundary intensification (PBI) was formed, which authors suggest that a flow channel formed from dayside reconnection, traveled to the nightside, and ultimately triggered localized nightside reconnection and flow bursts in the plasma sheet. PBIs are meso-scale auroral forms that occur near the nightside auroral oval poleward boundary and are associated with bursty bulk flows (BBFs) in the plasma sheet and streamers in the auroral oval (de la Beaujardière et al., 1994; Zesta et al., 2000, 2002, 2006; Lyons et al., 2011; Nishimura et al., 2010; Pitkänen et al., 2011, 2013; Shi et al., 2012). The question of patch formation therefore remains, or perhaps polar cap patches can be formed via both precipitation and nonprecipitation mechanisms. Lyons et al. (2016) summarize some of these unsolved questions.

4.2.2.3 Meso-scale conductance

In Section 4.2.1.3, we discussed how satellite overpasses can provide characteristic or average energy Eq. (4.14) and energy flux of a precipitating population, which can be used to determine conductance, especially on global scales. Average energy may be a better metric than characteristic energy because it is independent of the parent distribution. In addition to satellite overpasses, characteristic or average energy and energy flux can be indirectly measured from the ground by meridian scanning photometers (MSPs) and all-sky-imagers that record different wavelengths (colors) of the aurora (Strickland et al., 1989). Because more energetic particles travel farther and to lower altitudes before colliding with an atmospheric particle, the most energetic particles lead to a blue or purple emission thanks to the abundance of nitrogen at altitudes <60 km. Less energetic particles that deposit their energy at altitudes from 60 to 150 km interact with oxygen, resulting in green aurora. Finally, the least energetic particles that collide with oxygen molecules >150 km result in a red aurora. Therefore, if the blue to green emission ratio is large, the characteristic energy will be larger. If the red to green emission ratio is large, the characteristic energy will be lower.

Rees and Luckey (1974) modeled these relationships and presented figures that draw the relationship between color ratios and characteristic energy and energy flux. They did not, however, consider neutrals in the atmosphere. Neutrals were later included by Strickland et al. (1989) and Hecht et al. (2006), who considered the atmospheric composition when drawing the relationship between color ratios and characteristic energy, and between the blue line emission and characteristic energy to determine energy flux. Recently, Kaeppler et al. (2015) compared characteristic energy and energy flux estimates from PFISR observations to photometer observations using the ratio method, along with Scanning Doppler Imager (SDI)-derived characteristic energy and energy flux. The best agreement was found to be during a diffuse aurora, which was partially attributed to the fairly uniform intensity of visible emission for diffuse aurora over the SDI cell.

While satellite observations have provided a global perspective of precipitation-driven conductance enhancements, MSPs, ASIs, and ISRs provide a perspective of the conductance on meso-scales. The advantage of the ISR, MSP, and ASI measurements is that they can capture the time history of auroral evolution and precipitation-driven conductance enhancements.

ISRs are one of the best available instruments for estimating ionospheric conductances as they directly measure the altitude profiles of electron density (from backscattered power of ISR) and can estimate the electron and ion temperatures, which are crucial elements for calculating ionospheric conductivity. Extensive efforts have been made to derive the Pedersen and Hall conductivities from the ISR measurements. Methods were developed in some early works (e.g., Brekke et al., 1974; Horwitz et al., 1978; Vickrey et al., 1981) to demonstrate how the Hall and Pedersen conductivities can be derived from the Chatanika ISR measurements. Similar methods were later applied to the European Incoherent Scatter (EISCAT) radar (e.g., Schlegel, 1988; Senior, 1991; Brekke and Hall, 1988; Moen and Brekke, 1990), the Sondrestorm radar (e.g., Watermann et al., 1993; Thayer, 1998), and the Poker Flat ISR (PFISR) (e.g., Kaeppler et al., 2015; Lam et al., 2019) to estimate ionospheric conductivities/conductances over the polar cap and in the auroral zone.

Recently, Kaeppler et al. (2015) presented observations of conductivity associated with discrete and diffuse aurora at the Poker Flat Incoherent Scatter Radar (PFISR), see Fig. 4.12. PFISR made observations during an interval of a fairly stable east–west aligned auroral from 12:00 to 13:00 UT, which broke up into diffuse aurora in PFISR field-aligned look direction (marked by the white circle). Another equatorward moving east–west arc was observed near 13:30 UT, which modestly intensified and broke up into diffuse aurora after 14:42 UT. For this event, the PFISR-derived average energy agreed best during diffuse auroral intervals and lower average energies; after 14:00 UT, the average energy increased in the PFISR observations relative to the SDI. Good agreement was found for the energy flux between calibrated photometers for emission 427.8 nm and the PFISR-derived energy flux.

Since ISRs do not distinguish the different ion species nor do they typically measure any neutral properties, assumptions have to be made concerning the partition of ion species as well as neutral density, composition, and temperature. Static ion composition models (e.g., Brekke and Hall 1988; Bilitza and Reinisch, 2008) and empirical thermosphere models (e.g., Hedin et al., 1977; Picone et al., 2002) are commonly used to specify the ion-neutral collision frequencies needed for estimating the Pedersen and Hall conductivities.

Another limitation concerning ISRs is their temporal and spatial coverage. There is only a handful of ISRs operating around the world. Also, due to limited field-of-view (with a typical beam width of $\sim 1°$), ISRs essentially measure local ionospheric conductances, making it difficult to obtain global conductance maps from the ISR data alone. For example, many measurements are made along a single-point look direction that is notionally aligned with the local geomagnetic field, though some modes can scan across various elevation angles (e.g., Vondrak and Baron, 1976). Most recently, with the advancement of phased array ISRs, such as the advanced modular incoherent scatter radar, different radar beams can be formed to look in multiple directions in the sky (e.g., Kelly and Heinselman, 2009).

One can see the difficulty that exists in determining ionosphere conductance observationally. ISRs provide single-point observations, or, at best, a scan over multiple elevation angles but with a relatively limited field of view. Statistics of conductance can be obtained, but are limited to an annulus in magnetic latitude/magnetic local time. In addition, except for PFISR, ISR observations are generally not continuous

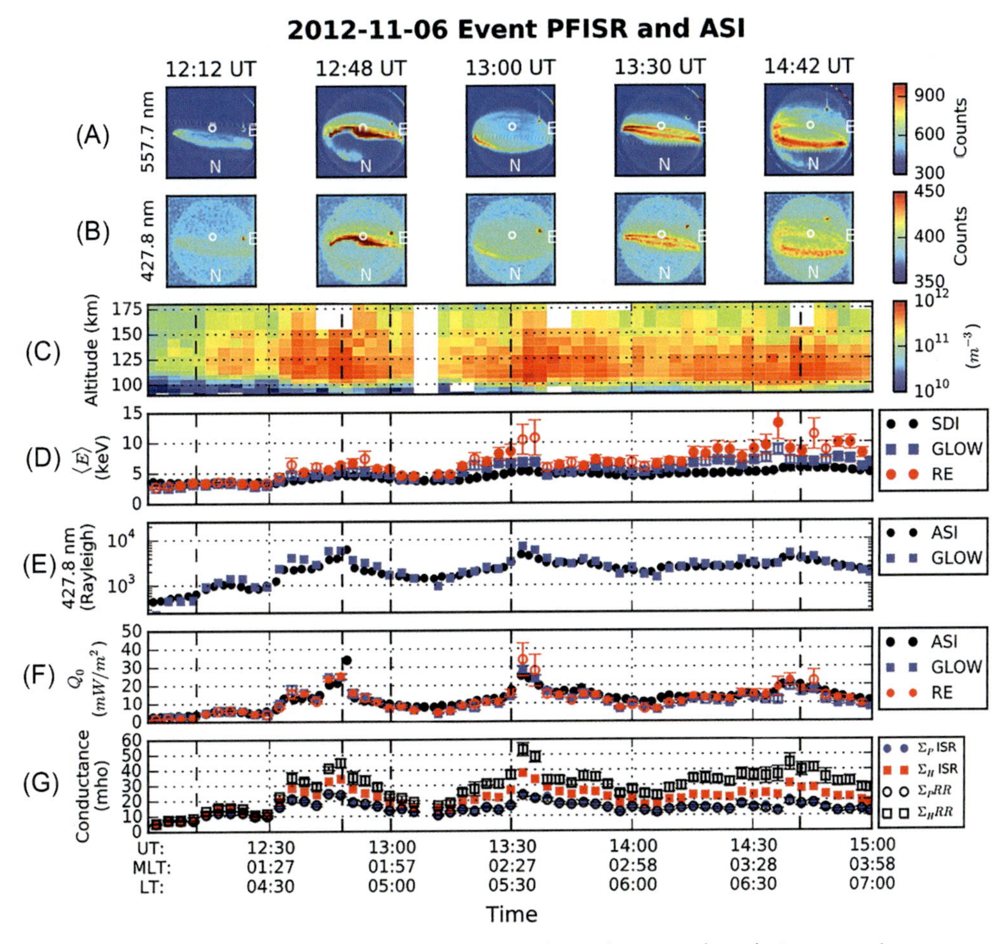

Fig. 4.12 (A) and (B) are included for context, to show the auroral evolution over time at two wavelengths (557.7 and 427.8 nm). PFISR has a look direction in the middle of the ASI field of view, marked with the *white circle*. (C) PFISR electron number density. (D) Average/characteristic energy determined from a scanning doppler imager (SDI) and estimates from the GLOW and range-energy (RE) models. (E) The 427.8-nm wavelength intensity as observed by the ASI and predicted by GLOW. (F) The energy flux derived from the 427.9 nm emission along with the predicted energy flux from GLOW and RE. (G) The Pederson (*circles*) and Hall (*square*) conductances as determined from ISR data (*filled in*) and the Robinson relation (*empty*). *From Kaeppler, S.R., Hampton, D.L., Nicolls, M.J., Strømme, A., Solomon, S.C., Hecht, J.H., Conde, M.G., 2015. An investigation comparing ground-based techniques that quantify auroral electron flux and conductance. J. Geophys. Res. Space Phys. 120, 9038–9056. https://doi.org/10.1002/2015JA021396.*

observations. For that reason, techniques have been developed to infer auroral conductances from other supplementary observations such as energetic particle precipitation by spacecraft-borne particle detectors and auroral emissions by space- and ground-based photometric imagers.

Meridian scanning photometers similarly provide a one-dimensional dataset over multiple latitudes but along a single longitude line. DMSP satellite overpasses provide two-dimensional swaths of energy flux and characteristic energy data, but only along the orbital track and only for the moment in time the satellite is overhead. These datasets can be used statistically to quantify conductance variations relative to some input parameter (e.g., geomagnetic activity indices, solar wind inputs, time of year, day, etc.). They struggle to provide adequate real-time data in order to study conductance variability for a particular substorm or storm, however.

Perhaps the best solution at present for event studies is to employ an array of all-sky-imagers that can measure multiple wavelength (color) intensities, as these provide 2D datasets that cover a particular region continuously. Until now, only a few all-sky-imagers were deployed, but with the initiation of the TREx Project (https://www.ucalgary.ca/aurora/projects/trex), these multiwavelength all-sky-imagers are being deployed in an array across all of Canada. Work is also being done to utilize single-wavelength or white-light all-sky-imagers to determine conductance (e.g., Kosch et al., 1998; Nishimura et al., 2019; Lam et al., 2019; Gabrielse et al., 2021). By comparing the Pedersen conductance values obtained from EISCAT with the intensities of auroral emission at 557.7 nm measured by a colocated all-sky-imager (ASI), Kosch et al. (1998) suggested a way to estimate auroral conductance using ASIs. A similar approach was taken by Lam et al. (2019) who derived an analytical formula to infer the Pedersen conductance directly from the white-light ASI measurements. These two studies, however, showed a 30% difference in their derived relationship between the Pedersen conductance and the emission intensity. One possible explanation for the discrepancy is that Kosch et al. (1998) used magnetic-field-aligned radar data and the 557.7-nm green-line auroral emissions over zenith whereas Lam et al. (2019) made use of nonmagnetic-field-aligned radar data and off-zenith white-light auroral emissions.

The above two studies focused only on the Pedersen conductance as both were based on auroral emissions from a single filter. The Hall conductance, on the other hand, is strongly dependent on the mean energy of precipitating particles (Vickrey et al., 1981; Brekke et al., 1989), and therefore requires measurements of simultaneous emissions from at least two different wavelengths (Rees and Luckey, 1974). A number of studies have shown that both the characteristic energy and energy flux of auroral precipitating electrons can be inferred from optical auroral images with a combination of emission lines at different wavelengths (e.g., Vondrak and Sears, 1978; Mende et al., 1984; Robinson et al., 1992; Lanchester et al., 2009; Kaeppler et al., 2015; Grubbs II et al., 2018). Using the auroral emission rates at the 427.8 and 630.0 nm wavelengths, together with the aid of the GLOW model, Adachi et al. (2017) derived the Pedersen and Hall conductances and compared them with the conductance values from the collocated EISCAT radar. Their study found that while the temporal variations of the photometrically derived conductances are generally in good agreement with the ISR observations, their magnitudes tend to be lower than the radar-derived conductance values, particularly for the Hall conductance. The authors attributed the discrepancy partially to the long lifetime of the 630.0-nm red-line emission that leads to an inaccurate estimate of the mean energy of the precipitating electrons and the assumption of the Maxwellian energy spectral distribution that becomes invalid for discrete aurora arcs (Lanchester et al., 2009; Kaeppler et al., 2015). Uncertainties associated with the specification of neutral compositions and collision cross section as well as the omission of proton precipitation may also affect the estimation of auroral conductances.

Besides auroral photometric imagers, riometer measurements of cosmic noise absorption (CNA) can also be used to infer the Pedersen and Hall conductances (Walker and Bhatnagar, 1989; Makarevitch et al., 2004; Senior et al., 2007). For example, by comparing the absorption of the imaging riometer at Kilpisjärvi with the simultaneous measurements by the EISCAT-Tromso radar, Senior et al. (2007) found that the CAN absorption is strongly correlated with the Hall conductance but the correlation with the Pedersen conductance is somewhat weaker. Their results also revealed that the relationship between the CAN absorption and the Hall and Pedersen conductance possessed a strong MLT dependence.

Additionally, Ahn et al. (1983, 1998) developed a conductance model based on ground magnetometer data. The model was built on the empirical relationship between the auroral conductances derived from the Chatanika IS radar and ground magnetic field perturbations observed in the same region. One advantage of the Ahn model is that it yields two-dimensional auroral conductance maps based on instantaneous ground magnetometer observations, and therefore provides a better depiction of the temporal variations associated with the aurora.

4.2.3 Small-scale precipitation and conductivity

"Convection" is not considered to occur on small scales, though electric fields related to other mechanisms exist on small scales. We, therefore, do not spend much time discussing small-scale phenomena here but point out their existence as their importance cannot be overlooked. Small-scale precipitation and conductivity changes result from small-scale mechanisms such as instabilities and waves. Rockets and auroral imagers are primary observation tools to study small (or fine) scale (10s km) ionospheric precipitation and conductivity.

For example, Evans et al. (1977) used a rocket experiment that flew over an auroral arc to determine the energy spectra of the precipitating electrons. They calculated the ion production rate by using the algorithm developed by Rees (1963). Then, the altitude profiles of electron density were obtained by applying altitude-dependent effective recombination coefficients. Finally, using the ion-neutral collision frequencies based on a neutral atmosphere model (Jones and Rees, 1973), the altitude-dependent Pedersen and Hall conductivities were computed.

Nishimura et al. (2010) used THEMIS all-sky-imagers to present direct evidence that a naturally occurring electromagnetic wave, lower-band chorus, can drive pulsating aurora. Later, Dahlgren et al. (2015) used ASK (Auroral Structure and Kinetics), a ground-based optical instrument, to find coexisting small-scale auroral features resulting from different high- and low-energy populations of precipitating electrons on the same field line. They found that the high-energy precipitation formed pulsating patches of 0.1 Hz with a 3-Hz modulation as well as nonpulsating discrete auroral filaments. They found low-energy precipitation on the same field line as the discrete filaments with no pulsation. The different structures drifted at different speeds in different directions. They proposed that the high- and low-energy electron populations were accelerated by different mechanisms, at different distances from Earth, with the small-scale structures caused by local instabilities above the ionosphere. See a study by Nishimura et al. (2020) for an in-depth review of diffuse and pulsating aurora.

Diffuse auroras also involve fine structures at a few to a few tens of km (Sergienko et al., 2008). These fine-scale structures can form along the western edges of eastward drifting pulsating auroral patches when the convection drift slows down. These fine-scale structures grow from Rayleigh–Taylor instability at the boundary of a drifting patch. Fukuda et al. (2016) reported expansion speeds of $\sim$10s km/s, comparable to Alfvén speeds, suggesting the mechanism behind the fine structures is slow- and fast-mode Alfvén waves.

Further, the fine-scale structure has been linked to energy-dispersed electron bursts accelerated by inertial Alfvén waves (Stasiewicz et al., 2000). Lynch et al. (1999) and Hallinan et al. (2001) combined optical and in situ rocket measurements to show that Alfvén waves accelerated electrons. Though out of the scope of this section, to give a proper review, we encourage the interested reader to consider these effects as well.

4.3 Simulating particle precipitation of magnetospheric origin in global models

4.3.1 Introduction

Particle precipitation from the magnetosphere, as discussed in Sections 4.1 and 4.2, can drive aurora via collisions within the upper atmosphere. The aurora is often categorized into two types according to their source, or observationally, according to their energy spectrogram: diffuse aurora and discrete aurora. The diffuse auroras are unstructured emissions due to particles with an energy of a few eV to tens of keV precipitating from the magnetosphere, while the discrete aurora is often well structured either with an inverted "V" shape in the energy spectrogram (or mono-energetic, with only 1–2 energy channels capturing the accelerated particles) or a broadband feature (with 3 or more energy channels observing accelerated particles) (Burch, 1991; Newell et al., 2009; Newell, 2000). The particles that create discrete aurora are accelerated from their source region in the magnetosphere into the ionosphere altitude. Such acceleration is associated with electric potential drop along magnetic field lines or dispersive Alfvén waves along the fields. On the other hand, the particles related to diffuse aurora can travel down to the atmosphere without additional energization. This is because the particle's pitch angle may be altered locally in such a way that the particle falls into the loss cone. As mentioned in Section 4.2, such pitch angle change is called "pitch angle scattering" or "pitch angle diffusion."

What causes the pitch angle scattering? As discussed in Section 4.2.1.2, pitch angle scattering is associated with the violation of the first adiabatic invariant $\mu = W_\perp / B$, where $W_\perp$ is the perpendicular kinetic energy of the particle and B is the magnitude of the magnetic field (see Eq. 4.1). That is, when the electric and magnetic fields experienced by the particle change significantly during one gyroperiod, the particle could not complete a closed gyroorbit (see Eqs. 4.1–4.3). The particle, therefore, exhibits chaotic motion. The first adiabatic invariant is not only violated when the fields change rapidly in time but also when the fields exhibit a large spatial gradient, in which case the particle also experiences a large change of fields within one gyroorbit (see Fig. 4.13B). For magnetospheric dynamics, two-pitch angle scattering mechanisms are proposed: wave-particle interactions and chaotic scattering in an inhomogeneous magnetic field.

4.3.1.1 Wave-particle interactions

There is a rich variety of electromagnetic waves in the magnetosphere with sufficiently high frequencies comparable to the gyration frequency of the trapped particle. The magnetically guided electromagnetic waves with a circularly polarized E-vector can resonate with the gyrofrequency of the particle, thus changing the value of the corresponding first adiabatic invariant. Here we take the electron interactions with whistler-mode waves as an example to illustrate the wave-particle interaction and associated pitch angle diffusion. Fig. 4.13A illustrates a right-hand circularly polarized electromagnetic wave with **E** and **B** wave vectors perpendicular to the background geomagnetic field, such as whistler-mode waves. The wave vectors rotate in the same sense as the electron gyration with a gyrofrequency of Ω_{ce}. The electron observes a constant Doppler-shifted electric field

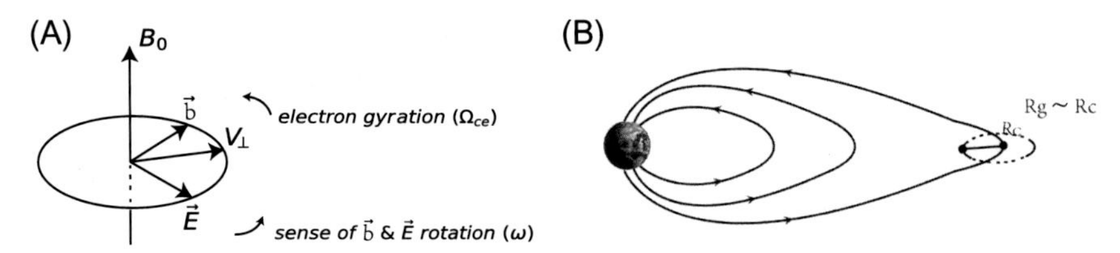

Fig. 4.13 (A) Right-hand circularly polarized electromagnetic wave represented by *E* and *b* vectors. The electron gyrates in the same sense around the background magnetic field B_0. (B) Near the central plasma sheet, the magnetic field lines are stretched largely, the curvature radius of the field lines at the equator is comparable to the particle's gyroradius.

such that changes in its energy will take place if the wave frequency ω satisfies the following resonance condition:

$$\omega - kv_{\|} = N\Omega_{ce}, \quad N = 0, \pm 1, \pm 2\ldots \tag{4.15}$$

where k is the wave number and $v_{\|}$ is the particle velocity in the parallel direction to the background magnetic field. This is referred to as *wave-particle resonance*, which can lead to wave growth or damping and particle diffusion in pitch angles. For an electron distribution uniformly distributed in the azimuthal directions perpendicular to the geomagnetic field, some electrons will be deflected to smaller pitch angles and some to larger pitch angles during the encounter with waves of finite length. The overall effect of many such encounters will be diffusion in pitch angle. Those electrons entering into the loss cone precipitate into the upper atmosphere.

The circularly polarized whistler and ion cyclotron waves appear to be the most important in scattering electrons and ions respectively, in the magnetosphere (Jordanova et al., 2001; Ni et al., 2008). For diffusive electron precipitation scattered from the plasma sheet or inner magnetosphere, three major plasma waves are found to be highly responsible: whistler-mode chorus waves, whistler-mode hiss waves, and electrostatic cyclotron harmonic (ECH) waves. The chorus waves are frequently observed in the low-density region outside the plasmapause and characteristically occurred in two bands separated around 0.5 f_{ce} (f_{ce} is the electron gyrofrequency) (Tsurutani and Smith, 1974), while broad-band hiss waves are usually detected inside the plasmapause. Recent studies found that the chorus waves are more effective than the ECH waves in scattering the plasma sheet electrons from a few hundred eV to tens of keV down to the auroral zone to produce intense diffuse auroral precipitation (Thorne et al., 2010; Ni et al., 2011a, b). The ECH waves are more dominant at higher latitudes, or in regions outside $L = 8$ (Ni et al., 2011a, b).

For diffusive ion precipitation, one responsible wave is the electromagnetic ion cyclotron (EMIC) wave which contains three wave bands below the hydrogen gyrofrequency: He^+ band, H^+ band, and O^+ band. The EMIC waves occur with a high possibility in the dusk sector (Min et al., 2012) and the dayside magnetosphere (Allen et al., 2015, 2016; Saikin et al., 2016). Therein, the EMIC wave can strongly resonate with protons at energy from a few keV to 100 keV (Ni et al., 2016), and thus scatter them efficiently.

4.3.1.2 Field-line curvature scattering

If the magnetic field exhibits large inhomogeneity in space such that it changes abruptly in a spatial scale comparable to the gyroradius of a charged particle, the first adiabatic invariant is violated and the particle then behaves chaotically. This process is referred to as *magnetic field-line curvature (FLC) scattering*. When the magnetic field lines are largely stretched, the curvature radius may become as small as the particle gyroradius as shown in Fig. 4.13B. The particle gyrating during its orbit experiences large changes in the magnetic field because the near-Earth magnetic field is much larger than that on the tail side. Its gyroorbit is thus not closed. The particle's first adiabatic invariant is broken and the particle suffers pitch angle changes while conserving its kinetic energy. Such condition usually appears in the central plasma sheet on the nightside, so such FLC scattering is also called *current sheet scattering*.

The parameter representing the degree of chaotic scattering due to the FLC effect is $\kappa = R_c/\rho$, where ρ is the particle gyroradius in the equatorial plane, and R_c is the field-line curvature radius. The scattering rate becomes stronger when the ratio is smaller. When the ratio is larger than a certain threshold level (e.g., $\kappa = 8$), the scattering can be neglected or considered weak (Sergeev et al., 1983). Due to the much larger gyroradius of ions, the FLC scattering loss is more efficient on ions than on electrons. Therefore, the FLC scattering process could be another important mechanism for pitch angle diffusing ions in addition to the EMIC waves as mentioned above.

4.3.2 Simulating electron precipitation in global geospace circulation models

Particle precipitation is not only a major loss mechanism for magnetospheric populations, but it also provides an important energy source to the upper atmosphere, which in turn provides feedback to magnetospheric dynamics. In global geospace circulation models (GGCM) (e.g., Raeder et al., 2001; Lyon et al., 2004; Tóth et al., 2005, 2012), to account for such coupling of the ionosphere-magnetosphere system, particle precipitation is the key element in bridging the two systems. Compared to ion precipitation, electron precipitation dominates the energy source over the top ionosphere, hence we will describe the modeling of electron precipitation first.

In general, there are two approaches in GGCM to determine the electron precipitation flux: (1) MHD approximation and (2) physics-based kinetic approach. The first method, simple and efficient, is adopted by MHD models that are incapable of resolving kinetic physics, while the second one is more appreciable in kinetic models that are able to solve distribution functions and particle flux in loss cones.

4.3.2.1 MHD parameterization

In global MHD models, the ionosphere serves as the inner boundary where it solves the electric potential needed in the magnetosphere. The electric potential solver is built over a sphere about 110 km above the Earth's surface and is controlled by two quantities: the height-integrated conductance and field-aligned currents. As the ionospheric aurora conductance is directly associated with the particle precipitation at the top of the ionosphere, particularly the dominant electron precipitation, the specification of the electron precipitation flux in these models is required prior to solving the electric potential.

MHD models are known to be incapable of capturing kinetic physics, such as distribution functions and loss cone flux. Therefore, an approximation is made in order to specify the electron precipitation in MHD models. Following the adiabatic kinetic theory (Knight, 1973; Lyons et al., 1979; Fridman and Lemaire, 1980), global models can specify the electron precipitation based on MHD parameters (e.g., Raeder et al., 2001; Tanaka, 2000; Zhang et al., 2015). We take the methodology used in Zhang et al. (2015) as an example. The electron thermal flux F_0 in the magnetosphere source region for precipitating electrons is described as follows:

$$F_0 = \beta \frac{N_e \sqrt{T_e}}{\sqrt{2\pi m_e}} \tag{4.16}$$

where N_e and T_e are the electron number density and temperature at the source region. Since MHD models only resolve the ion fluid quantities, the electron number density can be set to be equal to the proton number density, and the electron temperature is assumed to be a small fraction of the proton temperature according to plasma sheet observations. β represents the filling factor of the loss cone in the plasma sheet. It can range from 0 to 1 depending on location and geomagnetic activity (Zhang et al., 2015).

For the diffusive electron precipitation, the energy flux F_E and averaged energy $<E>$ are formulated as follows:

$$F_E = 2F_0 T_e$$
$$<E> = 2T_e \tag{4.17}$$

For discrete or mono-energetic electron precipitation that is associated with upward field-aligned currents $J_{||}$ (as electrons are accelerated down to the ionosphere), the energy flux F_E and averaged energy E are written as (Zhang et al., 2015)

$$F_E = \frac{J_{\|}}{e}\left[2T_e + eV\frac{1-e^{-eV/T_e(R_m-1)}}{1+\left(1-\frac{1}{R_m}\right)e^{-eV/T_e(R_m-1)}}\right]$$

$$< E >= 2T_e + eV\frac{1-e^{-eV/T_e(R_m-1)}}{1+\left(1-\frac{1}{R_m}\right)e^{-eV/T_e(R_m-1)}}$$

$$eV = T_e(R_m-1)\ \ln\frac{R_m-1}{R_m-\frac{J_{\|}}{eF_0}} \tag{4.18}$$

where R_m is the ratio of the magnetic field between the ionospheric footprint and the source region in the equatorial plane, and eV represents the energy required to accelerate the thermal electrons in the magnetosphere into the ionosphere altitude. When eV drops to zero, meaning that no electron acceleration occurs, or electron precipitation takes place from the magnetosphere without additional energization, the above energy flux and averaged energy become the same as the diffusive precipitation.

Fig. 4.14 shows the electron precipitation pattern determined from the global MHD model, Lyon-Fedder-Mobarry (LFM) model (Lyon et al., 2004; Wiltberger et al., 2004; Merkin and Lyon, 2010), during quiet time. The diffusive precipitation (illustrated in red) contributes the majority of the energy deposition on the nightside-to-dawnside and dominates the total power input at high latitudes. This relative relationship is consistent with observations (Newell et al., 2009).

4.3.2.2 *Kinetic method*

An alternative way of specifying the precipitating flux is based on kinetic physics that solves phase-space distribution functions and represents more realistically the micro-scale physics than MHD theory. In the kinetic ring current model, the net variation of the distribution f is due to convection around the Earth and various loss mechanisms along the drift path, such as the loss due to charge exchange (for ions only), Coulomb collision loss, pitch angle scattering due to interactions with waves or other scattering processes, drift loss out of the magnetopause boundary, and absorption of particles at low altitude or loss inside the loss cone. The governing equation in the kinetic models for the ring current particle distribution is the bounce-averaged Fokker-Plank equation:

$$<\frac{df}{dt}>=<\frac{\partial f}{\partial t}>_{loss} \tag{4.19}$$

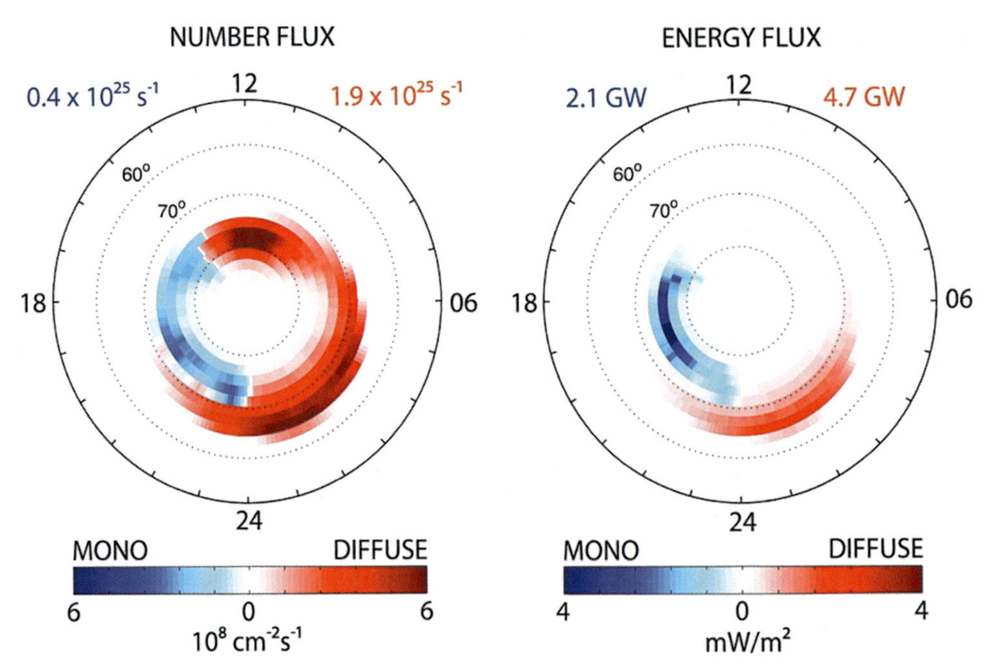

Fig. 4.14 The simulated electron number flux and energy flux due to both diffuse and mono-energetic precipitations at the ionospheric altitude. *Adapted from Zhang, B., Lotko, W., Brambles, O., Wiltberger, M., Lyon, J., 2015. Electron precipitation models in global magnetosphere simulations. J. Geophys. Res. Space Phys. 120, 1035–1056. https://doi.org/10.1002/2014JA020615.*

where the right-hand term represents the loss mechanisms experienced by particles in the inner magnetosphere. Of importance for the electron loss processes is the pitch angle scattering induced by waves, the efficiency of which can be quantified by loss rates, either the electron lifetimes or wave-induced pitch angle diffusion coefficients. The following sections will describe these two kinds of loss rates in detail.

Decay with lifetimes

The lifetime τ represents the e-folding time of exponentially decaying fluxes. Such lifetime is useful in kinetic modeling of electron distributions as the complexity of various loss processes can be represented by a single model parameter. With the lifetime, the loss term in the kinetic model can be written as:

$$< \frac{\partial f}{\partial t} >_{loss} = -\frac{f}{\tau} \tag{4.20}$$

The lifetime can be determined empirically (e.g., Chen and Schulz, 2001b; Albert, 1999; Claudepierre et al., 2020a, b). For example, from a long-term statistical database of differential electron fluxes from Van Allen Probes, Claudepierre et al. (2020a) calculated electron lifetimes from fitting the exponentially decaying flux with an exponential function. It should be noted that such empirical lifetimes that represent the total decay include effects due to a number of loss mechanisms.

The lifetimes can also be computed theoretically based on quasilinear diffusion rates. From using an integral expression (Albert and Shprits, 2009) on bounce-averaged pitch angle diffusion coefficients $D_{\alpha\alpha}$, the lifetime can be theoretically estimated as

$$\tau = \int_{\alpha c}^{\pi/2} (2 < D_{\alpha\alpha} > \tan(\alpha))^{-1} d\alpha \tag{4.21}$$

where α is the equatorial pitch angle, α_c is the loss cone angle. $D_{\alpha\alpha}$ is quasi-linear pitch angle diffusion coefficient induced by a specific scattering process, such as wave-particle interactions. Depending on specific waves that lead to pitch angle scattering, the lifetimes are applied in global models to account for the influence by the specific waves, such as chorus wave outside the plasmapause and hiss wave inside the plasmapause (e.g., Orlova and Shprits, 2014; Orlova et al., 2016).

Earlier lifetime models applied in kinetic codes to effectively compute electron losses assumed a loss rate with strong pitch angle scattering everywhere (Schulz, 1974, 1998; Chen & Schulz, 2001a; Harel et al., 1981; Toffoletto et al., 2003), meaning that the pitch angle distribution is essentially isotropic and that the electron loses its "memory" of its equatorial pitch angle before completing a bounce period (Chen & Schulz, 2001a). However, this approach tends to overestimate precipitation loss, resulting in unrealistically low trapped electron flux (Chen et al., 2015). Efforts have been continuously made on improving the loss models in order to better account for observations, such as electron lifetimes parameterized for whistler waves inside the plasmasphere from using quasi-linear diffusion theory (Albert, 1999), loss rates with a fraction of strong diffusion rate (Gkioulidou et al., 2012), a time-independent analytical electron lifetime model that included a smooth shift from weak diffusion (Albert, 1994) in the plasmasphere into strong diffusion (Schulz, 1974, 1998) in the plasma sheet with an MLT dependence (Chen and Schulz, 2001b), and a loss model with Kp, MLT, energy, and L-dependence of hiss (Orlova et al., 2014, 2016) inside the plasmasphere and chorus waves (Orlova and Shprits, 2014) outside the plasmasphere.

Chen et al. (2015) found that the static loss model with a smooth transition from weak to strong diffusion overestimates the observed electron flux at GEO on the morning and dayside, while the MLT and Kp parameterized loss rate model from Orlova et al. (2014, 2016) and Orlova and Shprits (2014) reproduced reasonably well the LANL/GEO-trapped electron flux. Perlongo et al. (2017) simulated the aurora electron precipitation using modified electron loss rates from Chen and Schulz (2001a, b) by introducing and changing the maximized lifetime, finding that the choice of lifetime models can influence the location of the aurora as well as electric potential at high latitudes.

Ferradas et al. (2019) compared three different types of electron loss models to examine their impact on the model capability of explaining the observed electron flux during the March 17, 2013 event, including (1) the loss model from Albert (1999) for hiss waves inside the plasmasphere and for diffusion not strong everywhere from Chen and Schulz (2001b) for chorus waves outside the plasmasphere; (2) the loss model from Orlova et al. (2016) and Orlova and Shprits (2014) for hiss and chorus waves respectively; (3) lifetime model obtained, by using Eq. (4.7), from bounce-averaged pitch angle diffusion coefficients derived by using wave properties obtained from Van Allen Probes observations (Li et al., 2015, 2016). Their comparisons indicated that the more recent loss models based on Van Allen Probes observations overall showed good agreement with observed trapped flux.

Diffusion loss with diffusion coefficient

The method of calculating lifetimes in Albert and Shprits (2009) and others gives a very good approximation to the exact lifetime. However, the exact lifetime calculated in Eq. (4.21) does not always lead to a good approximation to using the pitch angle diffusion coefficients, especially when considering pitch angle and local time dependence or during transient processes (e.g., substorm injections) before the distribution is settled down. Therefore, when the particle distribution is anisotropic (which is common during storms/substorms) or its gradient near the edge of the loss cone is pronounced, or the electron lifetime is longer than the loss timescale, it may be incorrect to use lifetime for the wave-induced scattering. For example, given a pitch angle distribution that monotonically increases with pitch angles, a diffusive process attempts to eliminate the sharp gradient toward isotropy. Fig. 4.15A illustrates an example with an initial distribution (t_0) with a positive gradient in pitch angle and a distribution after diffusion (t_1). Particles within the loss cone then precipitate into the upper atmosphere, as shaded for $\alpha<\alpha_c$. On the other hand, with a lifetime approach in Fig. 4.15B, as the decay time is independent of the pitch angle, the distribution at t_1 is decreased as a whole at all pitch angles exponentially at the same decay rate. The shaded precipitating flux within the loss cone is clearly lower than the case in Fig. 4.15A. Therefore, lifetime models can be applicable when the pitch angle dependence of the particle distributions is weak. But in some circumstances, diffusion coefficients as a function of pitch angle are more appropriate to quantify the efficiency of particle precipitation.

With pitch angle diffusion coefficients $D_{\alpha\alpha}$, the wave-induced scattering loss in the kinetic model is fulfilled through a diffusion equation by solving the loss term of Eq. (4.4).

$$< \frac{\partial f}{\partial t} >_{loss} = \frac{1}{h\mu} \frac{\partial}{\partial \mu} \left[h\mu < D_{\mu\mu} > \frac{\partial f}{\partial \mu} \right]$$
$$< D_{\mu\mu} > = \left(1-\mu^2\right) < D_{\alpha\alpha} > \tag{4.22}$$

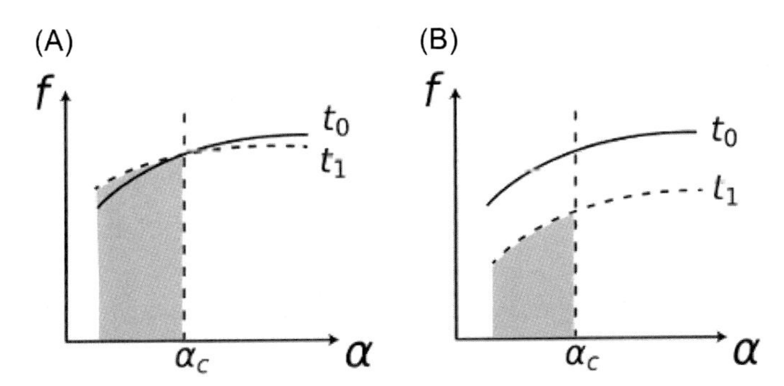

Fig. 4.15 Illustration of (A) diffusive loss using diffusion coefficients and (B) decay loss using lifetimes in altering the distribution functions and precipitating flux.

where $\mu = cos\alpha, h = \frac{1}{2R}\int_{s}^{s'} \frac{ds}{\sqrt{1-B/B_m}}$ is proportional to the bounce period between two magnetic mirror points represented by B_m.

The derivation of pitch angle diffusion coefficients $D_{\alpha\alpha}$ requires many basic inputs to the calculations. For example, for the wave-induced pitch angle scattering rates, detailed information of the waves is required, including the wave amplitude, wave spectrum, wave normal angle, and background magnetic field and plasma density. Lyons (1974) derived general expressions for the resonant diffusion coefficients with the quasi-linear theory of Kennel and Engelmann (1966) for any wave mode in any wave energy distribution and wave normal angle. The above-required wave information is usually parameterized from a rich pool of satellite measurements and can be applied to the general expression to quantitatively estimate the scattering rates.

Many studies have numerically quantified the scattering rates by various plasma waves. For example, Glauert and Horne (2005) built the PADIE code to determine the chorus wave scattering coefficients based on statistical observations of wave properties from CRRES. Based on more recent statistical wave parameters observed by the Van Allen Probes (Li et al., 2015, 2016; Ni et al., 2013), pitch angle diffusion coefficients that depend on energy, MLT, L, pitch angle, and substorm activity levels were also calculated (Li et al., 2015; Ma et al., 2016). Fig. 4.16 shows an example of bounce-averaged pitch angle diffusion coefficients associated with the chorus waves and ECH waves at $L=6$ during geomagnetically active times. The lower-band chorus wave can interact with electrons with $E > 1$ keV, while the upper-band chorus waves efficiently diffuse electrons with E of 0.1–1 keV at small pitch angles. On the contrary, ECH waves do not influence electrons very much.

Both the lifetime decay method and the diffusion loss method with pitch angle diffusion coefficients were investigated in Yu et al. (2016) to understand the ring current electron dynamics. It was found that both loss methods demonstrate similar temporal evolution of the trapped ring current electrons, indicating that the impact of using these two loss methods is small on the trapped electron population. However, the lifetime

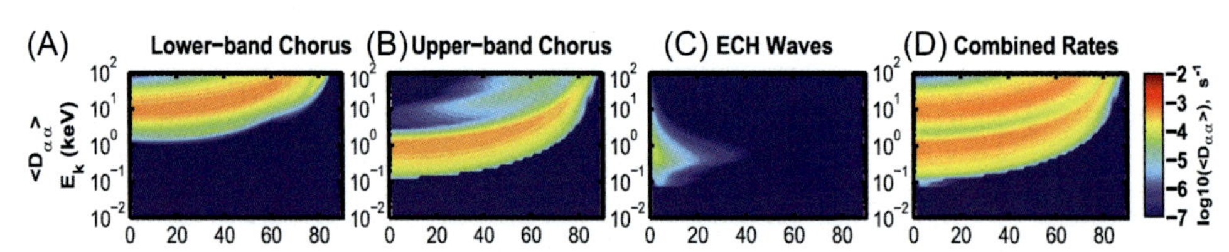

Fig. 4.16 Bounce-averaged diffusion coefficients$\langle D\alpha\alpha\rangle$in (equatorial pitch angle, electron kinetic energy) space for (A) lower band chorus, (B) upper band chorus, (C) ECH waves, and (D) combined diffusion at $L=6$ under geomagnetically active conditions ($AE^* > 300\,nT$). *Adapted from Ni, B., Thorne, R. M., Meredith, N.P., Horne, R.B., Shprits, Y.Y., 2011b. Resonant scattering of plasma sheet electrons leading to diffuse auroral precipitation: 2. Evaluation for whistler mode chorus waves. J. Geophys. Res. 116, A04219. https://doi.org/10.1029/2010JA016233.*

method hardly captured any observed electron precipitation in the large L shell (i.e., $4 < L < 6.5$) region, whereas the diffusion loss method produces much better agreement with NOAA/POES measurements of precipitating electron flux. The diffusion loss method with pitch angle diffusion coefficients, which carries more comprehensive information of the loss, is, therefore, advantageous for capturing electron precipitation in $4 < L < 6.5$.

4.3.3 Simulating ion precipitation in global geospace circulation models

Compared to electron precipitation, ion precipitation typically plays a secondary role in affecting the ionospheric electrodynamics, as the ions usually carry much less energy flux and contribute less to the ionization rate in the upper atmosphere due to their larger mass and shorter mean free path in the dense atmosphere. Nevertheless, ion precipitation may dominate over electrons in some regions, such as the dusk sector if electrons do not drift far enough. Studies showed that medium-energy (tens of keV) proton precipitation could carry more energy flux than electrons in most MLTs, such as $12 < MLT < 24$ and $0 < MLT < 3$, as shown in Fig. 4.17 (Tian et al., 2020), providing substantial energy deposition down to the ionosphere. The conductance resulting from ion precipitation in the subaruroal region can sometimes be on the order of several mhos (Galand and Richmond, 2001; Zou et al., 2014), and can distort the potential pattern (Khazanov et al., 2003). For the diffuse aurora that is formed mainly due to low-energy (a few keV) particle precipitation, statistical studies found that the relative roles of the precipitating electrons and ions in the global energy flux are roughly around 60% and 15%, respectively (Newell et al., 2009), suggesting that although it is secondary to the electron precipitation, the ion precipitation is still important.

Major mechanisms that drive ions to precipitate down to the upper atmosphere include (1) adiabatic loss when ions are transported toward Earth and experience widened loss cones, (2) the resonant interactions with EMIC waves, and (3) pitch angle diffusion due to increased field-line curvature radius. When only the adiabatic loss is included, the

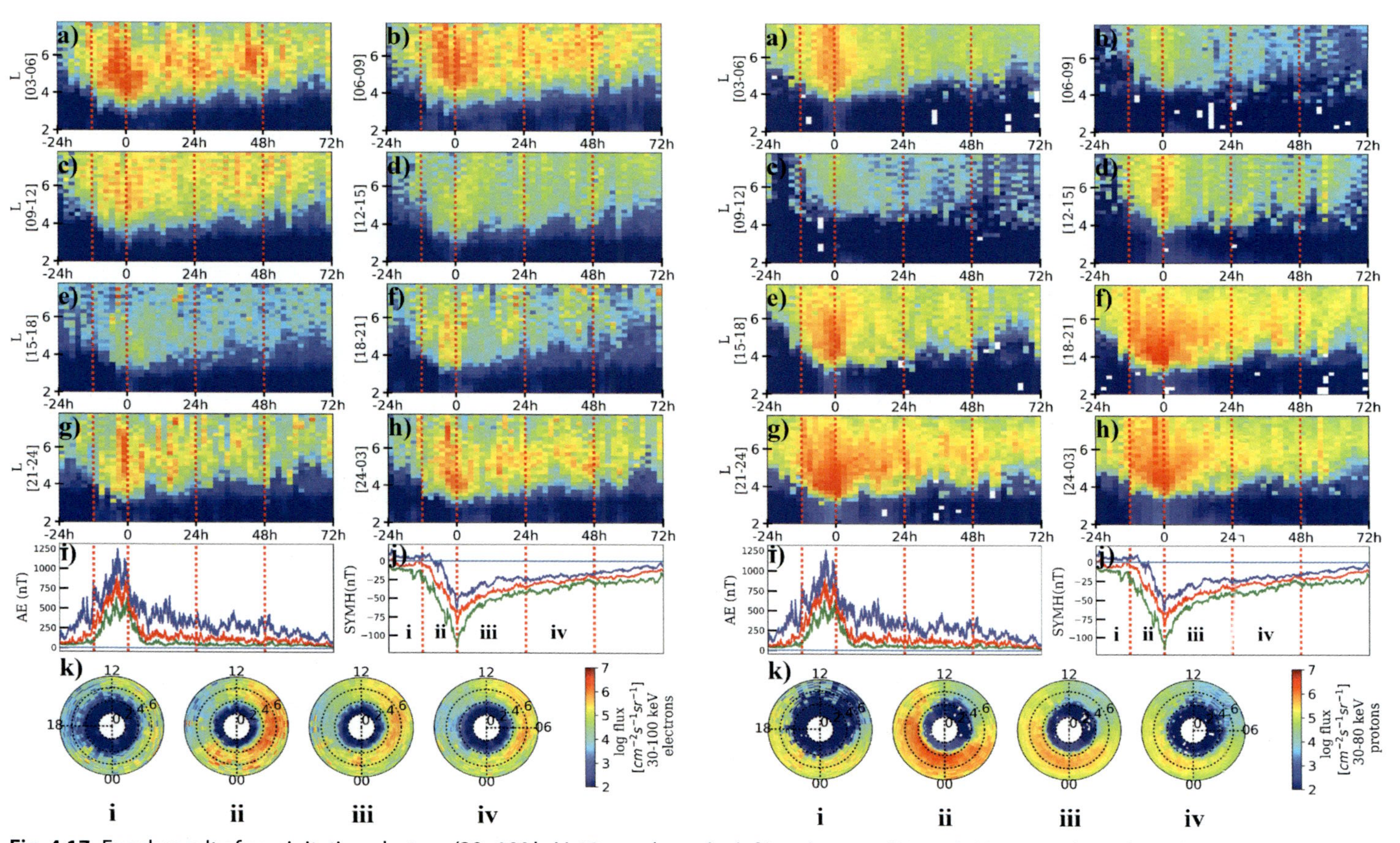

Fig. 4.17 Epoch result of precipitating electron (30–100 keV; 10 panels on the left) and proton (30–80 keV, 10 panels on the right) flux during storms. The precipitating proton flux on the right panels (D, E, F, G, H, representing MLTs from noon eastward to nightside and early morning) exceeds that of electrons in these local times. *Adapted from Tian, X., Yu, Y., Yue, C., 2020. Statistical survey of storm-time energetic particle precipitation, J. Atmos. Sol. Terr. Phys. 199 (1364–6826) 105204. https://doi.org/10.1016/j.jastp.2020.105204.*

ring current decay appears to be slower than reality (Ebihara et al., 2011). The intensity of the subsequent loss cone precipitating proton flux also significantly deviates from observations (Shreedevi et al., 2021), suggesting that loss mechanisms are further needed.

FLC scattering occurs when the particle gyroradius is comparable to the curvature radius of the magnetic fields. This condition is usually satisfied at times when the magnetic field is largely stretched and its curvature radius is narrowed. As discussed in Sections 4.2.1.2 and 4.3.1.2, ions of the same energy could meet this threshold more easily than electrons. Therefore, the FLC scattering process works more efficiently on ions. Similar to the electron loss, the pitch angle scattering loss of ions due to FLC can be quantified through either the lifetime or diffusion coefficients, via Eqs. (4.20) or (4.22). Chen et al. (2019) applied a simple loss rate model of FLC scattering in the ring current model, assuming an energy-dependent lifetime related to the FLC scattering. Their study indicated that the FLC scattering takes place in the nightside inner magnetosphere around $4 < L < 7$ where magnetic fields are less dipolar. Precipitation occurring in these regions, however, is sporadic and localized, and thus contributes sporadically to the ionosphere conductance in localized regions at auroral latitudes.

Ebihara et al. (2011) used pitch angle diffusion coefficients formulated by Young et al. (2008) to quantify the pitch angle diffusion of protons caused by FLC scattering. They examined the ring current decay rate and found that the ring current shows more rapid recovery with an e-folding time of 6 h when the FLC scattering of protons is included, as opposed to an e-folding time of 12 h when only charge exchange and adiabatic loss cone loss alone are included. Yu et al. (2020) further extended the investigation on the role of FLC scattering by including the diffusion of both protons and heavier ions (e.g., $O+$, $He+$). They found that under the FLC scattering process, the pitch angle diffusion is stronger in the night sector for $L > 4$–5 where the magnetic field configuration is more stretched and is more effective on ions above 10 keV, as shown in Fig. 4.18. However, it was also demonstrated that the FLC scattering alone cannot fully account for the widely observed precipitating proton flux in the inner magnetosphere, particularly in the inner zone of $L < 4$, indicating that more precipitating mechanisms are needed there, such as the EMIC waves.

Ions resonating with EMIC waves can be diffused stochastically in pitch angle, breaking the first adiabatic invariant and leading to loss cone ion precipitation. This process can also be quantified via the diffusion Eq. (4.8). Associated pitch angle diffusion coefficients for different bands of EMIC waves in Fig. 4.19 show that the He-band wave interacts with ions with higher energies (above 10 keV) than H-band. Considering this mechanism for scattering and precipitating ring current ions, previous studies have established the association of EMIC waves with the proton aurora or the localized energetic proton precipitation equatorward of the isotropy boundary (e.g., Jordanova et al., 2001, 2007; Nishimura et al., 2014a; Yahnina and Yahnin, 2014; Spasojevic and Frey, 2004; Fuselier et al., 2004; Chen et al., 2014). This thus provides additional energy source

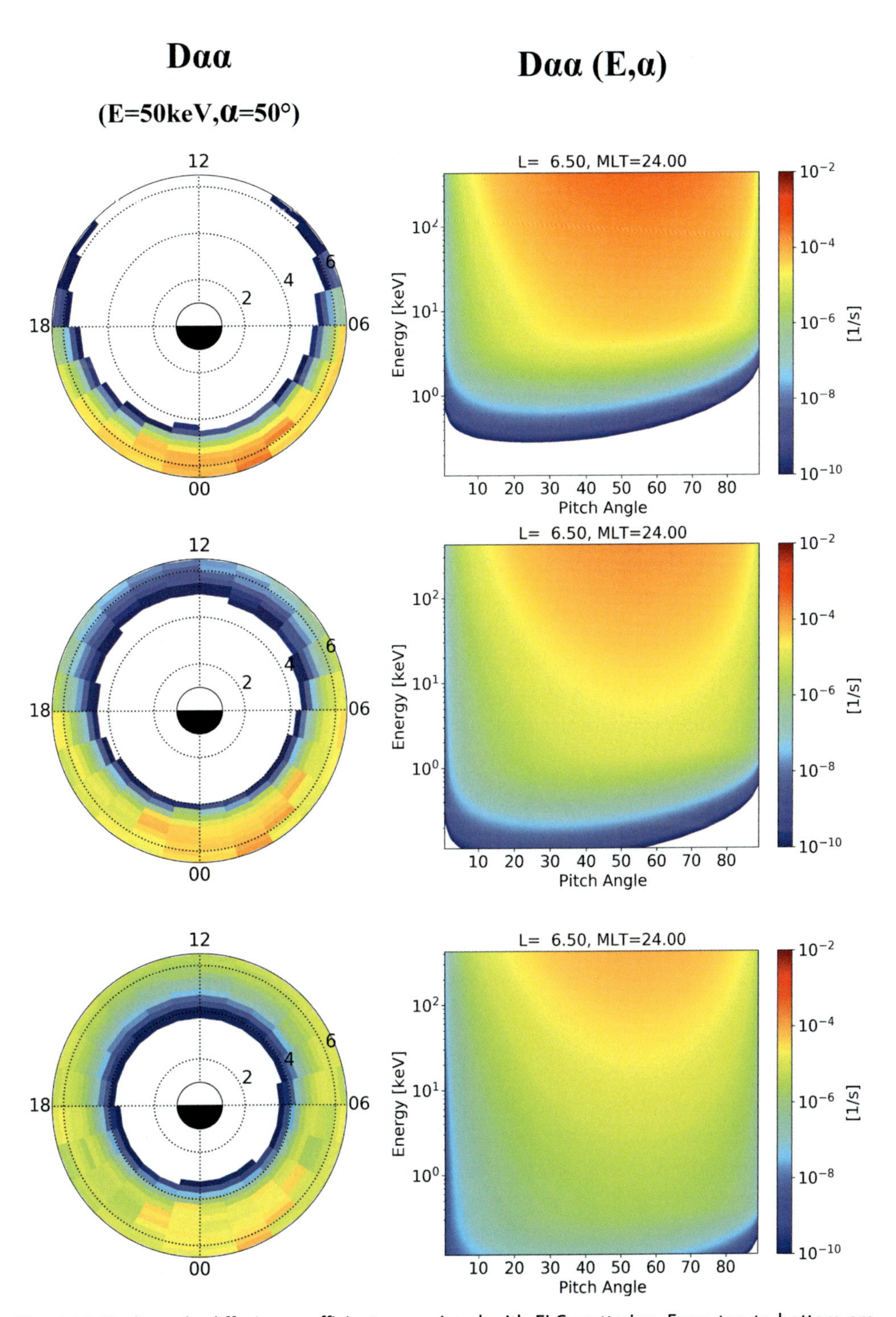

Fig. 4.18 Pitch angle diffusion coefficients associated with FLC scattering. From top to bottom are coefficients for H^+, He^+, and O^+ ions, respectively, in the equatorial plane for $E = 50\,\text{keV}$ and $\alpha = 50$ (left) and in the energy-pitch angle space (right). *Modified from Yu, Y., Tian, X.B., Jordanova, V.K., 2020. The effect of field line curvature (FLC) scattering on the ring current dynamics and isotropy boundary. J. Geophys. Res. Space Phys. https://doi.org/10.1029/2020JA027830.*

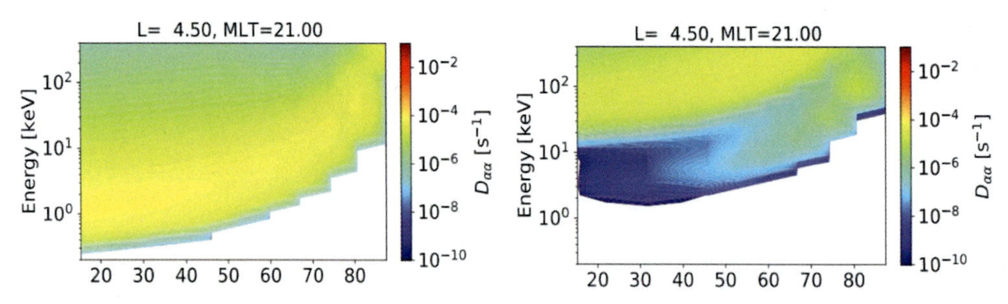

Fig. 4.19 Pitch angle diffusion coefficient as a function of energy and pitch angle for (left) H-band and (right) He-band EMIC waves, determined from statistical EMIC wave model by Saikin et al. (2016).

to the top ionosphere, specifically in the subauroral region. Recently, Shreedevi et al. (2021) reported a simulation study of proton precipitation enabled by both bands of EMIC waves, and showed good agreement with NOAA/POES observations during storm time, implying that EMIC waves are indeed a primary source for precipitating protons. Zhu et al. (2021) further investigated the relative role of FLC scattering and EMIC wave diffusion on the proton precipitation and revealed that the FLC scattering process is more effective at outer regions ($L > 5$) on the nightside and EMIC wave diffusion significantly contributes to the inner regions ($L < 5$) in the dusk sector.

4.3.4 Precipitation-associated ionospheric conductance

As the magnetospheric particle precipitation plays a critical role in altering the ionosphere-thermosphere dynamics, especially in changing the ionospheric conductivity, understanding this coupling process is the key to modeling the magnetosphere-ionosphere system. In GGCM, the ionosphere is often treated as a spherical shell at 100km above the ground surface and the electrodynamics is solved via the Poisson equation:

$$\nabla \cdot (\Sigma \cdot \nabla \Phi) = -J_{\parallel} sinI \tag{4.23}$$

provided the field-aligned current $J_{\parallel}$ from the global magnetospheric model along with a specified conductance Σ (i.e., height-integrated conductivity, usually over 90–500km), the electric potential Φ can be determined, which is an essential parameter in connecting the ionosphere and magnetosphere systems.

As discussed in Section 4.2, the first-principles calculation of ionospheric conductivity is not trivial because it requires important thermospheric and ionospheric characteristics in the calculation, such as the chemistry, reaction rates, and electron/neutral density and temperature. Therefore, empirical functions that parameterize the Hall and Pedersen conductances based on incident energetic electron precipitation were chosen to be used in GGCM. One such popular empirical function is from Robinson et al. (1987):

$$\Sigma_P^e = \frac{40 < E^e >}{16 + < E^e >^2} \sqrt{F_E^e}$$
$$\Sigma_H^e = 0.45 < E^e >^{0.85} \Sigma_P^e \tag{4.24}$$

where Σ_P^e, Σ_H^e are the Pedersen and Hall conductances induced by precipitating electrons, $< E^e >$, F_E^e are the average energy and energy flux of precipitating electrons. Though the above expressions assume a Maxwellian energy distribution in the precipitating flux, they have been shown to work well if correct average energy and energy flux are used (Robinson et al., 1987).

Recently, Khazanov et al. (2018) suggested that reflected electrons between two magnetic conjugate points bring in secondary superthermal ($E < 500$–$600\,\text{eV}$) electron fluxes, providing a new contribution to the auroral conductances. To include the secondary electron fluxes, the standard techniques for calculating the ionospheric conductance based on magnetospheric precipitation must be modified. The newly updated parametric relations are as follows:

$$\Sigma_P^K = K_P(< E >)\Sigma_P^e$$
$$\Sigma_H^K = K_H(< E >)\Sigma_H^e \tag{4.25}$$

The correction factors K_P and K_H, as functions of the mean energy of precipitating electrons' $< E >$, account for the superthermal multiple reflection processes of the secondary electrons (see details in Khazanov et al., 2018). With the reflected secondary electrons taken into account within the Rice Convection Model, Khazanov et al. (2019) discovered a significant increase in both Pedersen and Hall conductances, which further impact the MI coupling processes including subauroral dynamics.

While efficient, it should be noted that the above-parameterized formalisms bypass the complexity of the thermosphere-ionosphere system in determining the conductance, which apparently deviates from a self-consistent treatment of the magnetosphere-ionosphere system. The inclusion of the physics-based ionosphere-thermosphere processes, through degrading the simulation efficiency, is necessary for a more profound understanding of the coupled system. Such efforts have recently been made to elevate the physics-based self-consistency in global circulation models (e.g., Connor et al., 2016; Raeder et al., 2016; Wiltberger et al., 2017; Xi et al., 2016; Yu et al., 2018). For example, Raeder et al. (2016) applied the Open Global General Circulation Model (OpenGGCM) coupled with the Coupled Thermosphere Ionosphere Model (CTIM) to study the subauroral polarization streams (SAPS), with the CTIM receiving precipitating electron flux and subsequently calculating ionospheric conductivity globally. They found that in the region with low conductivity and coincident trough of electron density, the

SAPS are reproduced with a much larger flow speed than the earlier study of Yu et al. (2015) that does not include a self-consistent ionosphere-thermosphere model. Yu et al. (2018) coupled the kinetic ring current model with the electron transport code GLOW by providing the full spectrum of precipitating electrons from tens of eV to hundreds of keV to GLOW for determining the ionospheric ionization and conductivity. They found that energetic electron precipitation over 30 keV could drive enhanced ionization at low altitudes in the D region and create a secondary conductivity layer in addition to the primary one in the E region. Such a two-layer conductivity profile was also reported from observations (Hosokawa and Ogawa, 2010), suggesting that ionospheric electrodynamics is more sophisticated than previously thought.

4.4 Quantifying ionospheric conductances induced by auroral electron precipitation with empirical models and data assimilation techniques

4.4.1 Introduction

The ionosphere is embedded within the dense atmosphere where collisions between charged particles and neutrals cause the electrons and ions to move in different directions under the influence of Earth's magnetic field, forming electric currents in the ionosphere. The electrical conductivity of the ionosphere is composed of three components: the parallel conductivity along the direction of the magnetic field, the Pedersen conductivity in the direction of the electric field but perpendicular to the magnetic field, and the Hall conductivity that is perpendicular to both the electric field and magnetic field. Although the parallel conductivity is very large in magnitude, ionospheric electric currents are determined by the Pedersen and Hall conductivities since the electric field in the ionosphere is mainly perpendicular to the magnetic field.

The Pedersen and Hall conductivities are theoretically derived from the electron and ion momentum equations, and they are expressed in Eqs. (4.10) and (4.11). Since the Pedersen and Hall conductivities concentrate in a narrow altitude range, with the Pedersen conductivity peaking around 125 km and the Hall conductivity peaking around 110 km (Richmond, 1995), it is conceptually convenient to consider the height-integrated Pedersen and Hall conductivities, or Pedersen and Hall conductances, when analyzing the current-carrying capacity of the ionosphere:

$$\Sigma_P = \int_{z_1}^{z_2} \sigma_P dz \tag{4.26}$$

$$\Sigma_H = \int_{z_1}^{z_2} \sigma_H dz \tag{4.27}$$

where z_1 and z_2 are the bottom and top altitudes of the ionosphere.

Ionospheric conductivity is critically important in the electrodynamic coupling of the magnetosphere-ionosphere-thermosphere system. Momentum transfer between the magnetosphere and ionosphere is accomplished by magnetic field-aligned currents (FACs), which flow between the magnetosphere and the ionosphere, and are closed by ionospheric currents perpendicular to the magnetic field. When the neutral winds are absent, ionospheric horizontal currents are described by Ohm's law:

$$\vec{J}_{\perp} = \Sigma_P \vec{E}_{\perp} + \Sigma_H \vec{b} \times \vec{E}_{\perp} \tag{4.28}$$

where $\vec{E}_{\perp}$ is the electric field perpendicular to the magnetic field, and $\vec{b}$ is the unit vector of the magnetic field. Generally speaking, the magnetosphere-ionosphere coupling is analogous to current and voltage generators. In the case of a current generator, the total currents flowing into the ionosphere are fixed so that the electric field becomes inversely proportional to ionospheric conductances in order to maintain a constant current. Conversely, for a voltage generator, the electric potential drop (or the electric field) is fixed so that the currents increase when the conductances increase. The true coupling process between the magnetosphere and ionosphere lies somewhere in between these two extreme scenarios, and the partitioning of the electric field and currents is controlled by ionospheric conductances.

Ionospheric conductances are key parameters in global magnetic hydrodynamic (MHD) models in terms of magnetosphere-ionosphere coupling. Raeder et al. (2001) showed that the MHD simulation results are strongly dependent on the parameterization of auroral conductances. In particular, they found that the different conductance specifications determine whether the magnetosphere enters into a steady magnetospheric convection mode or develops a substorm. Other studies also suggested that magnetospheric convection is controlled by ionospheric conductances to a large extent (Wolf, 1970; Fedder and Lyon, 1987; Ridley et al., 2004). Siscoe et al. (2002) attributed ionospheric conductances as a controlling factor of polar-cap potential saturation by setting up nonlinear feedback between the region 1 FACs and magnetic reconnection at the dayside magnetopause. Ionospheric conductances are found to regulate the nightside magnetic reconnection as well, and the inhomogeneous distribution of the conductances leads to asymmetric flows in the plasma sheet that are faster in the premidnight sector than in the postmidnight sector (Lotko et al., 2014). Another important form of the magnetosphere-ionosphere coupling is via Alfvénic waves, which bounce back and forth along the magnetic field lines and modulate field-aligned currents and the precipitation of auroral energetic particles. The reflection of Alfvén waves at the ionospheric boundary is dependent on the Pedersen conductivity (Lysak, 1986, 1991).

Ionospheric conductances are also key parameters in energy and momentum transfer between the ionosphere and thermosphere. In addition to auroral precipitation, Joule heating is another important form of magnetospheric energy deposition in the ionosphere. Joule heating causes thermospheric upwelling and changes neutral composition due to the rising of molecular-rich air (Fuller-Rowell et al., 1997). Impulsive Joule heating launches large-scale gravity waves that propagate equatorward toward middle and low latitudes, altering the mean circulation of the thermosphere (Prölss, 1995). Joule heating is defined as $q_J = \sigma_P E'^2$, where $\vec{E}' = \vec{E}_\perp + \vec{V}_n \times \vec{B}$ and $\vec{V}_n$ is the neutral wind velocity. In cases when the wind speed is substantially smaller than plasma convection, height-integrated Joule heating can be written as $Q_J = \Sigma_P E_\perp{}^2$. Therefore, Joule heating is directly proportional to the Pedersen conductivity or conductance. The momentum coupling between the ionosphere and thermosphere is through Ampere acceleration resulting from ion-neutral collisions (Richmond, 1995):

$$\frac{\vec{J} \times \vec{B}}{\rho} = \frac{\sigma_P B^2}{\rho}\left(\vec{V}_E - \vec{V}_n\right)_\perp + \frac{\sigma_H B^2}{\rho} \vec{b} \times \left(\vec{V}_E - \vec{V}_n\right) \tag{4.29}$$

where ρ is neutral mass density, $\vec{V}_E = \vec{E}_\perp \times \vec{B} / B^2$, and $\sigma_P B^2/\rho$ and $\sigma_H B^2/\rho$ are called ion-drag coefficients. It is evident that both the Pedersen and Hall conductivities have important consequences on thermospheric dynamics.

The main sources of ionospheric conductivity are solar radiation of ultraviolet (UV) and extreme ultraviolet (EUV), and the precipitation of energetic particles from the magnetosphere. The calculation of solar-induced ionospheric conductances is relatively straightforward, which can be described as a function of solar radiation (commonly represented by the F10.7 index) and solar zenith angle (e.g., Rasmussen et al., 1988; Brekke and Hall, 1988; Moen and Brekke, 1990; Senior, 1991). Realistic estimation of ionospheric conductances by auroral precipitation, on the other hand, is very difficult due to inadequate global observations as well as the lack of knowledge concerning the energy spectral distributions of precipitating particles.

This section provides a brief overview of various techniques that have been developed to estimate auroral-induced ionospheric conductances using the data products described in Section 4.2. Section 4.4.2 presents some representative empirical and statistical conductance models. Section 4.4.3 describes how global conductance patterns can be obtained through data assimilation. Finally, some concluding remarks are given in Section 4.4.4.

4.4.2 Empirical and statistical conductance models

Because the ionization profile (and thus the electron density profile) is largely determined by the characteristics of the precipitating particles, several empirical formulas have been developed to parameterize auroral conductances using the mean energy and energy flux of the precipitating electrons (Wallis and Budzinski, 1981; Vickrey et al., 1981; Harel et al., 1981; Reiff, 1984; Robinson et al., 1987). Among them, the most widely used empirical formulas are the so-called Robinson's formulas (24), expressed through the average energy (in keV) and the energy flux (in ergs/cm^2/s) of precipitating electrons.

Robinson's formulas are based on Maxwellian energy spectra and are applicable for precipitating electrons with energies from 500 eV to 30 keV. Therefore, the contributions by soft precipitating electrons with energies less than 500 eV are not included in these formulas. In addition, the formulas were based on the results from three satellite passes over the Chatanika radar (Vondrak and Robinson, 1985). Despite these shortcomings, Robinson's formulas have been widely adopted by the research community, largely owing to their easy implementation. Robinson's formulas are commonly used in global MHD models for estimating ionospheric conductances (Fedder et al., 1995; Raeder et al., 2001; Ridley et al., 2004; Wiltberger et al., 2009), and they are also the basis for some of the conductance models (e.g., Hardy et al., 1987; Ahn et al., 1998).

Understanding the magnetosphere-ionosphere electrodynamic coupling requires the knowledge of ionospheric conductances globally. To provide such knowledge, a number of statistical conductance models have been developed based on the in situ measurements of auroral precipitation by various satellites. The first two-dimensional statistical auroral conductance model was developed by Wallis and Budzinski (1981) based on energetic particle measurements by the Isis-2 satellite that operated over the period of 1971–1974. The model has a spatial grid of 1° in magnetic latitude (MLAT) and 2 h in MLT. Because of the limited data from the satellite, the model represents only two activity ranges according to the planetary-K (*Kp*) index: $0 \leq Kp \leq 3$ and $3 < Kp \leq 9$. A similar statistical auroral conductance model was constructed by Spiro et al. (1982) based on measurements from the Atmosphere Explorer C and D satellites which flew from 1974 to 1976. This model has a spatial resolution of 1° in MLAT and 1 h in MLT, and are binned separately into four ranges of the hourly averaged auroral electrojet (AE) index (e.g., AE $\leq$ 100 nT, 100 nT < AE $\leq$ 300 nT, 300 nT < AE $\leq$ 600 nT, and AE > 600 nT), and five ranges of the *Kp* index (e.g., $Kp \leq 1$, $1 < Kp \leq 2$, $2 < Kp \leq 3$, $3 < Kp \leq 4$, and $Kp > 4$). To facilitate their use, the model was later fitted into functional forms by Reiff (1984).

These early statistical conductance models were hampered by insufficient data coverage due to the limited lifetime of the satellites, and consequently, they were binned into rather coarse geomagnetic activity levels. With the expansion of satellite missions and the longer spans of their data coverage, more sophisticated auroral conductance models become available. The Hardy model (Hardy et al., 1987) and the Fuller-Rowell and Evans model (Fuller-Rowell and Evans, 1987) are among the most popular ones.

The Hardy model is based on three polar-orbiting satellites (e.g., DMSP F2 and F4, and P78–1). The model has a spatial resolution of 1° in MLAT and 1 h in MLT, and is parameterized by the *Kp* index. The more extensive dataset allows the model to cover a broader range of geomagnetic activities from $Kp = 0$ to $Kp >= 6$ in a total of 7 discrete levels. In a parallel effort, Fuller-Rowell and Evans (1987) developed similar statistical models based on the National Oceanic and Atmospheric Administration (NOAA) Television and Infrared Observatory Satellite (TIROS) measurements of auroral precipitating particles by combining 6 years of data from two satellites (e.g., NOAA-6 and -7). Their model is parameterized by the 10-level hemispheric power index (HPI), and has a spatial resolution of 1° in MLAT and 2° in MLT.

Departing from the traditional statistical conductance models (e.g., Fuller-Rowell and Evans, 1987; Hardy et al., 1987) that simply sort the data according to selected driving parameters (e.g., HPI or *Kp*) and then average them over the MLAT-MLT bins, McGranaghan et al. (2015) employed empirical orthogonal functions (EOFs) to characterize the modes of the Pedersen and Hall conductances based on the spectrally resolved particle precipitation measured by the DMSP satellites. A total of 50 EOFs were constructed to describe the spatial variations of the Pedersen and Hall conductance patterns. The authors discussed the possible external drivers for the first four EOF modes, which together accounted for 50%–53% of the total conductance variabilities.

4.4.3 Global auroral conductance patterns derived from data assimilation

Since the coupling process of the solar wind-magnetosphere-ionosphere is highly nonlinear, no two geomagnetic storms are exactly the same. Consequently, the acceleration and dynamics of auroral precipitation vary drastically from one event to another. Statistical models represent the average distributions of ionospheric conductances that are very useful to climatological studies, they are often inadequate when applied to real events especially those associated with geomagnetic storms and substorms. In order to better represent the spatiotemporal variations of auroral conductances during real events, it is important to estimate the conductances by combining the different observations associated with the specific event using advanced data assimilation technology.

Green et al. (2007) demonstrated a technique to obtain the real-time ionospheric conductances. Their method involved separate data assimilations for the curl-free and divergence-free components of horizontal current based on satellite and ground magnetometer measurements, respectively. Together with the assimilated electric potential patterns based on the Super Dual Auroral Radar Network (SuperDARN) and DMSP ion drift data, the Pedersen and Hall conductances were obtained by applying Ohm's law. In a similar approach, Cousins et al. (2015) estimated ionospheric conductances based on the assimilated patterns of magnetic potential functions from AMPERE and electric potential from SuperDARN. Their results revealed that, although the conductances derived from this method are generally consistent with statistical models, the method also results in unphysical values owing to the uncertainties associated with sparse and often nonoverlapping coverage of the two datasets.

To illustrate the dynamical variations of auroral conductance, McGranaghan et al. (2016) carried out an assimilative analysis of the Hall and Pedersen conductance patterns over a week-long period associated with a moderate geomagnetic activity using an optimal interpolation (OI) technique. The authors demonstrated that the OI technique is capable of capturing the spatiotemporal variations in the derived conductance patterns. The fitted ion drifts and magnetic field perturbations show better agreement with the SuperDARN and AMPERE observations when using the OI-derived conductances compared to those using statistical conductance models. An important feature of the works by McGranaghan and colleagues is the elimination of any assumptions on the energy spectral distributions of precipitating electrons as their derivation of auroral conductances was based on the precipitating electron energy spectra directly measured by the DMSP particle detectors. This approach represents a significant advance over other methods, including the Assimilative Mapping of Ionospheric Electrodynamics (AMIE) procedure. Next generation particle precipitation model has been recently proposed by McGranaghan et al. (2020).

AMIE is specifically designed to obtain snapshots of ionospheric electrodynamic fields by synthesizing simultaneous observations from various ground- and space-based instruments. As described by Richmond and Kamide (1988) and Richmond (1992), the algorithm performs an optimally constrained, weighted least-squares fit of a set of coefficients to the observed data. The data fitting is carried out in two steps. In the first step, AMIE assimilates the Pedersen and Hall conductances by modifying statistical auroral conductance models (e.g., Fuller-Rowell and Evans, 1987) with available conductance-related data. The fitted Pedersen and Hall conductance patterns are then used to interrelate ionospheric electric fields and currents via Ohm's law. In the second step, AMIE combines all data related to electric fields (such as radar and satellite measurements of ion drifts) and currents (such as magnetic field perturbations observed on the ground and in space) to obtain an optimal estimation of electric potentials over the high-latitude ionosphere. Once the conductances and electric potentials are obtained, other electrodynamical parameters such as ionospheric equivalent currents, horizontal currents, field-aligned currents, and height-integrated Joule heating can be readily calculated.

Since AMIE fits the electric potential distribution simultaneously to the observations of electric fields (or ion drifts) and magnetic perturbations, it is important that the conductances obtained in the first step of the fitting process are as accurate as possible in order to ensure that the magnetic perturbations contribute properly to the fitted electric potential pattern. For that purpose, AMIE incorporates various direct and indirect information about the Pedersen and Hall conductance whenever available, such as in situ measurements of precipitating electrons from DMSP and NOAA satellites, conductance measurements from ISRs, auroral emissions observed by ground-based photometers and spacecraft auroral imagers, and magnetic perturbations by ground magnetometers. The inference of auroral conductances from ground magnetometers is based on the empirical relation derived by Ahn et al. (1998) as discussed earlier. It is worth noting that the conductance data injected into AMIE are inferred from the mean energy and energy flux of precipitating electrons measured by DMSP and NOAA satellites using Robinson's formulas, which are built on the premise of Maxwellian energy spectra.

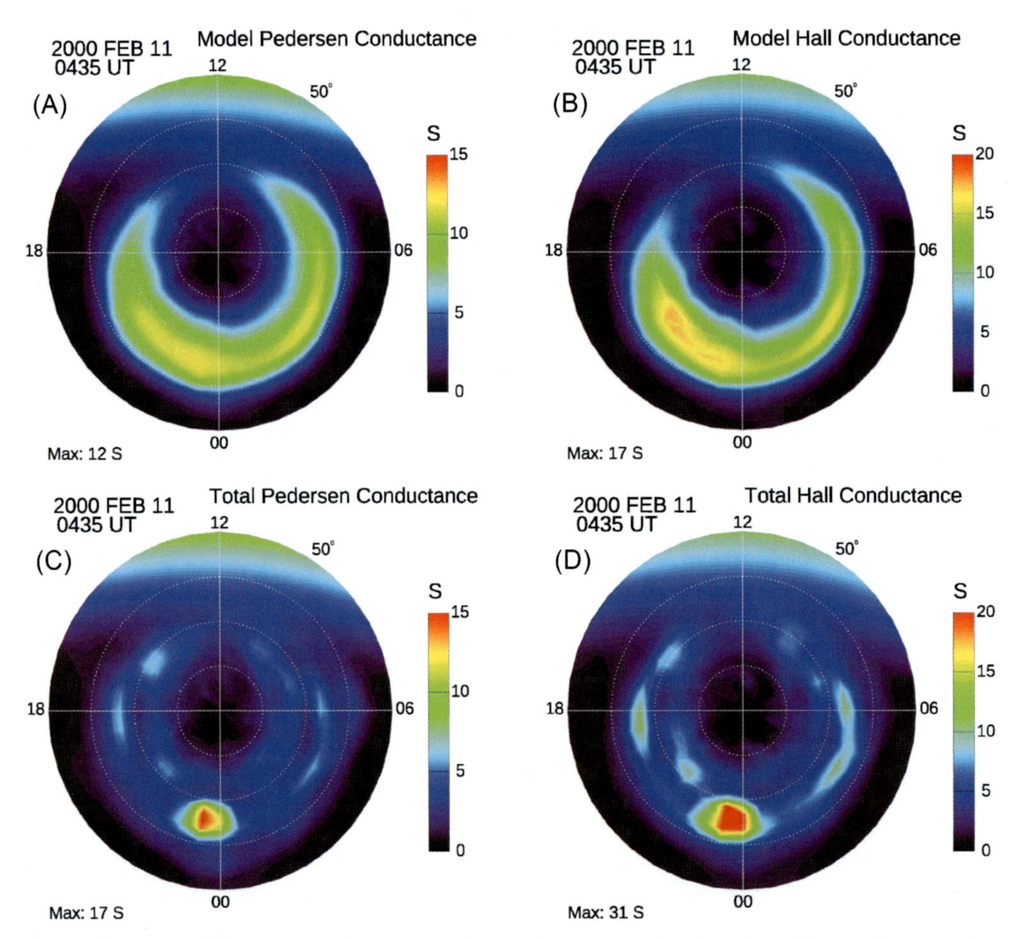

Fig. 4.20 (top row) Pedersen and Hall conductances from the Fuller-Rowell and Evans (1987) model, and (bottom row) the fitted Pedersen and Hall conductances from AMIE.

An example of how AMIE improves the estimates of Pedersen and Hall conductance based on various space- and ground-based observations is shown in Fig. 4.20. The top row presents the background Pedersen and Hall conductance models of Fuller-Rowell and Evans (1987) for the corresponding auroral activity at 04:35 UT on February 11, 2000, during the growth phase of an isolated substorm (Lu et al., 2002). The bottom row shows the fitted conductance patterns by synthesizing the observations of Polar UVI auroral images and ground magnetic perturbations. The model conductances show a relatively smooth distribution with enhanced conductances in the premidnight sector and some modest enhancement in the early morning sector. The fitted conductance patterns, on the other hand, depict a locally confined enhancement near local midnight.

The distributions of auroral conductances directly affect the derivation of ionospheric electric fields and currents in the AMIE procedure, and the effects are highlighted in Fig. 4.21. The top row shows the fitted patterns of electric potentials and field-aligned currents associated with the conductances derived from the Fuller-Rowell and Evans model, and the bottom row shows similar patterns but based on the fitted conductances from AMIE. Though the convection patterns associated with the model and fitted auroral conductances look very similar in terms of their large-scale configuration, noticeable differences can be seen near local midnight. In addition, the cross polar-cap potential drop is 59 kV using the model conductances, and 72 kV using the fitted conductance. The FAC patterns associated with the model and fitted conductances exhibit more

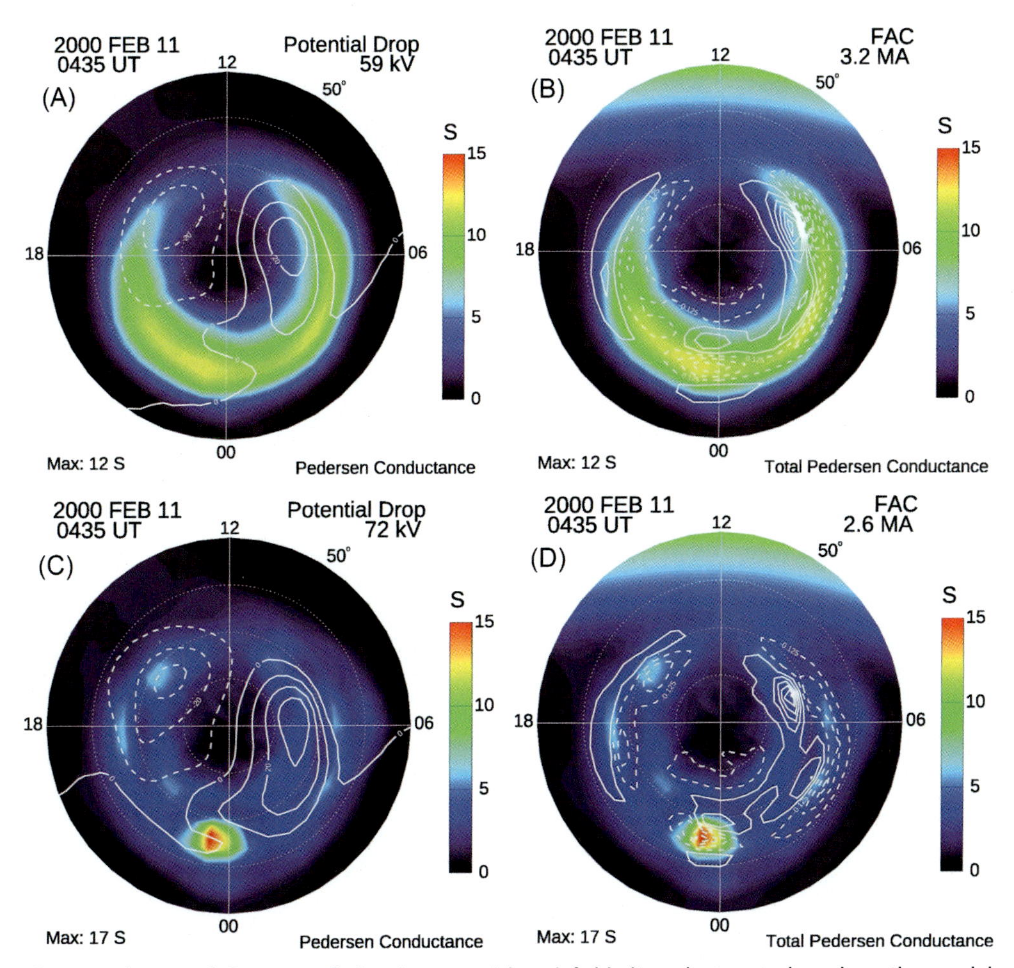

Fig. 4.21 (top row) Patterns of electric potential and field-aligned currents based on the model conductances. (bottom row) Patterns of electric potential and field-aligned currents based on the fitted conductances. The value of the cross-polar potential drop is shown in the upper right corner in the left panels, and the total downward/upward field-aligned currents in the right panels. The *dashed and solid contours* in the right panels represent the upward and downward field-aligned currents, respectively.

striking differences, especially on the nightside. The model conductances result in a nearly continuous band of upward FAC that connects the region 2 current on the dawnside and the duskside region 1 FAC on the duskside. In contrast, the fitted conductance results in a locally confined upward FAC structure near local midnight that is morphologically separated from the dawnside region 2 and the duskside region 1 FACs.

AMIE weights all observations according to the inverse square of their estimated errors so that less reliable data contributes less to the fitting. Fig. 4.22 illustrates how the estimation of Pedersen conductance is influenced by the different data inputs to

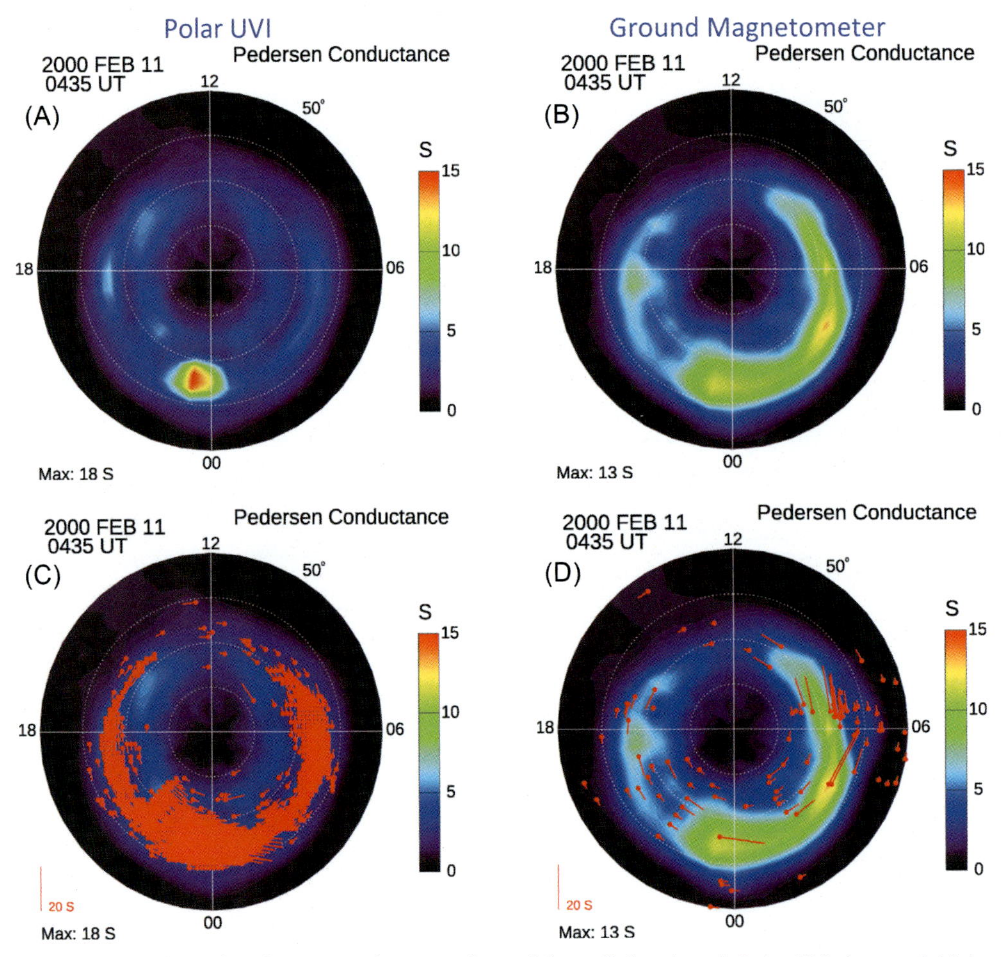

Fig. 4.22 Patterns of Pedersen conductance derived from (left column) Polar UVI data and (right column) ground magnetometer data, respectively. The corresponding input data are overlaid in the bottom panels.

AMIE. The upper left panel shows the fitted Pedersen conductance based on the Polar UVI auroral emissions, and the upper right panel shows the fitted Pedersen conductance inferred from ground magnetometer data. The corresponding input data for the fitted conductance patterns are highlighted in the bottom row. The fitted Pedersen conductance based on the Polar UVI data shows a localized enhancement near local midnight whereas the fitted Pedersen conductance inferred from ground magnetic perturbation indicates large-conductance values mostly on the dawnside. Since the conductances inferred from the Polar UVI data are considered more reliable than those inferred from ground magnetometer data, when these two datasets are combined, the Polar UVI data have a much larger influence than the magnetometer data in the fitted conductance patter. This is evident by comparing Fig. 4.22A (using Polar UVI data) and Fig. 4.22B (using magnetometer) with Fig. 4.20D (using both datasets).

4.4.4 Concluding remarks

Substantial progress has been made to improve the estimation of global ionospheric conductances. Many techniques have been developed to infer ionospheric conductances using measurements not only from ISRs but also from other space- and ground-based instruments. Our understanding of the global morphology of ionospheric conductance is also significantly improved through the development of empirical and statistical conductance models, and lately through the application of data assimilation. However, accurate specification of ionospheric conductances induced by auroral precipitation remains a great challenge to the research community since all currently available technologies for deriving ionospheric conductance require some form of assumptions and approximations, such as the energy spectral and pitch angle distributions of auroral precipitating particles, knowledge of neutral density, composition, and temperature, and the spatial distribution of major ion species throughout the ionosphere. A major breakthrough in this regard would require comprehensive measurements of all neutral and plasma properties simultaneously.

As noted before, Robinson's formulas are based on very limited observations and are applicable for precipitating electrons with Maxwellian energy spectra. A recent investigation by Kaeppler et al. (2015) indicated that modification to Robinson's formula concerning the ratio of Hall to Pedersen conductance is required in order to be more consistent with the ISR data for the Hall conductance. More systematic assessments are needed to further improve Robinson's formulas or develop new empirical conductance models that can be implemented as easily.

Robinson's formulas have been extensively used in global MHD and ring current kinetic models to calculate ionospheric conductances from the model-derived mean energies and energy fluxes of precipitating electrons from the magnetosphere. Recent work by Khazanov et al. (2018) showed that the secondary superthermal electrons produced by multiple reflections between the magnetically conjugate ionospheric locations can increase ionospheric conductances by 40%–100%, depending on the mean energy of the primary precipitating electrons originated in the magnetosphere. This important effect, however, is not considered by MHD and ring current kinetic models; consequently, direct applications of Robinson's formulas in these models can lead to erroneous simulation results. As demonstrated by Khazanov et al. (2019), the incorporation of the secondary superthermal electrons in the Rice Convection Model leads to significant increases of the Pedersen and Hall conductances, which in turn alter ring current energization, intensify subauroral polarization drifts, and suppress the plasma interchange instability. Furthermore, Wing et al. (2019) found that the secondary superthermal electrons are mainly responsible for the DMSP-observed soft precipitating electrons with energies less than a few 100s eV, which deviate from the Maxwellian distributions of the high-energy precipitating electrons in the diffuse auroral region. Although these soft precipitating electrons are not a major contributor to the height-integrated conductivities or conductances, they can significantly enhance the local conductivities at high altitudes above the E-region ionosphere.

While this section focuses on ionospheric conductances induced by precipitating electrons, the effects of energetic protons should not be overlooked. In fact, precipitating ions contribute about 5%–25% of the total auroral energy flux (Hardy et al., 1989; Hubert et al., 2002; Coumans et al., 2004, 2006; Newell et al., 2005, 2014). In addition, the location of proton aurora tends to offset with respect to the location of electron aurora such that proton precipitation can be a locally dominant ionization source, particularly in the dayside cusp and at the equatorward edge of the electron auroral oval in the premidnight sector (Hardy et al., 1987, 1989; Newell et al., 2005; Frey et al., 2002). To facilitate the estimation of Pedersen and Hall conductances associated with auroral proton precipitation, Galand and Richmond (2001) developed a simple parameterization that is analogous to Robinson's formulas for auroral electron precipitation. Their derivation was based on a proton transport code for computing the electron production rate under the assumption of a Maxwellian distribution for the incident proton energy fluxes. For given mean energy and energy flux of precipitating protons, such as statistical models of Hardy et al. (1989) and Newell et al. (2005, 2014), the simple parameterization can be readily applied to obtain the proton-produced auroral conductances. Based on the Doppler-shifted Lyman-α emission measured by the IMAGE FUV-SI12 camera, Coumans et al. (2004) were able to derive instantaneous global maps of Pedersen and Hall conductances induced by precipitating protons. However, since the FUV instrument had only one optical filter measuring proton emission, the mean energies of precipitating protons were adopted from the statistical model of Hardy et al. (1989) for the corresponding Kp index.

4.5 Open questions

There remain open questions on the topics of convection, precipitation, and conductance, only a few of which we could cover in this chapter. For example, what is the generation mechanism of polar cap arcs? Do they occur on open or closed field lines, or both? Are polar cap patches formed via a precipitation mechanism or a nonprecipitation mechanism? Or do they form from both types? What is the contribution of meso-scale auroral forms to the total energy flux precipitated into the ionosphere? And what is their contribution to changes in conductance? The limitations in our current datasets had made quantifying meso-scale contributions difficult, but there is a path forward. Combining two-dimensional all-sky-imager data, which provide a continuous view of the night sky, with satellite overpasses and well-understood ISR observations for calibration, may help answer some of these questions. A multisatellite mission with onboard cameras recording the auroral oval would also help answer these questions by providing auroral data regardless of cloud coverage, with the ability to link auroral forms to electric fields, field-aligned currents, and in situ particle data.

Additional questions also emerge as computational abilities improve. Recent model coupling that includes kinetic processes highlight effects we were previously missing, leading us to question what physics are we misunderstanding? Continuing to build self-consistently coupled physics-based models with more comprehensive physics and more self-consistent causes and effects and performing data-model comparisons will help elucidate what drives the MIT system. Furthermore, as the ion precipitation is increasingly paid more attention to, how the ionosphere-thermosphere system could respond is worth revisiting if the incident precipitating flux does not contain electrons only. Ion contribution to auroral conductance and further variations of ionospheric electrodynamics are nonnegligible as mentioned in Section 4.3.3, efforts are therefore also demanded toward this direction.

A common theme throughout the sections is the difficulty in determining conductance observationally since all methods lean on underlying assumptions. A major breakthrough in this regard would require comprehensive measurements of all neutral and plasma properties simultaneously.

Acknowledgments

We thank K.H. Glassmeier, U. Auster, and W. Baumjohann for THEMIS/FGM data, C.W. Carlson and J. P. McFadden for the THEMIS/ESA data, and D. Larson and R.P. Lin for the use of the THEMIS/SST data. We also thank K. Ogilvie at NASA GSFC and CDAWeb for providing the WIND/SWE data. We thank DMSP SSUSI team and Yongliang Zhang for SSUSI aurora image. We gratefully acknowledge the SuperMAG collaborators (http://supermag.jhuapl.edu/info/?page=acknowledgement).

Dr. Gabrielse's efforts were supported by National Aeronautics and Space Administration Grant 80NSSC20K0725 and AFOSR Grant FA9559-16-1-0364. The work by Dr. Lu was supported by National

Aeronautics and Space Administration under the Living with a Star program under grant 80NSSC17K071, the Heliophysics Supporting Research program under grant NNX17AI39G, and by AFOSR through award FA9559-17-1-0248. This material is based upon work supported by the National Center for Atmospheric Research (NCAR), which is a major facility sponsored by the National Science Foundation (NSF) under Cooperative Agreement No. 1852977. Dr. Yu's efforts were supported by NSFC Grants 41974192 and 41821003. Dr. Kaeppler's work was supported by AFOSR FA9550-19-1-0130 and NSF AGS 1853408. The work by Dr. Wang was supported by AFOSR FY2016 MURI and NASA 80NSSC20K0714.

References

Adachi, K., Nozawa, S., Ogawa, Y., et al., 2017. Evaluation of a method to derive ionospheric conductivities using two auroral emissions (428 and 630 nm) measured with a photometer at Tromsø (69.6°N). Earth Planets Space. https://doi.org/10.1186/s40623-017-0677-4.

Ahn, B.-H., R. M. Robinson, Y. Kamide, and S.-I. Akasofu (1983), Electric conductivities, electric fields and auroral particle energy injection rate in the auroral ionosphere and their empirical relation to the horizontal magnetic disturbances, Planet. Space Sci., 31, 641, doi:/https://doi.org/10.1016/0032-0633(83)90005-3.

Ahn, B.-.H., Richmond, A.D., Kamide, Y., Kroehl, H.W., Emery, B.A., de la Beaujardiére, O., Akasofu, S.-.I., 1998. An ionospheric conductance model based on ground magnetic disturbance data. J. Geophys. Res. 103 (A7), 14769–14780. https://doi.org/10.1029/97JA03088.

Aksnes, A., Stadsnes, J., Bjordal, J., Østgaard, N., Vondrak, R.R., Detrick, D.L., Rosenberg, T.J., Germany, G.A., Chenette, D., 2002. Instantaneous ionospheric global conductance maps duringan isolated substorm. Ann. Geophys. 20, 1181. https://doi.org/10.5194/angeo-20-1181-2002.

Aksnes, A., Stadsnes, J., Lu, G., Østgaard, N., Vondrak, R.R., De-trick, D.L., Rosenberg, T.J., Germany, G. A., Schulz, M., 2004. Effects of energetic electrons on the electrodynamics in the iono-sphere. Ann. Geophys. 22, 475. https://doi.org/10.5194/angeo-22-475-2004.

Albert, J.M., 1994. Quasi-linear pitch angle diffusion coefficients: retaining high harmonics. J. Geophys. Res. 99 (A12), 23,741–23,745. https://doi.org/10.1029/94JA02345.

Albert, J.M., 1999. Analysis of quasi-linear diffusion coefficients. J. Geophys. Res. 104, 2429–2442. https://doi.org/10.1029/1998JA900113.

Albert, J.M., Shprits, Y.Y., 2009. Estimates of lifetimes against pitch angle diffusion. J. Atmos. Sol. Terr. Phys. 71 (16), 1647–1652. https://doi.org/10.1016/J.JASTP.2008.07.004.

Allen, R.C., Zhang, J.-C., Kistler, L.M., Spence, H.E., Lin, R.-L., Klecker, B., Dunlop, M.W., André, M., Jordanova, V.K., 2015. A statistical study of EMIC waves observed by cluster: 1. Wave properties. J. Geophys. Res. Space Phys. 120, 5574–5592. https://doi.org/10.1002/2015JA021333.

Allen, R.C., Zhang, J.-C., Kistler, L.M., Spence, H.E., Lin, R.-L., Klecker, B., Dunlop, M.W., André, M., Jordanova, V.K., 2016. A statistical study of EMIC waves observed by cluster: 2. Associated plasma conditions. J. Geophys. Res. Space Phys. 121, 6458–6479. https://doi.org/10.1002/2016JA022541.

Amm, O., Pajunpaa, A., Brandstrom, U., 1999. Spatial distribution of conductances and currents associated with a north-south auroral form during a multiple-substorm period. Ann. Geophys. 17 (11), 1385–1396. https://doi.org/10.1007/s00585-999-1385-6.

Anderson, D.N., Buchau, J., Heelis, R.A., 1988. Origin of density enhancements in the winter polar cap ionosphere. Radio Sci. 23 (4), 513–519. https://doi.org/10.1029/RS023i004p00513.

Anderson, B.J., Korth, H., Waters, C.L., Green, D.L., Merkin, V.G., Barnes, R.J., Dryud, L.P., 2014. Development of large-scale Birkeland currents determined from the active magnetosphere and planetary electrodynamics response experiment. Geophys. Res. Lett. 41, 3017–3025. https://doi.org/10.1002/2014GL059941.

Angelopoulos, V., 2008. The THEMIS mission. Space Sci. Rev. 141, 5. https://doi.org/10.1007/s11214-008-9336-1.

Angelopoulos, V., 2011. The ARTEMIS mission. Space Sci. Rev. 165, 3–25. https://doi.org/10.1007/s11214-010-9687-2.

Angelopoulos, V., Baumjohann, W., Kennel, C.F., Coroniti, F.V., Kivelson, M.G., Pellat, R., Walker, R.J., Lühr, H., Paschmann, G., 1992. Bursty bulk flows in the inner central plasma sheet. J. Geophys. Res. 97 (A4), 4027–4039. https://doi.org/10.1029/91JA02701.

Angelopoulos, V., et al., 1993. Characteristics of ion flow in the quiet state of the inner plasma sheet. Geophys. Res. Lett. 20, 1711–1714.

Angelopoulos, V., et al., 1994b. Magnetotail flow bursts: association to global magnetospheric circulation, relationship to ionospheric activity and direct evidence for localization. Geophys. Res. Lett. 24, 2271–2274.

Angelopoulos, V., Kennel, C.F., Coroniti, F.V., Pellat, R., Kivelson, M.G., Walker, R.J., Russell, C.T., Baumjohann, W., Feldman, W.C., Gosling, J.T., 1994a. Statistical characteristics of bursty bulk flow events. J. Geophys. Res. 99 (A11), 21257–21280. https://doi.org/10.1029/94JA01263.

Antonova, E.E., 2006. Quasiturbulent transport and LLBL properties. Adv. Space Res. 37, 532–536. https://doi.org/10.1016/j.asr.2006.01.019.

Ashour-Abdalla, M., Frank, L.A., Paterson, W.R., Peroomian, V., Zelenyi, L.M., 1996. Proton velocity distributions in the magnetotail: theory and observations. J. Geophys. Res. 101, 2587–2598. https://doi.org/10.1029/95JA02539.

Bailey, S.M., Barth, C.A., Solomon, S.C., 2002. A model of nitric oxide in the lower thermosphere. J. Geophys. Res. 107 (A8), 1205. https://doi.org/10.1029/2001JA000258.

Bame, S.J., Asbridge, J.R., Felthauser, H.E., Hones, E.W., Strong, I.B., 1967. Characteristics of the plasma sheet in the Earth's magnetotail. J. Geophys. Res. 72 (1), 113–129. https://doi.org/10.1029/JZ072i001p00113.

Basu, B., Jasperse, J.R., Strickland, D.J., Daniell Jr., R.E., 1993. Transport-theoretic model for the electron-proton-hydrogen atom aurora. 1: theory. J. Geophys. Res. 98, 21,517–21,532. https://doi.org/10.1029/93JA01646.

Berkey, T., Cogger, L.L., Ismail, S., Kamide, Y., 1976. Evidence for a correlation between sun-aligned arcs and the interplanetary magnetic field direction. Geophys. Res. Lett. 3 (3), 145–147. https://doi.org/10.1029/GL003i003p00145.

Bilitza, D., Reinisch, B.W., 2008. International reference ionosphere 2007: improvements and new Parameters. Adv. Space Res. 42, 599–609. https://doi.org/10.1016/j.asr.2007.07.048.

Birn, J., Priest, E.R. (Eds.), 2007. Reconnection of Magnetic Fields: Magnetohydrodynamics and Collisionless Theory and Observations. Cambridge University Press, Cambridge.

Birn, J., Raeder, J., Want, Y.L., Wolf, R.A., Hesse, M., 2004. On the propagation of bubbles in the geomagnetic tail. Ann. Geophys. 22, 1773–1786.

Borovsky, J., Elphic, R.C., Funsten, H.O., Thomsen, M.F., 1997. The Earth's plasma sheet as a laboratory for flow turbulence in high-[beta] MHD. Aust. J. Plant Physiol. 57, 1–34. https://doi.org/10.1017/S0022377896005259.

Borovsky, J.E., Thomsen, M.F., Elphic, R.C., 1998. The driving of the plasma sheet by the solar wind. J. Geophys. Res. 103, 17,617–17,639. https://doi.org/10.1029/97JA02986.

Boström, R., 1974. Ionosphere-magnetosphere coupling. In: McCormac, B.M. (Ed.), Magnetospheric Physics. Astrophysics and Space Science Library (A Series of Books on the Recent Developments of Space Science and of General Geophysics and Astrophysics Published in Connection with the Journal Space Science Reviews). vol. 44. Springer, Dordrecht, https://doi.org/10.1007/978-94-010-2214-9_4.

Brekke, A., Doupnik, J.R., Banks, P.M., 1974. Incoherent scatter measurements of *E* region conductivities and currents in the auroral zone. J. Geophys. Res. 79 (25), 3773–3790. https://doi.org/10.1029/JA079i025p03773.

Brekke, A., Hall, C., 1988. Auroral ionospheric quiet summer time conductances. Ann. Geophys. 6 (4), 361–375.

Brekke, A., Hall, C., Hansen, T.L., 1989. Auroral ionospheric conductances during disturbed conditions. Ann. Geophys. 7, 269–280.

Burch, J.L., 1991. Diagnosis of auroral acceleration mechanisms by particle measurements. In: Meng, C.-I., Rycroft, M.J., Frank, L.A. (Eds.), Auroral Physics. Cambridge University Press, Cambridge, pp. 97–109.

Carlson, H.C., Moen, J., Oksavik, K., Nielsen, C., McCrea, I.W., Pedersen, T., Gallop, P., 2006. Direct observations of injection events of subauroral plasma into the polar cap. Geophys. Res. Lett. 33, L05103. https://doi.org/10.1029/2005GL025230.

Chen, Y.-J., Heelis, R.A., 2018. Mesoscale plasma convection perturbations in the high-latitude ionosphere. J. Geophys. Res. Space Phys. 123, 7609–7620. https://doi.org/10.1029/2018JA025716.

Chen, M.W., Schulz, M., 2001a. Simulations of storm time diffuse aurora with plasmasheet electrons in strong pitch angle diffusion. J. Geophys. Res. 106 (February), 1873–1886. https://doi.org/10.1029/2000JA000161.

Chen, M.W., Schulz, M., 2001b. Simulations of diffuse aurora with plasma sheet electrons in pitch angle diffusion less than everywhere strong. J. Geophys. Res. 106 (December), 28,949–28,966. https://doi.org/10.1029/2001JA000138.

Chen, C.X., Wolf, R.A., 1993. Interpretation of high-speed flows in the plasma sheet. J. Geophys. Res. 98 (21), 409–21,419.

Chen, C.X., Wolf, R.A., 1999. Theory of thin-filament motion in Earth's magnetotail and its application to bursty bulk flows. J. Geophys. Res. 104, 14613–14626.

Chen, L., Jordanova, V.K., Spasojevic, M., Thorne, R.M., Horne, R.B., 2014. Electromagnetic ion cyclotron wave modeling during the geospace environment modeling challenge event. J. Geophys. Res. Space Phys. 119 (4), 2963–2977.

Chen, M.W., Lemon, C.L., Orlova, K., Shprits, Y., Hecht, J., Walterscheid, R.L., 2015. Comparison of simulated and observed trapped and precipitating electron fluxes during a magnetic storm. Geophys. Res. Lett. 42, 8302–8311. https://doi.org/10.1002/2015GL065737.

Chen, M.W., Lemon, C.L., Hecht, J., Sazykin, S., Wolf, R.A., Boyd, A., Valek, P., 2019. Diffuse auroral electron and ion precipitation effects on RCM-E comparisons with satellite data during the 17 March 2013 storm. J. Geophys. Res. Space Phys., 124. https://doi.org/10.1029/2019JA026545.

Christon, S.P., Mitchell, D.G., Williams, D.J., Frank, L.A., Huang, C.Y., Eastman, T.E., 1988. Energy spectra of plasma sheet ions and electrons from −50 eV/e to −1 MeV during plasma temperature transitions. J. Geophys. Res. 93, 2562–2572.

Claudepierre, S.G., Ma, Q., Bortnik, J., O'Brien, T.P., Fennell, J.F., Blake, J.B., 2020a. Empirically estimated electron lifetimes in the Earth's radiation belts: Van Allen Probe observations. Geophys. Res. Lett. 47, e2019GL086053. https://doi.org/10.1029/2019GL086053.

Claudepierre, S.G., Ma, Q., Bortnik, J., O'Brien, T.P., Fennell, J.F., Blake, J.B., 2020b. Empirically estimated electron lifetimes in the Earth's radiation belts: comparison with theory. Geophys. Res. Lett. 47. e2019GL086056. https://doi.org/10.1029/2019GL086056.

Coley, W.R., Heelis, R.A., 1995. Adaptive identification and characterization of polar ionization patches. J. Geophys. Res. 100 (A12), 23,819–23,827. https://doi.org/10.1029/95JA02700.

Connor, H.K., Zesta, E., Fedrizzi, M., Shi, Y., Raeder, J., Codrescu, M.V., Fuller-Rowell, T.J., 2016. Modeling the ionosphere-thermosphere response to a geomagnetic storm using physics-based magnetospheric energy input: OpenGGCM-CTIM results. J. Space Weather Space Clim. 6, A25. https://doi.org/10.1051/swsc/2016019.

Coumans, V., Gérard, J.-C., Hubert, B., Meurant, M., Mende, S.B., 2004. Global auroral conductance distribution due to electron and proton precipitation from IMAGE-FUV observations. Ann. Geophys. https://doi.org/10.5194/angeo-22-1595-2004.

Coumans, V., Gérard, J.-.C., Hubert, B., Meurant, M., 2006. Global auroral proton precipitation observed by IMAGE-FUV: noon and midnight brightness dependence on solar wind characteristics and IMF orientation. J. Geophys. Res. 111. https://doi.org/10.1029/2005JA011317, A05210.

Cousins, E.D.P., Shepherd, S.G., 2010. A dynamical model of high-latitude convection derived from SuperDARN plasma drift measurements. J. Geophys. Res. 115. https://doi.org/10.1029/2010JA016017, A12329.

Cousins, E.D.P., Matsuo, T., Richmond, A.D., 2015. Mapping high-latitude ionospheric electrodynamics with SuperDARN and AMPERE. J. Geophys. Res. Space Phys. 120, 5854–5870. https://doi.org/10.1002/2014JA020463.

Cowley, S.W.H., 1981. Magnetospheric asymmetries associated with the y-component of the IMF. Planet. Space Sci. 29, 79.

Cowley, S.W.H., Lockwood, M., 1992. Excitation and decay of solar wind-driven flows in the magnetosphere-ionosphere system. Ann. Geophys. 10, 103–115.

Craven, J.D., Kamide, Y., Frank, L.A., Akasofu, S.-I., Sugiura, M., 1984. Distribution of aurora and ionospheric currents observed simultaneously on a global scale. In: Potemra, T.A. (Ed.), Magnetospheric Currents. Geophys. Monogr. Ser. 28. AGU, Washington, DC.

Dahlgren, H., Lanchester, B.S., Ivchenko, N., 2015. Coexisting structures from high- and low-energy precipitation in fine-scale aurora. Geophys. Res. Lett. 42, 1290–1296. https://doi.org/10.1002/2015GL063173.

de la Beaujardière, O., Lyons, L.R., Friis-Christensen, E., 1991. Sondrestrom radar measurements of the reconnection electric field. J. Geophys. Res. 96 (A8), 13,907–13,912. https://doi.org/10.1029/91JA01174.

de la Beaujardière, O., Lyons, L.R., Ruohoniemi, J.M., Friis-Christensen, E., Danielsen, C., Rich, F.J., Newell, P.T., 1994. Quiet-time intensificationsalong the poleward auroral boundary near midnight. J. Geophys. Res. 99 (A1), 287–298. https://doi.org/10.1029/93JA01947.

Donovan, E., Jackel, B., Klumpar, D., Strangeway, R., 2003. Energy dependence of the isotropy boundary latitude. In: Proceedings of Atmospheric Studies by Optical Methods. vol. 92. Sodankylä Geophysical Observatory Publications, Finland, pp. 11–14.

Dungey, J.W., 1963. The structure of the exosphere or adventures in velocity space. In: DeWitt, C., Hieblot, J., Lebeau, A. (Eds.), Geophysics, the Earth's Environment. Gordon and Breach, New York, pp. 503–550.

Ebihara, Y., Fok, M.-C., Immel, T.J., Brandt, P.C., 2011. Rapid decay of storm time ring current due to pitch angle scattering in curved field line. J. Geophys. Res. Space Phys. 116 (A3). https://doi.org/10.1029/2010JA016000.

Evans, D.S., Maynard, N.C., Trøim, J., Jacobsen, T., Egeland, A., 1977. Auroral vector electric field and particle comparisons, 2, electrodynamics of an arc. J. Geophys. Res. 82 (16), 2235–2249. https://doi.org/10.1029/JA082i016p02235.

Fang, X., Randall, C.E., Lummerzheim, D., Solomon, S.C., Mills, M.J., Marsh, D.R., Jackman, C.H., Wang, W., Lu, G., 2008. Electron impact ionization: a new parameterization for 100 eV to 1 MeV electrons. J. Geophys. Res. 113. https://doi.org/10.1029/2008JA013384, A09311.

Fang, X., Randall, C.E., Lummerzheim, D., Wang, W., Lu, G., Solomon, S.C., Frahm, R.A., 2010. Parameterization of monoenergetic electron impact ionization. Geophys. Res. Lett. 37. https://doi.org/10.1029/2010GL045406, L22106.

Fedder, J.A., Lyon, J.G., 1987. The solar wind–magnetosphere–ionosphere current–voltage relationship. Geophys. Res. Lett. 14, 880.

Fedder, J.A., Slinker, S.P., Lyon, J.G., Elphinstone, R.D., 1995. Global numerical simulation of the growth phase and the expansion onset for a substorm observed by Viking. J. Geophys. Res. 100 (A10), 19083–19093. https://doi.org/10.1029/95JA01524.

Ferradas, C.P., Jordanova, V.K., Reeves, G.D., Larsen, B.A., 2019. Comparison of electron loss models in the inner magnetosphere during the 2013 St. Patrick's Day geomagnetic storm. J. Geophys. Res. Space Phys. 124, 7872–7888. https://doi.org/10.1029/2019JA026649.

Frank, L.A., Craven, J.D., Burch, J.L., Winningham, J.D., 1982. Polar views of the Earth's aurora with dynamics explorer. Geophys. Res. Lett. 9, 1001–1004. https://doi.org/10.1029/GL009i009p01001.

Frey, H.U., Meade, S.B., Immel, T.J., Fuselier, S.A., Claflin, E.S., Gérard, J.-.C., Hubert, B., 2002. Proton aurora in the cusp. J. Geophys. Res. 107 (A7). https://doi.org/10.1029/2001JA900161.

Fridman, M., Lemaire, J., 1980. Relationship between auroral electrons fluxes and field aligned electric potential difference. J. Geophys. Res. 85, 664–670. https://doi.org/10.1029/JA085iA02p00664.

Fuller-Rowell, T.J., Evans, D.S., 1987. Height-integrated Pedersen and Hall conductivity patterns inferred from the TIROS-NOAA satellite data. J. Geophys. Res. 92 (A7), 7606–7618. https://doi.org/10.1029/JA092iA07p07606.

Fuller-Rowell, T.J., Codrescu, M.V., Roble, R.G., Richmond, A.D., 1997. How does the thermosphere and ionosphere react to a geomagnetic storm? In: Tsurutani, B.T., Gonzalez, W.D., Kamide, Y., Arballo, J.K. (Eds.), Magnetic Storms., https://doi.org/10.1029/GM098p0203.

Fukuda, Y., Kataoka, R., Miyoshi, Y., Katoh, Y., Nishiyama, T., Shiokawa, K., Ebihara, Y., Hampton, D., Iwagami, N., 2016. Quasi-periodic rapid motion of pulsating auroras. Polar Sci. 10 (3), 183–191. https://doi.org/10.1016/j.polar.2016.03.005.

Fuselier, S.A., Gary, S.P., Thomsen, M.F., Clain, E.S., Hubert, B., Sandel, B.R., Immel, T., 2004. Generation of transient dayside subauroral proton precipitation. J. Geophys. Res. Space Phys. 109 (A12). https://doi.org/10.1029/2004JA010393.

Gabrielse, C., Angelopoulos, V., Runov, A., Kepko, L., Glassmeier, K.H., Auster, H.U., McFadden, J., Carlson, C.W., Larson, D., 2008. Propagation characteristics of plasma sheet oscillations during a small storm. Geophys. Res. Lett. 35, L17S13. https://doi.org/10.1029/2008GL033664.

Gabrielse, C., Nishimura, Y., Lyons, L., Gallardo-Lacourt, B., Deng, Y., Donovan, E., 2018. Statistical properties of mesoscale plasma flows in the nightside high-latitude ionosphere. J. Geophys. Res. Space Phys. 123, 6798–6820. https://doi.org/10.1029/2018JA025440.

Gabrielse, C., Pinto, V., Nishimura, Y., Lyons, L., Gallardo-Lacourt, B., Deng, Y., 2019. Storm time mesoscale plasma flows in the nightside high-latitude ionosphere: a statistical survey of characteristics. Geophys. Res. Lett. 46, 4079–4088. https://doi.org/10.1029/2018GL081539.

Gabrielse, C, Nishimura, T., Chen, M., Hecht, J.H., Kaeppler, S.R., Gillies, D.M., Reimer, A.S., Lyons, L.R., Deng, Y., Donovan, E., Evans, J.S., 2021. Estimating precipitating energy flux, average energy, and hall auroral conductance from THEMIS all-sky-imagers with focus on mesoscales. Front. Phys. 9. https://doi.org/10.3389/fphy.2021.744298. https://www.frontiersin.org/article/10.3389/fphy.2021.744298.

Galand, M., Richmond, A.D., 2001. Ionospheric electrical conductances produced by auroral proton precipitation. J. Geophys. Res. 106 (A1), 117–125. https://doi.org/10.1029/1999JA002001.

Gallardo-Lacourt, B., Nishimura, Y., Lyons, L.R., Zou, S., Angelopoulos, V., Donovan, E., et al., 2014. Coordinated SuperDARN THEMIS ASI observations of meso-scale flow bursts associated with auroral streamers. J. Geophys. Res. Space Phys. 119, 142–150. https://doi.org/10.1002/2013JA019245.

Ganushkina, N.Y., Pulkkinen, T.I., Kubyshkina, M.V., Sergeev, V.A., Lvova, E.A., Yahnina, T.A., Yahnin, A.G., Fritz, T., 2005. Proton isotropy boundaries as measured on mid- and low-altitude satellites. Ann. Geophys. 23, 1839–1847.

Germany, G.A., Parks, G.K., Brittnacher, M.J., Cumnock, J., Lummerzheim, D., Spann, J.F., Chen, L., Richards, P.G., Rich, F.J., 1997. Remote determination of auroral energy characteristic during substorm activity. Geophys. Res. Lett. 24 (8), 995–998. https://doi.org/10.1029/97GL00864.

Gjerloev, J.W., 2012. The SuperMAG data processing technique. J. Geophys. Res. 117. https://doi.org/10.1029/2012JA017683, A09213.

Gkioulidou, M., Wang, C.-P., Wing, S., Lyons, L.R., Wolf, R.A., Hsu, T.-S., 2012. Effect of an MLT dependent electron loss rate on the magnetosphere-ionosphere coupling. J. Geophys. Res. 117. https://doi.org/10.1029/2012JA018032, A11218.

Glauert, S.A., Horne, R.B., 2005. Calculation of pitch angle and energy diffusion coefficients with the PADIE code. J. Geophys. Res. 110. https://doi.org/10.1029/2004JA010851, A04206.

Goertz, C.K., Nielsen, E., Korth, A., Glassmeier, K.H., Haldoupis, C., Hoeg, P., Hayward, D., 1985. Observations of a possible ground signature of flux transfer events. J. Geophys. Res. 90 (A5), 4069–4078. https://doi.org/10.1029/JA090iA05p04069.

Golovchanskaya, I.V., Maltsev, Y.P., 2005. On the identification of plasma sheet flapping waves observed by cluster. Geophys. Res. Lett. 32. https://doi.org/10.1029/2004GL021552, L02102.

Golovchanskaya, I.V., Kullen, A., Maltsev, Y.P., Biernat, H., 2006. Ballooning instability at the plasma sheet-lobe interface and its implications for polar arc formation. J. Geophys. Res. 111, A11216 (2006) https://doi.org/10.1029/2005JA011092.

Green, D.L., Waters, C.L., Korth, H., Anderson, B.J., Ridley, A.J., Barnes, R.J., 2007. Technique: large-scale ionospheric conductance estimated from combined satellite and ground-based electromagnetic data. J. Geophys. Res. 112. https://doi.org/10.1029/2006JA012069, A05303.

Grubbs II, G., Michell, R., Samara, M., Hampton, D., Jahn, J.-.M., 2018. Predicting electron population characteristics in 2-D using multispectral ground-based imaging. Geophys. Res. Lett. 44. https://doi.org/10.1002/2017GL075873.

Gussenhoven, M.S., Hardy, D.A., Heinenmann, N., Burkhardt, R.K., 1984. Morphology of the polar rain. J. Geophys. Res. 89 (A11), 9785–9800.

Haaland, S.E., Paschmann, G., Forster, M., Quinn, J.M., Torbert, R.B., McIlwain, C.E., et al., 2007. High-latitude plasma convection from cluster EDI measurements: method and IMF-dependence. Ann. Geophys. 25 (1), 239–253. https://doi.org/10.5194/angeo-25-239-2007.

Haerendel, G., Paschmann, G., Sckopke, N., Rosenbauer, H., Hedgecock, P.C., 1978. J. Geophys. Res. 83, 3195–3216.

Hallinan, T.J., Kimball, J., Stenbaek-Nielsen, H.C., Lynch, K., Arnoldy, R., Bonnell, J., Kintner, P., 2001. Relation between optical emissions, particles, electric fields, and Alfvén waves in a multiple rayed arc. J. Geophys. Res. 106 (A8), 15,445–15,454.

Hardy, D.A., Burke, W.J., Gussenhoven, M.S., 1982. DMSP optical and electron measurements in the vicinity of polar cap arcs. J. Geophys. Res. 87, 2413–2430. https://doi.org/10.1029/JA087iA04p02413.

Hardy, D.A., Gussenhoven, M.S., Holeman, E., 1985. A statistical model of auroral electron precipitation. J. Geophys. Res. 90 (A5), 4229–4248. https://doi.org/10.1029/JA090iA05p04229.

Hardy, D.A., Gussenhoven, M.S., Raistrick, R., McNeil, W.J., 1987. Statistical and functional representations of the pattern of auroral energy flux, number flux, and conductivity. J. Geophys. Res. 92, 12,275.

Hardy, D.A., Gussenhoven, M.S., Brautigam, D., 1989. A statistical model of auroral ion precipitation. J. Geophys. Res. 94 (A1), 370–392. https://doi.org/10.1029/JA094iA01p00370.

Harel, M., Wolf, R.A., Reiff, P.H., Spiro, R.W., Burke, W.J., Rich, F.J., Smiddy, M., 1981. Quantitative simulation of a magnetospheric substorm 1. Model logic and overview. J. Geophys. Res. 86 (A4), 2217–2241. https://doi.org/10.1029/JA086iA04p02217.

Hecht, J.H., Strickland, D.J., Conde, M.G., 2006. The application of ground-based optical techniques for inferring electron energy deposition and composition change during auroral precipitation events. J. Atmos. Sol. Terr. Phys. 68, 1502–1519.

Hedin, A.E., et al., 1977. A global thermospheric model based on mass spectrometer and incoherent scatter data MSIS, 1. N_2 density and temperature. J. Geophys. Res. 82 (16), 2139–2147. https://doi.org/10.1029/JA082i016p02139.

Henderson, M.G., 2012. Auroral substorms, poleward boundary activations, auroral streamers, omega bands, and onset precursor activity. In: Keiling, A., Donovan, E., Bagenal, F., Karlsson, T. (Eds.), Auroral Phenomenology and Magnetospheric Processes: Earth and Other Planets. https://doi.org/10.1029/2011GM001165.

Henderson, M.G., Reeves, G.D., Murphree, J.S., 1998. Are north–south structures an ionospheric manifestation of bursty bulk flows? Geophys. Res. Lett. 25, 3737–3740.

Heppner, J.P., 1977. Empirical models of high-latitude electric fields. J. Geophys. Res. 82, 1115–1125.

Heppner, J.P., Maynard, N.C., 1987. Empirical high-latitude electric field models. J. Geophys. Res. 92, 4467–4489.

Hietala, H., Drake, J.F., Phan, T.D., Eastwood, J.P., McFadden, J.P., 2015. Ion temperature anisotropy across a magnetotail reconnection jet. Geophys. Res. Lett. 42 (18), 7239–7247. https://doi.org/10.1002/2015GL065168.

Hori, T., Maezawa, K., Saito, Y., Mukai, T., 2000. Average profile of ion flow and convection electric field in near-earth plasma sheet. Geophys. Res. Lett. 27, 1623–1626.

Horwitz, J.L., Doupnik, J.R., Banks, P.M., 1978. Chatanika Radar observations of the latitudinal distributions of auroral zone electric fields, conductivities, and currents. J. Geophys. Res. 83 (A4), 1463–1481. https://doi.org/10.1029/JA083iA04p01463.

Hosokawa, K., Ogawa, Y., 2010. Pedersen cur- rent carried by electrons in auroral D-region. Geophys. Res. Lett. 37. https://doi.org/10.1029/2010GL044746, L18103.

Hosokawa, K., Kashimoto, T., Suzuki, S., Shiokawa, K., Otsuka, Y., Ogawa, T., 2009. Motion of polar cap patches: a statistical study with all-sky airglow imager at Resolute Bay, Canada. J. Geophys. Res. 114. https://doi.org/10.1029/2008JA014020, A04318.

Hosokawa, K., Moen, J.I., Shiokawa, K., Otsuka, Y., 2011. Motion of polar cap arcs. J. Geophys. Res. 116. https://doi.org/10.1029/2010JA015906, A01305.

Hosokawa, K., Kullen, A., Milan, S., et al., 2020. Aurora in the polar cap: a review. Space Sci. Rev. 216, 15. https://doi.org/10.1007/s11214-020-0637-3.

Hubert, B., Gérard, J.-C., Evans, D.S., Meurant, M., Mende, S.B., Frey, H.U., Immel, T.J., 2002. Total electron and proton energy input during auroral substorms: remote sensing with IMAGE-FUV. J. Geophys. Res. 107 (A8). https://doi.org/10.1029/2001JA009229.

Hwang, K.-J., 2015. Magnetopause waves controlling the dynamics of Earth's magnetosphere. J. Astron. Space Sci. 32 (1), 1–11. https://doi.org/10.5140/JASS.2015.32.1.1.

Ismail, S., Wallis, D.D., Cogger, L.L., 1977. Characteristics of polar cap sun-aligned arcs. J. Geophys. Res. 82 (29), 4741–4749. https://doi.org/10.1029/JA082i029p04741.

Jones, R.A., Rees, M.H., 1973. Time dependent studies of the aurora—I. ion density and composition. Planet. Space Sci. 21 (4), 537–557. https://doi.org/10.1016/0032-0633(73)90069-X.

Jordanova, V.K., Farrugia, C.J., Thorne, R.M., Khazanov, G.V., Reeves, G.D., Thomsen, M.F., 2001. Modeling ring current proton precipitation by electromagnetic ion cyclotron waves during the May 14–16, 1997, storm. J. Geophys. Res. Space Phys. 106 (A1), 7–22. https://doi.org/10.1029/2000JA002008.

Jordanova, V.K., Spasojevic, M., Thomsen, M.F., 2007. Modeling the electromagnetic ion cyclotron wave-induced formation of detached subauroral proton arcs. J. Geophys. Res. Space Phys. 112 (A8). https://doi.org/10.1029/2006JA012215.

Kaeppler, S.R., Hampton, D.L., Nicolls, M.J., Strømme, A., Solomon, S.C., Hecht, J.H., Conde, M.G., 2015. An investigation comparing ground-based techniques that quantify auroral electron flux and conductance. J. Geophys. Res. Space Phys. 120, 9038–9056. https://doi.org/10.1002/2015JA021396.

Kaeppler, S.R., Sanchez, E., Varney, R.H., Irvin, R.J., Marshall, R.A., Bortnik, J., Reimer, A.S., Reyes, P., 2020. Chapter 6 - Incoherent scatter radar observations of 10–100 keV precipitation: review and outlook. In: Jaynes, A., Usanova, M. (Eds.), The Dynamic Loss of Earth's Radiation Belts: From Loss in the Magnetosphere to Particle Precipitation in the Atmosphere., https://doi.org/10.1016/B978-0-12-813371-2.00006-8.

Kamide, Y., Craven, J.D., Frank, L.A., Ahn, B.-.H., Akasofu, S.-.I., 1986. Modeling substorm current systems using conductivity distributions inferred from DE auroral images. J. Geophys. Res. 91 (A10), 11235–11256. https://doi.org/10.1029/JA091iA10p11235.

Kaufmann, R.L., Ball, B.M., Paterson, W.R., Frank, L.A., 2001. Plasma sheet thickness and electric currents. J. Geophys. Res. 106, 6179–6193.

Kaufmann, R.L., Paterson, W.R., Frank, L.A., 2004. Pressure, volume, density relationships in the plasma sheet. J. Geophys. Res. 109. https://doi.org/10.1029/2003JA010317, A08204.

Kauristie, K., Sergeev, V.A., Kubyshkina, M., Pulkkinen, T.I., Angelopoulos, V., Phan, T., Lin, R.P., Slavin, J.A., 2000. Ionospheric current signatures of transient plasma sheet flows. J. Geophys. Res. 105, 10,677–10,688.

Kelly, J.D., Heinselman, C.J., 2009. Initial results from poker flat incoherent scatter radar (PFISR). J. Atmos Solar-Terr. Phys. 71 (6), 635. https://doi.org/10.1016/j.jastp.2009.01.009.

Kennel, C., Engelmann, F., 1966. Velocity space diffusion from weak plasma turbulence in a magnetic field. Phys. Fluids 9 (12), 2377–2389.

Khazanov, G.V., Liemohn, M.W., Newman, T.S., Fok, M., Spiro, R.W., 2003. Self-consistent magnetosphere-ionosphere coupling: theoretical studies. J. Geophys. Res. 108 (A3), 1122. https://doi.org/10.1029/2002JA009624.

Khazanov, G.V., Tripathi, A.K., Sibeck, D., Himwich, E., Glocer, A., Singhal, R.P., 2015. Electron distribution function formation in regions of diffuse aurora. JGR 120 (11), 9891–9915. https://doi.org/10.1002/2015JA021728.

Khazanov, G.V., Robinson, R.M., Zesta, E., Sibeck, D.G., Chu, M., Grubbs, G.A., 2018. Impact of precipitating electrons and magnetosphere-ionosphere coupling processes on ionospheric conductance. Space Weather 16. https://doi.org/10.1029/2018SW001837.

Khazanov, G.V., Chen, M.W., Lemon, C.L., Sibeck, D.G., 2019. The magnetosphere-ionosphere electron precipitation dynamics and their geospace consequences during the 17 march 2013 storm. J. Geophys. Res. Space Phys. 124, 6504–6523. https://doi.org/10.1029/2019JA026589.

Kiehas, S.A., Runov, A., Angelopolos, V., Hietala, H., Korovinksiy, D., 2018. Magnetotail fast flow occurrence rate and dawn-dusk asymmetry at XGSM ~ -60 RE. J. Geophys. Res. Space Phys. 123, 1767–1778. https://doi.org/10.1002/2017JA024776.

Kim, H.-J., Lyons, L.R., Zou, S., Boudouridis, A., Lee, D.-Y., Heinselman, C., McCready, M., 2009. Evidence that solar wind fluctuations substantially affect the strength of dayside ionospheric convection. J. Geophys. Res. 114. https://doi.org/10.1029/2009JA014280, A11305.

Knight, S., 1973. Parallel electric fields. Planet. Space Sci. 21, 741–750. https://doi.org/10.1016/0032-0633(73)90093-7.

Kosch, M.J., Hagfors, T., Schlegel, K., 1998. Extrapolating EISCAT Pedersen conductances to other parts of the sky using ground-based TV auroral images. Ann. Geophys. 16, 583–588. https://doi.org/10.1007/s00585-998-0583-y.

Kubyshkina, D.I., Sormakov, D.A., Sergeev, V.A., Semenov, V.S., Erkaev, N.V., Kubyshkin, I.V., Ganushkina, N.Y., Dubyagin, S.V., 2014. How to distinguish between kink and sausage modes in flapping oscillations? J. Geophys. Res. Space Phys. 119, 3002–3015. https://doi.org/10.1002/2013JA019477.

Lam, M.M., Freeman, M.P., Jackman, C.M., Rae, I.J., Kalmoni, N.M.E., Sandhu, J.K., Forsyth, C., 2019. How well can we estimate Pedersen conductance from the THEMIS white-light all-sky cameras? J. Geophys. Res. Space Phys. 124, 2920–2934. https://doi.org/10.1029/2018JA026067.

Lanchester, B.S., Ashrafi, M., Ivchenko, E., 2009. Simultaneous imaging of aurora on small scale in OI (777.4 nm) and $N_2$1P to estimate energy and flux of precipitation. Ann. Geophys. 27, 2881–2891. https://doi.org/10.5194/angeo-27-2881-2009.

Lassen, K., Danielsen, C., 1978. Quiet time pattern of auroral arcs for different directions of the interplanetary magnetic field in the Y-Z plane. J. Geophys. Res. 83, 5277. https://doi.org/10.1029/ja083ia11p05277.

Li, S.-S., Angelopoulos, V., Runov, A., Kiehas, S.A., 2014. Azimuthal extent and properties of midtail plasmoids from two-point ARTEMIS observations at the Earth-Moon Lagrange points. J. Geophys. Res. Space Phys. 119, 1781–1796. https://doi.org/10.1002/2013JA019292.

Li, W., Ma, Q., Thorne, R.M., Bortnik, J., Kletzing, C.A., Kurth, W.S., et al., 2015. Statistical properties of plasmaspheric hiss derived from Van Allen probes data and their effects on radiation belt electron dynamics. J. Geophys. Res. Space Phys. 120, 3393–3405. https://doi.org/10.1002/2015JA021048.

Li, W., Santolik, O., Bortnik, J., Thorne, R.M., Kletzing, C.A., Kurth, W.S., Hospodarsky, G.B., 2016. New chorus wave properties near the equator from Van Allen probes wave observations. Geophys. Res. Lett. 43, 4725–4735. https://doi.org/10.1002/2016GL068780.

Liang, J., Donovan, E., Ni, B., Yue, C., Jiang, F., Angelopoulos, V., 2014. On an energy-latitude dispersion pattern of ion precipitation potentially associated with magnetospheric EMIC waves. J. Geophys. Res. Space Phys. 119, 8137–8160. https://doi.org/10.1002/2014JA020226.

Lui, A.T.Y., Hamilton, D.C., 1992. Radial profiles of quiet time magnetospheric parameters. J. Geophys. Res. 97 (A12), 19325–19332. https://doi.org/10.1029/92JA01539.

Liu, J., Angelopoulos, V., Runov, A., Zhou, X.-Z., 2013. On the current sheets surrounding dipolarizing flux bundles in the magnetotail: the case for wedgelets. J. Geophys. Res. Space Phys. 118, 2000–2020. https://doi.org/10.1002/jgra.50092.

Lockwood, M., Carlson, H.C., 1992. Production of polar cap electron density patches by transient magnetopause reconnection. Geophys. Res. Lett. 19 (17), 1731–1734. https://doi.org/10.1029/92GL01993.

Lotko, W., Smith, R.H., Zhang, B., Ouellette, J.E., Brambles, O.J., Lyon, J.G., 2014. Ionospheric control of magnetotail reconnection. Science 345 (6193), 184–187. https://doi.org/10.1126/science.1252907.

Lu, G., Donovan, E.F., Nagai, T., Mukai, T., Lummerzheim, D., Parks, G.K., Frank, L.A., Singer, H.J., Moldwin, M.B., Posch, J.L., Engebretson, M.J., Watermann, J., 2002. Substorm developement as seen through coordinated multi-instrument observations. In: Winglee, R.M. (Ed.), Proceedings of Substorm-6, pp. 63–70.

Lummerzheim, D., Rees, M.H., Craven, J.D., Frank, L.A., 1991. Ionospheric conductances derived from DE-1 auroral images. J. Atmos. Terr. Phys. 53 (3–4), 281–292. https://doi.org/10.1016/0021-9169(91)90112-K.

Lynch, K.A., Pietrowski, D., Torbert, R.B., Ivchenko, N., Marklund, G., Primdahl, F., 1999. Multiple-point electron measurements in a nightside auroral arc: auroral turbulence II particle observations. Geophys. Res. Lett. 26 (22), 3361–3364.

Lyon, J.G., Fedder, J.A., Mobarry, C.M., 2004. The Lyon-Fedder- Mobarry (LFM) global MHD magnetospheric simulation code. J. Atmos. Sol.-Terr. Phys. 66, 1333–1350. https://doi.org/10.1016/j.jastp.2004.03.020.

Lyons, L., 1974. Electron diffusion driven by magnetospheric electrostatic waves. J. Geophys. Res. 79 (4).

Lyons, L.R., Speiser, T.W., 1982. Evidence for current sheet acceleration in the geomagnetic tail. J. Geophys. Res. 87, 2276.

Lyons, L.R., Evans, D.S., Lundin, R., 1979. An observed relation between magnetic field aligned electric fields and downward electron energy fluxes in the vicinity of auroral forms. J. Geophys. Res. 84 (A2), 457–461. https://doi.org/10.1029/JA084iA02p00457.

Lyons, L.R., Nagai, T., Blanchard, G.T., Samson, J.C., Yamamoto, T., Mukai, T., Nishida, A., Kokobun, S., 1999. Association between Geotail plasma flows and auroral poleward boundary intensifications observed by CANOPUS photometers. J. Geophys. Res. 104, 4485–4497.

Lyons, L.R., Zesta, E., Xu, Y., Sanchez, E.R., Samson, J.C., Reeves, G.D., Ruohomiemi, J.M., Sigwarth, J. B., 2002. Auroral poleward boundary intensifications and tail bursty flows: a manifestation of a large-scale ULF oscillation? J. Geophys. Res. 107 (A11), 1352. https://doi.org/10.1029/2001JA000242.

Lyons, L.R., Nishimura, Y., Kim, H.-J., Donovan, E., Angelopoulos, V., Sofko, G., Nicolls, M., Heinselman, C., Ruohoniemi, J.M., Nishitani, N., 2011. Possible connection of polar cap flows to pre- and post-substorm onset PBIs and streamers. J. Geophys. Res. 116. https://doi.org/10.1029/2011JA016850, A12225.

Lyons, L.R., Nishimura, Y., Zou, Y., 2016. Unsolved problems: mesoscale polar cap fow channels' structure, propagation, and effects on space weather disturbances. J. Geophys. Res. Space Phys. 121, 3347–3352. https://doi.org/10.1002/2016JA022437.

Lysak, R.L., 1986. Coupling of the dynamic ionosphere to auroral flux tubes. J. Geophys. Res. 91 (A6), 7047–7056. https://doi.org/10.1029/JA091iA06p07047.

Lysak, R.L., 1991. Feedback instability of the ionospheric resonant cavity. J. Geophys. Res. 96 (A2), 1553–1568. https://doi.org/10.1029/90JA02154.

Ma, Q., Li, W., Thorne, R.M., Bortnik, J., Reeves, G.D., Kletzing, C.A., et al., 2016. Characteristic energy range of electron scattering due to plasmaspheric hiss. J. Geophys. Res. Space Phys. 121, 11,737–11,749. https://doi.org/10.1002/2016JA023311.

Maezawa, K., Hori, T., 1998. The distant magnetotail: its structure, IMF dependence, and thermal properties. In: Nishida, A., Baker, D.N., Cowley, S.W.H. (Eds.), New Perspectives on the Earth's Magnetotail, Geophys. Monogr. Ser., Vol. 105. AGU, Washington, D. C, pp. 1–20.

Makarevitch, R.A., Honary, F., McCrea, I.W., Howells, V.S.C., 2004. Imaging riometer observations of drifting absorption patches in the morning sector. Ann. Geophys. 22, 3461–3478.

Makita, K., Meng, C.I., Akasofu, S.I., 1991. Transpolar auroras, their particle precipitation, and IMF By component. J. Geophys. Res. 96 (14), 085.

McEwen, D.J., Harris, D.P., 1996. Occurrence patterns of F layer patches over the north magnetic pole. Radio Sci. 31 (3), 619–628. https://doi.org/10.1029/96RS00312.

McGranaghan, R., Knipp, D.J., Matsuo, T., Godinez, H., Redmon, R.J., Solomon, S.C., Morley, S.K., 2015. Modes of high-latitude auroral conductance variability derived from DMSP energetic electron precipitation observations: empirical orthogonal function analysis. J. Geophys. Res. Space Phys. 120 (11), 013–11,031. https://doi.org/10.1002/2015JA021828.

McGranaghan, R., Knipp, D.J., Matsuo, T., Cousins, E., 2016. Optimal interpolation analysis of high-latitude ionospheric Hall and Pedersen conductivities: application to assimilative ionospheric electrodynamics reconstruction. J. Geophys. Res. Space Phys. 121, 4898–4923. https://doi.org/10.1002/2016JA022486.

McGranaghan, R., Ziegler, J., Bloch, T., et al., 2020. Next generation particle precipitation: mesoscale prediction through machine learning (a case study and framework for progress). Space Weather 19. 2011.10117.

McPherron, R.L., 2005. Magnetic pulsations: their sources and relation to solar wind and geomagnetic activity. Surv. Geophys. 26, 545. https://doi.org/10.1007/s10712-005-1758-7.

McPherron, R.L., Hsu, T.-S., Kissinger, J., Chu, X., Angelopoulos, V., 2011. Characteristics of plasma flows at the inner edge of the plasma sheet. J. Geophys. Res. 116, A00I33. https://doi.org/10.1029/2010JA015923.

Mende, S.B., Eather, R.H., Rees, M.H., Vondrak, R.R., Robinson, R.M., 1984. Optical mapping of ionospheric conductance. J. Geophys. Res. 89 (A3), 1755–1763. https://doi.org/10.1029/JA089iA03p01755.

Merkin, V.G., Lyon, J.G., 2010. Effects ofthe low-latitude inospheric boundary condition on the global magnetosphere. J. Geophys. Res. 115. https://doi.org/10.1029/2010JA015461, A10202.

Milan, S.E., Hubert, B., Grocott, A., 2005. Formation and motion of a transpolar arc in response to dayside and nightside reconnection. J. Geophys. Res. 110. https://doi.org/10.1029/2004JA010835, A01212.

Min, K., Lee, J., Keika, K., Li, W., 2012. Global distribution of EMIC waves derived from THEMIS observations. J. Geophys. Res. Space Phys. 117. https://doi.org/10.1029/2012JA017515, A05219.

Moen, J., Brekke, A., 1990. On the importance of ion composition to conductivities in the auroral ionosphere. J. Geophys. Res. 95 (A7), 10687–10693. https://doi.org/10.1029/JA095iA07p10687.

Moen, J., Brekke, A., 1990. The solar flux influence on quiet time conductances in the auroral ionosphere. Geophys. Res. Lett. 20 (10), 971–974.

Moen, J., Gulbrandsen, N., Lorentzen, D.A., Carlson, H.C., 2007. On the MLT distribution of F region polar cap patches at night. Geophys. Res. Lett. 34. https://doi.org/10.1029/2007GL029632, L14113.

Moen, J., Rinne, Y., Carlson, H.C., Oksavik, K., Fujii, R., Opgenoorth, H., 2008. On the relationship between thin Birkeland current arcs and reversed flow channels in the winter cusp/cleft ionosphere. J. Geophys. Res. 113. https://doi.org/10.1029/2008JA013061, A09220.

Mouikis, C.G., Kistler, L.M., Liu, Y.H., Klecker, B., Korth, A., Dandouras, I., 2010. H^+ and O^+ content of the plasma sheet at 15–19 Re as a function of geomagnetic and solar activity. J. Geophys. Res. 115, A00J16. https://doi.org/10.1029/2010JA015978.

Nagai, T., Machida, S., 1998. Magnetic reconnection in the near-earth magnetotail. In: Nishida, D.B.A., Cowley, S. (Eds.), New Perspectives on the Earth's Magnetotail. American Geophysical Union, Washington, DC, pp. 211–224, https://doi.org/10.1029/GM105p0211.

Nagata, D., Machida, S., Ohtani, S., Saito, Y., Mukai, T., 2007. Solar wind control of plasma number density in the near-Earth plasma sheet. J. Geophys. Res. 112. https://doi.org/10.1029/2007JA012284, A09204.

Nakamura, R., Baumjohann, W., Schödel, R., Brittnacher, M., Sergeev, V.A., Kubyshkina, M., Mukai, T., Liou, K., 2001. Earthward flow bursts, auroral streamers, and small expansions. J. Geophys. Res. 106, 10,791–10,804.

Newell, P.T., 2000. Reconsidering the inverted-V particle signature: relative frequency of large-scale electron acceleration events. J. Geophys. Res. 105, 15779. https://doi.org/10.1029/1999JA000051.

Newell, P.T., Burke, W.J., Sa'nchez, E.R., Meng, C.-I., Greenspan, M.E., Clauer, C.R., 1991a. The low-latitude boundary layer and the boundary plasma sheet at low altitude: prenoon precipitation regions and convection reversal boundaries. J. Geophys. Res. 96, 21,013.

Newell, P.T., Gjerloev, J.W., 2014. Local geomagnetic indices and the prediction of auroral power. J. Geophys. Res. Space Phys. 119. https://doi.org/10.1002/2014JA020524.

Newell, P.T., Meng, C.-I., 1988. The cusp and the cleft/boundary layer: low-altitude identific ations a nd statistical local time variation. J. Geophys. Res. 93, 14,549.

Newell, P.T., Meng, C.-I., 1998. Open and closed low latitude boundary layer. In: Moenetal, J. (Ed.), Polar Cap Boundary Phenomena. Kluwer Acad, Norwell, Mass, pp. 91–101.

Newell, P.T., Burke, W.J., Meng, C.-I., Sa'nchez, E.R., Greenspan, M.E., 1991b. Identification and observations of the plasma mantle at low altitude. J. Geophys. Res. 96, 35.

Newell, P.T., Feldstein, Y.I., Galperin, Y.I., Meng, C.-I., 1996. Morphology of nightside precipitation. J. Geophys. Res. 101 (A5), 10,737–10,748. https://doi.org/10.1029/95JA03516.

Newell, P.T., Ruohoniemi, J.M., Meng, C.-I., 2004. Maps of precipitation by source region, binned by IMF, with inertial convection streamlines. J. Geophys. Res. 109. https://doi.org/10.1029/2004JA010499, A10206.

Newell, P.T., Wing, S., Sotirelis, T., Meng, C.-.I., 2005. Ion aurora and its seasonal variations. J. Geophys. Res. 110. https://doi.org/10.1029/2004JA010743, A01215.

Newell, P.T., Sotirelis, T., Wing, S., 2009. Diffuse, mono-energetic, and broadband aurora: the global precipitation budget. J. Geophys. Res. 114. https://doi.org/10.1029/2009JA014326, A09207.

Newell, P.T., Liou, K., Zhang, Y., Sotirelis, T., Paxton, L.J., Mitchell, E.J., 2014. OVATION Prime-2013: extension of auroralnprecipitation model to higher disturbance levels. Space Weather 12, 368–379. https://doi.org/10.1002/201 4SW00105.

Ni, B., Thorne, R.M., Shprits, Y.Y., Bortnik, J., 2008. Resonant scattering of plasma sheet electrons by whistler-mode chorus: contribution to diffuse auroral precipitation. Geophys. Res. Lett. 35. https://doi.org/10.1029/2008GL034032, L11106.

Ni, B., Thorne, R.M., Horne, R.B., Meredith, N.P., Shprits, Y.Y., Chen, L., Li, W., 2011a. Resonant scattering of plasma sheet electrons leading to diffuse auroral precipitation: 1. Evaluation for electrostatic electron cyclotron harmonic waves. J. Geophys. Res. 116, A04218. https://doi.org/10.1029/2010JA016232.

Ni, B., Thorne, R.M., Meredith, N.P., Horne, R.B., Shprits, Y.Y., 2011b. Resonant scattering of plasma sheet electrons leading to diffuse auroral precipitation: 2. Evaluation for whistler mode chorus waves. J. Geophys. Res. 116, A04219. https://doi.org/10.1029/2010JA016233.

Ni, B., Bortnik, J., Thorne, R.M., Ma, Q., Chen, L., 2013. Resonant scattering and resultant pitch angle evolution of relativistic electrons by plasmaspheric hiss. J. Geophys. Res. Space Physics 118, 7740–7751. https://doi.org/10.1002/2013JA019260.

Ni, B., Thorne, R.M., Zhang, X., Bortnik, J., Pu, Z., Xie, L., Hu, Z.-j., Han, D., Shi, R., Zhou, C., Gu, X., 2016. Origins of the earth's diffuse auroral precipitation. Space Sci. Rev. 200 (1), 205–259. https://doi.org/10.1007/s11214-016-0234-7.

Nishimura, Y., Lyons, L., Zou, S., Angelopoulos, V., Mende, S., 2010. Substorm triggering by new plasma intrusion: THEMIS all-sky imager observations. J. Geophys. Res. 115. https://doi.org/10.1029/2009JA015166, A07222.

Nishimura, Y., Bortnik, J., Li, W., Lyons, L.R., Donovan, E.F., Angelopoulos, V., Mende, S.B., 2014a. Evolution of nightside subauroral proton aurora caused by transient plasma sheet. J. Geophys. Res. Space Phys. 119 (7), 5295–5304. https://doi.org/10.1002/2014JA020029.

Nishimura, Y., et al., 2014b. Day-night coupling by a localized flow channel visualized by polar cap patch propagation. Geophys. Res. Lett. 41, 3701–3709. https://doi.org/10.1002/2014GL060301.

Nishimura, et al., 2019. AGU Book on Solar/Heliosphere 3: Advances in Ionospheric Research. submitted.

Nishimura, Y., Lessard, M.R., Katoh, Y., et al., 2020. Diffuse and pulsating Aurora. Space Sci. Rev. 216, 4. https://doi.org/10.1007/s11214-019-0629-3.

Noja, M., Stolle, C., Park, J., Lühr, H., 2013. Long-term analysis of ionospheric polar patches based on CHAMP TEC data. Radio Sci. 48, 289–301. https://doi.org/10.1002/rds.20033.

Obara, T., Kitayama, M., Mukai, T., Kaya, N., Cogger, L., Murphree, S., 1988. Simultaneous observations of Sun-aligned polar cap arcs in both hemispheres by EXOS-C and Viking. Geophys. Res. Lett. 15 (7), 713–716. https://doi.org/10.1029/GL015i007p00713.

Ohtani, S., Wing, S., Ueno, G., Higuchi, T., 2009. Dependence of premidnight field-aligned currents and particle precipitation on solar illumination. J. Geophys. Res. 114. https://doi.org/10.1029/2009JA014115, A12205.

Oksavik, K., Moen, J.I., Rekaa, E.H., Carlson, H.C., Lester, M., 2011. Reversed flow events in the cusp ionosphere detected by SuperDARN HF radars. J. Geophys. Res. 116. https://doi.org/10.1029/2011JA016788, A12303.

Orlova, K., Shprits, Y., 2014. Model of lifetimes of the outer radiation belt electrons in a realisticmagnetic field using realistic chorus wave parameters. J. Geophys. Res.Space Phys. 119, 770–780. https://doi.org/10.1002/2013JA019596.

Orlova, K., Spasojevic, M., Shprits, Y., 2014. Activity-dependent global model of electron loss inside the plasmasphere. Geophys. Res. Lett. 41, 3744–3751. https://doi.org/10.1002/2014GL060100.

Orlova, K., Shprits, Y., Spasojevic, M., 2016. New global loss model of energetic and relativistic electrons based on Van Allen probes measurements. J. Geophys. Res. Space Physics 121, 1308–1314. https://doi.org/10.1002/2015JA021878.

Østgaard, N., Stadsnes, J., Bjordal, J., Vondrak, R.R., Cummer, S.A., Chenette, D.L., Schulz, M., Pronko, J. G., 2000. Cause of the localized maximum of X-ray emission in the morning sector: a comparison with electron measurements. J. Geophys. Res. 105 (A9), 20869–20883. https://doi.org/10.1029/1999JA000354.

Østgaard, N., Stadsnes, J., Bjordal, J., Germany, G.A., Vondrak, R.R., Parks, G.K., Cummer, S.A., Chenette, D.L., Pronko, J.G., 2001. Auroral Electron Distributions Derived from Comb.

Panov, E.V., Artemyev, A.V., Baumjohann, W., Nakamura, R., Angelopoulos, V., 2013. Transient electron precipitation during oscillatory BBF braking: THEMIS observations and theoretical estimates. J. Geophys. Res. Space Phys. 118, 3065–3076. https://doi.org/10.1002/jgra.50203.

Papitashvili, V.O., Rich, F.J., 2002. High-latitude ionospheric convection models derived from defense meteorological satellite program ion drift observations and parameterized by the interplanetary

magnetic field strength and direction. J. Geophys. Res. 107 (A8), 1198. https://doi.org/10.1029/2001JA000264.
Paxton, L.J., Schaefer, R.K., Zhang, Y., Kil, H., Hicks, J.E., 2018. SSUSI and SSUSI-lite: providing space situational awareness and support for over 25 years. J. Hopkins APL Tech. Dig. 34 (3).
Perlongo, N.J., Ridley, A.J., Liemohn, M.W., Katus, R.M., 2017. The effect of ring current electron scattering rates on magnetosphere-ionosphere coupling. J. Geophys. Res. Space Phys. 122. https://doi.org/10.1002/2016JA023679.
Picone, J.M., Hedin, A.E., Drob, D.P., Aikin, A.C., 2002. NRLMSISE-00 empirical model of the atmosphere: statistical comparisons and scientific issues. J. Geophys. Res. 107 (A12), 1468. https://doi.org/10.1029/2002JA009430.
Pitkänen, T., Aikio, A.T., Amm, O., Kauristie, K., Nilsson, H., Kaila, K.U., 2011. EISCAT-cluster observations of quiet-time near-earth magnetotail fast flows and their signatures in the ionosphere. Ann. Geophys. 29, 299–319. https://doi.org/10.5194/angeo-29-299-2011.
Pitkänen, T., Aikio, A.T., Juusola, L., 2013. Observations of polar cap flow channel and plasma sheet flowbursts during substorm expansion. J. Geophys. Res. Space Phys. 118, 774–784. https://doi.org/10.1002/jgra.50119.
Prölss, G.W., 1995. Ionospheric F-region storms. In: Volland, H. (Ed.), Handbook of Atmospheric Electrodynamics. CRC Press, Boca Raton, Fla, pp. 195–248.
Raeder, J., McPherron, R.L., Frank, L.A., Kokubun, S., Lu, G., Mukai, T., Paterson, W.R., Sigwarth, J.B., Singer, H.J., Slavin, J.A., 2001. Global simulation of the Geospace environment modeling substorm challenge event. J. Geophys. Res. 106 (A1), 381–395. https://doi.org/10.1029/2000JA000605.
Raeder, J., Cramer, W.D., Jensen, J., Fuller-Rowell, T., Maruyama, N., Toffoletto, F., Vo, H., 2016. Subauroral polarization streams: a complex interaction between the magnetosphere, ionosphere, and thermosphere. J. Phys. Conf. Ser. 767. https://doi.org/10.1088/1742-6596/767/1/012021, 012021.
Raj, A., Phan, R., Lin, R.P., Angelopoulos, V., 2002. Wind survey of high-speed bulk flows and field-aligned beams in the near-earth plasma sheet. J. Geophys. Res. 107, 1419. https://doi.org/10.1029/2001JA007547.
Rasmussen, C.E., Schunk, R.W., Wickwar, V.B., 1988. A photochemical equilibrium model for ionospheric conductivity. J. Geophys. Res. 93 (A9), 9831–9840. https://doi.org/10.1029/JA093iA09p09831.
Redmon, R.J., Denig, W.F., Kilcommons, L.M., Knipp, D.J., 2017. New DMSP database of precipitating auroral electrons and ions. J. Geophys. Res. Space Phys. 122, 9056–9067. https://doi.org/10.1002/2016JA023339.
Rees, M.F., 1963. Auroral ionization and excitation by incident energetic electrons. Planet. Space Sci. 11 (10), 1209–1218. https://doi.org/10.1016/0032-0633(63)90252-6.
Rees, D., 1989. Physics and Chemistry of the Upper Atmosphere. Cambridge Univ. Press, New York.
Rees, M.H., Luckey, D., 1974. Auroral electron energy derived from ratio of spectroscopic emissions. I—model computations. J. Geophys. Res. 79, 5181–5186. https://doi.org/10.1029/JA079i034p05181.
Rees, M.H., Lummerzheim, D., Roble, R.G., Winningham, J.D., Craven, J.D., Frank, L.A., 1988. Auroral energy deposition rate, characteristic electron energy, and ionospheric parameters derived from Dynamics Explorer 1 images. J. Geophys. Res. 93 (A11), 12841–12860. https://doi.org/10.1029/JA093iA11p12841.
Reiff, P.H., 1984. Models of auroral-zone conductances. In: Potemra, T.A. (Ed.), Magnetospheric Currents. https://doi.org/10.1029/GM028p0180.
Reiff, P.H., Burch, J.L., 1985. IMF By-dependent plasma flow and Birkeland currents in the dayside magnetosphere 2. A global model for northward and southward IMF. J. Geophys. Res. 90, 1595–1609.
Ren, J., Zou, S., Gillies, R.G., Donovan, E., Varney, R.H., 2018. Statistical characteristics of polar cap patches observed by RISR-C. J. Geophys. Res. Space Phys. 123, 6981–6995. https://doi.org/10.1029/2018JA025621.
Rezhenov, B.V., 1995. A possible mechanism for theta aurora formation. Ann. Geophys. 13, 698.
Rich, F., Hairston, M., 1994. Large-scale convection patterns observed by DMSP. J. Geophys. Res. 99 (A3), 3827–3844. pgs 4004–4006 https://doi.org/10.1029/93JA03296.

Richmond, A.D., 1992. Assimilative mapping of ionospheric electrodynamics. Adv. Space Res. 12, 59–68. https://doi.org/10.1016/0273-1177(92)90040-5.

Richmond, A., 1995. Ionospheric electrodynamics. In: Volland, H. (Ed.), Handbook of Atmospheric Electrodynamics, Vol. 2. CRC Press, Boca Raton, Fla, pp. 249–290.

Richmond, A.R., Kamide, Y., 1988. Mapping electrodynamic features of the high-latitude ionosphere from localized observations: technique. J. Geophys. Res. 93 (A6), 5741–5759. https://doi.org/10.1029/JA093iA06p05741.

Ridley, A., Gombosi, T., DeZeeuw, D., 2004. Ionospheric control of the magnetosphere: conductance. Ann. Geophys. 22 (2), 567–584. https://doi.org/10.5194/angeo-22-567-2004.

Rinne, Y., Moen, J., Oksavik, K., Carlson, H.C., 2007. Reversed flow events in the winter cusp ionosphere observed by the European incoherent scatter (EISCAT) Svalbard radar. J. Geophys. Res. 112. https://doi.org/10.1029/2007JA012366, A10313.

Robinson, R.M., Vondrak, R.R., Miller, K., Dabbs, T., Hardy, D., 1987. On calculating ionospheric conductances from the flux and energy of precipitating electrons. J. Geophys. Res. 92 (A3), 2565–2569. https://doi.org/10.1029/JA092iA03p02565.

Robinson, R., Dabbs, T., Vickrey, J., Eastes, R., Del Greco, F., Huffman, R., Meng, C., Daniell, R., Strickland, D., Vondrak, R., 1992. Coordinated measurements made by the Sondrestrom radar and the Polar Bear Ultraviolet Imager. J. Geophys. Res. 97 (A3), 2863–2871. https://doi.org/10.1029/91JA02803.

Robinson, R.M., Kaeppler, S.R., Zanetti, L., Anderson, B., Vines, S.K., Korth, H., Fitzmaurice, A., 2020. Statistical relations between auroral electrical conductances and field-aligned currents at high latitudes. J. Geophys. Res. Space Phys. 125. https://doi.org/10.1029/2020JA028008, e2020JA028008.

Roble, R.G., Rees, M.H., 1977. Time-dependent studies of the aurora: effects of particle precipitation on the dynamic morphology of ionospheric and atmospheric properties. Planet. Space Sci. 25 (11). https://doi.org/10.1016/0032-0633(77)90146-5.

Rodger, A.S., Pinnock, M., Dudeney, J.R., Baker, K.B., Greenwald, R.A., 1994. A new mechanism for polar patch formation. J. Geophys. Res. 99 (A4), 6425–6436. https://doi.org/10.1029/93JA01501.

Runov, A., et al., 2005. Electric current and magnetic field geometry in flapping magnetotail current sheets. Ann. Geophys. 23, 1391–1403. https://doi.org/10.5194/angeo-23-1391-2005.

Runov, A., Angelopoulos, V., Sitnov, M.I., Sergeev, V.A., Bonnell, J., McFadden, J.P., Larson, D., Glassmeier, K.-H., Auster, U., 2009. THEMIS observations of an earthward- propagating dipolarization front. Geophys. Res. Lett. 36. https://doi.org/10.1029/2009GL038980, L14106.

Ruohoniemi, J.M., Greenwald, R.A., 1996. Statistical patterns of high-latitude convection obtained from Goose Bay HF radar observations. J. Geophys. Res.-Atmos. 101 (A10), 21,743–21,763. https://doi.org/10.1029/96JA01584.

Ruohoniemi, J.M., Greenwald, R.A., 2005. Dependencies of high-latitude plasma convection: consideration of interplanetary magnetic field, seasonal, and universal time factors in statistical patterns. J. Geophys. Res. 110. https://doi.org/10.1029/2004JA010815, A09204.

Russell, C.T., Elphic, R.C., 1978. Space Sci. Rev. 22, 681–715.

Russell, C.T., Elphic, R.C., 1979. ISEE observations of flux transfer events at the dayside magnetopause. Geophys. Res. Lett. 6, 33.

Saikin, A.A., Zhang, J.-C., Smith, C.W., Spence, H.E., Torbert, R.B., Kletzing, C.A., 2016. The dependence on geomagnetic conditions and solar wind dynamic pressure of the spatial distributions of EMIC waves observed by the Van Allen Probes. J. Geophys. Res. Space Phys. 121 (5), 4362–4377. https://doi.org/10.1002/2016JA022523.

Saunders, M.A., Russell, C.T., Sckopke, N., 1984. Flux transfer events: scale size and interior structure. Geophys. Res. Lett. 11 (2), 131–134. https://doi.org/10.1029/GL011i002p00131.

Schlegel, K., 1988. Auroral zone E-region conductivities during solar minimum derived from EISCAT data. Ann. Geophys. 6, 129–137.

Schulz, M., 1974. Particle lifetimes in strong diffusion. Astrophys. Space Sci. 31 (1), 37–42. https://doi.org/10.1007/BF00642599.

Schulz, M., 1998. Particle drift and loss rates under strong pitch angle diffusion in Dungey's model magnetosphere. J. Geophys. Res. 103 (A1), 61–67. https://doi.org/10.1029/97JA02042.

Schunk, R., Nagy, A., 2009. Ionospheres: Physics, Plasma Physics, and Chemistry, second ed. Cambridge Atmospheric and Space Science Series, Cambridge University Press, Cambridge, https://doi.org/10.1017/CBO9780511635342.

Seki, K., Hirahara, M., Terasawa, T., Shinohara, I., Mukai, T., Saito, Y., Machida, S., Yamamoto, T., Kokubun, S., 1996. Coexistence of earth-origin O+ and solar wind-origin H^{+}/he^{++} in the distant magnetotail. Geophys. Res. Lett. 23 (9), 985–988. https://doi.org/10.1029/96GL00768.

Senior, C., 1991. Solar and particle contributions to auroral height-integrated conductivities from EISCAT data - a statistical study. Ann. Geophys. 9, 449–460.

Senior, A., Kavanagh, A.J., Kosch, M.J., Honary, F., 2007. Statistical relationships between cosmic radio noise absorption and ionospheric electrical conductances in the auroral zone. J. Geophys. Res. 112. https://doi.org/10.1029/2007JA012519, A11301.

Sergeev, V.A., Bösinger, T., 1993. Particle dispersion at the nightside boundary of the polar cap. J. Geophys. Res. 98 (A1), 233–241. https://doi.org/10.1029/92JA01667.

Sergeev, V.A., Sazhina, E., Tsyganenko, N., Lundblad, J., Soraas, F., 1983. Pitch-angle scattering of energetic protons in the magnetotail current sheet as the dominant source of their isotropic precipitation into the nightside ionosphere. Planet. Space Sci. 31 (10), 1147–1155.

Sergeev, V.A., Liou, K., Meng, C.-I., Newell, P.T., Brittnacher, M., Parks, G., Reeves, G.D., 1999. Development of auroral streamers in association with localized impulsive injections to the inner magnetotail. Geophys. Res. Lett. 26 (3), 417–420.

Sergeev, V.A., Sauvaud, J.-.A., Popescu, D., Kovrazhkin, R.A., Liou, K., Newell, P.T., et al., 2000. Multiple-spacecraft observation of a narrow transient plasma jet in the Earth's plasma sheet. Geophys. Res. Lett. 27 (6), 851–854.

Sergeev, V.A., Liou, K., Newell, P.T., Ohtani, S.-I., Hairston, M.R., Rich, F., 2004. Auroral streamers: characteristics of associated precipitation, convection and field-aligned currents. Ann. Geophys. 22 (2), 537–548. https://doi.org/10.5194/angeo-22-537-2004.

Sergeev, V.A., Chernyaev, I.A., Dubyagin, S.V., Miyashita, Y., Angelopoulos, V., Boakes, P.D., et al., 2012. Energetic particle injections to geostationary orbit: relationship to flow bursts and magnetospheric state. J. Geophys. Res. 117. https://doi.org/10.1029/2012JA017773, A10207.

Sergienko, T., Sandahl, I., Gustavsson, B., Andersson, L., Brändström, U., Steen, Å., 2008. A study of fine structure of diffuse aurora with ALIS-FAST measurements. Ann. Geophys. 26, 3185–3195.

Shi, Y., Zesta, E., Lyons, L.R., Xing, X., Angelopoulos, V., Donovan, E., McCready, M.A., Heinselman, C.J., 2012. Multipoint observations ofsubstorm pre-onset flows and time sequence in the ionosphere and magne-tosphere. J. Geophys. Res. 117. https://doi.org/10.1029/2011JA017185, A09203.

Shreedevi, P.R., Yu, Y., Ni, B., Saikin, A., Jordanova, V.K., 2021. Simulating the ion precipitation from the inner magnetosphere by H-band and He-band electro magnetic ion cyclotron waves. J. Geophys. Res.: Space Phys. 126. e2020JA028553. https://doi.org/10.1029/2020JA028553.

Siscoe, G.L., Erickson, G.M., Sonnerup, B.U.Ö., Maynard, N.C., Schoendorf, J.A., Siebert, K.D., Weimer, D.R., White, W.W., Wilson, G.R., 2002. Hill model of transpolar potential saturation: comparisons with MHD simulations. J. Geophys. Res. 107 (A6). https://doi.org/10.1029/2001JA000109.

Sojka, J.J., Bowline, M.D., Schunk, R.W., Decker, D.T., Valladares, C.E., Sheehan, R., et al., 1993. Modeling polar cap F-region patches using time varying convection. Geophys. Res. Lett. 20 (17), 1783–1786. https://doi.org/10.1029/93GL01347.

Sojka, J.J., Bowline, M.D., Schunk, R.W., 1994. Patches in the polar ionosphere: UT and seasonal dependence. J. Geophys. Res. 99 (A8), 14,959–14,970. https://doi.org/10.1029/93JA03327.

Solomon, S.C., 2017. Global modeling of thermospheric airglow in the far ultraviolet. J. Geophys. Res. Space Phys. 122, 7834–7848. https://doi.org/10.1002/2017JA024314.

Solomon, S.C., Abreu, V.J., 1989. The 630 nm dayglow. J. Geophys. Res. 94 (A6), 6817–6824. https://doi.org/10.1029/JA094iA06p06817.

Solomon, S.C., Hays, P.B., Abreu, V.J., 1988. The auroral 6300 Å emission: observations and modeling. J. Geophys. Res. 93 (A9), 9867–9882. https://doi.org/10.1029/JA093iA09p09867.

Sonnerup, B., 1980. Theory of the low-latitude boundary layer. J. Geophys. Res. 85 (A5), 2017–2026. https://doi.org/10.1029/JA085iA05p02017.

Southwood, D.J., 1985. Theoretical aspects of ionosphere-magnetospheresolar wind coupling. Adv. Space Res. 5 (4), 7–14. https://doi.org/10.1016/0273-1177 (85)90110-3.
Southwood, D.J., 1987. The ionospheric signature of flux transfer events. J. Geophys. Res. 92 (A4), 3207–3213. https://doi.org/10.1029/JA092iA04p03207.
Spasojevic, M., Frey, H.U., Thomsen, M.F., Fuselier, S.A., Gary, S.P., Sandel, B.R., Inan, U.S., 2004. The link between a detached subauroral proton arc and a plasmaspheric plume. Geophys. Res. Lett. 31 (4). https://doi.org/10.1029/2003GL018389.
Spiro, R.W., Reiff, P.H., Maher, L.J., 1982. Precipitating electron energy flux and auroral zone conductances-An empirical model. J. Geophys. Res. 87 (A10), 8215–8227. https://doi.org/10.1029/JA087iA10p08215.
Stasiewicz, K., Khotyaintsev, Y., Berthomier, M., Wahlund, J.-E., 2000. Identification of widespread turbulence of dispersive Alfvén waves. Geophys. Res. Lett. 27 (2), 173–176.
Strickland, D.J., Meier, R.R., Hecht, J.H., Christensen, A.B., 1989. Deducing composition and incident electron spectra from ground-based auroral optical measurements: theory and model results. J. Geophys. Res. 94, 13,527–13,539.
Strickland, D.J., Daniell Jr., R.E., Jasperse, J.R., Basu, B., 1993. Transport-theoretic model for the electron-proton-hydrogen atom aurora. 2: model results. J. Geophys. Res. 98, 21,533–21,548. https://doi.org/10.1029/93JA01645.
Strickland, D.J., Hecht, J.H., Christensen, A.B., Kelly, J., 1994. Relationship between energy flux Q and mean energy $\langle E \rangle$ of auroral electron spectra based on radar data from the 1987 CEDAR Campaign at Sondre Stromfjord, Greenland. J. Geophys. Res. 99, 19,467–19,473. https://doi.org/10.1029/94JA01901.
Tanaka, T., 2000. The state transition model of the substorm onset. J. Geophys. Res. 105, 21,081–21,096. https://doi.org/10.1029/2000JA900061.
Terasawa, T., Fujimoto, M., Mukai, T., Shinohara, I., Saito, Y., Yamamoto, T., Machida, S., Kokubun, S., Lazarus, A.J., Steinberg, J.T., Lepping, R.P., 1997. Solar wind control of density and temperature in the near-Earth plasma sheet: WIND/GEOTAIL collaboration. Geophys. Res. Lett. 24, 935–938. https://doi.org/10.1029/96GL04018.
Thayer, J.P., 1998. Height-resolved Joule heating rates in the high-latitude *E* region and the influence of neutral winds. J. Geophys. Res. 103 (A1), 471–487. https://doi.org/10.1029/97JA02536.
Thorne, R.M., Ni, B., Tao, X., Horne, R.B., Meredith, N.P., 2010. Scattering by chorus waves as the dominant cause of diffuse auroral precipitation. Nature 467, 943–946. https://doi.org/10.1038/nature09467.
Tian, X., Yu, Y., Yue, C., 2020. Statistical survey of storm-time energetic particle precipitation. J. Atmos. Sol. Terr. Phys. 199 (1364–6826). https://doi.org/10.1016/j.jastp.2020.105204, 105204.
Toffoletto, F., Sazykin, S., Spiro, R., Wolf, R., 2003. Inner magnetospheric modeling with the rice convection model. Space Sci. Rev. 107 (1/2), 175–196. https://doi.org/10.1023/A:1025532008047.
Tóth, G., et al., 2005. Space weather modeling framework: a new tool for the space science community. J. Geophys. Res. 110 (A12). https://doi.org/10.1029/2005JA011126, A12226.
Tóth, G., et al., 2012. Adaptive numerical algorithms in space weather modeling. J. Comput. Phys. 231, 870–903. https://doi.org/10.1016/j.jcp.2011.02.006.
Tsunoda, R.T., 1988. High-latitude F region irregularities: a review and synthesis. Rev. Geophys. 26 (4), 719–760. https://doi.org/10.1029/RG026i004p00719.
Tsurutani, B.T., Smith, E.J., 1974. Postmidnight chorus: a substorm phenomenon. J. Geophys. Res. 79, 118.
Tsyganenko, N.A., Mukai, T., 2003. Tail plasma sheet models derived from Geotail particle data. J. Geophys. Res. 108 (A3). https://doi.org/10.1029/2002JA009707.
Valladares, C.E., Carlson Jr., H.C., Fukui, K., 1994. Interplanetary magnetic field dependency of stable sun-aligned polar cap arcs. J. Geophys. Res. 99, 6247–6272. https://doi.org/10.1029/93JA03255.
van Eyken, A.P., Rishbeth, H., Willis, D.M., Cowley, S.W.H., 1984. Initial EISCAT observations of plasma convection at invariant latitudes 70°–77°. J. Atmos. Terr. Phys. 46, 635–641. https://doi.org/10.1016/0021-9169(84)90081-3.
Vasyliunas, V.M., 1970. Mathematical models of magnetospheric convec- tions and its coupling to the ionosphere. In: McCormac, B.M. (Ed.), Particles and Fields in the Magnetosphere. D. Reidel, Hingham, Mass, pp. 60–71, https://doi.org/10.1007/978-94-010-3284-1_6.

Vickrey, J.F., Vondrak, R.R., Matthews, S.J., 1981. The diurnal and latitudinal variation of auroral zone ionospheric conductivity. J. Geophys. Res. 86 (A1), 65–75. https://doi.org/10.1029/JA086iA01p00065.

Vondrak, R.R., Baron, M.J., 1976. Radar measurements of the lat- itudinal variation of auroral ionization. Radio Sci. 11 (11), 939. https://doi.org/10.1029/RS011i011p00939.

Vondrak, R., Robinson, R., 1985. Inference of high-latitude ionization and conductivity from AE-C measurements of auroral electron fluxes. J. Geophys. Res. 90 (A8), 7505–7512. https://doi.org/10.1029/JA090iA08p07505.

Vondrak, R.R., Sears, R.D., 1978. Comparison of incoherent scatter radar and photometric measurements of the energy distribution of auroral electrons. J. Geophys. Res. 83, 1665–1667.

Walker, J.K., Bhatnagar, V.P., 1989. Ionospheric absorption, typical ionization, conductivity, and possible synoptic heating parameters in the upper atmosphere. J. Geophys. Res. 94 (A4), 3713–3720. https://doi.org/10.1029/JA094iA04p03713.

Wallis, D.D., Budzinski, E.E., 1981. Empirical models of height integrated conductivities. J. Geophys. Res. 86 (A1), 125–137. https://doi.org/10.1029/JA086iA01p00125.

Walsh, A.P., Fazakerley, A.N., Forsyth, C., Owen, C.J., Taylor, M.G.G.T., Rae, I.J., 2013. Sources of electron pitch angle anisotropy in the magnetotail plasma sheet. J. Geophys. Res. Space Phys. 118, 6042–6054. https://doi.org/10.1002/jgra.50553.

Wang, C.-P., Xing, X., 2021. Solar wind entry into midtail current sheet via low-latitude mantle under dominant IMF By: ARTEMIS observation. J. Geophys. Res.: Space Phys. 126. https://doi.org/10.1029/2021JA029402. e2021JA029402.

Wang, C.-P., Lyons, L.R., Weygand, J.M., Nagai, T., McEntire, R.W., 2006. Equatorial distributions of the plasma sheet ions, their electric and magnetic drifts, and magnetic fields under different interplanetary magnetic field B_z conditions. J. Geophys. Res. 111. https://doi.org/10.1029/2005JA011545, A04215.

Wang, C.-P., Lyons, L.R., Nagai, T., Weygand, J.M., McEntire, R.W., 2007. Sources, transport, and distributions of plasma sheet ions and electrons and dependences on interplanetary parameters under northward interplanetary magnetic field. J. Geophys. Res. 112. https://doi.org/10.1029/2007JA012522, A10224.

Wang, C.-P., Lyons, L.R., Wolf, R.A., Nagai, T., Weygand, J.M., Lui, A.T.Y., 2009. Plasma sheet $PV^{5/3}$ and nV and associated plasma and energy transport for different convection strengths and AE levels. J. Geophys. Res. 114. https://doi.org/10.1029/2008JA013849, A00D02.

Wang, C.-P., Lyons, L.R., Nagai, T., Weygand, J.M., Lui, A.T.Y., 2010. Evolution of plasma sheet particle content under different interplanetary magnetic field conditions. J. Geophys. Res. 115. https://doi.org/10.1029/2009JA015028. A06210.

Wang, C.-.P., Gkioulidou, M., Lyons, L.R., Wolf, R.A., Angelopoulos, V., Nagai, T., Weygand, J.M., Lui, A.T.Y., 2011. Spatial distributions of ions and electrons from the plasma sheet to the inner magnetosphere: comparisons between THEMIS-Geotail statistical results and the rice convection model. J. Geophys. Res. 116. https://doi.org/10.1029/2011JA016809, A11216.

Wang, C.-P., Gkioulidou, M., Lyons, L.R., Angelopoulos, V., 2012a. Spatial distributions of the ion to electron temperature ratio in the magnetosheath and plasma sheet. J. Geophys. Res. 117. https://doi.org/10.1029/2012JA017658, A08215.

Wang, C.-P., Zaharia, S.G., Lyons, L.R., Angelopoulos, V., 2012b. Spatial distributions of ion pitch angle anisotropy in the near-Earth magnetosphere and tail plasma sheet. J. Geophys. Res. 118, 1–12. https://doi.org/10.1029/2012JA018275.

Wang, C.-P., Lyons, L.R., Angelopoulos, V., 2014a. Properties of low-latitude mantle plasma in the Earth's magnetotail: ARTEMIS obser- vations and global MHD predictions. J. Geophys. Res. Space Phys. 119. https://doi.org/10.1002/2014JA020060.

Wang, C.-P., Gkioulidou, M., Lyons, L.R., Xing, X., Wolf, R.A., 2014b. Interchange motion as a transport mechanism for formation of cold-dense plasma sheet. J. Geophys. Res. Space Phys. 119. https://doi.org/10.1002/2014JA020251.

Wang, B., Nishimura, Y., Lyons, L.R., Zou, Y., Carlson, H.C., Frey, H.U., Mende, S.B., 2016. Analysis of close conjunctions between dayside polar cap airglow patches and flow channels by all-sky imager and DMSP. Earth, Planets and Space 68 (1), 150. https://doi.org/10.1186/s40623-016-0524-z.

Wang, C.-P., Kim, H.-J., Yue, C., Weygand, J.M., Hsu, T.-S., Chu, X., 2017a. Effects of solar wind ultralow-frequency fluctuations on plasma sheet electron temperature: regression analysis with support vector machine. J. Geophys. Res. Space Phys. 122. https://doi.org/10.1002/2016JA023746.
Wang, C.-P., Merkin, V.G., Angelopoulos, V., 2017b. Mesoscale perturbations in midtail lobe/mantle during steady northward IMF: ARTEMIS observation and MHD simulation. J. Geophys. Res. Space Phys. 122. https://doi.org/10.1002/2017JA024305.
Wang, C.-P., Liu, Y.-H., Xing, X., Runov, A., Artemyev, A., 2020. An event study of simultaneous earthward and tailward bursty fast flows in the Earth's mid-tail. J. Geophys. Res. Space Phys. 125. https://doi.org/10.1029/2019JA027406, e2019JA027406.
Watermann, J., de la Beaujardiere, O., Rich, F.J., 1993. Comparison of ionospheric electrical conductances inferred from coincident radar and spacecraft measurements and photoionization models. J. Atmos. Terr. Phys. 55 (11–12). https://doi.org/10.1016/0021-9169(93)90127-K.
Weimer, D.R., 1995. Models of high-latitude electric potentials derived with a least error fit of spherical harmonic coefficients. J. Geophys. Res. 100 (A10), 19,595–19,607. https://doi.org/10.1029/95JA01755.
Welling, D., Andre, M., Dandouras, I., Delcourt, D., Fazakerley, A., Fontaine, D., Foster, J., Ilie, R., Kistler, L., Lee, J., Liemohn, M.W., Slavan, J., Wang, C.-P., Wiltberger, M., Yau, A., 2015. The Earth: plasma sources, losses, and transport processes. Space Sci. Rev. https://doi.org/10.1007/s11214-015-0187-2.
Wiltberger, M., Wang, W., Burns, A.G., Solomon, S.C., Lyon, J.G., Goodrich, C.C., 2004. Initial results from the coupled magnetosphere ionosphere thermosphere model: magnetospheric and ionospheric responses. J. Atmos. Sol.-Terr. Phys. 66, 1411–1423. https://doi.org/10.1016/j. jastp.2004.03.026.
Wiltberger, M., Weigel, R.S., Lotko, W., Fedder, J.A., 2009. Modeling seasonal variations of auroral particle precipitation in a global-scale magnetosphere-ionosphere simulation. J. Geophys. Res. 114. https://doi.org/10.1029/2008JA013108, A01204.
Wiltberger, M., Merkin, V., Zhang, B., Toffoletto, F., Oppenheim, M., Wang, W., Stephens, G.K., 2017. Effects of electrojet turbulence on a magnetosphere-ionosphere simulation of a geomagnetic storm. J. Geophys. Res. Space Phys. 122 (5), 5008–5027. https://doi.org/10.1002/2016JA023700.
Wing, S., Newell, P.T., 1998. Central plasma sheet ion properties as inferred from ionospheric observations. J. Geophys. Res. 103 (A4), 6785–6800. https://doi.org/10.1029/97JA02994.
Wing, S., Johnson, J.R., Chaston, C.C., Echim, M., Escoubet, C.P., Lavraud, B., Lemon, C., Nykyri, K., Otto, A., Raeder, J., Wang, C.-P., 2014. Review of solar wind entry into and transport within the plasma sheet. Space Sci. Rev. 184 (1–4), 33–86. https://doi.org/10.1007/s11214-014-0108-9.
Wing, S., Khazanov, G.V., Sibeck, D.G., Zesta, E., 2019. Low energy precipitating electrons in the diffuse aurorae. Geophys. Res. Lett. 46, 3582–3589. https://doi.org/10.1029/2019GL082383.
Winningham, J.D., Heikkila, W.J., 1974. Polar cap Electron fluxes observed with ISIS-l. Geophys. Res. 79, 949.
Wolf, R.A., 1970. Effects of ionospheric conductivity on convective flow of plasma in the magnetosphere. J. Geophys. Res. 75 (25), 4677–4698. https://doi.org/10.1029/JA075i025p04677.
Xi, S., Lotko, W., Zhang, B., Wiltberger, M., Lyon, J., 2016. Effects of auroral potential drops on plasma sheet dynamics. J. Geophys. Res. Space Phys. 121 (11), 11,129–11,144. https://doi.org/10.1002/2016JA022856.
Xing, X., Wolf, R.A., 2007. Criterion for interchange instability in a plasma connected to a conducting ionosphere. J. Geophys. Res. 112. https://doi.org/10.1029/2007JA012535. A12209.
Yahnina, T.A., Yahnin, A.G., 2014. Proton precipitation to the equator of the isotropic boundary during the geomagnetic storm on November 20-29, 2003. Cosm. Res. 52 (1), 79–85. https://doi.org/10.1134/S0010952514010092.
Young, S.L., Denton, R.E., Anderson, B.J., Hudson, M.K., 2008. Magnetic field line curvature induced pitch angle diffusion in the inner magnetosphere. J. Geophys. Res. Space Phys. 113 (A3). https://doi.org/10.1029/2006JA012133.
Yu, Y., Jordanova, V., Zou, S., Heelis, R., Ruohoniemi, M., Wygant, J., 2015. Modeling subauroral polarization streams during the 17 March 2013 storm. J. Geophys. Res. Space Phys. 120, 1738–1750. https://doi.org/10.1002/2014JA020371.

Yu, Y., Jordanova, V.K., Ridley, A.J., Albert, J.M., Horne, R.B., Jeffery, C.A., 2016. A new ionospheric electron precipitation module coupled with RAM-SCB within the geospace general circulation model. J. Geophys. Res. Space Phys. 121. https://doi.org/10.1002/2016JA022585.

Yu, Y., Jordanova, V.K., McGranaghan, R.M., Solomon, S.C., 2018. Self-consistent modeling of Electron precipitation and responses in the ionosphere: application to low-altitude energization during substorms. Geophys. Res. Lett. 45 (13), 6371–6381. https://doi.org/10.1029/2018GL078828.

Yu, Y., Tian, X.B., Jordanova, V.K., 2020. The effect of field line curvature (FLC) scattering on the ring current dynamics and isotropy boundary. J. Geophys. Res. Space Phys. https://doi.org/10.1002/2020/2020JA027830.

Yue, C., Wang, C.-P., Nishimura, Y., Murphy, K.R., Xing, X., Lyons, L., Henderson, M., Angelopoulos, V., Lui, A.T.Y., Nagai, T., 2015. Empirical modeling of 3-D force-balanced plasma and magnetic field structures during substorm growth phase. J. Geophys. Res. Space Phys. 120. https://doi.org/10.1002/2015JA021226.

Zesta, E., Lyons, L.R., Donovan, E., 2000. The auroral signature of earthward flow burst observed in the magnetotail. Geophys. Res. Lett. 27, 3241–3244. https://doi.org/10.1029/2000GL000027.

Zesta, E., Donovan, E., Lyons, L., Enno, G., Murphree, J.S., Cogger, L., 2002. The two-dimentional structure of auroral poleward boundary intensification (PBIs). J. Geophys. Res. 107 (A11), 1350. https://doi.org/10.1029/2001JA000260.

Zesta, E., Lyons, L., Wang, C.-.P., Donovan, E., Frey, H., Nagai, T., 2006. Auroral poleward boundary intensification (PBIs): their two-dimensional structure and associated dynamics in the plasma sheet. J. Geophys. Res. 111. https://doi.org/10.1029/2004JA010640, A05201.

Zhang, H., Zong, Q., 2020. Transient phenomena at the magnetopause and bow shock and their ground signatures. In: Zong, Q., Escoubet, P., Sibeck, D., Le, G., Zhang, H. (Eds.), Dayside Magnetosphere Interactions., https://doi.org/10.1002/9781119509592.ch2.

Zhang, Q.-.H., Zhang, B.C., Liu, R.Y., Dunlop, M.W., Lockwood, M., Moen, J., et al., 2011. On the importance of interplanetary magnetic field |by| on polar cap patch formation. J. Geophys. Res. 116. https://doi.org/10.1029/2010JA016287, A05308.

Zhang, B., Lotko, W., Brambles, O., Wiltberger, M., Lyon, J., 2015. Electron precipitation models in global magnetosphere simulations. J. Geophys. Res. Space Phys. 120, 1035–1056. https://doi.org/10.1002/2014JA020615.

Zhu, L., Schunk, R.W., Sojka, J.J., 1997. Polar cap arcs: a review. J. Atmos. Sol.-Terr. Phys. 59, 1087–1126. https://doi.org/10.1016/S1364-6826(96)00113-7.

Zhu, M., Yu, Y., Tian, X., Shreedevi, P.R., Jordanova, V.K., 2021. On the ion precipitation due to field line curvature (FLC) and EMIC wave scattering and their subsequent impact on ionospheric electrodynamics. J. Geophys. Res.: Space Phys. 126. https://doi.org/10.1029/2020JA028812. e2020JA028812.

Zou, S., et al., 2010. Identification of substorm onset location and pre-onset sequence using Reimei, THEMIS GBO, PFISR and Geotail. J. Geophys. Res. 115. https://doi.org/10.1029/2010JA015520, A12309.

Zou, S., Moldwin, M.B., Ridley, A.J., Nicolls, M.J., Coster, A.J., Thomas, E.G., Ruohoniemi, J.M., 2014. On the generation/decay of the storm-enhanced density plumes: role of the convection flow and field-aligned ion flow. J. Geophys. Res. Space Phys. 119, 8543–8559. https://doi.org/10.1002/2014JA020408.

Zou, Y., Nishimura, Y., Lyons, L.R., Shiokawa, K., 2017. Localized polar cap precipitation in association with nonstorm time airglow patches. Geophys. Res. Lett. 44, 609–617. https://doi.org/10.1002/2016GL071168.

Further reading

Brekke, A., Moen, J., 1993. Observations of high latitude ionospheric conductances. J. Atmos. Terr. Phys. 55 (11–12), 1493–1512. https://doi.org/10.1016/0021-9169(93)90126-J.

Chapman, S., 1956. The electrical conductivity of the ionosphere, a review. Nuovo Cimento 4 (Suppl., [10]), 1385.

Christensen, A.B., et al., 2003. Initial observations with the Global Ultraviolet Imager (GUVI) in the NASA TIMED satellite mission. J. Geophys. Res. 108 (A12), 1451. https://doi.org/10.1029/2003JA009918.

Crowley, G., 1996. Critical review of ionospheric patches and blobs. Rev. Radio Sci. 1993–1996, 619–648.

Dungey, J.W., 1961. Interplanetary magnetic fields and the auroral zones. Phys. Rev. Lett. 6, 47–48.

Foster, J.C., 1993. Storm time plasma transport at middle and high latitudes. J. Geophys. Res. 98 (A2), 1675–1689. https://doi.org/10.1029/92JA02032.

Frank, L.A., Craven, J.D., Gurnett, D.A., Shawhan, S.D., Burch, J.L., Winningham, J.D., et al., 1986. The theta aurora. J. Geophys. Res. 91, 3177–3224. https://doi.org/10.1029/JA091iA03p03177.

Fuller-Rowell, T.J., Codrescu, M.V., Roble, R.G., Richmond, A.D., 1997. How does the thermosphere and ionosphere react to a geo- magnetic storm? In: Magnetic Storm, Geophys. Monogr. Ser., Vol. 98. AGU, Washington, D. C, pp. 203–205.

Germany, G.A., Torr, D.G., Richards, P.G., Torr, M.R., John, S., 1994. Determination of ionospheric conductivities from FUV auroral emissions. J. Geophys. Res. 99 (A12), 23297–23305. https://doi.org/10.1029/94JA02038.

Gillies, R.G., van Eyken, A., Spanswick, E., Nicolls, M.J., Kelly, J., Greffen, M., et al., 2016. First observations from the RISR-C incoherent scatter radar. Radio Sci. 51, 1645–1659. https://doi.org/10.1002/2016RS006062.

Harel, M., Wolf, R., Reiff, P., Smiddy, M., 1979. Computer Modeling of Events in the Inner Magneto-sphere. American Geophysical Union (AGU), pp. 499–512, https://doi.org/10.1029/GM021p0499.

Hones Jr., E.W., Craven, J.D., Frank, L.A., Evans, D.S., Newell, P.T., 1989. The horse-collar aurora: a frequent pattern of the aurora in quiet times. Geophys. Res. Lett. 16, 37.

Hysell, D.L., 2015. The radar Aurora. In: Zhang, Y., Paxton, L.J. (Eds.), Auroral Dynamics and Space Weather., https://doi.org/10.1002/9781118978719.ch14.

Kivelson, M., Russell, C.T. (Eds.), 1995. Introduction to Space Physics. Cambridge Univ. Press, Cambridge, https://doi.org/10.1017/9781139878296.

Lassen, K., 1972. On the classification of high-latitude auroras. Geofys. Publ. 29, 87–104.

Lummerzheim, D., Lilensten, J., 1994. Electron transport and energy degradation in the ionosphere: evaluation of the numerical solution, comparison with laboratory experiments and auroral observations. Ann. Geophys. 12, 1039–1051. https://doi.org/10.1007/s00585-994-1039-7.

Maggiolo, R., Echim, M., Simon Wedlund, C., Zhang, Y., Fontaine, D., Lointier, G., Trotignon, J.-G., 2012. Polar cap arcs from the magnetosphere to the ionosphere: kinetic modelling and observations by cluster and TIMED. Ann. Geophys. 30, 283–302. https://doi.org/10.5194/angeo-30-283-2012.

Meng, C.-I., Akasofu, S.-I., 1976. The relation between the polar cap auroral arc and the auroral oval arc. J. Geophys. Res. 81 (22). https://doi.org/10.1029/JA081i022p04004.

Murphree, J.S., Cogger, L.L., 1981. Observed connections between apparent polar cap features and the instantaneous diffuse auroral oval. Planet. Space Sci. 29, 1143.

Nielsen, E., Craven, J.D., Frank, L.A., Heelis, R.A., 1990. Ionospheric flows associated with a transpolar arc. J. Geophys. Res. 95 (A12), 21,169–21,178. https://doi.org/10.1029/JA095iA12p21169.

Paterson, W.R., Frank, L.A., Kokubun, S., Yamamoto, T., 1998. Geotail survey of ion flow in the plasma sheet: observations between 10 and 50 R_E. J. Geophys. Res. 103, 11,811–11,825.

Saikin, A.A., 2018. The Spatial Distributions, Wave Properties, and Generation Mechanisms of Inner Magnetosphere EMIC Waves. Doctoral Dissertations, p. 2428. https://scholars.unh.edu/dissertation/2428.

Semeter, J., Doe, R., 2002. On the proper interpretation of ionospheric conductance estimated through satellite photometry. J. Geophys. Res. 107 (A8). https://doi.org/10.1029/2001JA009101.

Sitnov, M., Birn, J., Ferdousi, B., Gordeev, E., Khotyaintsev, Y., Merkin, V., Motoba, M., Otto, A., Panov, E., Pritchett, P., Pucci, F., Raeder, J., Runov, A., Sergeev, V., Velli, M., Zhou, X., 2019. Explosive magnetotail activity. Space Sci. Rev. 215 (4), 31. https://doi.org/10.1007/s11214-019-0599-5.

Trondsen, T.S., Cogger, L.L., 1998. A survey of small-scale spatially periodic distortions of auroral forms. J. Geophys. Res. 103, 9405–9416.

Weber, E.J., Buchau, J., Moore, J.G., Sharber, J.R., Livingston, R.C., Winningham, J.D., Reinisch, B.W., 1984. F layer ionization patches in the polar cap. J. Geophys. Res. 89, 1683–1694. https://doi.org/10.1029/JA089iA03p01683.

CHAPTER 5

Electromagnetic energy input and dissipation

Stephen R. Kaeppler[a], Delores J. Knipp[b], Olga P. Verkhoglyadova[c], Liam M. Kilcommons[d], and Weijia Zhan[a]
[a]Department of Physics and Astronomy, Clemson University, Clemson, SC, United States
[b]Smead Aerospace Engineering Department, University of Colorado, Boulder, CO, United States
[c]Jet Propulsion Laboratory, California Institute of Technology, Pasadena, CA, United States
[d]University of Colorado, Boulder, CO, United States

Chapter Outline

Cross-Scale Coupling and Energy Transfer in the Magnetosphere-Ionosphere-Thermosphere System
http://doi.org/10.1016/B978-0-12-821366-7.00006-8

5.1 Electromagnetic energy transfer to Earth's high-latitude upper atmosphere on multiple scales

The discovery of Earth's ionosphere and the beginning of the space age excited a curiosity about the sources, intensity, and distribution of energy available to the upper atmosphere. Three board categories of external energy sources were recognized: solar, particle, and electromagnetic energy. By the early 1930s, Chapman had worked out the basic theory of the effects of short-wave monochromatic radiation on Earth's ionosphere (Chapman, 1931). Thirty years later, a theory of auroral bombardment of the high-latitude upper atmosphere (Chamberlain, 1961) and an exposition of the role of electromagnetic energy transfer through Joule heating of the upper atmosphere (Cole, 1962) were published. Knipp et al. (2004) showed solar short-wave energy to be the largest overall energy contributor. However, they noted that a combination of auroral-particle energy deposition and electromagnetic energy transfer between the magnetosphere and ionosphere accounted for most of the variability and for extremes in ionospheric and thermospheric behavior.

There is a spectrum of temporal and spatial scales associated with electromagnetic energy transfer. Determining which (sub)ranges of these scales effectively contribute to energy that is dissipated in Earth's upper atmosphere is a major challenge in geospace science. Space-based low earth orbiting (LEO) platforms permit rapid latitudinal scans of energy-related measurements. However, when fields change rapidly (Alfvénic fluctuations), the motion of a LEO spacecraft may introduce measurement ambiguity due to Doppler effects. In general, fields that are slowly varying (quasistatic) can be measured by LEO spacecraft. However, insufficient local-time (orbit-plane) coverage and low revisit cadence within the orbit plane have thus far kept such energy-input measurements largely in the statistical realm.

5.1.1 Basic physics of electromagnetic energy transfer

The magnetosphere/solar wind dynamo is the source of most of the auroral particle (kinetic) and electromagnetic energy available to the high-latitude thermosphere and ionosphere. In the dynamo process, and especially during space weather disturbances, the conversion of solar wind energy into electrical energy strengthens electric fields. The electric fields in turn do work on charged particles and contribute to their redistribution. Highly conducting space plasmas carry currents. These currents generate their own disturbance magnetic fields, which superpose on Earth's dipole magnetic field. Electromagnetic energy transfer (the Poynting flux vector) is a key agent in this chain.

A conservation-of-energy relation for electromagnetic energy based on Maxwell's equation, called Poynting's theorem (Poynting, 1884), provides a continuity equation for electromagnetic-field energy density and flow in the presence of charged matter:

$$\frac{\partial W}{\partial t} + \nabla \cdot \mathbf{S} + \mathbf{j} \cdot \mathbf{E} = 0 \tag{5.1}$$

where W is the energy density of the electric and magnetic fields, **S** is the electromagnetic energy flux (Poynting flux vector) into/out of the considered volume, **j** is the electric current density, and **E** is the electric field. The left term represents time rate of change of the electromagnetic energy density in the volume, while the middle term quantifies divergence of the electromagnetic energy flux. The last term in Eq. (5.1) represents the volumetric energy exchange rate due to momentum transfer between charged particles and fields. In vacuum, the last term on the left-hand side equals zero.

For upper atmosphere applications, most of the downward Poynting flux absorbed at altitudes below about 2000 km goes into frictional heating of the thermosphere and ionosphere (including Joule and Ohmic heating), with a small fraction powering work done on the neutral gas by the Lorentz force (Matsuo and Richmond, 2008). Thus, the first term is usually set to zero under the assumption that energy is not being stored in the electric and magnetic fields. Equivalently, this means that neither the inductance nor capacitance of the volume changes. However, Alfvénic Poynting flux flowing in relatively narrow perpendicular channels (perpendicular wavelengths less than about 10 km) may power topside ionospheric ion upflows and outflows through collisionless wave-particle interactions and the Ponderomotive force (Chaston et al., 2006). See also Lotko (2004) for discussion of magnetic energy storage at the low-altitude boundary of the system. In a plasma, the possibility of energy storage in fields due to the absence, Eq. (5.1) becomes

$$-\nabla \cdot \mathbf{S} = \mathbf{E} \cdot \mathbf{j} \tag{5.2}$$

indicating that energy transferred to the charge carriers decreases the field energy flowing through the volume. The energy transfer ultimately produces Joule heating of the matter in the ionosphere (e.g., Cole, 1962, 1975) and/or transfer of mechanical energy between ions and neutrals transfer in the magnetosphere-ionosphere-thermosphere (MIT) system (Thayer, 1998a, 2000; Fujii et al., 1999; Thayer and Semeter, 2004).

The remainder of Sections 5.1 and 5.2 focus on divergence of the electromagnetic energy flux into the polar ionosphere. The transfer is efficient when, in the presence of strong horizontal electric fields, intense electric currents flow along geomagnetic field lines above the ionosphere and close horizontally within the ionosphere. This section will review aspects of the intensity and spatial-temporal distribution of the electromagnetic energy transfer from a space-based point of view. Section 5.3 discusses further partition and dissipation of the energy once it has arrived in the IT system.

5.1.2 Definitions and terminology

5.1.2.1 Poynting flux vector and Poynting's theorem

The formal terminology for transfer of incident electromagnetic energy flux through a volume is the Poynting flux vector, **S**, which is related to electric **E** and magnetic **B** fields as

$$\mathbf{S} = \mu_0^{-1}(\mathbf{E} \times \mathbf{B}) \tag{5.3}$$

where μ_0 is the permeability of free space. In Earth's polar regions, an electrostatic convection electric field is usually present. Further, as a result of MIT coupling, field-aligned currents (FACs) create a perturbation magnetic field, $\delta\mathbf{B}$, superposed on the background magnetic field, $\mathbf{B_0}$. The perturbation Poynting vector, S_p, describes the energy transfer in the MIT system:

$$\mathbf{S_p} = \mu_0^{-1}(\mathbf{E} \times \delta\mathbf{B}) \tag{5.4}$$

The perturbation Poynting vector excludes the contribution from $\mu_0^{-1}(\mathbf{E} \times \mathbf{B_0})$, which is nondivergent and, therefore, does not contribute to energy conversion (Kelley et al., 1991). Using the equations and definitions, above results in an expression of the perturbation Poynting vector parallel to the background magnetic field in a plasma as

$$\mathbf{S}_{\|} = \mu_0^{-1}(\mathbf{E} \times \delta\mathbf{B}) = -\mathbf{E} \cdot \mathbf{j} \tag{5.5}$$

In applications to near-Earth space, $\mathbf{E}$ is the large-scale electrostatic field from the magnetosphere, and $\delta\mathbf{B}$ is the perturbation magnetic field caused by FACs, which are also called as Birkeland currents. The quantity represents conversion of electromagnetic energy to heat and mechanical energy. In many instances, the (partial) conversion of $\mathbf{E} \cdot \mathbf{j}$ to mechanical energy is ignored and $\mathbf{j}$ is represented simply as the product of Pedersen conductance and the electrostatic E field, ($\sigma_p\mathbf{E}$). Assuming that $\mathbf{S}_{\|}$ does not exit the base of the ionosphere or leak out the sides of the considered volume and that the plasma is at rest, then the divergence of the Poynting flux is simplified to ($\sigma_p\mathbf{E}^2$). These simplifying assumptions are behind the assertion that field-aligned Poynting flux can be treated as the equivalent of Joule heating in the upper atmosphere. Foster et al. (1983) used these simplifications to produce the first global space-based estimates of Joule heating from the AE-C satellite.

If bulk motion, $\mathbf{U_n}$ of neutral gas is to be considered, then the RHS of Eq. (5.5) should be further expanded as

$$\mathbf{j} \cdot \mathbf{E}' = \mathbf{j} \cdot (\mathbf{E} + \mathbf{U_n} \times \mathbf{B}) + (\mathbf{j} \times \mathbf{B}) \cdot \mathbf{U_n} \tag{5.6}$$

where the primed electric field has an additional element $\mathbf{U_n} \times \mathbf{B}$. Here, $\mathbf{j} \cdot \mathbf{E}'$ represents the sum of Joule heating the (first term on RHS) and generation of bulk kinetic energy of the ionosphere/thermosphere medium (second term on RHS) by the $\mathbf{j} \times \mathbf{B}$ force. Both $\mathbf{E}$ and $\mathbf{U_n}$ are reference-frame dependent. For low Earth orbit (LEO), the frame is usually fixed with the rotating Earth.

Some words of caution should accompany the description above and interpretation of many discussion and figures in the literature. Only in the limiting case when magnetic flux tubes are bounded by equipotential surfaces, does Poynting flux deposit energy locally allowing one to use the Poynting flux instead of its divergence to estimate Joule heat (Richmond, 2010). Gradients in Pedersen and Hall conductances break the assumption that Poynting flux inflow at the top side of the ionosphere equals local Joule heat dissipation (either height-integrated or along magnetic flux tubes) (Vanhamaki et al., 2012). Further, as indicated in many modeling studies, electromagnetic energy travels throughout the ionosphere. Thus, despite Poynting flux deposition from the magnetosphere in high latitudes, some of the energy conversion may occur at lower latitudes.

5.1.2.2 Temporal variations

The Poynting flux vector, also called Poynting vector (and under specific conditions, Poynting flux) in various articles, is discussed in two forms in upper atmosphere electrodynamics: steady/quasistatic or perturbation/Alfvénic. For quasistatic situations, the time-dependent terms in the Maxwell equations governing ionospheric electrodynamics are neglected; the opposite is true for Alfvénic fluctuations in the fields.

5.1.2.3 Quasistatic Poynting flux

Designating $\mathbf{S}_{\parallel}$ as the geomagnetically field-aligned portion of $\mathbf{S_p}$ we have in Eqs. (5.4), (5.5), a description of electrostatic energy transfer as it is typically measured by LEO satellites. Many studies (see Table 5.1) have associated $\delta\mathbf{B}$ with large-scale FACs and a

Table 5.1 Summary of space-based observations of quasistatic (DC) Poynting flux (inspired by Table 1 in Hartinger et al., 2015).

Study	Observational platform/method	Location/event	Frequency	$S_{\parallel}$ (mW/m^2)
		Single pass and event studies		
Sugiura (1986) and Maynard et al. (1991)	DE 2 polar pass	Southern cusp, strong storm	Mix	200 (Max), otherwise 0–20
Knudsen (1990) and Kelley et al. (1991)	HILAT 2 polar passes	$\Lambda > 40$	Mix Mostly DC	5
Gary et al. (1994)	DE 2 polar pass	$\Lambda > 40$	Mix Mostly DC	1–10
Deng et al. (1995)	DE 2 polar pass	Polar cap and auroral zone Moderate storm	Mix Mostly DC	1–20
Huang and Burke (2004)	DMSP F13, F15 polar pass	Auroral zone Super storm	Mix Mostly DC	110 (Max)
Johansson et al. (2004)	Cluster	Southern hemisphere Evening sector	Mix Mostly DC	0.3 Cluster altitude ~33 at ionosphere

Continued

Table 5.1 Summary of space-based observations of quasistatic (DC) Poynting flux (inspired by Table 1 in Hartinger et al., 2015)—cont'd

Study	Observational platform/method	Location/event	Frequency	$S_{\|\|}$ (mW/m^2)
Waters et al. (2004)	Iridium and SuperDARN	$\Lambda > 60$, 2 moderate storms	DC	1–15
Korth et al. (2005)	Iridium and DMSP F13	Polar cap	Mix	50 (Max)
	and 15 polar passes	Strong IMF Bz+ superstorm	Mostly DC	
Burke et al. (2010)	DMSP F13, 15, and 16	$\Lambda > 40$ Mostly auroral	Mix	20–70 (Max)
		Superstorm	Mostly DC	
Huang et al. (2014)	DMSP 15, 16, and 17	$\Lambda > 60$	Mix	50–350 (Max)
Horvath and Lovell (2018a, b, c)	Numerous polar passes	Strong IMF Bz+ and By	Mostly DC	120 (Max)
		Intense storms		
Strangeway et al. (2000)	FAST	$\Lambda > 60$ Mostly cusp and	DC	120 (4000 km)
		Polar cap storm time		~500 ionosphere
		Strong IMF By		
		Survey studies		
Gary et al. (1995)	DE 2 polar pass and binning	$\Lambda > 40$	Mix	Avg. 1–10
		18-Month global survey	Mostly DC	
Olsson et al. (2004)	Astrid 2	$\Lambda > 40$	Mix	Avg. 1–10
	High-latitude binning	6-Month global survey	Mostly DC	
Strangeway et al. (2005)	FAST at 4000 km	$\Lambda > 50$	DC	1–100
	33 Polar passes	Survey of (near) cusp		Fast altitude
Knipp et al. (2011)	DMSP F15	$\Lambda > 50$	Mix	1–170 (Max)
	6 years	Dayside survey	Mostly DC	
		Strong IMF By		
Huang et al. (2016, 2017)	DMSP 15, 16, 17, and 18	$\Lambda > 50$	Mix	1–130
	Polar passes	High-latitude survey	Mostly DC	
		30 Storms		
Lu et al. (2018)	DMSP F15	Near-cusp survey	Mix	Avg. 1–18
	Low-latitude boundary	12,000 Near cusp	Mostly DC	13 (Cusp avg.)
	layer (LLBL) and cusp	Crossings		8 (LLBL avg.)

Table 5.1 Summary of space-based observations of quasistatic (DC) Poynting flux (inspired by Table 1 in Hartinger et al., 2015)—cont'd

Study	Observational platform/method	Location/event	Frequency	$S_{\parallel}$ (mW/m^2)
Rastaetter et al. (2016)	DMSP F15	$\Lambda > 50$	Mix	
	Polar passes	6 Event studies	Mostly DC	10 Auroral zone (AZ)
	6 Event studies	Mostly moderate storms		20 Active AZ
Knipp et al. (2021)	DMSP 15, 16, and 18	$\Lambda > 50$	Mix	Avg. 1–10
	Polar passes	High-latitude survey Seven satellite years *Modeling studies*	Mostly DC	
Weimer (2005)	DE-2 and spherical Harmonic analysis for high-latitude distribution	$\Lambda > 50$ Modeled from more than 2500 polar passes, 18-month survey	DC	0–20
Cosgrove et al. (2014)	FAST	$\Lambda > 60$	Mix	0–18
	and EOF analysis for high-latitude distribution	Modeled from >13,600 Orbits, 5-year survey		

geomagnetic field-aligned static *perturbation Poynting vector*. The downward-directed component of $\mathbf{S}_{\parallel}$ is often also referred to as *Poynting flux* in the literature. Richmond (2010) notes that temporal variations in $\delta\mathbf{B}$ with periods exceeding ~100 s ($f <$ 0.01 Hz) create only small induction electric field perturbations, which are usually ignored in electrostatic applications. We proceed using that terminology in our quest to describe when and how some of the energy flux is retained (converges) in the volume of geospace at high latitudes.

5.1.2.4 Alfvénic Poynting flux

Electric and magnetic fields may be electromagnetic rather than electrostatic or magnetostatic. If so, the Poynting vector be restated as the *Alfvénic Poynting vector*:

$$\mathbf{S}_{\parallel} = \mu_0^{-1}(\delta\mathbf{E} \times \delta\mathbf{B}) \tag{5.7}$$

where $\delta\mathbf{B}$ is determined by removing the local, average $\mathbf{B}$ value rather than the background value $\mathbf{B_0}$. Note the meaning of the symbols for perturbations between

Eqs. (5.4), (5.7) changes. In Eq. (5.4), the δ indicates a time stationary perturbation, while in Eq. (5.7), the δ indicates small-scale spatiotemporal variability.

In the Alfvénic (alternating current [AC]) approach, the magnetic induction term is retained. This approach naturally includes ultralow-frequency (ULF) wave solutions. Dynamic processes occurring at temporal scales from seconds to a minute fit in this category. Interestingly, variations occurring on multiminute timescales are also categorized as quasistatic in many studies. A long-running debate has centered on the relative amount and influence of electromagnetic energy transferred via quasistatic fields and currents (direct currents [DC]) versus higher frequency (Alfvénic [AC]) field variations. See discussion in Mann et al. (2020) and references therein.

5.1.3 Satellite estimates of the quasistatic Poynting flux vector

5.1.3.1 Satellite observations of earthward Poynting flux: Auroral zone

Theories and estimates of upper atmosphere energy deposition and conversion have a history that largely aligns with the beginning of the space age. An appendix in Sugiura (1986) lists over 20 journal references on the subject from the mid-1960s to mid-1980s. In the early 1980s, the low-altitude Dynamics Explorer-2 (DE-2) spacecraft hosted instruments that would start a legacy of space-based measurements of the Poynting vector (see Table 5.1—note that the $\mathbf{S}_{\|}$ frequency is listed as "mix." According to Burke et al. (2017) and Forsyth et al. (2017), most quasistatic measurements include an Alfvénic component). The DE-2 orbit was a highly inclined and elliptical orbit (300–1000 km). Sugiura (1984) noted that the DE-2 electric field and the perturbation magnetic field within the field-aligned current (FAC) regions were, as a rule, orthogonal to each other and highly correlated. He deduced that the downward Poynting vector calculated from the cross product of the fields was equivalent to the ionospheric energy dissipation (assuming that the height-integrated Hall current was divergence free). A subsequent NASA Technical Note on DE-2 measurements in the auroral oval reported the peak energy transfer to be a few tens of $\mathrm{mW/m^2}$ under moderately disturbed conditions (Sugiura, 1986).

Knudsen (1990) and Kelley et al. (1991) further developed the theory and methods related to determining the local vertical Poynting vector, which they called Poynting flux. They reported peak values of downward energy flux of $\sim$5 $\mathrm{mW/m^2}$ from HILAT spacecraft polar passes. In the auroral oval, the Poynting flux was almost exclusively downward. Similarly, Gary et al. (1994) argued that Poynting's theorem, applied to a magnetic flux tube segment bounded at the top by the satellite and at the bottom by the base of the ionosphere, allowed the satellite-measured $\mathbf{S}_{\|}$ to be equated to the rate of electromagnetic energy conversion taking place in the volume. They used DE-2's retarding potential analyzer (RPA) and ion drift meter (IDM) to calculate the horizontal electric field via $\mathbf{E} = -\mathbf{U_n} \times \mathbf{B_0}$. They corrected $\mathbf{E}$ for corotation and also baseline-corrected the magnetic measurement to produce $\delta\mathbf{B}$ and ultimately $\mathbf{S}_{||}$. In auroral regions, they reported the peak energy flux to be a few tens of mW/m^2 into the ionosphere under moderately disturbed conditions. Subsequent to northward turning of the IMF after a long interval of southward IMF, they reported an upward energy flux of a few mW/m^2 in the polar cap.

The DE-2 data provided an opportunity for early model-data comparisons. Deng et al. (1995) used $\mathbf{S}_{||}$ data and particle data from DE-2, as well as a vector spherical harmonic neutral atmosphere model, to show that the height-integrated energy conversion rate, the height-integrated Joule heating rate, and the DE-2 derived $\mathbf{S}_{||}$ were nearly equivalent at points along a DE-2 polar pass.

Gary et al. (1995) went on to calculate $\mathbf{S}_{||}$ for 576 orbits over the lifetime of DE-2. The data were displayed as bin averages in combined-hemisphere polar plots. For quiet conditions, the focus of the peak energy flux was near at the dayside, near-cusp region (top left figure on their page 3). For active geomagnetic conditions, the bin-average values near 65 MLT are $\sim$16 mW/m^2 (top right figure on their page 3). Similar patterns were derived by Olsson et al. (2004) from $\sim$6 months (3386 polar traverses) of Astrid-2 satellite data (see their Figs. 2 and 3). They too found persistent high levels of $\mathbf{S}_{||}$ in the near-noon region during quiet times.

A combined ground- and space-based approach was used in Waters et al. (2004) to provide event-based maps of the global Poynting flux with an approximately hourly cadence. They used spherical harmonic methods to merge the $\mathbf{E}$ data from Super Dual Auroral Radar Network (SuperDARN) with $\delta\mathbf{B}$ derived using magnetometer data from the Iridium satellite constellation for two storms. They reported maximum $\mathbf{S}_{||}$ in the dawn and dusk sectors of a few tens of mW/m^2 and average values in the high-latitude regions of $\sim$5 mW/m^2.

After 1999, the Defense Meteorological Satellite Program (DMSP) spacecraft carried boom mounted magnetometers. With these new data, event and storm studies became feasible. With less interference from spacecraft operations, the $\delta\mathbf{B}$ data became easier to interpret. A series of papers by staff and associates from the Air Force Research Laboratory reported $\mathbf{S}_{||}$ values for large storms. Table 5.1 shows maximum storm times $\mathbf{S}_{||}$ values exceeding 100 mW/m^2 (Huang and Burke, 2004; Korth et al., 2005; Burke et al., 2010; Huang et al., 2014, 2016, 2017; Horvath and Lovell, 2018a, b, c). Most of these are noncusp values (see the following section); however, some are reported as polar cap values during strong northward IMF.

To bring a current-day perspective on the statistics of quasistatic Poynting flux, we present a new analysis of average high-latitude DMSP $\mathbf{S}_{||}$ derived from seven satellite years of data in Figs. 5.1 and 5.2. We use only the highest quality ion drift data and detrended magnetometer data in the Poynting flux calculation. The top row of Fig. 5.1 shows $\mathbf{S}_{||}$ in both hemispheres for conditions roughly equivalent to those in Gary et al. (1995) for $Kp < 3$ but at higher spatial resolution. The bottom row shows

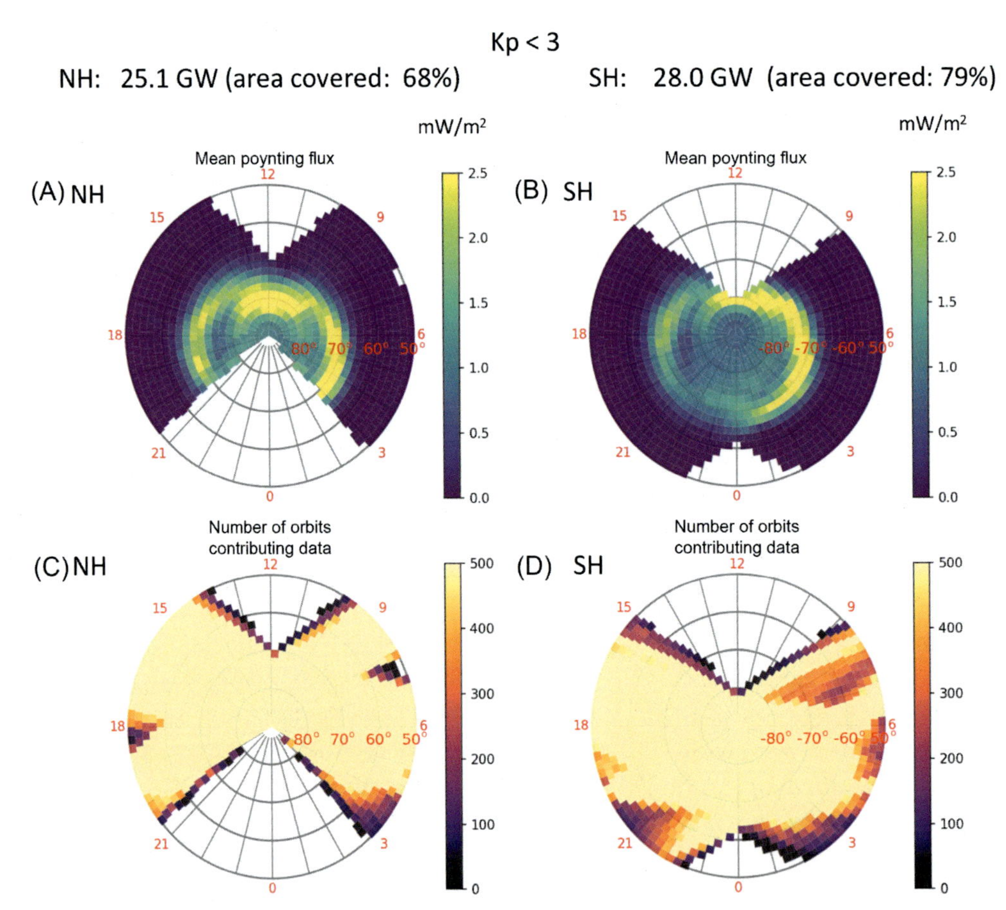

Fig. 5.1 Mean Earth-directed Poynting flux derived from seven satellite years of DMSP satellite measurements during intervals of Kp < 3. The top row (A and B) shows the Northern hemisphere distribution on the left and the Southern hemisphere distribution on the right. The percent of aerial cover with data in the equal-area bins is shown in parentheses at the top. The total power in GW is the sum over the bins with data. The Northern hemisphere is viewed from above and the Southern hemisphere is viewed from inside. Noon is at the top. The outer circle is 50 degrees magnetic latitude. The bottom row (C and D) shows the number of DMSP orbits contributing to the averages in the top row. The Northern hemisphere has better dayside coverage, while the Southern hemisphere has better nightside coverage.

the number of satellite passes with data in each bin. Fig. 5.2 shows $\mathbf{S}_{\|}$ for both hemispheres, roughly equivalent to those in Gary et al. (1995) for Kp $\geq$ 3. Note that the color bars in Figs. 5.1 and 5.2 are different.

For Kp < 3 conditions (~70% occurrence), the near-noon $\mathbf{S}_{\|}$ dominates the patterns. During Kp $\geq$ 3 conditions, the dayside Poynting flux patterns expand and intensify, and additional strong $\mathbf{S}_{\|}$ appears at dawn and dusk, presumably due to enhanced convection electric fields and storm-enhanced field-aligned currents in those locations.

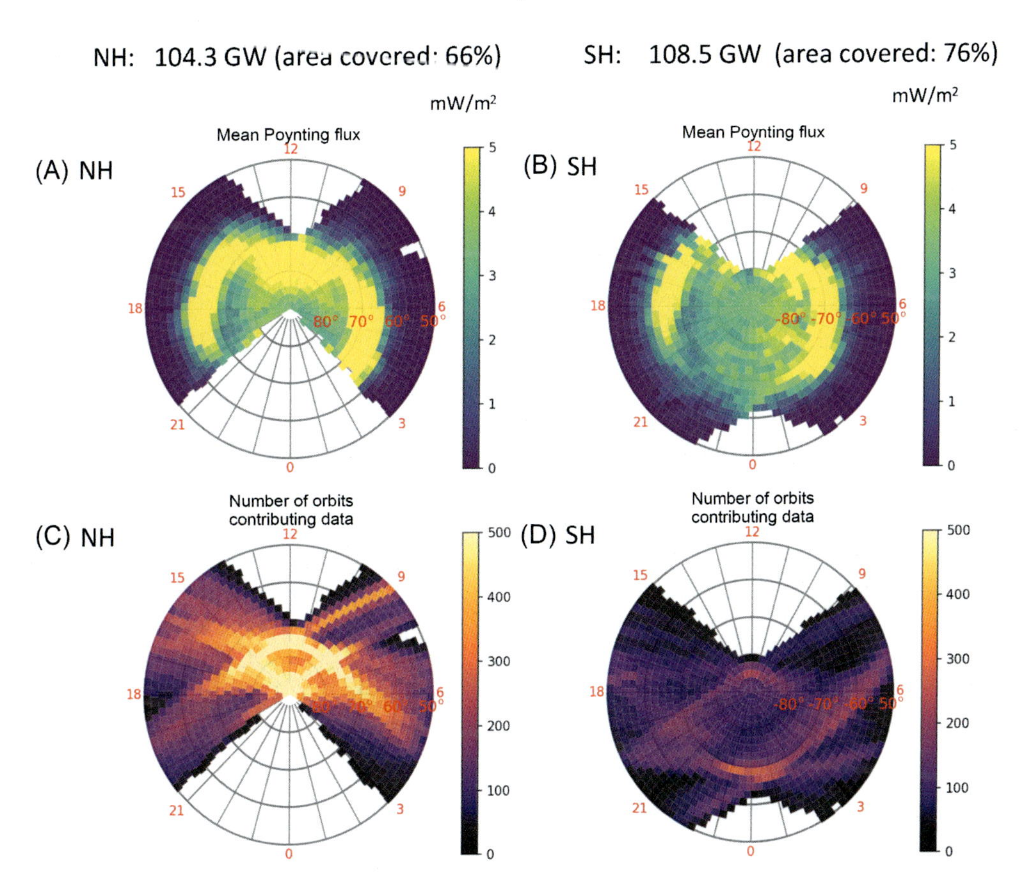

Fig. 5.2 Mean Earth-directed Poynting flux derived from seven satellite years of DMSP satellite measurements during intervals of Kp $\geq$ 3. The top row (A and B) shows the Northern hemisphere distribution on the left and the Southern hemisphere distribution on the right. Note the color bar has changed from Fig. 5.1 to accommodate the higher values of Poynting flux associated with higher levels of geomagnetic activity. The percent of aerial cover with data in the equal-area bins is shown in parentheses at the top. The total power in GW is the sum over the bins with data. The Northern hemisphere is viewed from above and the Southern hemisphere is viewed from inside. Noon is at the top. The outer circle is 50 degrees magnetic latitude. The bottom row (C and D) shows the number of DMSP orbits contributing to the averages in the top row. The Northern hemisphere has better dayside coverage, while the Southern hemisphere has better nightside coverage.

A further interesting feature of these patterns, and those in Knipp et al. (2021), is the hemispheric asymmetry. In both figures, the Northern hemisphere contains almost as much $\mathbf{S}_{\|}$ as the Southern hemisphere, but the northern coverage is substantially less than that in the Southern hemisphere. The implication is that the Northern hemisphere receives more $\mathbf{S}_{||}$ for all geomagnetic conditions. Recently, Pakhotin et al. (2021) showed a similar result for Alfvénic $\mathbf{S}_{||}$ derived from Swarm satellite day-to-night passes. They ascribed the asymmetry to the larger separation between the magnetic dipole and Earth's rotation axis in the Southern hemisphere, which in turn creates a different solar illumination in the two hemispheres.

5.1.3.2 Satellite observations: Earthward Poynting flux-nonauroral zone

In an often-overlooked NASA report, Sugiura (1986) noted that during the magnetic storm of September 6, 1982, the peak energy flux reached 200 $\mathrm{mW/m^2}$ in the cusp region (see Maynard et al. (1991) for event data and context related to the passage of the September 6, 1982 fast solar wind transient). Similarly, Strangeway et al. (2000) mapped an intense cusp $\mathbf{S}_{||}$ of 100 $\mathrm{mW/m^2}$ to ionospheric altitudes from Fast Auroral Snapshot (FAST) data, also during a strong storm with large IMF By conditions. In a subsequent survey, Strangeway et al. (2005) also reported large $\mathbf{S}_{||}$ mapped to the dayside throat region. A survey of DMSP $\mathbf{S}_{||}$ data revealed similar large near-noon enhancements for IMF By $> |10\mathrm{nT}|$ (Knipp et al., 2011). Li et al. (2011) using the Open Geospace General Circulation Model showed that such extreme Poynting flux is caused by high-latitude reconnection under conditions of large IMF clock angle and large IMF magnitude. Horvath and Lovell (2018a) reported extreme $\mathbf{S}_{||}$ during strong northward IMF in the summer hemisphere.

An interesting aspect of the Kp ≥ 3 maps is the amount of $\mathbf{S}_{||}$ in the polar cap regions. The patterns are qualitatively consistent with Gary et al. (1995) and with the Weimer (2005) maps for southward IMF, which show modest polar cap Poynting flux. Recently, Huang et al. (2014, 2016, 2017) and Horvath and Lovell (2018a, b) have shown large, but localized $\mathbf{S}_{||}$ in the polar cap during some intense storms with large IMF By and/or IMF Bz+ components. Our patterns may be consistent with an "averaged view" of those results. A rough estimate for average $\mathbf{S}_{||}$ of 4 $\mathrm{mW/m^2}$ covering the region above 75 MLT gives ~30 GW of energy transfer at polar cap latitudes. This is a substantial fraction of the globally integrated model values in Table 5.2. The contribution of Poynting flux to the polar cap regions remains an open research topic.

5.1.3.3 Poynting flux models from satellite observations and comparisons

Weimer (2005) used ~2500 DE-2 polar passes to derive magnetic and electric potential models as a function of IMF and solar wind plasma conditions. The potentials were further processed to provide electric field and magnetic perturbations and a model of $\mathbf{S}_{||}$, which by assumption was equated to energy dissipation or "Joule heating rate." This model provides output as a function of IMF and solar wind plasma conditions and dipole tilt angle. While the Weimer (2005) near-noon region energy dissipation is less prominent than in other mappings, the auroral region energy dissipation is significant. Weimer (2005) reported a localized $\mathbf{S}_{||}$ hotpot in the mid-afternoon region, which elongates in longitude during active conditions. An energy deposition model developed from

Table 5.2 Comparison of hemispherically integrated Poynting flux and location of maximum model $\mathbf{S}_{\parallel}$ by IMF clock angle from the Weimer (2005) and Cosgrove et al. (2014) models.

IMF clock angle	Weimer (2005) Integrated $S_{\parallel}$ (GW)	Max $S_{\parallel}$ location	Cosgrove et al. (2014) Integrated $S_{\parallel}$ (GW)	Max $S_{\parallel}$ location
0	7	80 degrees 16 MLT	33	80 degrees 13 MLT
45	16	80 degrees 09 MLT	57	79 degrees 13 MLT
90	49	72 degrees 16 MLT	107	68 degrees 16 MLT
135	106	76 degrees 16 MLT	159	67 degrees 16 MLT
180	137	75 degrees 17 MLT	175	67 degrees 16 MLT
225 (−135)	100	76 degrees 17 MLT	145	66 degrees 16 MLT
270 (−90)	40	78 degrees 13 MLT	91	68 degrees 16 MLT
315 (−45)	13	80 degrees 14 MLT	45	80 degrees 13 MLT
None	10	72 degrees 06 MLT	NA	NA

~13,600 polar passes of the Fast Auroral Snapshot (FAST) spacecraft shows a $\mathbf{S}_{\parallel}$ hotspot tilted to slightly after noon (Cosgrove et al., 2014) during northward IMF situations. A mid-afternoon $\mathbf{S}_{\parallel}$ hotspot is also prominent for intervals dominated by IMF By or Bz− and is present at ~16 MLT even for quiet (northward IMF) times. Both models provide a combined-hemisphere view and assume the IMF By effect is mirrored between the hemispheres. Table 5.2 provides a comparison of hemispherically integrated $\mathbf{S}_{\parallel}$ from these models. Billett et al. (2021) have also developed an energy deposition model by matching electric field observations from the Super Dual Auroral Radar Network (SuperDARN) with Active Magnetosphere and Planetary Electrodynamics Response Experiment (AMPERE) magnetic perturbation data. Although their energy deposition magnitudes are considerable smaller than those from Weimer (2005) and Cosgrove et al. (2014), the Billett et al. patterns provide more definition of IMF By effects on the dayside.

Recently, Rastätter et al. (2016) compared the Weimer (2005) and Cosgrove et al. (2014) empirical models described earlier, along with output from eight physics-based models, against orbit-integrated DMSP $\mathbf{S}_{\parallel}$. A six-event challenge that focused on quiet to moderate activity was run by the NASA Community Coordinated Modeling center. The physics-based models included three-dimensional (3D) models of the ionosphere/thermosphere and two-dimensional (2D) ionospheric electrodynamics modules of global magnetosphere MHD models. Rastätter et al. (2016) stated: "Overall, for each model tested there was a significant spread of yields across the six events ... a consistent relation between Poynting flux and Joule dissipation could not be established because the standard deviation in model outputs were comparable to the average yield." An additional benefit of the test was establishment of a methodology to describe uncertainty in DMSP Poynting flux calculations that account for quality of measurements and the often missing in-track ion drifts (cross-track electric field component). The uncertainties were estimated at 25% when only good quality data drift data were used. Higher uncertainty should be assigned if incorporating DMSP data with missing components or if lower quality data are used in the Poynting flux calculation, as has often been the case.

5.1.4 Satellite estimates of the Alfvénic Poynting vector (flux)

We provide Table 5.3 (expanded from a similar one in Hartinger et al., 2015) as a starting point for comparing the influences of quasistatic and Alfvénic Poynting flux. A quick look at the right-most columns of Table 5.3 shows that Alfvénic Poynting flux has a wide

Table 5.3 Summary of space-based observations of Alfvénic (AC) Poynting flux (inspired by Table 1 in Hartinger et al., 2015).

Study/effect	Observational platform/method	Location and/or mapping	Frequency/ scale size	$S_{\|\|}$ (mW/m^2)[a]
Osaki et al. (1998) MI coupling	Akebono	$\Lambda \sim 30$, 2 passes Plasmasphere	~20 mHz	0.002 at $\sim L = 3$
Vaivads et al. (1999)	Freja	1700 km; 2 passes	ELF	0.01–0.1
Chaston et al. (1999) Inertial Alfven Waves O+ heating and upflow	FAST	Dayside auroral zone Nightside auroral	~1 Hz	10
Angelopoulos et al. (2002) Auroral substorm	Polar/geotail	Plasmasheet	ULF	31–640
Keiling et al. (2002) Auroral generation	Polar	$\Lambda > 60$	>5.5 mHz	0.1–125
Wygant et al. (2002) Energy partition	Polar	Plasmasheet Boundary Layer	DC ULF	4 (DC) 100 (ULF)
Golovchanskaya and Maltsev (2004) Discrete auroral arcs	DE-2	$\Lambda > 60$	ELF	0.3
Chaston et al. (2005) Dayside auroral electron acceleration	Cluster/FAST	$70 > \Lambda > 80$ Low-lat boundary layer	1–50 mHz	1

Table 5.3 Summary of space-based observations of Alfvénic (AC) Poynting flux (inspired by Table 1 in Hartinger et al., 2015)—cont'd

Study/effect	Observational platform/method	Location and/or mapping	Frequency/ scale size	$S_{\|\|}(mW/m^2)$
Dombeck et al. (2005) Auroral electron Acceleration	Polar/FAST	$72 > \Lambda > 66$ Plasmasheet Boundary layer	0.005–4 Hz	2
Rae et al. (2007) Standing Alfvén waves	Polar	$76.5 > \Lambda > 66$ Dusk sector field lines	1.5 mHz	0.4–0.46
Lileo et al. (2008)	Cluster	6 RE and dusk Plasmasheet boundary layer	30 MHz	600 mW/m^2 upward
Nishimura et al. (2010) Storm SI response Convection change	Cluster	7.7 RE and dawn Global response		0.15 at 7.7 Re Upward and downward
Burke et al. (2010) Superstorm	DMSP	Dusk auroral zone	1 Hz	70
Hartinger et al. (2011) Joule heating	THEMIS	$71.5 > \Lambda > 70.5$ Closed field lines	5 mHz	0.7
Baddeley et al. (2017) Auroral arcs	DMSP	Auroral zone	1 Hz	0.7–2.5
Burke et al. (2017) Ionospheric reflection coefficients	DMSP	Polar cap Intense storm	1 Hz	20–200
Pakhotin et al. (2018, 2020) Quasistatic/Alfvénic comparison	Swarm	$\Lambda > 50$	0.2–16 Hz	1–40
		Survey studies		
Keiling et al. (2002) Auroral generation	Polar	$\Lambda > 60$	>5.5 mHz	0.1–125
Janhunen et al. (2005) Energy partition	Polar/astrid 2	$74 > \Lambda > 64$ Conjugate auroral oval	DC ULF	1–7 (DC) 0.1–1 (ULF)
Hartinger et al. (2015) Joule heating	THEMIS	$80 > \Lambda > 60$ Open/closed field lines	3–30 mHz	0.001–1

Continued

Table 5.3 Summary of space-based observations of Alfvénic (AC) Poynting flux (inspired by Table 1 in Hartinger et al., 2015)—cont'd

Study/effect	Observational platform/method	Location and/or mapping	Frequency/ scale size	$S_{\|\|}$ (mW/m^2)
Hatch et al. (2016, 2018) Ion outflow rates	FAST	$\Lambda > 60$	Inertial Alfvén waves 1–5 Hz	0.01–50 (Max)
Park et al. (2017) Ionospheric reflection coefficients	Swarm	$\Lambda > 60$	2 Hz	1
Ohtani (2019) Substorm phases	Geotail	Lobe 10–30 Re Open/closed field lines	ULF	0.01
Keiling et al. (2019) Auroral acceleration region	Swarm	$\Lambda > 60$	>5.5 mHz	0.1–125 500 (Max) for individual wave
Pakhotin et al. (2021) Hemispheric asymmetries	Polar	$80 > \Lambda > 60$	~2 Hz	0.1–1

[a]Projected to ionosphere unless noted.

range of values and several frequency bands. (See reviews by Stasiewicz et al. (2000) and Keiling (2009) for Alfvén wave descriptions.) Janhunen et al. (2005) reported that, based on an Astrid spacecraft survey, 60%–80% of energy flux near Earth (1–5 Re) was in the DC Poynting flux category, while less than 10% was in Alfvénic flux. The remainder was in particle energy flux. Despite the relatively low intensity compared to quasistatic Poynting flux, Alfvénic Poynting flux can have a multitude of influences on MIT coupling as noted in the "effect" notation of column 1 of Table 5.1.

Alfvén waves are related to auroral electron acceleration and conductance (Chaston et al., 2005, 2008), which in turn influences auroral kinetic energy deposition in the high-latitude ionosphere (see Keiling, 2009; Keiling et al., 2019 and references therein). Alfvénic waves and Poynting flux have roles in dayside and nightside auroral generation (e.g., Angelopoulos et al., 2002; Chaston et al., 2005) and auroral arcs in the polar cap (Golovchanskaya and Maltsev, 2004). More general discussion of the role of Alfvénic Poynting in auroral acceleration can be found in Keiling et al. (2003), Dombeck et al. (2005), and Baddeley et al. (2017). Vaivads et al. (1999) and Hatch et al. (2016) have further linked the Alfvénic Poynting flux to ion outflow on the nightside and dayside, respectively. Alfvénic waves potentially modify the efficiency of Joule heating from incident energy flux.

As noted by Yoshikawa (2002) and Nishimura et al. (2010), electrostatic mapping of **E** within the MIT system is well accepted for treating convection electric fields in the magnetosphere. However, electromagnetic coupling, including electromagnetic energy transport into/out of the magnetospheric dynamo as Poynting fluxes should be considered for understanding the energy source of convection **E**. Alfvénic waves are needed to modify quasisteady FACs in response to changes in the solar wind-MIT interactions. See discussion in Pakhotin et al. (2020) and Mann et al. (2020).

Incident Alfvén waves may be partially reflected from the ionosphere and/or may become trapped in the ionosphere wave guide. Several new studies have paid special attention to calculating the reflection coefficient, which is likely frequency dependent (e.g., Burke et al., 2017; Park et al., 2017). In addition, Lotko and Zhang (2018) reported that Alfvén waves have height-dependent heating rates in the thermosphere. Thus, the question of Alfvén wave contribution to upper atmospheric heating remains open. These limits are due in part to instrumentation and inability to monitor the vast regions of space where Alfvén waves are generated. The problem is further exacerbated by lack of agreement about an appropriate demarcation between DC and AC frequencies. Luhr et al. (2015) have suggested that large-scale FACs with scales of 150 km and with persistence periods of order 60 s or more are quasistationary, although this has been challenged by Pakhotin et al. (2018). Verkhoglyadova et al. (2018) estimated that Alfvén wave energy deposition in the frequency band from 0.5 Hz to several Hertz produced between 10% and 30% of the value of static Joule heating in localized regions.

5.1.5 Upward-directed Poynting flux

Thus, far our discussions have focused primarily on Earthward-directed Poynting flux. Low levels of upward-directed Poynting flux have been detected by LEO spacecraft. Gary et al. (1995) reported a few instances of upward DC Poynting flux, measured by the DE spacecraft, of $\sim$2.5 mW/m^2. Model output also suggests that large upward electromagnetic energy fluxes are rare, with the most likely occurrence near the high-latitude convection reversal boundaries were neutral winds act as a generator and contribute to mechanical energy transfer that is directed outward (Thayer et al., 1995). Thayer (1998a, b) used Sonderstrom incoherent scatter radar to investigate electromagnetic energy transfer at high latitudes for 48 h and found a short-lived instance of net upward transfer. The near-cusp region is another location of highly variable neutral winds and convection changes. Lu et al. (2018) analyzed 5 years of DMSP dayside data. Of 660 cusp crossings, 2% showed upward Poynting flux of 3 mW/m^2 or larger.

Beyond LEO, Lileo et al. (2008) reported a large upward Alfvénic Poynting flux ($\sim$600 mW/m^2) observed by a Cluster spacecraft at 5 RE in the evening auroral zone. They speculated that the large flux was due to a plasma instability. Upward/outward-directed AC Poynting flux is more likely at high altitudes since large fractions of magnetospherically generated AC Poynting flux is reflected from the ionosphere (Burke et al., 2010, 2017).

5.2 Summary and challenges for Poynting flux estimates

The fate of electromagnetic energy transferred from the magnetosphere to the upper atmosphere has been the subject of multidecadal studies and numerous individual and satellite constellations. Even so, the state of knowledge about the multitude of ways that this energy is manifest in the upper atmosphere is not well developed.

Summarizing the literature above related to direct current (DC) Poynting flux, we see that most measurements are made in LEO or just outside. At low levels of geomagnetic activity, the typical DC Poynting flux auroral values are 5 mW/m^2 or less. For moderate levels of activity, these values rise to 10–20 mW/m^2. Higher values, exceeding 100 mW/m^2, have been reported in dayside flow channels associated with large IMF By component values and intense reverse convection flow in the polar cap. This Poynting flux is dissipated as Joule heat and transferred to the mechanical energy (wind) of the ion-neutral medium. The relative balance of dissipation to transfer is an active area of investigation that requires combined modeling and observational efforts.

Alternating current (AC) Poynting flux is typically reported at much lower levels, ~0.1 mW/m^2, but may span five orders of magnitude (see Table 5.3, right column) depending on frequency and measurement location in the magnetosphere. AC Poynting flux is associated with precipitating particle acceleration, discrete aurora, and ion outflow from the ionosphere. The roles of AC Poynting flux on the thermosphere may be indirect, creating sharp conductivity gradients that feed electric fields that further accelerate flows in auroral and subauroral regions.

The following challenges relate to DC Poynting flux appear in the literature:

1. What is the timescale and source(s) of intense flow channels?
2. What is the role of conductivity gradients in facilitating Poynting flux?
3. How much Poynting flux is transferred to neutral wind motion?
4. Is there intense PF at small scales (scales smaller than can be measured at 1 s, which has been typical of LEO measurements)?
5. Does substorm onset contribute to Poynting flux deposition?
6. What is the instantaneous global 3D distribution of DC Poynting flux?
7. Several recent studies have suggested that a large fraction of total DC Poynting flux is deposited in the polar cap. The amount, source, and effects of this energy merit more attention.
8. In some regions where flows change rapidly, there may be a net upward Poynting flux. Thus, far the quoted values have been within the uncertainty of the measurements. A concerted effort should be made to determine the circumstances of upward PF generated and its magnitude.

The following challenges relate to AC Poynting flux:

1. Does AC Poynting flux have a direct role in Joule heating and if so what is the relative value compared to DC Poynting flux.

2. Does the amount of reflected Poynting flux vary with frequency and with level of geomagnetic activity?
3. What is the instantaneous global 3D distribution of AC Poynting flux? In terms of observations, there is not general agreement about the demarcation between quasistatic and Alfvénic Poynting flux, or even if a demarcation makes physical sense.

We have touched only briefly on modeling DC Poynting flux by comparing the IMF clock angles values from the empirical models of Weimer (2005) and Cosgrove et al. (2014). We did not attempt to summarize the multitude of individual modeling studies. Rather we note that in terms of modeling quasistatic Poynting flux, Rastätter et al. (2015) showed that models and data have large standard deviations in the yields and measurements, making model-data comparison very challenging.

As discussed in the following section, there is still not good understanding of the ultimate fate of the Poynting flux in terms of how much is dissipated as Joule heat and how much is transferred to mechanical energy of winds. The multimodal interactions of Alfvénic Poynting flux with the upper atmosphere need more investigation, measurements, and modeling. Within the archives of the satellite missions listed in Tables 5.1 and 5.3, there is likely a treasure of data that could provide conjugate studies of electromagnetic energy transfer from the outer to inner magnetosphere and then into the upper atmosphere. The brief review should serve as a motivator to begin the (re)discovery and to devise missions that address the previous questions.

5.3 Joule heating in the high-latitude ionosphere

The high-latitude E-region ionosphere is a highly variable interface between the magnetosphere above and the lower thermosphere below. Ions $E \times B$ drift in the F-region ionosphere, but with decreasing altitude and increasing thermospheric mass density, collisions between ions and neutrals become numerous. These collisions cause the ions to become decoupled from the local magnetic field and drift perpendicular to magnetic field lines. As a consequence of these collisions, cross-field currents flow within the E-region ionosphere, which close the electrical magnetosphere-ionosphere (MI) circuit (e.g., Boström, 1964). These closure currents enable a critical pathway for input energy, in the form of Poynting flux and direct particle precipitation, to be transferred from the magnetosphere and dissipated within the ionosphere-thermosphere (IT) through Joule heating (e.g., Richmond and Thayer, 2000; Thayer, 2000). Cole (1962) was one of the first to recognize the importance of Joule heating in the ionosphere-thermosphere.

Observing Joule heating at multiple scales remains a challenging problem. To quantify Joule heating requires simultaneous observations of plasma (i.e., electric field and conductivities) and thermospheric (i.e., neutral winds) parameters in the E-region, which is difficult to realize. The electric field can be estimated in situ, see the review by Mozer (2016) and references therein, and by incoherent scatter radar (e.g., Johnson, 1990;

Heinselman and Nicolls, 2008). Conductivity can also be estimated from satellite observations (e.g., Lummerzheim et al., 1991; Germany et al., 1994), rockets (e.g., Evans et al., 1977; Kaeppler et al., 2014), riometers (e.g., Senior et al., 2007), all-sky imagers (e.g., Kosch et al., 1998; Lam et al., 2019), and incoherent scatter radar (e.g., Brekke and Moen, 1993; Kaeppler et al., 2015), although it is important to note that conductivity is not a quantity that is directly measured with any technique (see Chapter 4.4 for a discussion on ionospheric conductivity). The neutral winds are an especially difficult parameter to accurately specify (e.g., Larsen, 2002), yet it has been shown through past studies that the influence of the neutral wind can significantly impact the magnitude of energy transfer to the ionosphere-thermosphere (e.g., Thayer and Vickrey, 1992; Lu et al., 1995; Fujii et al., 1999; Thayer, 2000). A particularly difficult regime to quantify energy input and dissipation is during active auroral conditions, which are associated with rapid spatial and temporal changes. Yet, these intervals can deposit significant energy locally to the thermosphere.

5.3.1 Basic physics of Joule heating

We start by presenting an overview of fundamental equations and concepts associated with Joule heating. Similar discussions can be found in Thayer (1998b), Thayer and Semeter (2004), and Aikio et al. (2012). Joule heating is a consequence of Poynting's theorem (5.1), where W is the total energy density contained in the electric and magnetic fields, $W = (B^2/2\mu_0) + (\epsilon_0 E^2/2)$, $\mathbf{S}$ is the Poynting flux vector, $\mathbf{S} = (\mathbf{E} \times \mathbf{B})/\mu_0$, and $\mathbf{j} \cdot \mathbf{E}$ is the Joule heating. Eq. (5.1) has units of W/m^2, and the permittivity and permeability of free space are represented as μ_0 and ϵ_0, respectively. The following sign convention is used to describe the flow of energy (e.g., Cowley, 1991) for $\mathbf{j} \cdot \mathbf{E} > 0$, energy is transferred from the electromagnetic fields into the plasma, while for $\mathbf{j} \cdot \mathbf{E} < 0$ the plasma transfers energy into the electromagnetic fields.

We derive expressions for Joule heating when the ionospheric electric field is defined in the frame of the neutral atmosphere. We define the perpendicular electric field in the neutral wind frame,

$$\mathbf{E}' = \mathbf{E} + \mathbf{U}_n \times \mathbf{B} \tag{5.8}$$

where $\mathbf{E}$ corresponds to the electric field in the plasma frame, $\mathbf{U_n}$ is the neutral wind, and $\mathbf{B}$ is the magnetic field vector. At high latitudes in the Northern hemisphere, the magnetic field is pointed downward toward the earth. Choosing to use the electric field in the frame of the neutral winds is a matter of convention and the interpretation of this choice is discussed in Section 5.1.1. Using Ohm's law, we define the current density, $\mathbf{j}$,

$$\mathbf{j} = \sigma \cdot (\mathbf{E} + \mathbf{U_n} \times \mathbf{B}) \tag{5.9}$$

where σ is the conductivity tensor that contains the Pedersen, Hall, and parallel conductivity. The Pedersen current density flows in the direction parallel to the perpendicular electric field ($\mathbf{j}_P || \mathbf{E}_\perp$), while the Hall current flows in the direction perpendicular to both the mean magnetic field and the perpendicular electric field ($\mathbf{j}_H \perp \mathbf{E}_\perp \perp \mathbf{B}$). The Pedersen and Hall currents can be decomposed as

$$\mathbf{j} = \mathbf{j_P} + \mathbf{j_H} = \sigma_P \mathbf{E}_\perp - \sigma_H (\mathbf{E}_\perp \times \mathbf{B}) \tag{5.10}$$

where σ_P and σ_H are the Pedersen and Hall conductivities with units of siemens/m or mho/m (note: mho is inverse of an Ohm). Evans et al. (1977) defines the Pedersen and Hall conductivities calculated locally at altitude z as

$$\sigma_P(z) = \frac{n_e(z)e}{B} \left[\sum_i C_i \sum_n \left[\frac{\nu_{in}/\Omega_i}{1 + \nu_{in}^2/\Omega_i^2} \right] + \frac{\nu_{en}/\Omega_e}{1 + \nu_{en}^2/\Omega_e^2} \right] \tag{5.11a}$$

$$\sigma_H(z) = \frac{n_e(z)e}{B} \left[\sum_i C_i \sum_n \left[\frac{1}{1 + \nu_{in}^2/\Omega_i^2} \right] - \frac{1}{1 + \nu_{en}^2/\Omega_e^2} \right] \tag{5.11b}$$

where $n_e(z)$ is the altitude-resolved electron density, e is the elementary electron charge, B is the magnetic field strength, C_i is the relative ion concentration which can be derived from models (e.g., Richards et al., 2009, 2010), Ω_i and Ω_e are the ion and electron cyclotron frequencies, and ν_{in} and ν_{en} are the ion-neutral and electron-neutral collision frequencies, respectively. To be consistent with convention, we define conductance as the altitude-integrated conductivity,

$$\Sigma_{P(H)} = \int \sigma_{P(H)}(z)dz \tag{5.12}$$

where z corresponds to altitude, notionally along the magnetic flux tube.

Joule heating defined in the frame of the neutral wind can be decomposed in the following way:

$$\mathbf{j} \cdot \mathbf{E}'_\perp = \sigma_P \mathbf{E}'_\perp \cdot \mathbf{E}'_\perp \tag{5.13}$$

$$= \mathbf{j} \cdot E_\perp - \mathbf{U_n} \cdot (\mathbf{j} \times \mathbf{B}) \tag{5.14}$$

$$= \sigma_P [E_\perp^2 + (\mathbf{U_n} \times \mathbf{B})^2 - 2\mathbf{U_n} \cdot (\mathbf{E} \times \mathbf{B})] \tag{5.15}$$

We can now define a number of important quantities using a consistent set of definitions from the previous investigations (e.g., Thayer, 1998b; Aikio et al., 2012). The Joule heating rate, which is the Joule heating including the neutral wind contribution, is defined as

$$q_j = \mathbf{j} \cdot \mathbf{E}'_\perp \tag{5.16}$$

the passive energy deposition rate is

$$q_j^E = \sigma_p E_\perp^2 \tag{5.17}$$

and the mechanical energy transfer rate is

$$q_m = \mathbf{U_n} \cdot (\mathbf{j} \times \mathbf{B}) \tag{5.18}$$

We also call Joule heating, $\mathbf{j} \cdot \mathbf{E}_\perp$, the electromagnetic energy transfer rate to be consistent with the previous investigations (e.g., Thayer, 1998b; Aikio et al., 2012). The passive energy deposition rate corresponds to case in which the ionosphere-thermosphere acts as a passive load given the magnetospheric electric field. Note that the passive energy deposition rate is a positive definite quantity. The electromagnetic energy transfer rate $\mathbf{j} \cdot \mathbf{E}_\perp$ corresponds to the energy that is transferred to or from the plasma, depending on the sign.

The mechanical energy transfer rate is particularly important because it is not positive definite and the direction of the neutral wind relative to the $\mathbf{E} \times \mathbf{B}$ direction becomes important for determining whether the neutral winds are an energy source or an energy sink. Lu et al. (1995) performed a modeling study using the Thermosphere-Ionosphere-Electrodynamics General Circulation Model (TIEGCM) driven by the Assimilative Mapping of Ionospheric Electrodynamics (AMIE) model to interpret observations obtained during an interval from March 28–29, 1992. From this modeling investigation, they found that only 6% of the energy went into the mechanical acceleration of the neutral species, while 94% of the electromagnetic energy was dissipated. As will be reviewed in Section 5.4.2, many ISR investigations have found that the contribution to the mechanical energy transfer term was larger, of the order of 20%–40% (e.g., Fujii et al., 1999; Thayer, 2000).

We can also gain further insight into the altitude dependence of the Joule heating rate by expanding Eq. (5.15) in the following way:

$$\mathbf{j} \cdot \mathbf{E}' = \sigma_p \mathbf{E}' \cdot \mathbf{E}' = \sum_i \frac{q_i n_i}{B_0} \left(\frac{\kappa_e}{1+\kappa_e^2} + \frac{\kappa_i}{1+\kappa_i^2} \right) \mathbf{E}' \cdot \mathbf{E}' \tag{5.19}$$

where $\kappa_{i(e)} = \Omega_{i(e)}/\nu_{i(e)n}$ is the ratio of the ion (electron) cyclotron frequency to the ion (electron) neutral collision frequency. The ion momentum equation including the Lorentz force and ion-neutral drag can be used to derive an expression for κ_i,

$$\kappa_i = \frac{\Omega_i}{\nu_{in}} = \frac{|\mathbf{V_i} - \mathbf{U_n}|}{\left| \frac{\mathbf{E}}{|\mathbf{B}|} + \frac{\mathbf{V_i} \times \mathbf{B}}{|\mathbf{B}|} \right|} \tag{5.20}$$

where $\mathbf{V_i}$ corresponds to the ion drift velocity. In the limit that $\kappa_i \ll 1$, $V_i \approx U_n$, and the ion-neutral collision frequency is large; this case corresponds to the ions being collisionally coupled to the neutral atmosphere motion. In the other extreme, $\kappa_i \gg 1$, the ions remain magnetized and the motion is consistent with the $\mathbf{E} \times \mathbf{B}$ drift. We define the altitude at which $\kappa_i = 1$ as the ion demagnetization altitude. At this altitude, the increase in the ion-neutral collision frequency results in the mean free path of the ions becomes comparable to the ion gyroradius resulting in collisions that occur while the ion is executing cyclotron motion. These collisions cause a randomizing of the ion's velocity, which results in a drift that is perpendicular to the local magnetic field. For the term $\kappa_i/(1+\kappa_i^2)$, we can derive the condition that the Pedersen conductivity maximizes at the altitude, where $\kappa_i = 1$.

5.3.2 Physical interpretation of the Joule heating rate

We have defined Joule heating in the frame of the neutral atmosphere, which we called the Joule heating rate, but we have not yet provided a physical interpretation of the terms in Eq. (5.15). Joule heating is the conversion of energy contained within the electromagnetic fields into heat (e.g., Cowley, 1991; Jackson, 1998). There has been some debate regarding the precise physical meaning of the Joule heating in the frame of the neutral winds (e.g., Vasyliunas and Song, 2005; Strangeway, 2012). Vasyliunas and Song (2005) argue that the "ionospheric Joule heating," as defined by Eq. (5.15), does not have the same physical meaning as Joule heating or Ohmic dissipation that is described in electrodynamics textbooks (e.g., Jackson, 1998). Instead, the authors state that Joule heating in the frame of the neutral atmosphere corresponds to frictional energy transfer between the plasma and neutrals. Therefore, Joule dissipation should correspond to the transfer of electromagnetic energy in the plasma reference frame, not the neutral reference frame (e.g., Strangeway, 2012). That the equations describing the frictional energy transfer have the same form as the Joule heating equation, $\mathbf{j} \cdot \mathbf{E}$, is coincidental (Vasyliunas and Song, 2005). Finally, for the mechanical energy transfer rate, $\mathbf{U_n} \cdot (\mathbf{j} \times \mathbf{B})$, additional detailed calculations are required to determine whether the work done changes the flow or the thermal energy of the neutrals (Aikio et al., 2012).

Strangeway (2012) expanded upon the work of Vasyliunas and Song (2005) by carefully deriving friction heating terms and the perpendicular currents that include the conductivities, effects from pressure gradients, and other forces. They estimated Joule heating in the frame of the plasma and neutrals. They found that the Joule dissipation in the frame of the ions was related to the electron-ion heating rate, the electron neutral heating rate, and proportional to the electron-neutral collision frequency, as shown in Eq. (35) in Strangeway (2012). Considering that these terms are small in the ionosphere, the Joule dissipation rate within the frame of the ions is very small, especially relative to the dissipation in the frame of the neutrals. The authors also demonstrated that in the absence of additional nonelectromagnetic forces, the Joule heating rate in the frame of the neutrals equals to the heating rate of individual fluids. They conclude that because $\mathbf{j} \cdot \mathbf{E} > 0$ corresponds to the conversion of electromagnetic energy into mechanical energy, which leads to enhanced heating, the term "Joule heating" is an appropriate term.

While Strangeway (2012) provides a detailed physical derivation, Appendix 1 of Thayer and Semeter (2004) provides a simplified derivation showing the connection between Joule heating as described by single fluid MHD and the frictional drag associated with individual species. We briefly review their derivation here. Thayer and Semeter (2004) consider both the neutral and plasma contributions together as a single fluid.

The thermosphere corresponds to the majority of the mass density in the MHD equation. Thayer and Semeter (2004) also derive expressions for the frictional heating term for the ions and neutrals, represented as Eqs. (A.1) and (A.2), respectively (and not repeated here for brevity). From the ion frictional heating equation, to good approximation below 400 km (e.g., Thayer and Semeter, 2004 and references therein)

$$3k_B(T_i - T_n) \approx m_n(\mathbf{V_i} - \mathbf{U_n})^2 \tag{5.21}$$

where k_b is the Boltzmann constant, and T_i and T_n correspond to the ion and neutral temperature, respectively. From the MHD equations, the Joule heating rate can be expressed as:

$$\mathbf{j} \cdot \mathbf{E}' = \sum_i q_i n_i(\mathbf{V_i} - \mathbf{V_e}) \cdot \mathbf{E}' \tag{5.22}$$

where we have used the definition of current density as $\mathbf{j} = \sum_i q_i n_i(\mathbf{V_i} - \mathbf{V_e})$, where $\mathbf{V_e}$ is the electron drift velocity, which to good approximation moves in the $E \times B$ direction down to 80 km (e.g., Boström, 1964; Kelley, 2009). Note, the current density is a frame independent quantity. Expressions can be derived for $\mathbf{V_i}$ and $\mathbf{V_e}$ by solving the steady-state ion and electron momentum equations, as shown in Eqs. (A.10) in Thayer and Semeter (2004). These expressions for $\mathbf{V_i}$ and $\mathbf{V_e}$ can be used in Eq. (5.22), to derive the following expression for the Joule heating rate:

$$\mathbf{j} \cdot \mathbf{E}' = \sum_i n_i m_i \nu_{in}(\mathbf{V_i} - \mathbf{U_n})^2 \tag{5.23}$$

Eq. (5.23) shows that the Joule heating rate is equal to the frictional heating rate when Eq. (5.21) is satisfied. As noted in Aikio et al. (2012), Eq. (5.23) is consistent with results derived in Vasyliunas and Song (2005), where it was found that the Joule heating rate was proportional to the square of the difference between the ion velocity and neutral winds.

To summarize, the most agreed upon interpretation for Joule heating in the frame of the neutral winds is that it corresponds to frictional drag between the ion and the neutrals, as described in Eq. (5.23). From Eq. (5.23), one would expect that maximum Joule heating rate occurs when the ion and the neutral drift toward each other. This situation would generate the most frictional drag because the neutrals and ions would collide more directly with each other.

5.4 Observational studies of Joule heating

5.4.1 Incoherent scatter radar observations

Incoherent scatter radars (ISRs) have been used to study E-region electrodynamics and neutral winds since the 1970s (e.g., Brekke et al., 1973; Brekke and Rino, 1978). Given that ISR can simultaneously estimate the plasma and neutral parameters in the E-region, our understanding of Joule heating has improved significantly as a result of ISR investigations. Before reviewing the key papers on the topic, we present an overview of how ISR observations of E-region electrodynamics and neutral winds are used to estimate the Joule heating rate.

5.4.1.1 Technique

In the early 1960s, a statistical theory was developed (e.g., Dougherty and Farley, 1960; Farley et al., 1961), which described how the power spectral density can be estimated from ionospheric state parameters, that is, electron density (n_e), ion and electron temperature (T_i, T_e), and line-of-sight velocity (v_{LOS}). Through the Wiener-Khinchin theorem, the power spectral density can be related to the autocorrelation function (ACF) through the Fourier transform. The ACF is estimated using voltage measurements obtained by a radar. The backscatter from ionospheric electrons occurs at the Bragg wavelength; however, these electrons are influenced by collective plasma motion that includes the effects of ions. The ionospheric state parameters are estimated as a function of range by fitting the modeled ISR ACF, using the aforementioned statistical theory, to the estimated ACF from the measured voltage samples. The modeled ACF includes pulse smearing effects as well. There are two basic types of radar coding schemes that are used in ISR experiments to produce altitude-resolved profiles of the ionospheric state parameters: so-called long-pulse modes, which are primary for the F-region, and alternating codes (Lehtinen and Haggstrom, 1987) which are used to resolve the E-region. Further information about the ISR technique can be found in the following resources (e.g., Evans, 1969; Kudeki and Milla, 2011).

To quantify the Joule heating rate, we need estimates of the Pedersen conductivity, the perpendicular current densities, the electric field, and the E-region neutral winds. The Pedersen conductivity is calculated using Eq. (5.11). The current density is:

$$\mathbf{j} = en_e(\mathbf{V_i} - \mathbf{V_e}) \tag{5.24}$$

where we have assumed quasineutrality, $en_e \approx \sum_i q_i n_i$. To good approximation (Boström, 1964), the electrons drift in the $\mathbf{E} \times \mathbf{B}$ direction to an altitude of ~80 km; therefore, to quantify the perpendicular electron drift requires an estimate of the electric field. If we consider a coordinate system of geomagnetic east, north, and up as x, y, and z, respectively, we can decompose Eq. (5.24) into the following components perpendicular to the local magnetic field:

$$j_x(z) = en_e(V_{ix}(z) + E_y/B) \tag{5.25}$$

$$j_y(z) = en_e(V_{iy}(z) - E_x/B) \tag{5.26}$$

where $\mathbf{B} = -B\hat{z}$, which is the magnetic field direction in the Northern hemisphere.

The perpendicular ion velocity (plasma drifts) in the F-region, to good approximation move in the $\mathbf{E} \times \mathbf{B}$ direction, therefore, enabling an estimate of the electric field. Heinselman and Nicolls (2008) showed how the line-of-sight (LOS) velocity measurements can be related to the parallel and perpendicular ion velocity in the F-region as a linear matrix problem,

$$\mathbf{v}_{LOS}(z) = \mathbf{A}(z)\mathbf{V}_i(z) + \mathbf{e}_{LOS}(z) \tag{5.27}$$

where the matrix $\mathbf{A}$ corresponds to a projection with matrix elements being defined in Eq. (10) in Heinselman and Nicolls (2008), $\mathbf{V}_i(z)$ corresponds to the components of the ion velocity perpendicular and parallel to the local magnetic field as a function of altitude (range), and $\mathbf{e}_{LOS}$ is the error of the LOS velocity measurements.

Estimating $\mathbf{V}_i(z)$ given $\mathbf{v}_{LOS}$ amounts to a matrix inversion of $\mathbf{A}(z)$. As reviewed in Johnson (1990), in the case of three-beam experiments the inversion can be direct, provided that the matrix $\mathbf{A}$ is well conditioned. However, as discussed in Heinselman and Nicolls (2008) for the case relevant to the advanced modular incoherent scatter radars (AMISR) (Kelly and Heinselman, 2009), the number of LOS look directions (i.e., V_{LOS}) is larger than the number of parameters being estimated per range, a different methodology is used to handle the overdetermined problem. Methods have been developed to estimate the perpendicular ion velocity in the F-region given different information to regularize the solution, that is, to provide additional information to enable a solution of the overdetermined problem. Heinselman and Nicolls (2008) used a Bayesian regularization in which information about the solutions (perpendicular ion velocity) can be incorporated into a covariance matrix. For the Bayesian methodology, the information used in the covariance matrices corresponds to some physical meaningful bounds on the perpendicular velocities. Nygrén et al. (2011) used similar mathematics as the method described in Heinselman and Nicolls (2008), but statistical methods were used to provide the additional information for regularization.

The neutral winds are also derived from ISR observations and the winds are needed for the Joule heating rate calculation. Brekke et al. (1973) developed a technique to estimate the E-region winds, which involved solving the steady-state ion momentum equation that included the ion-neutral drag term. The solution presented by Brekke et al. (1973) can be derived from Eq. (5.20) for the $\mathbf{U_n}$ term as a function of altitude, z,

$$\mathbf{U_n}(\mathbf{z}) = \mathbf{V_i}(z) - \frac{\Omega_i}{\nu_{in}|\mathbf{B}|}(\mathbf{E} + \mathbf{V}_i(z) \times \mathbf{B}) \tag{5.28}$$

Many ISR investigations have used this technique to quantify the E-region neutral winds (e.g., Brekke et al., 1994; Nozawa and Brekke, 1995, 1999a, b, 2000; Nozawa et al., 2005). Heinselman and Nicolls (2008) showed that Eq. (5.27) can be used to simultaneously estimate the E-region winds and the electric field, given alternating code and long-pulse data. The key difference was that the matrix $\mathbf{A}(\mathbf{z})$ is replaced with a different forward model, $\mathbf{A}(\mathbf{z})' = \mathbf{A}(\mathbf{z}) \cdot \mathbf{D}(z)$, where $\mathbf{D}(z)$ contains the ion-neutral physics, which is described in Eqs. (15)–(17) of their paper. The inversion is then conducted using the linear Bayesian inversion (e.g., Heinselman and Nicolls, 2008) or statistical methods (e.g., Nygrén et al., 2011).

5.4.2 Previous investigations

A few early attempts were made to quantify the Joule heating rate at the Chatanika ISR (e.g., Wickwar et al., 1975; Banks, 1977), while at the same time techniques were being developed to estimate the neutral winds in the E-region (Brekke et al., 1973, 1974). However, these early radar investigations used square long pulses with a range resolution of approximately 48 km (Brekke and Rino, 1978), which effectively smeared out the E-region. Rino et al. (1977) developed a single pulse technique that could achieve higher altitude resolution of ~24 km. Brekke and Rino (1978) used this single pulse mode and the neutral wind estimation technique to derive the first altitude-resolved Joule heating rate that included the neutral wind contribution. They found that the maximum Joule heating rate occurred at an altitude above the maximum Pedersen conductivity. The Joule heating rate was found to be positive, which suggested the ionosphere acted as a sink of magnetospheric energy.

The Chatanika ISR observations in Brekke and Rino (1978) had a range resolution of 24 km, which was still rather coarse for resolving the E-region (Fujii et al., 1998). With the advent of alternating codes (e.g., Lehtinen and Haggstrom, 1987; Huuskonen et al., 1996), the range resolution of ISR observations improved to the point that the E-region could be resolved. In the late 1990s and early 2000s, there were a few important case studies (e.g., Thayer, 1998b; Fujii et al., 1998) and statistical studies (e.g., Fujii et al., 1999; Thayer, 2000) that quantified Joule heating in the ionosphere-thermosphere (IT) system. These studies emphasized the impact that the neutral winds played in modulating the Joule heating rate.

Thayer (1998b) examined two events using the Sondrestrom ISR, August 5, 1993 and May 2, 1995, in which the altitude-resolved neutral wind and Joule heating rate were calculated. The altitude resolution of the observations was ~6 km in the E-region. These case studies were associated with moderate-to-strong geomagnetic activity, with the May 2, 1995 being the stronger event. They found the Joule heating rate was highly dependent upon the altitude structure of the neutral wind profile: at higher (lower) E-region altitudes the neutral winds supported an enhancement (reduction) of the Joule heating rate. The integrated Joule heating rate, which includes the neutral wind contribution, could be larger or smaller than the passive energy deposition rate, depending on the configuration of the neutral wind. For example, near 1800 UT on August 5, 1993, the ISR observations showed a 75% reduction and a 50% increase in the Joule heating rate in the upper and lower E-region altitudes, respectively. This resulted in a 40% reduction in the integrated Joule heating rate. The largest difference in the Joule heating rate was found during intervals when the electric field would change direction, while the neutral wind direction would remain steady as a result of the longer response time of the thermosphere. For these case studies, there was one interval with a 400% enhancement of the Joule heating rate, which occurred after a substorm.

Fujii et al. (1998) examined an event from May 3, 1988 using the EISCAT radars in a tristatic configuration after a disturbed interval with a prolonged interval of ~4 h of magnetometer perturbations. The long interval was important because the IT system was steadily driven by Lorentz forcing. The observations were provided at 101, 109, 119, and 132 km using data from the CP-1 radar mode. One of the key findings from this investigation was that the role of the mechanical heating term, $\mathbf{U_n} \cdot (\mathbf{j} \times \mathbf{B})$, was not negligible and at times can be as significant as the Joule heating rate, $\mathbf{j} \cdot \mathbf{E}'$. They also found the equivalent neutral wind-driven electric field, that is, $\mathbf{U_n} \times \mathbf{B}$, increased in magnitude at higher altitudes. For example, at 119 km altitude, the equivalent electric field magnitude was found to be ~50% of the electric field magnitude. In addition, the direction of $\mathbf{U_n} \times \mathbf{B}$ was generally opposite the electric field direction. Finally, the electromagnetic energy transfer rate, $\mathbf{j} \cdot \mathbf{E}$, was found to increase with altitude, which they attributed to the altitude dependence of the Pedersen conductivity.

Statistical studies using ISR observations produced results that supported the role that neutral winds have on the Joule heating rate (Fujii et al., 1999; Thayer, 2000). Fujii et al. (1999) examined 28 days of EISCAT data from 1989 to 1991, while the study by Thayer (2000) examined 95 h of Sondrestrom data from 1993 to 1998. Thayer (2000) found that on average the neutral winds reduced the Joule heating rate by up to 30%–50% relative to the passive energy deposition rate. This meant that the winds acted to reduce the total energy deposition rate relative to the energy deposition rate predicted by the electric field and conductivities alone. Fujii et al. (1999) found that at 117 km altitude, on average, ~90% of the energy was dissipated through Joule heating, while only ~10% was dissipated through the mechanical heating term. However, at 125 km, on average, 65% of energy was converted to Joule heat and the remaining 35% was dissipated through mechanical energy transfer. Fujii et al. (1999) and Thayer (2000) presented a significantly different perspective than model results by Lu et al. (1995), that only 6% of the energy was dissipated through the mechanical heating term.

Both statistical studies also found an asymmetry in the energy deposition rate between the dusk and dawn sectors. In addition, both studies also found noticeable reductions in the Joule heating rate during the noon and midnight magnetic local-time sector. Thayer (2000) concluded the dawn sector had a larger passive energy deposition rate relative to the dusk sector and that the neutral wind in the dawn MLT sector caused a larger reduction in the Joule heating rate relative to the dusk MLT sector. However, one important difference between both studies was the region of geospace that was sampled by the ISRs.

Thayer (2000) pointed out that the EISCAT radars used in the studies (Fujii et al., 1999) were located at ~66.4 degrees magnetic latitude, while Sondrestrom was located at 74.2 degrees magnetic latitude. Therefore, Sondrestrom sampled the auroral oval in the dawn, noon, and dusk sectors, while remaining in the polar cap in the midnight sector; while the EISCAT radars provided good coverage of the auroral oval in the midnight sector.

Aikio and Selkälä (2009) performed a statistical investigation of conductances, electric fields, and Joule heating obtained from an experiment spanning March 6, 2006 to April 6, 2006 using EISCAT Tromsø. In contrast to previous statistical investigations, these data were obtained from a continuous 1 month data collection interval versus data collected over multiple years and seasons. The analysis did not include the contribution from the neutral winds, but only the Joule heating associated with the electric field, namely, the passive energy deposition rate. The electric fields, conductance, and passive energy deposition rate were compared for $Kp = 0–2$, and $Kp > 3$, corresponding to quiet and active intervals. Dusk and dawn enhancements were observed in the Joule heating rate, similar to other observations (e.g., Fujii et al., 1999; Thayer, 2000). Aikio and Selkälä (2009) also found for $Kp > 3$ that the Joule heating observations in the dawn sector were larger than the dusk sector. They found in the dawn sector both the electric field and conductivity were enhanced, while in the dusk sector the electric field was more strongly enhanced relative to the conductivity. In addition, Aikio and Selkälä (2009) provided fits of Joule heating in the form $q_j = AE + BE^2$, where E corresponded to the electric field magnitude and A and B were fit coefficients.

Aikio et al. (2012) undertook an EISCAT-based statistical investigation of Joule heating that included the effects of neutral winds for two long experimental runs from September 6–30, 2005 and for an active interval on November 11–19, 2003. The focus of the investigation was to quantify the altitude-integrated Joule heating rate, passive energy deposition rate, and mechanical energy transfer rate. The observations were separated into three activity levels based on Kp, quiet ($0 < Kp < 2$), moderate ($3 < Kp < 4$), and active conditions ($Kp \geq 5$). During quiet condition, the authors found that the neutral winds were a significant contribution to the Joule heating rate. Aikio et al. (2012) found that the altitude-integrated mechanical heating rate (Q_m) was approximately a factor of 2 larger during active intervals relative to moderate intervals, and a factor of 4 larger in active versus quiet intervals. They also found that the mechanical heating rate was found to negative, which means that the neutral winds are doing work onto the plasma and acting as a generator. The electromagnetic energy input was found to be larger in the evening MLT sector versus the morning MLT sector; however, the Joule heating rate was largest in the morning sector because the neutral winds reduce the Joule heating rate in the evening sector. For moderate conditions, the neutral winds increased the Joule heating rate in the morning sector, but decrease the Joule heating rate in the evening sector.

While Aikio et al. (2012) focused on the altitude-integrated parameters, Cai et al. (2013) examined the altitude-resolved Joule heating rate and mechanical energy transfer rate. Cai et al. (2013) used the same dataset as Aikio et al. (2012) and binned the data by Kp in a similar way. They found that the peak altitude of the Joule heating rate was different than the maximum altitude of the Pedersen conductivity, similar to other observations (e.g., Brekke and Rino, 1978; Hurd and Larsen, 2016). During quiet conditions, Cai et al. (2013) found that the maximum energy dissipation occurs near the peak of the Pedersen conductivity, while the mechanical heating rate increases with altitude. For active conditions, they also showed a reduction in the eveningside Joule heating maximum was the result of the Joule heating altitude profile descending to 115 km and reducing in magnitude by 15%–45%.

5.4.3 Rocket observations

There have been few in situ observations that have directly investigated the effects the neutral winds have on the Joule heating rate. Two early rocket-based studies that calculated Joule heating included Evans et al. (1977) and Watanabe et al. (1991); however, neither study included the contribution to Joule heating from the neutral winds. Evans et al. (1977) conducted one of the first rocket experiments to quantify the Joule heating from the electric field only and particle energy flux near a discrete auroral arc, in which they found significant Joule heating equatorward of a discrete auroral arc. This observation may have been partially attributed to the relatively strong total electric field magnitude observed on the equatorward side of the auroral arc.

As part of the ERRRIS rocket mission, two rockets were launched into quiet conditions and one rocket into moderate-to-active conditions (Watanabe et al., 1991). They found that the Joule heating during the quiet conditions and the active conditions were comparable in magnitude. However, during active conditions, the peak of the Joule heating rate increased to 130 km altitude relative to 110 km altitudes during the quiet conditions. Moreover, they found that within an auroral arc, the electron heat flux exceeded the Joule heating rate by a factor of 5, while outside the arc, the Joule heating was expected to dominant. The winds were inferred using the rocket-based observations; therefore, it was not possible to determine the impact of the winds on the Joule heating.

Sangalli et al. (2009) presented results from the JOULE II rocket mission, which show some of the first in situ measurements of Joule heating that includes the neutral winds. Sangalli et al. (2009) analyzed data from the 21.138 and 41.065 payloads which launched at approximately 1245 UT on January 19, 2007 into pulsating aurora and a northward arc, which 15 min previously had been a relatively stable quiet auroral arc (Burchill et al., 2012). The authors combined in situ electric field, ion plasma velocity, and neutral wind data to produce estimates of the Joule heating rate with and without the inclusion of the neutral winds. Fig. 5.3 shows that neglecting the neutral wind resulted in an

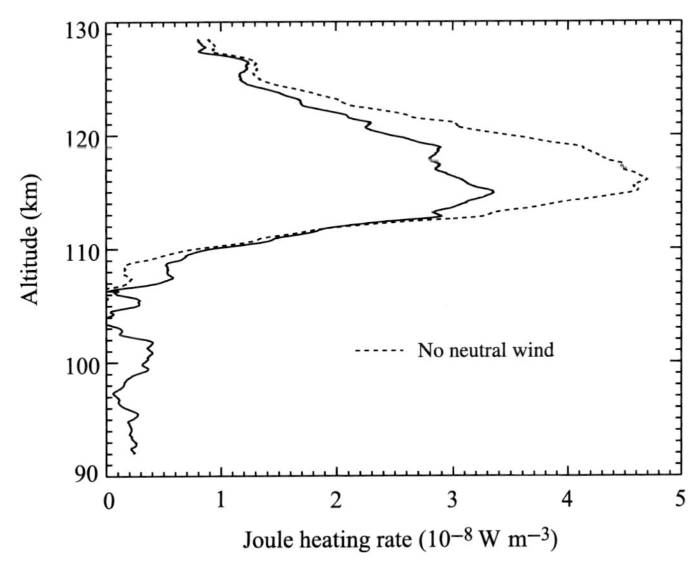

Fig. 5.3 Joule heating derived from JOULE II data (Sangalli et al., 2009), which include the contribution from the neutral winds as shown as the *solid line*.

overestimate of the Joule heating rate by 28% and the altitude-integrated Pedersen and Hall currents by +14% and −6%, respectively.

Hurd and Larsen (2016) reanalyzed data from three sounding rocket missions that used barium to measure the electric fields in the F-region, which were mapped to the E-region. One of the rockets included in this investigation was COPE-2 (e.g., Larsen et al., 1989), while the other rockets were from a campaign in the 1960s in Norway, as discussed by Wescott (e.g., Wescott et al., 1969; Hurd and Larsen, 2016 and references therein). For the COPE-2 mission, which was relatively quiet, Hurd and Larsen (2016) found the Joule heating rate was altered by up to 12% when including a relatively uniform neutral wind. They also reported that the Joule heating rate was largest just above the peak altitude of the Pedersen conductivity. For the three cases examined, the moderately active case had the largest Joule heating rate, due to large fluctuations in the electric field. In contrast, the most active case had a relatively uniform electric field, thus resulting in little Joule heating dissipation. The conclusion of the investigation was that in all three cases the local fluctuation in the Joule heating rate due to electric field variations were approximately a factor of 2–3 larger than the Joule heating rate associated with the mean electric field.

5.4.4 Coherent radar theory and observations

Coherent scatter radar provides a unique perspective of Joule heating by accessing both very small scales, but also providing spatial observations over mesoscales. However, investigations specifically focused on Joule heating from coherent radar observations are relatively limited compared to incoherent scatter radar.

Dimant and Oppenheim (2011a, b) have demonstrated through modeling investigations that small-scale Farley-Buneman waves in the auroral electrojet can at times significantly modify the global Pedersen conductance. Two mechanisms have been suggested that can cause up to a factor of 2 increase in the Pedersen conductance (e.g., Hysell, 2015). First, wave heating can reduce the recombination rates, therefore, increasing the background electron density, and second, wave turbulence can enhance the conductance (e.g., Dimant and Oppenheim, 2011a, b).

A recent investigation by Kiene et al. (2019) merged together multiple datasets to assess the Joule heating rate in the F-region. While the E-region has the largest Joule heating magnitude, the F-region has the largest Joule heating magnitude per unit mass (e.g., Thayer and Semeter, 2004 and references therein). Kiene et al. (2019) combined ion drifts that were derived from SuperDARN observations using local divergence-free fitting (Bristow et al., 2016) at ~50 km^2 resolution, Poker Flat Incoherent Scatter Radar (PFISR) observations, and Scanning Doppler Imager (SDI) data that produce estimates of the neutral winds (Conde and Smith, 1995, 1998). These data were combined to produce a mesoscale (~1000 km) perspective of Joule heating and ion temperature enhancements in the F-region associated with aurora. They found that during particularly intense periods, the neutral winds reduced the overall Joule heating rate significantly, by more than a factor of 5 during some time intervals. They also quantified the large-scale ion temperature structure, and they found some patchy temperature enhancements associated with specific auroral forms. The work by Kiene et al. (2019) provides a useful methodology for understanding Joule heating on mesoscales.

5.5 Recent observations from the Poker Flat Incoherent Scatter Radar

The Poker Flat Incoherent Scatter Radar (PFISR) is an advanced modular incoherent scatter radar, which is a monostatic electronically steerable phased-array radar that is capable of beam steering on a pulse-to-pulse basis (Kelly and Heinselman, 2009) and PFISR is located near Poker Flat, Alaska, the United States (Longitude: 147.47°W, Longitude: 65.12°N). Since 2007, PFISR has been operating in a four-beam 1% duty cycle background mode called the "IPY mode," which has provided nearly continuous observations of the E- and F-region ionosphere. Higher duty cycle modes are also run during dedicated experiments. Alternating codes are used to resolve the E-region ionospheric state parameters (Lehtinen and Haggstrom, 1987). PFISR experiments use a 16-baud randomized strong alternating code (e.g., Lehtinen and Haggstrom, 1987) with 30 μs (4.5 km) bauds. The data are oversampled at 10 μs and processed using fractional lag processing (e.g., Huuskonen et al., 1996). For the F-region, a long-pulse experiment using a

480 or 330 μs uncoded pulses are gated to have a spacing of 36 and 24.5 km with a range resolution of 72 and 49 km, respectively.

We present results of Joule heating derived from PFISR observations. The full database of neutral winds and Joule heating parameters are calculated from March 2010 to June 2019; we present a small, but representative, set of examples from these data. The estimation of the electric field and neutral winds used methods described in Heinselman and Nicolls (2008) and discussed in Section 5.4.1.1. For the neutral winds, estimates of the geographic zonal, meridional, and vertical winds were produced as a function of time and on a uniform altitude grid that spanned from 90 to 130 km in 5 km increments. The electron density observed by PFISR was used to calculate the Hall and Pedersen conductivity using Eq. (5.11). As described in Section 5.3.1, the perpendicular current density, Pedersen conductivity, electric field, and neutral winds were used in Eqs. (5.15), (5.17), and (5.18) to calculate the Joule heating rate, the passive energy deposition rate, and the mechanical energy transfer rate. These values were then integrated with respect to altitude to produce the integrated Joule heating rate, integrated passive energy deposition rate, and the integrated mechanical energy transfer rate.

In Fig. 5.4, we present an example of the 2 days of PFISR observations for June 18, 2017 and June 19, 2017 which show a day with a Joule heating event and a typical quiet day, respectively. Fig. 5.4A and B shows the passive energy deposition rate and the Joule heating rate as a function of altitude, respectively. Fig. 5.4C shows the percent difference between the passive energy deposition rate and the Joule heating rate, normalized to Joule heating rate; −100% corresponds to the limiting case, where only the neutral winds are contributing to the Joule heating rate. This ratio gives an indication as to the effect of the electric field on the Joule heating rate. Fig. 5.4D shows the meridional and zonal electric fields as orange and blue, respectively. Note that the left column and right column have different magnitudes. Fig. 5.4E and F shows the altitude-resolved zonal and meridional neutral winds, respectively. All of these data are plotted with respect to magnetic local time. At PFISR, magnetic midnight occurs at ~11 UT, solar local time (LT) is ~UT-9.8 h, and LT is +1.2 h ahead of MLT.

On June 18, 2017, there are two maxima in the passive energy deposition rate and the Joule heating rate, one that occurs in the evening sector, 1600–2100 MLT, and the other which occurs in the morning sector, 0000–0400 MLT. There is a minimum which occurs near magnetic midnight, which has been reported in the previous observational papers (e.g., Fujii et al., 1999; Thayer, 2000; Aikio et al., 2012; Cai et al., 2013) and has been attributed to the reversal of plasma drifts associated with the Harang discontinuity (e.g., Aikio et al., 2012; Cai et al., 2013). The Joule heating rate on the morningside extends down to lower altitudes, relative to the eveningside; this effect could be could be partially attributed to the harder auroral precipitation observed in the morningside associated with diffuse and pulsating aurora (e.g., Jones et al., 2009; Hosokawa and Ogawa, 2015). For June 18, 2017, Fig. 5.4D shows that the meridional electric field

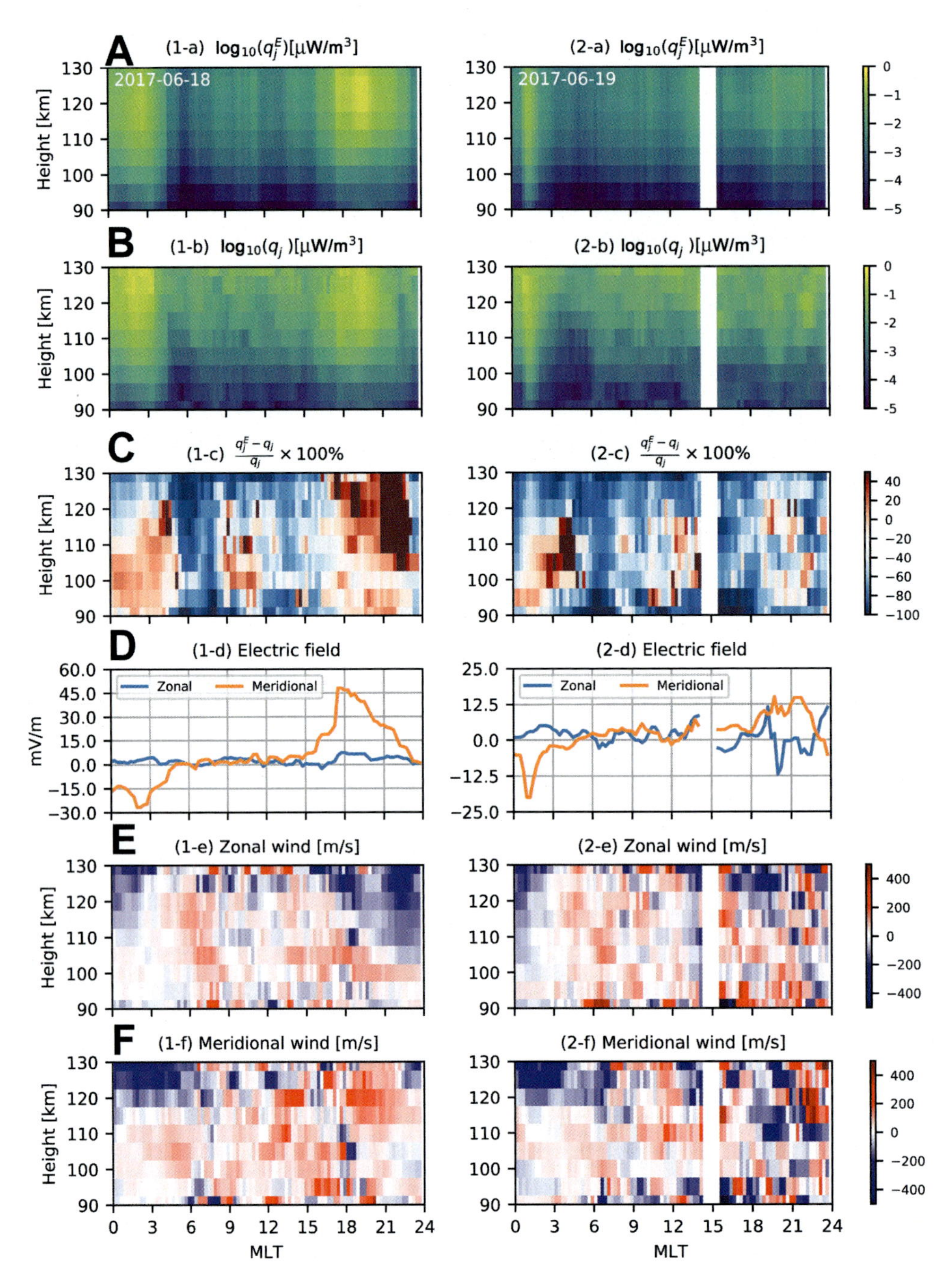

Fig. 5.4 An example of 2 days that include a Joule heating event and a typical quiet day. The left and right column are June 18, 2017 and June 19, 2017, which correspond to the active and quiet conditions, respectively. Rows A and B are the passive energy deposition rate and the Joule heating rate, respectively. Row C is the percent difference between the passive energy deposition rate and the Joule heating rate. Row D is the zonal and meridional electric fields shown as *blue* and *orange*, respectively. Rows E and F correspond to the altitude-resolved zonal and meridional winds, respectively.

has a larger magnitude in the evening sector relative to the morning sector. Although on the morningside, Fig. 5.4C shows the electric field has a stronger impact on the Joule heating rate at lower altitudes in the dawn sector, as indicated by the positive magnitude, relative to the evening sector. Fig. 5.4C suggests that the winds may reduce the Joule heating rate on the morningside. On the eveningside, near 2100–2300 MLT, there is ~2–3 h delay in MLT between the peak meridional electric field and the largest percent difference in the Joule heating, as shown in Fig. 5.4C.

For the E-region neutral winds, a 1-h running average window was applied to the data that are presented. During the interval 0600–1800 MLT which corresponds to the daytime, the winds are somewhat noisy, which is mostly attributed to the low duty cycle IPY mode, relative to the higher duty cycle modes run during the nightside MLT sector.

June 19, 2017 shows a typical quiet day. There is an isolated morningside event near 0200 MLT; however, the remainder of the day shows small passive energy deposition and Joule heating rates. In these cases, the Joule heating rate does not simply cease, but due to the neutral winds and the weak electric field, some energy transfer occurs, even if the magnitude is small. As shown on June 19, 2017, not every day has an associated Joule heating event, such as the event shown on June 18, 2017.

Fig. 5.5 shows statistical results for the whole month of June 2017. Approximately 400 h of data were used to generate Fig. 5.5. The data are subdivided by columns into geomagnetically quiet and active conditions, corresponding to $AE \leq 200$ and $AE > 200$, respectively. Fig. 5.5A and B shows the occurrence rates for the passive energy deposition rate and the Joule heating rate, respectively. The occurrence rate is a percentage relative to the total number of hours for June 2017 per altitude and is represented as a discrete, normalized color bar. We see that for the active conditions, the occurrence rates show the evening and dawnside enhancements, while for quiet conditions, there is an eveningside enhancement that peaks at 2100 MLT. For active conditions, the eveningside enhancement is also broader in MLT and extends to lower altitudes relative to the quiet times.

Fig. 5.5C–E shows the median passive energy deposition rate, Joule heating rate, and the mechanical energy transfer rate for quiet and active conditions as columns, respectively. The passive energy deposition rate and the Joule heating rate are plotted on a logarithmic scale to show the full range. For quiet conditions, there is an enhancement in the passive energy transfer rate associated with the electric fields near ~2100 MLT, which correlates with the larger occurrence rate. However, we see the median Joule heating rate, which includes the neutral wind contribution, has a contribution during all MLT sectors down to 110 km, and to slightly lower altitudes near ~2100 MLT.

During the active interval, the dusk and dawn enhancements in the median passive energy transfer rate and the Joule heating rate are clear. The mechanical energy transfer rate shown in Fig. 5.5E shows that during quiet times, at altitudes above 110 km, the sign of the mechanical energy transfer rate is negative which means that the neutral winds are

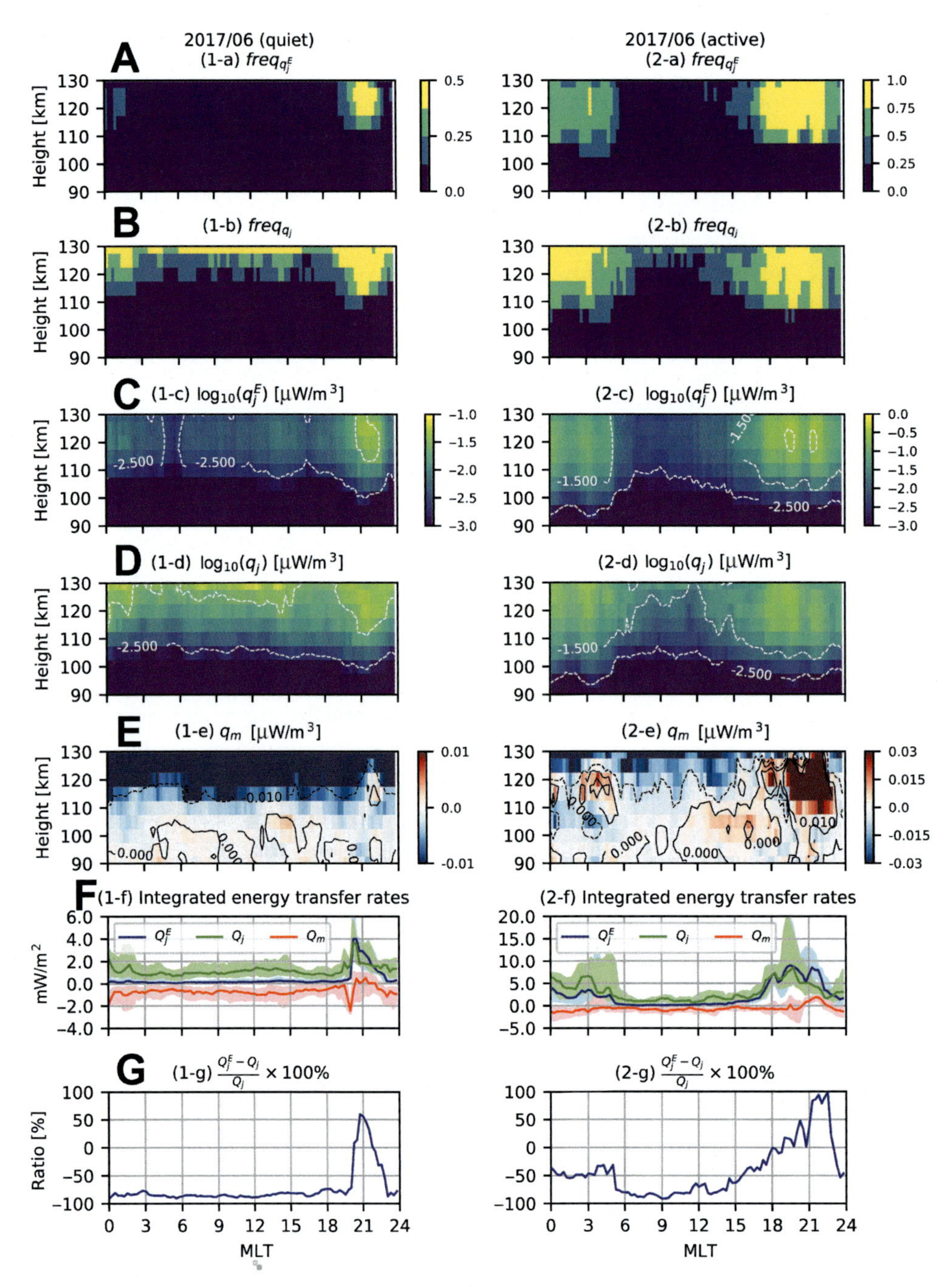

Fig. 5.5 Median results are presented for the month of June 2017. Rows A and B show the occurrence rate for the passive energy deposition rate and the Joule heating rate, respectively. Rows C, D, and E, are the median altitude-resolved passive energy deposition rate, Joule heating rate, and mechanical energy transfer rate. Row F shows the median integrated passive energy deposition rate, Joule heating rate, and mechanical energy transfer rate as *blue*, *green*, and *red*, respectively. The *shaded regions* correspond to the 25% and 75% quartiles. Row G shows the percent difference between the integrated passive energy deposition rate relative to the Joule heating rate. More details can be found in the text.

doing the work on the plasma. For active conditions, especially between ~1800 and 2200 MLT, at altitudes above 110 km, the work is being done by the plasma onto the neutral wind, which is acting as an energy sink.

Fig. 5.5F shows the median integrated passive energy deposition rate, Joule heating rate, and mechanical energy transfer rate as blue, green, and red, respectively. The shaded region corresponds to the 25% and 75% quartiles. For quiet times, this illustrates that the neutral winds are responsible for the Joule heating rate for most MLT sectors, except for an interval near ~2100 MLT where the electric field does provide some contribution to the Joule heating rate. The mechanical energy transfer term, as shown as the red line, is negative indicating that the winds are doing work and thus acting as an energy source. The median integrated Joule heating rate provided by the winds is of the order of 1–2 mW/m^2 during quiet intervals. During active conditions, the median integrated Joule heating rate and the integrated passive energy deposition rate are approximately equal in the evening sector, while in the morning sector the median Joule heating rate is larger than the passive energy deposition rate. The median Joule heating rate and passive energy transfer rate are a factor of ~2–5 larger during active intervals relative to quiet times. Also, during active intervals, the mechanical energy transfer rate becomes positive for a small interval near 2100 MLT, which indicates that the winds are acting as an energy sink.

Fig. 5.5G shows the percent difference between the integrated passive energy deposition rate and the Joule heating rate. Again, for interpretation, if the ratio is −100% that indicates that the neutral winds are the only factor contributing to the integrated Joule heating rate. In this case, we see that for quiet conditions, the neutrals are the primary energy dissipation mechanism, with the electric field contributing a very small magnitude, expect near ~2100 MLT when the electric field produces an appreciable contribution to the energy deposition. For the active interval, the neutral winds are the dominant in the morning sector from 0000 to 0500 MLT, while in the evening sector the integrated Joule heating rate and the integrated passive energy deposition rate are approximately equal from 1500 to 2400 MLT, except near 2100 MLT. Near 2100 MLT, the energy deposition is dominated by the ionospheric electric field, as indicated by the ~100% enhancement in the passive energy deposition rate.

5.6 Modeling approaches to energy budget estimates

Energy is an important physical quantity that characterizes general state of a physical system. It is one of the integrals of motion of the IT system (Schunk and Nagy, 2009). Changes in energy budget (input and dissipation) can provide crucial insights into interaction with neighboring regions. In our case, energy change in the IT system may indicate coupling with the heliosphere, including the solar wind, magnetosphere, and middle atmosphere. Analysis of energy partitioning, for example, auroral heating or nitric oxide (NO)-based thermospheric cooling, helps to separate and study specific pathways of such coupling. This section focuses on modeling of dissipation (mostly Joule heating) in the IT. It was shown that Joule heating is one of the main routes of energy deposition in the IT system, accounting for ~50%–70% of energy dissipation during geomagnetic storms (Knipp et al., 2004; Turner et al., 2009). Several approaches to estimation of energy-related parameters using empirical modeling and first-principles models will be briefly overviewed. Impacts of temporal and spatial resolutions in observations and models on electromagnetic energy budget estimation will be discussed. Uncertainties in the definitions of Joule heating across models will be pointed out. Challenges and future research directions will be outlined.

Definitions and the key observation-based approaches to Joule heating and Poynting flux estimations were discussed in the previous sections. Observational and modeling sources for Joule heating and Poynting flux that we will overview are summarized in Table 5.4. Here, we distinguish between empirical models, first-principles models, and measurements from ground and satellite-based instruments. JH_N denotes the height-integrated total Joule heating estimated over the Northern hemisphere (Knipp et al., 2005) and W05 refers to the Weimer05 model (Weimer, 2005). We will describe different modeling approaches below in some detail starting with empirical models.

5.6.1 Empirical models of Joule heating and their applications

Empirical models are derived from statistical analysis of observations performed over a certain period of time. Physical parameter of interest is estimated based on a simple formula (often using a polynomial regression) and is dependent on a number of parameters, for example, time, geomagnetic activity index. Empirical formulae or models are very useful to evaluate or predict typical values of physical parameters that are difficult to measure globally or during a specific time interval. These models are often utilized to define boundary conditions for the first-principles IT models. For instance, the Weimer05 model (W05) (Weimer, 2005) is widely used to specify electric field potential in the high-latitude ionosphere. The first empirical model of the Joule heating in the Northern hemisphere (Knipp et al., 2005) is parametrized by Polar Cap index (PC) and Dst index and is

Table 5.4 Sources for PF and JH estimations.

Energy channel	Empirical models	First-principles models	Measurements and data assimilation
Joule heating	JH_N	GITM	AMIE+
	W05	TIEGCM	Radars, ISRs
	McHarg et al. (2005)		Sounding rockets
Poynting flux	Weimer (2005) and Cosgrove et al. (2014)	–	Satellite measurements
			Sounding rockets

comprised of four sets of coefficients for different seasons determined by the best fit to a Joule heating dataset constructed using AMIE. AMIE is the data assimilation technique that incorporates available measurements from AMPERE, ground magnetometers, DMSP/SSUSI, and SuperDARN radars (Richmond and Kamide, 1988; Lu et al., 1996; McHarg et al., 2005). McHarg et al. (2005) used AMIE results to derive an empirical model of Joule heating depending on the interplanetary field clock angle. Joule heating model W05 (Weimer, 2005) is parametrized by the solar wind inputs (density, velocity, and magnetic field components) and optional AL index (Weimer, 2005; Rastaetter et al., 2016). The CEJH empirical model was proposed by Zhang et al. (2005).

Community Coordinated Modeling Center (CCMC) hosts several empirical models that can be run instantly or on request (https://ccmc.gsfc.nasa.gov/models/models_at_glance.php). The drawback of using an empirical model is that the underlying dataset is constructed over a limited time interval (usually several years) and thus could be biased toward certain solar or geomagnetic activity conditions (a certain solar cycle phase or lack of large geomagnetic storms during the measurement interval). Measurements could have a limited spatial coverage, for example, do not cover all local times. For instance, W05 utilized the Dynamic Explorer 2 measurements of electric and magnetic field performed in a polar orbit in 1981–83. Moreover, by statistical nature of empirical modeling, a model may not be sufficiently accurate for detailed case studies. Then, direct observations or first-principle models are utilized. However, empirical models can be successfully applied for statistical analysis of overall energy budget and its partitioning at large spatial scales (e.g., Turner et al., 2009; Hajra et al., 2014; Verkhoglyadova et al., 2017).

5.6.2 Joule heating in first-principles models

First-principles models solve the 3D momentum, energy, and continuity equations and provide time series of main physical parameters (e.g., density, ion velocity) on a spatial grid for several neutral and ion species. Thermosphere Ionosphere Electrodynamics General Circulation Model (TIEGCM) (Richmond et al., 1992; Roble and Ridley, 1994) and Global Ionosphere Thermosphere Model (GITM) (Ridley et al., 2006) are examples of widely used global first-principles models of the Earth's ionosphere-thermosphere. There have been a number of studies on Joule heating estimations based on these two models (Lu et al., 1995, 1998, 2010; Thayer et al., 1995; Deng and Ridley, 2007; Deng et al., 2009, 2011; Zhu and Ridley, 2016; Verkhoglyadova et al., 2016, 2017) and we briefly introduce the models here. These models take solar wind parameters and F10.7 values (as the proxy for solar EUV radiation) as inputs. The altitude range of TIEGCM and GITM is from approximately 97 to 500 km, and from approximately 90 to 600 km, respectively. Horizontal resolution varies depending on version of a model and is typically about 2.5–5 degrees. These models utilize empirical models or assimilative models to determine particle precipitation fluxes in certain energy bands and electric field at the upper boundary as their external "drivers." Typically, particle precipitation is specified either by the Fuller-Rowell and Evans model (Fuller-Rowell and Evans, 1987) or Oval Variation, Assessment, Tracking, Intensity, and Online Nowcasting (OVATION) Prime model (Newell et al., 2009). The particle precipitation can be also determined from AMIE (Richmond and Kamide, 1988). The W05 model is often used as an upper boundary condition for electric field potential in the high-latitude ionosphere. TIEGCM incorporates diurnal and semidiurnal tidal driving at the lower model boundary through empirical models (e.g., Hagan and Forbes, 2002). Empirical models, such as Mass Spectrometer and Incoherent Scatter (MSIS) (Hedin, 1991) and Horizontal Wind Model (HWM) (Drob et al., 2008), set up the initial IT state. This brief description emphasizes importance of empirical models to initiate and drive the first-principles models.

GITM evaluates Joule heating through frictional heating due to relative drift between ions and neutrals (Eq. 5.23) and calculates it through contributions from several species (Ridley et al., 2006; Zhu and Ridley, 2016). TIEGCM defines Joule heating as Ohmic heating through electric field and local conductivity, for example, Eq. (5.17) (Cole, 1962; Lu et al., 1998). The same definition is used in AMIE-based evaluation of Joule heating. Reference-frame dependence of electric field and neutral wind impact (Thayer et al., 1995) are often not accounted for (e.g., see Fig. 5.3). These definitions are consistent under certain conditions, but they are not always met (see Thayer and Semeter, 2004 and the previous section for detailed discussion). Thus, these inconsistencies in definitions of Joule heating across first-principles models make it difficult to intercompare modeling results (Verkhoglyadova et al., 2016, 2017).

Joule heating estimates are sensitive to the high-altitude electric field specification. Comparison of such estimates based on W05 and several AMIE-based data assimilation procedures showed up to 2.5 times differences (Huang et al., 2012a). Accurate specification of altitude dependence of Joule heating has important physical consequences. It is shown that Joule heating peaks below 150 km in altitude (Huang et al., 2012b). However, a small energy deposition in F-layer around 300 km in altitude can have a major effect on neutral density (Deng et al., 2011).

Assuming the total energy balance in the IT system, one can suggest proxies for estimating difficult to obtain energy channels. Since the NO cooling is one of the dominant cooling mechanisms in the IT during geomagnetically disturbed conditions (Mlynczak et al., 2003; Lin et al., 2018) and Joule heating is often the dominant heating channel, there is a correlation between the NO cooling and Joule heating albeit with a time lag (Lu et al., 2010) that can be explored further. Potentially Joule heating estimates could be used to approximate total cooling in the IT system.

By definition, Joule heating depends on specification of ionospheric conductivity which in turn depends on particle precipitation inputs. Self-consistent estimation of high-latitude convective electric field and particle precipitation for model driving is an ongoing challenge. Modeling study of Sheng et al. (2020) showed that lack of consistency between the drivers strongly affects resultant Joule heating estimates. Current development of AMGeO (or Assimilative Mapping of Geospace Observations, https://amgeo.colorado.edu) aims to unravel interconnection of ionospheric electrodynamic parameters, including electrostatic potential and ionospheric conductances at multiscales using ground-based and satellite measurements. Custom temporal resolution in AMGeO is 5 min. However, the assimilative model can run at a higher temporal cadence.

Verkhoglyadova et al. (2016, 2017) performed comparative study of IT energy budget estimates using GITM, empirical models, and observations. Their analysis confirmed that Joule heating is controlled by external driving during geomagnetic storm intervals. The global integrated Joule heating power reached the value of ~1500 GW in AMIE estimates, while JH_N was tracking AMIE at lower values with the peak of ~700 GW. GITM estimates were lower. However, we would like to point out that these models utilized different Joule heating definitions, which make it difficult to cross-compare estimates.

5.7 Characterization of Joule heating across different spatial and temporal scales

Joule heating estimation depends on a theoretical definition used (see the previous section), electric field and plasma parameters inferred either from model or measurements and spatiotemporal scale of the energy deposition. One cause of the scale-dependent estimate of Joule heating is its dependence on spatiotemporal resolution of electric field. Typically, first-principles models are driven by large-scale and relatively steady electric fields (Richmond, 2010). It was first demonstrated in Codrescu et al. (1995) that variable component of electric field contributes to energy deposition. It is established that resolved variability and structuring of the electric field is an important factor in determining modeled ionospheric Joule heating, and can improve our understanding of the IT energy budget (Crowley and Hackert, 2001; Deng and Ridley, 2007; Matsuo and Richmond, 2008; Deng et al., 2009; Cosgrove and Codrescu, 2009; Cosgrove et al., 2009). To address field variability, one can include it in the first-principles model driver either through empirical model, for example, based on DE-2 measurements (Matsuo et al., 2003; Deng et al., 2009; Zhu et al., 2018), or through regional drivers derived from measurements and blended with large-scale electric field forcing, for example, using ISR measurements in a multibeam mode (Ozturk et al., 2020), or SuperDARN measurements (Liuzzo et al., 2015). Chapter 1.3 provides detailed discussion of high-latitude electric field variability and its impact on Joule heating.

Another cause of dependence of Joule heating and Poynting flux estimates on a scale is in underlying physical mechanisms of energy input and dissipation at different scales. While DC processes at the interface between magnetosphere and high-latitude IT are intrinsically large-scale and quasistatic, AC processes are dynamic, mesoscale and can manifest themselves as Alfvén waves (Lotko, 2004, 2007; Lysak and Song, 2006). For instance, Lyons et al. (2016) showed importance of dynamic mesoscale flow structures in storm-time IT dynamics. Data analysis by McGranaghan et al. (2017) identified numerous mesoscale (at 150–250 km in the horizontal scale) high-latitude FAC systems with spatial distribution different from that of large-scale FACs. Pakhotin et al. (2018, 2020) analyzed Swarm data in the high-latitude ionosphere and showed a continuum of electromagnetic-field fluctuations with frequencies up to 8 Hz. Their analysis of a FAC system crossing revealed evidence for large-amplitude magnetic structures at 10 s timescales. Multiinstrument analysis of 530 Swarm auroral crossing events (Wu et al., 2020) demonstrated that Alfvénic fluctuations occur within FAC boundaries and downward current sheets over a wide range of locations and conditions. Observations by Swarm and the enhanced Polar Outflow Probe (e-POP) (Miles et al., 2018) indicated that Alfvén waves (in the frequency range from 1 to 10 Hz) are important for discrete auroral arc dynamics. Analysis of Cluster and FAST observations showed increased IT heating in the daytime auroral oval in connection with Alfvén waves (Chaston et al., 2005). FAST data analysis (Hatch et al., 2016) demonstrated that the downward Poynting flux carried by Alfvén waves at and above auroral latitudes increases by a factor of 4–5 during geomagnetic storms. Detailed overview of Alfvénic Poynting vector estimations from satellite measurements is presented in Chapter 5.1.4 and Table 5.3.

There are yet no strict physics-based definitions of characteristic scale ranges based on the IT physics. These scale ranges are likely different in reference to different parameters and processes (particle fluxes, currents and plasma convection, fields). Brief summary of observational and modeling capabilities at different spatial and temporal scales (based on current understanding) is shown in Table 5.5. Joule heating estimates at spatial large and mesoscales can be provided by all three sources, direct measurements, empirical models based on measurements from satellite constellations, and high-resolution first-principles modeling. Small-scale spatial features can be captured in rocket experiments, but these are local measurements. Currently, there is a lack of understanding meso- and small-scale drivers and computing limitations to perform adequate modeling of small spatial structures. Clayton et al. (2021) is a successful application of high-resolution regional modeling to interpret rocket observations.

Situation is more complicated in the temporal domain. Modeling at temporal mesoscale can be performed for specific events, while campaigns can provide measurements to estimate Joule heating with necessary temporal resolution in selected regions. Empirical models can in principle capture temporal variability of electric and magnetic fields at mesoscale (e.g., Deng et al., 2009), but they best perform in statistical studies. Robust approaches for determining mesoscale drivers of first-principles models are an active field of research. Sounding rocket experiments provide important measurements of electric and magnetic fields including Alfvén wave processes at subsecond resolution. Among notable experiments were the Sounding of the Ion Energization Region: Resolving

Table 5.5 Sources for JH estimation across the scales.

Source	Spatial scales			Temporal scales		
	Large (>500 km)	Meso (150–500 km)	Small (<150 km)	Large (>15 min)	Meso (1–10 min)	Small (10s sec-1 min)
Measurements	X	X	X (Local)	X	X (Campaigns)	X
Empirical models	X	X		X	X (Statistical)	
First-principles Models	X	X	Drivers?	X	X (Event-based)	Dynamic Effects

Here, "data" stands for individual measurements or data assimilation.

Ambiguities (SIERRA) multiple-payload experiment (Klatt et al., 2005), JOULE II (Sangalli et al., 2009), Cascades-2 (Lundberg et al., 2012), and the Magnetosphere-Ionosphere Coupling in the Alfvén Resonator (MICA) (Lynch et al., 2015) and the Ionospheric Structuring: In Situ and Ground-based Low Altitude Studies (ISINGLASS) mission (Clayton et al., 2019). First-principles models that can reproduce dynamic estimates of energy deposition in the IT by Alfvén wave processes are under development.

5.8 Summary and outlook on challenges in Joule heating estimates

The observations of Joule heating at high latitudes have illustrated a few common features:

1. In magnetic local time, there is an dusk/evening and morning/dawn peak in the Joule heating, while there is a minimum in Joule heating near magnetic midnight. The ISR observations suggest that the dayside Joule heating is modest; however, more observations under the aurora oval on the dayside are required for this to be conclusive.
2. The peak altitude of the Joule heating occurs at a higher altitude relative to the peak of the Pedersen conductivity.
3. The neutral winds undoubtedly play an important role in the modulation of energy transfer in the high latitudes. However, there is some uncertainty regarding during which MLT sectors the neutral winds in fact enhance or reduce the overall Joule heating rate.

Energy transport and deposition at high latitudes are scale dependent. One of the most outstanding challenges is *understanding the spatial and temporal characteristics of energy deposition*. We need multipoint measurements that will ensure needed high- and mesoscale resolution of electric and magnetic field measurements that are not limited to a specific region (ISR) or a limited time interval (sounding rockets). A first step toward addressing this are more coordinated observational campaigns of ISRs and analysis of previously observed storms when multiple ISRs were making observations. The observations by Thayer (2000), Fujii et al. (1999), Aikio et al. (2012), and results within this chapter suggest some differences in the Joule heating rate as a function of magnetic latitude. Having simultaneous and nearly continuous observations from the ISRs can provide a perspective on where energy is flowing.

While the ISRs have provided a statistical perspective, ultimately they make observations in a ring in MLT and magnetic latitude, with very little knowledge of what is happening outside of that ring. Not denying critical importance of the above observational platforms, we would like to advocate for distributed measurements from a satellite platform, possibly a small satellite constellation (e.g., Verkhoglyadova et al., 2020). Several NASA missions are in development: Dione CubeSat Mission (Zesta et al., AGU Fall Meeting, SA24A-03, 2021) and Auroral Reconstruction CubeSwarm (ARCS, see Burleigh et al. (AGU Fall Meeting, SA24A-05, 2021)). A four-dimensional (4D) knowledge of energy distribution in the IT system is frontier topic within the larger community and one of the goals of flagship missions, such as the Geospace Dynamics Constellation (GDC) and Daedalus (Sarris et al., 2020).

The other challenge is to *understand the dissipation of energy between multiple scales*. While the ISR observations provide a view at one scale size, coherent scatter radar may provide observations at 1–10 m scale sizes. It is important to understand how much power is dissipated at those scale sizes and whether small-scale dissipation can actually account for more dissipation than has previously been predicted.

JH is indirectly inferred from measurements. Development and refining of *independent techniques to determine Joule heating* are needed for validation of both AMIE estimates and future modeling results. Different definitions of Joule heating result in a major uncertainty in the total budget estimations and complicate cross-comparisons. Discussion of benefits and limitations of the adopted methods can benefit the science community and will chart a path forward to more accurate evaluation of energy budget for the IT system.

Substantial progress has been achieved in understanding Joule heating dynamics in the IT system with first-principles models. However, further efforts are needed to *improve self-consistent specification of both electric field and particle precipitation drivers at multiscales* (see also Chapter 5.3). It has been convincingly demonstrated that better resolved high-latitude drivers affect estimation of IT energy budget. Moreover, physics-based definition of scale ranges for the IT system will benefit understanding of underlying energy deposition processes. Role of Alfvén waves in local heating is needed to be further quantified and modeled. Understanding energy budget of the coupled magnetosphere-IT system at the multiscales and improving its estimates are important for current space weather forecasting efforts.

Acknowledgments

The contributions by SRK and WZ were supported by the National Science Foundation AGS-1853408. This material is based on the work supported by the Poker Flat Incoherent Scatter Radar, which is a major facility funded by the National Science Foundation through cooperative agreement AGS-1840962 to SRI International. Monthly PFISR data files for the month of March 2010–June 2019 that were used in this chapter are archived at http://doi.org/10.5281/zenodo.3885547. Version v0.6.2.2020.07.01 was used for analysis in this chapter. Portions of OV research were performed at the Jet Propulsion Laboratory, California Institute of Technology, under a contract with NASA. DJK was partially supported by a grant from the Air Force Office of Scientific Research (AFOSR), Award No. FA9550-17-1-0258. DJK and LMK were also partially supported by AFOSR through a Multi-University Research Initiative (MURI) award FA9559-16-1-0364 led by the University of Texas Arlington. DMSP Poynting flux data processing is partially supported by NASA Award 80NSSC20K1784.

References

Aikio, A.T., Selkälä, A., 2009. Statistical properties of joule heating rate, electric field and conductances at high latitudes. Ann. Geophys. 27 (7), 2661–2673. https://doi.org/10.5194/angeo-27-2661-2009.

Aikio, A.T., Cai, L., Nygrén, T., 2012. Statistical distribution of height-integrated energy exchange rates in the ionosphere. J. Geophys. Res. Space Phys. 117 (A10). https://doi.org/10.1029/2012JA018078.

Angelopoulos, V., Chapman, J.A., Mozer, F.S., Scudder, J.D., Russell, C.T., Tsuruda, K., Mukai, T., Hughes, T.J., Yumoto, K., 2002. Plasma sheet electromagnetic power generation and its dissipation along auroral field lines. J. Geophys. Res. 107 (A8), 1181. https://doi.org/10.1029/2001JA900136.

Baddeley, L.J., Lorentzen, D.A., Partamies, N., Denig, M., Pilipenko, V.A., Oksavik, K., Chen, X., Zhang, Y., 2017. Equatorward propagating auroral arcs driven by ULF wave activity: multipoint ground- and space-based observations in the dusk sector auroral oval. J. Geophys. Res. Space Phys. 122. https://doi.org/10.1002/2016JA023427.

Banks, P., 1977. Observations of joule and particle heating in the auroral zone. J. Atmos. Terr. Phys. 39 (2), 179–193. https://doi.org/10.1016/0021-9169(77)90112-X.

Billett, D.D., Perry, G.W., Clausen, L.B.N., Archer, W.E., McWilliams, K.A., Haaland, S., et al., 2021. The relationship between large scale thermospheric density enhancements and the spatial distribution of Poynting flux. Journal of Geophysical Research: Space. Physics 126. https://doi.org/10.1029/2021JA029205. e2021JA029205.

Boström, R., 1964. A model of the auroral electrojects. J. Geophys. Res. 69, 4983–4999. https://doi.org/10.1029/JZ069i023p04983.

Brekke, A., Moen, J., 1993. Observations of high latitude ionospheric conductances. J. Atmos. Terr. Phys. 55, 1493–1512.

Brekke, A., Rino, C.L., 1978. High-resolution altitude profiles of the auroral zone energy dissipation due to ionospheric currents. J. Geophys. Res. Space Phys. 83, 2517–2524. https://doi.org/10.1029/JA083iA06p02517.

Brekke, A., Doupnik, J.R., Banks, P.M., 1973. A preliminary study of the neutral wind in the auroral E region. J. Geophys. Res. Space Phys. 78, 8235–8250. https://doi.org/10.1029/JA078i034p08235.

Brekke, A., Doupnik, J.R., Banks, P.M., 1974. Observations of neutral winds in the auroral E region during the magnetospheric storm of August 3–9, 1972. J. Geophys. Res. (1896–1977) 79 (16), 2448–2456. https://doi.org/10.1029/JA079i016p02448.

Brekke, A., Nozawa, S., Sparr, T., 1994. Studies of the E region neutral wind in the quiet auroral ionosphere. J. Geophys. Res. Space Phys. 99, 8801–8825. https://doi.org/10.1029/93JA03232.

Bristow, W.A., Hampton, D.L., Otto, A., 2016. High-spatial-resolution velocity measurements derived using local divergence-free fitting of SuperDARN observations. J. Geophys. Res. Space Phys. 121 (2), 1349–1361. https://doi.org/10.1002/2015JA021862.

Burchill, J.K., Clemmons, J.H., Knudsen, D.J., Larsen, M., Nicolls, M.J., Pfaff, R.F., Rowland, D., Sangalli, L., 2012. High-latitude E region ionosphere-thermosphere coupling: a comparative study using in situ and incoherent scatter radar observations. J. Geophys. Res. Space Phys. 117, A02301. https://doi.org/10.1029/2011JA017175.

Burke, W.J., Huang, C.Y., Weimer, D.R., Wise, J.O., Wilson, G.R., Lin, C.S., Marcos, F.A., 2010. Energy and power requirements of the global thermosphere during the magnetic storm of November 10, 2004. J. Atmos. Sol. Terr. Phys. 72, 309–318. https://doi.org/10.1016/j.jastp.2009.06.005.

Burke, W.J., Kilcommons, L.M., Hairston, M.R., 2017. Storm time coupling between the magnetosheath and the polar ionosphere. J. Geophys. Res. Space Phys. 122, 7541–7554. https://doi.org/10.1002/2017JA024101.

Cai, L., Aikio, A.T., Nygrén, T., 2013. Height-dependent energy exchange rates in the high-latitude E region ionosphere. J. Geophys. Res. Space Phys. 118 (11), 7369–7383. https://doi.org/10.1002/2013JA019195.

Chamberlain, J.W., 1961. Theory of auroral bombardment. Astrophys. J. 134, 401–424.

Chapman, S., 1931. Absorption and dissociative or ionising effects of monochromatic radiation in an atmosphere on a rotating earth. Proc. Phys. Soc. 43, 1047–1055.

Chaston, C.C., et al., 2005. Energy deposition by Alfvén waves into the dayside auroral oval: cluster and FAST observations. J. Geophys. Res. 110, A02211. https://doi.org/10.1029/2004JA010483.

Chaston, C.C., Carlson, C.W., Peria, W.J., Ergun, R.E., McFadden, J.P., 1999. FAST observations of inertial Alfven waves in the dayside aurora. Geophys. Res. Lett. https://doi.org/10.1029/1998GL900246.

Chaston, C.C., Genot, V., Bonnell, J.W., Carlson, C.W., McFadden, J.P., Ergun, R.E., Strangeway, R.J., Lund, E.J., Hwang, K.J., 2006. Ionospheric erosion by Alfvén waves. J. Geophys. Res. 111, A03206. https://doi.org/10.1029/2005JA011367.

Chaston, C.C., Salem, C., Bonnell, J.W., Carlson, C.W., Ergun, R.E., Strangeway, R.J., McFadden, J.P., 2008. The turbulent Alfvenic aurora. Phys. Rev. Lett. 100, 175003. https://doi.org/10.1103/PhysRevLett.100.175003.

Clayton, R., Lynch, K., Zettergren, M., Burleigh, M., Conde, M., Grubbs, G., et al., 2019. Two-dimensional maps of in situ ionospheric plasma flow data near auroral arcs using auroral imagery. J. Geophys. Res. Space Phys. 124, 3036–3056. https://doi.org/10.1029/2018JA026440.

Clayton, R., Burleigh, M., Lynch, K.A., Zettergren, M., et al., 2021. Examining the aururalionosphere in three dimensions using reconstructed 2D maps of auroral data to drive the 3D GEMINI model. J. Geophys. Res. Space Phys. e2021JA029749. https://doi.org/10.1029/2021JA029749.

Codrescu, M.V., Fuller-Rowell, T.J., Foster, J.C., 1995. On the importance of E-field variability for joule heating in the high-latitude thermosphere. Geophys. Res. Lett. 22, 2393–2396.

Cole, K.D., 1962. Joule heating of the upper atmosphere. Aust. J. Phys. 15, 223–235. https://doi.org/10.1071/PH620223.

Cole, K.D., 1975. Energy deposition in the thermosphere caused by the solar wind. J. Atmos. Terr. Phys. 37, 939. https://doi.org/10.1016/0021-9169(75)90008-2.

Conde, M., Smith, R.W., 1995. Mapping thermospheric winds in the auroral zone. Geophys. Res. Lett. 22 (22), 3019–3022. https://doi.org/10.1029/95GL02437.

Conde, M., Smith, R.W., 1998. Spatial structure in the thermospheric horizontal wind above Poker Flat, Alaska, during solar minimum. J. Geophys. Res. Space Phys. 103 (A5), 9449–9471. https://doi.org/10.1029/97JA03331.

Cosgrove, R.B., Codrescu, M., 2009. Electric field variability and model uncertainty: a classification of source terms in estimating the squared electric field from an electric field model. J. Geophys. Res. 114, A06301. https://doi.org/10.1029/2008JA013929.

Cosgrove, R.B., Lu, G., Bahcivan, H., Matsuo, T., Heinselman, C.J., McCready, M.A., 2009. Comparison of AMIE-modeled and Sondrestrom-measured Joule heating: a study in model resolution and electric field-conductivity correlation. J. Geophys. Res. 114, A04316. https://doi.org/10.1029/2008JA013508.

Cosgrove, R.B., et al., 2014. Empirical model of Poynting flux derived from FAST data and a cusp signature. J. Geophys. Res. 119, 411–430. https://doi.org/10.1002/2013JA019105.

Cowley, S.W.H., 1991. Acceleration and heating of space plasmas: concepts. Ann. Geophys. 9, 176–187.

Crowley, G., Hackert, C.L., 2001. Quantification of high latitude electric field. Geophys. Res. Lett. 28 (14), 2783–2786.

Deng, Y., Ridley, A.J., 2007. Possible reasons for underestimating Joule heating in global models: E field variability, spatial resolution and vertical velocity. J. Geophys. Res. 112, A09308. https://doi.org/10.1029/2006JA012006.

Deng, W., Killeen, T.L., Burns, A.G., Johnson, R.M., Emery, B.A., Roble, R.G., Winningham, J.D., Gary, J.B., 1995. One-dimensional hybrid satellite track model for the Dynamics Explorer 2 (DE 2) satellite. J. Geophys. Res. 100 (A2), 1611–1624. https://doi.org/10.1029/94JA02075.

Deng, Y., Maute, A., Richmond, A.D., Roble, R.G., 2009. Impact of electric field variability on Joule heating and thermospheric temperature and density. Geophys. Res. Lett. 36, L08105. https://doi.org/10.1029/2008GL036916.

Deng, Y., Huang, Y., Lei, J., Ridley, A.J., Lopez, R., Thayer, J., 2011. Energy input into the upper atmosphere associated with high speed solar wind streams in 2005. J. Geophys. Res. 116, A05303. https://doi.org/10.1029/2010JA016201.

Dimant, Y.S., Oppenheim, M.M., 2011a. Magnetosphere-ionosphere coupling through E region turbulence: 1. Energy budget. J. Geophys. Res. Space Phys. 116 (A9). https://doi.org/10.1029/2011JA016648.

Dimant, Y.S., Oppenheim, M.M., 2011b. Magnetosphere-ionosphere coupling through E region turbulence: 2. Anomalous conductivities and frictional heating. J. Geophys. Res. Space Phys. 116 (A9). https://doi.org/10.1029/2011JA016649.

Dombeck, J., Cattell, C., Wygant, J.R., Keiling, A., Scudder, J., 2005. Alfvén waves and Poynting flux observed simultaneously by Polar and FAST in the plasma sheet boundary layer. J. Geophys. Res. 110, A12S90. https://doi.org/10.1029/2005JA011269.

Dougherty, J.P., Farley, D.T., 1960. A theory of incoherent scattering of radio waves by a plasma. Proc. R. Soc. Lond. Ser. A 259, 79–99. https://doi.org/10.1098/rspa.1960.0212.
Drob, D.P., et al., 2008. An empirical model of the Earth's horizontal wind fields: HWM07. J. Geophys. Res. 113, A12304. https://doi.org/10.1029/2008JA013668.
Evans, J.V., 1969. Theory and practice of ionosphere study by Thomson scatter radar. Proc. IEEE 57 (4), 496–530.
Evans, D.S., Maynard, N.C., Trøim, J., Jacobsen, T., Egeland, A., 1977. Auroral vector electric field and particle comparisons, 2, electrodynamics of an arc. J. Geophys. Res. (1896–1977) 82 (16), 2235–2249. https://doi.org/10.1029/JA082i016p02235.
Farley, D.T., Dougherty, J.P., Barron, D.W., 1961. A theory of incoherent scattering of radio waves by a plasma II. Scattering in a magnetic field. Proc. R. Soc. Lond. Ser. A 263, 238–258. https://doi.org/10.1098/rspa.1961.0158.
Forsyth, C., Rae, I.J., Mann, I.R., Pakhotin, I.P., 2017. Identifying intervals of temporally invariant field-aligned currents from Swarm: assessing the validity of single-spacecraft methods. J. Geophys. Res. Space Phys. 122, 3411–3419. https://doi.org/10.1002/2016JA023708.
Foster, J.C., St.-Maurice, J.-P., Abreu, V.J., 1983. Joule heating at high latitudes. J. Geophys. Res. 88 (A6), 4885–4897. https://doi.org/10.1029/JA088iA06p04885.
Fujii, R., Nozawa, S., Matuura, N., Brekke, A., 1998. Study on neutral wind contribution to the electrodynamics in the polar ionosphere using EISCAT CP-1 data. J. Geophys. Res. Space Phys. 103, 14731–14740. https://doi.org/10.1029/97JA03687.
Fujii, R., Nozawa, S., Buchert, S.C., Brekke, A., 1999. Statistical characteristics of electromagnetic energy transfer between the magnetosphere, the ionosphere, and the thermosphere. J. Geophys. Res. Space Phys. 104, 2357–2366. https://doi.org/10.1029/98JA02750.
Fuller-Rowell, T., Evans, D., 1987. Height-integrated Pedersen and Hall conductivity patterns inferred from the TIROS-NOAA satellite data. J. Geophys. Res. 92, 7606. https://doi.org/10.1029/JA092iA07p07606.
Gary, J.B., Heelis, R.A., Hanson, W.B., Slavin, J.A., 1994. Field aligned Poynting flux observations in the high-latitude ionosphere. J. Geophys. Res. 99 (A6), 11417–11427. https://doi.org/10.1029/93JA03167.
Gary, J.B., Heelis, R.A., Thayer, J.P., 1995. Summary of field-aligned Poynting flux observations from DE2. Geophys. Res. Lett. 22 (14), 1861–1864. https://doi.org/10.1029/95GL00570.
Germany, G.A., Torr, D.G., Richards, P.G., Torr, M.R., John, S., 1994. Determination of ionospheric conductivities from FUV auroral emissions. J. Geophys. Res. Space Phys. 99 (23). https://doi.org/10.1029/94JA02038.
Golovchanskaya, I.V., Maltsev, Y.P., 2004. On the direction of the Poynting flux related to the mesoscale electromagnetic turbulence at high latitudes. J. Geophys. Res. 109, A10203. https://doi.org/10.1029/2004JA010432.
Hagan, M.E., Forbes, J.M., 2002. Migrating and nonmigrating diurnal tides in the middle and upper atmosphere excited by tropospheric latent heat release. 107 (D24), 4754. https://doi.org/10.1029/2001JD001236.
Hajra, R., Echer, E., Tsurutani, B.T., Gonzalez, W.D., 2014. Solar wind-magnetosphere energy coupling efficiency and partitioning: HILDCAAs and preceding CIR storms during solar cycle 23. J. Geophys. Res. Space Phys. 119, 2675–2690. https://doi.org/10.1002/2013JA019646.
Hartinger, M., Angelopoulos, V., Moldwin, M.B., Glassmeier, K.-H., Nishimura, Y., 2011. Global energy transfer during a magnetospheric field line resonance. Geophys. Res. Lett. 38, L12101. https://doi.org/10.1029/2011GL047846.
Hartinger, M.D., Moldwin, M.B., Zou, S., Bonnell, J.W., Angelopoulos, V., 2015. ULF wave electromagnetic energy flux into the ionosphere: Joule heating implications. J. Geophys. Res. Space Phys. 120, 494–510. https://doi.org/10.1002/2014JA020129.
Hatch, S.M., Chaston, C.C., Labelle, J., 2016. Alfvén wave-driven ionospheric mass outflow and electron precipitation during storms. J. Geophys. Res. 121, 7828–7846. https://doi.org/10.1002/2016JA022805.
Hatch, S.M., LaBelle, J., Chaston, C.C., 2018. Storm phase-partitioned rates and budgets of global Alfvénic energy deposition, electron precipitation, and ion outflow. J. Atmos. Sol. Terr. Phys. 167, 1–12. https://doi.org/10.1016/j.jastp.2017.08.009.

Hedin, A., 1991. Extension of the MSIS thermosphere model into the middle and lower atmosphere. J. Geophys. Res. 96, 1159–1172. https://doi.org/10.1029/90JA02125.

Heinselman, C.J., Nicolls, M.J., 2008. A Bayesian approach to electric field and E-region neutral wind estimation with the Poker Flat Advanced Modular Incoherent Scatter Radar. Radio Sci. 43, RS5013. https://doi.org/10.1029/2007RS003805.

Horvath, I., Lovell, B.C., 2018a. Polar cap energy deposition events during the 5–6 August 2011 magnetic storm. J. Geophys. Res. Space Phys. 123. https://doi.org/10.1002/2017JA025102.

Horvath, I., Lovell, B.C., 2018b. Investigating high-latitude energy deposition events occurring during the 17 January 2005 geomagnetic storm. J. Geophys. Res. Space Phys. 123, 6760–6775. https://doi.org/10.1029/2018JA025465.

Horvath, I., Lovell, B.C., 2018c. Polar ion temperature variations during the 22 January 2012 magnetic storm. J. Geophys. Res. Space Phys. 123, 7806–7824. https://doi.org/10.1029/2018JA025727.

Hosokawa, K., Ogawa, Y., 2015. Ionospheric variation during pulsating aurora. J. Geophys. Res. Space Phys. 120 (7), 5943–5957. https://doi.org/10.1002/2015JA021401.

Huang, C.Y., Burke, W.J., 2004. Transient sheets of field-aligned current observed by DMSP during the main phase of a magnetic superstorm. J. Geophys. Res. 109, A06303. https://doi.org/10.1029/2003JA010067.

Huang, Y., Deng, Y., Lei, J., Ridley, A., Lopez, R., Allen, R.C., MacButler, B., 2012a. Comparison of joule heating associated with highspeed solar wind between different models and observations. J. Atmos. Sol. Terr. Phys. 75–76, 5–14. https://doi.org/10.1016/j.jastp.2011.05.013.

Huang, Y., Richmond, A.D., Deng, Y., Roble, R., 2012b. Height distribution of Joule heating and its influence on the thermosphere. J. Geophys. Res. 117, A08334. https://doi.org/10.1029/2012JA017885.

Huang, C.Y., Su, Y.-J., Sutton, E.K., Weimer, D.R., Davidson, R.L., 2014. Energy coupling during the August 2011 magnetic storm. J. Geophys. Res. Space Phys. 119, 1219–1232. https://doi.org/10.1002/2013JA019297.

Huang, C.Y., Huang, Y., Su, Y.-J., Sutton, E.K., Hairston, M.R., Coley, W.R., 2016. Ionosphere-thermosphere (IT) response to solar wind forcing during magnetic storms. J. Space Weather Space Clim. 6, A4. https://doi.org/10.105/swsc/2015041.

Huang, C.Y., Huang, Y., Su, Y.-J., Hairston, M.R., Sotirelis, T., 2017. DMSP observation of high latitude Poynting flux during magnetic storm. J. Atmos. Sol. Terr. Phys. 164, 294–307. https://doi.org/10.1016/j.jastp.2017.09.005.

Hurd, L.D., Larsen, M.F., 2016. Small-scale fluctuations in barium drifts at high latitudes and associated Joule heating effects. J. Geophys. Res. Space Phys. 121, 779–789. https://doi.org/10.1002/2015JA021868.

Huuskonen, A., Lehtinen, M.S., Pirttilä, J., 1996. Fractional lags in alternating codes: improving incoherent scatter measurements by using lag estimates at noninteger multiples of baud length. Radio Sci. 31 (2), 245–261. https://doi.org/10.1029/95RS03157.

Hysell, D.L., 2015. The Radar Aurora (Chapter 14). In: American Geophysical Union (AGU), pp. 191–209. https://doi.org/10.1002/9781118978719.ch14.

Jackson, J.D., 1998. Classical Electrodynamics, third ed., Wiley and Sons, New York.

Janhunen, P., Olsson, A., Tsyganenko, N.A., Russell, C.T., Laakso, H., Blomberg, L.G., 2005. Statistics of a parallel Poynting vector in the auroral zone as a function of altitude using Polar EFI and MFE data and Astrid-2 EMMA data. Ann. Geophys. 23, 1797–1806.

Johansson, T., Figueiredo, S., Karlsson, T., Marklund, G., Fazakerley, A., Buchert, S., Lindqvist, P.-A., Nilsson, H., 2004. Intense high-altitude auroral electric fields—temporal and spatial characteristics. Ann. Geophys. 22, 2485–2495. https://doi.org/10.5194/angeo-22-2485-2004.

Johnson, R.M., 1990. Lower-thermospheric neutral winds at high latitude determined from incoherent scatter measurements—a review of techniques and observations. Adv. Space Res. 10, 261–275. https://doi.org/10.1016/0273-1177(90)90259-3.

Jones, S., Lessard, M., Fernandes, P., Lummerzheim, D., Semeter, J., Heinselman, C., Lynch, K., Michell, R., Kintner, P., Stenbaek-Nielsen, H., Asamura, K., 2009. PFISR and ROPA observations of pulsating aurora. J. Atmos. Sol. Terr. Phys. 71 (6), 708–716. https://doi.org/10.1016/j.jastp.2008.10.004.

Kaeppler, S.R., Nicolls, M.J., Strømme, A., Kletzing, C.A., Bounds, S.R., 2014. Observations in the E region ionosphere of kappa distribution functions associated with precipitating auroral electrons and discrete aurorae. J. Geophys. Res. Space Phys. 119 (12), 10164–10183. https://doi.org/10.1002/2014JA020356.

Kaeppler, S.R., Hampton, D.L., Nicolls, M.J., Strømme, A., Solomon, S.C., Hecht, J.H., Conde, M.G., 2015. An investigation comparing ground-based techniques that quantify auroral electron flux and conductance. J. Geophys. Res. Space Phys. 120, 9038–9056. https://doi.org/10.1002/2015JA021396.

Keiling, A., 2009. Alfvén waves and their roles in the dynamics of the Earth's Magnetotail: a review. Space Sci. Rev. 142, 73–156. https://doi.org/10.1007/s11214-008-9463-8.

Keiling, A., Wygant, J.R., Cattell, C., Peria, W., Parks, G., Temerin, M., Mozer, F.S., Russell, C.T., Kletzing, C.A., 2002. Correlation of Alfvén wave Poynting flux in the plasma sheet at 4–7 RE with ionospheric electron energy flux. J. Geophys. Res. 107 (A7), 1132. https://doi.org/10.1029/2001JA900140.

Keiling, A., Wygant, J.R., Catell, C.A., Mozer, F.S., Russell, C.T., 2003. The global morphology of wave Poynting flux: powering the aurora. Science 299, 383–386.

Keiling, A., Thaller, S., Wygant, J., Dombeck, J., 2019. Assessing the global Alfvén wave power flow into and out of the auroral acceleration region during geomagnetic storms. Sci. Adv. 5 (6), eaav8411. https://doi.org/10.1126/sciadv.aav8411.

Kelley, M., 2009. The Earth's Ionosphere: Plasma Physics and Electrodynamics. International Geophysics, Elsevier Science.

Kelley, M.C., Knudsen, D.J., Vickrey, J.F., 1991. Poynting flux measurements on a satellite: a diagnostic tool for space research. J. Geophys. Res. 96 (A1), 201–207. https://doi.org/10.1029/90JA01837.

Kelly, J.D., Heinselman, C.J., 2009. Initial results from Poker Flat Incoherent Scatter Radar (PFISR). J. Atmos. Sol. Terr. Phys. 71 (6), 635. https://doi.org/10.1016/j.jastp.2009.01.009.

Kiene, A., Bristow, W.A., Conde, M.G., Hampton, D.L., 2019. High-resolution local measurements of *F* region ion temperatures and Joule heating rates using SuperDARN and ground-based optics. J. Geophys. Res. Space Phys. 124 (1), 557–572. https://doi.org/10.1029/2018JA025997.

Klatt, E.M., Kintner, P.M., Seyler, C.E., Liu, K., MacDonald, E.A., Lynch, K.A., 2005. SIERRA observations of Alfvénic processes in the topside auroral ionosphere. J. Geophys. Res. 110, A10S12. https://doi.org/10.1029/2004JA010883.

Knipp, D., Kilcommons, L., Hairston, M., Coley, W.R., 2021. Hemispheric asymmetries in Poynting flux derived from DMSP spacecraft. Geophys. Res. Lett. 48. https://doi.org/10.1029/2021GL094781. e2021GL094781.

Knipp, D.J., Tobiska, W.K., Emery, B.A., 2004. Direct and indirect thermospheric heating sources for the solar cycle 21–23. Sol. Phys. 224, 495–505. https://doi.org/10.1007/s11207-005-6393-4.

Knipp, D.J., Welliver, T., McHarg, M.G., Chun, F.K., Tobiska, W.K., Evans, D., 2005. Climatology of extreme upper atmospheric heating events. Adv. Space Res. 36, 2506–2510.

Knipp, D., Eriksson, S., Kilcommons, L., Crowley, G., Lei, J., Hairston, M., Drake, K., 2011. Extreme Poynting flux in the dayside thermosphere: examples and statistics. Geophys. Res. Lett. 38, L16102. https://doi.org/10.1029/2011GL048302.

Knudsen, D.J., 1990. Alfven Waves and Static Fields in Magnetosphere/Ionosphere Coupling: In-Situ Measurements and a Numerical Model (Ph.D. thesis). Cornell University, Ithaca, NY. https://apps.dtic.mil/sti/pdfs/ADA356871.pdf.

Korth, H., et al., 2005. High-latitude electromagnetic and particle energy flux during an event with sustained strongly northward IMF. Ann. Geophys. 23, 1295–1310. https://doi.org/10.5194/angeo-23-1295-2005.

Kosch, M.J., Hagfors, T., Schlegel, K., 1998. Extrapolating EISCAT Pedersen conductances to other parts of the sky using ground-based TV auroral images. Ann. Geophys. 16 (5), 583–588. https://doi.org/10.1007/s00585-998-0583-y.

Kudeki, E., Milla, M.A., 2011. Incoherent scatter spectral theories—part I: a general framework and results for small magnetic aspect angles. IEEE Trans. Geosci. Remote Sens. 49 (1), 315–328.

Lam, M.M., Freeman, M.P., Jackman, C.M., Rae, I.J., Kalmoni, N.M.E., Sandhu, J.K., Forsyth, C., 2019. How well can we estimate Pedersen conductance from the Themis white-light all-sky cameras? J. Geophys. Res. Space Phys. 124 (4), 2920–2934. https://doi.org/10.1029/2018JA026067.

Larsen, M.F., 2002. Winds and shears in the mesosphere and lower thermosphere: results from four decades of chemical release wind measurements. J. Geophys. Res. Space Phys. 107, 1215. https://doi.org/10.1029/2001JA000218.

Larsen, M., Mikkelsen, I.S., Meriwether, J.W., Niciejewski, R., Vickery, J., 1989. Simultaneous observations of neutral winds and electric fields at spaced locations in the dawn auroral oval. J. Geophys. Res. Space Phys. 94 (A12), 17235–17243. https://doi.org/10.1029/JA094iA12p17235.

Lehtinen, M.S., Haggstrom, I., 1987. A new modulation principle for incoherent scatter measurements. Radio Sci. 22, 625–634. https://doi.org/10.1029/RS022i004p00625.

Li, W., Knipp, D., Lei, J., Raeder, J., 2011. The relation between dayside local Poynting flux enhancement and cusp reconnection. J. Geophys. Res. https://doi.org/10.1029/2011JA016566.

Lileo, S., Marklund, G.T., Karlsson, T., Johansson, T., Lindqvist, P.-A., Marchaudon, A., Fazakerley, A., Mouikis, C., Kistler, L.M., 2008. Magnetosphere-ionosphere coupling during periods of extended high auroral activity: a case study. Ann. Geophys. 26, 583–591. https://doi.org/10.5194/angeo-26-583-2008.

Lin, C.Y., Deng, Y., Venkataramani, K., Yonker, J., Bailey, S.M., 2018. Comparison of the thermospheric nitric oxide emission observations and the GITM simulations: ensitivity to solar and geomagnetic activities. J. Geophys. Res.: Space Phys. 123. https://doi.org/10.1029/2018JA025310. 10,239–10,253.

Liuzzo, L.R., Ridley, A.J., Perlongo, N.J., Mitchell, E.J., Conde, M., Hampton, D.L., Bristow, W.A., Nicolls, M.J., 2015. High-latitude ionospheric drivers and their effects on wind patterns in the thermosphere. J. Geophys. Res. Space Phys. 120, 715–735. https://doi.org/10.1002/2014JA020553.

Lotko, W., 2004. Inductive magnetosphere-ionosphere coupling. J. Atmos. Sol. Terr. Phys. 66, 1443–1456.

Lotko, W., 2007. The magnetosphere-ionosphere system from the perspective of plasma circulation: a tutorial. J. Atmos. Sol. Terr. Phys. 69, 191–211.

Lotko, W., Zhang, B., 2018. Alfvénic heating in the cusp ionosphere-thermosphere. J. Geophys. Res. Space Phys. 123, 10368–10383. https://doi.org/10.1029/2018JA025990.

Lu, Y., Deng, Y., Sheng, C., Kilcommons, L., Knipp, D.J., 2018. Poynting flux in the dayside polar cap boundary regions from DMSP F15 satellite measurements. J. Geophys. Res.: Space Phys. 123, 6948–6956. https://doi.org/10.1029/2018JA025309.

Lu, G., Richmond, A.D., Emery, B.A., Roble, R.G., 1995. Magnetosphere-ionosphere-thermosphere coupling: effect of neutral winds on energy transfer and field-aligned current. J. Geophys. Res. 100 (A10), 19643. https://doi.org/10.1029/95JA00766.

Lu, G., et al., 1996. High-latitude ionospheric electrodynamics as determined by the assimilative mapping of ionospheric electrodynamics procedure for the conjunctive SUNDIAL/ATLAS 1/GEM period of March 28–29. J. Geophys. Res. 101 (A12), 26697–26718. https://doi.org/10.1029/96JA00513.

Lu, G., et al., 1998. Global energy deposition during the January 1997 magnetic cloud event. J. Geophys. Res. 103 (A6), 11685–11694. https://doi.org/10.1029/98JA00897.

Lu, G., Mlynczak, M.G., Hunt, L.A., Woods, T.N., Roble, R.G., 2010. On the relationship of joule heating and nitric oxide radiative cooling in the thermosphere. J. Geophys. Res. 115, A05306. https://doi.org/10.1029/2009JA014662.

Luhr, H., Park, J., Gjerloev, J.W., Rauberg, J., Michaelis, I., Merayo, J.M.G., Brauer, P., 2015. Field-aligned currents' scale analysis performed with the Swarm constellation. Geophys. Res. Lett. 42, 1–8. https://doi.org/10.1002/2014GL062453.

Lummerzheim, D., Rees, M.H., Craven, J.D., Frank, L.A., 1991. Ionospheric conductances derived from DE-1 auroral images. J. Atmos. Terr. Phys. 53, 281–292. https://doi.org/10.1016/0021-9169(91)90112-K.

Lundberg, E.T., Kintner, P.M., Lynch, K.A., Mella, M.R., 2012. Multipayload measurement of transverse velocity shears in the topside ionosphere. Geophys. Res. Lett. 39, L01107. https://doi.org/10.1029/2011GL050018.

Lynch, K.A., et al., 2015. MICA sounding rocket observations of conductivity-gradient-generated auroral ionospheric responses: small-scale structure with large-scale drivers. J. Geophys. Res. Space Phys. 120, 9661–9682. https://doi.org/10.1002/2014JA020860.

Lyons, L.R., Nishimura, Y., Zou, Y., 2016. Unsolved problems: mesoscale polar cap flow channels' structure, propagation, and effects on space weather disturbances. J. Geophys. Res. 121, 3347–3352. https://doi.org/10.1002/2016JA022437.

Lysak, R.L., Song, Y., 2006. Magnetosphere-ionosphere coupling by Alfvén waves: beyond current continuity. Adv. Space Res. 38 (8), 1713.

Mann, I.R., Pakhotin, I.P., Rae, J., Murphy, K.R., Ozeke, L.G., Knudsen, D.J., Kale, A., Milling, D.K., 2020. Magnetosphere-ionosphere coupling and ionospheric dynamics during storms and substorms: role of Alfven waves (Chapter 15). In: Mandea, M., Corte, M., Yau, A., Petrovsky, E. (Eds.), Geomagnetism, Aeronomy and Space Weather, A Journey From the Earth's Core to the Sun, Special Publications of the International Union of Geodesy and Geophysics Series. Cambridge University Press. https://doi.org/10.1017/9781108290135.

Matsuo, T., Richmond, A.D., 2008. Effects of high-latitude ionospheric electric field variability on global thermospheric Joule heating and mechanical energy transfer rate. J. Geophys. Res. 113, A07309. https://doi.org/10.1029/2007JA012993.

Matsuo, T., Richmond, A.D., Hensel, K., 2003. High-latitude ionospheric electric field variability and electric potential derived from DE-2 plasma drift measurements: dependence on IMF and dipole tilt. J. Geophys. Res. Space Phys. 108 (A1), 1005. https://doi.org/10.1029/2002JA009429.

Maynard, N.C., Aggson, T.L., Basinska, E.M., Burke, W.J., Craven, P., Peterson, W.K., Sugiura, M., Weimer, D.R., 1991. Magnetospheric boundary dynamics: DE 1 and DE 2 observations near the magnetopause and cusp. J. Geophys. Res. 96 (A3), 3505–3522. https://doi.org/10.1029/90JA02167.

McGranaghan, R.M., Mannucci, A.J., Forsyth, C., 2017. A comprehensive analysis of multi-scale field aligned currents: characteristics, controlling parameters, and relationships. J. Geophys. Res. 122, 11,931–11,960, https://doi.org/10.1002/2017JA024742.

McHarg, M., Chun, F., Knipp, D., Lu, G., Emery, B., Ridley, A., 2005. High-latitude Joule heating response to IMF inputs. J. Geophys. Res. 110, A08309. https://doi.org/10.1029/2004JA010949.

Miles, D.M., Mann, I.R., Pakhotin, I.P., Burchill, J.K., Howarth, A.D., Knudsen, D.J., Lysak, R.L., Wallis, D.D., Cogger, L.L., Yau, A.W., 2018. Alfvénic dynamics and fine structuring of discrete auroral arcs: Swarm and e-POP observations. Geophys. Res. Lett. 45. https://doi.org/10.1002/2017GL076051.

Mlynczak, M., et al., 2003. The natural thermostat of nitric oxide emission at 5.3 μm in the thermosphere observed during the solar storms of April 2002. Geophys. Res. Lett. 30 (21), 2100. https://doi.org/10.1029/2003GL017693.

Mozer, F.S., 2016. DC and low-frequency double probe electric field measurements in space. J. Geophys. Res. Space Phys. 121 (11), 10942–10953. https://doi.org/10.1002/2016JA022952.

Newell, P.T., Sotirelis, T., Wing, S., 2009. Diffuse, monoenergetic, and broadband aurora: the global precipitation budget. J. Geophys. Res. 114, A09207. https://doi.org/10.1029/2009JA014326.

Nishimura, Y., Kikuchi, T., Shinbori, A., Wygant, J., Tsuji, Y., Hori, T., Ono, T., Fujita, S., Tanaka, T., 2010. Direct measurements of the Poynting flux associated with convection electric fields in the magnetosphere. J. Geophys. Res. 115, A12212. https://doi.org/10.1029/2010JA015491.

Nozawa, S., Brekke, A., 1995. Studies of the E region neutral wind in the disturbed auroral ionosphere. J. Geophys. Res. Space Phys. 100, 14717–14734. https://doi.org/10.1029/95JA00676.

Nozawa, S., Brekke, A., 1999a. Studies of the auroral E region neutral wind through a solar cycle: Quiet days. J. Geophys. Res. Space Phys. 104, 45–66. https://doi.org/10.1029/1998JA900013.

Nozawa, S., Brekke, A., 1999b. Seasonal variation of the auroral E-region neutral wind for different solar activities. J. Atmos. Sol. Terr. Phys. 61, 585–605. https://doi.org/10.1016/S1364-6826(99)00016-4.

Nozawa, S., Brekke, A., 2000. A case study of the auroral E region neutral wind on a quiet summer day: comparison of the European Incoherent Scatter UHF radar for deriving the E region wind. Radio Sci. 35, 845–863.

Nozawa, S., Brekke, A., Maeda, S., Aso, T., Hall, C.M., Ogawa, Y., Buchert, S.C., Röttger, J., Richmond, A.D., Roble, R., Fujii, R., 2005. Mean winds, tides, and quasi-2 day wave in the polar lower thermosphere observed in European Incoherent Scatter (EISCAT) 8 day run data in November 2003. J. Geophys. Res. Space Phys. 110 (A12). https://doi.org/10.1029/2005JA011128.

Nygrén, T., Aikio, A.T., Kuula, R., Voiculescu, M., 2011. Electric fields and neutral winds from monostatic incoherent scatter measurements by means of stochastic inversion. J. Geophys. Res. Space Phys. 116 (A5). https://doi.org/10.1029/2010JA016347.

Ohtani, S., 2019. Substorm energy transport from the magnetotail to the nightside ionosphere. J. Geophys. Res. Space Phys. 124, 8669–8684. https://doi.org/10.1029/2019JA026964.

Olsson, A., Janhunen, P., Karlsson, T., Ivchenko, N., Blomberg, L.G., 2004. Statistics of Joule heating in the auroral zone and polar cap using Astrid-2 satellite Poynting flux. Ann. Geophys. 22, 4133–4142. https://doi.org/10.5194/angeo-22-4133-2004.

Osaki, H., Takahashi, K., Fukunishi, H., Nagatsuma, T., Oya, H., Matsuoka, A., Milling, D.K., 1998. Pi2 pulsations observed from the Akebono satellite in the plasmasphere. J. Geophys. Res. 103 (A8), 17605–17615. https://doi.org/10.1029/97ja03012.

Ozturk, D.S., Meng, X., Verkhoglyadova, O.P., Varney, R., Reimer, A.S., Semeter, J.L., 2020. A new framework to incorporate high-latitude input for mesoscale electrodynamics. J. Geophys. Res. Space Phys. 125. e2019JA027562.

Pakhotin, I.P., Mann, I.R., Lysak, R.L., Knudsen, D.J., Gjerloev, J.W., Rae, I.J., Forsyth, C., Murphy, K. R., Miles, D.M., Ozeke, L.G., Balasis, G., 2018. Diagnosing the role of Alfvén waves in magnetosphere-ionosphere coupling: Swarm observations of large amplitude nonstationary magnetic perturbations during an Interval of Northward IMF. J. Geophys. Res. 123. https://doi.org/10.1002/2017JA024713.

Pakhotin, I.P., Mann, I.R., Knudsen, D.J., Lysak, R.L., Burchill, J.K., 2020. Diagnosing the role of Alfvén waves in global field-aligned current system dynamics during southward IMF: Swarm observations. J. Geophys. Res. Space Phys. 125. https://doi.org/10.1029/2019JA027277. e2019JA027277.

Pakhotin, I.P., Mann, I.R., Xie, K., et al., 2021. Northern preference for terrestrial electromagnetic energy input from space weather. Nat. Commun. 12, 199. https://doi.org/10.1038/s41467-020-20450-3.

Park, J., Luhr, H., Knudsen, D.J., Burchill, J.K., Kwak, Y.-S., 2017. Alfvén waves in the auroral region, their Poynting flux, and reflection coefficient as estimated from Swarm observations. J. Geophys. Res. Space Phys. 122, 2345–2360. https://doi.org/10.1002/2016JA023527.

Poynting, J.H., 1884. On the transfer of energy in the electromagnetic field. Philos. Trans. 175, 343–361.

Rae, I.J., Watt, C.E.J., Fenrich, F.R., Mann, I.R., Ozeke, L.G., Kale, A., 2007. Energy deposition in the ionosphere through a global field line resonance. Ann. Geophys. 25, 2529–2539. https://doi.org/10.5194/angeo-25-2529-2007.

Rastätter, L., Shim, J.S., Kuznetsova, M.M., Kilcommons, L.M., Knipp, D.J., Codrescu, M., Fuller-Rowell, T., Emery, B., Weimer, D.R., Cosgrove, R., et al., 2016. GEM-CEDAR challenge: Poynting flux at DMSP and modeled Joule heat. Space Weather 14, 113–135. https://doi.org/10.1002/2015SW001238.

Rastaetter, L., et al., 2016. GEM-CEDAR challenge: Poynting flux at DMSP and modeled joule heat. Space Weather 14, 113–135. https://doi.org/10.1002/2015SW001238.

Richards, P.G., Nicolls, M.J., Heinselman, C.J., Sojka, J.J., Holt, J.M., Meier, R.R., 2009. Measured and modeled ionospheric densities, temperatures, and winds during the international polar year. J. Geophys. Res. Space Phys. 114 (A12). https://doi.org/10.1029/2009JA014625.

Richards, P.G., Bilitza, D., Voglozin, D., 2010. Ion density calculator (IDC): a new efficient model of ionospheric ion densities. Radio Sci. 45 (5). https://doi.org/10.1029/2009RS004332.

Richmond, A.D., 2010. On the ionospheric application of Poynting's theorem. J. Geophys. Res. 115, A10311. https://doi.org/10.1029/2010JA015768.

Richmond, A., Kamide, Y., 1988. Mapping electrodynamic features of the high-latitude ionosphere from localized observations: technique. J. Geophys. Res. 93, 5741–5759. https://doi.org/10.1029/JA093iA06p05741.

Richmond, A.D., Thayer, J.P., 2000. Tutorial: Ionospheric Electrodynamics: A Tutorial. Geophysical Monograph Series, vol. 118 American Geophysical Union, Washington, DC, p. 131, https://doi.org/10.1029/GM118p0131.

Ridley, A.J., Deng, Y., Toth, G., 2006. The global ionosphere-thermosphere model. J. Atmos. Sol. Terr. Phys. 68. https://doi.org/10.1016/j.jastp.2006.01.008.

Rino, C.L., Brekke, A., Baron, M.J., 1977. High-resolution auroral zone E region neutral wind and current measurements by incoherent scatter radar. J. Geophys. Res. (1896–1977) 82 (16), 2295–2304. https://doi.org/10.1029/JA082i016p02295.

Richmond, A.D., Ridley, E.C., Roble, R.G., 1992. A thermosphere/ionosphere general circulation model with coupled electrodynamics. Geophys. Res. Lett. 19, 369. https://doi.org/10.1029/92GL00401.

Roble, R.G., Ridley, E.C., 1994. A thermosphere-ionosphere-mesosphere-electrodynamics general circulation model (TIME-GCM): equinox solar cycle minimum simulations (30–500 km). Geophys. Res. Lett. 21 (417). https://doi.org/10.1029/93GL03391.

Sangalli, L., Knudsen, D.J., Larsen, M.F., Zhan, T., Pfaff, R.F., Rowland, D., 2009. Rocket-based measurements of ion velocity, neutral wind, and electric field in the collisional transition region of the auroral ionosphere. J. Geophys. Res. Space Phys. 114, A04306. https://doi.org/10.1029/2008JA013757.

Sarris, T.E., Talaat, E.R., Palmroth, M., Dandouras, I., Armandillo, E., Kervalishvili, G., Buchert, S., Tourgaidis, S., Malaspina, D.M., Jaynes, A.N., Paschalidis, N., Sample, J., Halekas, J., Doornbos, E., Lappas, V., Jørgensen, T.M., Stolle, C., Clilverd, M., Wu, Q., Sandberg, I., Pirnaris, P., Aikio, A., 2020. Daedalus: a low-flying spacecraft for in situ exploration of the lower thermosphere-ionosphere. Geosci. Instrum. Methods Data Syst. 9 (1), 153–191. https://doi.org/10.5194/gi-9-153-2020.

Schunk, R.W., Nagy, A.F., 2009. Ionospheres: Physics, Plasma Physics, and Chemistry, second ed. Cambridge University Press, Cambridge, UK.

Senior, A., Kavanagh, A.J., Kosch, M.J., Honary, F., 2007. Statistical relationships between cosmic radio noise absorption and ionospheric electrical conductances in the auroral zone. J. Geophys. Res. Space Phys. 112 (A11). https://doi.org/10.1029/2007JA012519.

Sheng, C., Deng, Y., Zhang, S.-R., Nishimura, Y., Lyons, L.R., 2020. Relative contributions of ion convection and particle precipitation to exciting large-scale traveling atmospheric and ionospheric disturbances. J. Geophys. Res. Space Phys. 125 (2). https://doi.org/10.1029/2019JA027342. e2019JA027342.

Stasiewicz, K., Bellan, P., Chaston, C., et al., 2000. Small scale Alfvénic structure in the aurora. Space Sci. Rev. 92, 423–533. https://doi.org/10.1023/A:1005207202143.

Strangeway, R.J., 2012. The equivalence of Joule dissipation and frictional heating in the collisional ionosphere. J. Geophys. Res. Space Phys. 117, A02310. https://doi.org/10.1029/2011JA017302.

Strangeway, R.J., Russell, C.T., Carlson, C.W., McFadden, J.P., Ergun, R.E., Temerin, M., Klumpar, D. M., Peterson, W.K., Moore, T.E., 2000. Cusp field-aligned currents and ion outflows. J. Geophys. Res. 105 (A9), 21129–21141. https://doi.org/10.1029/2000JA900032.

Strangeway, R.J., Ergun, R.E., Su, Y.-J., Carlson, C.W., Elphic, R.C., 2005. Factors controlling ionospheric outflows as observed at intermediate altitudes. J. Geophys. Res. 110, A03221. https://doi.org/10.1029/2004JA010829.

Sugiura, M., 1984. A fundamental magnetosphere-ionosphere coupling mode involving field-aligned currents as deduced from DE-2 observations. Geophys. Res. Lett. 11, 877–880. https://doi.org/10.1029/GL011i009p00877.

Sugiura, M., 1986. Joule heating and field-aligned currents: preliminary results from DE-2. https://ntrs.nasa.gov/archive/nasa/casi.ntrs.nasa.gov/19860019864.pdf.

Thayer, J.P., 1998a. Radar measurements of the electromagnetic energy rates associated with the dynamic ionospheric load/generator. Geophys. Res. Lett. 25, 469–472. https://doi.org/10.1029/97GL03660.

Thayer, J.P., 1998b. Height-resolved Joule heating rates in the high-latitude E region and the influence of neutral winds. J. Geophys. Res. Space Phys. 103, 471–487. https://doi.org/10.1029/97JA02536.

Thayer, J.P., 2000. High-latitude currents and their energy exchange with the ionosphere-thermosphere system. J. Geophys. Res. Space Phys. 105, 23015–23024. https://doi.org/10.1029/1999JA000409.

Thayer, J.P., Semeter, J., 2004. The convergence of magnetospheric energy flux in the polar atmosphere. J. Atmos. Sol. Terr. Phys. 66, 807–824. https://doi.org/10.1016/j.jastp.2004.01.035.

Thayer, J.P., Vickrey, J.F., 1992. On the contribution of the thermospheric neutral wind to high-latitude energetics. Geophys. Res. Lett. 19, 265–268. https://doi.org/10.1029/91GL02868.

Thayer, J.P., Vickrey, J.F., Heelis, R.A., Gary, J.B., 1995. Interpretation and modeling of the high-latitude electromagnetic energy flux. J. Geophys. Res. 100 (A10), 19715–19728. https://doi.org/10.1029/95JA01159.

Turner, N.E., Cramer, W.D., Earles, S.K., Emery, B.A., 2009. Geoefficiency and energy partitioning in CIR-driven and CME-driven storms. J. Atmos. Sol. Terr. Phys. 71, 1023–1031. https://doi.org/10.1016/j.jastp.2009.02.005.

Vaivads, A., Andre, M., Norqvist, P., Oscarsson, T., Ronnmark, K., Blomberg, L., Clemmons, J.H., Santolik, O., 1999. Energy transport during O+ energization by ELF waves observed by the Freja satellite. J. Geophys. Res. 104 (A2), 2563–2572. https://doi.org/10.1029/1998JA900114.

Vanhamaki, H., Yoshikawa, A., Amm, O., Fujii, R., 2012. Ionospheric Joule heating and Poynting flux in quasi-static approximation. J. Geophys. Res. 117, A08327. https://doi.org/10.1029/2012JA017841.

Vasyliunas, V.M., Song, P., 2005. Meaning of ionospheric joule heating. J. Geophys. Res. Space Phys. 110 (A2). https://doi.org/10.1029/2004JA010615.

Verkhoglyadova, O., Meng, X., Mannucci, A.J., Tsurutani, B.T., Hunt, L.A., Mlynczak, M.G., Hajra, R., Emery, B.A., 2016. Estimation of energy budget of ionosphere-thermosphere system during two CIR-HSS events: observations and modeling. J. Space Weather Space Clim. 6 (A20). https://doi.org/10.1051/swsc/2016013.

Verkhoglyadova, O.P., Meng, X., Mannucci, A.J., Mlynczak, M.G., Hunt, L.A., Lu, G., 2017. Ionosphere thermosphere energy budgets for the ICME storms of March 2013 and 2015 estimated with GITM and observational proxies. Space Weather J. 15. https://doi.org/10.1002/2017SW001650.

Verkhoglyadova, O.P., Meng, X., Manucci, A.J., McGranaghan, R.M., 2018. Semianalytical estimation of energy deposition in the ionosphere by monochromatic Alfvén waves. J. Geophys. Res. Space Phys. 123, 5210–5222. https://doi.org/10.1029/2017JA025097.

Verkhoglyadova, O.P., Bussy-Virat, C.D., Caspi, A., Jackson, D.R., Kalegaev, V., Klenzing, J., Nieves-Chinchilla, J., Vourlidas, A., 2020. Addressing gaps in space weather operations and understanding with small satellites. Space Weather J., SWE21089. https://doi.org/10.1029/2020SW002566.

Watanabe, S., Whalen, B.A., Wallis, D.D., Pfaff, R.F., 1991. Observations of ion-neutral collisional effects in the auroral E region. J. Geophys. Res. Space Phys. 96, 9761–9771. https://doi.org/10.1029/91JA00561.

Waters, C.L., Anderson, B.J., Greenwald, R.A., Barnes, R.J., Ruohoniemi, J.M., 2004. High-latitude Poynting flux from combined Iridium and SuperDARN data. Ann. Geophys. 22, 2861–2875. https://doi.org/10.5194/angeo-22-2861-2004.

Weimer, D.R., 2005. Improved ionospheric electrodynamic models and application to calculating Joule heating rates. J. Geophys. Res. 110. https://doi.org/10.1029/2004JA010884.

Wescott, E.M., Stolarik, J.D., Heppner, J.P., 1969. Electric fields in the vicinity of auroral forms from motions of barium vapor releases. J. Geophys. Res. 74 (14), 3469–3487. https://doi.org/10.1029/JA074i014p03469.

Wickwar, V.B., Baron, M.J., Sears, R.D., 1975. Auroral energy input from energetic electrons and joule heating at Chatanika. J. Geophys. Res. 80 (31), 4364–4367. https://doi.org/10.1029/JA080i031p04364.

Wu, J., Knudsen, D.J., Gillies, D.M., Burchill, J.K., 2020. Swarm survey of Alfvénic fluctuations and their relation to nightside field-aligned current and auroral arc systems. J. Geophys. Res. 125 (3). https://doi.org/10.1029/2019JA027220. e2019JA027220.

Wygant, J.R., et al., 2002. Evidence for kinetic Alfvén waves and parallel electron energization at 4–6 RE altitudes in the plasma sheet boundary layer. J. Geophys. Res. 107 (A8), 1201. https://doi.org/10.1029/2001JA900113.

Yoshikawa, A., 2002. Excitation of a Hall-current generator by field-aligned current closure, via an ionospheric, divergent Hall-current, during the transient phase of magnetosphere-ionosphere coupling. J. Geophys. Res. 107 (A12), 1445. https://doi.org/10.1029/2001JA009170.

Zhang, X.X., Wang, C., Chen, T., Wang, Y.L., Tan, A., Wu, T.S., Germany, G.A., Wang, W., 2005. Global patterns of Joule heating in the high-latitude ionosphere. J. Geophys. Res. 110, A12208. https://doi.org/10.1029/2005JA011222.

Zhu, J., Ridley, A.J., 2016. Investigating the performance of simplified neutral-ion collisional heating rate in a global IT model. J. Geophys. Res. Space Phys. 121, 578–588. https://doi.org/10.1002/2015JA021637.

Zhu, Q., Deng, Y., Richmond, A., Maute, A., 2018. Small-scale and mesoscale variabilities in the electric field and particle precipitation and their impacts on Joule heating. J. Geophys. Res.: Space Phys. 123, 9862–9872. https://doi.org/10.1029/2018JA025771.

CHAPTER 6

Kinetics, ionization and electromagnetic waves

Chapter outline

6.1 Kinetic processes and their feedback to larger scales

Matthew A. Young[a], William J. Longley[b,c], Meers M. Oppenheim[d], and Yakov S. Dimant[d]

[a]Space Science Center, University of New Hampshire, Durham, NH, United States

[b]Department of Physics and Astronomy, Rice University, Houston, TX, United States

[c]University Corporation for Atmospheric Research, Boulder, CO, United States

[d]Center for Space Physics, Boston University, Boston, MA, United States

Cross-Scale Coupling and Energy Transfer in the Magnetosphere-Ionosphere-Thermosphere System
https://doi.org/10.1016/B978-0-12-821366-7.00005-6

6.1.1 Introduction

Kinetic plasma processes are fundamental to magnetosphere-ionosphere-thermosphere (M-I-T) coupling because they accurately describe the dynamics of a system out of equilibrium. However, they can also seem esoteric or excessive because they are not as intuitive as the fluid description. The two viewpoints that a kinetic description is both fundamental and less intuitive than a fluid description are not incompatible, and it is the objective of this section to bring them into harmony.

There are often situations in which we can assume that the velocities of a collection of particles follow a *Maxwellian distribution* (see Section 6.1.3.1), which is the unique particle distribution that describes a gas in *thermal equilibrium*. In these cases, we can describe the plasma as a *fluid*. In other situations, kinetic effects may drive the distribution away from Maxwellian, thereby rendering the fluid distribution less useful or even misleading. By *kinetic effects* or kinetic plasma processes, we mean the motions and interactions of ions, electrons, and neutral particles that constitute a plasma, without any assumptions about the distributions of their positions or velocities. Even when a distribution of particle velocities deviates from strictly Maxwellian, it is often still possible to approximate its behavior as fluidic within a certain range of parameters and either to incorporate kinetic effects in an ad hoc fashion or to simply avoid the regimes of parameter space in which kinetic effects are significant.

6.1.2 Glossary of terms

This section defines some terms relevant to this section, and which may be unfamiliar to the reader. This list is by no means exhaustive; rather, our aim is to make clear what we mean when we use each of these terms in this section.

- *Phase space*. The six-dimensional (6D) space of particle states defined by three position coordinates, $\vec{r} = (r_x, r_y, r_z)$, and three velocity coordinates, $\vec{v} = (v_x, v_y, v_z)$. A curve through phase space gives the complete history of a particular particle subject to a known set of forces.
- *Distribution function*. The density of particles at each point in phase space, at a particular time. Because phase space has three position dimensions and three velocity dimensions, the distribution function has units of (particle number) $\times$ (unit length)$^{-3}$ $\times$ (unit speed)$^{-3}$.
- *Mean free path*. The average distance which a particle in a distribution can expect to travel before experiencing a collision.
- *Thermal equilibrium*. The state of a particle distribution in which the particles no longer spontaneously exchange energy through collisions.
- *Maxwellian distribution*. The distribution of velocities of particles in thermal equilibrium.

- *Fluid*. A collection of particles which can be modeled as a continuum, with well-defined density, bulk velocity, and temperature. The latter requires a Maxwellian particle distribution.
- *Kinetic effects*. The results of processes that require a microscopic description of particle dynamics in phase space, rather than a macroscopic fluid description.
- *Boltzmann equation*. The equation which describes the evolution of a particle distribution over time. The Boltzmann equation simplifies to the Vlasov equation when collisions are negligible and when acceleration is due to the Lorentz force, see Section 6.1.3.4 for more detail.
- *Magnetohydrodynamics (MHD)*. A single-fluid description of a plasma that follows the average mass density, current density, and center-of-mass velocity under the influence of magnetic forces.
- *BGK collision operator*. A simplified form of the collision integral in the Boltzmann equation, representing the collisional relaxation of the distribution to a Maxwellian; see Section 6.1.3.5.
- *Moments of a distribution*. Average quantities derived by multiplying a particle distribution function by an integer power of velocity, then integrating over all velocity coordinates.
- *Wave-particle interactions*. Interactions between charged particles and electromagnetic waves in which particles gain energy from waves or vice versa.
- *Landau damping*. The collisionless damping of plasma waves through a fully kinetic wave-particle interaction.
- *Pitch angle*. The angle which a particle's velocity vector makes with the magnetic field.
- *Gyrofrequency*. The angular frequency at which a charged particle gyrates about the magnetic field.

6.1.3 Key concepts

6.1.3.1 Maxwellian distribution

The Maxwellian distribution describes a gas in *thermal equilibrium*. It is the distribution of velocities to which a uniform collection of particles relaxes when they are allowed to collide with each other while outside forces are negligible.

In terms of a single independent velocity coordinate—the speed along the x-axis, v_x, for example—the Maxwellian distribution is

$$f(v_x) = n\left(\frac{m}{2\pi k_B T}\right)^{1/2} \exp\left[-\frac{mv_x^2}{2k_B T}\right] \tag{6.1}$$

where m is the species mass, T is the temperature, n is the number density, and k_B is Boltzmann's constant. If the distribution has relaxed to equilibrium but is embedded in a large-scale flow, we simply need to replace v_x with $v_x - v_{0x}$ in the exponential,

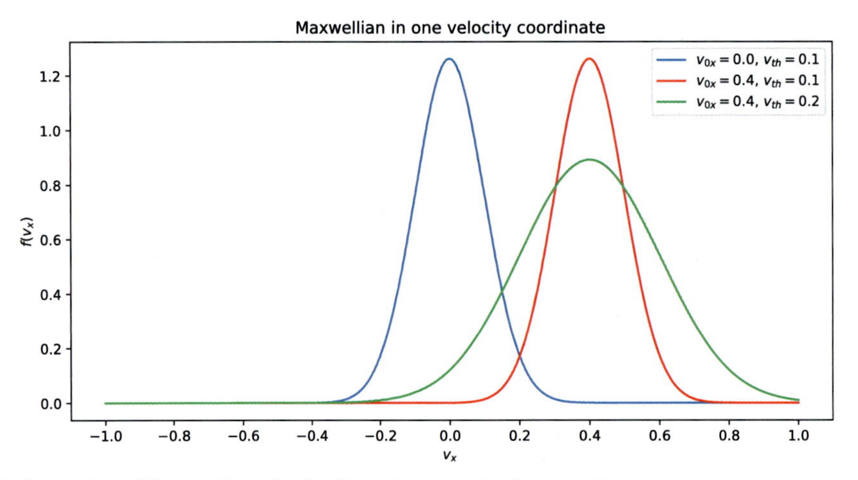

Fig. 6.1 Examples of Maxwellian distributions in one velocity coordinate.

$$f(v_x) = n\left(\frac{m}{2\pi k_B T}\right)^{1/2} \exp\left[-\frac{m}{2k_B T}(v_x - v_{0x})^2\right] \tag{6.2}$$

where v_{0x} is the speed of the large-scale flow. Instead of defining the Maxwellian in terms of temperature, we can define it in terms of the thermal velocity,[a] $v_{th} \equiv \sqrt{k_B T/m}$,

$$f(v_x) = n\left(\frac{1}{2\pi v_{th}^2}\right)^{1/2} \exp\left[-\frac{1}{2}\left(\frac{v_x - v_{0x}}{v_{th}}\right)^2\right] \tag{6.3}$$

Fig. 6.1 shows examples of one-dimensional (1D) Maxwellian distributions with different values of flow and thermal speeds.

Of course, there are many situations that require more than one velocity coordinate to describe the system. As shown in Fig. 6.2, the two-dimensional (2D) Maxwellian distribution in v_x and v_y is

$$f(v_x, v_y) = n\left(\frac{m}{2\pi k_B T}\right) \exp\left[-\frac{m\left(v_x^2 + v_y^2\right)}{2k_B T}\right] \tag{6.4}$$

and the three-dimensional (3D) Maxwellian distribution in v_x, v_y, and v_z is

$$f(v_x, v_y, v_z) = n\left(\frac{m}{2\pi k_B T}\right)^{3/2} \exp\left[-\frac{m\left(v_x^2 + v_y^2 + v_z^2\right)}{2k_B T}\right] \tag{6.5}$$

[a] There are many conventions for how thermal velocity is defined, differing by a factor on the order of unity.

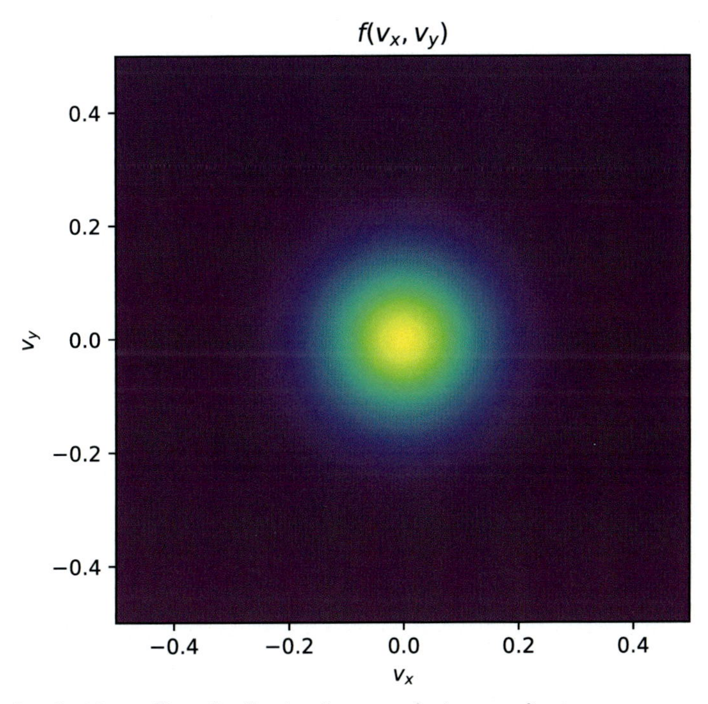

Fig. 6.2 Example of a Maxwellian distribution in two velocity coordinates.

Finally, we may only be interested in the distribution of speeds of particles in a volume of gas. While it may be tempting to multiply Eq. (6.1) by 3 to account for the other two components that would not give the correct expression. Instead, we need to put Eq. (6.5) in spherical coordinates with radius $v \equiv \sqrt{v_x^2 + v_y^2 + v_z^2}$, polar angle θ, and azimuthal angle ϕ. Then, the distribution of speeds is found by integrating over the polar and azimuthal angles, as shown in Fig. 6.3:

$$f(v) = \int\int f(v_x, v_y, v_z)\, v^2 \sin\theta \, d\theta \, d\phi \tag{6.6}$$

Since the 3D Maxwellian distribution is spherically symmetric, the angular integrals simplify to 4π, and we obtain

$$f(v) = 4\pi n v^2 \left(\frac{m}{2\pi k_B T}\right)^{3/2} \exp\left[-\frac{mv^2}{2k_B T}\right] \tag{6.7}$$

6.1.3.2 Other distributions

Power-law distributions are another common type of distribution. They arise in situations in which data points tend to cluster at a rate proportional to the number of data points that already exist in a particular region. For example, Fig. 6.4 shows the population in 1147 largest cities at the start of 2020 (https://worldpopulationreview.com/world-cities/), on

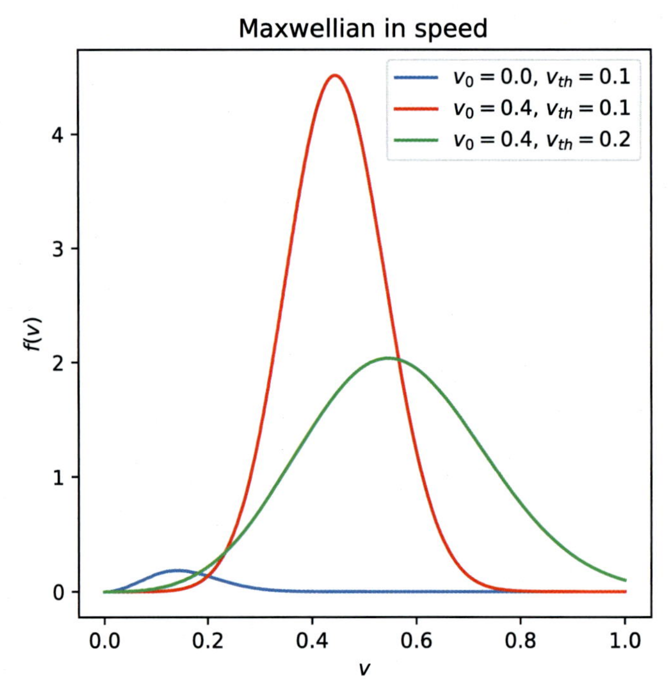

Fig. 6.3 Examples of Maxwellian distributions in speed.

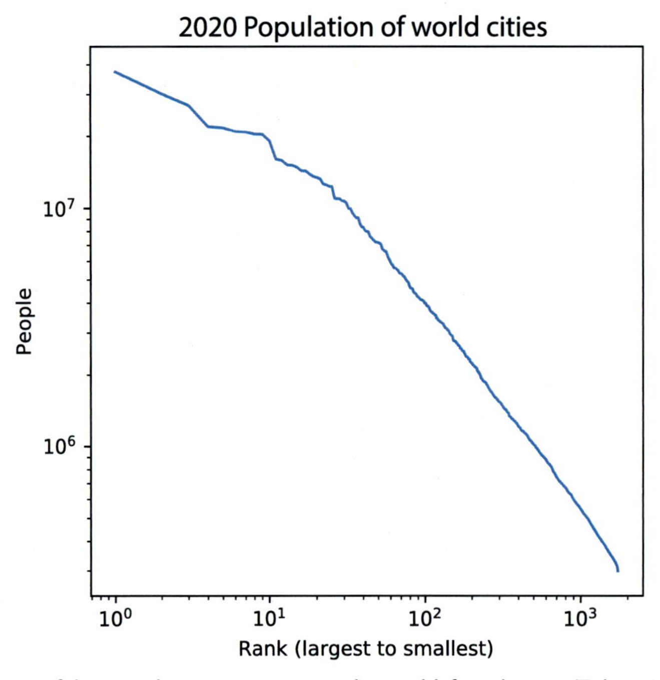

Fig. 6.4 Populations of the 1147 largest cities across the world, from largest (Tokyo, Japan) to smallest (Ziyang, China). *(Data available at: https://worldpopulationreview.com/world-cities/.)*

a log-log scale. The slope of the distribution starting around the $30th$ largest city is a nearly straight line in log-log space, meaning that the distribution can be written in the form

$$\frac{d\log_{10}(\text{People})}{d\log_{10}(\text{Rank})} \propto k$$

where k is the slope of the distribution of people in cities across the world. Rearranging terms, integrating, and exponentiating both sides tells us that the number of people in a particular city is proportional to its rank raised to the power k. Since a city's rank in this data set is a reflection of how many people live in that city, this relationship tells us that people tend to cluster in places where there are already a lot of people, which is not a surprise.

More generally, a power-law distribution has the form

$$f(v) = Cv^{-\lambda} \tag{6.8}$$

where neither C nor λ depends on v. Some physical systems exhibit behavior that does not cluster around a mean value, unlike in a Maxwellian distribution. Such systems tend to have power-law distributions, in which the amplitude at any one value is proportional to all the other values. This trait is called *scale invariance*. Sizes of eddies in classical fluid turbulence, connections in social networks (Kuby et al., 2009), seismic fault lengths (Scholz, 2007), and the fluence of energetic particles in the heliosphere (Mewaldt et al., 2001) are just a few examples of a phenomenon, which exhibit power-law distributions. In some systems (e.g., turbulence), kinetic effects become important only outside the power-law regime; in others (e.g., cosmic rays), kinetic effects directly produce the power law. Fig. 6.5 shows examples of power-law distributions with various values of C and λ. Note that many works cite $\lambda > 0$ as the *power-law index*, implying the negative sign shown in Eq. (6.8).

The *kappa* (κ) *distribution* is a generalization of the Maxwellian distribution, first proposed by Vasyliunas (1968) for the study of electron distributions in the magnetosphere:

$$f(v) = C\left[1 + \frac{1}{\kappa}\left(\frac{v^2}{v_{th}^2}\right)\right]^{-\kappa-1} \tag{6.9}$$

This distribution has the useful properties that it behaves like a Maxwellian at low energies and power law at high energies. In addition, the limit as $\kappa \rightarrow \infty$ is a Maxwellian distribution. Fig. 6.6 shows examples of kappa distributions for various values of C, κ, and v_{th}.

For examples of kappa distributions in space-physics research (not necessarily limited to the M-I-T system); see Schwadron et al. (2010), Livadiotis and McComas (2009), and Pierrard and Lazar (2010).

There are many situations in which a beam of particles passes through a background Maxwellian distribution. We can write the total distribution, comprising both the

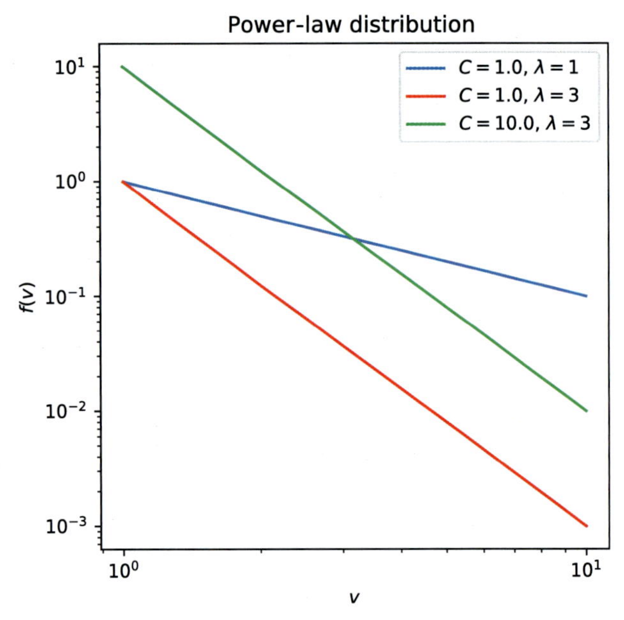

Fig. 6.5 Examples of power-law distributions.

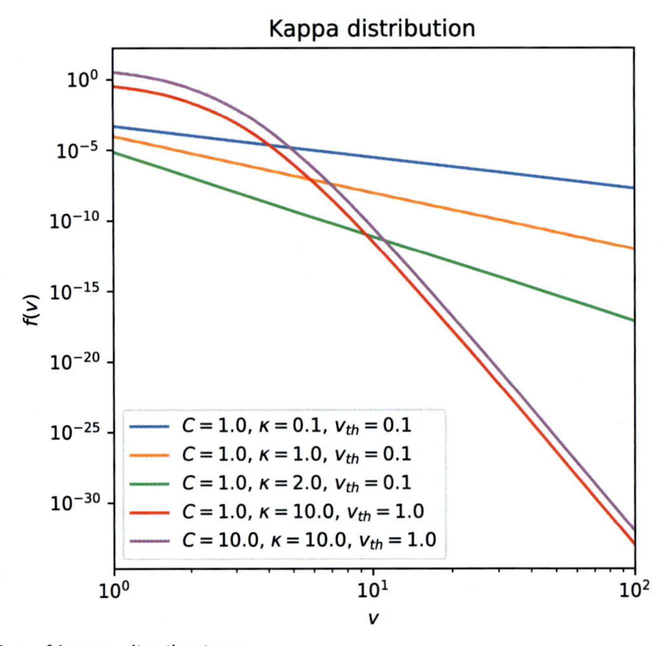

Fig. 6.6 Examples of kappa distributions.

Maxwellian and the beam, as the sum of two distributions, $f(\vec{v}) = f_a(\vec{v}) + f_b(\vec{v})$, where $f_a(\vec{v})$ is the background Maxwellian and $f_b(\vec{v})$ is the beam. It is quite likely that the beam itself is a Maxwellian distribution with its own thermal speed, $v_{th,\,b}$, and density, n_b. Then, the total distribution is

$$\begin{aligned} f(\vec{v}) &= n_a \left(\frac{1}{2\pi v_{th,\,a}}\right)^{3/2} \exp\left[-\frac{1}{2}\left(\frac{|\vec{v}_a - \vec{v}_{0a}|}{v_{th,\,a}}\right)^2\right] \\ &+ n_b \left(\frac{1}{2\pi v_{th,\,b}}\right)^{3/2} \exp\left[-\frac{1}{2}\left(\frac{|\vec{v}_b - \vec{v}_{0b}|}{v_{th,\,b}}\right)^2\right] \end{aligned}$$

We can typically assume that the beam comprises only a small fraction of the total distribution and that the velocities of the particles in the beam do not deviate much from their average value. The first assumption implies that the beam density is much smaller than the background density, $n_b \ll n_a$, and the second assumption implies that the beam's thermal speed is smaller than the background thermal speed, $v_{th,b} \ll v_{th,a}$.

This is an example of a *bump-on-tail distribution*. Fig. 6.7 shows a bump-on-tail distribution composed of a background Maxwellian distribution with $v_{0x} = 0$ and a beam

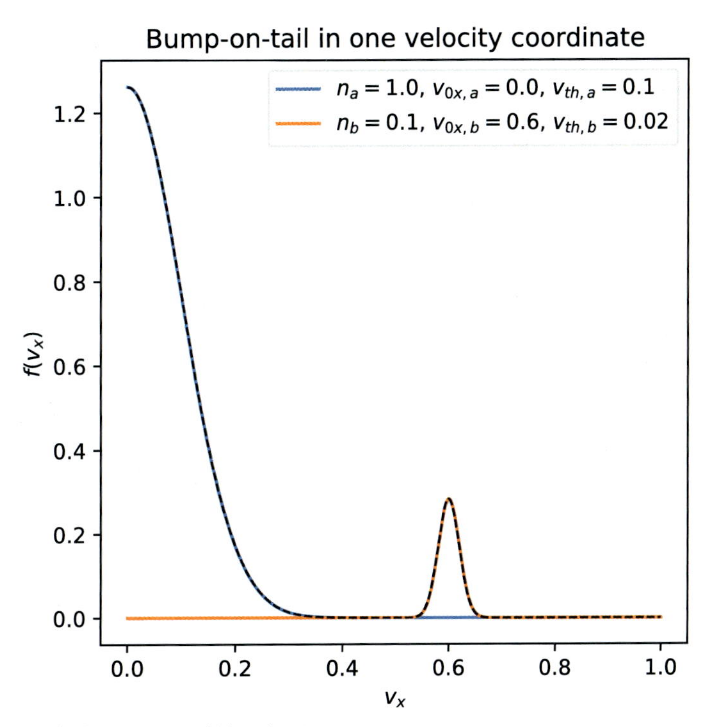

Fig. 6.7 Example of a bump-on-tail distribution.

Maxwellian distribution with $v_{0x} > 0$. The beam density is 10% the background density and the beam thermal speed is 20% the background thermal speed. The blue line shows the background Maxwellian, $f_a(v)$; the orange line shows the beam Maxwellian, $f_b(v)$; and the black dashed line shows the total distribution, $f(v) = f_a(v) + f_b(v)$. Bump-on-tail distributions often lead to kinetic instabilities, such as the upper-hybrid and electron cyclotron instabilities in the lower ionosphere (Basu et al., 1982; Longley et al., 2020).

6.1.3.3 Phase space

We can fully describe a particle distribution with three spatial coordinates, three velocity coordinates, and a temporal coordinate. The specific variable names are arbitrary, though there are some conventions. This section will use r_x, r_y, and r_z for spatial coordinates, v_x, v_y, and v_z for velocity coordinates, and t for the temporal coordinate. It is also common to see x, y, and z in use as spatial coordinates; we have chosen the r_j form to preserve symmetry with the v_j form of velocity coordinates.

Since a given distribution occupies a 6D *phase space* at a particular time, it is difficult to visualize. Furthermore, spatial and velocity coordinates come in pairs, so not even the available schemes for projecting three spatial dimensions onto a 2D image will help. One approach is to plot one spatial coordinate against one velocity coordinate.

For example, imagine a particle drifting in either the $+\hat{r}$ or $-\hat{r}$ direction at a constant speed. We can represent it as a straight line parallel to the r axis because the value of its position changes, while the value of its velocity does not. Another approach is to plot multiple velocity coordinates and use the result to draw conclusions about how a distribution moves through a known or assumed volume of space. This is a common way of analyzing particle data from satellites.

6.1.3.4 Boltzmann and Vlasov equations

Plasma theory is fundamentally a statistical description of the position and velocity of a collection of charged particles as it evolves over time. One cubic meter of plasma in the E-region ionosphere contains tens of billions of particles, each with their own 6D phase-space trajectories; the task of analyzing the dynamics of each of these individual particles quickly becomes intractable. Instead of following every particle in a plasma, we can consider all the particles in a small volume of phase space bounded by $(r_{x0}, r_{y0}, r_{z0}, v_{x0}, v_{y0}, v_{z0})$ and $(r_{x0} + dr_x, r_{y0} + dr_y, r_{z0} + dr_z, v_{x0} + dv_x, v_{y0} + dv_y, v_{z0} + dv_z)$, or d^3rd^3v. The *distribution function* of species s is defined as $f_s(t, \vec{r}, \vec{v}) = N_s(t, \vec{r}, \vec{v})/d^3rd^3v$—in other words, it is the density of particles of species s in the volume d^3rd^3v. We would like to know the temporal evolution of f_s through phase space in order to understand the kinetic dynamics of species s. Since $\vec{r} = \vec{r}(t)$ and $\vec{v} = \vec{v}(t)$, the total time derivative of $f_s(t, \vec{r}, \vec{v})$ is

$$\frac{df_s(t, \vec{r}, \vec{v})}{dt} = \frac{\partial f_s}{\partial t} + \frac{d\vec{r}}{dt} \cdot \nabla_r f_s + \frac{d\vec{v}}{dt} \cdot \nabla_v f_s$$

where $\nabla_r \equiv \partial/\partial_{r_x}\hat{r}_x + \partial/\partial_{r_y}\hat{r}_y + \partial/\partial_{r_z}\hat{r}_z$ is the familiar gradient in coordinate space and $\nabla_v \equiv \partial/\partial_{v_x}\hat{v}_x + \partial/\partial_{v_y}\hat{v}_y + \partial/\partial_{v_z}\hat{v}_z$ is the analogous gradient in velocity space.

So far, this description of a plasma has not included any way to add or remove particles from the phase-space volume d^3rd^3v, so $df_s/dt = 0$ by Liouville's theorem. Production and loss of charged particles are one way to add and remove particles that are physically intuitive, since they involve creating a charged particle from a neutral particle or vice versa. If we were only thinking about the plasma, the charged particle has effectively appeared in or disappeared from some region of the atmosphere—if a particle appears in or disappears from a region of physical space, it also appears in or disappears from the phase space of that species. Another way to add or remove particles from a certain region in phase space that is relevant to ionospheric plasma is through collisions, which disrupt the smooth phase-space trajectories described by Liouville's theorem. We can wrap these and similar phenomena up into a single "collision" term that balances df_s/dt and accounts for particles that move discontinuously into and out of regions of phase space. The resultant equation that describes the evolution of a particle distribution is (Nicholson, 1983)

$$\frac{\partial f_s}{\partial t} + \frac{d\vec{r}}{dt} \cdot \nabla_r f_s + \frac{d\vec{v}}{dt} \cdot \nabla_v f_s = \left[\frac{\delta f_s}{\delta t}\right]_c \tag{6.10}$$

This is the *Boltzmann equation*. In this notation, ∇_r represents a gradient in spatial coordinates and ∇_v represents a gradient in velocity coordinates.

We can drop the collision term in situations when collisions are negligible and when the change in velocity is due to a combination of external forces and self-consistent interparticle forces. For many plasmas of interest in the solar system, the external forces consist of the external electric and magnetic fields, while the interparticle forces consist of the self-consistent electric and magnetic fields due to particle motion. These assumptions transform Eq. (6.10) into

$$\frac{\partial f_s}{\partial t} + \frac{d\vec{r}}{dt} \cdot \nabla_r f_s + \frac{q_s}{m_s}\left(\vec{E} + \vec{v} \times \vec{B}\right) \cdot \nabla_v f_s = 0 \tag{6.11}$$

where both $\vec{E}$ and $\vec{B}$ comprise external and interparticle fields. Eq. (6.11) is the *Vlasov equation*. See Nicholson (1983) for a first-principles derivation of the Boltzmann and Vlasov equations.

The assumption that electric and magnetic fields are the only relevant forces is valid for a broad range of ionospheric phenomena. Although gravity plays a prominent role in many fluid processes, the kinetic phenomena to which this analysis applies occur on spatial scales smaller than a kilometer and temporal scales shorter than an hour—often much smaller and shorter—rendering the effects of gravity negligible. The assumption that collisions are negligible holds for most geospace phenomena above roughly 130 km, but fails when it comes to plasma instabilities in the E-region ionosphere, which can play a significant role in M-I-T coupling. Therefore, we need some appropriate form of the collision operator $[\delta f_s/\delta t]_c$ in order to fully understand the situation.

6.1.3.5 Collisions

The standard method in determining the collision operator for a given problem is through the use of the Boltzmann collision integral. The Boltzmann collision integral considers the probability of a particle being scattered into a new region of phase space based, weighted by the distribution of the target particles

$$\left[\frac{\delta f_1}{\delta t}\right]_c = \int\int |\vec{v}_1 - \vec{v}_2| \sigma(\Omega) \left(f_1' f_2' - f_1 f_2\right) d^3 v_2 \, d\Omega \tag{6.12}$$

where $f_1 \equiv f(\vec{v}_1)$ is the distribution for the scattered particles, $f_2 \equiv f(\vec{v}_2)$ is the distribution of the target particles, and $f_1' \equiv f\left(\vec{v}_1'\right)$ and $f_2' \equiv f\left(\vec{v}_2'\right)$ are those distributions modified by collisions. The cross section, σ, is a function of the solid angle, Ω. The $d^3 v_2$ integral captures the total effect that particles in f_2 have on particles in f_1 during collisions. The $d\Omega$ integral accounts for particles in f_1 begin scattered into all possible solid angles. For a more detailed explanation, see Appendix G of Schunk and Nagy (2004), Shkarofsky et al. (1966), Gurevich (1978), and Khazanov (2011).

We are now prepared to understand why the Maxwellian distribution describes a gas in thermal equilibrium. The left-hand side of Eq. (6.10) vanishes for a steady-state ($\partial/\partial t \rightarrow 0$), homogeneous ($\nabla_r \rightarrow 0$) gas when external forces are negligible ($d\vec{v}/dt = 0$). That means that the right-hand side must also vanish. According to Eq. (6.12), the only way for that to happen is if $f_1 f_2 = f_1' f_2'$. Taking the natural log of both sides yields $\ln f_1 + \ln f_2 = \ln f_1' + \ln f_2'$, which has the form of a quantity that is conserved during elastic binary collisions (i.e., collisions between two particles).

The other quantities that are conserved during elastic binary collisions are the three components of linear momentum and energy. For a gas comprising a single species of particles, the mass drops out of the conservation relations and the conserved quantities become v_x, v_y, v_z, and $|v|^2$. This suggests that we can express $\ln f(\vec{v})$ as a linear combination of v_x, v_y, v_z, and $|v|^2$, or $\ln f(\vec{v}) = c_0 + c_{1x} v_x + c_{1y} v_y + c_{1z} v_z + c_2 |v|^2$. Remember that we are trying to determine what sort of distribution satisfies this conservation relation, so we cannot yet assume anything about $\ln f(\vec{v})$. However, the c_i are just constants, so we can redefine them any way we want as long as they do not end up depending on the conserved variables. For convenience,[b] let us rewrite $c_0 \equiv \ln f_0 - \beta |\vec{v}_0^2|$, $\vec{c}_1 \equiv 2\beta \vec{v}_0$, and $c_2 \equiv -\beta$. The reasons will become clear shortly.

[b] Statements like this often hide the fact that the author has wrestled with the problem for a while before stumbling upon a useful result. Do not get the impression that this substitution should be obvious at first sight.

Now, we have

$$\ln f(\vec{v}) = \ln f_0 - \beta|\vec{v}_0|^2 + 2\beta\vec{v}\cdot\vec{v}_0 - \beta|\vec{v}|^2$$

$$\ln\left[\frac{f(\vec{v})}{f_0}\right] = -\beta\left|\vec{v}-\vec{v}_0\right|^2$$

$$f(\vec{v}) = f_0\exp\left(-\beta\left|\vec{v}-\vec{v}_0\right|^2\right)$$

The final line has the same form as the Maxwellian distributions in Section 6.1.3.1, which suggests that $f_0 \equiv n\left(2\pi v_{th}^2\right)^{-1/2}$ and $\beta \equiv \left(2v_{th}^2\right)^{-1}$. This is not a rigorous derivation, of course, but the interested reader can find additional details in ter Haar (1954), Hill (1956), Huang (1963), Choudhuri (1998), and Appendix H of Schunk and Nagy (2004).

There are many solutions to the Boltzmann collision integral depending on the particles involved and the forces exerted during the collision. We will discuss the form of two of the most prominent collision operators: the BGK operator, which is used primarily for collisions between charged and neutral species, and the Fokker-Planck operator, which describes collisions between two charged species.

The BGK operator (Bhatnagar et al., 1954) is the simplest approximation to the Boltzmann collision integral. The distribution f_1' in Eq. (6.12) is approximated as a constant, zeroth-order distribution f_0, and the distribution of target particles is approximated as constant during the collisions, that is, $f_2' = f_2$. The result of these harsh approximations is Struchtrup (1997)

$$\left[\frac{\delta f_s}{\delta t}\right]_c = -\nu_s\left(f_s - \frac{n_s}{n_{s0}}f_{s0}\right)$$

The integrals over $d^3v_2 d\Omega$ define the collision frequency as $\nu_s \equiv \int|\vec{v}_1 - \vec{v}_2|\sigma(\Omega) f_2' d^3v_2\, d\Omega$. Physically, the BGK operator represents the collisional relaxation of the distribution over a timescale of $1/\nu_s$ to the zeroth-order distribution f_{s0}, which is often assumed to be Maxwellian. Some works use an even simpler collision operator, which results from dimensional analysis: $\left[\frac{\delta f_s}{\delta t}\right]_c = -\nu_s f_s$. Integrating this simple operator over time shows $f_s(t) = \exp(-\nu_s t)$, and therefore this operator does not conserve particle density. This means the particle conserving term f_{s0} in the BGK operator is necessary for any realistic problem.

The BGK operator is convenient and can be used for a wide range of problems, where the exact form of collisions is not important. However, in many problems involving Coulomb collisions, where two charged particles collide through the Coulomb force, the more sophisticated Fokker-Planck operator is needed. The general form of the Fokker-Planck operator is

$$\left[\frac{\delta f_s}{\delta t}\right]_c = -\frac{\partial}{\partial\vec{v}}\cdot\left[f_s\vec{F}_D\right] + \frac{1}{2}\frac{\partial}{\partial\vec{v}}\frac{\partial}{\partial\vec{v}} : \left[f_s D\right] \tag{6.13}$$

The term $\vec{F}_D$ represents the frictional drag force on the particles and the term D represents a stochastic diffusion process due to the random nature of collisions. Both the drag force vector and the diffusion tensor depend strongly on the velocity of the particle. An equivalent collision frequency for the Fokker-Planck collision operator is

$$\nu_s = \frac{ne^4 \ln\Lambda}{2\pi\epsilon_0^2\mu^2 v^3} \tag{6.14}$$

The Coulomb collision frequency, therefore, depends on the speed of the scattering particle, the Coulomb force ($\propto e^2/\epsilon_0$), density of the target particles, reduced mass, μ, and Coulomb logarithm, $\ln\Lambda$. The Coulomb logarithm represents the ratio of the number of small angle, glancing collisions that occur for each large angle Coulomb collision (more specifically the ratio is $8 \ln\Lambda \approx 100$ for most plasmas). A more detailed description of the Fokker-Planck operator and its drag and diffusion coefficients is in Bellan (2008).

6.1.3.6 Moments of a distribution

Now that we have sketched the fundamentals of kinetic theory, let us see how it connects to a fluid description of plasma. Since the function $f_s(t, \vec{r}, \vec{v})$ describes the position and velocity of a collection of particles, integrating over all velocities must yield the number density:

$$\int_{-\infty}^{+\infty} f_s(t, \vec{r}, \vec{v})\, d^3v = n_s(t, \vec{r}) \tag{6.15}$$

As we previously saw, the Boltzmann equation describes how a particle distribution evolves over time throughout phase space. Based on Eq. (6.15), integrating the first term on the left-hand side (LHS) of Eq. (6.10) over all velocities should yield an expression for how the number density evolves in time:

$$\int_{-\infty}^{+\infty} \frac{\partial f_s}{\partial t}\, d^3v = \frac{\partial}{\partial t}\int_{-\infty}^{+\infty} f_s\, d^3v = \frac{\partial n_s}{\partial t}$$

Integrating the remaining terms should also yield information about how the plasma evolves throughout phase space, and that expression should complement the explicit temporal evolution that the first term contains. The second LHS term of the Boltzmann equation (6.10) gives us

$$\begin{aligned}
\int_{-\infty}^{+\infty} \frac{d\vec{r}}{dt}\cdot\nabla_r f_s\, d^3v &= \int_{-\infty}^{+\infty} \nabla_r\cdot\left(f_s\frac{d\vec{r}}{dt}\right) d^3v - \int_{-\infty}^{+\infty} f_s\nabla_r\cdot\frac{d\vec{r}}{dt}\, d^3v \\
&= \nabla_r\cdot\int_{-\infty}^{+\infty} f_s\frac{d\vec{r}}{dt}\, d^3v - \int_{-\infty}^{+\infty} f_s\frac{d}{dt}\nabla_r\cdot\vec{r}\, d^3v \\
&= \nabla_r\cdot\int_{-\infty}^{+\infty} \vec{v} f_s\, d^3v - 0 \\
&= \nabla_r\cdot\left(n_s\vec{V}_s\right)
\end{aligned}$$

The second RHS term vanishes because $\nabla_r \cdot \vec{r}$ is constant with respect to time, and the transition between the last two lines defines the average velocity:

$$\vec{V}_s \equiv \langle \vec{v} \rangle = \frac{1}{n_s} \int_{-\infty}^{+\infty} \vec{v} f_s(t, \vec{r}, \vec{v}) d^3v$$

Integrating the third LHS term of Eq. (6.10) yields

$$\int_{-\infty}^{+\infty} \frac{d\vec{v}}{dt} \cdot \nabla_v f_s \, d^3v = \int_{-\infty}^{+\infty} \nabla_v \cdot \left(f_s \frac{d\vec{v}}{dt} \right) d^3v - \int_{-\infty}^{+\infty} f_s \nabla_v \cdot \frac{d\vec{v}}{dt} \, d^3v$$

The first RHS term here is an exact integral,

$$\int_{-\infty}^{+\infty} \nabla_v \cdot \left(f_s \frac{d\vec{v}}{dt} \right) d^3v = \left. \frac{d\vec{v}}{dt} \right|_{+\infty} f_s(t, \vec{r}, \vec{v} \to +\infty) - \left. \frac{d\vec{v}}{dt} \right|_{-\infty} f_s(t, \vec{r}, \vec{v} \to -\infty) = 0$$

which vanishes under the very reasonable assumption that no particles have infinite velocities as long as $d\vec{v}/dt \neq 0$. The other term is

$$\int_{-\infty}^{+\infty} f_s \nabla_v \cdot \frac{d\vec{v}}{dt} \, d^3v = \int_{-\infty}^{+\infty} f_s \frac{d}{dt} \nabla_v \cdot \vec{v} \, d^3v = 0$$

which vanishes because $\nabla_v \cdot \vec{v}$ is constant with respect to time.

Thus, the full integral of the LHS of Eq. (6.10) is

$$\frac{\partial n_s}{\partial t} + \nabla_r \cdot \left(n_s \vec{V}_s \right)$$

The result of integrating the RHS of Eq. (6.10) depends on the particular collision term. Using the BGK collision operator,

$$\begin{aligned} \int_{-\infty}^{+\infty} \left[\frac{\delta f_s}{\delta t} \right]_c d^3v &= \int_{-\infty}^{+\infty} -\nu_s \left[f_s - \left(\frac{n_s}{n_{s0}} \right) f_{s0} \right] d^3v \\ &= -\nu_s \left[n_s - \left(\frac{n_s}{n_{s0}} \right) n_{s0} \right] \\ &= 0 \end{aligned}$$

In a spatially homogeneous system, this version implies that the density remains constant. This is a more physically realistic result for collision operators that do not model production or loss.

The results of integrating Eq. (6.10) over all velocities are called the *continuity equation*. In the absence of any processes that create particles (also called "sources"), or processes that destroy particles (also called "sinks"), it reads,

$$\frac{\partial n_s}{\partial t} + \nabla_r \cdot \left(n_s \vec{V}_s \right) = 0 \tag{6.16}$$

Some texts write this in terms of the particle flux, $\vec{\Gamma}_s \equiv n_s \vec{V}_s$, which has units of number of particles per unit area per unit time.

Eq. (6.16) describes the evolution of n_s once we know $\vec{V}_s$. Certain situations may involve a simple expression for $\vec{\Gamma}_s$ or $\vec{V}_s$ that we can plug in, but a self-consistent approach requires deriving a more general expression.

By integrating Eq. (6.10) over all velocities to get Eq. (6.16), we implicitly appealed to a general scheme to derive *moments* of a distribution by averaging some quantity over the distribution. The *n*th velocity moment of a phase-space distribution, $f\left(t, \vec{r}, \vec{v}\right)$, is defined as

$$\langle v^n \rangle = \frac{\int_{-\infty}^{+\infty} v^n f\left(t, \vec{r}, \vec{v}\right) d^3 v}{\int_{-\infty}^{+\infty} f\left(t, \vec{r}, \vec{v}\right) d^3 v}$$

where $\langle v^n \rangle$ may be a function of the position coordinates and time, but will not be a function of the velocity coordinates.

Notice that the *zeroth* moment of Eq. (6.10), which we just derived by effectively multiplying the entire equation by $v^0 = 1$ and integrating over all velocities, contains a term proportional to the *first* moment of the particle distribution, $\vec{V}_s \equiv \langle v^1 \rangle$. We will need to derive the first moment of Eq. (6.10) in order to know how $\vec{V}_s$ evolves. To do so, we will multiply the entire equation by the particle momentum, $m_s \vec{v}$, and integrate over all velocities. We could simply multiply by the velocity, $\vec{v}$, but including the mass lets us connect this to a familiar conserved quantity and makes some of the terms a little easier to work with. It will also be useful to write $d\vec{r}/dt = \vec{v}$ and $m_s d\vec{v}/dt = \vec{F}$, where $\vec{F}$ represents the sum of all external and intraparticle forces relevant to the system of interest.

Multiplying Eq. (6.10) by $m_s \vec{v}$ and integrating the first LHS term by parts yields

$$\begin{aligned}
\int_{-\infty}^{+\infty} m_s \vec{v} \left(\frac{\partial f_s}{\partial t} \right) d^3 v &= \int_{-\infty}^{+\infty} \frac{\partial \left(m_s \vec{v} f_s \right)}{\partial t} d^3 v - \int_{-\infty}^{+\infty} f_s \left(\frac{\partial m_s \vec{v}}{\partial t} \right) d^3 v \\
&= \frac{\partial}{\partial t} \int_{-\infty}^{+\infty} m_s \vec{v} f_s \, d^3 v - 0 \\
&= \frac{\partial \left(n_s m_s \vec{V}_s \right)}{\partial t} = \frac{m_s \partial \vec{\Gamma}_s}{\partial t}
\end{aligned}$$

The second term in the top line vanishes because the phase-space coordinates do not explicitly depend on time.

Integrating the second LHS term in Eq. (6.10) by parts gives us

$$\begin{aligned}\int_{-\infty}^{+\infty} m_s \vec{v}\left(\vec{v}\cdot\nabla_r f_s\right) d^3v &= \int_{-\infty}^{+\infty} \nabla_r \cdot \left(m_s \vec{v} f_s \vec{v}\right) d^3v - \int_{-\infty}^{+\infty} m_s \vec{v} f_s \nabla_r \cdot \vec{v}\, d^3v \\ &\quad - \int_{-\infty}^{+\infty} f_s m_s \vec{v}\cdot\left(\nabla_r \vec{v}\right) d^3v \\ &= \nabla_r \cdot \int_{-\infty}^{+\infty} m_s f_s \vec{v}\vec{v}\, d^3v - 0 - 0 \\ &= \nabla_r \cdot \left(n_s m_s \langle \vec{v}\vec{v}\rangle\right)\end{aligned}$$

The terms containing ∇_r acting on $\vec{v}$ in the top line vanish because $\vec{v}$ and $\vec{r}$ are independent. The second-rank dyadic tensor, $\vec{v}\vec{v}$, in the remaining term contains all nine possible products of the three velocity components. This term is easier to understand if we first split the velocity into an average component, $\vec{V}_s$, and a component that represents only fluctuations about the average, $\delta\vec{v}$. The product $\langle\vec{v}\vec{v}\rangle$ thus expands to $\left\langle\left(\vec{V}_s - \delta\vec{v}\right)\left(\vec{V}_s - \delta\vec{v}\right)\right\rangle = \langle \vec{V}_s\vec{V}_s + \delta\vec{v}\,\delta\vec{v} - 2\vec{V}_s\delta\vec{v}\rangle$, which we can write as $\vec{V}_s\vec{V}_s + \langle\delta\vec{v}\,\delta\vec{v}\rangle - 2\vec{V}_s\langle\delta\vec{v}\rangle$ by virtue of the fact that the average of an average is still the average. Finally, the term $2\vec{V}_s\langle\delta\vec{v}\rangle$ vanishes since the average of the fluctuating term is zero by our definition of $\delta\vec{v}$.

Returning to the second LHS term in Eq. (6.10), we now have

$$\int_{-\infty}^{+\infty} m_s \vec{v}\left(\vec{v}\cdot\nabla_r f_s\right) d^3v = \nabla_r \cdot \left(n_s m_s \vec{V}_s \vec{V}_s\right) + \nabla_r \cdot \left(n_s m_s \langle\delta\vec{v}\,\delta\vec{v}\rangle\right)$$

The quantity $n_s m_s \langle\delta\vec{v}\,\delta\vec{v}\rangle \equiv \mathsf{P}_s$ is defined as the fluid stress tensor.

The third LHS term in Eq. (6.10) gives us

$$\begin{aligned}\int_{-\infty}^{+\infty} \vec{v}\left(\vec{F}\cdot\nabla_v f_s\right) d^3v &= \int_{-\infty}^{+\infty} \nabla_v \cdot \left(\vec{v} f_s \vec{F}\right) d^3v - \int_{-\infty}^{+\infty} \vec{v} f_s \nabla_v \cdot \vec{F}\, d^3v \\ &\quad - \int_{-\infty}^{+\infty} f_s \vec{F}\cdot\left(\nabla_v \vec{v}\right) d^3v\end{aligned}$$

We can use the divergence theorem to transform the first of these terms into a surface integral at $|\vec{v}| = \infty$. If we make the very reasonable assumption that f_s has finite energy, then $f_s \to 0$ faster than $v^{-2} \to 0$ as $v \to \infty$. As a result, the integral over a spherical surface at $|\vec{v}| = \infty$ vanishes regardless of whatever form $\vec{F}$ takes.

The second of these terms contains $\nabla_v \cdot \vec{F}$. Here is where the form of $\vec{F}$ is crucial. When applying Eq. (6.10) to problems in space plasma physics, we can reasonably assume that the Lorentz force, $\vec{F}=q_s\left(\vec{E}+\vec{v}_s\times\vec{B}\right)$, is the dominant force on charged particles. This applies to both the external forces they are likely to encounter in the M-I-T system as well as the forces that they exert on each other. On large enough spatial scales or long enough timescales, $\vec{F}$ may need to include gravity. Substituting the Lorentz force,

$$\begin{aligned}\int_{-\infty}^{+\infty} \vec{v} f_s \nabla_v \cdot \vec{F}\, d^3v &= \int_{-\infty}^{+\infty} \vec{v} f_s \nabla_v \cdot q_s\left(\vec{E}+\vec{v}_s\times\vec{B}\right) d^3v \\ &= q_s \int_{-\infty}^{+\infty} \vec{v} f_s\left[\nabla_v \cdot \vec{E} + \nabla_v \cdot \left(\vec{v}_s\times\vec{B}\right)\right] d^3v\end{aligned}$$

The electric field is not a function of velocity, so $\nabla_v \cdot \vec{E}=0$. The term $\vec{v}_s\times\vec{B}$ *is* a function of velocity but each component of $\vec{v}_s\times\vec{B}$ is orthogonal to the corresponding component of ∇_v. Thus, $\nabla_v \cdot \left(\vec{v}_s\times\vec{B}\right)=0$ as well.

At this point, the third LHS term in Eq. (6.10) has reduced to

$$\begin{aligned}\int_{-\infty}^{+\infty} \vec{v}\left(\vec{F}\cdot\nabla_v f_s\right) d^3v &= \int_{-\infty}^{+\infty} f_s\left(\vec{F}\cdot\nabla_v\right)\vec{v}\, d^3v \\ &= q_s \int_{-\infty}^{+\infty} f_s\left(\vec{E}+\vec{v}_s\times\vec{B}\right)\cdot\left(\nabla_v \vec{v}\right) d^3v\end{aligned}$$

Since $\nabla_v \vec{v}$ is the identity tensor, the second line becomes simply the average over $\vec{E}+\vec{v}_s\times\vec{B}$ and we are left with

$$\int_{-\infty}^{+\infty} \vec{v}\left(\vec{F}\cdot\nabla_v f_s\right) d^3v = q_s n_s\langle\vec{E}+\vec{v}_s\times\vec{B}\rangle = q_s n_s\left(\langle\vec{E}\rangle+\vec{V}_s\times\langle\vec{B}\rangle\right)$$

All that remains is to evaluate the collision integral on the RHS of Eq. (6.10). The first moment of the BGK operator is

$$\begin{aligned}\int_{-\infty}^{+\infty} -\nu_s\left[f_s-\left(\frac{n_s}{n_{s0}}\right)f_{s0}\right] m_s\vec{v}\, d^3v &= -\nu_s\left[\int_{-\infty}^{+\infty} m_s\vec{v} f_s\, d^3v - \left(\frac{n_s}{n_{s0}}\right)\int_{-\infty}^{+\infty} m_s\vec{v} f_{s0}\, d^3v\right] \\ &= -\nu_s n_s m_s\vec{V}_s - \nu_s\left(\frac{n_s}{n_{s0}}\right)\int_{-\infty}^{+\infty} m_s\vec{v} f_{s0}\, d^3v\end{aligned}$$

If we assume f_{s0} is a Maxwellian, the second term contains an odd integrand integrated over an even interval, and therefore vanishes identically. The first moment of this collision operator describes the drag force on the fluid. See Chapter 4 of Schunk and Nagy (2004) for the moments of other collision operators.[c]

Combining all the first-moment terms gives us

$$\frac{\partial\left(n_s m_s \vec{V}_s\right)}{\partial t} + \nabla_r \cdot \left(n_s m_s \vec{V}_s \vec{V}_s\right) + \nabla_r \cdot \mathsf{P}_s - q_s n_s\left(\langle\vec{E}\rangle + \vec{V}_s \times \langle\vec{B}\rangle\right) = -\nu_s n_s m_s \vec{V}_s$$

In cases when there are no particle sources or sinks, such that Eq. (6.16) holds, we can further simplify this equation. The first step is to rewrite the quantity $\nabla_r \cdot \left(n_s m_s \vec{V}_s \vec{V}_s\right)$ in a more useful form by expanding both vectors into components, writing out all nine terms in the product, and rearranging. The process is not trivial, but with some patience, you can convince yourself that $\nabla_r \cdot \left(n_s m_s \vec{V}_s \vec{V}_s\right) = \left(n_s m_s \vec{V}_s \cdot \nabla_r\right)\vec{V}_s + \vec{V}_s\left[\nabla_r \cdot \left(n_s m_s \vec{V}_s\right)\right]$.

The next step is to expand the first two LHS terms and use Eq. (6.16):

$$\begin{aligned}
\frac{\partial\left(n_s m_s \vec{V}_s\right)}{\partial t} + \nabla_r \cdot \left(n_s m_s \vec{V}_s \vec{V}_s\right) &= n_s m_s \frac{\partial \vec{V}_s}{\partial t} + m_s \vec{V}_s \frac{\partial n_s}{\partial t} + \left(n_s m_s \vec{V}_s \cdot \nabla_r\right)\vec{V}_s \\
&\quad + \vec{V}_s\left[\nabla_r \cdot \left(n_s m_s \vec{V}_s\right)\right] \\
&= n_s m_s\left[\frac{\partial \vec{V}_s}{\partial t} + \left(\vec{V}_s \cdot \nabla_r\right)\vec{V}_s\right] \\
&\quad + m_s \vec{V}_s\left[\frac{\partial n_s}{\partial t} + \nabla_r \cdot \left(n_s \vec{V}_s\right)\right] \\
&= n_s m_s\left[\frac{\partial \vec{V}_s}{\partial t} + \left(\vec{V}_s \cdot \nabla_r\right)\vec{V}_s\right] + 0
\end{aligned}$$

It is common to move all the terms containing pressures, forces, and collisions to the RHS, to indicate that these terms drive the change in momentum. Finally, the first moment of Eq. (6.10) yields the *momentum equation*:

$$n_s m_s\left[\frac{\partial \vec{V}_s}{\partial t} + \left(\vec{V}_s \cdot \nabla_r\right)\vec{V}_s\right] = q_s n_s\left(\langle\vec{E}\rangle + \vec{V}_s \times \langle\vec{B}\rangle\right) - \nabla_r \cdot \mathsf{P}_s - \nu_s n_s m_s \vec{V}_s \tag{6.17}$$

[c] Schunk and Nagy (2004) also show how the first moment of the BGK operator is $-\nu_s n_s m_s(\vec{V}_s - \vec{V}_t)$ when the target species t has an average velocity $\vec{V}_t$.

As with Eq. (6.16), Eq. (6.17) contains a term with the next highest moment of f_s—in this case, $\nabla_r \cdot \mathsf{P}_s = \nabla_r \cdot \left(n_s m_s \langle \delta \vec{v}\, \delta \vec{v} \rangle\right)$. Just as we multiplied Eq. (6.10) by $m_s\, \vec{v}$ to derive an expression for the time dependence of $n_s \vec{V}_s$ when $\nabla_r \cdot \left(n_s \vec{V}_s\right)$ arose in Eq. (6.16), we can multiply Eq. (6.10) by $m_s\, \vec{v}\,\vec{v}$ to derive an expression for the time dependence of P_s. We can also multiply Eq. (6.10) by $m_s v^2$ to learn how the energy of the distribution evolves. Of course, the resultant second-moment equations will all contain third moments and so on. This process will never produce a closed system of equations, since the equation for the evolution of the nth moment will always contain a term proportional to the $(\mathrm{n}+1)$th moment. We must always close the system with some relationship based on our understanding of the physics. In many cases, a simple form of the pressure tensor, $\mathsf{P}_s\left(t, \vec{r}\right) = n_s\left(t, \vec{r}\right) k_B T_s \mathsf{I}$, where T_s is a constant temperature and I is the identity tensor, will adequately describe the system. See Chapter 3 of Schunk and Nagy (2004) for a more detailed discussion of deriving a closed system of fluid transport equations from a kinetic starting point.

There are many systems that we can describe as a single fluid because frequent collisions drive the distribution function back to equilibrium whenever something perturbs it. Conversely, there are plenty of systems for which only a kinetic treatment will suffice (Khazanov et al., 2012). There are also intermediate cases. For one example, we may find that describing a plasma consisting of three different species as three separate but interacting fluids explains a particular phenomenon with sufficient accuracy. For another example, we may find that we only care about the kinetic effects of one species while treating the other species as fluids. The hybrid simulation model makes use of this assumption.

This section has shown how to formally pass from a kinetic description to a fluid description, with the caveat that we will need some additional physical assumptions in order to truncate the series of moment equations. When the particle distribution is Maxwellian, second- and higher-order moments either vanish or take on simple forms, such as the one shown above for the pressure tensor. This is one reason why the Maxwellian distribution is so useful. It also represents a potential analytical trap: prematurely truncating the moment equations despite the fact that the particle distribution under study does not behave like a Maxwellian leads to an incomplete analysis and may exclude the most interesting physics!

6.1.3.7 Linearization and perturbation theory

The Boltzmann equation accurately describes the kinetic behavior of a plasma evolving in time. However, this equation is a nonlinear partial differential equation, or even a nonlinear integrodifferential equation when including the collision integral, and does not have a general, analytic solution. The process of linearization is a standard method for obtaining analytic solutions of the Boltzmann equation by assuming the system is a linear combination of plane waves. Linearization is done by taking the Laplace transform in time and the Fourier transform in space of the Boltzmann equation, and solving the resulting algebraic equation. The Laplace transform of a function g from the time domain t to the frequency domain ω is defined as

$$g(\omega) = \int_0^{\infty} g(t) e^{i(\omega - i\gamma)t} dt \tag{6.18}$$

and the Fourier transform from the spatial domain to the wave number (k) domain is

$$g(\vec{k}) = \int_{-\infty}^{\infty} g(\vec{r}) e^{-i\vec{k}\cdot\vec{r}} d\vec{r} \tag{6.19}$$

Note that other sign conventions exist, but do not change the physics of the problem when interpreted consistently.

The Laplace transform is a generalization of the typical Fourier transform. The Fourier transform is recovered from Eq. (6.18) by taking $\gamma = 0$ and extending the integration bounds[d] from $-\infty$ to ∞. While the Fourier transform seeks to express the function $g(t)$ in an infinite series of sine and cosine functions, the Laplace transform expresses $g(t)$ as an infinite series of *exponentially damped* sine and cosine functions. From a physical view of plasmas, we expect solutions for waves to have some kind of damping source, which at the kinetic level can be provided by collisions (Section 6.1.3.5) or Landau damping (Section 6.1.3.8). Wave growth and instabilities can also be described by the Laplace transform as long as the wave amplitude grows exponentially, which is referred to as the linear growth phase.

The utility of the Fourier-Laplace transform comes from how it transforms the Boltzmann equation. Many standard texts detail the mathematics of the Laplace transform (Arfken et al., 2012), but the primary rules of interest are that the time and spatial derivatives of the distribution f transform as

$$\frac{\partial}{\partial t} f(t, \vec{r}, \vec{v}) \rightarrow -i(\omega - i\gamma) f(\omega - i\gamma, \vec{r}, \vec{v}) - if(t=0, \vec{r}, \vec{v}) \tag{6.20}$$

$$\vec{v} \cdot \nabla_r f(t, \vec{r}, \vec{v}) \rightarrow i(\vec{k} \cdot \vec{v}) f(t, \vec{k}, \vec{v}) \tag{6.21}$$

Applying the Fourier-Laplace transform to the Vlasov equation yields (Drummond and Pines, 1962)

[d] From a signal processing point of view, the integration starting at $t = 0$ in Eq. (6.18) is equivalent to invoking causality.

$$-i(\omega - i\gamma)f_s(\omega - i\gamma, \vec{k}, \vec{v}) - if_s(t=0, \vec{k}, \vec{v}) + i(\vec{k} \cdot \vec{v})f_s(\omega - i\gamma, \vec{k}, \vec{v}) \\ = +\frac{q_s}{m_s}\int_0^\infty dt\, e^{i(\omega - i\gamma)t}\int d\vec{\xi}\; \vec{E}(t, \vec{k} - \vec{\xi}) \cdot \nabla_v f_s(t, \vec{\xi}, \vec{v}) \tag{6.22}$$

The Fourier-Laplace transform of the Vlasov equation no longer contains time or spatial derivatives and is in principle easier to solve. However, the integral term represents the nonlinear coupling of different wave modes ($\vec{k}$ and $\vec{\xi}$). The process of *linearization* expresses the distribution as the sum of a zeroth-order part that is constant in time and space, and a first-order perturbation:

$$f_s(\omega, \vec{k}, \vec{v}) = F_{0,s}(\vec{v}) + f_{1,s}(\omega, \vec{k}, \vec{v})$$

This expansion of the distribution can be substituted into Eq. (6.22). Retaining only first-order terms, the linearized Vlasov equation is then

$$-i(\omega - i\gamma)f_{1,s}(\omega, \vec{k}, \vec{v}) - if_{1,s}(t=0, \vec{k}, \vec{v}) + i(\vec{k} \cdot \vec{v})f_{1,s}(\omega, \vec{k}, \vec{v}) \\ = \frac{q_s}{m_s}\vec{E}(\omega, \vec{k}) \cdot \nabla_v F_{0,s}(\vec{v}) \tag{6.23}$$

The linearization of the Boltzmann equation follows the same method once the collision operator is specified.

The benefit of linearizing the Vlasov equation is that a nonlinear differential equation can now be solved as a linear algebraic equation. Furthermore, many steady-state solutions such as wave dispersion relations do not require the initial condition term, $f_s(t = 0)$, and it can be dropped. However, for some problems, the process of linearization is too harsh as it neglects the wave coupling term, $(q_s/m_s)\int d\vec{\xi}\,\vec{E}(\vec{k} - \vec{\xi}) \cdot \nabla_v f_s(\omega - i\gamma, \vec{\xi}, \vec{v})$, in Eq. (6.22). This limitation on linear theory is typically violated when the wave electric field has a high amplitude, such as whistler waves in the radiation belts. Quasilinear theory is a method for solving such problems where the wave coupling term is retained—see Drummond and Pines (1962) and Nicholson (1983).

6.1.3.8 Landau damping

Landau damping is a kinetic plasma phenomenon that does not have a fluid or neutral gas analog since it involves the interaction of charged particles with wave electric fields. A common viewpoint[e] of Landau damping is that particles near the phase speed of a wave ($v = \omega/k$) see a nearly constant electric field, which accelerates or decelerates particles toward the phase speed. This can be seen from the solution for Langmuir waves, which

[e] This viewpoint breaks down when the finite amplitude of the wave electric field is considered; see Drummond (2004).

solve Eq. (6.23) with only an electron distribution. The resulting dispersion relation is Nicholson (1983)

$$1 = \frac{\omega_p^2}{k^2} \int_{-\infty}^{\infty} \frac{\nabla_v F_0(v)}{v - (\omega - i\gamma)/k} dv \tag{6.24}$$

The solution for the dispersion relation requires integrating over velocity, which runs into a problem from the singularity at

$$v = \frac{\omega - i\gamma}{k} \tag{6.25}$$

If the Fourier transform was used instead of the Laplace transform, the singularity would be on the path of integration at $v = \omega/k$. The mathematical solution of Landau damping arises from how this singularity is handled. The correct way of solving the integral is to deform the integration contour around the singularity, and then use the Plemelj theorem. This results in a complex-valued dispersion relation, which in a simplified 1D geometry is (Chen et al., 2015)

$$\omega = \omega_p \left(1 + i \frac{\pi \omega_p^2}{2 k^2} \left[\frac{\partial f_0}{\partial v} \right]_{v = \omega_p/k} \right) \tag{6.26}$$

where $f_0 = F_0/n_0$, and the plasma frequency is defined as $\omega_p^2 = ne^2/m_e\epsilon_0$.

The imaginary component of ω is the Landau damping term, and is directly proportional to the gradient of $F_0(v)$ evaluated at the phase speed of the wave. Different wave modes such as the ion-acoustic mode or whistler mode lead to different dispersion relations, but the Landau damping is always dependent on the gradient of $F_0(v)$ evaluated at the phase speed of the wave. For a Maxwellian distribution, this derivative is negative and the wave experiences a net damping force. However, an instability can occur when a bump-on-tail distribution is present, such as shown in Fig. 6.7. The bump-on-tail distribution has a positive derivative near the bump, and waves will be generated by electrons moving at the phase speed. This process is called inverse Landau damping.

The wave-particle interaction of Landau damping is a fundamental result of kinetic theory. For a magnetized plasma, Landau damping is further generalized to the case of cyclotron damping, where particles moving at the speed

$$v = \frac{\omega - n\Omega}{k} \tag{6.27}$$

can damp or enhance a wave. In this case, an integer multiple of the gyrofrequency (Ω) matches the rotation of the electric field, causing the particle to see a nearly constant electric field. In the equatorial ionosphere inverse cyclotron damping can generate an instability that is seen by radars as 150 km echoes (Longley et al., 2020).

Landau and cyclotron damping is one of the most important phenomenon neglected when moving from a kinetic to fluid description of the plasma. In most fluid cases, the distributions are assumed to be Maxwellian, in which case inverse Landau and cyclotron damping cannot occur. Since Landau damping occurs at velocities proportional to the wavelength ($v = \omega/k \propto \lambda$), the fluid description means only long wavelength behavior can be studied. This is an important caveat for phenomena such as the Farley-Buneman instability in the E-region ionosphere (Section 6.1.4.1).

6.1.4 Example phenomena

This section briefly describes four phenomena within the scope of M-I-T coupling in which kinetic effects play a meaningful role. The phenomena are: (1) the Farley-Buneman instability, which can significantly heat E-region electrons during geomagnetic storms; (2) dispersive Alfvén waves, which can accelerate electrons along magnetic field lines; (3) ion upflow and outflow, which transports ions against gravity and potentially out of the ionosphere; and (4) precipitation, which increases ionospheric conductivity.

6.1.4.1 Farley-Buneman instability

The Farley-Buneman instability (FBI), or "modified two-stream instability" (Farley, 1963a, b; Buneman, 1963), as well as the gradient drift instability (GDI) (Simon, 1963; Hoh, 1963; Maeda et al., 1963) and thermal instabilities (Dimant and Sudan, 1997; Kagan and Kelley, 2000; Dimant and Oppenheim, 2004), occur at altitudes between roughly 80 and in the ionosphere. In this highly dissipative region of the lower ionosphere, electrons are strongly magnetized ($\Omega_e \gg \nu_{en}$), while ions are mostly unmagnetized ($\nu_{in} \gg \Omega_i$) due to frequent collisions with the dense neutral atmosphere. Here, we use Ω_s to stand for the gyrofrequency of species s and ν_{sn} to represent the frequency of collisions between species s and neutral particles.

The major driving force behind all these instabilities is a sufficiently strong DC electric field $\vec{E}_0$ perpendicular to the geomagnetic field, $\vec{B}_0$. Due to the high electron mobility along $\vec{B}_0$, there is almost no component of $\vec{E}_0$ parallel to $\vec{B}_0$ above an altitude of 75 km. At E-region altitudes, where the electrons and ions have dramatically different magnetization properties, this DC field leads to a strong $\vec{E}_0 \times \vec{B}_0$-drift of magnetized electrons, while unmagnetized ions are collisionally tied to the motion of neutrals. The net result of this disparity is the formation of intense currents, called *electrojets*, in the $\vec{E}_0 \times \vec{B}_0$ direction.

Electrojet currents often develop waves through a number of instability processes. The FBI occurs when the electron velocity becomes supersonic and the ions can no longer quickly cancel any electric fields that develop due to small perturbations in the density. These perturbations grow into fully turbulent waves that disrupt radio signals and cause large radar echoes.

All collisional E-region instabilities result in excitation of low-frequency waves in which quasineutral plasma density perturbations travel parallel to the wave vector, $\vec{k}$. Perturbation wavevectors are nearly perpendicular to $\vec{B}_0$, but small components along $\vec{B}_0$ may be important to the nonlinear instability saturation. Small parallel components may play a crucial role in an effect known as anomalous electron heating (AEH), which has been observed by radars since the early 1980s (e.g., St.-Maurice et al., 1981). More recent observations (e.g., Foster and Erickson, 2000; Bahcivan, 2007) have shown that the FBI can heat electrons by an order of magnitude through AEH. This electron heating can indirectly increase the ionospheric conductance by reducing recombination. In addition, the FBI turbulence directly modifies the ionospheric conductance through the effect of nonlinear current and anomalous transport (see Dimant and Oppenheim, 2011). Particle-in-cell simulations by Oppenheim and Dimant (2013) showed AEH on a kinetic scale, and MHD simulations by Wiltberger et al. (2017) showed that the attendant effects can play a significant role in M-I coupling.

Much of the FBI behavior can be described as a fluid, since the typical wavelengths are often much greater than the characteristic ion-neutral *mean free path*. However, the fact that ion Landau damping efficiently quenches shorter-wavelength perturbations means that a full description requires kinetic theory. An analysis by Schmidt and Gary (1973) showed that the instability growth rate based on a fluid treatment increases in proportion to k^2, whereas the instability growth rate based on a kinetic treatment reaches a maximum determined by local physical parameters. In other words, the fluid growth rate implies that infinitesimally small waves grow instantaneously, whereas the kinetic growth rate predicts waves with wavelengths in a certain range will grow most quickly, while others either fail to grow or grow more slowly. The former is nonphysical and the latter is consistent with observations.

There are two major areas where sufficiently strong DC fields perpendicular to $\vec{B}_0$ form electrojets and drive the E-region instabilities: (1) around the geomagnetic equator and (2) at high latitudes. The origin of the strong DC field in the equatorial and high-latitude areas is vastly different. At the magnetic equator, a moderate DC field (typically in the range of 10–20 mV/m) appears as a result of neutral atmospheric winds generating a collisional dynamo (Kelley, 2009). At the high latitudes, typically a much stronger field (often within the 20–200 mV/m range) maps down the magnetic field lines from the Earth's magnetosphere. This usually happens during magnetic storms or other perturbing events. The high-latitude E region is very important for M-I coupling because field-aligned currents (FACs) close there, affecting the entire near-Earth plasma environment.

At mid-latitudes, quiet-time electric fields are much weaker and usually cannot drive E-region instabilities. However, FBI-like plasma turbulence has been observed by radars at certain locations around sporadic-E layers, where the weak ambient electric field can be amplified due to electric polarization of dense plasma structures (Haldoupis et al., 1996; Shalimov and Haldoupis, 2005). Electric fields associated with a geomagnetic disturbance phenomenon called subauroral polarization streams (SAPS) can easily reach the FBI threshold (Foster et al., 2004), and although increased electron mobility along the geomagnetic field at mid-latitudes (cf. Kelley, 2009) may cancel the electric field before producing measurable FBI irregularities, disturbed mid-latitude electric fields can lead to a range of other density perturbations (Eltrass et al., 2014; Zakharenkova and Cherniak, 2020).

The peak wavelength for FBI growth in the E-region ionosphere is a few meters, which is small compared to typical length scales involved in M-I-T coupling. However, there is compelling evidence that the FBI couples to the GDI when the latter creates isolated electric fields strong enough to drive the former (Pfaff et al., 1987a, b; Kudeki et al., 1987; Young et al., 2019). Once initiated, the patches of FBI heat the plasma and move it across magnetic field lines, thereby increasing the local conductivity and subsequently reducing the GDI electric field. Although the GDI is itself a purely fluid instability, this coupling to the FBI makes kinetic considerations important.

6.1.4.2 Dispersive Alfvén waves

The fluid-based formalism of MHD includes three fundamental wave modes: the fast magnetosonic mode, the slow magnetosonic mode, and the Alfvén mode. The first two are analogous to sound waves in a neutral gas and the third is the magnetic-field-line analog of a wave on a string. MHD has proven to be a powerful tool when analyzing plasmas that evolve on large spatial or long temporal scales, but its assumptions ignore any potential kinetic effects. This section will describe how Alfvén waves differ from their MHD form in some of those cases. However, the reader should bare in mind that compressive magnetic-field perturbations, which are characteristic of magnetosonic modes, are also present in Alfvén waves when kinetic effects are important. This suggests that a complete analysis should consider all three modes (Hollweg, 1999).

One type of non-MHD Alfvén wave is the inertial Alfvén wave (IAW), for which the electron thermal speed, $V_{th,e}$, is less than the Alfvén wave speed, V_A. Another type is the kinetic Alfvén wave (KAW), for which $V_{th,e} > V_A$. Stasiewicz et al. (2000) group IAW and KAW under the term dispersive Alfvén waves (DAW). Both IAW and KAW extend the field-aligned (or shear) Alfvén waves of MHD to include an electric field parallel to the mean magnetic field. They are therefore associated with electron acceleration above auroral structures. In the IAW case, the parallel electric field arises when the wavelength perpendicular to the mean magnetic field is on the order of the electron inertial length (or skin depth), $d_e \equiv c/\omega_{pe}$, which is a measure of extent to which a low-frequency electromagnetic wave can penetrate a collisionless plasma. In the KAW case, the parallel electric field is a result of the nonnegligible gradient in electron pressure parallel to the mean magnetic field. Despite the different parameter regimes, Lysak and Lotko (1996) provide a dispersion relation of which IAW and KAW are limiting cases.

There have been numerous observations of DAW and their effects in M-I coupling. The *Polar* spacecraft observed a correlation between electromagnetic energy flux carried by Alfvén waves (i.e., Alfvén Poynting flux) in the magnetosphere and UV emission in the auroral ionosphere (Wygant et al., 2002). A statistical study of Alfvén-wave Poynting flux and energy flux of field-aligned electrons by Chaston et al. (2003) showed that over the altitude range covered by the *FAST* satellite, Poynting flux decreased with altitude in a way consistent with a picture of Alfvén waves accelerating electrons into the ionosphere.

Because simultaneously measuring accelerated electrons and the accelerating Alfvén wave in the auroral ionosphere is extremely challenging, Schroeder et al. (2017) devised a laboratory experiment in which they launched a series of IAWs through an isolated plasma column and measured the electron distribution function. They were able to definitively verify that the parallel electric field in the IAW accelerated electrons to energies consistent with those measured in the auroral ionosphere. Their analysis included a technique for simplifying Eq. (6.10) by averaging over the gyroperiod of traveling electrons by justifying the assumption that the phase of each electron's gyration about the mean magnetic field did not significantly affect their results.

A combined fluid/kinetic model by Watanabe (2014) of M-I coupling showed how KAWs help to couple the magnetosphere to the ionosphere by carrying FACs. When the perpendicular wavelength of shear Alfvén waves becomes short enough, kinetic effects on the spatial scale of the ion gyroradius and temporal scale of the electron inertial response become important. According to their model, parallel electric fields in magnetospheric KAWs accelerate electrons into the ionosphere as ionospheric perturbations in density, FAC, and perpendicular electric field develop. These perturbations give way to a feedback instability, which self-consistently enhances electron precipitation. In turn, the precipitating electrons enhance instability growth by ionizing neutral particles.

6.1.4.3 Ion upflow and outflow

Ion upflow and outflow are important components in transporting heavy ions (mostly O^+) from the ionosphere to the magnetosphere, mass loading the solar wind, and possibly even atmospheric evolution (Yamauchi, 2019). The community has historically divided ion upflow into two phenomenological types (cf. Wahlund et al., 1992): Type-1 ion upflow occurs in regions of downward field-aligned currents (FAC) adjacent to auroral arcs, where ions develop a greater temperature perpendicular to the magnetic field than parallel ($T_\perp > T_\parallel$). This temperature anisotropy leads to pressure gradients that give the ions enough energy to flow upward, against the current. Note that, unlike in a neutral gas, we can think of the temperature of charged species as having "directions" because the magnetic force only acts perpendicular to the magnetic field. This is a distinctly kinetic characteristic. Type-2 ion upflow occurs above auroral arcs, when precipitating electrons heat ionospheric electrons, after which energized electrons travel upward along magnetic field lines and temporarily leave the ions behind. This separation sets up an ambipolar electric field, which eventually drags the ions upward to preserve the quasineutral state of the plasma. Here, electron precipitation is the relevant kinetic process; it arises in the treatment of the collision integral and becomes a source term in the fluid continuity equation.

Wave-particle interactions can create non-Maxwellian ion distributions by preferentially energizing a portion of the ions perpendicular to the magnetic field. These distributions are anisotropic because they are not the same in all directions in velocity space. Such anisotropic distributions can arise when the electric field within low-frequency broadband waves preferentially accelerate ions transverse to the magnetic field (Kintner et al., 1992, 1996). André et al. (1994) showed non-Maxwellian velocity distributions observed by the Freja satellite during an ion heating event. Specifically, distributions of both H^+ and O^+ became elongated in the $v_\perp$ dimension, indicating a mechanism that preferentially increases ion energy perpendicular to the magnetic field. The authors attributed the observed heating to a two-step process in which waves with frequencies at or near the ion cyclotron frequency gradually heat the ions to the point where lower hybrid waves, which they also observed (Eriksson et al., 1994; Eliasson et al., 1994), could take over and more efficiently accelerate the ions.

Two other examples of anisotropic upflowing ion distributions occur in the related phenomena of ion beams and ion conics (Welling et al., 2015). Ion beams comprise ions with energies from 10s of eV to a few keV flowing out of the ionosphere along the magnetic field, and tend to occur at altitudes between 5000 and 8000 km. Ion conics also comprise ions with energies from 10s of eV to a few keV traveling from the ionosphere toward the magnetosphere, but unlike ion beams, their pitch-angle distributions peak at an angle to the magnetic field and have a minimum along it. Shelley et al. (1976) and Sharp et al. (1977) were the first to report conclusive observations of ion beams and conics, respectively, in satellite data. They attributed ion beams to a process that preferentially accelerates ions in the direction parallel to the magnetic field, and attributed ion conics to a process that preferentially accelerates ions in the plane perpendicular to the magnetic field. Ion beans and conics are both common, but their occurrence frequencies differ in both latitude and altitude (Yau and André, 1997).

The kinetic processes that accelerate ions in the high-latitude ionosphere act as a source of plasma for the magnetosphere, and even the solar wind. This runs contrary to the popular image of M-I-T coupling in which the solar wind provides particles and energy to the magnetosphere, which the magnetosphere stores until its time to create the aurora. However, this outward flux of plasma and its significant implications to the M-I-T system has been studied since the 1960s (see Moore et al., 1999) and continues to be a topic of interest (see Welling et al., 2015).

6.1.4.4 Precipitation

Charged-particle precipitation from the magnetosphere into the ionosphere forms a core component of the coupled M-I system. Weak electron and ion precipitation from the magnetosphere into the ionosphere creates a diffuse band, several hundred kilometers wide, around the entire auroral oval (Roeder and Koons, 1989). This is known as the diffuse aurora. Despite lacking the isolated spikes in energy density that accompany FACs and auroral arcs, precipitating electrons contribute the majority of the diffuse aurora, which itself accounts for 71%–84% of the incoming ionospheric energy flux (Newell et al., 2009; Ni et al., 2012).

Whereas field-aligned potential drops accelerate electrons to produce the familiar discrete auroral arcs, a different process drives the diffuse-auroral precipitation. Magnetospheric phenomena that cause electrons to leave the plasmasphere and travel along magnetic field lines into the ionosphere must interact with electrons on the temporal and spatial scales of electron motion, while doing so over hundreds of kilometers in longitude (not to mention the considerably longer distances electrons must travel along a field line). The major consequence of electrons precipitating into the ionosphere is a change in the ionospheric conductivity, which in turn affects a multitude of M-I coupling processes (Khazanov et al., 2003). In addition to increasing the conductivity in one region of ionosphere, precipitating electrons can produce secondary electrons that reflect back along closed magnetic field lines and redistribute energy to the magnetosphere (Khazanov et al., 2014).

What causes the increased flux of electrons into the ionosphere? The electrons that comprise diffuse-auroral precipitation come from the central plasma sheet in the magnetosphere, where they would normally be destined to drift adiabatically in Earth's magnetic field. However, those electrons share the magnetosphere not only with ions, but also with a variety of wave modes. There are two types of waves in particular that can effectively scatter a plasma-sheet electron into the ionosphere by changing its *pitch angle*. Waves of the first type are called electromagnetic whistler-mode chorus, or often simply "chorus," waves. This so-called chorus is actually a collection of right-hand circularly polarized waves—the same sense as the electron gyration about the magnetic field—with frequencies below the local *gyrofrequency*. Waves of the second type are called electron cyclotron harmonic (ECH) waves. They are electrostatic waves with frequencies between harmonics of the electron gyrofrequency.

Thorne et al. (2010) presented results from a kinetic diffusion analysis that showed that chorus waves play the dominant role in pitch-angle scattering plasma-sheet electrons into the ionosphere. A subsequent pair of reports analyzed in detail the contributions of chorus (Ni et al., 2011b) and ECH waves (Ni et al., 2011a) to diffuse-auroral precipitation. They found that ECH played a negligible role in scattering electrons from the magnetosphere compared to chorus waves, making chorus waves the dominant producer of diffuse precipitating electrons from around geostationary orbit. Farther out in the magnetosphere, ECH can play a more central role (Ni et al., 2012).

6.1.5 Summary

This section has motivated the need for kinetic description of a plasma, outlined the connection between the kinetic and fluid descriptions of a plasma, and provided some examples of kinetic phenomena in magnetosphere-ionosphere-thermosphere coupling. The authors hope that it has not only encouraged the reader to look beyond the fluid description of whatever phenomenon they are studying, but also shown them that kinetic physics is not merely the purview of the theoretician. Although the kinetic framework begins by considering the smallest elements of a plasma—the constituent electrons, ions, and neutrals—examples such as enhanced conductivity due to the Farley-Buneman instability or ion outflow due to temperature anisotropy illustrate how important kinetic effects can be to large-scale processes. This section is by no means exhaustive; rather, it should provide a starting point from which the interested researcher can further their understanding about kinetic processes in magnetosphere-ionosphere-thermosphere coupling. We especially hope that the key concepts in Section 6.1.3 will provide a useful reference to common topics in kinetic theory for anyone interested in learning more about where and when kinetic phenomena matter to their research.

6.2 Fast calculation of particle-impact ionization from precipitating energetic electrons and protons in the Earth's atmosphere

Xiaohua Fang

Laboratory for Atmospheric and Space Physics, University of Colorado, Boulder, CO, United States

6.2.1 Introduction

Energetic particle precipitation globally and continuously takes place within the Earth's atmosphere. Precipitating particles at different horizontal and vertical locations have different origins with different energies and types (primarily electrons and protons). The source regions of incident particles include the Earth's magnetosphere (due to convection and scattering loss of plasma sheet, radiation belt, and ring current particles), the Sun and heliosphere (solar energetic particles [SEPs]), and outside of the solar system (galactic cosmic rays [GCRs]). Particle precipitation of the magnetospheric origin focuses on high-latitude auroral regions and is extended to subauroral and mid-latitude regions, constituting a major part of particle impact of interest to the magnetosphere-ionosphere-thermosphere (MIT) coupling community (e.g., Lyons, 1997). The precipitating energy flux of these particles is carried mostly by electrons (e.g., Hardy et al., 1987; Newell et al., 2009) and in a part by protons (e.g., Hardy et al., 1989; Fang et al., 2007b; Emery et al., 2008), with a single particle energy of 100s eV up to 100s keV. SEPs are another category of precipitating particles, which are dominated by protons and get accelerated either in association with solar flare eruptions on the Sun or by shocked solar winds particularly driven by fast coronal mass ejections (e.g., Reames, 1999; Desai and Giacalone, 2016). The SEP energy is of the order of MeV and is extended to keV at the low-energy end and occasionally as high as GeV during ground-level enhancement events. The precipitation of these particles that are external to the magnetosphere is subject to the shielding of geomagnetic fields. The SEP bombardment is concentrated in the polar regions (e.g., Jackman et al., 2008), where magnetic field lines are open to the interplanetary space and the associated cutoff rigidity is relatively low in comparison with that at low latitudes (e.g., Smart et al., 2000). GCRs, which are dominated by energetic protons, represent an isotropic population of extrasolar origin (e.g., Gaisser 1990). With a high energy from $\sim$1 MeV up to $\sim 10^3$ GeV, precipitating GCRs are present everywhere from the ground up to the stratosphere (e.g., Usoskin and Kovaltsov, 2006). Because of the deep penetration and involvement of various hadronic, electromagnetic, and muonic processes, GCRs have a negligible impact on the MIT system and are not discussed here.

Over the penetration course within the atmosphere, incident energetic electrons and protons suffer a series of complex collisional processes with ambient neutrals, including elastic and inelastic (such as ionization, excitation, dissociation) collisions. Precipitating particles are subject to energy cascading and angular scattering as a result of collisional interactions with the atmosphere, whose occurrence frequency increases dramatically as the density of the encountered atmosphere exponentially increases with increasing penetration depth. Unlike electron transport, other types of inelastic collisions for impinging protons take place with charge change involved. Incident protons may be neutralized by capturing an electron from the background atmosphere and become fast hydrogen atoms: $H^+ + M \rightarrow H + M^+$ (charge exchange). Here, H^+ and H indicate an energetic proton and hydrogen atom, respectively. M stands for an atmospheric neutral species. Newly created hydrogens may subsequently turn back into protons through $H + M \rightarrow H^+ + M + e^-$ (electron stripping). Frequent charge exchange and electron-stripping collisions, below 300 km altitude for auroral protons (e.g., Fang et al., 2004), require a simultaneous solution of the tightly coupled H^+-H transport for precipitating protons. Moreover, there is another complexity for the proton precipitation problem. Unlike charged particles of electrons and protons that are constrained by the Lorentz force to spiral around local magnetic field lines, the generated hydrogen atoms retain most of the proton energy but are able to freely move across magnetic field lines. The resulting spreading effect is unique to precipitating proton beams (Davidson, 1965), requiring a special attention to relatively less energetic auroral protons (Fang et al., 2004, 2005).

One of the central focuses of particle precipitation studies is on the quantification of the particle-impact atmospheric ionization, which is directly related with the intensification of ionospheric densities and electrical conductivities. These properties are well known to have a profound global impact, for example, on ionospheric electrodynamics, field-aligned currents, Joule heating, magnetic reconnection, and magnetospheric convection (e.g., Wolf, 1970; Ridley et al., 2004). As a result, being one of the first links in the MIT interplay chain, particle precipitation plays a critical role and requires an accurate description. There are several ways that particle-impact ionization happens, in which neutral atmospheric species are positively charged through the loss of an electron. One ionizing source is direct ionization collisions over the course of electron and proton (and newly generated hydrogen) transport. The other is through ionization collisions by energetic secondary electrons, which are produced in either direct ionization collisions or electron-stripping reactions of hydrogen atoms during the proton transport. Moreover, a part of atmospheric ionization is produced through charge exchange collisions of precipitating protons with neutral constituents. There is another ionizing source for high-energy electron precipitation. The deceleration of energetic electrons by the atmosphere is accompanied by a process called bremsstrahlung (from "braking radiation" in German), in which part of the kinetic energy of slowed-down electrons are converted to electromagnetic radiation in a wavelength range near X-rays. The secondary ionization by bremsstrahlung photons occurs at significantly lower altitudes and with an intensity of two to four orders of magnitude smaller than the main particle-impact ionization peak (Rees, 1964; Frahm et al., 1997; Xu et al., 2018). The bremsstrahlung-induced ionization is minimal in the MIT system and also insignificant for the nearby atmospheric chemistry according to Xu et al. (2018). The bremsstrahlung effect is neglected in our discussions.

6.2.2 Particle-impact ionization calculation

Given the aforementioned complexity of energetic particle interactions with the atmosphere, the transport and the resulting ionization cannot be analytically obtained. Numerous computational models have been developed with different sophistications under various approximations and assumptions, which may be roughly grouped into three categories: range calculation under a continuous-slowing-down approximation (CSDA), numerical solution of theoretical transport equations, and Monte Carlo simulation of collisional processes.

6.2.2.1 CSDA range calculation

In a CSDA model, an energetic particle (electron or proton) is assumed to continuously lose its energy passing through the atmospheric matter, during which collisional details are not considered and the particle keeps its moving direction without being deflected from its incident angle. The residual range R of the particle (in units of g/cm^2) measures the encountered atmospheric column mass integrated along the penetration path. The range is energy dependent and is given by (e.g., Spencer, 1959)

$$R(E) \equiv \int_{z_{\min}(E)}^{z} \rho(z') \frac{dz'}{\mu} = \int_{0}^{E} \frac{dE'}{S(E')} \tag{6.28}$$

where E' and z' denote the intermediate energy and altitude during the transport, respectively; ρ is the atmospheric mass density (g/cm^3); $z_{\min}(E)$ is the stopping altitude, depending on the energy; $\mu = \cos\theta$, where θ is the incident angle with respect to the vertical direction; $S(E)$ is the energy-dependent mass stopping power (in units of keV/[g/cm^2]), which describes the average rate of energy loss per unit column mass of travel. S and R can be derived from each other by

$$S(E) = \left(\frac{dR(E)}{dE}\right)^{-1} \tag{6.29}$$

Fig. 6.8 presents the energy dependence of R and S for electrons and protons as established by the National Institute of Standards and Technology (NIST) (Berger et al., 2017). It is seen from Eqs. (6.28), (6.29) that a linear $\log(R)$-$\log(E)$ relationship within an energy range would lead to a linear $\log(S)$-$\log(E)$ relationship, and vice versa. This has been applied in Fig. 6.8 for the extension of the data at the low-energy end. The monotonic increase of the range with energy for electrons and protons determines that a more energetic particle is able to penetrate deeper (cf. Fang et al., 2010, 2013). As demonstrated in Fig. 6.8, previous range approximations adopted by early studies (with one exception) agree well with the modern knowledge from the NIST. The direct comparison here is also helpful to infer energy limits in the application of the previous works, whose accuracy in part depends on the accuracy of the adopted analytic approximations on $R(E)$.

The energy dissipation rate (in units of keV/cm^3 / s) can be obtained through

$$\eta(z) = 2\pi\rho(z)\int_{\mu=0}^{1}\int_{E=0}^{\infty} I(z, E, \mu)S(E)dEd\mu \tag{6.30}$$

where $I(z, E, \mu)$ is the differential particle intensity (cm^{-2} s^{-1} keV^{-1} sr^{-1}) within the atmosphere, describing the local energy and angular distribution at z. The angular integration in Eq. (6.30) is over the downward hemisphere given the neglect of backward scattering. $I(z, E, \mu)$ can be solved together with the energy loss history starting from the topside precipitating condition of $I_0(z_{\max}, E, \mu)$ down to the altitude of interest. Obliquely incident particles encounter more atmospheric matters than vertically incident particles over a given penetration altitude, and thus lose energy faster. As a result, $I(z, E, \mu)$ deviates from $I_0(z_{\max}, E, \mu)$ more noticeably at oblique angles and dramatically near the end of penetration (where the range is close to $R(E)$). Given that $R(E)$ is more conveniently measurable than $S(E)$ and are convertible into each other, such a method is often called range calculation. Using empirical $R(E)$ analytic expressions, Eq. (6.30) may be further simplified (e.g., Jackman et al., 1980).

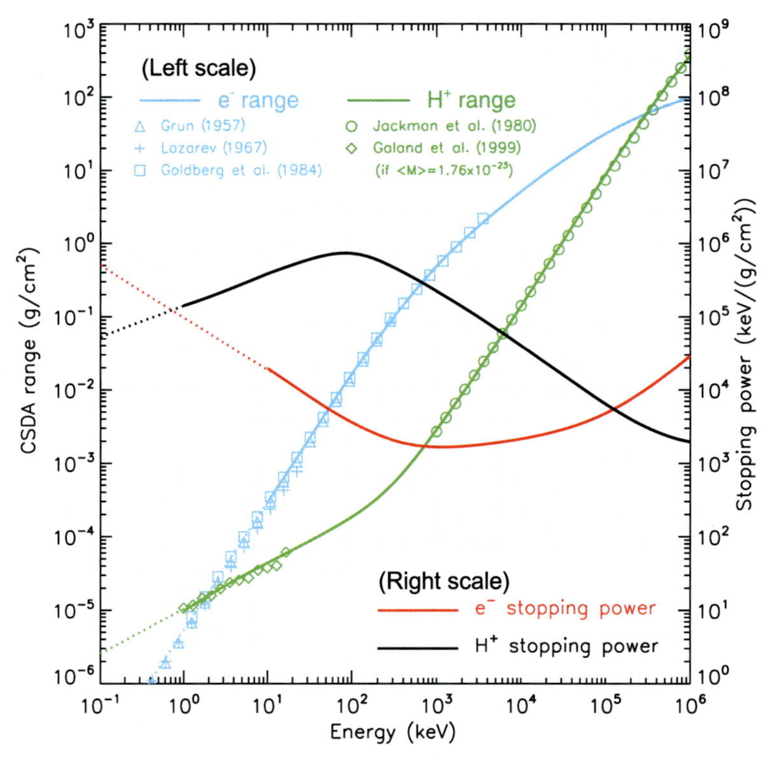

Fig. 6.8 The NIST-established electron and proton ranges (*left axis*) and stopping powers (*right axis*) as a function of energy in the dry air at sea level on Earth. The *dashed lines* are the extension of the NIST data to low energies assuming a linear trend on the log-log scale. The range-energy relationships from early results are superposed in symbols for comparison, in which two corrections have been made in order to have a good agreement with NIST. For Goldberg et al. (1984), one coefficient in their formula for $E > 295$ eV is changed from 0.295 to 0.469 to correct an apparent discontinuity problem. For Galand et al. (1999), the parameter of mean atmospheric molecular mass in their formula is set to be 1.76×10^{-23} g here, which, however, is ~60% lower than that at 120 km altitude predicted by the Mass Spectrometer Incoherent Scatter (MSIS-90) model (Hedin, 1991). It is implied that the range used by Galand et al. (1999) is a factor of ~2.5 larger than the NIST data.

The particle-impact ionization rate ($cm^{-3}s^{-1}$) is readily yielded by dividing $\eta(z)$ by the value of an average energy loss in producing an ion pair ($\Delta\epsilon$), that is,

$$q(z) = \frac{\eta(z)}{\Delta\epsilon} \tag{6.31}$$

For electron and proton precipitation in the Earth's atmosphere, $\Delta\epsilon$ are conventionally assumed to be around 3.5×10^{-2} keV (or 35 eV). More discussions on $\Delta\epsilon$ will be given later in Section 6.2.3.

6.2.2.2 Theoretical transport calculation

The accuracy of the CSDA range calculation method becomes questionable when the precipitating particle energy is low, where the neglect of angular scattering is problematic. This is the case for the precipitation of relatively low-energy photoelectrons and auroral electrons and protons. An alternative model category is to take into account the details of energy and directional changes during elastic and inelastic collisions by solving theoretical transport equations, which are essentially based on the Boltzmann equation:

$$\left(\frac{\partial}{\partial t} + \mathbf{v} \cdot \nabla + \mathbf{a} \cdot \nabla_v\right) f(\mathbf{r}, \mathbf{v}, t) = \frac{\delta f}{\delta t} \tag{6.32}$$

where $f(\mathbf{r}, \mathbf{v}, t)$ (in units of $cm^{-6}\ s^3$) describes the phase-space distribution of a particle species (e^-, H^+, or newly generated H) at position $\mathbf{r}$, velocity $\mathbf{v}$, and time t. The acceleration term due to external forces, such as the Lorentz and gravitational forces, is represented by $\mathbf{a}$. The term on the right-hand side, $\frac{\delta f}{\delta t}$, is the Boltzmann collisional integral, which covers relevant collisional effects on particle distributions. In practice, simplified transport equations have been derived from the Boltzmann equation by replacing the phase-space distribution function $f(\mathbf{r}, \mathbf{v}, t)$ with a differential number flux $\phi(l, E, \mu)$ under a steady-state approximation, where l is the distance along the magnetic field line and μ is the cosine of pitch angle. Numerous variants of the simplification of the Boltzmann equation have been developed to solve the particle transport in the atmosphere: for electron precipitation (Spencer, 1959; Nagy and Banks, 1970; Banks and Nagy, 1970; Banks et al., 1974; Strickland et al., 1976; Solomon et al., 1988; Lummerzheim et al., 1989; Link, 1992; Lummerzheim and Lilensten, 1994) and for proton precipitation (Jasperse and Basu, 1982; Basu et al., 1987, 1993; Strickland et al., 1993; Jasperse, 1997; Galand et al., 1997). The formulations of the transport equations in these works look significantly different due to the use of different techniques and approximations. The formulation details are not presented, except that we briefly outline representative transport equations here.

For electron precipitation, the transport equations under a classic two-stream approximation are simplified as (Schunk and Nagy, 2009):

$$\langle\mu\rangle\frac{\partial\Phi^+}{\partial l} = -\sum_s n_s\sigma_s^t\Phi^+ + \frac{\sum_s n_s\sigma_s^e}{2}(\Phi^+ + \Phi^-) + \frac{Q_{sec}}{2} + Q_{casc}^+ \tag{6.33}$$

$$-\langle\mu\rangle\frac{\partial\Phi^-}{\partial l} = -\sum_s n_s\sigma_s^t\Phi^- + \frac{\sum_s n_s\sigma_s^e}{2}(\Phi^+ + \Phi^-) + \frac{Q_{sec}}{2} + Q_{casc}^- \tag{6.34}$$

where n_s is the number density of atmospheric species s, and σ_s^t and σ_s^e are the total and elastic scattering cross sections between e^- and s, respectively. Φ^+ and Φ^- are the particle streams integrated over the upward (+) and downward (−) hemispheres, respectively. They are defined, under the assumption of azimuthal angular symmetry, as

$$\Phi^+(l,E) \equiv 2\pi \int_0^1 \phi(l,E,\mu)d\mu \tag{6.35}$$

$$\Phi^-(l,E) \equiv 2\pi \int_{-1}^0 \phi(l,E,\mu)d\mu \tag{6.36}$$

It is worth noting that Φ^+ and Φ^- are not a net flux (which is the parameter most often measured), but half the omnidirectional flux. There is no μ term in the integral of Eqs. (6.35), (6.36), which is otherwise required to calculate the perpendicular component to a horizontal unit area for a net flux. Eqs. (6.33), (6.34) are numerically solved over energy bins in a monotonically descending order, given that energy cascading happens from high to low energies. Q_{sec} and Q_{casc} cover secondary-electron productions and energy cascading effects, respectively. It has been assumed that newly produced secondary electrons have an isotropic angular distribution and on average the forward and backward scattering probabilities in elastic collisions are equal. The average pitch angles for the upward and downward streams are assumed such that $\langle\mu\rangle = 1/2$.

For proton precipitation, the problem turns to the simultaneous solution of the transport of protons and hydrogen atoms, the latter of which are produced in charge exchange collisions suffered by protons. Since charge state transition collisions are sufficiently frequent in the deep layers of the atmosphere, their transports are tightly coupled and cannot be solved individually or independently. The coupled H^+-H transport equations, given by Jasperse and Basu (1982) and Basu et al. (2001), are written as

$$\begin{aligned}
&\left(\mu\frac{\partial}{\partial z} + \sum_s n_s(z)(\sigma_{s,\mathrm{P}}(E) + \sigma_s^{10}(E))\right)\phi_{\mathrm{P}}(z,E,\mu) \\
&= \sum_s n_s(z) \int\int \sum_J \sigma_{s,\mathrm{P}}^J(E',\mu' \to E,\mu)\phi_{\mathrm{P}}(z,E',\mu')dE'd\mu' \\
&+ \sum_s n_s(z) \int\int \sigma_s^{01}(E',\mu' \to E,\mu)\phi_{\mathrm{H}}(z,E',\mu')dE'd\mu'
\end{aligned} \tag{6.37}$$

$$\begin{aligned}
&\left(\mu\frac{\partial}{\partial z} + \sum_s n_s(z)(\sigma_{s,\mathrm{H}}(E) + \sigma_s^{01}(E))\right)\phi_{\mathrm{H}}(z,E,\mu) \\
&= \sum_s n_s(z) \int\int \sum_J \sigma_{s,\mathrm{H}}^J(E',\mu' \to E,\mu)\phi_{\mathrm{H}}(z,E',\mu')dE'd\mu' \\
&+ \sum_s n_s(z) \int\int \sigma_s^{10}(E',\mu' \to E,\mu)\phi_{\mathrm{P}}(z,E',\mu')dE'd\mu'
\end{aligned} \tag{6.38}$$

where the subscripts P and H indicate protons and hydrogen atoms, respectively. The term $\sigma^J_{s,\,\mathrm{A}}(E', \mu' \rightarrow E, \mu)$ (with A replaced by either P or H) describes differential cross sections for any state transition from (E', μ') to (E, μ). The superscript J denotes the type of collisions between projectiles and atmospheric species s, including elastic, ionization, excitation, charge exchange ("10"), and electron-stripping ("01") collisions.

The resulting ionization rates can be derived from particle flux distributions using

$$q(z) = 2\pi \sum_s n_s(z) \int \int P^I_s(z, E, \mu) dE d\mu \tag{6.39}$$

where P^I_s denotes the ionization efficiency of the atmospheric species s.

For electron precipitation,

$$P^I_s(z, E, \mu) = \sigma^I_{s,e}(E) \phi(z, E, \mu) \tag{6.40}$$

where $\sigma^I_{s,e}$ represents the electron-impact ionization cross section.

For proton precipitation, more ionizing sources need to be considered. P^I_s is then given by

$$\begin{aligned} P^I_s(z, E, \mu) &= \sigma^I_{s,\mathrm{P}}(E) \phi_{\mathrm{P}}(z, E, \mu) + \sigma^I_{s,\mathrm{H}}(E) \phi_{\mathrm{H}}(z, E, \mu) \\ &\quad + \sigma^{10}_{s,\mathrm{P}}(E) \phi_{\mathrm{P}}(z, E, \mu) + \sigma^I_{s,e}(E) \phi_e(z, E, \mu) \end{aligned} \tag{6.41}$$

The four terms on the right-hand side of Eq. (6.41) correspond to various ionizing processes due to proton- and hydrogen-impact ionization collisions, charge exchange collisions, and secondary-electron-impact ionization collisions, respectively.

6.2.2.3 Monte Carlo transport calculation

The Monte Carlo approach is a numerical technique of using repeated random number sampling to predict the occurrence and consequence of processes that follow probability distributions. The precipitating particle transport and collisions with the background atmosphere are a stochastic process that is amenable to Monte Carlo modeling on a collision-by-collision basis. The Monte Carlo calculation is concerned with the distance from point of departure to the next collision, and when a collision happens, turns to the determination of the nature of target hit and the type of process involved. The collision probability for a particle moving through the atmosphere is given by (e.g., Cashwell and Everett, 1959)

$$\int P_r(z') dz' = \int \exp(-\tau) d\tau \tag{6.42}$$

where τ is the classic "optical" depth defined by

$$d\tau = \sum_s \sum_J n_s(z) \sigma^J_s(E) \frac{dz}{\mu} \tag{6.43}$$

where $\sigma_s^J(E)$ represents the energy-dependent cross section between the projectile and the atmospheric species s for a specific collision type J. When the magnetic field is vertically inclined or is not considered (e.g., for newly produced hydrogen atoms), $\mu = \cos\theta$. Otherwise, $\mu = \cos\theta \sin\gamma$, where γ is the inclination of the magnetic field to the horizontal.

There are two ways to predict where the next collision would happen. One is to directly infer the probable path length from Eq. (6.42) using

$$\Delta\tau = -\ln(r_n) \tag{6.44}$$

where r_n is a random number uniformly distributed between 0 and 1. Alternatively, the cumulative collision probability over a finite path (Δz) is obtained by

$$\Delta P_r = 1 - \exp(-\Delta\tau(\Delta z)) \tag{6.45}$$

A collision happens if ΔP_r is greater than a random number. If not, then the projectile continues moving another finite distance until a collision occurs in accordance with the above procedure. The latter approach is advantageous in the case that the particle motion is subject to external forces (e.g., magnetic mirroring forces, background electric fields) or the guiding center approximation is invalid.

If a collision with ambient neutrals occurs, a further Monte Carlo decision is made to determine the collision type and the target neutral species. The random choice is weighted by products of cross sections and neutral densities at the determined collision point. That is, the probability is proportional to the ratio of $n_s\sigma_s^J / \sum_s \sum_J n_s\sigma_s^J$. The next step after the specification of the collision is to assess collision outcomes such as energy degradation and angular deflection. Depending on the reaction nature, energetic secondary electrons may be generated or charge state changes (for proton precipitation) may happen. Their transport and collisional consequences will be further considered either as a similar Monte Carlo process or using a fluid approach as described in Section 6.2.2.2. The bookkeeping process is followed until the incident test particle and energetic by-products (such as secondary, tertiary, and subsequent generations) lose a majority of the energy to the atmosphere such that they are unable to move significantly from their positions. When a cutoff energy limit is met, the particle is regarded as being thermalized. That test particle is removed from further consideration with all of the leftover energy going to the background heating. The Monte Carlo procedure is then restarted for another precipitating primary particle from the topside boundary. By tracing a sufficiently large number of test particles that are initially distributed according to the energy and angular distributions, a variety of atmospheric effects (including the impact ionization) are recorded and accumulated along particle trajectories and finally are statistically analyzed.

The accuracy of the Monte Carlo approach greatly relies on an accurate understanding of the details of all relevant collisional processes. Therefore, the approach is subject to the limitation on availability and accuracy of differential cross sections ($\sigma_s^J(E)$) as well as probability distributions in energy loss and angular scattering. There are uncertainties due to limited or even lack of lab measurements or theoretical calculations for relevant processes. Moreover, Monte Carlo is a brute force approach; the energy cascading and angular deflection can only be reliably realized in statistics by tracing a large number of test particles. As a result, Monte Carlo models are computationally intensive and consume considerably more time than the other approaches.

Numerous models have been developed for the implementation of Monte Carlo approaches to calculate the atmospheric effects from precipitating electrons (Berger, 1963; Solomon, 1993; Lehtinen et al., 1999; Xu et al., 2018) and from protons (Porter and Green, 1975; Kozelov, 1993; Decker et al., 1996; Synnes et al., 1998; Lorentzen, 2000; Gérard et al., 2000; Solomon, 2001; Fang et al., 2004, 2005, 2007a, c). Recently, general-purpose radiation transport packages such as GEometry ANd Tracking (Geant4) (Agostinelli et al., 2003; Allison et al., 2016) and Monte Carlo N-Particle (MCNP) extensions (Pelowitz et al., 2005; Werner et al., 2017), which are originally designed for numerical experiments of high-energy nuclear physics, have been adopted for space and planetary sciences (e.g., Schröter et al., 2006; Wissing and Kallenrode, 2009; Artamonov et al., 2017; Mesick et al., 2018; Banjac et al., 2019). However, precaution must be taken when applying the radiation transport packages to the study of particle precipitation in the atmosphere, as geomagnetic field effects are usually neglected and model uncertainties at the low-energy end are unclear.

6.2.3 Fast calculation methods

The time line of the development of the models for particle transport and impact ionization calculation, as described in Section 6.2.2, follows the trend of the focus shifting from a simplistic scaling of lab measurements toward an in-depth investigation of the details and characteristics of collisional impact. However, there is a dilemma in the model application as an increase of model sophistication levels generally comes with more demands for computing resources. There are two peaks in history on particular demands for fast calculation methods. One is in the early era, when computer capabilities barely allowed extensive computations. The other is in the present era, when large-scale calculations are favored to integrate individual space environment elements into a tightly coupled system. As part of the direct links in the MIT coupling, the atmospheric effects from precipitating particles are routinely considered but usually not the end products. It is extremely computational expensive and thus practically infeasible, even with supercomputers, to couple the aforementioned transport calculations into global models. An offline transport calculation using prescribed atmospheres has been attempted, which, however, is an unfavorable option due to the compromised accuracy from the use of inconsistent atmospheric conditions. Therefore, an accurate and fast method that is able to directly and self-consistently estimate the ionization rate within any desired atmospheres without having to actually solve particle transport is much needed by the science community, which has been termed "parameterization." A parameterization method essentially is to use empirically derived simple functions to replicate results from complex model calculations.

Fig. 6.9 lists the published parameterization methods of particle-impact ionization calculation. Although the based models are very different and have various accuracies, nearly all the parameterizations share a similar appearance and inherent physics in their formulations, for both electron and proton precipitations. Let us first take Lazarev (1967) as an example. For monoenergetic, vertically incident electrons, the resulting ionization rate is given by Lazarev (1967) as

$$q_L(z) = \frac{Q_0}{\Delta\epsilon_L}\frac{\rho(z)}{R_{L0}} D_L(\chi_L) \tag{6.46}$$

where χ_L is a dimensionless parameter as a proxy of altitude. It stands for the fractional penetration, which is defined to be the ratio of the penetrated column mass to the expected total range as

$$\chi_L(z) \equiv \frac{\int_z^{z_{max}} \rho(z')dz'}{R_{L0}} \tag{6.47}$$

R_{L0} is empirically specified to be $R_{L0} = 4.6 \times 10^{-6} E_0^{1.65}$, where E_0 is the incident energy in keV. Note a good agreement of this empirical $R_{L0} - E_0$ relationship with the NIST data near auroral electron energies as seen in Fig. 6.8. In addition, Q_0 in Eq. (6.46) is the total incident energy flux at the topside boundary (z_{max}) in units of keV/cm^2 / s. $\Delta\epsilon_L$ (in keV) is a constant of the mean energy loss per ion-pair production as noted in Fig. 6.9. $D_L(\chi_L)$ is an empirically derived, dimensionless energy dissipation function solely dependent on χ_L,

$$D_L(\chi_L) = 4.2\chi_L \exp(-{\chi_L}^2 - \chi_L) + 0.48 \exp(-17.4{\chi_L}^{1.37}) \tag{6.48}$$

Eq. (6.46) can be simply understood as follows. The term $Q_0/\Delta\epsilon_L$ amounts to the altitude-integrated total ionization expected from the topside energy input. The term $\rho(z)/R_{L0}$ has the dimension of reciprocal length. In an isothermal atmosphere with a scale height of H, the term is approximately equal to $1/H$ times $\rho(z)/\rho(z_{min})$. Therefore, $Q_0/\Delta\epsilon_L \cdot \rho(z)/R_{L0}$ can be roughly regarded as the altitude-averaged ionization rate (i.e., total ionization divided by the characteristic altitude scale of H), modified by an altitude-varying factor of $\rho(z)/\rho(z_{min})$. More weights, as implied by the modification factor, are put at the lower altitudes because $\rho(z_{min}) > \rho(z)$. This is consistent with the fact that most of precipitating auroral energies are deposited at low altitudes, where more

		Features			Fast calculation application					
	Reference	Models	Collisions	Secondary effects	Precipitating angular distribution	Precipitating energy spectrum	Precipitating energy (keV)[a]	$\Delta\epsilon$ (eV)[b]	Format	Error analysis
e^-	Spencer (1959)	Theoretical transport calculation	No collisional details (except for angular scattering from elastic collisions)		isotropic, vertical	Monoenergetic	$25 \le E_{mono} \le 10^4$	N/A[c]	Tabular	✗
	Rees (1963)	Experimental work of Gnur (1957)	No collisional details		isotropic, vertical	Monoenergetic	$0.4 \le E_{mono} \le 300$	35	Parameterized[e]	✗
	Lazarev (1967)	Empirical calculation	No collisional details		Vertical	Monoenergetic	$0.1+ \le E_{mono} \le 32$	34 (air) 31 (O_2) 35 (N_2)	Parameterized	✗
	Roble and Ridley (1987)	Empirical calculation from Lazarev (1967)	No collisional details		Vertical	Maxwellian	$0.1+ \le E_c \le 32$	35	Parameterized	✗
	Lummerzheim (1992)	Extension of Spencer (1959) with modification from Barret and Hays (1976)	No collisional details		isotropic, specific angular distributions	Monoenergetic	$25 \le E_{mono} \le 10^4$	35	Tabular	✗
		extension of Roble and Ridley (1987)	No collisional details		Isotropic	Maxwellian	$0.1+ \le E_c \le 10^2$	35	Parameterized	✗
	Fang et al. (2008)	Theoretical (two-stream/multistream) transport calculation	Elastic and inelastic collisions	Secondary electrons included; no bremsstrahlung X-ray	Isotropic	Maxwellian	$0.1 \le E_c \le 10^3$	N/A[d]	Parameterized	✓
	Fang et al. (2010)	Theoretical (two-stream/multistream) transport calculation	Elastic and inelastic collisions	Secondary electrons included; no bremsstrahlung X-ray	Isotropic	Monoenergetic	$0.1 \le E_{mono} \le 10^3$	N/A[d]	Parameterized	✓
	Artamonov et al. (2017)	Monte Carlo calculation (based on GEANT4)	Elastic and inelastic collisions	Secondary electrons included; bremsstrahlung X-ray included	Isotropic, directional	Monoenergetic	$10^2 \le E_{mono} \le 5x10^5$	35	Tabular	✗
H^+	Rees (1982)	Monte Carlo calculation of Reme (1969)	Elastic and inelastic collisions	No secondary electrons	Isotropic	Monoenergetic	$0.2 \le E_{mono} \le 60$	36	Parameterized[e]	✗
	Lummerzheim (1992)	Extension of Roble and Ridley (1987) using Jackman et al. (1980)	No collisional details		Isotropic	Maxwellian	$10^3 \le E_c \le 10^6$	35	Parameterized	✗
	Galand et al. (1999)	Theoretical (multistream) transport calculation	Elastic and inelastic collisions	No secondary electrons	Isotropic	Maxwellian	$1 \le E_c \le 20$	20–28 (energy dependent)	Parameterized	✗
	Fang et al. (2013)	Monte Carlo calculation (primary H^+ and coupled H); theoretical (multistream) transport calculation (secondary e^-)	Elastic and inelastic collisions	Secondary electrons included	Isotropic	Monoenergetic	$0.1 \le E_{mono} \le 10^3$	N/A[d]	Parameterized	✓

[a] E_{mono} and E_c stand for the monoenergy and charactersitic energy of precipitating particles, respectively.

[b] $\Delta\epsilon$ stands for the mean energy loss per ion pair production, which has been used to convert the energy dissipation rate into the ionization rate.

[c] It is the energy dissipation rate that was directly calculated by Spencer [1959], which has been adopted by other researchers to convert into the ionization rate.

[d] The ionization rate is directly calculated by the models. There is no need for an assumption on the mean energy loss per ion pair production ($\Delta\epsilon$).

[e] The normalized energy dissipation functions in Rees [1963] and Rees [1982] are not parameterized, which, instread, are described by graphs.

Fig. 6.9 List of fast methods for energetic electron- and proton-impact ionization calculations.

collisions take place when particles are stopped by the denser atmosphere. Finally, the dimensionless term $D_L(\chi)$ provides a further correction, also taking into account the fact that the ionization rate rapidly drop beneath the ionization peak layer.

Note that the incident energy dependence of ionization rate altitude profiles is partially allowed for through the consideration of E_0 in R_{L0} (and subsequently into χ_L and D_L). E_0 is taken as an input parameter, which describes the initial precipitating energy and thus holds unchanged in the ionization calculation. For any given energy spectrum $\phi(E)$ of precipitating electrons (in units of $\text{cm}^{-2}\text{s}^{-1}\,\text{keV}^{-1}$), the ionization rate can be obtained by making the following integration:

$$q_L(z) = \frac{\rho(z)}{\Delta\epsilon_L} \int \frac{E\phi(E)}{R_{L0}(E)} D_L(\chi_L)\,dE \tag{6.49}$$

That is, the incident energy spectrum is decomposed into contiguous monoenergetic components, and the resulting ionization rate is obtained by applying the parameterization to calculate the contributions from individual components and finally summing them up. This is achieved by replacing Q_0 in the parameterization formula with $E\phi(E)dE$ and then integrating over the incident energy.

Roble and Ridley (1987) extend the parameterization of Lazarev (1967) to apply to incident nonmonoenergetic electrons with a Maxwellian spectrum, which often is a good approximation for auroral precipitation. Using the kinetic theory of gases on the Maxwellian-Boltzmann velocity distribution within a volume, we have the Maxwellian energy distribution of the number flux precipitating per unit area as

$$\phi_M(E) = \frac{Q_0}{2E_c^3} E \exp\left(-\frac{E}{E_c}\right) \tag{6.50}$$

where $E_c = k_B T$ is the characteristic energy in units of keV (k_B and T being the Boltzmann constant and charged-particle temperature, respectively). That is, ϕ_M peaks at $E = E_c$. It is inferred from Eq. (6.50) that the total incident energy is $\int E\phi_M(E)dE = Q_0$, and the average energy is $\langle E\rangle = \int E\phi_M(E)dE / \int \phi_M(E)dE = 2E_c$.

According to Roble and Ridley (1987), the impact ionization from vertically incident electrons in a Maxwellian energy distribution is obtained from

$$q_R(z) = \frac{Q_0}{2\Delta\epsilon} \frac{1}{H(z)} D_R(\chi_R) \tag{6.51}$$

where $\Delta\epsilon = 3.5 \times 10^{-2}$ keV, H is the atmospheric density scale height in units of cm by $H = \text{kT}/(\text{mg})$. The fractional penetration χ_R is defined as

$$\chi_R(z) \equiv \left(\frac{\rho(z)H(z)}{4\times 10^{-6} E_c^{1.65}}\right)^{1/1.65} \tag{6.52}$$

and the dimensionless energy dissipation rate D_R is empirically fit by

$$D_{\mathrm{R}}(\chi_{\mathrm{R}}) = \sum_{m=0}^{M} C_{(4m+1)} \chi_{\mathrm{R}}^{C_{(4m+2)}} \exp\left(-C_{(4m+3)} \chi_{\mathrm{R}}^{C_{(4m+4)}}\right) \qquad (6.53)$$

where $M = 1$ and the eight parameters (C_{4m+k}, $m = 0, 1; k = 1, \ldots, 4$) are constants. Considering that $\rho(z)H(z)$ is a good approximation of the column mass above altitude z, the definitions of χ_{R} and χ_{L} are similar except for an additional exponent of 0.606 in χ_{R}. Lummerzheim (1992) extends the work of Roble and Ridley (1987) to electron precipitation with a Maxwellian energy and isotropically angular distribution, adopting the identical equations except for a different set of constant C_{4m+k} coefficients.

It should also be pointed out that χ_{L} and χ_{R} nominally, but not strictly, represent the fractional penetration. Actually, their values may exceed 1 particularly near the end of the penetration. This is because complex physical processes (including energy cascading, angular scattering, and secondary effects) are involved during the particle transport. The net effect after passing through an atmospheric layer not only is determined by the penetrated mass within the layer, but also, and more importantly, varies with the density structure over the course. It is thus not surprising to see that the range data, which are obtained for a particle passing through a uniform medium (see Fig. 6.8), solely are unable to yield an accurate description of particle motion and the induced collisional effects within the atmosphere. The denominators in Eqs. (6.47), (6.52) are close to but not strictly the maximum column mass that precipitating particles is able to penetrate within a realistic atmosphere. Nevertheless, this is not necessarily problematic as far as empirical functions are concerned here. It is underscored that the accuracy of the based calculation models greatly limits that of the empirical formulas, which, however, is often unclear. Another compelling need is for understanding the energy limits on the application of the parameterizations. This is particularly important because large-scale community models evolve to take into account precipitating particles of various origins and thus spanning a very broad energy range (e.g., from 100s eV soft electrons up to MeV SEPs). There has been a lack of error analysis of how accurately the parameterization methods work for various energies and under different atmospheres.

As discussed in Section 6.2.2, efforts have been made to develop more sophisticated (and more accurate) theoretical and Monte Carlo models beyond simplistic CSDA range calculation models. This facilitates taking advantage of the improved transport models to update the old empirical laws. Such a revisit was initiated for isotropically incident electrons with a Maxwellian energy distribution by Fang et al. (2008), with the aim to improve and extend the Roble and Ridley (1987) method using theoretical transport models. Two validated models have been employed: one from Lummerzheim et al. (1989) and Lummerzheim and Lilensten (1994) for the characteristic energy E_c from 100 eV to 10 keV and the other from Solomon et al. (1988) and Solomon and Abreu (1989) for higher E_c up to 1 MeV. It is confirmed by Fang et al. (2008) that the Roble and Ridley (1987) normalization functions manage to account for the sensitivity of ionization altitude profiles to background atmospheric conditions. By converting direct physical quantities of q_{R} and z into dimensionless quantities of D_{R} and χ_{R}, respectively (see Eqs. 6.51, 6.52), the $D_{\mathrm{R}}(\chi_{\mathrm{R}})$ profiles become closely clustered with each other, even though the original $q_{\mathrm{R}}(z)$ profiles vary significantly, and in some circumstances dramatically, with atmospheric conditions. However, using the more accurate model results, it is noticed by Fang et al. (2008) that the $D_{\mathrm{R}} - \chi_{\mathrm{R}}$ relationship manifests itself differently for various incident characteristic energies. As a consequence, the fixed, energy-independent functional form in Eq. (6.53) works fine near E_c=10 keV but causes errors in other scenarios. For example, it is found that the entire ionization altitude profile would have shifted upward by up to 40 km in altitude for $E_c < 1$ keV and downward by up to ~10 km for $E_c > 100$ keV. The predicted peak ionization rate would have been generally underestimated, by −15% at $E_c = 0.1$ keV and −30% at $E_c = 10^3$ keV. The large errors at low energies are due to the limitations of the based range calculation model, particularly the neglect of angular scattering. Moreover, the validity of the parameterization becomes considerably problematic at high energies near 1 MeV, where the adopted range approximation of $R = 4 \times 10^{-6} E^{1.65}$ (as shown in Eq. 6.52) is invalid (see Fig. 6.8). These shortcomings are overcome in Fang et al. (2008) by adopting the same Eqs. (6.51)–(6.53) but allowing the C coefficients in Eq. (6.53) to vary with incident energy. The new coefficients are expressed as

$$C_i(E_c) = \exp\left(\sum_{j=0}^{3} P_{ij}\left(\ln(E_c)\right)^j\right) \tag{6.54}$$

where C_i ($i = 1, \ldots, 8$) corresponds to C_{4m+k} ($m = 0, 1$; $k = 1, \ldots, 4$) in Eq. (6.53), and P_{ij} are constant parameters determined through a fit to the theoretical transport model results. An extensive error analysis has been performed for the validation of the new method by Fang et al. (2008). A remarkable agreement is observed between the results of the new parameterization and the transport models, under thousands of atmospheric conditions from the MSIS-90 model. An excellent reproduction of the ionization altitude profile is achieved: the altitudes of the peak and e-folding ionization rates are accurately captured with an error significantly less than 5 km. The errors in the peak ionization rate and the altitude-integrated total ionization rate generally fall well within ±5%. These highlight the capability of the updated parameterization method to conduct fast particle-impact ionization calculation for precipitating Maxwellian electrons over a broad energy range of $0.1-10^3$ keV, and at the same time, to maintain a satisfactory accuracy level as the theoretical transport models.

While a Maxwellian energy approximation of particle precipitation as in Eq. (6.50) is often reasonable, which has been widely adopted by community global models, exceptions are also observed (e.g., Rees, 1969; Frahm et al., 1997). A parameterization for monoenergetic inputs has the flexibility to deal with any given energy spectrum using a method similar to that described in Eq. (6.49). Following Fang et al.'s (2008) work, Fang et al. (2010, 2013) developed fast ionization calculation methods for isotropically incident, monoenergetic electrons and protons, respectively. A wide incident energy range, spanning four orders of magnitude, from 0.1 to 10^3 keV is covered by both methods. Empirical functions were derived from the fit to extensive calculation results from sophisticated transport models. In the electron-impact study of Fang et al. (2010), the same theoretical transport models are employed as those in Fang et al. (2008). In the proton-impact study of Fang et al. (2013), a coupled model framework is developed. The primary ionization from incident protons (and the generated hydrogen atoms) is calculated using the Monte Carlo model of Fang et al. (2004, 2005, 2007a, c); the additional ionization from secondary electrons during the proton/hydrogen transport is calculated using the transport model of Lummerzheim et al. (1989) and Lummerzheim and Lilensten (1994).

The Fang et al. (2010, 2013) parameterizations have a similar formulation for calculating the ionization rate from a monoenergetic input of E_0, written as

$$\chi_F(z) \equiv \left(\frac{\rho(z)H(z)}{\alpha E_0^{1/\beta}}\right)^{\beta} \tag{6.55}$$

$$q_F(z) = \frac{Q_0}{\Delta\epsilon}\frac{1}{H(z)}D_F(\chi_F) \tag{6.56}$$

The fit parameters for electrons are $\alpha = 2.23 \times 10^{-6}$ and $\beta = 0.7$, and those for protons are $\alpha = 1.07 \times 10^{-5}$ and $\beta = 0.9$. The dimensionless energy dissipation function D_F shares the same formula as D_R in Eq. (6.53) with the following exceptions. First, χ_R is replaced by χ_F as specified by Eq. (6.55). Second, M equals 1 and 2 for electrons and protons, respectively. Third, the energy dependence of C_{4m+k} ($m=0,\ldots,M$; $k=1,\ldots,4$) is described by Eq. (6.54), where E_c is changed to E_0. Moreover, specific sets of fit parameters P_{ij} are derived separately for electron and proton precipitations. The two parameterizations have been validated by an excellent agreement with sophisticated transport model results, to which these empirical functions are derived to fit. It is confirmed through extensive error analyses that the agreement is maintained under various conditions of precipitating energies and background atmospheres. There is another computational advantage of using $1/H(z)$ in Eq. (6.56) in place of $\rho(z)/R_{L0}$ in Eq. (6.46) for monoenergetic particle precipitation. For a realistic application such as that shown in Eq. (6.49), the energy dependence of R_{L0} leads to extra computations inside the integration over incident energy bins. In contrast, this is not the case for the use of $1/H(z)$, which results in enhanced computational efficiency.

It should be pointed out that although $\Delta\epsilon = 3.5 \times 10^{-2}$ keV explicitly appears in the equations of the modern parameterizations of Fang et al. (2008, 2010, 2013), it no longer carries the meaning of the average energy loss per ion-pair formation like in CSDA range calculation (Eq. 6.31) or in the other parameterizations (see Fig. 6.9 and Eq. 6.46). The transport models, on which the new parameterizations are based, directly use relevant ionizing cross sections to make the ionization calculation. This is very different from those earlier works, in which the mean energy loss approximation is critical to convert energy deposition into the ion-pair production. Therefore, the uncertainties/errors with the assumption on the mean energy loss per ion-pair production are a concern for old parameterizations but become irrelevant for the new ones. The inclusion of $\Delta\epsilon$ in the modern parameterizations is in part for a historical reason to have a similar formula appearance in consistence with earlier works, and in part to make the energy dissipation function D_F dimensionless.

As a key parameter in old range calculation models and parameterizations, the mean energy loss per ion-pair production approximation has been widely adopted, as noted in Fig. 6.9. Reid (1961) adopted the value of $\Delta\epsilon(H^+) = 36$ eV for proton-impact ionization according to the work of Bailey (1959). $\Delta\epsilon(e^-) = 35$ eV was used by Rees (1963) for electron precipitation. Porter et al. (1976) reported that $\Delta\epsilon$ had a nearly stable value of close to 35 eV over a wide precipitating energy range of 10^{-2}–10^3 keV for both electrons and protons, except that a rapid increase was suggested when the incident electron energy decreased to lower than 100 eV. Similar energy dependence of $\Delta\epsilon(e^-)$ was reported by Green et al. (1977), in which its dependence on the atmospheric composition was found to be important. For example, $\Delta\epsilon(e^-)$ for 2-keV electrons within the gas of $N_2/O_2/O$ was estimated to be 35.4/31.0/27.3 eV (Green et al., 1977) and 34.3/27.8/26.4 eV (Wedlund et al., 2011). The energy dependence of $\Delta\epsilon$ (H^+) within the Earth's atmosphere has also been discussed by Fang et al. (2013, and references therein), suggesting an important variability: $\Delta\epsilon$ (H^+) nearly monotonically increases from 22 to 33 eV when incident energy increases from 100 eV to 1 MeV. It is clear that $\Delta\epsilon$ is not strictly a constant but depends on incident energy and penetration altitude to some degree. The altitude dependence of $\Delta\epsilon$ comes from the fact that the atmospheric composition changes over the course of particle transport and the particle energy monotonically decreases due to the deceleration from the atmosphere particularly near the end of the penetration. However, due to historical reasons, the ratio of the energy deposition rate to the net ionization rate has been conveniently approximated as a constant of 35 eV for both $\Delta\epsilon$ (e^-) and $\Delta\epsilon$ (H^+), regardless of precipitating energy or altitude. The concern for the use of a constant $\Delta\epsilon = 35$ eV is worth careful consideration, which is also applicable to the very recent revisit of the electron-impact ionization calculation by Artamonov et al. (2017) particularly for energies lower than 1 MeV.

As an application example, Fig. 6.10 shows the resulting ionization rates from isotropically incident monoenergetic electrons and protons using the parameterizations of Fang et al. (2010, 2013), respectively. Two representative background atmospheres from the MSIS-90 model are used: one in a quiet-time case ($A_p = 5$, $F_{10.7} = 50$) and the other in a geomagnetic and solar active case ($A_p = 250$, $F_{10.7} = 300$). The color contour plots show

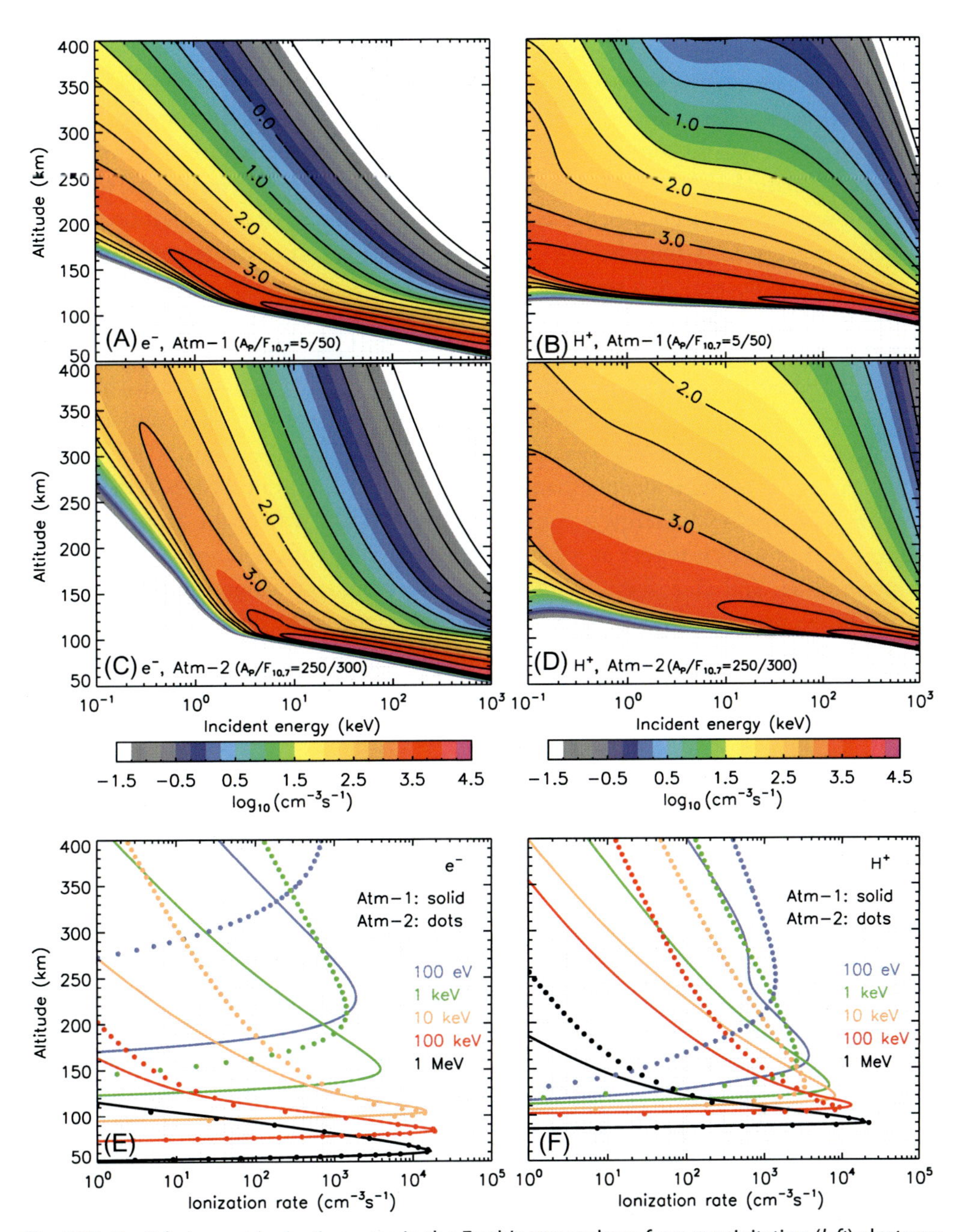

Fig. 6.10 Particle-impact ionization rates in the Earth's atmosphere from precipitating (*left*) electrons and (*right*) protons. Particles are assumed to be monoenergetic and have an isotropic angular distribution, with a normalized total energy flux of 1 erg/cm^2/s. Two background MSIS atmospheric conditions are used for comparison, which are representative of atmospheric structures during quiet times (panels A and B) and during geomagnetic and solar active periods (panels C and D), respectively. The comparisons for five representative energies (100 eV/1 keV/10 keV/100 keV/1 MeV, in *blue/green/yellow/red/black*) are highlighted in panels (E) and (F). The *solid lines* and *dot symbols* indicate the ionization altitude profiles under quiet and disturbed atmospheric conditions, respectively.

the dependence of the particle-impact ionization rate as a function of incident energy and altitude. The altitude profiles for five representative incident energies are also shown. A straightforward observation is that the altitude profile of the ionization rate (or approximately energy deposition/loss rate) is sensitive to the vertical structure of atmospheric densities. In accordance with the enhancement of ambient neutral densities due to atmospheric upwelling under disturbed conditions, more energy loss (and thus more ionization) takes place at high altitudes, which weakens the amount of the energy flux that is able to penetrate deeper. As a result, the ionization peak is generally weaker under disturbed atmospheric conditions. This is particularly prominent for <10 keV electrons and <100 keV protons, whose ionizations peak above $\sim$100-km altitude.

It is of interest to compare the penetration capability of energetic electrons and protons within a specific atmosphere, that is, between Fig. 6.10A and B and between Fig. 6.10C and D. Generally speaking, when the incident energy is lower than $\sim$2 keV, protons penetrate deeper than electrons, and the opposite occurs at the greater energies. This threshold energy is consistent with Fig. 6.8, where the range curves of electrons and protons intersect at $\sim$2 keV. A smaller value of the range corresponds to less penetration, and vice versa. Within the incident energy range of 10^{-1}–10^{3} keV, Fig. 6.8 shows that electrons have a more dramatic change in the penetrated atmospheric column mass, spanning >6 orders of magnitude. In contrast, proton ranges vary considerably less, by about three orders of magnitude. Such a difference of the penetration range between electrons and protons manifests itself in the penetration altitude difference as presented in Fig. 6.10.

The dependence of the ionization altitude profile on the background atmosphere, which has been demonstrated in Fig. 6.10, is further elaborated in Fig. 6.11. The figure shows the variability in the vertical distributions of the densities of atmospheric major species, total column masses, and the resulting ionization rates from isotropically incident, monoenergetic electrons and protons (using Fang et al., 2010, 2013, respectively). The variation of the atmospheric conditions is obtained by driving the empirical MSIS-90 model with various parameters in order to cover a broad range of timescales (including a diurnal cycle and a 11-year solar cycle), geographic locations (0–360 degrees in longitude and 55–75 degrees in latitude), solar activities (50–300 in $F_{10.7}$), and geomagnetic activities (10–250 in A_p). The shaded areas in Fig. 6.11 indicate the range of variation, that is, the minimum and maximum values as a result of the numerical experiments. Immediately apparent is the large variability in the atmospheric density distribution at high altitudes, particularly above 100 km. As less energetic particles are liable to be significantly decelerated at higher altitudes where more energy is lost for ionization, more prominent variability is shown there. For example, in response to atmospheric changes, the ionization rate from 100-eV electrons exhibits more than three orders of magnitude variation within the altitudes of $\sim$200–300 km. The bumpy structures seen in the particle-impact ionization near 110–120 km altitudes (Fig. 6.11C and D) are attributed

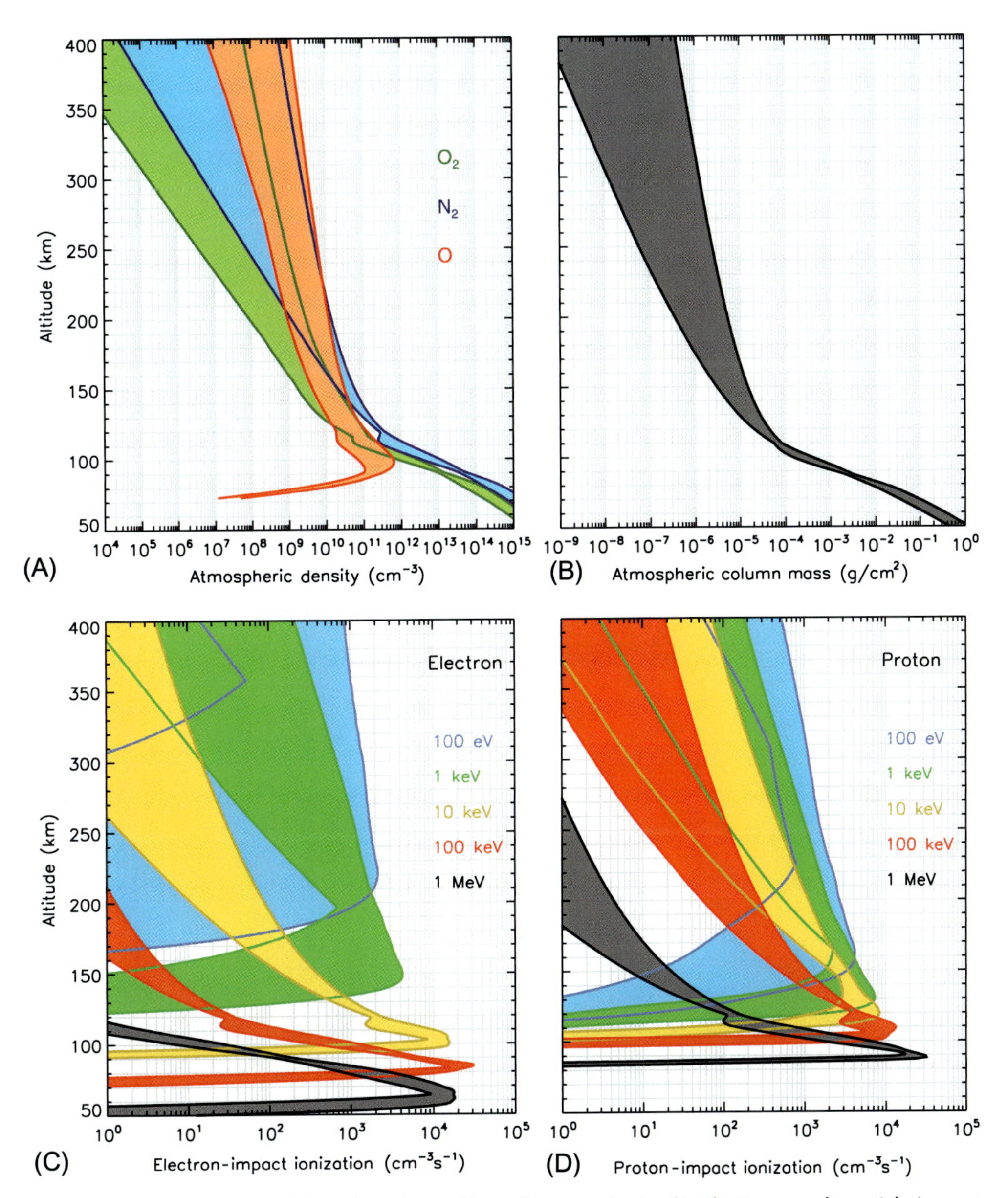

Fig. 6.11 The variability of the altitude profiles of atmospheric distributions and particle-impact ionization rates. Panels (A) and (B) show the MSIS-90 model-calculated number densities of major atmospheric species O_2/N_2/O (*green/blue/red*) and total atmospheric column mass, respectively. Panels (C) and (D) show the variability of the resulting ionization from incident electrons and protons, respectively, for five representative energies. Particles are assumed to be monoenergetic and have an isotropic angular distribution, with a normalized total energy flux of 1 erg/cm^2/s. The *shaded areas* in the panels indicate the range of variation in response to the MSIS-90 model outputs that are driven by various parameters.

to similar features in the MSIS-predicted atmospheric profiles (Fig. 6.11A and B), which are found to be associated with disturbed conditions due to geomagnetic activities. It is unclear whether this is physically meaningful or due to some problem with the empirical model of MSIS-90 in accounting for atmospheric responses during extreme geomagnetic events, which, nevertheless, does not affect our discussions. The high degree of variability as illustrated in Fig. 6.11 clearly underscores the necessity and importance of implementing self-consistent particle-impact ionization calculation within any global models, which are facilitated by the availability of the accurate and computationally efficient parameterization models. It should be pointed out that none of the empirical ionization formulae that have been reviewed in this section can be directly extended to nonterrestrial atmospheres, considering that different atmospheric compositions would lead to different interaction processes between charged particles and atmospheric neutrals and thus different collisional effects.

6.3 Electromagnetic fields of magnetospheric disturbances in the conjugate ionospheres: Current/voltage dichotomy

Vyacheslav A. Pilipenko[a, b], Mark J. Engebretson[c], Michael D. Hartinger[d], Evgeny N. Fedorov[b], and Shane Coyle[d]

[a]Space Research Institute, Moscow, Russia

[b]Institute of Physics of the Earth, Moscow, Russia

[c]Augsburg University, Minneapolis, MN, United States

[d]Virginia Tech, Blacksburg, VA, United States

6.3.1 Introduction: Current/voltage dichotomy

When considering a circuit analogy of the magnetosphere-ionosphere current systems it is physically intuitive to distinguish between generators which deliver a fixed current, and those in which the voltage is fixed (Lysak, 1985). This qualitative consideration may be visualized as a magnetospheric generator with an internal resistance and an ionospheric load. If the internal resistance is small compared to the load on the circuit, the ionosphere will receive a fixed voltage, whereas if the load is small compared to the internal resistance, a fixed current will be delivered to the ionosphere (Lysak, 1990).

Magnetosphere-ionosphere transient current systems with typical timescales in the lowest-frequency portion of the ULF band (1–15 min) are also often described using this electrical circuit analogy (Hartinger et al., 2017). A magnetospheric process can generate a potential difference that maps along magnetic field lines to the ionosphere, where it drives electric fields. On the other hand, magnetospheric processes can generate divergent currents perpendicular to the background magnetic field which in the ionosphere are closed via field-aligned currents (FACs). In the case of a current generator, the magnetic effect on the ground must be nearly the same under sunlit or dark ionospheres, whereas in the case of a voltage generator the ground magnetic response must be much larger under a highly conductive ionosphere (Sibeck et al., 1996; Lam and Rodger, 2004). The dependence of the ULF response to magnetospheric driving on the ionospheric conductance may be an important observational feature of a coupled magnetosphere-ionosphere system. An adequate incorporation of the interhemispheric differences and their effects on magnetosphere-ionosphere coupling is vitally important for observations and modeling/simulations (Zesta et al., 2016).

There are several possibilities to examine the dependence of ground magnetic response to a magnetospheric driver on the ionospheric conductance. The first one is to study the daily variations of magnetic disturbance amplitude at a selected station. However, the intensity of a magnetospheric driver may vary from hour to hour. The second is to examine the seasonal variation of the disturbance amplitude. Upon examining the seasonal variations, for a magnetospheric FAC generator the local ionospheric resistance plays the role of a load resistance, whereas the Alfvén wave resistance and the resistance of conjugate ionospheres play the role of an internal source resistance (Pilipenko et al., 2019). Such a combination of several poorly known parameters makes an unambiguous interpretation of seasonal variations difficult. Finally, the current/voltage paradigm can be tested in a straightforward way with observations at conjugate stations under strongly asymmetric ionospheres.

An important aspect of the ULF magnetospheric disturbance interaction with the ionosphere which should be taken into account when considering the current/voltage dichotomy is that FACs may interact with the ionosphere in a different way in regimes of forced driving or excitation of resonant field line oscillations (Pilipenko et al., 2019). Here, we compare the results of observations of various ULF phenomena, such as TCVs, Pc5 waves, SCs, and MPEs at conjugate magnetometer arrays in Greenland and Antarctica from the viewpoint of the voltage/current generator dichotomy. But first, theoretical predictions of a simple "plasma box" model of the magnetosphere with asymmetric conjugate ionospheres driven by magnetospheric FAC are summarized.

6.3.2 Model of a magnetosphere with asymmetric ionospheres

The "plasma box" model of the magnetosphere (Fig. 6.12) mimics the magnetosphere at middle and high latitudes. Magnetospheric straight field lines with length $2L$ are terminated by thin ionospheric layers with height-integrated Pedersen and Hall conductances Σ_P and Σ_H, respectively. The coordinate x corresponds to the Earthward radial direction, the z-axis is along the geomagnetic field $\mathbf{B}_0$, and the y-axis corresponds to the azimuthal direction (the system is homogeneous in this direction).

Let at some magnetic shell a transient pulse of FAC $j_z(t)$ be excited by an external transverse current $j_x^{(e)}$. The radial direction of an external current corresponds to an

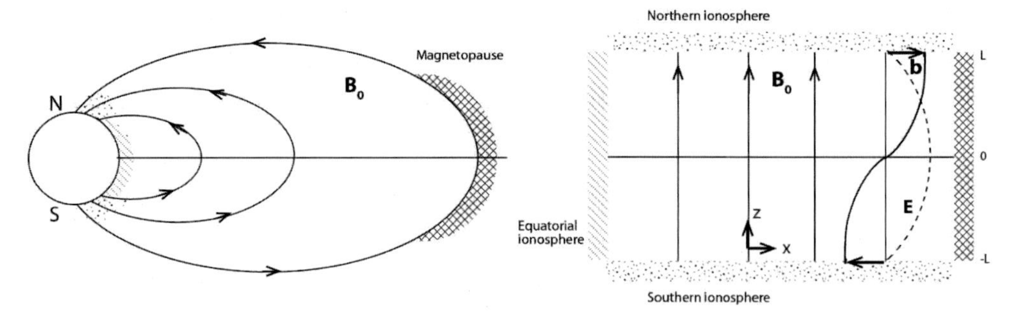

Fig. 6.12 Field-aligned currents excitation in a simple "magnetospheric box" model: homogeneous plasma with constant Alfvén velocity V_A immersed in a straight magnetic field B_o. The field lines with length $2l$ are terminated by conjugate dayside Southern (S) and Northern (N) ionospheres with conductances $\Sigma_P^{(S)}$ and $\Sigma_P^{(N)}$. The oscillations are driven by an external transverse current $j_x^{(d)}$.

azimuthally (along the Y-axis) large-scale disturbance. The electric field of the disturbance $\mathbf{E}_\perp = (E_x, E_y)$ is described by the MHD wave equation

$$(\partial_z^2 - V_A^{-2}\partial_t^2)E_x(t,z) = \mu_0 \partial_t j_x^{(e)}(t,z) \tag{6.57}$$

The magnetic component of a disturbance can be found from Maxwell's equations as $\partial_t B_y = -\partial_z E_x$. Eq. (6.57) describes an Alfvénic-type disturbance carrying a nonsteady FAC $j_z = \Sigma_A \nabla E_\perp$, where $\Sigma_A = (\mu_o V_A)^{-1}$ is the magnetospheric Alfvén conductance determined by the Alfvén velocity V_A. The fundamental Alfvén wave period in the magnetospheric box model is $T_A = 4L/V_A$.

Eq. (6.57) should be augmented by the impedance-type boundary conditions at conjugate Northern (N) and Southern (S) ionospheres ($z = \mp L$) as follows (Newton et al., 1978):

$$B_y = \mp \mu_0 \Sigma_P E_x \tag{6.58}$$

Let the driving current be localized at the magnetospheric equatorial plane ($z = 0$), namely, $j_x^{(e)}(z) = J_e(t)\delta(z)$. Integrating Maxwell's equations across the area occupied by a driver, the conditions on a jump of transverse magnetic and electric fields can be obtained

$$\{E_x\}|_{z=0} = 0, \quad \{B_y\}|_{z=0} = -\mu_0 J_e(t) \tag{6.59}$$

where $\{\ldots\}$ denote a jump across the source region. The interaction of an Alfvén wave with the ionosphere is characterized by the reflection coefficient

$$R^{(S,N)} = \frac{\overline{\Sigma}_P^{(S,N)} - 1}{\overline{\Sigma}_P^{(S,N)} + 1} \tag{6.60}$$

determined by the ratio between the ionospheric Pedersen conductance and wave conductance, $\overline{\Sigma}_P^{(S,N)} = \Sigma_P^{(S,N)}/\Sigma_A$.

Steady-state solutions above the ionospheres for $E_x(z) = ik\varphi$ and $B_y(z) = -ik\psi$ can be obtained via the electric and magnetic potentials

$$\begin{aligned}\varphi^{(S,N)}(z) &= \mp V_A\left[\exp\{\mp ik_A(z\pm L)\} - R^{(S,N)}\exp\{\pm ik_A(z\pm L)\}\right]\\ \psi_{S,N}(z) &= \exp\{\mp ik_A(z\pm L)\} + R^{(S,N)}\exp\{\pm ik_A(z\pm L)\}\end{aligned} \tag{6.61}$$

Here, the upper/lower signs correspond to S/N hemispheres, and $k_A = \omega/V_A$ is the Alfvén wave number. These solutions satisfy the boundary conditions at the ionosphere (6.58) and describe the field-aligned structure of a disturbance. The field-aligned distribution of electromagnetic field above and below the source of disturbances is found with account for the merging condition (6.59). Then, the ratio between electric $E^{(N)}/E^{(S)} = E_x(L)/E_x(-L)$ and magnetic $B^{(N)}/B^{(S)} = B_y(L)/B_y(-L)$ fields in the conjugate ionospheres can be derived as follows:

$$\frac{E^{(N)}}{E^{(S)}} = \frac{1 - i\overline{\Sigma}_P^S \tan(k_A L)}{1 - i\overline{\Sigma}_P^N \tan(k_A L)} \tag{6.62}$$

$$\frac{B^{(N)}}{B^{(S)}} = -\frac{\overline{\Sigma}_P^N}{\overline{\Sigma}_P^S}\frac{1 - i\overline{\Sigma}_P^S \tan(k_A L)}{1 - i\overline{\Sigma}_P^N \tan(k_A L)} \tag{6.63}$$

The magnetic B_x disturbance on the ground is related to an azimuthal magnetic disturbance B_y above the ionosphere by the well-known formula (Hughes and Southwood, 1976; Alperovich and Fedorov, 2007)

$$B_x^{(S,N)} = B_y^{(S,N)}\frac{\Sigma_H^{(S,N)}}{\Sigma_P^{(S,N)}}\exp(-kh) \tag{6.64}$$

where h is the height of the ionospheric conductive sheet and k is the horizontal wave vector of the disturbance. The ratio between the ground magnetic responses at the conjugate points is

$$\frac{B_x^{(N)}}{B_x^{(S)}} = \frac{B^{(N)}}{B^{(S)}}\frac{\Sigma_H^{(N)}}{\Sigma_P^{(N)}}\frac{\Sigma_P^{(S)}}{\Sigma_H^{(S)}} \tag{6.65}$$

Different situations are possible, as follows from these relationships (6.62 and 6.65), depending on the timescale of the disturbance τ and the local field-line resonant period T_A.

6.3.2.1 Relationship between ground magnetic disturbances at conjugate sites

In the case of nonresonant quasi-DC forced driving, when $\omega \ll \Omega_A$ (or $k_A L \ll 1$), from Eqs. (6.62), (6.65), the ratio between ground magnetic disturbances at conjugate points follows:

$$\frac{B_x^{(N)}}{B_x^{(S)}} = \frac{\overline{\Sigma}_H^{(N)}}{\overline{\Sigma}_H^{(S)}} \tag{6.66}$$

Thus, upon a quasi-DC driving the magnetic response must be larger under a higher local ionospheric conductivity, which corresponds to a voltage generator regime.

In the case of resonant driving, when the frequency of an oscillatory driver matches the local field line frequency of a magnetic shell, $\omega \to \Omega_A$ (or $k_A L = \pi/2$), the ratio of magnetic responses at conjugate points can be obtained from Eqs. (6.62), (6.65)

$$\frac{B_x^{(N)}}{B_x^{(S)}} \simeq \frac{\overline{\Sigma}_H^{(N)} \overline{\Sigma}_P^{(S)}}{\overline{\Sigma}_P^{(N)} \overline{\Sigma}_H^{(S)}} \tag{6.67}$$

The ratio of ground magnetic responses at conjugate points does not depend on variations of the ionospheric conductance, which corresponds to the current generator regime. The change from voltage to current generator regime upon transfer from nonresonant to resonant driving is similar to the antireflective coating in optics, where the input impedance of a multilayered system changes dramatically when the wavelength of incident light is a multiple of the layer width.

6.3.2.2 Field-aligned magnetosphere-ionosphere currents

The relationship between the FACs flowing into conjugate ionospheres $j_z \equiv j_\parallel$ and a driver current J_e can be found using Ampere's law $j_z = \mu_0^{-1} ikB_y$ as follows $j_\parallel^{(N)}/j_\parallel^{(S)} = B^{(N)}/B^{(S)}$. In the case of nonresonant quasi-DC driving, the FAC ratio is

$$\frac{j_\parallel^{(N)}}{j_\parallel^{(S)}} = -\frac{\Sigma_P^{(N)}}{\Sigma_P^{(S)}} \tag{6.68}$$

This relationship predicts that a larger FAC flows into a more-conductive ionosphere. In the regime of resonant driving, the magnitudes of FACs above the conjugate ionospheres are the same, namely

$$j_\parallel^{(N)} = -j_\parallel^{(S)} = kJ_e \frac{\overline{\Sigma}_P^{(N)} \overline{\Sigma}_P^{(S)}}{\overline{\Sigma}_P^{(S)} + \overline{\Sigma}_P^{(N)}} \tag{6.69}$$

In the case of strongly asymmetric ionospheres, for example, $\overline{\Sigma}_P^{(N)} \gg \overline{\Sigma}_P^{(S)}$, from the above relationships it follows that FACs into both ionospheres are proportional to Pedersen conductivity of the less-conductive hemisphere, $j_{\|}^{(N)} = j_{\|}^{(S)} \propto \overline{\Sigma}_P^{(S)}$.

6.3.2.3 *Ionospheric electric fields in conjugate ionospheres*

HF radar sounding of the ionosphere provides the possibility to monitor the ionospheric Doppler plasma velocities and consequently the electric field. The dense array of SuperDARN radars in both Northern and Southern hemispheres makes it possible to measure the ionospheric electric field response in conjugate ionospheres. The relevant magnitudes of electric fields in conjugate ionospheres are shown to be as follows.

In two limiting cases of quasi-DC forced driving and resonant excitation of Alfvén eigenoscillations, we arrive from Eq. (6.62) at the following ratios between electric disturbances at conjugate ionospheres:

$$\frac{E_N}{E_S} \simeq 1 \text{ when } \tau \gg T_A, \quad \frac{E_N}{E_S} \simeq \frac{\overline{\Sigma}_P^{(S)}}{\overline{\Sigma}_P^{(N)}} \text{ when } \tau \simeq T_A \tag{6.70}$$

These relationships may be verified by comparing the amplitudes of ionospheric electric field disturbances detected simultaneously by SuperDARN radars in the Northern and Southern hemispheres.

6.3.3 Spatial structure of resonant ULF waves and ionospheric conductivity

In the previous consideration, it has been assumed that a disturbance generated in the magnetosphere is a localized stream of FACs. However, the resonant ULF waves, such as Pc4-5 waves, are in fact a coupled MHD mode. So, upon interhemispheric comparison, their latitudinal structure should be taken into account. MHD disturbances from remote parts of the magnetosphere propagate inside the magnetosphere and, through a mode transformation, excite standing Alfvén oscillations. The process of the mode transformation is most effective in the vicinity of the resonant shells, where the local eigenfrequency $f_A(x)$ of Alfvén field line oscillations coincides with the frequency f of an external source (Chen and Hasegawa, 1974). The mathematical description of the spatial structure of the field perturbation in the magnetosphere can be expressed in the form of asymptotic expansion in the vicinity of a resonant field line (Kivelson and Southwood, 1986). The expansion of the azimuthal magnetic component $B_y(x,f)$ has a singularity at a resonant shell ($x \to x_A(f)$). At the same time, the radial $B_x(x, f)$ component has just a weak logarithmic singularity near the resonance, so the resonant behavior of this component would hardly be noticeable. The spectral MHD theory (e.g., Krylov et al., 1979) states that the resonant frequency and field-aligned structure of the coupled MHD mode are described by ordinary differential 1D equations, identical to equations for the uncoupled Alfvén modes in a homogeneous plasma (e.g., plasma box model).

Upon transmission through the ionosphere, the horizontal structure of ULF waves is distorted. This distortion can be analytically described for the Alfvén wave with a resonant Lorentz-type spatial structure transmitting through the "thin" ionospheric layer above an infinitely conducting ground (Hughes and Southwood, 1976; Alperovich and Fedorov, 2007): (a) the $\pi/2$ rotation of the wave polarization ellipse takes place, $B_y^{(m)} \rightarrow B_x^{(g)}$ (superscripts m and g denote field above the ionosphere and on the ground); (b) the width δ_g of the resonance peak, as observed at the ground, is wider as compared with the width δ_m above the ionosphere, $\delta_g = \delta_m + h$. The leading term which describes the amplitude of the north-south ULF wave component at the ground $B_x^{(g)}$ in the vicinity of a resonant magnetic shell can be written as

$$|B_x^{(g)}(x,f)| = B_o(f)\frac{\delta_m}{\sqrt{[x - x_A(f)]^2 + \delta_g^2}} \tag{6.71}$$

According to Eq. (6.71), the latitudinal structure of the ULF field can be qualitatively represented as the combination of a "source" spectrum $B_o(f)$ and a magnetospheric Alfvénic resonance response. The "source" part is related to a disturbance transported by a large-scale fast compressional wave and has a weak dependence on the x coordinate. The resonant magnetospheric response related to the Alfvén waves excitation is strongly localized and it causes rapid enhancement of amplitude at a resonant shell. The Alfvénic part is dominant in the vicinity of the resonant peak, $|x - x_A(f)| \leq \delta_g$. The width δ_g is determined by the dominant damping mechanism (Yumoto et al., 1995). One of the main damping mechanisms is ionospheric Joule dissipation (Newton et al., 1978). In order to drive a resonant system to the saturation level, a driver must be applied continuously during a time period of $\sim T_A Q$ (where Q is the quality factor of the resonant system). A realistic value for the dayside Alfvénic resonator in the Pc5 band is $Q \simeq 5-10$.

6.3.4 Alfvénic pulse incident on conjugate ionospheres

The transient Alfvénic response of the high-latitude magnetosphere to an interplanetary shock is a short impulse with duration τ much less than the Alfvén wave eigenperiod T_A of a field line, $\tau \ll T_A$. In this case, after excitation Alfvénic pulses propagate independently without attenuation toward Northern and Southern hemispheres away from the magnetospheric equatorial plane. At $t < T_A/4$, only incident (i) pulses can be seen in both hemispheres above the ionospheres. In this case, they are described as follows:

$$E_x^{(i)}(t,z) = \frac{1}{2\Sigma_A}\begin{cases} J_e(t + z/V_A) & \text{at } z < 0 \\ J_e(t - z/V_A) & \text{at } z > 0 \end{cases} \tag{6.72}$$

The Alfvénic pulses reflected from the ionospheres (r) can be presented as

$$E_x^{(r)}(t,z) = -\frac{1}{2\Sigma_A}\left[R^{(S)}J_e\left(t-\frac{z+2L}{V_A}\right)+R^{(N)}J_e\left(t+\frac{z-2L}{V_A}\right)\right] \tag{6.73}$$

The total field structure around the ionosphere ($z \simeq \pm L$) is formed by incident (6.72) and reflected (6.73) pulses. The magnetic disturbance B_x on the Earth's surface produced by the Hall current induced by the impulse total field in the ionosphere $J_y = -\Sigma_H E_x$ is

$$B_x = -\mu_0 \frac{\overline{\Sigma}_H}{\overline{\Sigma}_P + 1} J_e(t - L/V_A) \tag{6.74}$$

In respect to the magnitude of the ionospheric conductance, the dependence of the ground response changes from a voltage generator regime ($\overline{\Sigma}_P \ll 1$) to a current regime ($\overline{\Sigma}_P \gg 1$). The ground magnetic response in the Northern and Southern hemispheres from Eq. (6.74) is as follows:

$$\frac{B_x^{(N)}}{B_x^{(S)}} = \frac{\overline{\Sigma}_H^{(N)}}{\overline{\Sigma}_H^{(S)}} \frac{\overline{\Sigma}_P^{(S)} + 1}{\overline{\Sigma}_P^{(N)} + 1} \tag{6.75}$$

In general, this relationship does not correspond directly to either the current or voltage regimes.

6.3.5 Resolving the current-voltage dualism with conjugate observations

The theoretical model predicts that upon the excitation of FAC in the magnetospheric resonator the dependence on the ionospheric conductance may look like a voltage or current generator, depending on the parameters of a magnetospheric disturbance, in particular on timescale of the driver regime τ and the local eigenperiod of the magnetospheric resonator T_A (Pilipenko et al., 2019). Here, we compare the results of various types of ULF disturbances at conjugate sites in the Arctic and Antarctic.

6.3.5.1 *Database and models*

We use data from the conjugate Antarctica-Greenland magnetometer arrays. Virginia Tech (http://mist.nianet.org) deployed an autonomous adaptive low-power instrument platform (AAL-PIP) network with sampling cadence 1 s in Antarctica. These sites are designed to be magnetically conjugate to the Greenland West Coast magnetometer chain along the $\Lambda \simeq 40$degrees magnetic meridian operated by the Technical University of Denmark (https://www.space.dtu.dk). Fig. 6.13 shows the locations of the AAL-PIP stations and the geomagnetic conjugate points of their counterparts in Greenland: PG0-UPN, PG1-UMQ, PG2-GDH, PG3-ATU, PG4-SKT, and PG5-GHB. Conjugate pair coordinates are given in Table 6.1. All stations are equipped with three-axes

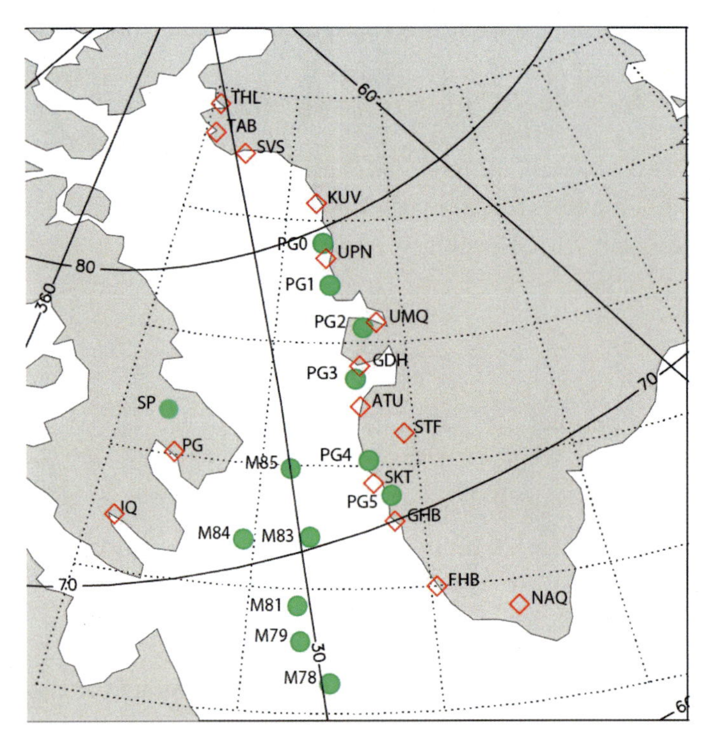

Fig. 6.13 Map of Arctic Canada and Greenland, showing stations in the Northern hemisphere (*diamonds*) and the conjugate mapped locations of Southern hemisphere stations (*green circles*). *Solid lines* show corrected geomagnetic coordinates.

fluxgate magnetometers. The sensor axes are oriented along local magnetic north (B_x), local magnetic east (B_y), and vertically down (B_z). The AAL-PIP profile is augmented by the latitudinal profile of automated low-powered British Antarctic Survey (BAS) stations.

Because of the lack of direct information about the ionospheric conductivities, one has to use information from empirical models. Here, the height-integrated (80–1000 km) ionospheric conductivities have been estimated using the online resource of the Kyoto University (http://wdc.kugi.kyoto-u.ac.jp/ionocond/sigcal), based on the IRI-2016 model (Bilitza et al., 2017). The difference in ionospheric conductances in opposite hemispheres is controlled not only by the difference in solar illumination, but also auroral electron precipitation as well. A source of information on the auroral precipitation is the OVATION-prime (OP) model driven by solar wind and IMF parameters (http://sourceforge.net/projects/ovation-prime/). The OP model is based on energetic particle measurements from the Defense Meteorological Satellite Program (DMSP) satellites (Newell et al., 2009). We applied a combined model which accounts for two main

Table 6.1 Geographic and geomagnetic locations of the Greenland stations and conjugate points of Antarctic stations based on the IGRF for epoch 2015 (https://omniweb.gsfc.nasa.gov/vitmo/cgm.html).

Greenland Antarctica	Code	Geographic Lat.	Location lang.	Geomagnetic lat.	Conjugate lang.
	PG0	−83.67	88.68	78.7	38.2
	PG1	−84.50	77.20	77.3	37.3
	PG2	−84.42	57.95	75.5	39.1
	PG3	−84.81	37.63	73.8	36.6
	PG4	−83.34	12.25	71.1	36.1
	PG5	−81.96	05.71	69.7	37.0
Thule	THL	77.47	290.77	84.4	27.5
Savissivik	SVS	76.02	294.90	82.7	31.2
Kullorsuaq	KUV	74.57	302.82	80.4	40.3
Upernavik	UPN	72.78	303.85	78.6	38.7
Umanaq	UMQ	70.68	307.87	76.0	41.2
Godhavn	GDH	69.25	306.47	74.8	38.2
Attu	ATU	67.93	306.43	73.5	37.1
S. Stromfjord	STF	67.02	309.28	72.1	40.0
Sukkertoppen	SKT	65.42	307.10	70.9	36.4
Godthab	GHB	64.17	308.27	69.5	37.1
Frederikshab	FHB	62.00	310.32	66.9	38.4

contributors: solar photo-ionization and auroral electron precipitation. A contribution of electron precipitation is computed using the empirical model developed by Robinson et al. (1987) that relates particle flux and energy output to conductance.

Here, we compare the results of conjugate observations of such types of ULF disturbances as TCVs, Pc5 waves, SCs, and MPEs at high latitudes.

6.3.5.2 Magnetospheric global FAC systems

Numerous studies of large-scale magnetospheric FACs were performed using magnetic field gradients measured by low-orbiting satellites (Christiansen et al., 2002). The statistical results obtained correspond to a quasisteady (as compared with the Alfvénic time) case. Typically, empirical modeling of quasisteady magnetosphere-ionosphere current system indicated that these FACs are driven by a voltage generator. Here, we mention just a few. The study of Haraguchi et al. (2004) statistically examined the dependence of the intensities of dayside large-scale FACs on the ionospheric conductance using DMSP-F7 satellite data. In the dayside region, the intensity of R1 FAC (current flows into/away the ionosphere in the prenoon/postnoon sector) and cusp (R0) currents had a high correlation with ionospheric conductivity, suggesting that these FACs are driven by a voltage-like source. Shi et al. (2010) statistically investigated features of the FAC distribution in the plasma-sheet boundary layers in the magnetotail using the curlometer technique to calculate the current from four-point magnetic field measurements. The occurrence and polarities of FACs in the Northern hemisphere were found to be different from those in the Southern hemisphere. The interhemispheric difference between the FAC densities suggests that a source of these currents must be a voltage generator. These observational results match the model prediction of nonresonant driving of FACs.

6.3.5.3 Traveling convection vortices

Traveling convection vortices (TCVs) are specific daytime high-latitude structures driven by localized FACs (Friis-Christensen et al., 1988; Engebretson et al., 2013). The terrestrial manifestation of a TCV is an isolated magnetic impulse event (MIE)—a perturbation of the geomagnetic field with a duration of ~5–10 min and amplitude of ~100 nT (Vorob'ev, 1993). The typical double vortex TCV structure of Hall currents is formed around a pair of upward and downward FACs between the ionosphere and the magnetosphere (McHenry and Clauer, 1987). The physical mechanism of excitation of a TCV is not uniquely determined. Various driving mechanisms were claimed, such as the pulsed reconnection of the interplanetary and magnetospheric magnetic fields (Lanzerotti et al., 1990), pulsed variations in the dynamic pressure of the solar wind (Glasmeier, 1992), sporadic regions of hot plasma in the ion foreshock (hot flow anomalies) (Murr and Hughes, 2003), etc.

As an example, we consider the transient TCV events on January 19, 2013 at 1430 and 1730 UT triggered by a weak solar wind pressure increase (Kim et al., 2015). In Fig. 6.14, one can see a remarkable amplitude-phase similarity between magnetic variations in Antarctica and Greenland. A bipolar vortex-like structure is clearly seen in the ionospheric convection flow in both hemispheres. Fig. 6.15 shows the latitudinal profile of TCV amplitude determined as the extreme value of Max$B_x(t)$-Min$B_x(t)$ during the time interval 1420–1450 UT. The maximal intensity in the Northern hemisphere, ~140 nT, is observed at $\Phi = 72$ degrees (STF). A maximum in the Southern hemisphere is hard to evaluate because of the very limited number of stations. Anyway, the difference is not more than 20%. At the same time, the contrast between conductances is about an order of magnitude. Thus, interhemispheric properties of this TCV event rather correspond to the current generator regime.

Earlier observational studies (Kim et al., 2013) of interhemispheric conjugate behavior of MIE/TCVs at high latitudes attempted to reveal how a difference in ionospheric conductivity can play a role in creating asymmetry in TCV structures. The statistical study of Lanzerotti et al. (1991), which used data from the Iqaluit-South Pole conjugate pair, found that TCVs/MIEs were of similar magnitude in the two hemispheres. A study of TCVs at the same conjugate pair by Murr et al. (2002) showed that the amplitudes of the magnetic perturbations were similar in the two hemispheres in the sunlit and dark ionospheres. Lam and Rodger (2004) also observed that conjugate TCVs were of similar intensity in both hemispheres regardless of any difference in conductivity. They found no

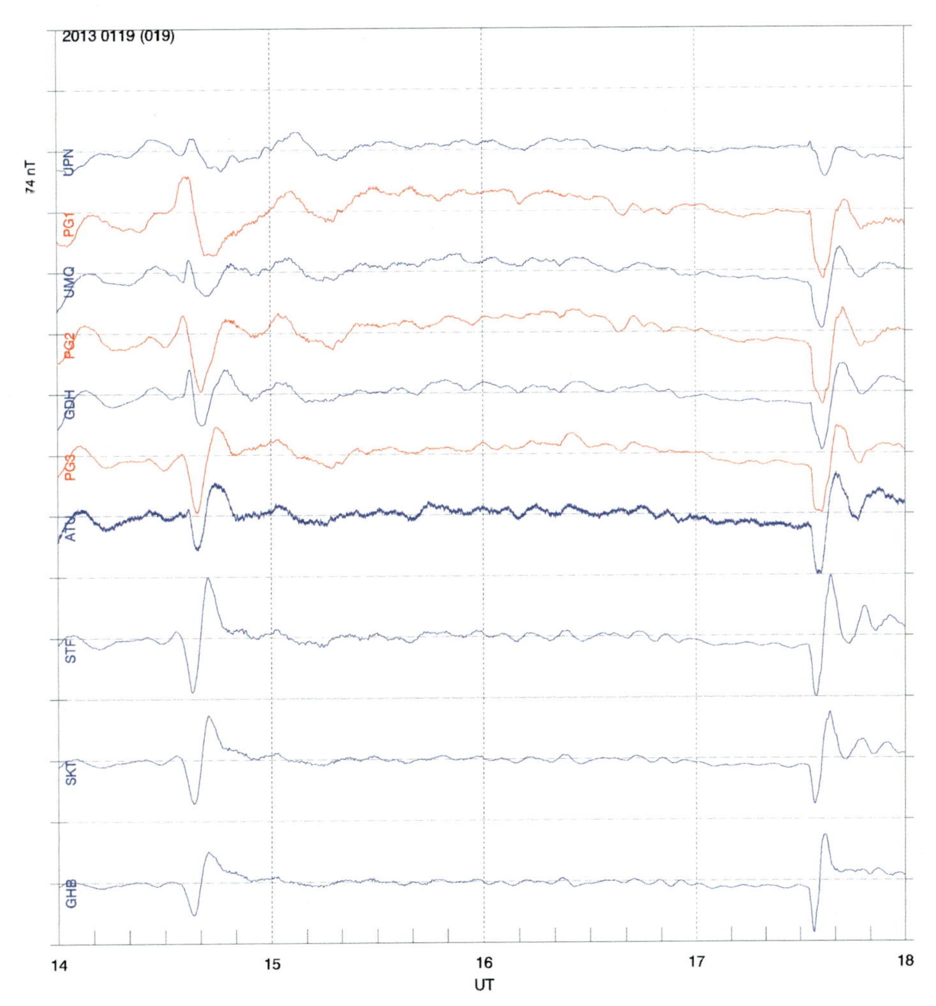

Fig. 6.14 Stacked magnetograms (B_x component) from conjugate pairs of stations in Antarctica (*red lines*) and Greenland (*blue lines*) on January 19, 2013.

statistical difference between events occurring during conditions when one hemisphere was dark and other events when both were dark or light. Based on these observational results, it seems likely that TCVs are to be associated with a current generator.

Thus, according to the model, TCVs are to be excited at a resonant field line. However, TCV waveforms appear rather as impulsive or heavily damped oscillations. Nonetheless, it is possible that resonant effects play a significant role in the TCV formation. Though TCVs are generated by some extra-magnetospheric transients, the "epicenters" of TCV-associated FACs are found well inside the magnetosphere (Moretto and Yahnin, 1998), but not at the magnetopause. Numerical 3D modeling also showed

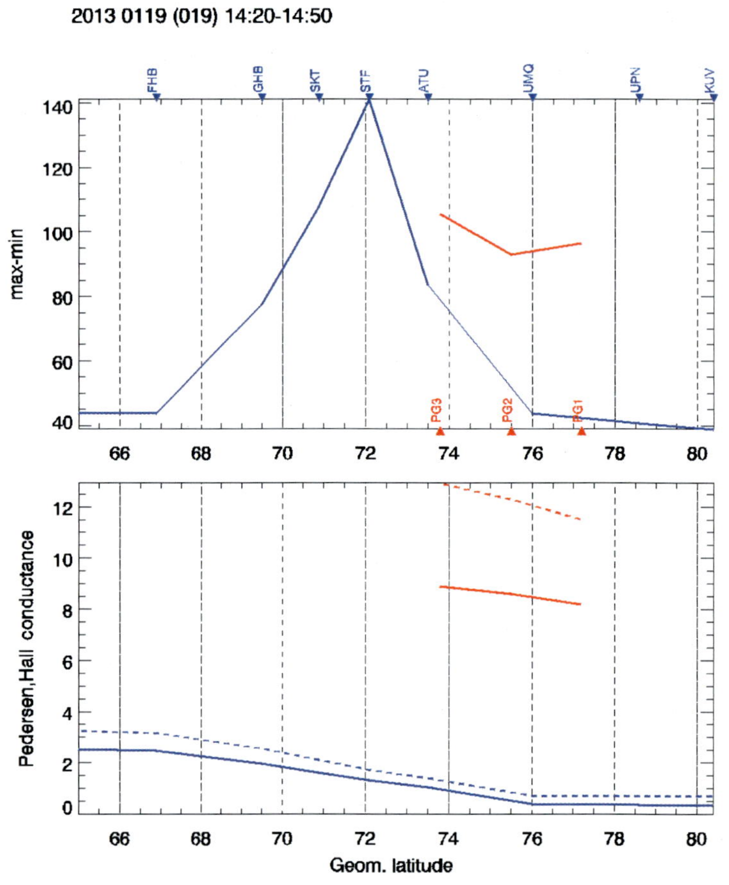

Fig. 6.15 (*Top panel*) The latitudinal profile of TCV amplitude determined as the extreme value of Max$B_x(t)$-Min$B_x(t)$ during the time interval 1420–1450 UT, January 19, 2013 in Antarctica (*red lines*) and Greenland (*blue lines*). (*Bottom panel*) The latitudinal profile of Pedersen (*solid lines*) and Hall (*dotted lines*) conductances in Antarctica (*red lines*) and Greenland (*blue lines*).

that the generation of TCVs and the associated FACs by an interplanetary tangential discontinuity originated well inside the magnetopause (Chen et al., 2000). A possible way to reconcile the intermagnetospheric TCV response with a magnetosheath source is the assumption that the FACs which drive the TCV are produced in the Alfvén resonance region, where the initial impulse duration matches the field line eigenperiod, that is, $\tau \simeq T_A$. The actual driving time of a TCV is less than the time to reach the resonance saturation, $T_A Q$, but a TCV still can be excited because the driver intensity is much stronger than that of monochromatic Pc5 wave drivers. Thus, TCVs are likely a spatially localized phenomenon, and most probably associated with a nearly standing resonant Alfvénic mode. Therefore, the match of its conjugate features to the current generator agrees with the theoretical predictions for resonant driving of the magnetospheric Alfvén resonator.

6.3.5.4 Pc5 waves

The consideration of the current/voltage dichotomy for ULF waves (Pc3-5 pulsations) should be done with great care. In general, the field of ULF waves is composed of a fast magnetosonic (compressional) mode and an Alfvén mode. The transmission mechanisms of the compressional mode and Alfvén mode through the ionosphere are very different (Pilipenko et al., 2011). A FAC carried by an Alfvén wave cannot penetrate into the insulating atmosphere and instead spreads over the ionosphere, whereas the ground magnetic response is produced by the Hall currents. The magnetic field compression transported by a fast mode wave "feels" the ionosphere only weakly, and directly penetrates to the Earth's surface. Both modes, azimuthally large scale, produce a main ground response in the same B_x component. The contribution of the Alfvénic part becomes dominant in the vicinity of the peak of the resonant structure.

The north-south asymmetry of the amplitude of Pc5 pulsations was studied by Obana et al. (2005) using magnetic field data from one pair of the conjugate Kotzebue-Macquarie Island stations at $L \simeq 5.4$. The power ratio showed an "offset," probably caused by a regular bias of the statistical position of the wave resonant peak and the station location. The relative stability of the offset during seasonal variations may provide indirect evidence of the independence of the ground magnetic signal on the ionospheric conductance. A specific difficulty in analysis of conjugate studies of Pc5 pulsations is that resonant frequency-dependent amplification occurs in a small latitudinal region (about a few hundred kilometers). Conjugate observation results are strongly influenced by uncertainties in the difference between the pulsation resonant peak and an observation site. Therefore, any conclusion on Pc5 asymmetry demands a thorough preliminary determination of resonant wave structure in both hemispheres.

An example of this approach from Pilipenko et al. (2021) is given below. Spectral analysis of magnetometer data is made in a running 30-min window. The spectra of horizontal components for all stations in both hemispheres were calculated, and the central frequency band of recorded Pc5 waves has been identified. Latitudinal plots of $B_x(\Phi)$ spectral power at the central frequency of the selected band depending on geomagnetic latitude Φ were constructed. These plots enable us to compare the amplitudes of spatial peaks at two conjugate stations, where the maximum was observed. At the same time, from the OP model the ionospheric conductances were estimated. Using these values, we compare the observed ratio with the theoretical predictions. Here, we present contrasting Pc5 events during the Northern summer and Northern winter.

Stacked magnetograms of the B_x component from conjugate pairs of stations during the Northern winter event on January 25, 2016 are shown in Fig. 6.16. Quasimonochromatic Pc5 waves appeared during 06–20 UT in the recovery phase of a weak substorm with onset at ~07 UT. Signatures of field line resonance can be seen even from visual inspection: the localized amplitude of magnetic variations at GDH-PG3 stations, apparent poleward propagation, and dominance of the B_x component over the B_y component (not shown).

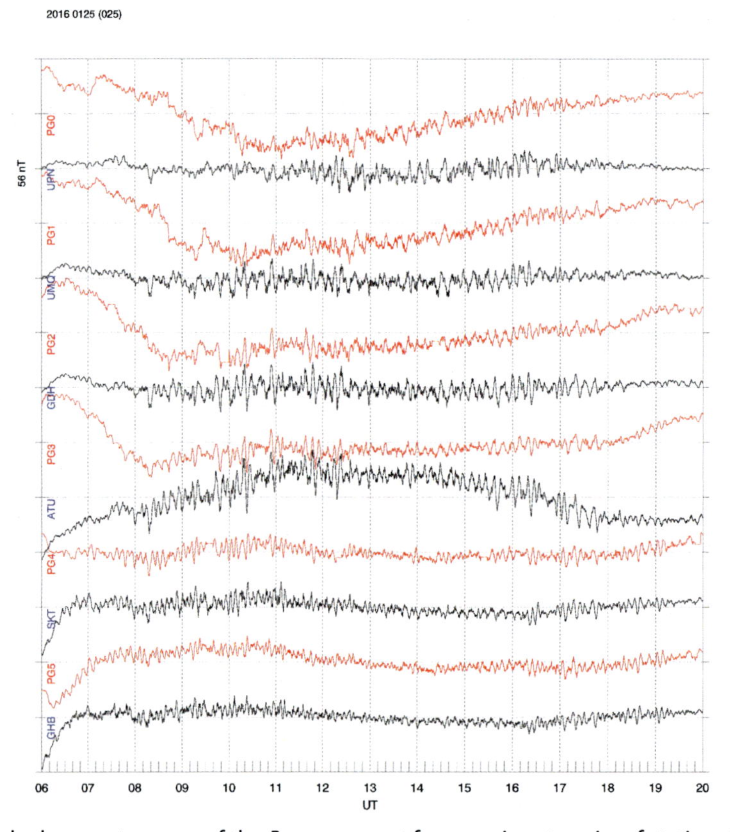

Fig. 6.16 Stacked magnetograms of the B_x component from conjugate pairs of stations in Antarctica (*red lines*) and Greenland (*blue lines*) during the Northern winter event on January 25, 2016.

Spectra of the B_x component (not shown) demonstrate the occurrence of a spectral peak in the band between 2.0 and 2.5 mHz at most stations in both hemispheres, so the latitudinal structure of the spectral power of monochromatic Pc5 waves is examined at this frequency. The latitudinal distribution of spectral power along the profile has a maximum at $\Phi \simeq 75$ degrees in both hemispheres (Fig. 6.17). Spectral power in the Northern hemisphere is larger than that in the Southern hemisphere, but just weakly, ~20%–40%. The predicted ionospheric height-integrated conductances are around $\Sigma^{(S)} \simeq 5-6$ S, and $\Sigma^{(N)} \simeq 1$ S.

Stacked magnetograms of the B_x component from conjugate pairs of stations for the Northern summer event on June 13, 2016 are shown in Fig. 6.18. Quasimonochromatic Pc5 waves appeared between 09 and 15 UT. Even from a visual inspection of magnetograms, one can see a regular increase of the dominant frequency from 2.5 mHz at high latitudes (PG2-UMQ) toward 3.5 mHz at lower latitudes (PG5-SKT). During the 13–14 UT interval, the latitudinal peaks of spectral power at $f = 2.0-2.5$ mHz are at the same

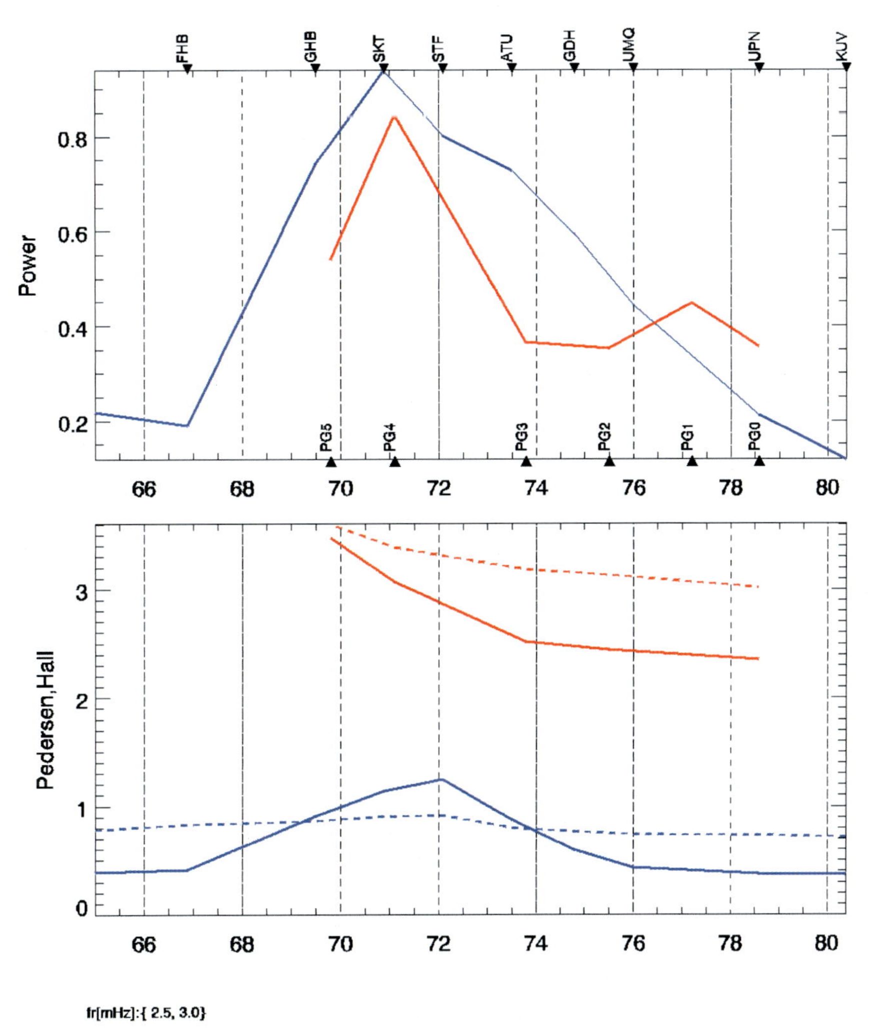

Fig. 6.17 (*Top panel*) The latitudinal distribution of spectral power (*X* component) in the 2.0–2.5 mHz band along the profile in the Northern (*blue lines*) and Southern (*red lines*) hemispheres during the Northern winter event on January 25, 2016. (*Bottom panel*) The predicted ionospheric height-integrated Pedersen (*solid lines*) and Hall (*dashed lines*) conductances in Antarctica (*red lines*) and Greenland (*blue lines*).

geomagnetic latitude in both hemispheres, around 74 degrees (ATU/PG3) (Fig. 6.19). The interhemispheric contrast between ionospheric conductances is rather substantial, $\Sigma_H^{(N)} \simeq 11$ S, whereas $\Sigma_H^{(S)} \simeq 0.8$ S. The profile of the ionospheric conductance is weakly dependent on latitude, and strong gradients are absent. Despite the strong asymmetry of the ionospheric conductance, the magnetic field spectral power is nearly the same in both

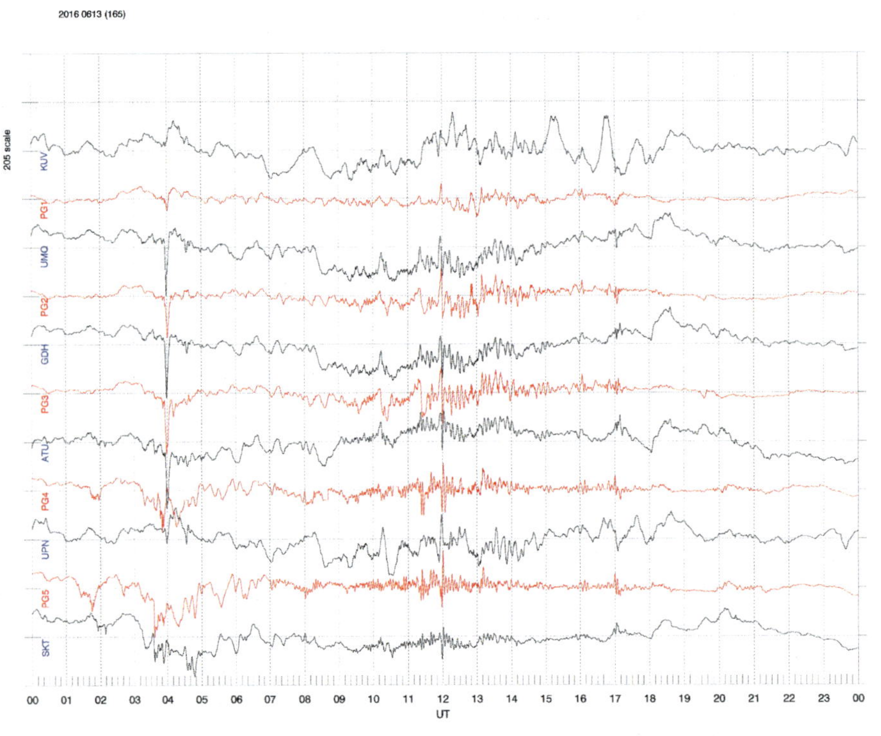

Fig. 6.18 Stacked magnetograms of the B_x component from conjugate pairs of stations in Antarctica (*red lines*) and Greenland (*blue lines*) for the Northern summer event on June 13, 2016.

hemispheres, $B^{(N)} \simeq B^{(S)}$. Thus, the interhemispheric contrast between the Pc5 amplitudes is much less than the contrast between ionospheric conductivities. The Pc5 wave excitation is much closer to the current generator regime than to the voltage generator regime.

6.3.5.5 High-latitude SC impulses

A sudden commencement (SC) is a complicated large-scale transient response of the magnetosphere-ionosphere system to an interplanetary shock. It consists of a magnetic field compression transported to the ground by a fast compressional mode wave and a global vortex-like disturbance produced by an FAC system at the magnetopause. The SC-associated compression of the magnetosphere can excite damped Psc5 pulsations in the magnetosphere localized at a latitudinally narrow resonant magnetic shell where the impulse duration $\tau \simeq T_A$. Beyond this narrow region, no periodic resonant response to the SC pulse is observed. Thus, the SC main impulse is a transient FAC stimulated by an interplanetary shock and predominantly it corresponds either to the condition $\tau \gg T_A$ (at low latitudes, where $T_A < 1$ min) or $\tau \ll T_A$ (at high latitudes, where $T_A \simeq$ 5–10 min). Formally, low latitude SC observations are to be treated as an example of

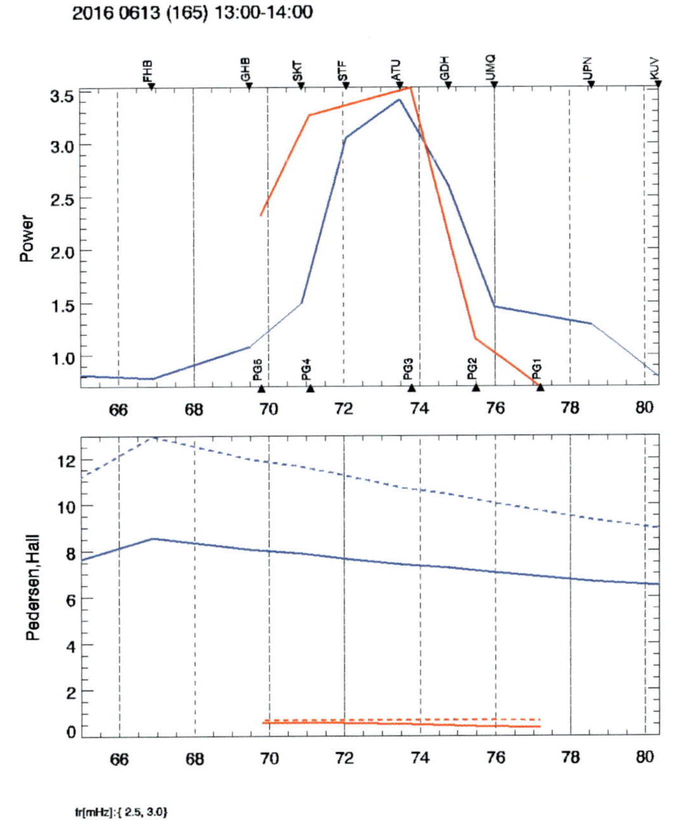

Fig. 6.19 (*Top panel*) The latitudinal distribution of spectral power (*X* component) in the 2.0–2.5 mHz band along the profile in the Northern (*blue lines*) and Southern (*red lines*) hemispheres during the Northern summer event on June 13, 2016. (*Bottom panel*) The predicted ionospheric height-integrated Pedersen (*solid lines*) and Hall (*dashed lines*) conductances in Antarctica (*red lines*) and Greenland (*blue lines*).

a quasi-DC excitation. Studies at low latitudes ($L < 3$) indeed revealed an interhemispheric difference in SC signatures: their amplitude was significantly larger in the summer hemisphere than in the winter one (Yumoto et al., 1996). Shinbori et al. (2012) also found that the size of the diurnal variation of the SC main impulse increased significantly during the summer, compared with that during the winter. Based on the seasonal and interhemispheric variations of SCs at low latitudes, it may be concluded that the SC in this region is caused by a voltage generator rather than a current generator. The regime of a voltage generator is natural for a nonresonant quasi-DC disturbance (Pilipenko et al., 2019).

However, in contrast to low latitudes, for an SC pulse at high latitudes ($L > 6$), the inverse condition takes place, $\tau \ll T_A$. Here, we present typical SC events on a quiet background during summer and winter, and determine from conjugate station data the ratio between amplitudes of pulses $B_x^{(N)}/B_x^{(S)}$ at conjugate pairs of stations in Antarctica and Greenland. The magnetospheric Alfvén conductance is supposed not to differ considerably throughout the magnetosphere, $\Sigma_A^{(N)} \simeq \Sigma_A^{(S)} \simeq 1$ S.

We first present the Northern winter event that occurred on January 7, 2014, 15 UT (Fig. 6.20, top panel). According to the IRI-2016 model, for the central PG3 station in the Southern hemisphere $\Sigma_P^{(S)} \simeq 5.0$ S and $\Sigma_H^{(S)} \simeq 6.0$ S. For the conjugate station ATU in the Northern hemisphere $\Sigma_P^{(N)} \simeq 2.4$ S, and $\Sigma_H^{(N)} \simeq 2.1$ S. In this event, the contrast between the Northern and Southern conductances is not so strong. There is an additional factor that may obscure the observed relationship between the ground magnetic response and the ionospheric conductance, which should be taken into account. Though the model considered in Section 6.3.4 is homogeneous, in reality the plasma density N_e in the upper ionosphere at the sunlit end of a field line probably should be higher than at the dark end. As a result, the Alfvén wave conductance, $\Sigma_A \propto \sqrt{N_e}$, must be higher at the sunlit end. Thus, the contrast in the ratio Σ_P/Σ_A, which determines the reflection condition and ground response, is to be less distinct between (N) and (S) ionospheres. Here, as a proxy of plasma density in the upper ionosphere, N_e, total electron content (TEC) maps from GPS receivers have been used (http://vt.superdarn.org). Based on very sparse observations in Greenland and Antarctica, TEC in the Northern hemisphere is ~12 TECu, while in the Southern hemisphere TEC is ~15 TECu. Then, Eq. (6.75) gives us $B_x^{(N)}/B_x^{(S)} \simeq 0.5$. The correction for the asymmetry of $\Sigma_A^{(N)}/\Sigma_A^{(S)} \simeq \sqrt{TEC^{(N)}/TEC^{(S)}} \simeq 0.9$ does not modify this estimate noticeably. Thus, the model predicts that by the order of magnitude the amplitudes of pulses in conjugate ionospheres are to be about the same. The observed amplitudes of the SC pulse at station pairs GDH-PG2, ATU-PG3, and SKT-PG4 are indeed about the same within 15%.

Then, we consider the typical Northern summer event on June 7, 2014, 17 UT (Fig. 6.20, bottom panel). For the central PG3 station in the Southern hemisphere, the ionospheric conductances estimated with the use of the IRI-2016 model are as follows: $\Sigma_P^{(S)} \simeq 0.2$ S and $\Sigma_H^{(S)} \simeq 0.2$ S. For the conjugate station ATU in the Northern hemisphere, the ionospheric conductances are: $\Sigma_P^{(N)} \simeq 6.8$ S, and $\Sigma_H^{(N)} \simeq 8.8$ S. Therefore, it is reasonable to assume that $\overline{\Sigma}_P^{(N)} \gg 1$ and $\overline{\Sigma}_P^{(S)} \ll 1$. According to the TEC maps for this event, the TEC in the Northern hemisphere is ~15 TECu, while in the Southern hemisphere it is below 5 TECu. Then, Eq. (6.75) gives us $B_x^{(N)}/B_x^{(S)} \simeq 6.5$. The asymmetry of $\Sigma_A^{(N)}/\Sigma_A^{(S)} \simeq \sqrt{TEC^{(N)}/TEC^{(S)}} \simeq 1.7$ decreases this estimate down to ~3.8. This prediction matches surprisingly well the observed ratio between magnetic responses in conjugate ionospheres: for the ATU-PG3 pair the ratio is ~3.8.

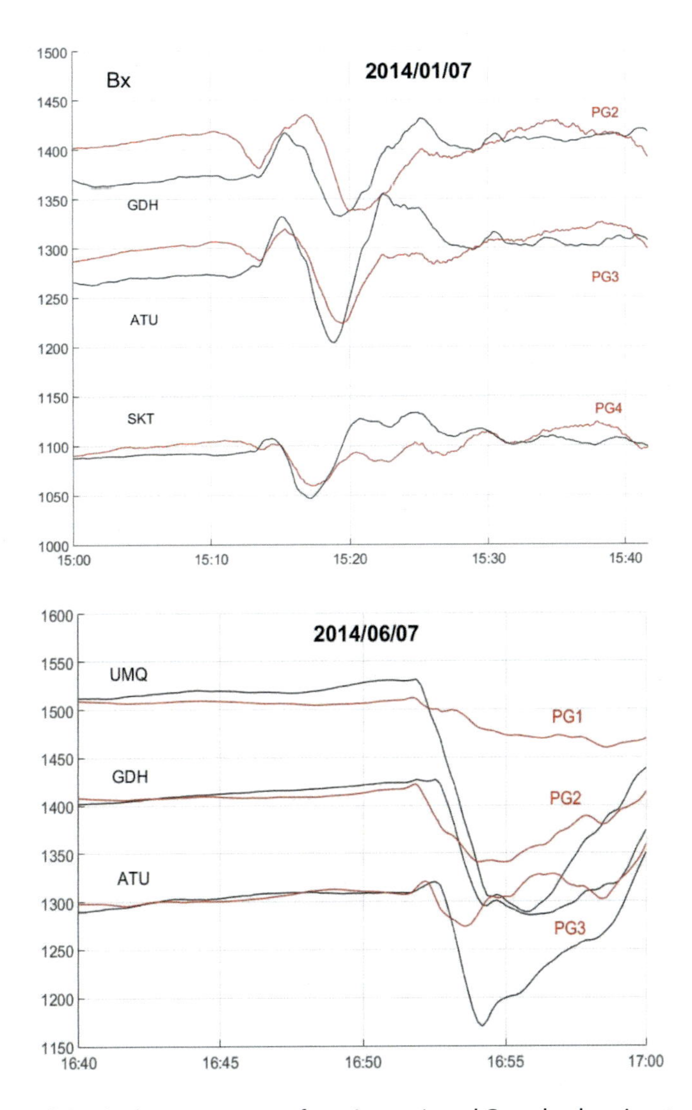

Fig. 6.20 (*Top panel*) Stacked magnetograms from Antarctic and Greenland conjugate pairs of stations during the Northern summer SC event on June 7, 2014. (*Bottom panel*) Stacked magnetograms from Antarctic and Greenland conjugate pairs of stations during the Northern winter SC event on January 7, 2014.

6.3.5.6 Magnetic perturbation events

When the auroral oval expands to subauroral latitudes, impulsive MPEs with duration ~5–15 min occur during even nonstorm times and at up to 78 degrees magnetic latitude (Engebretson et al., 2019a). The MPEs appear roughly simultaneously at near

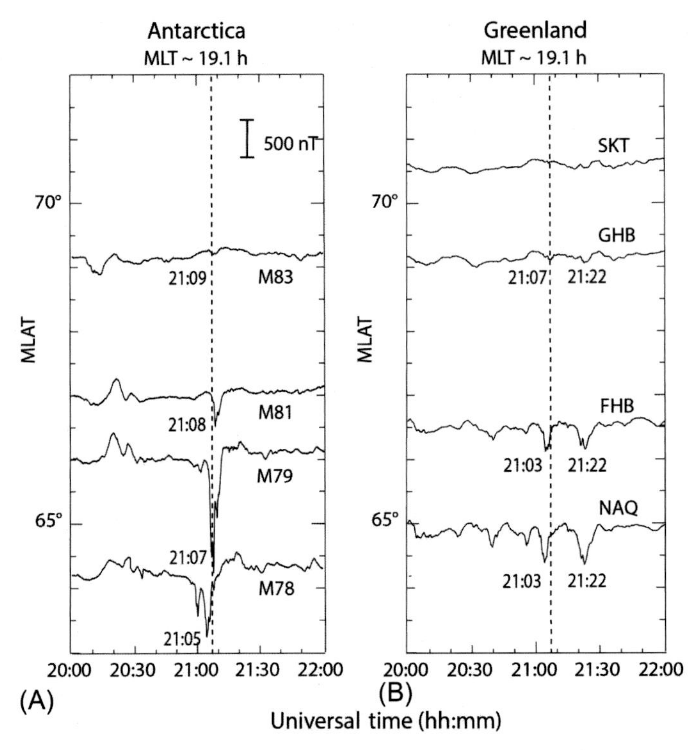

Fig. 6.21 Magnetograms from stations in the Northern hemisphere (A) and at magnetically conjugate locations in Antarctica (B) during May 8, 2016.

magnetically conjugate locations in each hemisphere (Engebretson et al., 2019b). Here, we present data from a 2D set of ground-based magnetometers in the Northern hemisphere and magnetometers at magnetically conjugate locations in Antarctica during May 8, 2016 (Fig. 6.21). This MPE interval (2102–2130 UT) includes an extremely large $dB/dt \simeq 37.7$ nT/s at M79, but MPEs appear only from 64 to 69 degrees. Multiple substorm onsets occurred ~1–3 h before the beginning of the MPE interval. Magnetograms from BAS LPM M79 and FHB on this day show that a single B_x minimum at M79 appeared at 21:07 UT, and two B_x minima appeared at FHB at 21:04 and 21:22 UT, respectively. The largest $|dB_x/dt|$ excursion at FHB was substantial (6.7 nT/s), but was a factor of ~5 smaller than that at M79.

The amplitude and location data for the MPE during this interval can be used to estimate its latitudinal and longitudinal scale size. Both B_x and the derivatives in each component at the Antarctic stations were highly localized in MLAT: the maximum $|dB_x/dt|$ value decreased to less than half its maximal value within 1–1.7 degrees. The latitudinal distances ranged from 100 to 200 km, and for the two somewhat less closely spaced West Greenland stations at nearly the same magnetic longitude, the latitudinal distance was

~200 km. These latitudinal distances are roughly comparable with the ~275 km 2D half-amplitude radius calculated for several events in Arctic Canada using the SECs technique by Engebretson et al. (2019a).

The first MPE observed in Greenland occurred within ~3 min of the much larger MPE observed in Antarctica, and conversely there was no evidence of the second Greenland MPE at any of the Antarctic stations. Thus, for both MPEs there was an apparent lack of conjugacy. However, at least some of this lack of conjugacy might be attributed to longitudinal localization of both MPEs: BAS LPM stations M79 and M81 were located ~9 degrees in magnetic longitude west of the conjugate point of FHB, at distances of ~ 430 km.

During the interval under study, both perturbation and derivative amplitudes were consistently larger by a factor of ~3 in the Southern hemisphere than in the Northern hemisphere at comparable latitudes (Fig. 6.22A and B). Fig. 6.22C shows the ionospheric conductances calculated using an updated AMIE procedure based on an empirical model parameterized by solar zenith angle and the solar radio flux index, F10.7 (Cousins et al., 2015). This augmented model contributed only negligible additional conductances because the modeled auroral zone was located below the latitude range of the available stations.

It has been suggested that MPEs are driven by localized FACs, which transfer an image of a "spacequake" in the magnetotail to the high-latitude nightside ionosphere. However, the exact mechanism of FAC generation and specific channel of the disturbance propagation have not been established. Current/voltage generator properties may be helpful to reveal this mechanism. Inverse interhemispheric patterns are evident during northern summer events (not shown here): magnetic perturbations and derivatives were mostly larger in the Southern hemisphere, and conductances were much larger in the Northern hemisphere. Northern hemisphere conductances based on the AMIE model increased relatively smoothly with MLT, while Southern hemisphere conductances were nearly constant. These relations indicate that for MPEs the voltage generator model is not applicable. On the other hand, the difference between Northern and Southern amplitudes is too large to fit the current generator regime, even under the known uncertainties in the modeled conductances.

6.3.6 Discussion: Additional factors influencing ground magnetometer observations

The simple consideration in this review necessarily neglected some, hopefully of secondary importance, factors. The factors that may complicate the examination of ionospheric conductance effect in magnetically conjugate points are (a) the differing effects on ground conductivity of coastlines and oceans in the north versus ice sheets in the south and (b) the fact that Antarctic stations are situated at 13–18 degrees higher geographic latitude than Arctic stations at similar MLAT.

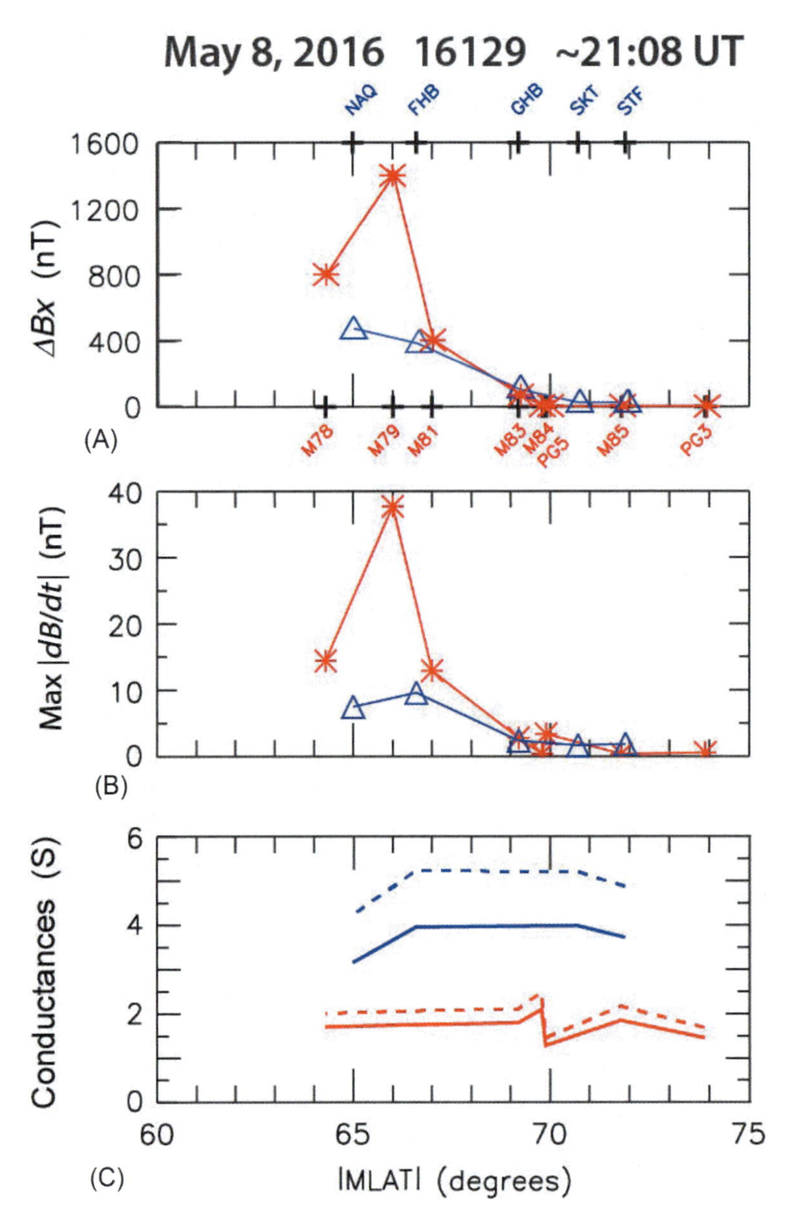

Fig. 6.22 Latitudinal profile of MPE amplitudes (A) and derivative amplitudes (B) in the Southern and Northern hemispheres during the MPE of May 8, 2016. (C) Ionospheric conductances calculated using an updated AMIE procedure. *Solid lines* and *dashed lines* denote Pedersen and Hall conductances, respectively.

The conclusion that Pc5 wave interhemispheric properties correspond to the current generator regime does not mean that Pc5 excitation efficiency in the magnetosphere does not depend on the ionospheric conductivity. By contrast, it seems reasonable that a low damping of Alfvén field line oscillations in the regions with elevated ionospheric conductance is favorable for the Pc5 wave excitation (Sutcliffe and Rostoker, 1979).

At high latitudes, the presence of the auroral acceleration region (AAR) must be taken into account upon consideration of the magnetospheric driver's internal resistance and the load resistance. A potential drop along auroral field lines can be modeled with the Ohm-type nonlocal current-voltage relationship (Fedorov et al., 2001). A special consideration is needed to reveal the influence of the mirror resistance along auroral field lines on interhemispheric properties of disturbances at these latitudes.

The model considered here has a passive ionosphere, whose conductances do not evolve due to the current/voltage imposed on the ionosphere. However, a very intense FAC can modify the ionosphere via plasma transport by Alfvén wave (Belakhovsky et al., 2016) or modulated precipitation of particles (Spanswick et al., 2005). An SC impulse can trigger sporadic precipitation of energetic electrons (Rosenberg et al., 1980), which may locally modify the ionospheric conductance. However, during daytime, such precipitation is not strong enough to modify considerably the ionospheric global parameters. Such extreme events of the ionospheric conductance modification by a magnetospheric driver are not considered in the linear theory. A steep gradient of the ionospheric conductance, for example, at the auroral oval boundary or cusp boundary, also neglected in the above consideration, can substantially distort the ground magnetic response (Glasmeier et al., 1984).

The above consideration is an element of the more global problem of the significance of the ionosphere for the a coupled magnetosphere-ionosphere system. Though it is generally thought that the occurrence of substorms is externally controlled by some instability in the magnetotail, Liou et al. (2018) provided a strong argument that the ionosphere plays an active role in the occurrence of substorms. Statistical analysis of UV images indicated that substorm onsets occurred more frequently when the ionosphere was dark, and in regions where the Earth's magnetic field is largest. There is mounting evidence suggesting a relationship between solar illumination and the occurrence frequency of auroral acceleration events, which produce discrete arcs (Newell et al., 1996). These facts suggest that auroral substorms occur more frequently when the ionospheric conductivity is lower.

6.3.7 Conclusion

Here, we have highlighted an important aspect of the magnetosphere-ionosphere interaction, which should be taken into account when considering the ground response to magnetospheric FAC driving. A simple box model of the magnetosphere predicts that nonsteady FACs interact with the ionosphere in a different way in cases of a forced driving or resonant excitation. A quasi-DC driving of FAC corresponds to a voltage generator regime, when the ground magnetic response is proportional to the ionospheric Hall conductance. A resonant excitation corresponds to the current generator regime, when the ground magnetic response practically does not depend on the ionospheric conductance. According to the suggested conception, quasi-DC nonresonant disturbances such as global magnetospheric FAC systems and SCs at low latitudes correspond to a voltage generator. Such magnetospheric phenomena as TCVs and Pc5 waves should be considered as a resonant response of magnetospheric field lines and correspond to a current generator. For a short SC pulse at high latitudes when a conjugate ionosphere does not influence the pulse interaction with the ionosphere of interest, the interhemispheric asymmetry of pulse amplitude in the general does not correspond either to the current or voltage generator. The typical ULF events shown here support the outline of the classification scheme that enabled us to resolve the current/voltage dichotomy. The aspects of the current/voltage dichotomy presented here should be taken into account when considering the magnetosphere-ionosphere interaction, and can be applied to improve global MHD simulations.

Acknowledgments

This work was supported by NASA LWS Grant #80NSSC19K0080 and NSF Award AGS-1755350. This work used the XSEDE and TACC computational facilities, supported by NSF grant ACI-1053575. This study is supported by a grant 20-05-00787 from the Russian Fund for Basic Research (ENF), and the US National Science Foundation grants PLR-1744828 to Virginia Tech (MDH and SC) and AGS-2013648 to the Augsburg University (VAP and MJE).

References

Agostinelli, S., et al., 2003. GEANT4-a simulation toolkit. Nucl. Instrum. Methods A 506 (25), 250–303.

Allison, J., Amako, K., Apostolakis, J., Arce, P., Asai, M., Aso, T., Bagli, E., Bagulya, A., Banerjee, S., Barrand, G.J.N.I., et al., 2016. Recent developments in Geant4. Nucl. Instrum. Methods Phys. Res. A 835, 186–225.

Alperovich, L.S., Fedorov, E.N., 2007. Hydromagnetic Waves in the Magnetosphere and the Ionosphere. Astrophysics and Space Science Library, vol. 353, XXIV, Springer Netherlands, 418 pp. https://doi.org/10.1007/978-1-4020-6637-5.

André, M., Norqvist, P., Vaivads, A., Eliasson, L., Norberg, O., Eriksson, A.I., Holback, B., 1994. Transverse ion energization and wave emissions observed by the Freja satellite. Geophys. Res. Lett. 21 (17), 1915–1918. https://doi.org/10.1029/94GL00699.

Arfken, G.B., Weber, H.J., Harris, F.E., 2012. Mathematical Methods for Physicists, seventh. Academic Press, Boston, MA, https://doi.org/10.1016/B978-0-12-384654-9.00032-3.

Artamonov, A., Mironova, I., Kovaltsov, G., Mishev, A., Plotnikov, E., Konstantinova, N., 2017. Calculation of atmospheric ionization induced by electrons with non-vertical precipitation: updated model CRAC-EPII. Adv. Space Res. 59 (9), 2295–2300.

Bahcivan, H., 2007. Plasma wave heating during extreme electric fields in the high-latitude E region. Geophys. Res. Lett. 34 (15), L15106. https://doi.org/10.1029/2006GL029236.

Bailey, D.K., 1959. Abnormal ionization in the lower ionosphere associated with cosmic-ray flux enhancements. Proc. IRE 47 (2), 255–266.

Banjac, S., Herbst, K., Heber, B., 2019. The atmospheric radiation interaction simulator (ATRIS): description and validation. J. Geophys. Res. Space Phys. 124 (1), 50–67.
Banks, P.M., Nagy, A.F., 1970. Concerning the influence of elastic scattering upon photoelectron transport and escape. J. Geophys. Res. 75 (10), 1902–1910.
Banks, P.M., Chappell, C.R., Nagy, A.F., 1974. A new model for the interaction of auroral electrons with the atmosphere: spectral degradation, backscatter, optical emission, and ionization. J. Geophys. Res. 79 (10), 1459–1470.
Basu, B., Chang, T., Jasperse, J.R., 1982. Electrostatic plasma instabilities in the daytime lower ionosphere. Geophys. Res. Lett. 9 (1), 68–71. https://doi.org/10.1029/GL009i001p00068.
Basu, B., Jasperse, J.R., Robinson, R.M., Vondrak, R.R., Evans, D.S., 1987. Linear transport theory of auroral proton precipitation: a comparison with observations. J. Geophys. Res. Space Phys. 92 (A6), 5920–5932.
Basu, B., Jasperse, J.R., Strickland, D.J., Daniell Jr, R.E., 1993. Transport-theoretic model for the electron-proton-hydrogen atom aurora: 1. Theory. J. Geophys. Res. Space Phys. 98 (A12), 21517–21532.
Basu, B., Decker, D.T., Jasperse, J.R., 2001. Proton transport model: a review. J. Geophys. Res. Space Phys. 106 (A1), 93–105.
Belakhovsky, V., Pilipenko, V., Murr, D., Fedorov, E., Kozlovsky, A., 2016. Modulation of the ionosphere by Pc5 waves observed simultaneously by GPS/TEC and EISCAT. Earth Planets Space 68 (1), 1–13.
Bellan, P.M., 2008. Fundamentals of Plasma Physics. Cambridge University Press.
Berger, M.J., 1963. Monte Carlo calculation of the penetration and diffusion of fast charged particles. Methods Comput. Phys. 1, 135–215.
Berger, M.J., Coursey, J.S., Zucker, M.A., Chang, J., 2017. ESTAR, PSTAR, and ASTAR: Computer Programs for Calculating Stopping-Power and Range Tables for Electrons, Protons, and Helium Ions (version 2.0.1). National Institute of Standards and Technology, Gaithersburg, MD. http://physics.nist.gov/Star.
Bhatnagar, P.L., Gross, E.P., Krook, M., 1954. A model for collision processes in gases. I. Small amplitude processes in charged and neutral one-component systems. Phys. Rev. 94, 511–525. https://doi.org/10.1103/PhysRev.94.511.
Bilitza, D., Altadill, D., Truhlik, V., Shubin, V., Galkin, I., Reinisch, B., Huang, X., 2017. International reference ionosphere 2016: from ionospheric climate to real-time weather predictions. Space Weather 15 (2), 418–429.
Buneman, O., 1963. Excitation of field aligned sound waves by electron streams. Phys. Rev. Lett. 10, 285.
Cashwell, E.D., Everett, C.J., 1959. A Practical Manual on the Monte Carlo Method for Random Walk Problems. Pergamon, New York
Chaston, C.C., Bonnell, J.W., Carlson, C.W., McFadden, J.P., Ergun, R.E., Strangeway, R.J., 2003. Properties of small-scale Alfvén waves and accelerated electrons from FAST. J. Geophys. Res. Space Phys. 108 (A4). https://doi.org/10.1029/2002JA009420.
Chen, L., Hasegawa, A., 1974. A theory of long-period magnetic pulsations. 1. Steady state excitation of field line resonance. J. Geophys. Res. 79, 1024–1032.
Chen, G.X., Lin, Y., Cable, S., 2000. Generation of traveling convection vortices and field-aligned currents in the magnetosphere by response to an interplanetary tangential discontinuity. Geophys. Res. Lett. 27 (21), 3583–3586.
Chen, M.W., Lemon, C.L., Guild, T.B., Keesee, A.M., Lui, A., Goldstein, J., Rodriguez, J.V., Anderson, P. C., 2015. Effects of modeled ionospheric conductance and electron loss on self-consistent ring current simulations during the 5–7 April 2010 storm. J. Geophys. Res. Space Phys. 120 (7), 5355–5376. https://doi.org/10.1002/2015JA021285.
Choudhuri, A.R., 1998. The Physics of Fluids and Plasmas: An Introduction for Astrophysicists. Cambridge University Press, Cambridge.
Christiansen, F., Papitashvili, V.O., Neubert, T., 2002. Seasonal variations of high-latitude field-aligned currents inferred from Ørsted and Magsat observations. J. Geophys. Res. Space Phys. 107 (A2), SMP-5.
Cousins, E.D.P., Matsuo, T., Richmond, A.D., 2015. Mapping high-latitude ionospheric electrodynamics with SuperDARN and AMPERE. J. Geophys. Res. Space Phys. 120 (7), 5854–5870.

Davidson, G.T., 1965. Expected spatial distribution of low-energy protons precipitated in the auroral zones. J. Geophys. Res. 70 (5), 1061–1068.

Decker, D.T., Kozelov, B.V., Basu, B., Jasperse, J.R., Ivanov, V.E., 1996. Collisional degradation of the proton-H atom fluxes in the atmosphere: a comparison of theoretical techniques. J. Geophys. Res. Space Phys. 101 (A12), 26947–26960.

Desai, M., Giacalone, J., 2016. Large gradual solar energetic particle events. Living Rev. Solar Phys. 13 (1), 1–132.

Dimant, Y.S., Oppenheim, M.M., 2004. Ion thermal effects on E-region instabilities: linear theory. J. Atmos. Solar Terr. Phys. 66, 1639–1654. https://doi.org/10.1016/j.jastp.2004.07.006.

Dimant, Y.S., Oppenheim, M.M., 2011. Magnetosphere-ionosphere coupling through E-region turbulence: 2. Anomalous conductivities and frictional heating. J. Geophys. Res. 116 (A09), A09304. https://doi.org/10.1029/2011JA016649.

Dimant, Y.S., Sudan, R.N., 1997. Physical nature of a new cross-field current-driven instability in the lower ionosphere. J. Geophys. Res. 102, 2551–2564. https://doi.org/10.1029/96JA03274.

Drummond, W.E., 2004. Landau damping. Phys. Plasmas 11 (2), 552–560. https://doi.org/10.1063/1.1628685.

Drummond, W.E., Pines, D., 1962. Non-linear stability of plasma oscillations. Nucl. Fusion Suppl. 2, 1047–1059. https://www-thphys.physics.ox.ac.uk/people/AlexanderSchekochihin/notes/PlasmaClassics/drummond_pines62.pdf.

Eliasson, L., André, M., Eriksson, A., Norqvist, P., Norberg, O., Lundin, R., Holback, B., Koskinen, H., Borg, H., Boehm, M., 1994. Freja observations of heating and precipitation of positive ions. Geophys. Res. Lett. 21 (17), 1911–1914. https://doi.org/10.1029/94GL01067.

Eltrass, A., Mahmoudian, A., Scales, W.A., de Larquier, S., Ruohoniemi, J.M., Baker, J.B.H., Greenwald, R.A., Erickson, P.J., 2014. Investigation of the temperature gradient instability as the source of midlatitude quiet time decameter-scale ionospheric irregularities: 2. Linear analysis. J. Geophys. Res. Space Phys. 119 (6), 4882–4893. https://doi.org/10.1002/2013JA019644.

Emery, B.A., Coumans, V., Evans, D.S., Germany, G.A., Greer, M.S., Holeman, E., Kadinsky-Cade, K., Rich, F.J., Xu, W., 2008. Seasonal, Kp, solar wind, and solar flux variations in long-term single-pass satellite estimates of electron and ion auroral hemispheric power. J. Geophys. Res. Space Phys. 113 (A6), A06311. https://doi.org/10.1029/2007JA012866.

Engebretson, M.J., Yeoman, T.K., Oksavik, K., Søraas, F., Sigernes, F., Moen, J.I., Johnsen, M.G., Pilipenko, V.A., Posch, J.L., Lessard, M.R., et al., 2013. Multi-instrument observations from Svalbard of a traveling convection vortex, electromagnetic ion cyclotron wave burst, and proton precipitation associated with a bow shock instability. J. Geophys. Res. Space Phys. 118 (6), 2975–2997.

Engebretson, M.J., Pilipenko, V.A., Ahmed, L.Y., Posch, J.L., Steinmetz, E.S., Moldwin, M.B., Connors, M.G., Weygand, J.M., Mann, I.R., Boteler, D.H., et al., 2019a. Nighttime magnetic perturbation events observed in Arctic Canada: 1. Survey and statistical analysis. J. Geophys. Res. Space Phys. 124 (9), 7442–7458.

Engebretson, M.J., Steinmetz, E.S., Posch, J.L., Pilipenko, V.A., Moldwin, M.B., Connors, M.G., Boteler, D.H., Mann, I.R., Hartinger, M.D., Weygand, J.M., et al., 2019b. Nighttime magnetic perturbation events observed in Arctic Canada: 2. Multiple-instrument observations. J. Geophys. Res. Space Phys. 124 (9), 7459–7476.

Eriksson, A.I., Holback, B., Dovner, P.O., Boström, R., Holmgren, G., André, M., Eliasson, L., Kintner, P. M., 1994. Freja observations of correlated small-scale density depletions and enhanced lower hybrid waves. Geophys. Res. Lett. 21 (17), 1843–1846. https://doi.org/10.1029/94GL00174.

Fang, X., Liemohn, M.W., Kozyra, J.U., Solomon, S.C., 2004. Quantification of the spreading effect of auroral proton precipitation. J. Geophys. Res. Space Phys. 109 (A4), A04309. https://doi.org/10.1029/2003JA010119.

Fang, X., Liemohn, M.W., Kozyra, J.U., Solomon, S.C., 2005. Study of the proton arc spreading effect on primary ionization rates. J. Geophys. Res. Space Phys. 110 (A7), A07302. https://doi.org/10.1029/2004JA010915.

Fang, X., Liemohn, M.W., Kozyra, J.U., Evans, D.S., 2007a. Global 30–240 keV proton precipitation in the 17–18 April 2002 geomagnetic storms: 2. Conductances and beam spreading. J. Geophys. Res. Space Phys. 112 (A5), A05302. https://doi.org/10.1029/2006JA012113.

Fang, X., Liemohn, M.W., Kozyra, J.U., Evans, D.S., DeJong, A.D., Emery, B.A., 2007b. Global 30–240 keV proton precipitation in the 17–18 April 2002 geomagnetic storms: 1. Patterns. J. Geophys. Res. Space Phys. 112 (A5), A05301. https://doi.org/10.1029/2006JA011867

Fang, X., Ridley, A.J., Liemohn, M.W., Kozyra, J.U., Evans, D.S., 2007c. Global 30–240 keV proton precipitation in the 17–18 April 2002 geomagnetic storms: 3. Impact on the ionosphere and thermosphere. J. Geophys. Res. Space Phys. 112 (A7), A07310. https://doi.org/10.1029/2006JA012144.

Fang, X., Randall, C.E., Lummerzheim, D., Solomon, S.C., Mills, M.J., Marsh, D.R., Jackman, C.H., Wang, W., Lu, G., 2008. Electron impact ionization: a new parameterization for 100 eV to 1 MeV electrons. J. Geophys. Res. Space Phys. 113 (A9), A09311. https://doi.org/10.1029/2008JA013384.

Fang, X., Randall, C.E., Lummerzheim, D., Wang, W., Lu, G., Solomon, S.C., Frahm, R.A., 2010. Parameterization of monoenergetic electron impact ionization. Geophys. Res. Lett. 37 (22), L22106. https://doi.org/10.1029/2010GL045406.

Fang, X., Lummerzheim, D., Jackman, C.H., 2013. Proton impact ionization and a fast calculation method. J. Geophys. Res. Space Phys. 118 (8), 5369–5378.

Farley, D.T., 1963a. A plasma instability resulting in field-aligned irregularities in the ionosphere. J. Geophys. Res. 68, 6083.

Farley, D.T., 1963b. Two-stream plasma instability as a source of irregularities in the ionosphere. Phys. Rev. Lett. 10 (7), 279–282.

Fedorov, E., Pilipenko, V., Engebretson, M.J., 2001. ULF wave damping in the auroral acceleration region. J. Geophys. Res. Space Phys. 106 (A4), 6203–6212.

Foster, J., Erickson, P., 2000. Simultaneous observations of E-region coherent backscatter and electric field amplitude at F-region heights with the Millstone Hill UHF Radar. Geophys. Res. Lett. 27 (19), 3177–3180. https://doi.org/10.1029/2000GL000042.

Foster, J.C., Erickson, P.J., Lind, F.D., Rideout, W., 2004. Millstone Hill coherent-scatter radar observations of electric field variability in the sub-auroral polarization stream. Geophys. Res. Lett. 31 (21), L21803. https://doi.org/10.1029/2004GL021271.

Frahm, R.A., Winningham, J.D., Sharber, J.R., Link, R., Crowley, G., Gaines, E.E., Chenette, D.L., Anderson, B.J., Potemra, T.A., 1997. The diffuse aurora: a significant source of ionization in the middle atmosphere. J. Geophys. Res. Atmos. 102 (D23), 28203–28214.

Friis-Christensen, E., Vennerstrøm, S., Clauer, C.R., McHenry, M.A., 1988. Irregular magnetic pulsations in the polar cleft caused by traveling ionospheric convection vortices. Adv. Space Res. 8 (9–10), 311–314.

Gaisser, T.K., 1990. Cosmic Rays and Particle Physics. Cambridge University Press, Cambridge.

Galand, M., Lilensten, J., Kofman, W., Sidje, R.B., 1997. Proton transport model in the ionosphere: 1. Multistream approach of the transport equations. J. Geophys. Res. Space Phys. 102 (A10), 22261–22272.

Galand, M., Roble, R.G., Lummerzheim, D., 1999. Ionization by energetic protons in thermosphere-ionosphere electrodynamics general circulation model. J. Geophys. Res. Space Phys. 104 (A12), 27973–27989.

Gérard, J.-C., Hubert, B., Bisikalo, D.V., Shematovich, V.I., 2000. A model of the Lyman-α line profile in the proton aurora. J. Geophys. Res. Space Phys. 105 (A7), 15795–15805.

Glasmeier, K.-H., 1992. Traveling magnetospheric convection twin-vortices-observations and theory. Ann. Geophys. 10 (8), 547–565.

Glasmeier, K.H., et al., 1984. On the influence of ionospheres with non-uniform conductivity distribution on hydromagnetic waves. J. Geophys. 54 (1), 125–137.

Goldberg, R.A., Jackman, C.H., Barcus, J.R., Søraas, F., 1984. Nighttime auroral energy deposition in the middle atmosphere. J. Geophys. Res. Space Phys. 89 (A7), 5581–5596.

Green, A.E.S., Jackman, C.H., Garvey, R.H., 1977. Electron impact on atmospheric gases, 2. Yield spectra. J. Geophys. Res. 82 (32), 5104–5111.

Gurevich, A.V., 1978. Nonlinear Phenomena in the Ionosphere. Springer, New York, https://doi.org/10.1007/978-3-642-87649-3.

Haldoupis, C., Schlegel, K., Farley, D.T., 1996. An explanation for type 1 radar echoes from the midlatitude E-region ionosphere. Geophys. Res. Lett. 23 (1), 97–100. https://doi.org/10.1029/95GL03585.

Haraguchi, K., Kawano, H., Yumoto, K., Ohtani, S., Higuchi, T., Ueno, G., 2004. Ionospheric conductivity dependence of dayside region-0, 1, and 2 field-aligned current systems: statistical study with DMSP-F7. Ann. Geophys. 22 (8), 2775–2783.

Hardy, D.A., Gussenhoven, M.S., Raistrick, R., McNeil, W.J., 1987. Statistical and functional representations of the pattern of auroral energy flux, number flux, and conductivity. J. Geophys. Res. Space Phys. 92 (A11), 12275–12294.

Hardy, D.A., Gussenhoven, M.S., Brautigam, D., 1989. A statistical model of auroral ion precipitation. J. Geophys. Res. Space Phys. 94 (A1), 370–392.

Hartinger, M.D., Xu, Z., Clauer, C.R., Yu, Y., Weimer, D.R., Kim, H., Pilipenko, V., Welling, D.T., Behlke, R., Willer, A.N., 2017. Associating ground magnetometer observations with current or voltage generators. J. Geophys. Res. Space Phys. 122 (7), 7130–7141.

Hedin, A.E., 1991. Extension of the MSIS thermosphere model into the middle and lower atmosphere. J. Geophys. Res. Space Phys. 96 (A2), 1159–1172.

Hill, T.L., 1956. Statistical Mechanics: Principles and Selected Applications. Dover Books on Physics, McGraw-Hill, ISBN: 9780486653907.

Hoh, F.C., 1963. Instability of penning-type discharges. Phys. Fluids 6, 1184–1191. https://doi.org/10.1063/1.1706878.

Hollweg, J.V., 1999. Kinetic Alfvén wave revisited. J. Geophys. Res. 104 (A7), 14811–14820. https://doi.org/10.1029/1998JA900132.

Huang, K., 1963. Statistical Mechanics. Wiley. ISBN 9780471417602.

Hughes, W.J., Southwood, D.J., 1976. The screening of micropulsation signals by the atmosphere and ionosphere. J. Geophys. Res. 81 (19), 3234–3240.

Jackman, C.H., Frederick, J.E., Stolarski, R.S., 1980. Production of odd nitrogen in the stratosphere and mesosphere: an intercomparison of source strengths. J. Geophys. Res. Oceans 85 (C12), 7495–7505.

Jackman, C.H., Marsh, D.R., Vitt, F.M., Garcia, R.R., Fleming, E.L., Labow, G.J., Randall, C.E., López-Puertas, M., Funke, B., von Clarmann, T., et al., 2008. Short- and medium-term atmospheric constituent effects of very large solar proton events. Atmos. Chem. Phys. 8 (3), 765–785.

Jasperse, J.R., 1997. Transport theoretic solutions for the beam-spreading effect in the proton-hydrogen aurora. Geophys. Res. Lett. 24 (11), 1415–1418.

Jasperse, J.R., Basu, B., 1982. Transport theoretic solutions for auroral proton and H atom fluxes and related quantities. J. Geophys. Res. Space Phys. 87 (A2), 811–822.

Kagan, L.M., Kelley, M.C., 2000. A thermal mechanism for generation of small-scale irregularities in the ionospheric E region. J. Geophys. Res. 105, 5291–5302. https://doi.org/10.1029/1999JA900415.

Kelley, M.C., 2009. The Earth's Ionosphere: Plasma Physics and Electrodynamics. Academic Press, Ithaca, New York.

Khazanov, G.V., 2011. Kinetic Theory of the Inner Magnetospheric Plasma. Springer, New York, https://doi.org/10.1007/978-1-4419-6797-8.

Khazanov, G.V., Liemohn, M.W., Newman, T.S., Fok, M.C., Spiro, R.W., 2003. Self-consistent magnetosphere-ionosphere coupling: theoretical studies. J. Geophys. Res. Space Phys. 108 (A3), 1122. https://doi.org/10.1029/2002JA009624.

Khazanov, G.V., Khabibrakhmanov, I.K., Glocer, A., 2012. Kinetic description of ionospheric outflows based on the exact form of Fokker-Planck collision operator: electrons. J. Geophys. Res. Space Phys. 117 (A11), A11203. https://doi.org/10.1029/2012JA018082.

Khazanov, G.V., Glocer, A., Himwich, E.W., 2014. Magnetosphere-ionosphere energy interchange in the electron diffuse aurora. J. Geophys. Res. Space Phys. 119 (1), 171–184. https://doi.org/10.1002/2013JA019325.

Kim, H., Cai, X., Clauer, C.R., Kunduri, B.S.R., Matzka, J., Stolle, C., Weimer, D.R., 2013. Geomagnetic response to solar wind dynamic pressure impulse events at high-latitude conjugate points. J. Geophys. Res. Space Phys. 118 (10), 6055–6071.

Kim, H., Clauer, C.R., Engebretson, M.J., Matzka, J., Sibeck, D.G., Singer, H.J., Stolle, C., Weimer, D.R., Xu, Z., 2015. Conjugate observations of traveling convection vortices associated with transient events at the magnetopause. J. Geophys. Res. Space Phys. 120 (3), 2015–2035.

Kintner, P.M., Vago, J., Chesney, S., Arnoldy, R.L., Lynch, K.A., Pollock, C.J., Moore, T.E., 1992. Localized lower hybrid acceleration of ionospheric plasma. Phys. Rev. Lett. 68 (16), 2448–2451. https://doi.org/10.1103/PhysRevLett.68.2448.

Kintner, P.M., Bonnell, J., Arnoldy, R., Lynch, K., Pollock, C., Moore, T., 1996. SCIFER-transverse ion acceleration and plasma waves. Geophys. Res. Lett. 23 (14), 1873–1876. https://doi.org/10.1029/96GL01863.

Kivelson, M.G., Southwood, D.J., 1986. Coupling of global magnetospheric MHD eigenmodes to field line resonances. J. Geophys. Res. Space Phys. 91 (A4), 4345–4351.

Kozelov, B.V., 1993. Influence of the dipolar magnetic field on transport of proton-H atom fluxes in the atmosphere. Ann. Geophys. 11 (8), 697–704.

Krylov, A.L., Lifshitz, A.E., Fedorov, E.N., 1979. About resonant properties of a plasma in a curvilinear magnetic field. Doklady AN SSSR 247, 1056–1061.

Kuby, M.J., Roberts, T.D., Upchurch, C.D., Tierney, S., 2009. Network analysis. In: Rob, K., Nigel, T. (Eds.), International Encyclopedia of Human Geography. Elsevier, Oxford, pp. 391–398.

Kudeki, E., Fejer, B.G., Farley, D.T., Hanuise, C., 1987. The Condor equatorial electrojet campaign-radar results. J. Geophys. Res. 92, 13561–13577. https://doi.org/10.1029/JA092iA12p13561.

Lam, M.M., Rodger, A.S., 2004. A test of the magnetospheric source of traveling convection vortices. J. Geophys. Res. 109, A02204, https://doi.org/10.1029/2003JA010214.

Lanzerotti, L.J., Wolfe, A., Trivedi, N., Maclennan, C.G., Medford, L.V., 1990. Magnetic impulse events at high latitudes: magnetopause and boundary layer plasma processes. J. Geophys. Res. Space Phys. 95 (A1), 97–107.

Lanzerotti, L.J., Konik, R.M., Wolfe, A., Venkatesan, D., Maclennan, C.G., 1991. Cusp latitude magnetic impulse events: 1. Occurrence statistics. J. Geophys. Res. Space Phys. 96 (A8), 14009–14022.

Lazarev, V.I., 1967. Absorption of the energy of an electron beam in the upper atmosphere. Geomagn. Aeron. 7, 219.

Lehtinen, N.G., Bell, T.F., Inan, U.S., 1999. Monte Carlo simulation of runaway MeV electron breakdown with application to red sprites and terrestrial gamma ray flashes. J. Geophys. Res. Space Phys. 104 (A11), 24699–24712.

Link, R., 1992. Feautrier solution of the electron transport equation. J. Geophys. Res. Space Phys. 97 (A1), 159–169.

Liou, K., Sotirelis, T., Mitchell, E.J., 2018. North-South asymmetry in the geographic location of auroral substorms correlated with ionospheric effects. Sci. Rep. 8 (1), 1–6.

Livadiotis, G., McComas, D.J., 2009. Beyond kappa distributions: exploiting Tsallis statistical mechanics in space plasmas. J. Geophys. Res. Space Phys. 114 (A11), A11105. https://doi.org/10.1029/2009JA014352.

Longley, W.J., Oppenheim, M.M., Pedatella, N.M., Dimant, Y.S., 2020. The photoelectron-driven upper hybrid instability as the cause of 150-km echoes. Geophys. Res. Lett. 47 (8). https://doi.org/10.1029/2020GL087391. e2020GL087391.

Lorentzen, D.A., 2000. Latitudinal and longitudinal dispersion of energetic auroral protons. Ann. Geophys. 18 (1), 81–89.

Lummerzheim, D., 1992. Comparison of Energy Dissipation Functions for High Energy Auroral Electron and Ion Precipitation. Geophysical Institute, University of Alaska, Fairbanks.

Lummerzheim, D., Lilensten, J., 1994. Electron transport and energy degradation in the ionosphere: evaluation of the numerical solution, comparison with laboratory experiments and auroral observations. Ann. Geophys. 12 (10), 1039–1051.

Lummerzheim, D., Rees, M.H., Anderson, H.R., 1989. Angular dependent transport of auroral electrons in the upper atmosphere. Planet. Space Sci. 37 (1), 109–129.

Lyons, L.R., 1997. Magnetospheric processes leading to precipitation. Space Sci. Rev. 80, 109–132.

Lysak, R.L., 1985. Auroral electrodynamics with current and voltage generators. J. Geophys. Res. Space Phys. 90 (A5), 4178–4190.

Lysak, R.L., 1990. Electrodynamic coupling of the magnetosphere and ionosphere. Space Sci. Rev. 52 (1), 33–87.

Lysak, R.L., Lotko, W., 1996. On the kinetic dispersion relation for shear Alfvén waves. J. Geophys. Res. 101 (A3), 5085–5094. https://doi.org/10.1029/95JA03712.

Maeda, K., Tsuda, T., Maeda, H., 1963. Theoretical interpretation of the equatorial sporadic *E* layers. Phys. Rev. Lett. 11 (9), 406–407. https://doi.org/10.1103/PhysRevLett.11.406.
McHenry, M.A., Clauer, C.R., 1987. Modeled ground magnetic signatures of flux transfer events. J. Geophys. Res. Space Phys. 92 (A10), 11231–11240.
Mesick, K.E., Feldman, W.C., Coupland, D.D.S., Stonehill, L.C., 2018. Benchmarking Geant4 for simulating galactic cosmic ray interactions within planetary bodies. Earth Space Sci. 5 (7), 324–338.
Mewaldt, R.A., Mason, G.M., Gloeckler, G., Christian, E.R., Cohen, C.M.S., Cummings, A.C., Davis, A. J., Dwyer, J.R., Gold, R.E., Krimigis, S.M., Leske, R.A., Mazur, J.E., Stone, E.C., von Rosenvinge, T. T., Wiedenbeck, M.E., Zurbuchen, T.H., 2001. Long-term fluences of energetic particles in the heliosphere. In: Wimmer-Schweingruber, R.F. (Ed.), American Institute of Physics Conference Series. Joint SOHO/ACE Workshop "Solar and Galactic Composition", November, vol. 598, pp. 165–170.
Moore, T.E., Lundin, R., Alcayde, D., André, M., Ganguli, S.B., Temerin, M., Yau, A., 1999. Chapter 2—Source processes in the high-latitude ionosphere. Space Sci. Rev. 88, 7–84. https://doi.org/10.1023/A:1005299616446.
Moretto, T., Yahnin, A., 1998. Mapping travelling convection vortex events with respect to energetic particle boundaries. Ann. Geophys. 16 (8), 891–899.
Murr, D.L., Hughes, W.J., 2003. Solar wind drivers of traveling convection vortices. Geophys. Res. Lett. 30 (7).
Murr, D.L., Hughes, W.J., Rodger, A.S., Zesta, E., Frey, H.U., Weatherwax, A.T., 2002. Conjugate observations of traveling convection vortices: the field-aligned current system. J. Geophys. Res. Space Phys. 107 (A10), 1306. https://doi.org/10.1029/2002JA009456.
Nagy, A.F., Banks, P.M., 1970. Photoelectron fluxes in the ionosphere. J. Geophys. Res. 75 (31), 6260–6270.
Newell, P.T., Meng, C.I., Lyons, K.M., 1996. Discrete aurorae are suppressed in sunlight. Nature 381 (6585), 766–767.
Newell, P.T., Sotirelis, T., Wing, S., 2009. Diffuse, monoenergetic, and broadband aurora: the global precipitation budget. J. Geophys. Res. Space Phys. 114 (A9), A09207. https://doi.org/10.1029/2009JA014326.
Newton, R.S., Southwood, D.J., Hughes, W.J., 1978. Damping of geomagnetic pulsations by the ionosphere. Planet. Space Sci. 26 (3), 201–209.
Ni, B., Thorne, R.M., Horne, R.B., Meredith, N.P., Shprits, Y.Y., Chen, L., Li, W., 2011a. Resonant scattering of plasma sheet electrons leading to diffuse auroral precipitation: 1. Evaluation for electrostatic electron cyclotron harmonic waves. J. Geophys. Res. Space Phys. 116 (A4). https://doi.org/10.1029/2010JA016232.
Ni, B., Thorne, R.M., Meredith, N.P., Horne, R.B., Shprits, Y.Y., 2011b. Resonant scattering of plasma sheet electrons leading to diffuse auroral precipitation: 2. Evaluation for whistler mode chorus waves. J. Geophys. Res. Space Phys. 116 (A4). https://doi.org/10.1029/2010JA016233.
Ni, B., Liang, J., Thorne, R.M., Angelopoulos, V., Horne, R.B., Kubyshkina, M., Spanswick, E., Donovan, E.F., Lummerzheim, D., 2012. Efficient diffuse auroral electron scattering by electrostatic electron cyclotron harmonic waves in the outer magnetosphere: a detailed case study. J. Geophys. Res. Space Phys. 117 (A1), A01218. https://doi.org/10.1029/2011JA017095.
Nicholson, D.R., 1983. Introduction to Plasma Theory. Wiley, New York.
Obana, Y., Yoshikawa, A., Olson, J.V., Morris, R.J., Fraser, B.J., Yumoto, K., 2005. North-South asymmetry of the amplitude of high-latitude Pc 3–5 pulsations: observations at conjugate stations. J. Geophys. Res. Space Phys. 110 (A10).
Oppenheim, M.M., Dimant, Y.S., 2013. Kinetic simulations of 3-D Farley-Buneman turbulence and anomalous electron heating. J. Geophys. Res. Space Phys. 118, 1306–1318. https://doi.org/10.1002/jgra.50196.
Pelowitz, D.B., et al., 2005. MCNPX User's Manual Version 2.5. 0. Los Alamos Natl. Lab. 76, 473.
Pfaff, R.F., Kelley, M.C., Kudeki, E., Fejer, B.G., Baker, K.D., 1987a. Electric field and plasma density measurements in the strongly driven daytime equatorial electrojet. 1. The unstable layer and gradient drift waves. J. Geophys. Res. 92, 13578–13596. https://doi.org/10.1029/JA092iA12p13578.

Pfaff, R.F., Kelley, M.C., Kudeki, E., Fejer, B.G., Baker, K.D., 1987b. Electric field and plasma density measurements in the strongly driven daytime equatorial electrojet. 2. Two-stream waves. J. Geophys. Res. 92, 13597–13612. https://doi.org/10.1029/JA092iA12p13597.

Pierrard, V., Lazar, M., 2010. Kappa distributions: theory and applications in space plasmas. Solar Phys. 267, 153–174. https://doi.org/10.1007/s11207-010-9640-2.

Pilipenko, V., Fedorov, E., Heilig, B., Engebretson, M.J., Sutcliffe, P., Lühr, H., 2011. ULF waves in the topside ionosphere: satellite observations and modeling. In: Liu, W., Fujimoto, M. (Eds.), The Dynamic Magnetosphere, IAGA Special Sopron Book Series, vol. 3, Springer, Dordrecht, pp. 257–269. https://doi.org/10.1007/978-94-007-0501-2 (Chapter 14).

Pilipenko, V.A., Fedorov, E.N., Hartinger, M.D., Engebretson, M.J., 2019. Electromagnetic fields of magnetospheric ULF disturbances in the ionosphere: current/voltage dichotomy. J. Geophys. Res. Space Phys. 124 (1), 109–121.

Pilipenko, V.A., Martines-Bedenko, V.A., Coyle, S., Fedorov, E.N., Hartinger, M.D., Engebretson, M.J., Edwards, T.R., 2021. Conjugate properties of magnetospheric Pc5 waves: Antarctica-Greenland comparison. J. Geophys. Res. Space Phys. 126, e2020JA028048. https://doi.org/10.1029/2020JA028048.

Porter, H.S., Green, A.E.S., 1975. Comparison of Monte Carlo and continuous slowing-down approximation treatments of 1-keV proton energy deposition in N2. J. Appl. Phys. 46 (11), 5030–5038.

Porter, H.S., Jackman, C.H., Green, A.E.S., 1976. Efficiencies for production of atomic nitrogen and oxygen by relativistic proton impact in air. J. Chem. Phys. 65 (1), 154–167.

Reames, D.V., 1999. Particle acceleration at the Sun and in the heliosphere. Space Sci. Rev. 90 (3), 413–491.

Rees, M.H., 1963. Auroral ionization and excitation by incident energetic electrons. Planet. Space Sci. 11 (10), 1209–1218.

Rees, M.H., 1964. Ionization in the Earth's atmosphere by aurorally associated bremsstrahlung X-rays. Planet. Space Sci. 12 (11), 1093–1108.

Rees, M.H., 1969. Auroral electrons. Space Sci. Rev. 10 (3), 413–441.

Rees, M.H., 1982. On the interaction of auroral protons wit the Earth's atmosphere. Planet. Space Sci. 30 (5), 463–472.

Reid, G.C., 1961. A study of the enhanced ionization produced by solar protons during a polar cap absorption event. J. Geophys. Res. 66 (12), 4071–4085.

Ridley, A.J., Gombosi, T.I., DeZeeuw, D.L., 2004. Ionospheric control of the magnetosphere: conductance. Ann. Geophys. 22 (2), 567–584.

Robinson, R.M., Vondrak, R.R., Miller, K., Dabbs, T., Hardy, D., 1987. On calculating ionospheric conductances from the flux and energy of precipitating electrons. J. Geophys. Res. Space Phys. 92 (A3), 2565–2569.

Roble, R.G., Ridley, E.C., 1987. An auroral model for the NCAR thermospheric general circulation model (TGCM). Ann. Geophys. 5, 369–382.

Roeder, J.L., Koons, H.C., 1989. A survey of electron cyclotron waves in the magnetosphere and the diffuse auroral electron precipitation. J. Geophys. Res. Space Phys. 94 (A3), 2529–2541. https://doi.org/10.1029/JA094iA03p02529.

Rosenberg, T.J., Siren, J.C., Lanzerotti, L.J., 1980. High time resolution riometer and X-ray measurements of conjugate electron precipitation from the magnetosphere. Nature 283 (5744), 278–280.

Schmidt, M.J., Gary, S.P., 1973. Density gradients and the Farley-Buneman instability. J. Geophys. Res. 78, 8261–8265. https://doi.org/10.1029/JA078i034p08261.

Scholz, C.H., 2007. 6.10-Fault mechanics. In: Gerald, S. (Ed.), Treatise on Geophysics. Elsevier, Amsterdam, pp. 441–483.

Schroeder, J.W.R., Skiff, F., Howes, G.G., Kletzing, C.A., Carter, T.A., Dorfman, S., 2017. Linear theory and measurements of electron oscillations in an inertial Alfvén wave. Phys. Plasmas 24 (3), 032902. https://doi.org/10.1063/1.4978293.

Schröter, J., Heber, B., Steinhilber, F., Kallenrode, M.B., 2006. Energetic particles in the atmosphere: a Monte-Carlo simulation. Adv. Space Res. 37 (8), 1597–1601.

Schunk, R.W., Nagy, A.F., 2004. Ionospheres. Cambridge University Press, Cambridge, p. 570.

Schunk, R., Nagy, A., 2009. Ionospheres: Physics, Plasma Physics, and Chemistry. Cambridge University Press, Cambridge.

Schwadron, N.A., Dayeh, M.A., Desai, M., Fahr, H., Jokipii, J.R., Lee, M.A., 2010. Superposition of Stochastic Processes and the Resulting Particle Distributions. Astrophys. J. 713, 1386–1392. https://doi.org/10.1088/0004-637X/713/2/1386.
Shalimov, S., Haldoupis, C., 2005. E-region wind-driven electrical coupling of patchy sporadic-E and spread-F at midlatitude. Ann. Geophys. 23 (6), 2095–2105. https://doi.org/10.5194/angeo-23-2095-2005.
Sharp, R.D., Johnson, R.G., Shelley, E.G., 1977. Observation of an ionospheric acceleration mechanism producing energetic (keV) ions primarily normal to the geomagnetic field direction. J. Geophys. Res. 82 (22), 3324. https://doi.org/10.1029/JA082i022p03324.
Shelley, E.G., Sharp, R.D., Johnson, R.G., 1976. Satellite observations of an ionospheric acceleration mechanism. Geophys. Res. Lett. 3 (11), 654–656. https://doi.org/10.1029/GL003i011p00654.
Shi, J.K., Cheng, Z.W., Zhang, T.L., Dunlop, M., Liu, Z.X., Torkar, K., Fazakerley, A., Lucek, E., Rème, H., Dandouras, I., et al., 2010. South-North asymmetry of field-aligned currents in the magnetotail observed by Cluster. J. Geophys. Res. Space Phys. 115 (A7).
Shinbori, A., Tsuji, Y., Kikuchi, T., Araki, T., Ikeda, A., Uozumi, T., Baishev, D., Shevtsov, B.M., Nagatsuma, T., Yumoto, K., 2012. Magnetic local time and latitude dependence of amplitude of the main impulse (MI) of geomagnetic sudden commencements and its seasonal variation. J. Geophys. Res. Space Phys. 117 , A08322. https://doi.org/10.1029/2012JA018006.
Shkarofsky, I.P., Johnston, T.W., Bachynski, M.P., 1966. The Particle Kinetics of Plasmas. Addison-Wesley, Reading, MA.
Sibeck, D.G., Greenwald, R.A., Bristow, W.A., Korotova, G.I., 1996. Concerning possible effects of ionospheric conductivity upon the occurrence patterns of impulsive events in high-latitude ground magnetograms. J. Geophys. Res. Space Phys. 101 (A6), 13407–13412.
Simon, A., 1963. Instability of a partially ionized plasma in crossed electric and magnetic fields. Phys. Fluids 6, 382–388. https://doi.org/10.1063/1.1706743.
Smart, D.F., Shea, M.A., Flückiger, E.O., 2000. Magnetospheric models and trajectory computations. Space Sci. Rev. 93, 305–333.
Solomon, S.C., 1993. Auroral electron transport using the Monte Carlo method. Geophys. Res. Lett. 20 (3), 185–188.
Solomon, S.C., 2001. Auroral particle transport using Monte Carlo and hybrid methods. J. Geophys. Res. Space Phys. 106 (A1), 107–116.
Solomon, S.C., Abreu, V.J., 1989. The 630 nm dayglow. J. Geophys. Res. Space Phys. 94 (A6), 6817–6824.
Solomon, S.C., Hays, P.B., Abreu, V.J., 1988. The auroral 6300 Å emission: observations and modeling. J. Geophys. Res. Space Phys. 93 (A9), 9867–9882.
Spanswick, E., Donovan, E., Baker, G., 2005. Pc5 modulation of high energy electron precipitation: particle interaction regions and scattering efficiency. Ann. Geophys. 23 (5), 1533–1542.
Spencer, L.V.C., 1959. Energy Dissipation by Fast Electrons. National Bureau of Standards Monograph 1, US Department of Commerce, National Bureau of Standards, Washington, D.C
St.-Maurice, J.P., Schlegel, K., Banks, P.M., 1981. Anomalous heating of the polar E region by unstable plasma waves. II—theory. J. Geophys. Res. 86, 1453–1462. https://doi.org/10.1029/JA086iA03p01453.
Stasiewicz, K., Bellan, P., Chaston, C., Kletzing, C., Lysak, R., Maggs, J., Pokhotelov, O., Seyler, C., Shukla, P., Stenflo, L., Streltsov, A., Wahlund, J.E., 2000. Small scale Alfvénic structure in the aurora. Space Sci. Rev. 92, 423–533.
Strickland, D.J., Book, D.L., Coffey, T.P., Fedder, J.A., 1976. Transport equation techniques for the deposition of auroral electrons. J. Geophys. Res. 81 (16), 2755–2764.
Strickland, D.J., Daniell Jr, R.E., Jasperse, J.R., Basu, B., 1993. Transport-theoretic model for the electron-proton-hydrogen atom aurora: 2. Model results. J. Geophys. Res. Space Phys. 98 (A12), 21533–21548.
Struchtrup, H., 1997. The BGK-model with velocity-dependent collision frequency. Contin. Mech. Thermodyn. 9 (1), 23–31. https://doi.org/10.1007/s001610050053.
Sutcliffe, P.R., Rostoker, G., 1979. Dependence of Pc5 micropulsation power on conductivity variations in the morning sector. Planet. Space Sci. 27 (5), 631–642.

Synnes, S.A., Søraas, F., Hansen, J.P., 1998. Monte-Carlo simulations of proton aurora. J. Atmos. Sol.-Terr. Phys. 60 (17), 1695–1705.

ter Haar, D., 1954, Elements of Statistical Mechanics. Rinehart and Company, New York.

Thorne, R.M., Ni, B., Tao, X., Horne, R.B., Meredith, N.P., 2010. Scattering by chorus waves as the dominant cause of diffuse auroral precipitation. Nature 467 (7318), 943–946. https://doi.org/10.1038/nature09467.

Usoskin, I.G., Kovaltsov, G.A., 2006. Cosmic ray induced ionization in the atmosphere: full modeling and practical applications. J. Geophys. Res. Atmos. 111 (D21).

Vasyliunas, V.M., 1968. A survey of low-energy electrons in the evening sector of the magnetosphere with OGO 1 and OGO 3. J. Geophys. Res. 73, 2839–2884. https://doi.org/10.1029/JA073i009p02839.

Vorob'ev, V.G., 1993. Dynamics of Hall vortices in a day-time high-latitude region. Geomagn. Aeron. 33 (5), 58–68.

Wahlund, J.E., Opgenoorth, H.J., Haggstrom, I., Winser, K.J., Jones, G.O.L., 1992. EISCAT observations of topside ionospheric ion outflows during auroral activity: revisited. J. Geophys. Res. (A3), 3019–3037. https://doi.org/10.1029/91JA02438.

Watanabe, T.H., 2014. A unified model of auroral arc growth and electron acceleration in the magnetosphere-ionosphere coupling. Geophys. Res. Lett. 41 (17), 6071–6077. https://doi.org/10.1002/2014GL061166.

Wedlund, C.S., Gronoff, G., Lilensten, J., Ménager, H., Barthélemy, M., 2011. Comprehensive calculation of the energy per ion pair or w values for five major planetary upper atmospheres. Ann. Geophys. 29 (1), 187–195.

Welling, D.T., André, M., Dandouras, I., Delcourt, D., Fazakerley, A., Fontaine, D., Foster, J., Ilie, R., Kistler, L., Lee, J.H., Liemohn, M.W., Slavin, J.A., Wang, C.-P., Wiltberger, M., Yau, A., 2015. The Earth: plasma sources, losses, and transport processes. Space Sci. Rev. 192 (1–4), 145–208. https://doi.org/10.1007/s11214-015-0187-2.

Werner, C.J., et al., 2017. MCNP User's Manual-Code Version 6.2. Los Alamos National Laboratory, LA-UR-17-29981.

Wiltberger, M., Merkin, V., Zhang, B., Toffoletto, F., Oppenheim, M., Wang, W., Lyon, J.G., Liu, J., Dimant, Y., Sitnov, M.I., Stephens, G.K., 2017. Effects of electrojet turbulence on a magnetosphere-ionosphere simulation of a geomagnetic storm. J. Geophys. Res. Space Phys. 122 (5), 5008–5027. https://doi.org/10.1002/2016JA023700.

Wissing, J.M., Kallenrode, M.B., 2009. Atmospheric ionization module Osnabrück (AIMOS): a 3-D model to determine atmospheric ionization by energetic charged particles from different populations. J. Geophys. Res. Space Phys. 114 (A6).

Wolf, R.A., 1970. Effects of ionospheric conductivity on convective flow of plasma in the magnetosphere. J. Geophys. Res. 75 (25), 4677–4698.

Wygant, J.R., Keiling, A., Cattell, C.A., Lysak, R.L., Temerin, M., Mozer, F.S., Kletzing, C.A., Scudder, J.D., Streltsov, V., Lotko, W., Russell, C.T., 2002. Evidence for kinetic Alfvén waves and parallel electron energization at 4–6 R_E altitudes in the plasma sheet boundary layer. J. Geophys. Res. Space Phys. 107 (A8), 1201. https://doi.org/10.1029/2001JA900113.

Xu, W., Marshall, R.A., Fang, X., Turunen, E., Kero, A., 2018. On the effects of bremsstrahlung radiation during energetic electron precipitation. Geophys. Res. Lett. 45 (2), 1167–1176.

Yamauchi, M., 2019. Terrestrial ion escape and relevant circulation in space. Ann. Geophys. 37 (6), 1197–1222. https://doi.org/10.5194/angeo-37-1197-2019.

Yau, A.W., André, M., 1997. Sources of ion outflow in the high latitude ionosphere. Space Sci. Rev. 80, 1–25. https://doi.org/10.1023/A:1004947203046.

Young, M.A., Oppenheim, M.M., Dimant, Y.S., 2019. Simulations of secondary Farley-Buneman instability driven by a kilometer-scale primary wave: anomalous transport and formation of flat-topped electric fields. J. Geophys. Res. Space Phys. 124, 734–748. https://doi.org/10.1029/2018JA026072.

Yumoto, K., Pilipenko, V., Fedorov, E., Kurneva, N., Shiokawa, K., 1995. The mechanisms of damping of geomagnetic pulsations. J. Geomagn. Geoelectr. 47 (2), 163–176.

Yumoto, K., Matsuoka, H., Osaki, H., Shiokawa, K., Tanaka, Y., Kitamura, T.I., Tachihara, H., Shinohara, M., Solovyev, S.I., Makarov, G.A., et al., 1996. North/south asymmetry of SC/SI magnetic variations observed along the 210° magnetic meridian. J. Geomagn. Geoelectr. 48 (11), 1333–1340.

Zakharenkova, I., Cherniak, I., 2020. When plasma streams tie up equatorial plasma irregularities with auroral ones. Space Weather 18 (2), e02375. https://doi.org/10.1029/2019SW002375.

Zesta, E., Boudouridis, A., Weygand, J.M., Yizengaw, E., Moldwin, M.B., Chi, P., 2016. Interhemispheric asymmetries in magnetospheric energy input. In: Ionospheric Space Weather: Longitude Dependence and Lower Atmosphere Forcing, Geophysical Monograph Series, 220, pp. 1–20.

CHAPTER 7

Ionosphere-thermosphere interaction

Chapter Outline

Cross-Scale Coupling and Energy Transfer in the Magnetosphere-Ionosphere-Thermosphere System
https://doi.org/10.1016/B978-0-12-821366-7.00003-2

7.1 Ionosphere-thermosphere interaction: Theoretical aspects

Jiuhou Lei[a, b, c] and Tong Dang[a, b, c]

[a]CAS Key Laboratory of Geospace Environment, School of Earth and Space Sciences, University of Science and Technology of China, Hefei, China

[b]Mengcheng National Geophysical Observatory, University of Science and Technology of China, Hefei, China

[c]CAS Center for Excellence in Comparative Planetology, Hefei, China

7.1.1 Introduction

The thermosphere is a layer of the Earth's atmosphere from 100 km to around 1000 km, with high atmospheric temperature and large variability in response to variations in solar extreme ultraviolet (EUV) radiation and geomagnetic activity. Molecular diffusion dominates the thermosphere and thus species separate out with altitude according to their atomic or molecular weights. In the lower thermosphere, the major components are N_2 and O_2, but above about 200 km the major component is O (Fig. 7.1A). Mostly

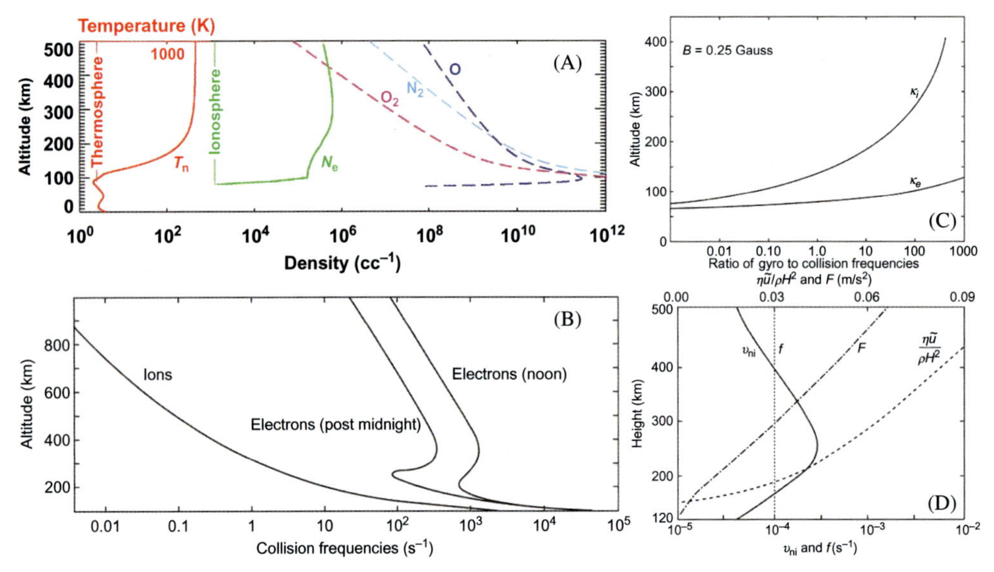

Fig. 7.1 Altitude profiles of thermospheric and ionospheric characteristics. (A) Temperature, thermospheric species, and ionospheric electron density. (B) Typical electron neutral plus electron ion collision frequency along with the ion-neutral collision frequency at a high sunspot number (Kelley, 2009). (C) Typical values for κ_e and κ_i in the equatorial ionosphere for a magnetic field of $0.25G = 2.5 \times 10^{-5}$ T (Kelley, 2009). (D) Neutral acceleration time constants as well as viscous and pressure accelerations for mid-day at sunspot minimum. The upper scale applies to the acceleration due to the pressure gradient, *F* (*dash and dot*), and the normalized kinematic viscosity parameter, $\eta\tilde{U}/(\rho H^2)$ (*long dashed*). The lower scale refers to the ion drag parameter, ν_{ni} (*solid*), and the Coriolis parameter for a latitude of 45 degrees, *f* (*short dashed*) (Rishbeth, 1972; Kelley, 2009). *((A) Courtesy of Alan Burns.)*

embedded in the thermosphere, the ionosphere is formed by photoionization by solar X-rays and EUV radiations and further divided as D, E, and F layers. The ionosphere can reflect waves in different bands and is important for communications and navigations.

Due to ion-neutral collisions, the ionosphere and thermosphere are closely coupled. The neutral atmosphere dominates the ionospheric changes and the charged particles also exert significant feedback on neutrals. These ion-neutral interactions occur all the times (during both quiet and storm time), at all regions (at polar and mid-low latitudes), and also global and local scales. This makes the ionosphere and thermosphere a complex coupled system, which is of critical importance to both space weather applications such as navigation and satellite operation and also scientific understanding.

Great efforts were made to explore the ionosphere-thermosphere (IT) interaction processes from both observations and modeling. In this section, we briefly describe the physics and studies of IT coupling processes from the modeling perspective. We present how the thermosphere drives the ionosphere and the ionospheric feedback, during both quiet and disturbed time. The IT coupling in solar eclipses and solar flares as well as in ionospheric irregularities are also discussed. Note that here only some of relevant IT coupling studies of modeling are listed, and no attempt is made to provide a complete compilation of references. Nevertheless, in this section, we discuss several fundamental IT coupling concepts that may be applied to understanding multiscale upper atmospheric processes and phenomena, some of which are dealt with in this section and additionally in Section 7.2 for large-scale IT structures, Section 7.3 for large-scale traveling ionospheric disturbances, and Section 7.4 for medium-scale traveling ionospheric disturbances. These multiscale processes develop independently but sometimes interact with each other, leading to a complicated dynamical IT system.

7.1.2 Thermospheric effects on the ionosphere

7.1.2.1 Photochemical influences

As the background of the ionosphere (Fig. 7.1A), the thermosphere has significant impact on the behavior of the ion and electron distribution in the ionosphere.

The ion continuity equation can be expressed as (most of the equations in this section can be found in Kelley, 2009; Rishbeth, 1972; Kosch et al., 2001, etc.)

$$\partial N_i/\partial t = P_i - L_i - \nabla \cdot (N_i \boldsymbol{V}_i) \tag{7.1}$$

where i is the ion species, N_i is the ion number density, and $\boldsymbol{V}_i$ is the ion velocity. The right-hand side of Eq. (7.1) represents the advection term, ion production (P_i) and loss (L_i), respectively. The O^+, which is the major ion in the F-region ionosphere, is mainly produced through photoionization of O.

$$O + h\nu \rightarrow O^+ + e \tag{7.2}$$

The major loss process of O^+ is mainly through the dissociative recombination with N_2 and O_2, predominantly with N_2,

$$O^+ + N_2 \rightarrow NO^+ + N \tag{7.3}$$

Therefore, the ionospheric density in the F2 layer strongly depends on the O/N_2 ratios,

$$N_{O^+} \propto [O]/[N_2] \tag{7.4}$$

Fig. 7.2 is the storm-time O/N_2 global distributions. The top and bottom panels are the Global Ultraviolet Imager (GUVI)-measured and Coupled Magnetosphere-Ionosphere-Thermosphere model (CMIT)-simulated O/N_2 during the April 2004 magnetic storm at a fixed local time of 9.3 h. Universal time goes from right to left, and UT day is shown at the bottom of the figure. Both observations and simulations show that high O/N_2 between about 0800 and 1800 UT in a latitudinal band

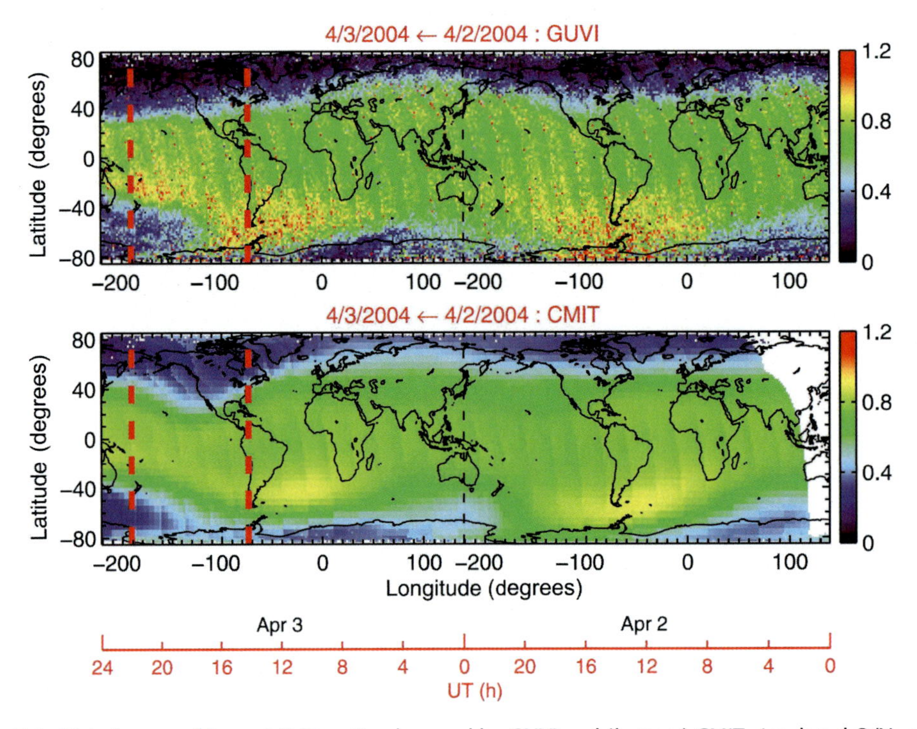

Fig. 7.2 Global maps of (*upper*) O/N_2 ratio observed by GUVI and (*bottom*) CMIT-simulated O/N_2 ratio on April 2 and 3, 2004 in UT day (time from right to left) for a nearly constant local time of 9.3 h. The two *vertical red lines* indicate the time interval of the initial phase. *(From Wang, W., Lei, J., Burns, A.G., Solomon, S.C., Wiltberger, M., Xu, J., Zhang, Y., Paxton, L., Coster, A., 2010. Ionospheric response to the initial phase of geomagnetic storms: common features. J. Geophys. Res. Space Phys.115 (A7).)*

around 40 degrees in the Southern hemisphere. The model also reproduced the observed O/N_2 depletion at later UTs in the Southern hemisphere. The O/N_2 changes during the initial phase of the storm will also affect ionospheric electron densities and TEC (see Wang et al., 2010). Section 7.2.6 provides further discussion on O/N_2 climatological variations and effects.

7.1.2.2 Neutral wind effect

The thermospheric wind generates from the inhomogeneity of the thermosphere. The simulated distribution of thermospheric temperature and wind in Fig. 7.3 generally corresponds to the solar-induced day-night variations. At high latitudes, the winds are greatly influenced by the momentum transfer from the convecting ions to the neutrals.

The neutral wind can significantly affect the behavior of ionospheric ions. This is expected on the basis of the momentum equation of ionospheric ion

$$N_i m_i d\mathbf{V_i}/dt = N_i m_i \mathbf{g} - \nabla P_i + N_i e(\mathbf{E} + \mathbf{V_i} \times \mathbf{B}) - N_i m_i \nu_{in}(\mathbf{V_i} - \mathbf{U}) \tag{7.5}$$

where m_i is the ion mass, $\mathbf{g}$ is gravity, P_i is the ion pressure, e is the electric charge, $\mathbf{V}_i$ is the ion velocity, $\mathbf{U}$ is the neutral velocity, $\mathbf{E}$ is the electric field, $\mathbf{B}$ is the magnetic field, and ν_{in} is the ion-neutral collision frequency. Fig. 7.1B shows the typical ion-neutral collision frequency ν_{in} and electron-ion/neutral collision frequency ν_e at solar maximum. As shown in this figure, the ion-electron collisions are important in the upper ionosphere where ν_e is greater than ν_{in} (say above 200 km). However, the ions and electrons move with the same velocity in the direction perpendicular to B in this region. Thus, we neglect the ion-electron collision term in Eq. (7.5).

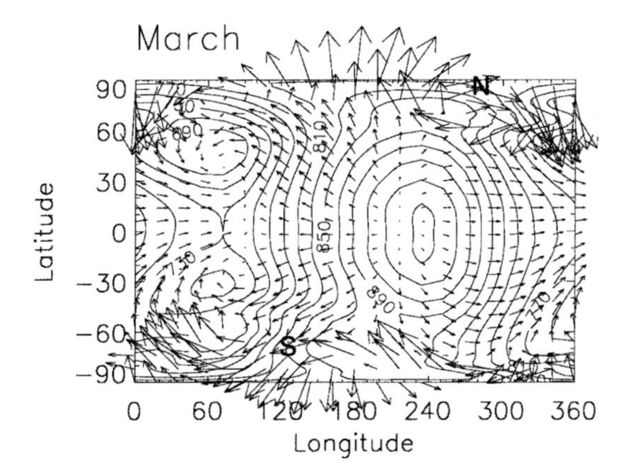

Fig. 7.3 Temperature contours and wind vector as a function of latitude and longitude at 0000 UT near 300 km, corresponding to average solar cycle and quiet geomagnetic conditions (Rishbeth et al., 2000a).

If we set $d\mathbf{V}_i/dt = 0$ in Eq. (7.5) for the steady state and neglect the gravity, pressure-gradient, and electron-ion collisions, the Eq. (7.5) can be simplified as

$$0 = N_i e(\mathbf{E} + \mathbf{V_i} \times \mathbf{B}) - N_i m_i \nu_{in}(\mathbf{V_i} - \mathbf{U}) \tag{7.6}$$

In a geomagnetic east, north, and vertical Cartesian coordinate system $\mathbf{U}$ (U_E, U_S, 0) and $\mathbf{B}$ (0, $B\cos I$, $-B\sin I$) (I is the magnetic dip angle) for the Northern hemisphere, the effective vertical ion drift V_z (positive upwards) can be solved from Eq. (7.6):

$$V_z = \frac{\kappa}{(1+\kappa^2)} U_E \cos I + \frac{\kappa}{(1+\kappa^2)} U_S \cos I \sin I \tag{7.7}$$

where

$$\kappa = \nu_{\text{in}}/\Omega \tag{7.8}$$

where Ω is the ion gyrofrequency and ν_{in} is the ion-neutral collision frequency.

The absolute values of κ_e and κ_i are plotted in Fig. 7.1C for an equatorial ionosphere with $B = 2.5 \times 10^{-5}$ T (0.25 G). The transition from a molecular ion plasma (NO^+ and $O_2{}^+$) to an atomic ion plasma (O^+) has been included in the calculation of κ. The absolute value of κ_e passes through unity near 75 km, while κ_i does so at 130 km.

An example of the neutral wind effect on the ions is the formation of the mid-latitude sporadic E (Es). The formation mechanism is associated with the wind shear, which was first proposed by Whitehead (1961) and Axford (1963). The first and second terms in the right-hand side of Eq. (7.7) define two processes of vertical ion convergence, which is associated with the vertical shears in zonal (U_E) and meridional (U_S) winds, respectively, as illustrated by Fig. 7.4 (Haldoupis, 2011). For instance, the zonal wind shear mechanism shown in the top panel involves the effect of the magnetic field horizontal component $B_H = B\cos I$ on the E-region zonal wind. The zonal wind has a vertical wind shear with a westward wind above and an eastward wind below the Es-layer to be formed. As a result, the ions converge, accumulate, and form the Es layer. Both zonal and meridional neutral wind shears can lead to the convergence of ion drift and the accumulation of plasma density in the metal layer.

In the ionospheric F-region, the thermospheric winds can directly blow the F2 region plasma vertically and horizontally along the geomagnetic lines by neutral-ion collisions. Fig. 7.5 displays the time evolution of the ionospheric electron density profile at a fixed longitude sector during a geomagnetic storm, and the meridional wind at that longitude (red lines in Fig. 7.5). It is seen that the enhancement of meridional wind owing to the energy injected from the magnetosphere into the high-latitude upper atmosphere can effectively push the electrons to move equatorward and can even accumulate the F2 region plasma at the equator. As a result, the F-region ionospheric peak density and its height at low-to-middle latitudes would also change significantly because of the neutral-ion collisions in the F2 region.

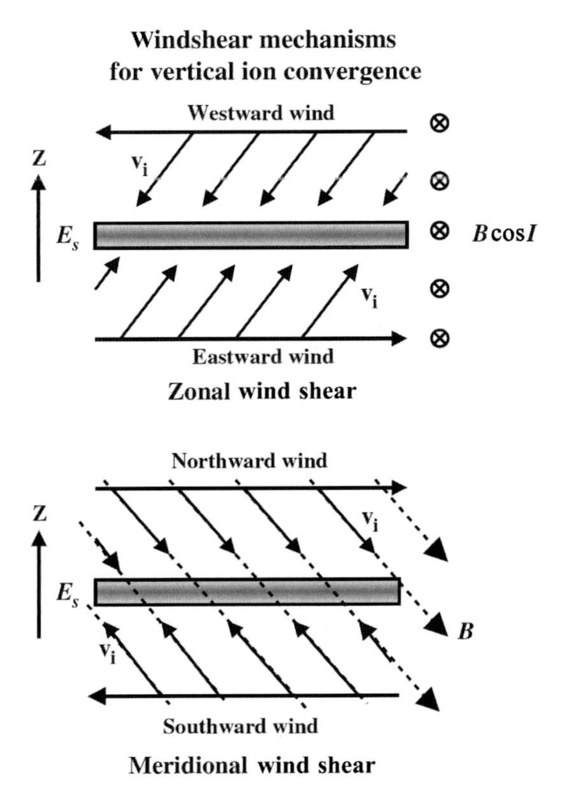

Fig. 7.4 Exemplifying sketches of the zonal (*top*) and meridional (*bottom*) wind shear mechanisms for vertical ion convergence into a thin ionization layer forming at the wind shear velocity null. More details on the two mechanisms are given in the text (Haldoupis, 2011).

Another example of the neutral wind effect on the equatorial and low-latitude ionosphere, as shown in Fig. 7.6, gives a series of Sheffield University Plasmasphere Ionosphere Model (SUPIM) simulations of the TEC during a geomagnetic storm (Lin et al., 2005a). Three case-controlled runs were carried out with the storm-generated equatorward winds only (bold gray line), with the storm-generated $\mathbf{E} \times \mathbf{B}$ drifts only (thin black line), and with both the storm-generated $\mathbf{E} \times \mathbf{B}$ drifts and the equatorward winds (bold black). It can be clearly seen in Fig. 7.6 that the TEC enhancement in the northern EIA peak due to the storm-generated neutral wind (bold gray line) is larger than that due to the storm-generated $\mathbf{E} \times \mathbf{B}$ drifts (thin black line). This indicates a significant contribution from neutral winds to produce a TEC enhancement during a storm.

Meanwhile, the motion of neutral gas can drive the ionospheric plasma across the geomagnetic field in the upper atmosphere, leading to ionospheric currents and electric fields. This is named as the neutral wind dynamo process (e.g., Richmond, 1979). The control equation relating the electric fields, currents, and neutral wind is Ohm's law.

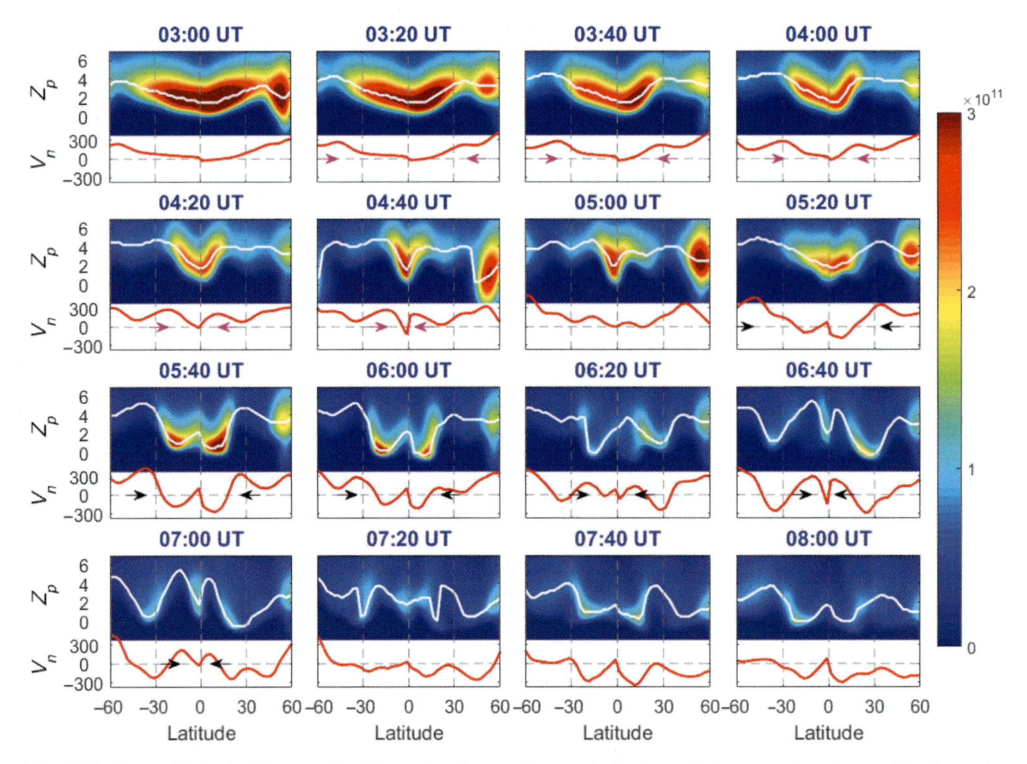

Fig. 7.5 Neutral wind effect on the F2 region ionosphere. Variation of electron density profile (in units of electrons m^{-3}) as a function of pressure level (Zp) and geographic latitude at the Sao Luis longitude sector during 03:00–08:00 UT on October 2; the hmF2 is shown by the *white line*. The meridional winds V_n (positive is equatorward, in units of $m\ s^{-1}$) at the altitude of 250 km is shown in the *bottom panels*, and the equatorward surges of winds associated with the TADs are given by *magenta* and *black arrows*, respectively (Ren and Lei, 2017).

$$\mathbf{J} = \sigma_{\|}\mathbf{E}_{\|} + \sigma_P(\mathbf{E}_{\perp} + \mathbf{U} \times \mathbf{B}) + \sigma_H \mathbf{b} \times (\mathbf{E}_{\perp} + \mathbf{U} \times \mathbf{B}) \tag{7.9}$$

where $\mathbf{J}$ is the current density; $\mathbf{E}_{\|}$ and $\mathbf{E}_{\perp}$ are the components of the electric field parallel and perpendicular to the geomagnetic field $\mathbf{B}$; $\mathbf{U}$ is the neutral wind velocity; $\mathbf{b}$ is a unit vector in the direction of $\mathbf{B}$; $\sigma_{\|}$ is the conductivity parallel to $\mathbf{B}$; σ_P is the Pedersen conductivity; and σ_H is the Hall conductivity. In addition, a closure or divergenceless of the ionospheric current is required:

$$\nabla \cdot \mathbf{J} = 0 \tag{7.10}$$

From Eqs. (7.9), (7.10), the ionospheric electric fields and currents are dominated by the neutral wind and ionospheric conductance variations.

The neutral winds cause a relative motion between ions and electrons, which causes charge separation and the current divergence and generates polarization electric fields. One can note a major contribution from neutral winds in Fig. 7.7 to the quiet-time zonal

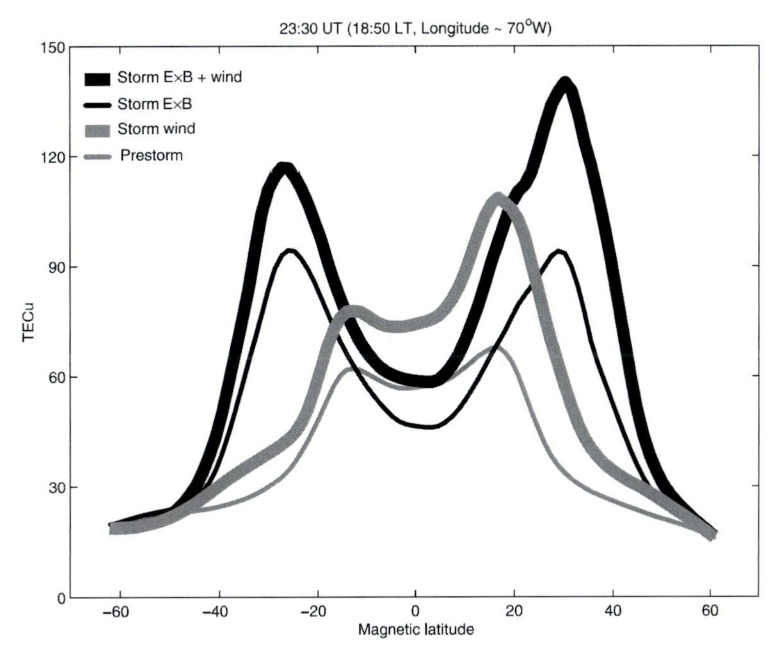

Fig. 7.6 The total electron content (TEC) between altitudes 100 and 2000 km from the SUPIM results at 2330 UT (1850 LT) at −70 degrees geographic longitude on the prestorm day (*thin gray line*), and the storm day with the storm-generated equatorward winds only (*bold gray line*), with the storm-generated $E \times B$ drifts only (*thin black line*), and with both the storm-generated $\mathbf{E} \times \mathbf{B}$ drifts and the equatorward winds (*bold black line*) (Lin et al., 2005a).

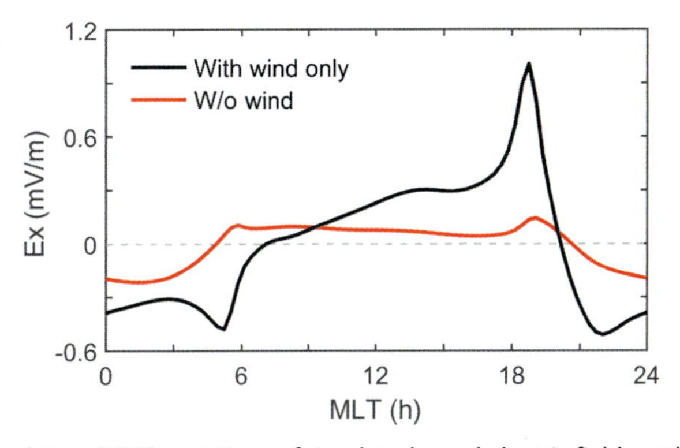

Fig. 7.7 Magnetic local time (MLT) variations of simulated zonal electric fields at the magnetic equator at 12 UT at pressure level 2 on March equinox during the geomagnetic quiet ($Kp = 2$) periods with moderate solar activity ($F10.7 = 130$). The *black line* shows the Ex generated by neutral wind dynamo. The *red line* indicates the Ex in the simulation without neutral winds, which is the PPEF from the high-latitude convection pattern.

electric field E_x at all local times. The equatorial E_x generated by the wind dynamo usually is eastward during the daytime and turns to the west at night. Specifically, there is often an eastward enhancement near and/or after sunset, which is called prereversal enhancement (Eccles, 1998; Fejer et al., 1991; Heelis et al., 1974; Richmond, 1979). Here the residual equatorial E_x without neutral winds (red lines) is from prompt penetration electric fields. In addition, the electric fields determine the $\mathbf{E} \times \mathbf{B}$ drifts, which will further modulate the ionospheric plasma distributions. Section 7.2.5 provides some additional discussions ultimately on the neutral wind dynamo effects.

The neutral wind also significantly contributes to electric field disturbances during storm times. During geomagnetic disturbances, the thermosphere is greatly disturbed. Both polar thermospheric temperature and neutral winds are greatly enhanced, due to the huge injection of solar wind energy (e.g., Fuller-Rowell et al., 1994). The disturbed winds further induce changes of the dynamo process including the global ionospheric electric field and currents, which refers to disturbance wind dynamo. The simulations by Richmond et al. (2003) (not shown here) clearly demonstrate that disturbance wind effects have a tendency to offset the effect of electric field disturbance associated with steady-state direct penetration.

7.1.3 Ionospheric imprints on the thermosphere

At the same time, although the ionospheric plasma density is a few orders less than the thermospheric species, the ionosphere can exert significant feedback on the variations of the thermosphere via the accumulative effects over time.

The horizontal momentum equation for neutral gas is

$$\frac{d\mathbf{U}}{dt} = -\frac{1}{\rho}\nabla P - 2\boldsymbol{\Omega} \times \mathbf{U} + \frac{1}{\rho}\nabla \cdot (\mu \nabla \mathbf{U}) - \nu_{ni}(\mathbf{U} - \mathbf{V}_i) \tag{7.11}$$

where $\mathbf{U}$ is the horizontal neutral wind, P is the pressure, and $\boldsymbol{\Omega}$ is the Earth's angular velocity with the vector direction along the Earth's axis of rotation toward the geographic north pole. If we neglect the force of pressure gradient, Coriolis force, and viscosity, the equation with only the ion drag effect becomes

$$\partial U / \partial t = \nu_{ni}(V_i - U) \tag{7.12}$$

Then solve the first-order, partial differential equation assuming constant coefficients

$$U(t) = U_0 e^{-\nu_{ni} t} + V_i(1 - e^{-\nu_{ni} t}) \tag{7.13}$$

Hence, we have the time constant of ion drag

$$\tau \approx 1/\nu_{ni} \tag{7.14}$$

Here it should be pointed out that the neutral-ion collision frequency ν_{ni} is different from the ion-neutral collision frequency ν_{in} by

$$\nu_{ni} = \frac{n_i m_n}{n_n m_n} \nu_{in} \tag{7.15}$$

In the ionospheric F-region, O^+ and O are the dominant ion and neutral species, respectively. The ion-neutral collision frequency ν_{in} can be obtained from an empirical formula (Schunk and Nagy, 2009)

$$\nu_{O^+-O} = 5.9 \times 10^{-17} T^{0.5} (1 - 0.096 \log T)^2 [O] \tag{7.16}$$

Therefore, the higher the electron density, the faster the neutral winds will respond to the ion drag force. For the sunlit ionospheric F-region peak, the time constant is usually 1 h, while the time constant can be 3–6 h for the sunlit E-region peak. For solar maximum, the typical time constant determined by Ponthieu et al. (1988) was about 1 h on the dayside and approximately 3 h on the nightside. Fig. 7.1D gives a quantitative comparison between the various terms in the neutral momentum equation for 45 degrees latitude. The ionosphere plays an important role in altering neutral wind variations (see Rishbeth, 1971). Fig. 7.8 displays a comparison of different terms of neutral wind momentum equation (Lei et al., 2007). The results indicate that the magnitude of viscosity force, pressure gradient, Coriolis force, and ion drag during equinox are comparable at

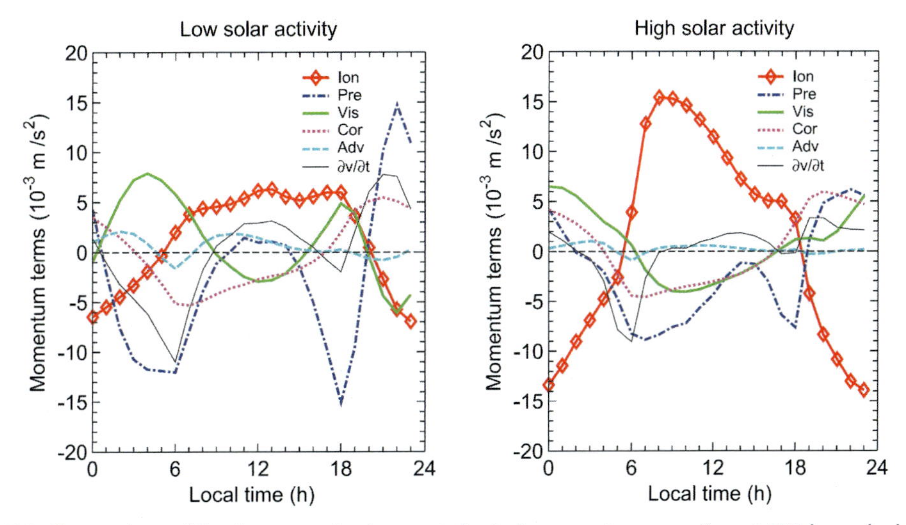

Fig. 7.8 Comparison of forcing terms in the neutral wind momentum equation at 300 km calculated from the TIE-GCM during equinox under (*left*) low and (*right*) high solar activity. The plotted terms are ion drag (Ion, *line with diamonds*), pressure gradient (Pre, *dash-dotted line*), viscosity force (Vis, *solid line*), Coriolis force (Cor, *dotted line*), and nonlinear momentum advection (Adv, *dashed line*), respectively. The net wind acceleration term (*∂v/∂t, thin solid line*) is also plotted (Lei et al., 2007).

low solar activity, except for much larger pressure gradients during sunrise and sunset period, whereas ion drag is much larger than the other forcing terms at high solar activity. As a result, the faster increase of ion drag than other momentum forces retards to neutral gas motion, leading to smaller neutral winds in low and middle latitudes at high solar activity.

7.1.3.1 Ionospheric impact on high-latitude winds and composition

Plasma convection is a common feature in the high-latitude ionosphere associated with the solar wind-magnetospheric-ionosphere interaction. Usually, the ionospheric convection conforms to a well-known two-cell pattern during the southward IMF, characterized by antisunward convection over the highest latitudes of the polar cap and sunward return flow along the dawn and dusk flanks at auroral latitudes (e.g., Heppner and Maynard, 1987; Ruohoniemi and Greenwald, 1996), leading to complex polar ionospheric structures such as the tongue of ionization (e.g., Foster et al., 2005; Dang et al., 2019).

At high latitudes, the pressure gradient of the neutral gas produced by solar EUV heating is generally balanced by the ion-neutral collision forcing in addition to Coriolis and centrifugal forcings. Thermospheric winds at high latitudes are strongly influenced by collisions with ions that are rapidly drifting in the electric and magnetic fields (Axford and Hines, 1961; Fuller-Rowell and Rees, 1980; Rees and Fuller-Rowell, 1989). The effect can be observed in neutral composition and temperature as well (e.g., Thayer et al., 1987). In addition, especially during storm times, the thermosphere is greatly disturbed by the changes of ionospheric convection and the associated conductivities, through the effects of Joule or frictional heating processes. The polar thermospheric neutral wind measured by DE-2 satellite (Fig. 7.9) illustrate that ion-drag and Coriolis forces play a significant role in the neutral wind in the polar region. Vortex formation of the nondivergent wind component is consistently more dominant in the dusk sector than in the dawn sector. The imprint of the ionospheric convection can be seen in the simulated winds at high latitudes, even during quiet time (see Fig. 7.3).

Due to the ion-drag effect, neutral winds also exhibit distinct a vortex structure in the lower thermosphere, depending on IMF orientations (Richmond et al., 2003; Deng and Ridley, 2006; Kwak and Richmond, 2007). The Thermosphere-Ionosphere-Electrodynamics General Circulation Model (TIE-GCM) simulated wind patterns in Fig. 7.10 show good agreement with the WINDII observations (Kwak et al., 2007). At around the 140 km altitude, there is a clear anticyclonic (clockwise) vortex on the dusk side, and an equatorward flow in the early morning hours, which are most pronounced when the IMF equation B_z is negative. The winds for positive B_y around 140 km have a two-cell pattern. At 111 km, the effects associated with ion drag appear to be weak, relative to the global-scale background winds associated with atmospheric

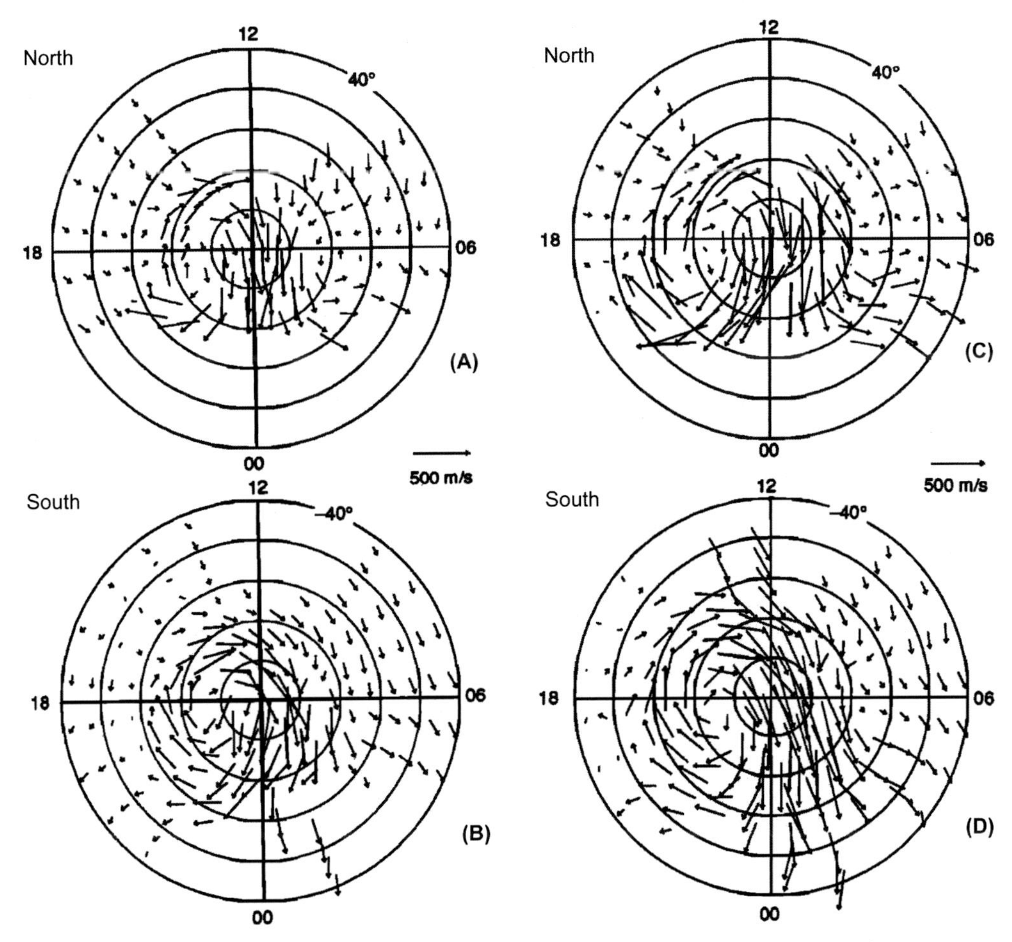

Fig. 7.9 Averaged DE 2 thermospheric neutral wind maps for December solstices, solar maximum conditions in geomagnetic coordinates during periods of Kp $<$ 3 for (A) northern and (B) southern high-latitude regions and periods of 3+ $<$ Kp $<$ 6 for (C) northern and (D) southern high-latitude regions (Thayer and Killeen, 1993).

tides present at all latitudes. The results indicate that the wind patterns show considerable similarity with ionospheric convection patterns, depending on the IMF conditions.

Zhang et al. (2015) reported a significant poleward surge in thermospheric winds at subauroral and mid-latitudes following the March 17–18, 2015 great geomagnetic storm. Further model calculation indicated that the observed neutral wind variations as driven by SubAuroral Polarization Stream (SAPS; Foster and Burke, 2002), through a scenario where strong ion flows cause a westward neutral wind, subsequently establishing a poleward wind surge due to the poleward Coriolis force on that westward wind.

The thermospheric species could be also redistributed by the plasma convection. As shown in Fig. 7.11, a "tongue" of neutral composition at middle latitudes in the daytime

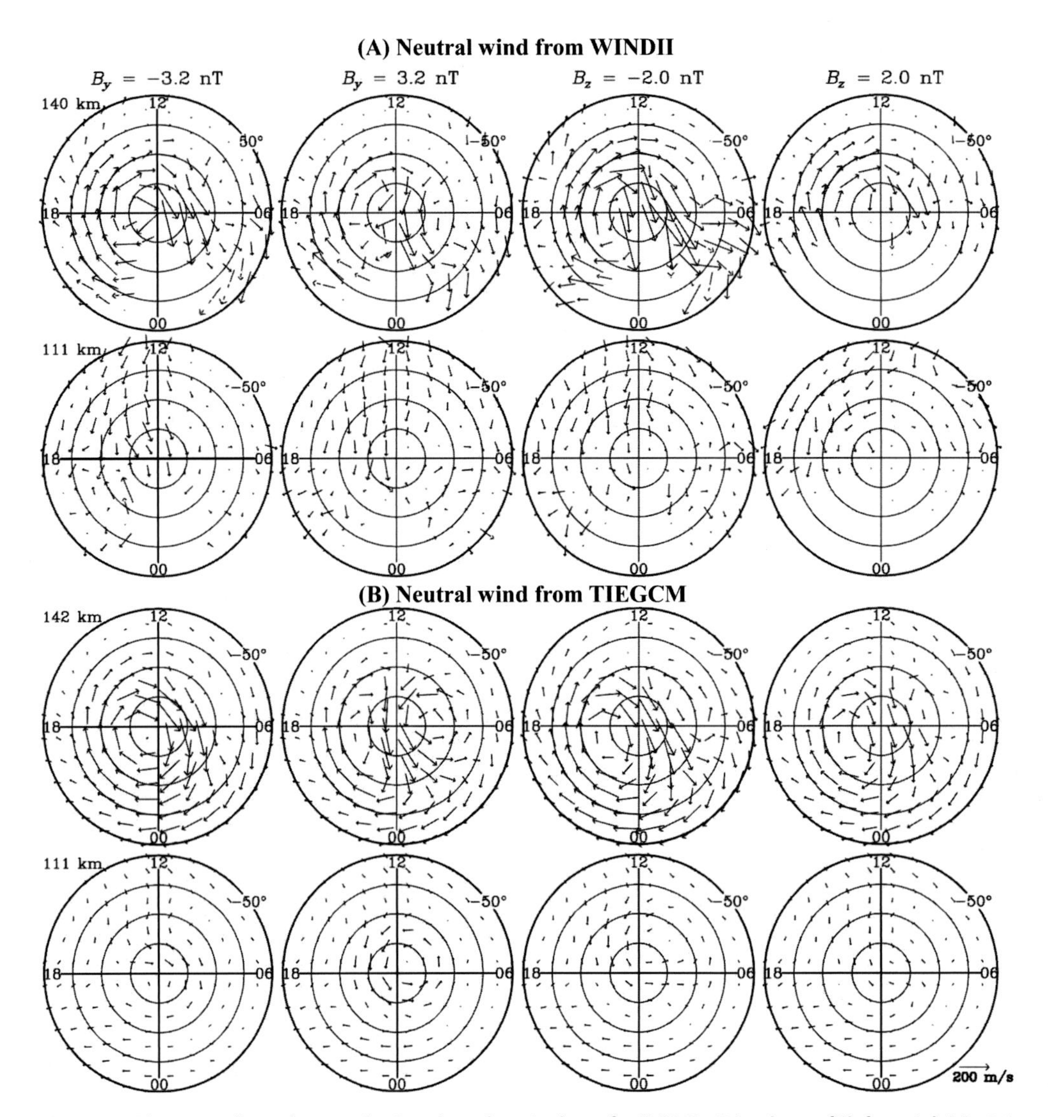

Fig. 7.10 The neutral winds over the Southern hemisphere for IMF (B_y, B_z) values of (*left to right*) (−3.2, 0.0), (+3.2, 0.0), (0.0, −2.0), and (0.0, +2.0) nT from (A) WINDII data and (B) the TIE-GCM run, respectively. These vectors are plotted against a fixed length of *arrow* of 200 ms^{-1} (Kwak et al., 2007).

is formed in much the same way that the tongue of ionization is (Burns et al., 2004). During B_z southward conditions parcels of air, which are rich in O/N_2, are drawn from the dayside by the antisunward winds associated with the neutral convection pattern and transported across the polar cap toward the night side auroral oval. This O/N_2-rich air can only be transported from a small region on the dayside due to the geometry of the transport, so the neutral "tongue" is narrow.

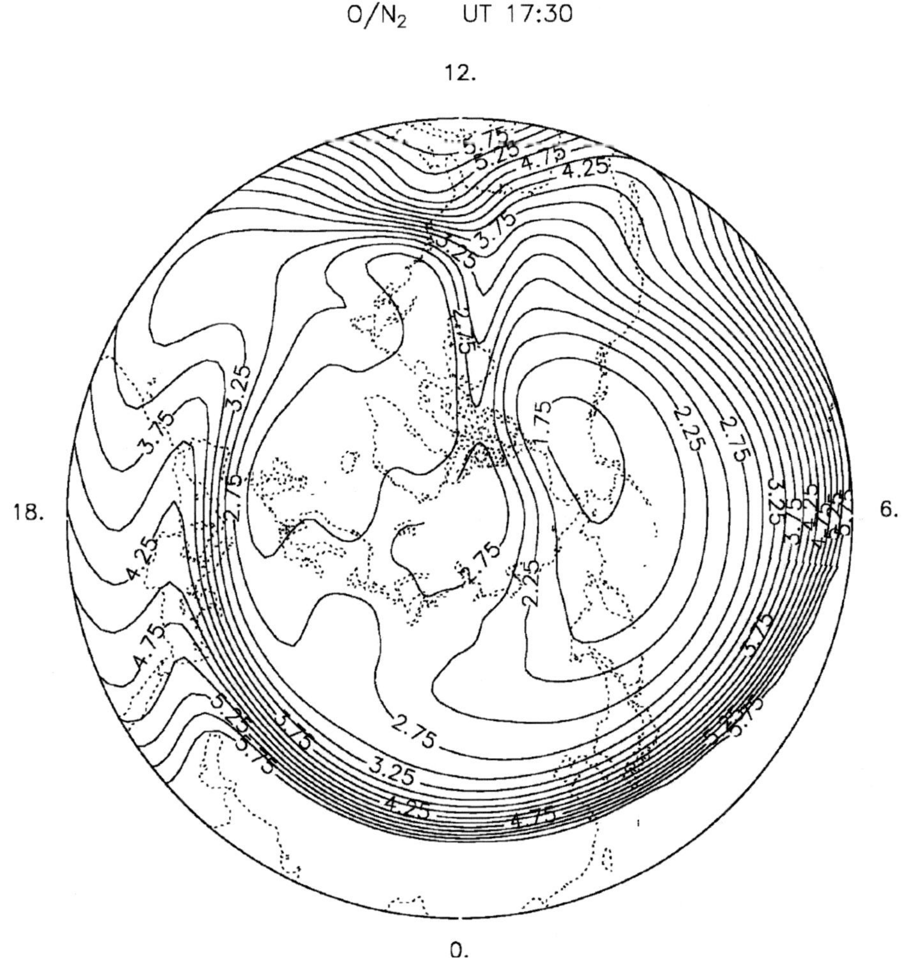

Fig. 7.11 A plot of the Thermosphere Ionosphere Nested Grid (TING) model calculation of O/N_2 on the $z = 2$ pressure surface (at an altitude of about 400 km in this case) for 1730 UT of the July 15, 2000. The plot is in geographic coordinates of local time and latitude. The lowest latitude plotted is 20°N (Burns et al., 2004).

7.1.3.2 EIA effects on the low-latitude thermosphere

Besides the polar region, the low-latitude thermosphere is also greatly impacted by the ionospheric behaviors through ion drag effect. The ion drag force per unit mass, F_{ni}, in the momentum equations of the neutral gas can be described in directions perpendicular and parallel to the magnetic field as follows:

$$F_{ni} = \frac{\mathbf{J} \times \mathbf{B}}{\rho} - \nu_{ni}(\mathbf{U}_{\parallel} - \mathbf{V}_{\parallel}) = \frac{\mathbf{J} \times \mathbf{B}}{\rho} + \nu_{ni}\mathbf{W}_{d\parallel} \tag{7.17}$$

where **J** is the electric current density; **B** is the magnetic field; ρ is the neutral mass density; ν_{ni} is the collision frequency of momentum transfer for the neutral gas with ions; $\mathbf{U}_{||}$, $\mathbf{V}_{||}$, and $\mathbf{W}_{d||}$ are the field-aligned components of neutral wind, ion velocity, and ion diffusion velocity, respectively. In each form of the right-hand side of Eq. (7.17), the first and second terms represent the perpendicular and field-aligned components of ion drag, respectively.

The equatorial ionization anomaly (EIA) is a well-known phenomenon of the ionosphere, which is characterized by a trough at the dip equator and two crests at about ±15 degrees geomagnetic latitudes (e.g., Namba and Maeda, 1939; Appleton, 1946; Lin et al., 2005b; Dang et al., 2016). The EIA is mainly caused by the so-called "equatorial fountain effect" (Duncan, 1960; Moffett and Hanson, 1965). The plasma over the equatorial region moves upward due to $\mathbf{E} \times \mathbf{B}$ drift then diffuses downward along the magnetic field lines into both hemispheres under the force of gravity and pressure gradient and results in the EIA features. Due to the ion drag effect, the thermosphere also exhibits a similar structure, named as equatorial thermospheric anomaly (ETA) (Philbrick and McIsaac, 1972; Lei et al., 2012b). During severe geomagnetic storms, this anomaly can be significantly enhanced by the so-called "super fountain." The anomalous thermospheric variation has been identified in the mass density (equatorial mass density anomaly [EMA]) and neutral temperature. Moreover, observations have indicated that eastward zonal winds prevail at low to middle latitudes (Liu et al., 2009a; Kondo et al., 2011). These strong winds, or equatorial wind jet (EWJ), maximize actually at the Earth's dip equator, instead of the geographic equator, as another evidence of EIA-related ion drag effect. Some of these ionospheric effects on the formation of ETA, especially EMA and EWJ structures, will be discussed in detail in Sections 7.2.2 and 7.2.3.

In addition to the EIA ion drag dynamic effect on ETA, Lei et al. (2012a) simulated the ion-neutral collision heating effect. This heating process transfers the ionospheric energy from thermal electrons and ions to the ambient neutral gas through collisions due to their temperature differences, and makes a major contribution to the ETA two neutral temperature peaks in the topside ionosphere aside the magnetic equator.

The EIA structure also influences the global thermospheric circulation via ion-neutral collisions. The summer-to-winter wind is significantly impacted by EIA through plasma-neutral collisional heating (see Fig. 7.12), which changes the summer-to-winter pressure gradient, and ion drag. Consequently, the wind is suppressed in the summer hemisphere as it encounters the EIA but accelerates after it passes the EIA in the winter hemisphere (Qian et al., 2016).

7.1.3.3 Ionospheric impacts on heating and TAD

The IT interactions take place through not only momentum transfer but also energy coupling between neutrals and ions. For the perspective of thermospheric energy, the governing temperature equation is

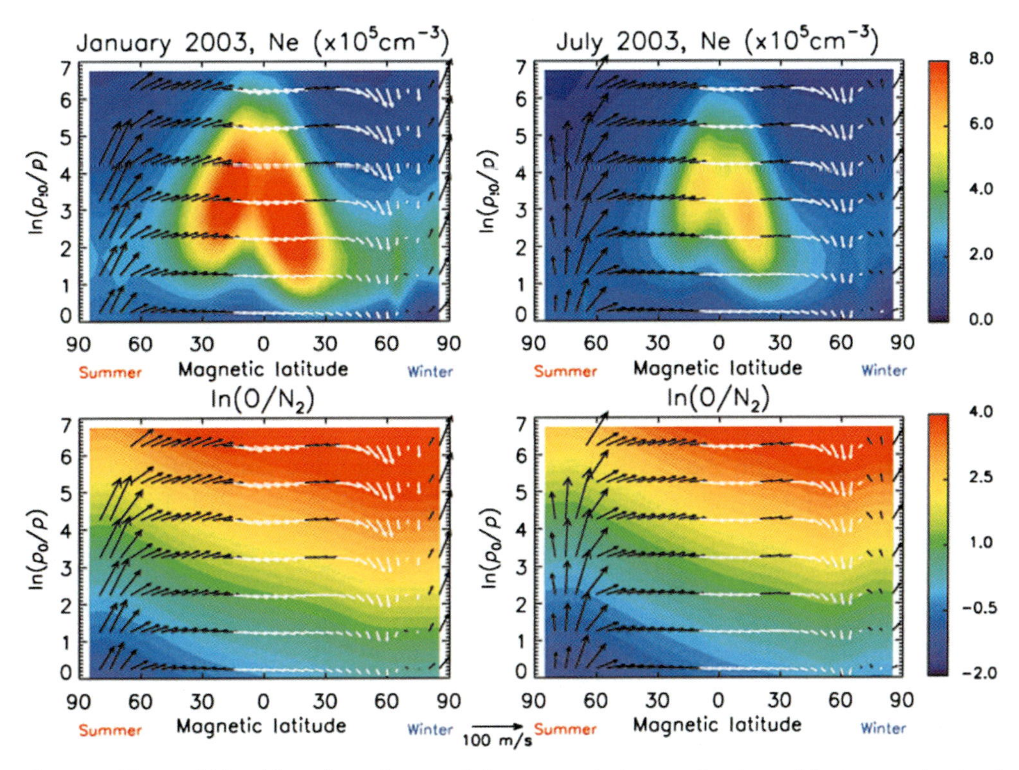

Fig. 7.12 (*Top row*) Monthly and zonal mean, daily averaged, electron density and (*bottom row*) natural logarithm of (O/N_2). Overplotted are vectors of the meridional wind and vertical wind, with the vertical wind multiplied by 100 to be compatible with the meridional wind in magnitude. *Black wind vectors* indicate upward vertical wind, whereas *white wind vectors* indicate downward vertical wind. (*Left column*) January 2003 and (*right column*) July 2003. The *y* axes are model pressure levels, the *x* axes are equivalent to geomagnetic latitudes, but with the summer hemisphere on the left-hand side and the winter hemisphere on the right-hand side (Qian et al., 2016).

$$\frac{\partial T}{\partial t} = -\mathbf{U}\cdot\nabla T - \frac{RT}{C_V\overline{m}}(\nabla\cdot\mathbf{U}) + \frac{1}{\rho C_V}\nabla\cdot[(\lambda+\rho C_P K_E)\nabla T + \rho\mathbf{g}K_E] + \frac{(Q-L)}{\rho C_V} \tag{7.18}$$

with neutral temperature T. $\mathbf{U}$ is the horizontal neutral velocity, R is the universal gas constant, $\overline{m}$ is mean atmospheric molar mass, $\mathbf{g}$ is gravity, K_E is the molecular thermal conductivity, λ is molecular heat conductivity, C_V is the specific heat at constant volume per unit mass, C_P is the specific heat at constant pressure per unit mass, K_E is the eddy diffusion coefficient, ρ is the atmospheric mass density, and Q and L are the heating and cooling rates, respectively. Thermospheric heating (Q) includes solar radiation heating, Joule heating, collisional heat transfers, and heat release by chemical reactions. Energy loss (L) process is primarily radiation cooling (NO, $O(^3P)$, and CO_2). It is well established that the contribution to high-latitude thermospheric heating by particle precipitation and Joule dissipation of electric fields often dominates the direct solar EUV input. The Joule heating can be written as

$$Q_J = \sigma_P(\mathbf{E} + \mathbf{U} \times \mathbf{B})^2 \tag{7.19}$$

where σ_P is the local Pedersen conductivity, **E** is the convection electric field, **U** is the neutral wind, and **B** is the ambient magnetic field. See Chapters 1 and 5 for detailed discussions.

The energy supplied to the magnetosphere from the solar wind is dissipated by deposition into the atmosphere via particle precipitation and Joule heating. This energy deposition by particles and electric fields, which can be highly structured both spatially and temporally, can perturb the atmospheric electron density, ion composition, electron and ion temperatures, and neutral air motion (e.g., Hays et al., 1973; Banks, 1977; Banks et al., 1981; Vickrey et al., 1982; Wilson et al., 2006). The comparison of particle heating rate and Joule heating rate calculated based on DE-2 measurements in Fig. 7.13 indicates that the particle heating rate is considerably smaller than the Joule heating rate in most regions below the satellite orbit but is important in the auroral regions where it can even exceed the Joule heating rate below 100 km (Deng et al., 1995).

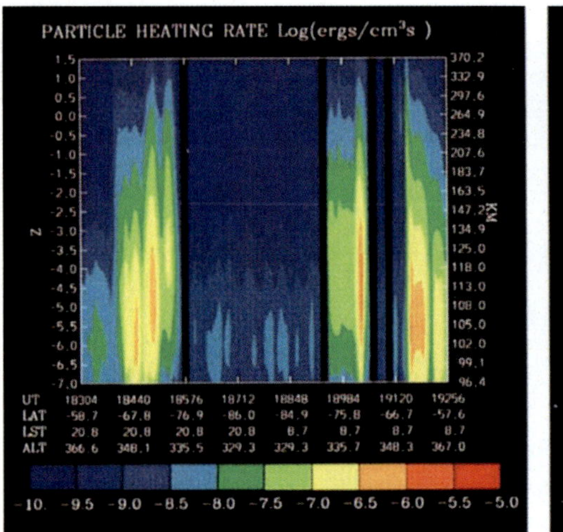

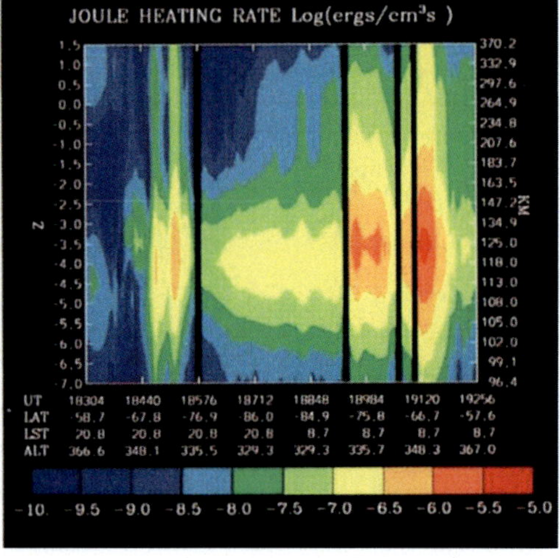

Fig. 7.13 The model calculated parameters below satellite orbit 1222 for (*left*) particle heating rate in log (ergs cm^{-3} s^{-1}), and (*right*) Joule heating rate in log (ergs cm^{-3} s^{-1}). *(Modified from Deng, W., Killeen, T.L., Burns, A.G., Johnson, R.M., Emery, B.A., Roble, R.G., Winningham, J.D., Gary, J.B., 1995. One-dimensional hybrid satellite track model for the dynamics explorer 2 (DE 2) satellite. J. Geophys. Res. Space Phys.100 (A2), 1611–1624.)*

It is known that the injection of magnetospheric energy at high latitudes can induce gravity waves in the thermosphere (e.g., Richmond, 1978; Forbes et al., 1995; Lu et al., 2012, 2020). Large-scale waves (greater than ~1000 km wavelength) can propagate over large horizontal distances and evolves to large-scale traveling atmospheric disturbances (TADs) in the thermosphere (Hocke and Schlegel, 1996; Balthazor and Moffett, 1999; Bruinsma and Forbes, 2010). The storm-induced TADs have also been well reproduced by numerical model (see Fig. 7.14 for a simulation case of the TADs that presented by the thermospheric winds and temperature). Furthermore, due to the availability of dense observational network since the start of GNSS era, some traveling waves in total electron content (TEC), referred as traveling ionospheric disturbances (TIDs), are commonly observed (e.g., Saito et al., 1998; Tsugawa et al., 2004). These TIDs, however, may or may not be manifestation of TADs, as described in details in Sections 7.3 and 7.4.

Besides generating large-scale wave structures, the Joule and particle heating are believed to have a great influence on the mesosphere and lower thermosphere (MLT) region. Using TIMED/SABER observations, Chang et al. (2009) found a 9-day period in global temperature in the MLT region, driven by Joule and particle energy deposition

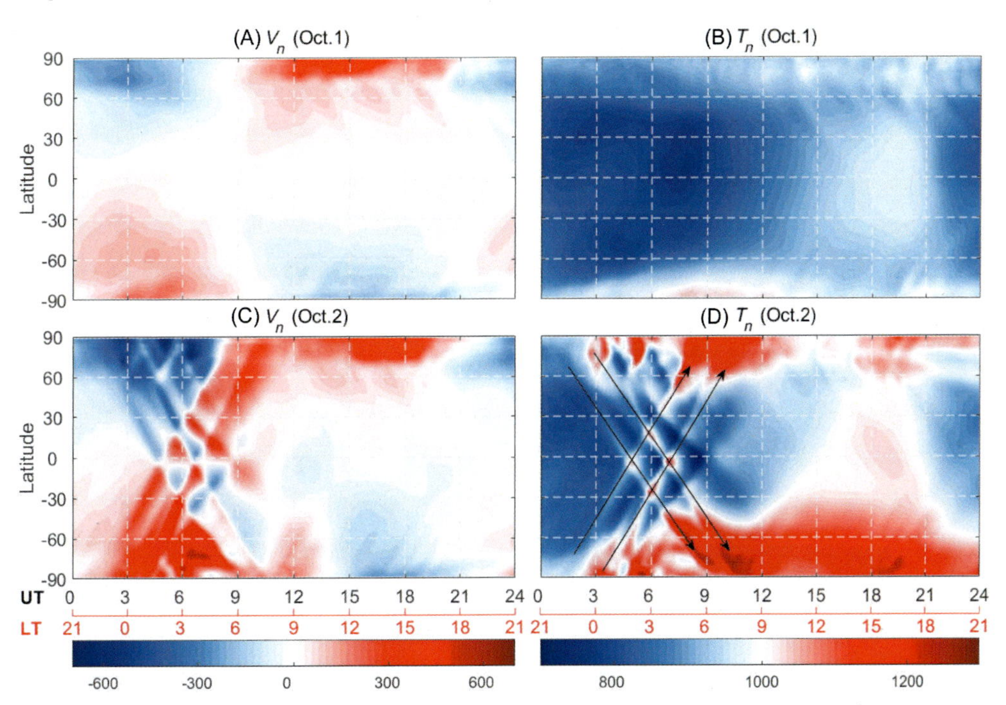

Fig. 7.14 Variations of meridional winds Vn (positive is northward, in units of m s^{-1}) and neutral temperature Tn (in units of K) at 250 km as a function of geographic latitude and universal time at the Sao Luis longitude sector on (A, B) quiet and (C, D) storm days. The storm time TADs are marked by *black arrows* (Ren and Lei, 2017).

from the 9-day recurrent geomagnetic activity forced by corotating interaction regions in the solar wind. Moreover, during geomagnetic storms, observations from LiDARs and satellites have shown that large neutral temperature increases and decreases occur in the middle and low latitudes of the mesosphere and lower thermosphere region (e.g., Biondi and Meriwether Jr., 1985; Nesse Tyssøy et al., 2010; Yuan et al., 2015). Prominent temperature responses to geomagnetic disturbances can be noted in the MLT region in Fig. 7.15 (Li et al., 2018), with an earlier temperature changes at night than during

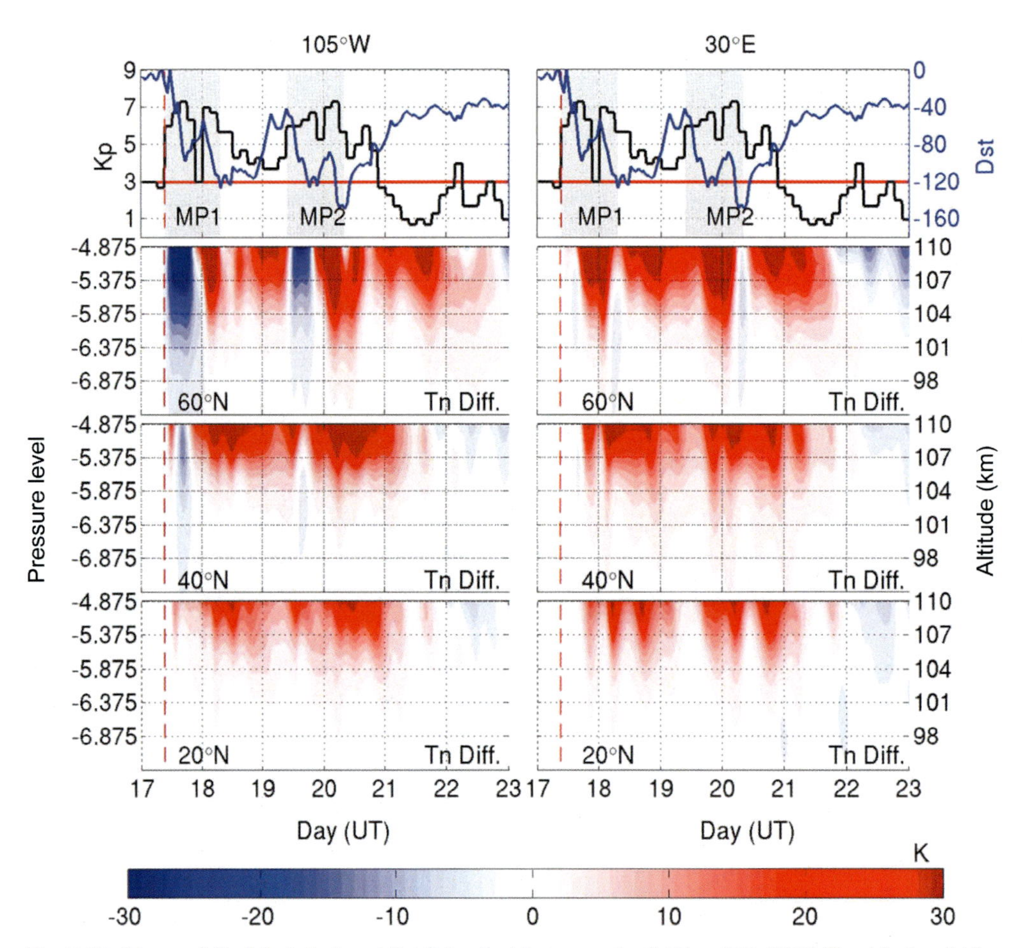

Fig. 7.15 (*Top row*) Kp (*black line*) and Dst (*blue line*) between April 17 and 23, 2002. The *black solid line* shows the real Kp during the storm, and the *red line* is a constant Kp value of 3.0 for nondisturbed conditions. The *second to bottom rows* are the Tn differences at 60°N, 40°N, and 20°N from 94 to 110 km between TIMEGCM simulations driven by the real Kp and by a constant Kp value of 3.0. The *left column* is for the longitude of 105°W where the storm began at 02:00 LT. The *right column* is for the longitude of 30°E and the onset of the storm was at 11:00 LT. The *red-dashed lines* indicate the beginning of the geomagnetic storm (Li et al., 2018).

daytime. The analysis in Li et al. (2018) has further illustrated that the temperature changes were produced mainly by adiabatic heating/cooling that was associated with vertical winds resulting from general circulation changes, with additional contributions from vertical heat advection.

7.1.4 Transient solar flux events and ionospheric irregularities

In the previous sections, we mainly discuss the climatology of IT coupling during both geomagnetically quiet and disturbed time. As compared to the geomagnetic activity, the solar eclipses and solar flares can influence the IT system via transient solar radiation flux changes. A solar eclipse occurs when the Moon passes between the Sun and the Earth. A solar eclipse provides a unique opportunity of active experiment of solar radiation effects on the coupled IT system and has been extensively studied by a number of works (e.g., Anastassiadis and Matsoukas, 1969; Davis et al., 2000; Ding et al., 2010; Le et al., 2009; Salah et al., 1986). It is realized that the short duration decrease of solar radiation over the Moon's shadow region which reduces both the heating to the upper atmosphere and the ionization can produce significant ionospheric and thermospheric responses on both local and global scales (Dang et al., 2018a, b; Lei et al., 2018a). The rapid cooling in the eclipse zone creates atmospheric pressure gradients and convergent horizontal winds. These disturbances can propagate as TADs in the direction determined by the eclipse path as well as the background conditions. In the ionosphere, responses result from not only the photochemistry but also dynamic forcing including disturbance winds, induced dynamo electric fields, plasma diffusion, and the ionosphere-plasmasphere exchange. TEC enhancements, for example, could arise even after the eclipse is over due to the eclipse induced large-scale TADs that propagated worldwide (Fig. 7.16).

Different from the solar eclipse, a solar flare is a sudden brightening of radiation in solar active regions. Solar flares occur frequently in solar maxima years. During solar flares, huge energy is abruptly released from the Sun, inducing enhancements with varying amplitudes at radio, visible light, and XUV wavelengths with durations ranging from about 10 min to hours. Explosive radiation propagates at light speed and strikes the Earth after 8 min. The flares abruptly change the structure and state of the ionosphere and thermosphere in the sunlit hemisphere, causing sudden ionospheric and thermospheric disturbances. Qian et al. (2019) conducted model simulations and data analysis to examine solar flare effects on the coupled thermosphere and ionosphere system in connection with flare location effects and to investigate the occurrences of large-scale TADs due to flares and storms. They found that large-scale TADs occurred when there were both flares and storms. The presence of the flares changed the magnitudes and propagation speeds of the large-scale TADs. However, these TIE-GCM simulations provided no evidence that large-scale TADs occurred when there were only flares and not storms, indicating that solar flares alone were unlikely sufficient to excite large-scale TADs. Some of the GITM

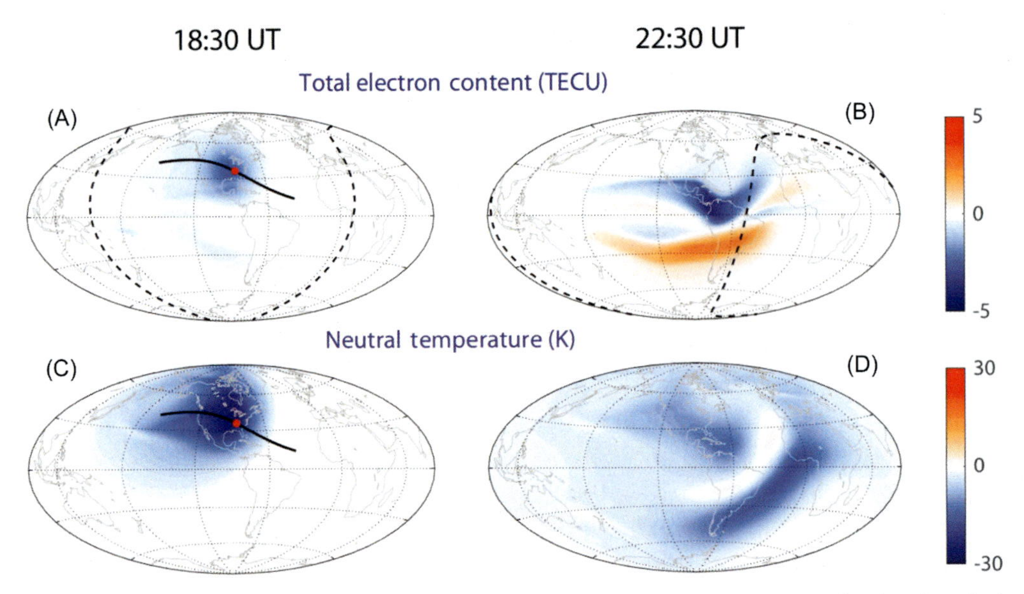

Fig. 7.16 Global maps of differential TEC and neutral temperature at pressure level 2 (300 km) between the TIE-GCM simulations with and without eclipse (with eclipse-without eclipse) for two selected UTs (18:30 and 22:30 UT). The *solid circle* indicates the location of the totality at 18:30 UT. The *solid and dashed lines* show the umbra path and solar terminator, respectively. *(Adapted from Dang, T., Lei, J., Wang, W., Zhang, B., Burns, A., Le, H., et al., 2018b. Global responses of the coupled thermosphere and ionosphere system to the August 2017 Great American Solar Eclipse. J. Geophys. Res. 123, 7040–7050. 2018JA025566.)*

simulations, however, provided a different scenario where flares can excite TADs (e.g., Pawlowski and Ridley, 2008). During significant solar flares, rapid heating appears significant in the thermosphere, and the ionospheric heating and upwelling are substantial (Mendillo et al., 2018; Zhang et al., 2019a). Flare-induced conductivity changes on the sunlit ionosphere in both high and equatorial latitudes may cause electromagnetic disturbances.

Besides the large-scale structures described earlier, the ion-neutral coupling is also important in meso- and small-scale processes. The ionospheric irregularities known as equatorial spread F (ESF) were first observed by Booker and Wells (1938) using ionosondes. It is now known that ionospheric irregularities were produced by Rayleigh-Taylor-like instability, and the scales of the irregularities vary from tens of centimeters to tens of kilometers. ESF is driven by plasma instability while is tightly coupled with the neutral atmosphere. For example, the neutral wind in the upper atmosphere generates a global electric field, dominant in the low- to mid-latitude ionosphere during quiet times, which plays a critical role in the development of ESF (Huang and Kelley, 1996; Huba and Joyce, 2007, 2010; Huba et al., 2009; Yokoyama et al., 2014; Yokoyama,

2017). Additionally, gravity waves originating in the troposphere, which propagate into the ionosphere, can act as "seeds" to trigger the Rayleigh-Taylor instability.

7.1.5 Summary

The IT coupling has continued to be an active research field during the past decades benefiting from the advanced observation and simulation tools. Due to ion-neutral collisions, the ionosphere and thermosphere make up a complex system associated with chemical, dynamic, and electrodynamic (sometimes energetic coupling) processes.

The thermospheric composition (O/N_2) affects ionospheric density variations through photochemical processes. The movement of ionospheric ions can be driven by the neutral wind along the magnetic field line. Moreover, the neutral gas flow takes ions going across the magnetic field lines and induces the ionospheric electric field and currents, which is referred as "neutral wind dynamo."

The ionospheric ions drag the motion of neutral gas, especially in the polar region. This ion drag effect leads to complex thermospheric features such as vortex patterns in the polar neutral winds, equatorial thermosphere anomaly, and the equatorial wind jet. The energy deposition from magnetosphere-solar interactions into the ionosphere can excite thermospheric large-scale TADs propagating globally. The transient solar radiation changes, including solar eclipses and solar flares, can affect the coupled IT system in complex ways. Furthermore, in a meso- and small-scale, the ion-neutral coupling plays a significant role in the formation and evolution of ionospheric irregularities.

We have demonstrated that the IT communities have made great progresses in understanding the IT two-way coupling from theoretical aspects. The interplay of the forcings from below and above (e.g., Pedatella and Liu, 2018; Lei et al., 2018b) could further reinforce the complex nature of the IT interaction. Nevertheless, the coordinated satellite in situ and ground-based observations of the thermosphere and ionosphere are still sparse and particularly needed to further improve our understanding of the ion-neutral coupling in the upper atmosphere.

7.2 Large-scale structures and ion-neutral coupling

Huixin Liu[a] and Scott England[b]

[a]Department of Earth and Planetary Science, Kyushu University, Fukuoka, Japan

[b]Department of Aerospace and Ocean Engineering, Virginia Polytechnic Institute and State University, Blacksburg, VA, United States

7.2.1 Introduction

The Earth's ionosphere-thermosphere (IT) system, being the transition regime from meteorology to outer space, is strongly forced via energy and momentum input by the Sun, the magnetosphere, and lower atmosphere. In response, it undergoes a variety of spatial and temporal changes: some are regular, such as diurnal, seasonal, and solar cycle changes: some are dramatic and impulsive, such as changes during solar flares and geomagnetic storms; and still some are neither regular nor dramatic, but occurs frequently, which is referred to as day-to-day variability. These variabilities form important parts of space weather, which affects the infrastructure of our modern high-tech society, for example, global communication and positioning system, satellite orbit control, and reentry of space vehicles.

There have been many extensive reviews on various aspects of the IT system and readers are referred to a study by Schunk (2014) for the current status and problems in the IT physics. Much of the large-scale regular structures of the IT system, such as variations with the latitude, local time, season, and solar cycle, are described by empirical models of the thermosphere and ionosphere, for example, the MSIS (Picone et al., 2002), the HWM (Drob et al., 2015), and IRI (Bilitza, 2018) models. Section 7.1 provides several fundamental IT coupling concepts. In this section, we describe some of the nonregular structures pertaining to ion-neural coupling processes in the upper atmosphere under quiet geomagnetic conditions.

7.2.2 The equatorial mass density anomaly

The first global distribution of the thermospheric density is provided by the CHAMP satellite (Liu et al., 2005), which revealed many interesting features unknown before the mission. Liu et al. (2017) reviewed various nonregular features in the thermosphere density, which are driven by different sources from the magnetosphere and the lower atmosphere. In this section, we pick up the equatorial thermospheric anomaly (ETA), which includes the equatorial mass density anomaly (EMA) and equatorial temperature anomaly, to illustrate the close thermosphere-ionosphere coupling.

The EMA is one of the major discoveries of the decade-long CHAMP mission between 2000 and 2010 (Stolle and Liu, 2014). It is a fundamental structure of the thermosphere in tropical regions during daytime, with its equinox configuration consisting of an air mass density trough near the Earth's dip equator flanked by density crests around ± 25 degrees magnetic latitude (see Fig. 7.17). Similar latitudinal structures were observed by the DE satellite in the thermosphere composition and temperature, which are collectively referred to as the ETA (Philbrick and McIsaac, 1972; Hedin and Mayr, 1973; Raghavarao et al., 1991). The EMA structure closely resembles the well-known EIA (Namba and Maeda, 1939; Appleton, 1946) in the ionosphere but with a ~ 10 degrees latitude shift in their crest location. They also share similar climatological features like the seasonal and solar cycle variations (Liu et al., 2007a). Note that Fig. 7.17 also shows the well-known wave-4 longitudinal structure, which is mainly driven by lower atmosphere DE3 tides (Liu et al., 2009b).

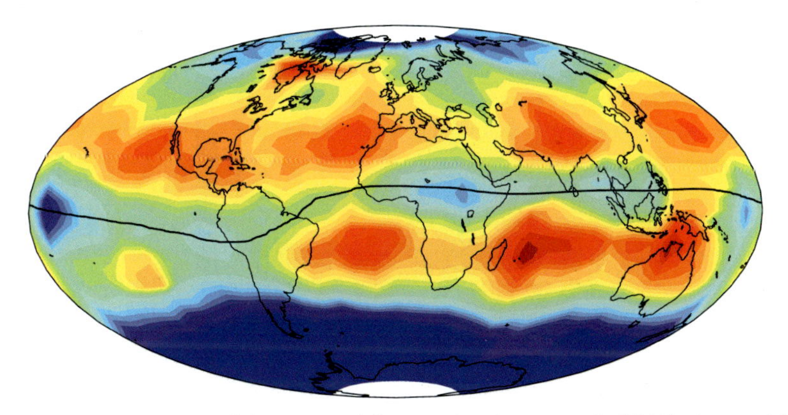

Fig. 7.17 The spatial structure of the equatorial mass density anomaly (EMA) observed by CHAMP satellite at 400-km near equinox in 2002.

The EMA/ETA reflects a significant magnetic control of the thermosphere via comprehensive ion-neutral coupling processes. Simulations by TIEGCM show that the ETA trough and crests evolve from different processes. The crests are produced mainly by the ion-neutral collision heating (Lei et al., 2012a), while the trough is mainly due to the field-aligned ion drag (Maruyama et al., 2003; Miyoshi et al., 2011; Lei et al., 2012b). Fig. 7.18 from the TIEGCM simulations by Lei et al. (2012b) demonstrate the trough neutral temperature reduction at the magnetic equator caused by introducing the field-aligned ion drag forcing in the simulation.

The heat transport due to zonal winds (Hedin and Mayr, 1973; Mayr et al., 1974) and chemical heating fueled by charge exchange between O^+ and O_2 or N_2 (Fuller-Rowell et al., 1997) are found to produce small contributions to the ETA formation. The same may apply to the formation of EMA, though one should keep in mind that thermosphere mass density at a fixed altitude can be influenced by changes not only in temperature but also in the mixing ratios of various gas constituents. Furthermore, the terdiurnal tides also significantly contribute to the formation of EMA trough, thus indicating partial contribution to EMA from the lower atmosphere.

It is unclear how the EMA extends in altitudes. Due to the limited height coverage of CHAMP, it could observe the EMA only within a small range from 360 to 410 km, which is less than one scale height (see Fig. 7.19). It would be instructive to see how EMA evolves vertically, that is, further down to the bottomside F-region and further up to the topside ionosphere, and how the interplay of different processes changes with height.

7.2.3 Thermosphere-ionosphere coupling via ion drag: The equatorial wind jet

Thermospheric wind plays a pivot role in the IT coupling through dynamo effects and field-aligned transportation of the plasma. However, it is currently one of the least

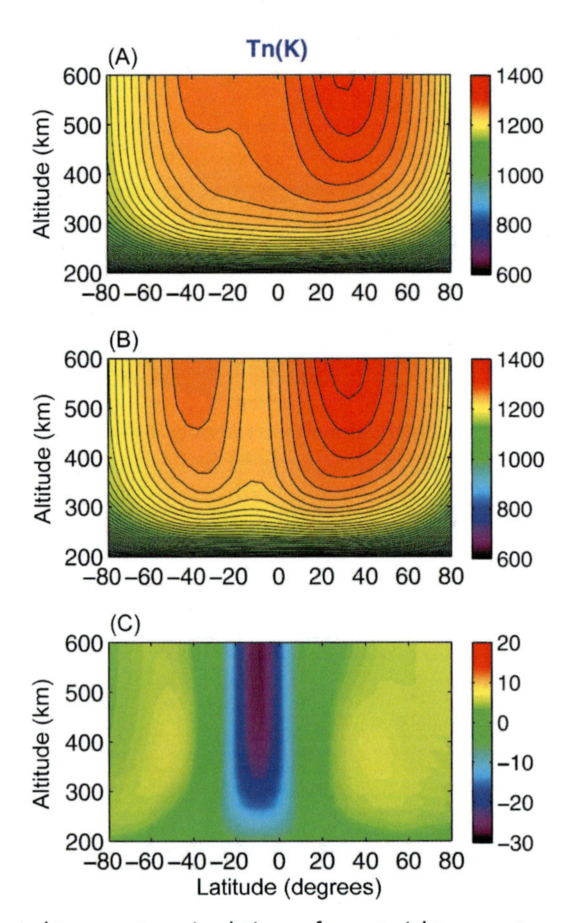

Fig. 7.18 TIEGCM neutral temperature simulations of equatorial temperature anomaly for the 60°W longitude at 14:00 LT under the influence of field-align ion drag. (A) Without and (B) with field-aligned ion drag in the momentum equations of the neutral gas. The corresponding differences between the two simulations are displayed in (C). Note that the magnetic equator lies at a geographic latitude of around 11°S at this longitude. *(From Lei, J., Thayer, J.P., Wang, W., Richmond, A.D., Roble, R., Luan, X., Dou, X., Xue, X., Li, T., 2012b. Simulations of the equatorial thermosphere anomaly: field-aligned ion drag effect. J. Geophys. Res. Space Phys.117 (A1).)*

observed quantitied in the thermosphere. The most widely used technique to measure wind is ground-based Fabry-Perot Interferometers, which provide observations at about 250 km but only at night. The accelerometer CHAMP is among the few missions that provide a decade-long global and homogenous measurement of the zonal wind. Liu et al. (2006, 2009a) provide a global picture of the wind, with its local time, season and solar cycle variations. Here we choose the equatorial wind jet to demonstrate the strong effect of ion-neutral coupling on thermosphere dynamics.

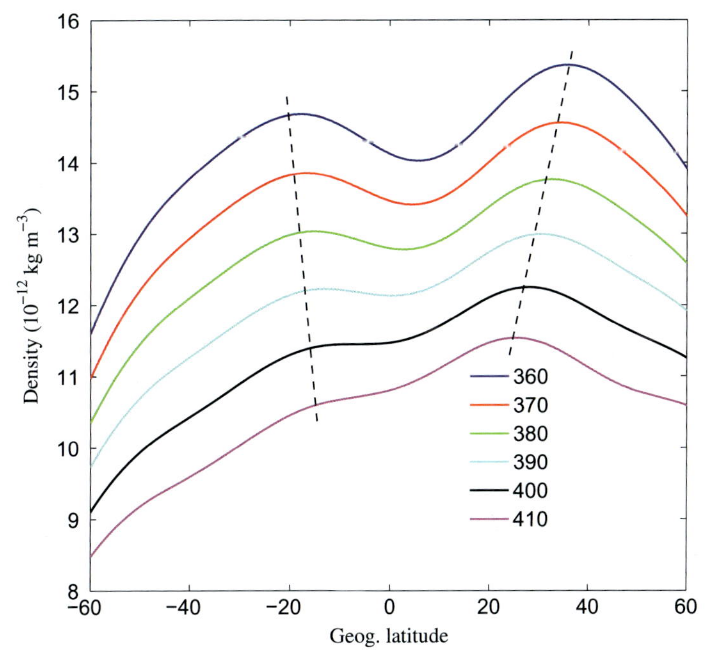

Fig. 7.19 Latitudinal profile of EMA at different altitudes around March equinox at 14 LT at high solar flux level $P = 180$ produced by the CHAMP model. *(From Liu, H., Hirano, T., Watanabe, S., 2013. Empirical model of the thermospheric mass density based on CHAMP satellite observations. J. Geophys. Res.118, 843–848. 10.1002/jgra.50144.)*

Given that thermosphere wind is largely driven by the pressure gradient in the neutral atmosphere to first order, we would expect the zonal wind structure to be organized in geographic coordinates. However, satellite measurements have revealed its close alignment with the dip equator (see Fig. 7.20). The zonal wind forms a strong wind jet over the dip equator during local times when the EIA structure exists, being particularly prominent around noon (westward) and after sunset (eastward). During non-EIA local times (about 05–08 LT), the dip equator becomes a host for weak wind (the blue belt in the right panel of Fig. 7.20). This remarkable feature of magnetic control was first scarcely spotted by DE-2 satellite (Raghavarao et al., 1991), and later confirmed by extensive observations from CHAMP near 400 km altitude (Liu et al., 2009a) and GOCE near 250 km (Liu et al., 2016).

The mechanism responsible for this dip-equator-aligned wind structure is the latitudinal distribution of the zonal ion drag, as confirmed by both TIEGCM (Kondo et al., 2011) and GAIA (Miyoshi et al., 2012) models. These work demonstrates that due to the latitudinal distribution of the plasma in the EIA region, the zonal ion drag minimizes at the dip equator where the plasma density is low. On the other hand, the pressure gradient

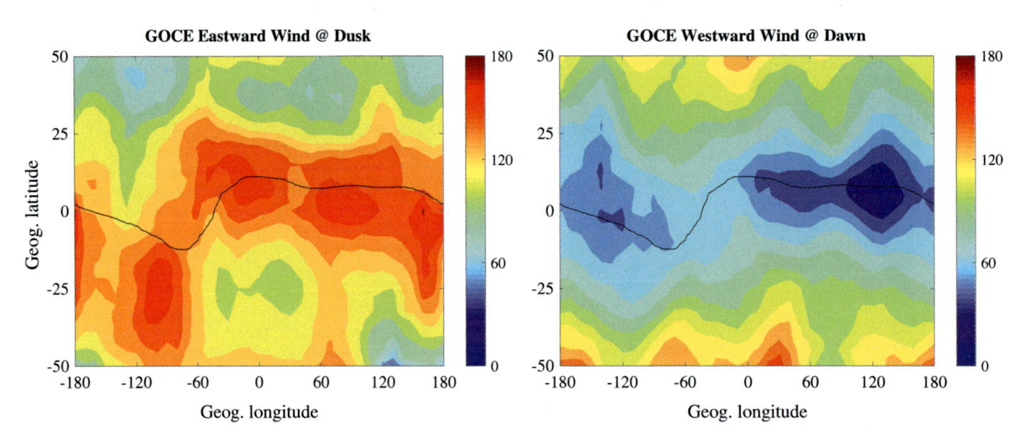

Fig. 7.20 Zonal winds at about 250 km altitude observed by GOCE. *Left*: eastward wind at dusk 19 LT; *right*: westward wind at dawn 07 LT. Note that the color coding represent the speed of the wind, not direction.

does not vary much between the EIA trough and crest regions. These factors combine together to give a stronger wind at the dip equator. During non-EIA periods (~05–08 LT), the reverse occurs and the ion drag maximizes at the dip equator due to slightly higher plasma density, thus weakening the wind (Fig. 7.21). It is noted that fast wind jet also exists at the dip equator in the topside ionosphere, where no EIA exists. This is a result of the direct upward extension of the F-region wind jet via strong viscosity (Kondo et al., 2011).

The regulation of ion drag on the zonal wind leads to a net eastward wind at magnetic latitudes below ±35 degrees as derived from CHAMP observations (see Fig. 7.22). This means the upper atmosphere rotates faster than the Earth, which is a phenomenon termed superrotation (Rishbeth, 2002). The superrotation speed is not constant but varies with season and solar cycle, due to corresponding changes in the ion drag (Liu et al., 2006). The average superrotation speed at the dip equator is about 22 m s^{-1} for moderate solar flux levels, but increases to 63 m s^{-1} for high solar flux levels.

7.2.4 Thermosphere-ionosphere coupling via wind-driven field-aligned motions at middle latitudes: MSNA in the ionosphere

The Mid-latitude Summer Night Anomaly (MSNA) refers to enhanced night-time ionospheric electron density at summer middle latitudes (Thampi et al., 2009; Liu et al., 2010). It is the generalized name for the so-called Weddell Sea Anomaly (in the Southern hemisphere) to capture the global nature of the phenomenon in both hemispheres.

The MSNA phenomenon was first discovered in the ionosonde *foF2* observations in the Southern hemisphere at Faraday (65.2°S, 64.6°W) near the Weddell Sea and was named the Weddell Sea Anomaly (WSA) (Bellchambers and Piggott, 1958; Dungey,

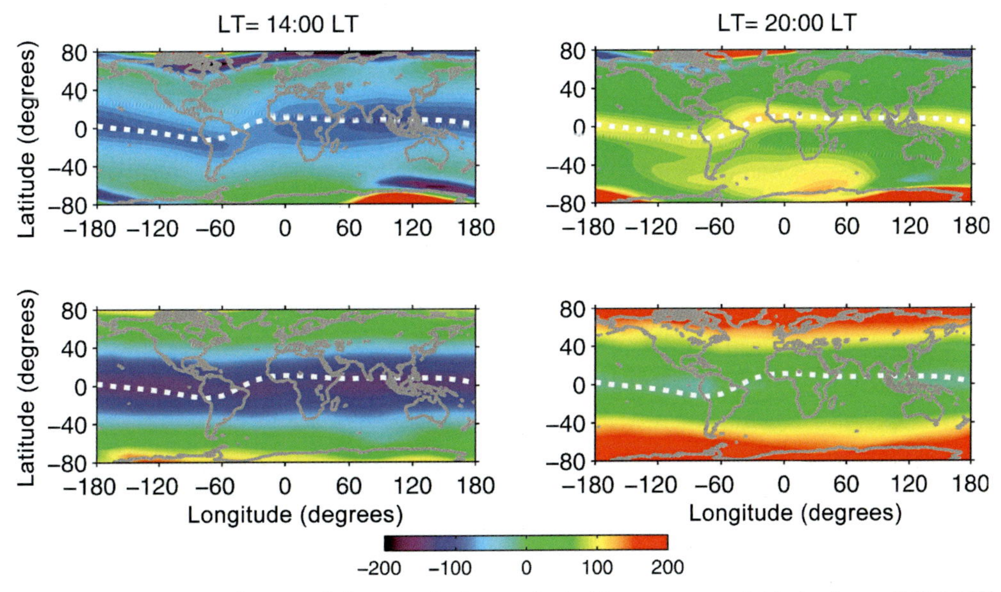

Fig. 7.21 TIEGCM simulations of thermospheric zonal wind jets at equatorial latitudes at (*left*) 14 LT and (*right*) 20 LT at about 400 km. (*Top*) The default model run and (*bottom*) the run without the ion drag forcing. The *white-dashed line* is the dip equator. Positive wind values are eastward. *(From Kondo, T., Richmond, A.D., Liu, H., Lei, J., Watanabe, S., 2011. On the formation of a fast thermospheric zonal wind at the magnetic dip equator. Geophys. Res. Lett.38 (10).)*

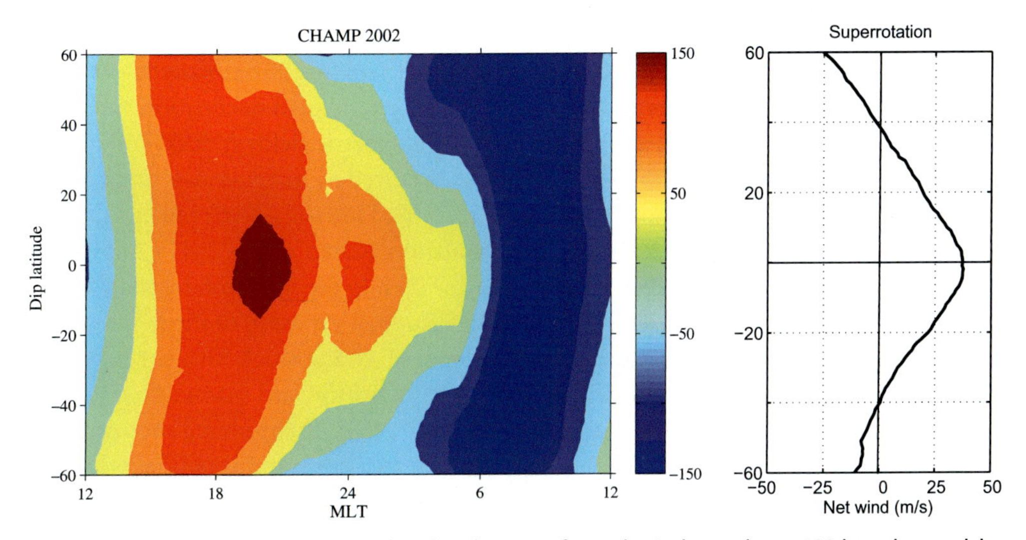

Fig. 7.22 *Left*: Dip latitude versus MLT distribution of zonal winds at about 400 km observed by CHAMP averaged over year 2002. *Right*: Net wind averaged over 24 LTs. Positive net wind indicates superrotation, while negative one indicates subrotation. *(Modified from Liu, H., Watanabe, S., Kondo, T., 2009a. Fast thermospheric wind jet at the Earth's dip equator. Geophys. Res. Lett.36. 10.1029/2009GL037377.)*

1961). Similar phenomenon was found to occur at Port Lockroy (65.8°S, 63.5°W) (Rastogi, 1960), and also in the Northern hemisphere at Millstone Hill (42.6°N, 71.5°W) (Evans, 1965). Global plasma density observations from the CHAMP satellite revealed WSA-like features in both hemispheres at middle latitudes (Liu et al., 2007b). The existence of this anomaly in the Northern hemisphere was further confirmed by the COSMIC satellite observations and ground ionosphere tomographic observations (Lin et al., 2009; Thampi et al., 2009). The MSNA can be seen in many ionospheric physical parameters such as the total electron content (TEC), plasma density, the F2 region peak density NmF2 and the peak height hmF2, and the 136.5 nm airglow from the O^{+}+e recombination (Horvath and Lovell, 2009; Burns et al., 2008; Lin et al., 2009; Jee et al., 2009; He et al., 2009; Liu et al., 2010; Hsu et al., 2011, and references therein), with MSNA in the plasma density at a fixed altitude (e.g., 400 km) being more prominent than in NmF2 or TEC (Liu et al., 2010).

Fig. 7.23 shows a global map of the MSNA regions observed by CHAMP at 400 km altitude. They include three distinct regions: the East Asian (EA) region centered around (53°N, 150°E), the Northern Atlantic (NA) region centered around (45°N, 50°W), and the South Pacific (SP) region centered around (60°S, 110°W). The Weddell Sea Anomaly (Bellchambers and Piggott, 1958; Dungey, 1961) falls inside the SP region Fig. 7.24 shows the climatology of MSNA in each region. We see that the anomaly occurs during March–August in EA and NA regions, and between August–March in SP region. The MSNA is more pronounced in the Southern hemisphere than in the Northern hemisphere (larger magnitudes and longer persistence months), and more pronounced at the solar minimum than at the solar maximum.

Many mechanisms have been proposed to explain the nighttime enhancement, such as the neutral wind effects in conjunction with the magnetic field declination and inclination, photoionization, plasmaspheric fluxes at night, electric fields at dusk, and neutral composition. Among these, the first one has been supported by many studies using observations (e.g., Horvath and Lovell, 2009; Jee et al., 2009; He et al., 2009; Liu et al., 2010, and references therein), and SAMI2 model simulation (Chen et al., 2011). However, Richards et al. (2017) simulated the MSNA in NmF2 using the FLIP model and reached opposite conclusions. They found that it is the longitudinal variation of the neutral wind itself, along with longitudinal variation of the neutral composition that are the most important factors for MSNA formation. They also found upward plasma flux at night in the summer hemisphere rather than downward flux, thus disapproving the contribution of downward plasma flux. It seems necessary to have continuous neutral wind and composition observations across the day-to-night transition period at middle latitudes covering all longitudes to confirm these findings. Contributions from the dusk electric field in the conjugate hemisphere is proposed by Burns et al. (2011). Several extended discussions, other MSNA aspects, and relevant references can be found in Zhang (2021).

In discussing the mechanisms, Liu et al. (2010) took a different perspective in terms of diurnal variation and zonal wave numbers, rather than focusing only on the nighttime

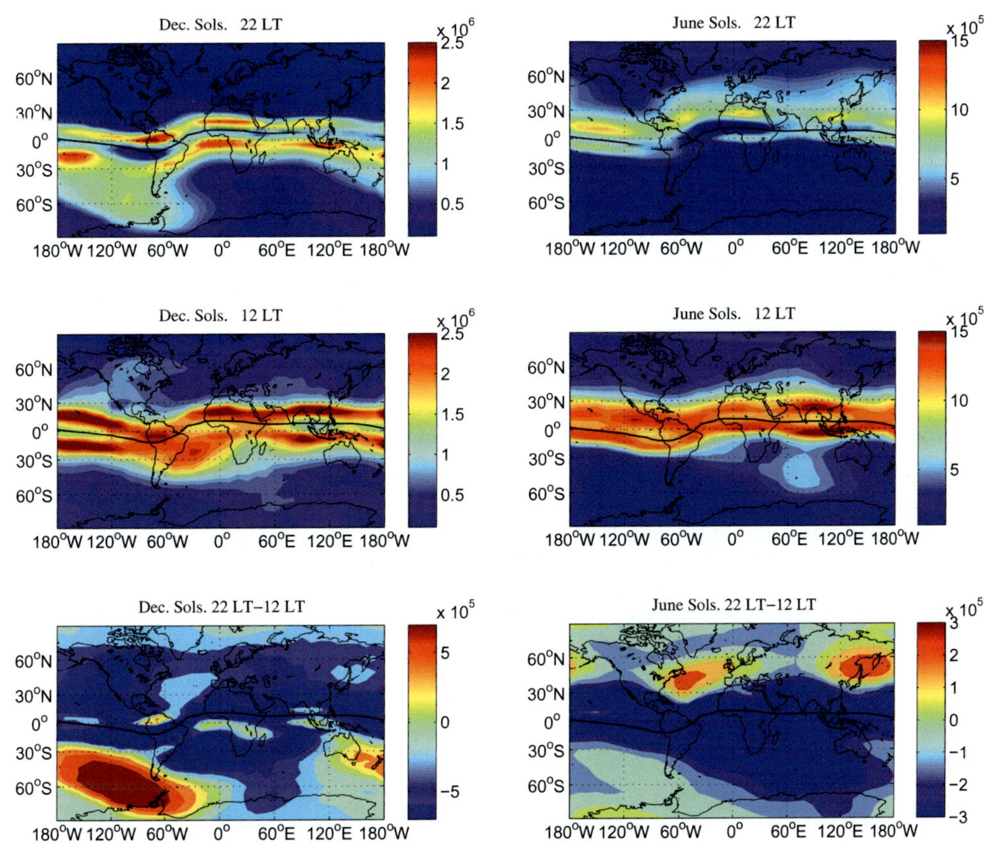

Fig. 7.23 Global distribution of the electron density observed by CHAMP at 400 km altitude. *Left*: For December solstice; *right*: for June solstice. We see the electron density at night (22 LT) is larger than at 12 LT around December solstice in the southern pacific centered at 60°S, 110°W (*lower left panel*), but around June solstice in East Asian (EA) region centered around (53°N, 150°E), Northern Atlantic (NA) region centered around (45°N, 50°W) (*lower right panel*). These regions are the MSNA regions. *(From Liu, H., Thampi, S., Yamamoto, M., 2010. Phase reversal of the diurnal cycle in the midlatitude ionosphere. J. Geophys. Res.115. 10.1029/2009ja014689.)*

enhancement done in most studies. They view the MSNA as a phase reversal of the diurnal cycle, which has a spatial zonal wave-1 structure in the Southern hemisphere and wave-2 structure in the Northern hemisphere driven by neutral winds projected in the global geomagnetic field frame (see Fig. 5 of Liu et al., 2010). These same wave patterns were also discussed in Zhang et al. (2011, Fig. 4) and Zhang et al. (2012). This tidal perspective was adopted by Xiong and Lühr (2014), who successfully identified the major underlying tidal components. Using CHAMP and GRACE plasma density observations, they found that the nonmigrating tidal component D0 (diurnal standing) is dominantly responsible for MSNA in the Southern hemisphere, while DE1 (eastward wave number 1) for the Northern hemisphere (see Fig. 7.25). Chang et al. (2015) also applied tidal

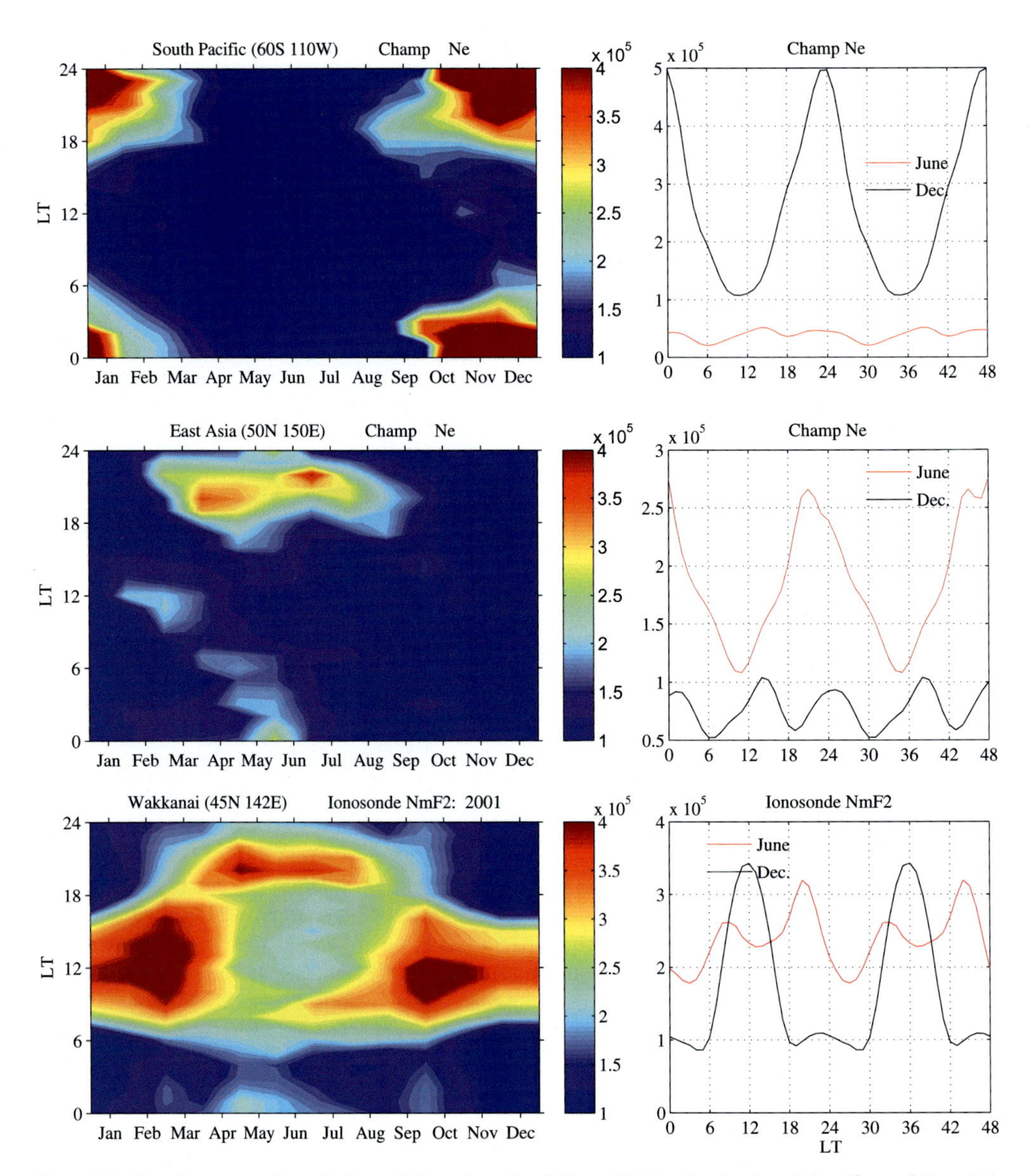

Fig. 7.24 Month-to-month variation of diurnal cycle of Ne at 400 km in the South Pacific and East Asia regions, along with NmF2 observed by an Ionosonde at Wakkanai. The solar flux level is F10.7 $\sim$ 90. Note that a phase reversal of the diurnal cycle occurs around local equinoxes and summer, while the electron density maximizes at night instead of at noon. *(From Liu, H., Thampi, S., Yamamoto, M., 2010. Phase reversal of the diurnal cycle in the midlatitude ionosphere. J. Geophys. Res.115. 10.1029/2009ja014689.)*

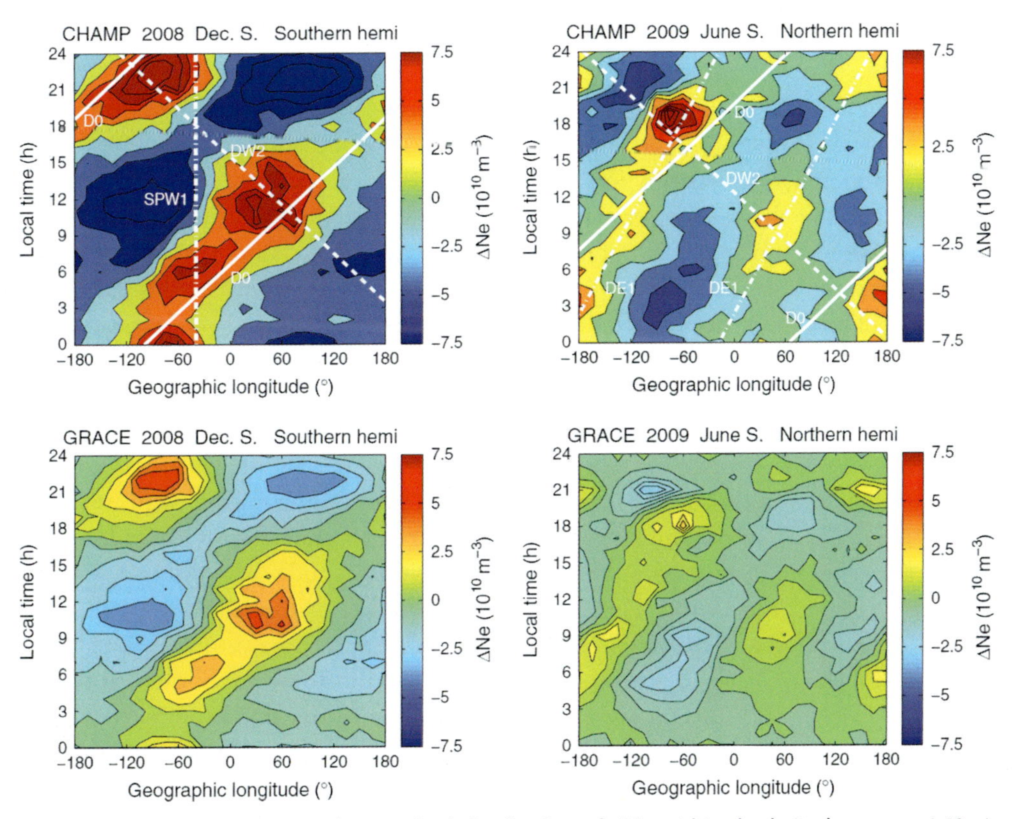

Fig. 7.25 The local time versus longitudinal distribution of ΔNe within the latitude ranges ±40–± 60 degrees MLAT for the *(left)* Southern and *(right)* Northern hemispheres during local summer from *(top)* CHAMP and *(bottom)* GRACE observations. *White lines* indicate the wave crests of the dominant tidal components. *(From Xiong, C., Lühr, H., 2014. The midlatitude summer night anomaly as observed by CHAMP and GRACE: interpreted as tidal features. J. Geophys. Res. Space Phys.119 (6), 4905–4915. 10.1002/2014JA019959.)*

analysis to the TEC and electron density observations from COSMIC and found that DW2 (westward wave number 2) and SPW1 (stationary planetary wave 1) also play a role in addition to D0 and DE1. Using the GAIA model, they further investigated the generation of these nonmigrating tides and showed they are likely due to in situ photoionization or plasma transport along magnetic field lines driven by neutral winds.

7.2.5 Thermosphere-ionosphere coupling via the wind-driven dynamo: The ionosphere wave-4 pattern

One of the most fundamental means of coupling between the atmosphere and ionosphere at low-to-mid latitudes is via the wind-driven dynamo in the E- and F-regions. The E-region, located between ~100 and 180 km, is primarily made up of molecular ion species (O_2^+ and NO^+), whereas the F-region, located between ~180 and 1000 km, is primarily made up of atomic O^+ (e.g., Johnson, 1966). The difference in the ratio between ion transport and lifetime time constants between these two regions means that to first order the E-region can be understood in terms of photo equilibrium, whereas the F-region is strongly impacted by ion-transport processes, as is clearly seen in the difference between the altitude of peak ion production at ~180 km and the plasma density peak at ~250–350 km (e.g., Matuura, 1966). This altitude difference results in-part from vertical transport, including diffusion and $\mathbf{E} \times \mathbf{B}$ drift. At low latitudes, where the geomagnetic field is close to horizontal, the $\mathbf{E} \times \mathbf{B}$ drift is the result of horizontal electric fields generated by the E- and F-region dynamos. Both the E- and F-region dynamos result from collisions between neutral thermospheric particles and the charged ionosphere. Owing to the difference in the collision cross-sections between neutral particles and ions compared to neutral particles and electrons, such collisions tend to impact ions more than electrons and can thus lead to a larger departure from gyromotion for ions than electrons. In the presence of a neutral wind, this can lead to a current and an associated conductivity, Hall or Pedersen which both play a role at E-region altitudes and where Pedersen is dominant at F-region altitudes (e.g., Heelis, 2004). Such currents can lead to electric fields where gradients in the winds or conductivity are present, the conditions for which vary with location and time. The E-region dynamo produces the dominant horizontal electric field during most of the day and night, whereas the F-region dynamo produces the dominant horizontal electric field around sunset (Heelis, 2004). Both of these dynamos are driven by neutral winds associated with atmospheric tides (Kato, 1957; Tarpley, 1970; Rishbeth, 1971; Richmond, 1979). Together, the uplift these electric fields create, along with the subsequent motion parallel to the magnetic field from pressure and gravity (Hanson and Moffett, 1966) produce the EIA. We choose to focus on the impact of nonmigrating atmospheric tides on the low-latitude ionosphere because these waves are known to impact the E-region dynamo and because their impacts are both clear in observations and well documented.

The observed impact of nonmigrating atmospheric tides modulating the E- and F-region dynamos is a longitudinal variation in the density of the EIA (TEC or plasma density at a fixed altitude; e.g., Lin et al., 2007; Scherliess et al., 2008; Wan et al., 2008; Liu and Watanabe, 2008). Airglow observations have revealed that this variation in TEC is primarily the result of variations in the F-region O^+ plasma density (e.g., Sagawa et al., 2005; Henderson et al., 2005; Immel et al., 2006; see Fig. 7.26). Corresponding changes in the strength of E-region equatorial-electrojet derived from magnetometer measurements have indicated that the horizontal electric field produced by the E-region dynamo varies with the same or similar longitudinal pattern to the F-region ion density response (e.g., England et al., 2006b; Lühr et al., 2008; Alken, 2009). In close relation to the electric fields in the E-region, the $\mathbf{E} \times \mathbf{B}$ drift measurements made at F-region altitudes show a similar longitudinal pattern (e.g., Hartman and Heelis, 2007; Kil et al., 2007; Ren et al.,

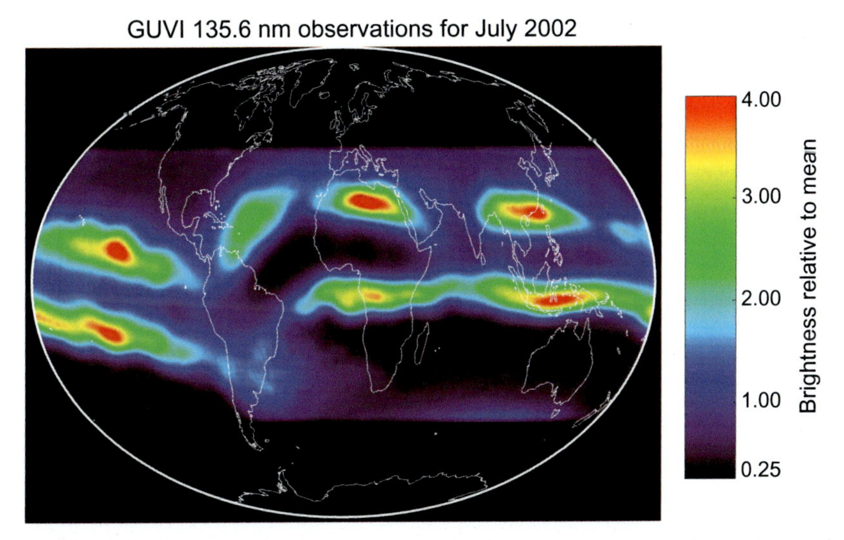

Fig. 7.26 TIMED-GUVI 135.6 nm observations for July 2002 during magnetically quiet periods. Image represents the morphology of O^+ at 20:30 local time. Brightnesses are relative to the mean at each magnetic latitude. *(Adapted from England, S.L., Zhang, X., Immel, T.J., Forbes, J.M., DeMajistre, R., 2009. The effect of non-migrating tides on the morphology of the equatorial ionospheric anomaly: seasonal variability. Earth Planets Space61, 493–503.)*

2009). Given that the altitude of the F-region density peak is related to the time-integral of the $\mathbf{E} \times \mathbf{B}$ drift at F-region altitudes throughout the day, variations in the height of the F-region peak has been observed to have the same kind of longitudinal pattern (seen in airglow observations, e.g., England et al., 2006a; McDonald et al., 2008). The wavenumber patterns created in the F-region O^+ density are seen to be both not purely symmetric about the magnetic equator (seen in airglow observations e.g., McDonald et al., 2008) and to vary with season, with a wave-4 structure being dominant throughout much of the year, but wavenumbers 1–3 being present as well (e.g., Scherliess et al., 2008; England et al., 2009; Xiong and Lühr, 2014).

The similarity of the longitudinal structures seen in the observed E-region electric fields, F-region $\mathbf{E} \times \mathbf{B}$ drifts, F-region peak altitudes and densities is certainly indicative of a mechanism by which tides perturb the E-region dynamo, and this produces the subsequent effects seen as originally suggested by Immel et al. (2006). However, the details of the coupling mechanism, including the role of winds at different altitudes, Pedersen and Hall conductivities, the source of hemispheric asymmetries, etc., are not apparent from observations alone and have been investigated with a variety of numerical models. Using an ionospheric model with an imposed $\mathbf{E} \times \mathbf{B}$ drift, England et al. (2008) were able to show that the observed change in the E-region electric field was sufficient in magnitude to create the observed changes in the F-region ion density. Using the same model, England et al. (2010) showed that other coupling mechanisms, including the F-region dynamo and neutral winds at F-region altitudes may also play a role in producing the observed changes in the F-region ionosphere. Using an ionospheric model with a self-consistent dynamo calculation, and observed winds from TIMED-TIDI, Ren et al. (2010) were able to show that hemispherically symmetric winds associated with the diurnal eastward wavenumber-3 tide (DE3) in the lower E-region driving currents in the Hall layer were perhaps the most important source of the coupling between tides and the F-region ionosphere, at least in producing the common wavenumber-4 structure, which was in agreement with the results from the coupled atmosphere-ionosphere models used by Jin et al. (2008) and Wan et al. (2012). Simulations by Hagan et al. (2009) show that other tides and the stationary planetary wave with wavenumber-4 may also provide key contributions to the generation of the wavenumber-4 signature in the ionosphere. Interestingly, recent work by Jiang et al. (2018) has also revealed that the hemispherically antisymmetric component of the DE3 tide may produce a wavenumber-3 pattern in the ionosphere, again via perturbations to the E-region dynamo. With the recent launch of NASA's Ionospheric CONnection Explorer (ICON; Immel et al., 2017), it is possible that any role neutral winds at F-region altitudes may play in shaping the ionospheric response to atmospheric tides will also be further elucidated.

It is noted that the EMA described in Section 7.2.2 also exhibits a wave-4 longitudinal structure when seen at a fixed local time (see Fig. 7.17). However, its formation mechanism is found to be quite different from that of the EIA. Using simultaneous thermosphere and ionosphere observations from the CHAMP satellite, Liu et al. (2009b) quantitatively compared the phase structures of the longitudinal variation in both EMA and EIA and revealed that the EMA's wave-4 is mainly driven by direct penetration of nonmigrating tides (e.g., DE3, SE2, SPW4) to the F-region heights, in contrast to the excitation via E-region dynamo for EIA.

7.2.6 Thermosphere-ionosphere coupling via changes in thermospheric composition

The composition of the thermosphere is one of the key elements in determining the density of the ionosphere. This is perhaps obvious in the light of the basic concept that the ionosphere is produced from the thermosphere and the thermosphere in turn controls important loss processes for ions, which are dependent on thermospheric composition. Nonetheless, the details of this and its implications for thermosphere-ionosphere coupling are important. The composition of the lower thermosphere—below the homopause—is essentially constant with time and location, but above this altitude diffusive equilibrium dominants, allowing for composition to vary with altitude according to species mass, production and loss of reactive species. This leads to a thermosphere that is dominated by atomic O between ~180 km and the exobase, with N_2 playing a secondary role over most of this altitude range (e.g., Hedin et al., 1977). While the E-region ionosphere is primarily made up of molecular ion species, the F-region ionosphere is dominated by O^+, and makes up by far majority of the whole ionosphere in terms of density (e.g., Johnson, 1966). The need to consider the relationship between thermospheric composition and O^+ is further highlighted by the consideration that the thermospheric composition is far more variable at F-region altitudes than at E-region altitudes. O^+ production depends on the density of O (and to a lesser extent the density of O_2), either via direct ionization or charge exchange with N_2^+, which is produced at a higher rate than O^+ (e.g., Torr and Torr, 1985). The total O^+ density is governed by this production rate, transport, and also loss. Aspects of F-region transport are discussed in Section 7.2.5. The primary loss mechanism for O^+ is via charge exchange with N_2 (e.g., Schunk, 1983). Thus, we choose to examine the impacts of thermospheric composition, especially the ratio of O to N_2, as understanding this is vital to understanding the impact of thermosphere-ionosphere coupling on ionospheric density.

Global-mean variations in both thermospheric mass density and composition have been observed on annual and semiannual time scales. On an annual time scale, the mass density is seen to vary by around 10% at 400 km altitude (e.g., Emmert and Picone, 2010), and the semiannual oscillation is observed to be larger at around 15% at 400 km altitude (e.g., Emmert, 2015). Along with the change in mass density seen, a change in the O/N_2 ratio is seen.

Early work by Jacchia (1965) suggested that changes in thermospheric temperature could account for this change in mass, but such changes were noted to be insufficient in magnitude and the magnitude of the semiannual oscillation has been observed to be anticorrelated with solar EUV (e.g., Yue et al., 2019). An alternative mechanism proposed by Fuller-Rowell (1998) suggested that the semiannual oscillation is produced by an enhancement in the mixing of thermospheric constituents at solstice, relative to equinox, associated with an increase in vertical and meridional circulation. The enhancement in this circulation is may originate in the larger interhemispheric temperature gradients present at solstice. The associated upwelling, in the presence of changing composition with altitude, leads to an increase in N_2 at higher altitudes and a reduction in the O/N_2 ratio (e.g., Rishbeth et al., 2000a). An alternative mechanism was proposed by Qian et al. (2009), who argued that seasonal changes in the eddy diffusion coefficient, which is in turn related to mixing in the thermosphere, could explain the semiannual oscillation in the O/N_2 ratio. Further, model simulations have shown that while much of the semiannual oscillation is related to some form of thermospheric mixing, as much as 30% may come from other processes such as the tilt of the Earth's rotation axis (e.g., Jones Jr. et al., 2018).

Along with the thermospheric variations, ionospheric density variations have been seen to display annual and semiannual cycles (e.g., Burkard, 1965). Rishbeth et al.

(2000b) argued that the semiannual variation in the O/N_2 ratio at low-to-mid latitudes produced by interhemispheric transport create a semiannual variation in the density of the F-layer peak (which is predominantly O^+). Wu et al. (2017) presented model simulations that showed the seasonally varying eddy mixing proposed by Qian et al. (2009) could reproduce the semiannual ionospheric variation observed by COSMIC.

Atmospheric waves that propagate up from the lower and middle atmosphere can modify thermospheric circulation as they dissipate, leading to increased eddy mixing and changing the thermospheric O/N_2 ratio (e.g., Jones Jr. et al., 2014). Recent observations from the GOLD mission of opportunity have provided a synoptic picture of thermospheric O/N_2 on a global scale. Using these data, Oberheide et al. (2020) have reported a global-scale response in thermospheric composition to a sudden stratospheric warming (SSW; Fig. 7.27), believed to be the result of an increase in atmospheric wave activity associated with the SSW, which may play an important role in determining how SSWs can impact the ionosphere.

7.2.7 Thermosphere-ionosphere coupling: Day-to-day variability

Day-to-day variabilities, or short-term variabilities, generally refer to variabilities with time scales above 1 day but below 1 month, to distinguish from longer timescale variabilities, such as intraseasonal, seasonal, annual, interannual, solar cycle, or longer. Day-to-day variability is an intricate but seriously underexplored aspect of the ionosphere and thermosphere (IT) system. Its ubiquitous but stochastic presence makes it difficult to track down and hard to predict, presenting a serious challenge for both theoretical understanding and space weather forecasting. Liu et al. (2020) gives a detailed review about our current knowledge on day-to-day variability and many open questions. Here we only summarize the main points to avoid repetition.

The sources of day-to-day IT variabilities can be categorized into three types: solar radiation, solar wind and geomagnetic storms, and meteorological forcing. The variabilities they produce can be impulsive with large magnitudes (e.g., during solar flares, eclipse, and severe geomagnetic storms), or periodic with periods of a few days (e.g., driven by planetary waves), or random (e.g., driven by tides, gravity waves) but still with significant magnitude to affect the global positioning system (GPS) and communications. Their relative contribution to the total observed variability depends not only on local time, latitude, height, season, and solar cycle, but also on which physical parameter we examine.

Fig. 7.28 presents an example of the day-to-day variation of the equatorial electrojet observed at Tatuoca in Brazil and Ponape in the Federated States of Micronesia in December 2009. This period is characterized with extremely quiet geomagnetic activity, with the maximum Kp value of 3 occurred only once on December 14. However, the EEJ

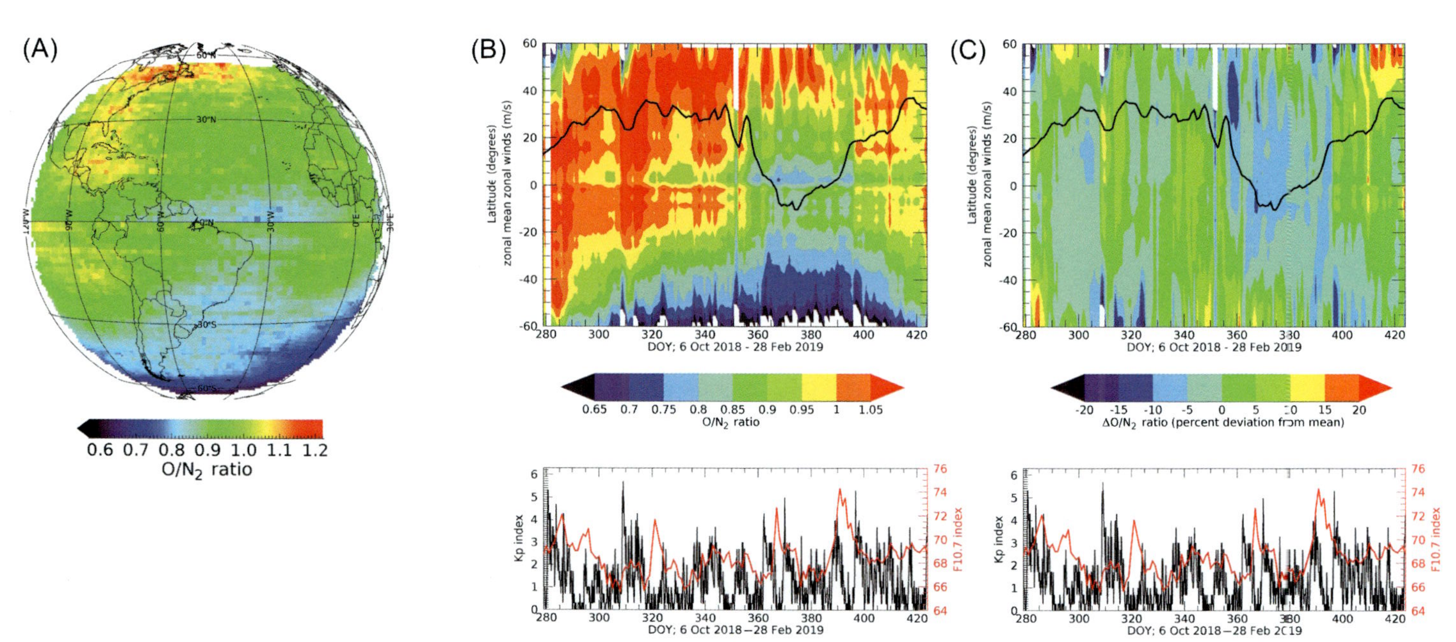

Fig. 7.27 (A) GOLD O/N_2 ratios for December 15, 2018 (Day 349, geomagnetic quiet, solar zenith angle <80 degrees, 113°W to 19°E), averaged over all 30-min cadence images taken during daytime. (B) Time series of O/N_2 daytime means averaged over all observed longitudes as a function of geographic latitude (*color contours*), along with 10-hPa mean zonal winds in the stratosphere at 60°N from MERRA-2 (*black line*). A SSW occurred around Day 370 with onset on Day 355. The *lower panel* shows the 3-h K_p index (*black*) and the F10.7 cm radio flux. (C) Same as (B) but with mean, annual, and semiannual components removed. Plotted as percent deviation. *(After Oberheide, J., Pedatella, N.M., Gan, Q., Kumari, K., Burns, A.G., Eastes, R.W., 2020. Thermospheric composition O/N response to an altered meridional mean circulation during sudden stratospheric warmings observed by GOLD. Geophys. Res. Lett.47 (1), e2019GL086313. 10.1029/2019GL086313.)*

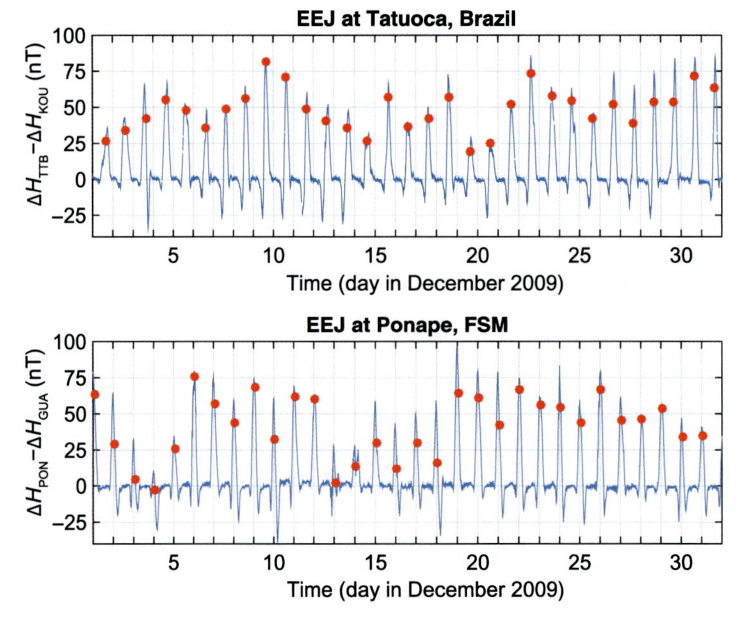

Fig. 7.28 Horizontal magnetic field perturbations associated with the EEJ at Tatuoca (1°S, 49°W) in Brazil and at Ponape (7°N, 158°E) in the Federated States of Micronesia during the quiet month of December 2009. The mid-day values are highlighted by the *red dots*. *(After Liu, H., Lei, J., Yamazaki, Y., 2020. Day-to-day variability of the thermosphere and ionosphere. In: Wang, W., Zhang, Y. (Eds.) Geophysical Monograph Series: Upper Atmosphere Dynamics and Energetics. American Geophysical Union.)*

still undergoes considerable day-to-day variations, indicating meteorological forcing from the lower atmosphere. On the other hand, high-speed solar wind streams (HSSs) can also cause day-to-day variation in the ionosphere and thermosphere with a period which is normally subharmonics of the solar rotation (27 days) (see Liu et al., 2020, for detailed review).

Our knowledge on the day-to-day variability is highly limited, and a systematic picture of such variability across an array of major IT parameters (e.g., plasma density and drift, thermosphere density, wind, and temperature) is yet to be achieved. For that, it is necessary to establish an "aeronometrics" to guide systematic quantification of the day-to-day IT variabilities, including its spatial and temporal dependence across a set of key IT parameters. It is important to distinguish the relative contributions to IT short-term variability from different drivers.

7.2.8 Atmosphere-ionosphere coupling on other planets

Clear signatures of atmosphere-ionosphere coupling are seen at other planets. Here we choose Mars as an example to contrast and compare with Earth, because despite being smaller, less massive, and having a thinner atmosphere than Earth, its thermosphere has a number of similarities to Earth's. Mars' thermosphere extends from approximately 100 to 200 km altitude, overlaps with a relatively tenuous ionosphere made up of both atomic and molecular ions and the planet rotates at almost the same angular rate as the Earth, thus is capable of supporting a similar set of tides and Rossby waves as Earth. A central difference between Earth and Mars in terms of atmosphere-ionosphere interactions is Mars' lack of a global magnetic field. Instead, Mars has more localized magnetic field features that vary strongly across its surface (Acuña et al., 1999). Estimate of the wind-driven dynamo effects at Mars thus varies strongly in relation to the surface magnetic field features (e.g., Riousset et al., 2014), but are nonetheless believed to be important (e.g., Lillis et al., 2019).

The impacts of atmosphere waves on the ionosphere are seen at both global and small scales at Mars. At a global scale, longitudinal perturbations of electron density have been interpreted as signature of atmospheric tides (e.g., Bougher et al., 2004). On smaller scales, modification of ion density and velocity have been seen, which are believed to be associated with atmospheric gravity waves (e.g., England et al., 2017; Collinson et al., 2019).

In regions where the localized magnetic fields are weak, the coupling between the atmosphere and the ionosphere is the result of advection by neutral winds—predominantly vertical motion and compression and expansion. Vertical advection, believed to be associated with long-wavelength gravity waves, is clearly seen to affect ion species differently, depending on their own scale heights (England et al., 2017). In the presence of a background magnetic field, evidence of currents resulting from dynamo driven winds are seen (Collinson et al., 2019).

7.2.9 Concluding remarks

We have briefly reviewed the thermosphere-ionosphere coupling processes that are particularly related to ion drag, wind-driven dynamo, composition, and field-aligned motion driven by neutral winds. Though many studies have been done on these topics, there remain many open questions. Among these, especially the day-to-day variability remains to be explored. What are the relative contributions to the day-to-day variability from different drivers? What is the day-to-day variability in upward propagating atmospheric waves? How to systematically quantify day-to-day variability (e.g., magnitude and its dependence on latitudes, longitudes, altitudes, season, solar cycle)? How common or different are these IT processes we observe on Earth in comparison to other planets like Venus and Mars? With the concurrent datasets from SWARM, COSMIC-2, ICON, GOLD, and TIMED that are currently available, there is good reason to believe many of these questions will be addressed and new ones be revealed.

7.3 Large-scale traveling ionospheric disturbances

Shun-Rong Zhang
MIT Haystack Observatory, Westford, MA, United States

7.3.1 Introduction

Ionospheric disturbances are a short-term deviation from regular climatology (such as diurnal variations). These disturbances can be local, regional, and sometimes global. A large category of these disturbances has characteristic wave properties in propagation and periodicity and is specifically termed as traveling ionospheric disturbances (TIDs). TIDs were first reported in the earlier days of the ionospheric sounding (e.g., Munro, 1948). A major milestone was achieved in 1950–60 when many TIDs were ascribed to atmospheric waves (Hines, 1959), the acoustic gravity waves (AGWs). For a long time in history, TIDs have been considered as AGW manifestation, and thus significant attention is paid to AGW excitation and propagation in the thermosphere to explain TID observations. In the literature, AGWs sometimes stand also for Atmospheric Gravity Waves, without referring to acoustic waves, the higher frequency branch of the atmospheric waves. The neutral atmospheric counterpart of TID is TAD. However, TAD as a collective terminology often designates disturbances associated with geomagnetic storms or substorms, and gravity waves (GWs) are a specific type of TADs where the gravitational force and the buoyancy force act to restore the hydrostatic equilibrium of the atmosphere.

Based on observations, atmospheric gravity waves manifested as TIDs are classified according to wave length, period, and propagation speed (Hunsucker, 1982). The large-scale GWs and large-scale TIDs (LSTIDs) have very large wavelengths at 1000 km or larger, fast horizontal propagation speed between 400 and 1000 m s^{-1}, long periods between 30 min and 3 h. These are predominately geomagnetic storm-related and propagate globally away from the auroral source region. Medium-scale GWs and medium-scale TIDs (MSTIDs) as well as small-scale GWs and small-scale TIDs (SSTIDs) have small wavelengths of several hundred kilometers, short periods from tens of minutes to 1 h (MSTIDs) and minutes (SSTIDs), and typically slower horizontal speeds (than that of large-scale waves). Earlier results suggested the MSTID connection to GWs that are excited by high-latitude processes or by lower atmospheric disturbances. Advances in recent decades indicate, however, the mid- and low-latitude MSTIDs, being characteristic for the night occurrence and specific propagation direction, are connected to electromagnetic processes of the ionospheric plasma (Perkins, 1973; Kelley and Fukao, 1991) and have nothing to do directly with atmospheric waves manifestation. Section 7.4 provides detailed discussion on MSTIDs.

Classification of TIDs based on the wave scale size and periodicity is meaningful because a specific type of TIDs likely corresponds to a given AGW excitation mechanism. Some of the known AGW mechanisms include (1) solar-terrestrial energy, momentum, and material depositions to the upper atmosphere at high latitudes. Auroral heating processes, for instance, yield primarily LSTIDs that can propagate equatorward to lower latitudes, dispersing the solar disturbance energy from the auroral zone globally. Some disturbances related to the geospace storm or substorm can also excite medium-scale TIDs that have regional impacts. (2) Immense energy releases, often impulsively, from natural processes below the upper atmosphere can excite upward propagating AGWs. They land in the upper atmosphere either directly, or indirectly through secondary waves (Vadas and Fritts, 2002); the origin of these disturbance processes ranges from the troposphere (deep convective events, tornadoes, and hurricanes or typhoons; Azeem et al., 2015; Nishioka et al., 2013) to the Earth's surface (earthquakes, tsunamis, etc.; see e.g., Liu et al., 2000; Komjathy et al., 2016). (3) Other transient solar-terrestrial phenomena (e.g., eclipses and perhaps solar flares; see Zhang et al., 2017a, 2019a), man-made space disturbances (e.g., rocket launches and nuclear expositions; see Lin et al., 2017). These latter two types typically excite MSTIDs. Numerous studies have devoted to understanding TIDs using observation and theoretical approaches. We provide at the end of this chapter only a small subset of papers from these researches.

By definition, identification of TIDs requires three-dimensional (3D) spatial information of the ionospheric disturbances. In recent years, ionospheric and thermospheric remote sensing in a 2D horizontal plane domain has made significant progress due to the use of dense receiver networks for Global Navigation Satellite Systems (GNSS) for the total electron content (TEC) measurement (Saito et al., 1998), All-Sky Imagers for detecting the atmospheric emission, as well as other observational systems. They can resolve TID/TAD features over a large spatial extent and therefore making substantial new insights. This section will describe our fundamental understanding of LSTIDs and the associated atmospheric disturbances, with considerable emphasis on new results for the TID characterization. Comprehensive reviews by Hunsucker (1982), Yeh and Liu (1974), Francis (1975), and Hocke and Schlegel (1996) provide great details and valuable resources for earlier work and should be consulted for interested readers.

7.3.2 Gravity waves in the thermosphere

GWs in the thermosphere as the main cause of TID propagation obey the fundamental conservation laws that form the base of the hydrodynamic equations.

7.3.2.1 Internal wave dispersion and polarization relations

Assuming the atmospheric perturbations can be linearized, the disturbance parameters neutral density, pressure, temperature, and motion can be solved for under an isothermal background atmosphere. The resulting dispersion relation for GWs as plane waves (Hines, 1960; Yeh and Liu, 1974) can be expressed as

$$k_h^2(1-\omega_b^2/\omega^2)+k_z^2=k_0^2(1-\omega_a^2/\omega^2) \tag{7.20}$$

where ω is the angular frequency; $\omega_a = c_0/2H$ is the acoustic cut-off frequency; $c_0=\sqrt{\gamma Hg}$ is the speed of sound; γ (~1.4) is the ratio of specific heats; H is the scale height; g is the gravitational acceleration; $\omega_b=\sqrt{(\gamma-1)}g/c_0$ is the Brunt-Väisälä frequency (or GW cut-off frequency). k_h and k_z are horizontal and vertical wavenumbers so that $k^2=k_h^2+k_z^2$ with $k_h^2=k_x^2+k_y^2$ and $k_0=\omega/c_0$. This dispersion equation represents two branches of internal wave propagation for which ω and k are real, that is, acoustic waves and gravity waves (Fig. 7.29). Acoustic waves, as a compression wave, have frequencies higher than the acoustic cut-off frequency ω_b ($\omega > \omega_a$). Their phase speed $v_{ph}=\omega/k$ is always higher than the sound speed. GWs have frequencies lower than the Brunt-Väisälä frequency ω_b and the gravitation is the major restoring forcing of the GWs. The GW wavenumber k is always larger than the wavenumber of the sound wave k_0, thus the GW phase speed is always less than the speed of sound. Low-frequency GWs exist for a larger range of horizontal wavelengths whereas high-frequency GWs exist mostly for shorter wavelengths (larger horizontal wavenumber, $\lambda = 2\pi/k$). A special case with $k_h=k_0$ represented by the 45 degrees line in Fig. 7.29 is the Lamb waves which propagate at the phase speed of the sound and experience no vertical disturbance speed.

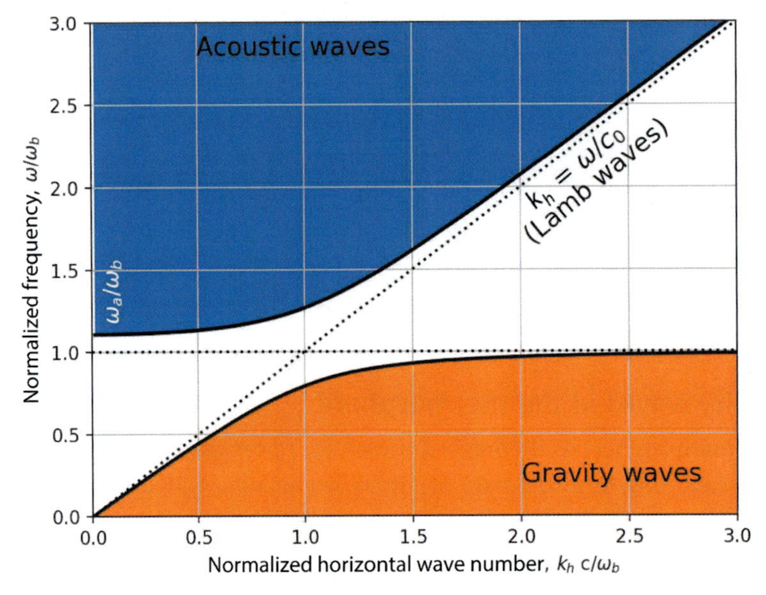

Fig. 7.29 Regions of propagation for the gravity branch and the acoustic branch in an isothermal atmosphere. The bounding curves are given by $k_z=0$. The regions that correspond to finite real k_z values are shown by the *shaded area. (This figure is based on Fig. 3 in Yeh, K.C., Liu, C.H., 1974. Acoustic-gravity waves in the upper atmosphere. Rev. Geophys.12 (2), 193–216.)*

More realistic dissipation properties can be obtained by using a nonhydrostatic and compressible theoretical scheme with primary damping mechanisms for high-frequency GWs with large vertical wavelengths, kinematic viscosity, and thermal diffusivity (Vadas, 2007). Fig. 7.30 shows sample dissipation relations of GWs launched from ground ($z_i = 0$) and at 120 km ($z_i = 120$ km) in an atmosphere whose exospheric temperature is ~1000 K among horizontal and vertical wavelengths λ_h and λ_z, horizontal and vertical dissipation distances x_{diss} and z_{diss}, maximum vertical wavelength $\lambda_z(z_{\mathrm{max}})$, intrinsic horizontal phase speeds c_{IH}, dissipation time τ_{diss}. To consider the background neutral wind induced Doppler shift, an intrinsic frequency is defined as

$$\omega_{Ir} = \omega_r - k_x U - k_y V \tag{7.21}$$

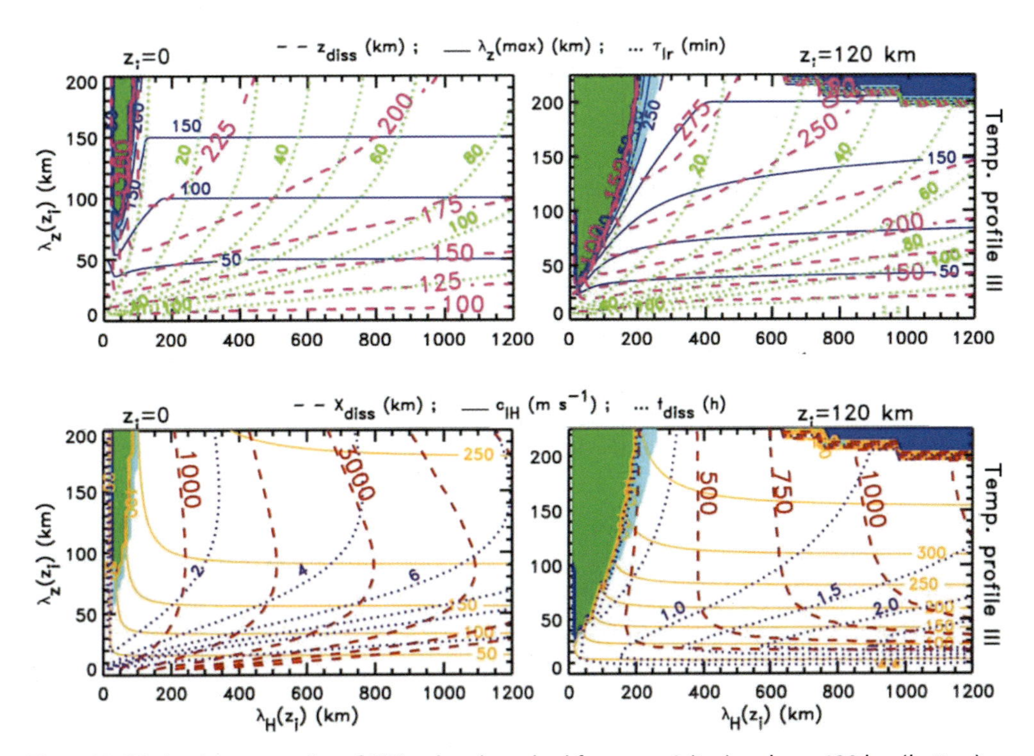

Fig. 7.30 Dissipation properties of GWs when launched from $z_i = 0$ (*top*) and $z_i = 120$ km (*bottom*) as functions of wavelengths λ_h (marked as in λ_H the figure) in the horizontal direction and $\lambda_z(zi)$ in the vertical direction. *Pink-dash lines* show GW dissipation altitudes, z_{diss}, in 25-km intervals. *Blue solid lines* show maximum vertical wavelengths prior to dissipation, λ_z(zmax), in 50-km intervals. *Green dot lines* show the GW intrinsic periods, τ_r, at 10, 20, 30, 40, 50, 60, 80, 100, and 180 min. *(From Vadas, S.L., 2007. Horizontal and vertical propagation and dissipation of gravity waves in the thermosphere from lower atmospheric and thermospheric sources. J. Geophys. Res. Space Phys. (1978–2012)112 (A6).)*

where U and V are horizontal background winds measured on the ground in x- and y-- directions, respectively, and $\omega_r = 2\pi/\tau_r$ is the angular frequency corresponding to the wave period τ_r measured on the ground.

The GW dissipation relation indicates that medium- and small-scale waves launched from ground ($z_i = 0$) with $\lambda_z > 50$ km and λ_h 100–400 km dissipate at $z_{\text{diss}} \sim$ 150−250 km vertical distance; these waves have intrinsic period $\tau_{\text{Ir}} < 60$ min period and propagate at $c_{\text{IH}} \geq 100$ m s^{-1} for less than ~2000 km horizontally. This suggests that if GWs from a tropospheric source propagate into the thermosphere as primary GWs, the corresponding MSTIDs are unlikely observed more than ~2000 km away from this source. However, it should be noted that these upward propagating GWs can be broken in the lower thermosphere and excite secondary or tertiary GWs which arrive in the thermosphere (e.g., $z_i = 120$ km) and exhibit very different wave characteristics from primary waves (e.g., Vadas and Fritts, 2002).

GWs launched from $z_i = 120$ km with $\lambda_z > 100$ km and λ_h 150–600 km dissipate at the highest altitudes. These GWs with large λ_h and λ_z travel the largest horizontal distances; GWs with larger intrinsic period τ_{Ir} tend to propagate larger horizontal distances X_{diss}. It is worthwhile noting that GW global propagation tends to follow the Earth's curvature, largely due to the restoring force of GWs being the gravitational force, is pointed toward the Earth's center. However, acoustic waves with frequency $\omega > \omega_a$ tends to propagate out into space because of the lack of the Earth-center pointed restoring force.

The group velocity of the atmospheric waves $\mathbf{v_g}$ can be expressed as

$$\mathbf{v_g} = \left[\mathbf{x}k_x(\omega^2 - \omega_b^2) + \mathbf{y}k_y(\omega^2 - \omega_b^2) + \mathbf{z}k_z\omega^2\right]\omega c_0^2/(\omega^4 - \omega_b^2 k_x^2 c_0^2) \tag{7.22}$$

where $\mathbf{x}$, $\mathbf{y}$, and $\mathbf{z}$ are directional unit vectors, and k_x, k_y, and k_z are wavenumbers in these directions. For acoustic waves, $\mathbf{v}_g$ in all directions have the same sign as corresponding k_x, k_y, and k_z. For gravity waves, however, $\mathbf{v}_g$ in the horizontal direction have the same sign as corresponding k_x and k_y, but the $\mathbf{v_g}$ vertical component has an opposite sign to k_z, that is, the GW vertical propagation in phase is opposite to that in the air parcel energy. The following equations demonstrate the polarization relationship between perturbations in pressure p', mass density ρ', air velocity in the horizontal directions v_x', v_y' and in the vertical direction v_z'.

$$\begin{aligned} p' &= v_z'(\omega^2 - \omega_b^2)/\omega[k_z + i\omega_a c_0(\gamma - 2)/\gamma] \\ p' &= v_x'/k_x = v_y'/k_y \\ p' &= \rho' c_0^2(\omega^2 - \omega_b^2)/[\omega^2 - g(\gamma - 1)(-ik_z + H/2)] \end{aligned} \tag{7.23}$$

7.3.2.2 GW propagation modes

Finally, we can summarize the propagation modes of GW landing primarily in the thermosphere (Mayr et al., 1990). Mode (1): when the disturbance is initiated in the lower thermosphere (~100–150 km), the direct wave from the lower thermosphere dominates near the source; the direct wave from further below (the lower atmosphere, <100 km) or the secondary wave the direct wave produces in the lower thermosphere can also reach the thermosphere; Mode (2): the direct wave can be important in the thermosphere at large distances away from its lower thermospheric source; Mode (3): the wave reflected from Earth's surface is important near the source; Mode (4): the ducted wave can also dominate at large distances away from its original source. The wave modes (1) and (2) are primarily affected by the properties of the thermosphere or upper atmosphere and are called the "upper modes." The wave modes (3) and (4) are primarily affected by the properties of the lower atmosphere and are called the "lower modes."

7.3.3 Plasma responses to neutral perturbations

Ionospheric density variation is governed by the continuity equations. For ions,

$$\partial N_i/\partial t = Q_i - L_i - \nabla \cdot (N_i \mathbf{u_i})$$

where the temporal change rate in ion number density N_i at time t is caused by the ion production rate Q_i, chemical loss rate L_i, and gradients of dynamic ion flux associated with ion velocity $\mathbf{u}_i$ and N_i. AGWs drive ionospheric variations through their influences on these three terms.

7.3.3.1 Influences on ionospheric dynamics

Ion velocity $\mathbf{u}_i$ is determined by the momentum equation,

$$N_i m_i (\mathbf{u_i} \cdot \nabla)\mathbf{u_i} = N_i e_i (\mathbf{E} + \mathbf{u_i} \times \mathbf{B}) + N_i m_i \nu_{in} (\mathbf{u} - \mathbf{u_i}) - \nabla (N_i k_B Ti) + N_i m_i \mathbf{g}$$

which is derived under several simplification assumptions, with quantities ion charge e_i, electric field $\mathbf{E}$, magnetic field $\mathbf{B}$, ion mass m_i, neutral winds $\mathbf{u}$, the ion-neutral momentum transfer frequency ν_{in}, gravitational acceleration $\mathbf{g}$, and ion pressure gradient involving ion temperature T_i and Boltzmann's constant k_B. The linearization for the ion continuity equation, to the first-order approximation, leads to,

$$\partial N_i'/\partial t = Q_i' - L_i' - \nabla \cdot (N_i^0 \mathbf{u_i'}) - \nabla \cdot (N_i' \mathbf{u_{i0}}) \tag{7.24}$$

with disturbance quantities marked by superscript prime "$'$" and nondisturbance quantities marked by superscript "0". The ion motion $\mathbf{u}_i = \mathbf{u}_i^0 + \mathbf{u}_i{}'$ is induced by a combined effect of neutral winds, electric fields, and ambipolar diffusion. The role of GWs for the ionospheric dynamics is to (1) modify neutral winds directly, (2) induce electric fields by neutral wind disturbance and plasma density inhomogeneity, and (3) impact plasma ambipolar diffusion by fluctuations in plasma and neutral densities and their temperatures. In particular, ambipolar diffusion in the vertical direction $\mathbf{u}_{\mathbf{id}}$ can be expressed as

$$\mathbf{u_{id}} = -\frac{k_B T \sin I}{m_i \nu_i n}\left[\frac{1}{N}\frac{\partial N_i}{\partial z} + \frac{1}{T_i}\frac{\partial T_i}{\partial z} + \frac{1}{T_e}\frac{\partial T_e}{\partial z} + \frac{m_i g \sin I}{k_B T}\right]$$

where T_e and T_i are electron and ion temperatures, and I is the magnetic dip angle. The GW influences on diffusion was believed primarily to introduce a phase shift from that due to dynamic effect alone (Clark et al., 1971). The diffusion term can be ignored only for those waves with periods less than about 5 min. The GW-induced change in plasma ambipolar diffusion is estimated as a few percentage (Hooke, 1968).

GW effects through neutral wind disturbances $\mathbf{u}'$, $\nabla \cdot (N_i^0 \mathbf{u_i'})$ in Eq. (7.24), has two origins which can be expressed as $N_i \nabla^{\|} \cdot \mathbf{u}'^{\|} + \mathbf{u}'^{\|} \cdot \nabla^{\|} N_i$ (here the neutral wind component perpendicular to the magnetic field line is assumed not to contribute to the plasma motion in the F region.) The first term $N_i \nabla^{\|} \cdot \mathbf{u}'^{\|}$ is the divergence term related to the vertical gradient of field-aligned motion induced by the horizontal winds. Thus the vertical phase variation of the horizontal wind disturbance projected along field lines should be normally nonzero except for that at the magnetic pole with $I = 90°$ or for zonal disturbances that transverse the magnetic field line. Clearly, the variation of the neutral gas horizontal motion along the field lines $\nabla^{\|} \cdot \mathbf{u}'^{\|}$ results in actual compression or rarefaction of the ionization. Other than the meridional component, the neutral wind **vertical** component is also height dependent and correlated to neutral density and pressure disturbances as shown in the GW polarization (Eq. 7.23). This component makes a direct contribution to this divergence term, which could be important at high latitudes with $I \sim 90$ degrees.

The second term $\mathbf{u}'^{\|} \cdot \nabla^{\|} N_i$ is convection term and will be generally nonzero except (1) at the magnetic pole where the horizontal wind disturbance is perpendicular to $\mathbf{B}$, (2) for zonal disturbances with $\mathbf{u}'^{\|} \sim 0$, (3) near the F2 peak with $(\nabla N_i^0 \cdot \mathbf{b} \sim 0)$, or (4) at the magnetic equator where ∇N_i^0 is pointed nearly vertically.

Apparently, the neutral wind disturbance component parallel to the magnetic field lines will generally have a direct contribution to the ion density disturbances (except for at the F2 peak). This is the case for mid-latitudes. At magnetic poles or for zonal disturbances, no ionospheric disturbances are directly induced through field-aligned plasma motion. Under these scenarios, other processes become important, including potentially more efficient diffusion (if $I = 0$), vertical wind disturbance (if $I = 0$), GW-induced electric field, and photochemical effects.

7.3.3.2 Influences on ionospheric photochemistry

The photoionization rate for a given height is proportional to the local number density of neutral gases and solar irradiation arriving at this altitude, which results from attenuation due to photoabsorption along the entire ray path of the sunlight. Thus a neutral density increase enhances not only the number density of parent neutrals to be ionized but also the attenuation and photoabsorption along the ray path. On the one hand, it is the local neutral gas number density fluctuations that predominate in determining changes in the production rate at altitudes several scale heights above the peak production height. On the other hand, at or below this altitude, it may be the local fluctuations in the value of the arriving irradiation flux which predominate in determining changes in the production rate. Thus the ionospheric production will strongly depend on altitude. In general, the presence of acoustic-gravity waves will have a large impact on the production in the lower F1 region (Hooke, 1968).

GW influences on the ion loss rate are through modifying the neutral density as well as plasma and neutral temperatures, thus affecting the charge exchange and dissociation reaction rates.

Given all these GW effects, ionospheric variations as TIDs can appear very complicated. Numerical simulations of ionospheric perturbations due to the passage of large-scale GWs can be conducted to demonstrate the characteristics of resulting TIDs (Kirchengast et al., 1996). Fig. 7.31 shows such a simulation using a first-principle ionosphere model (Zhang and Huang, 1995) that solves the ion continuity and momentum

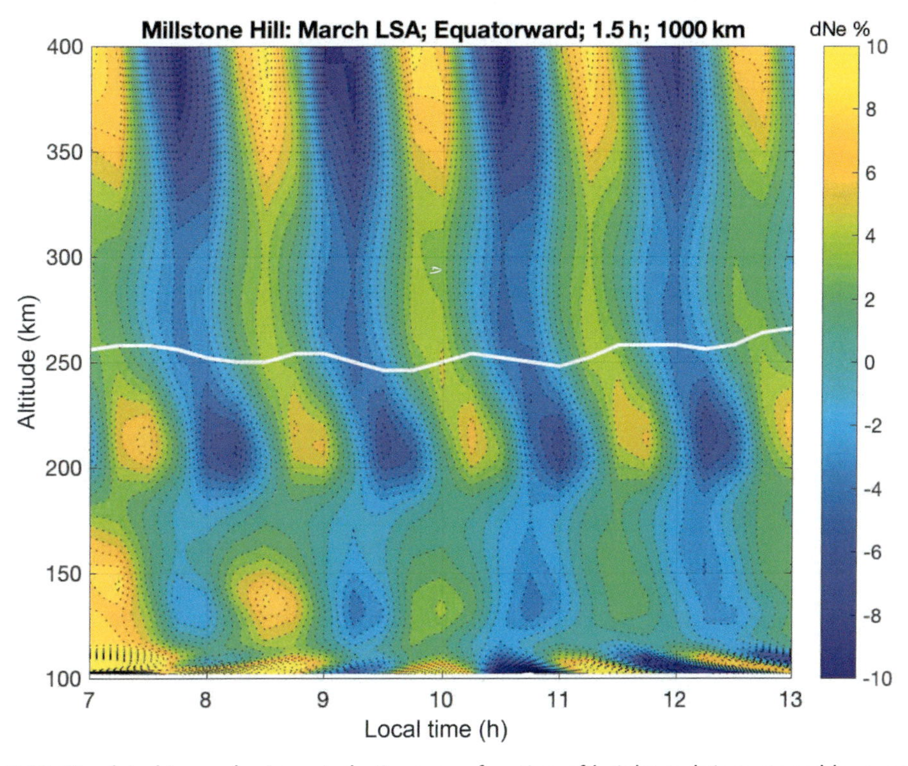

Fig. 7.31 Simulated ionospheric perturbations as a function of height and time caused by a model gravity wave at Millstone Hill. The gravity wave propagates equatorward with 1.5-h period and 1000 km horizontal wavelength in March at low solar activity. The model is specified by using the polarization relationship in Eq. (7.23).

equations for ion densities within 100–500 km altitudes. The classic GW theory with the polarization Eq. (7.23) is used to specify the wave proprieties. This model GW has a 1.5-h period and a 1000-km horizontal wavelength and propagate equatorward into Millstone Hill (42.6°N, 288.5°E, $I = 69$) in summer at low solar activity. The GWs cause $\pm 10\%$ amplitude TIDs predominantly in the F-region above 200 km. TIDs show downward phase progression. The TIDs appear to have a $\sim$150 km vertical wavelength. The F2 peak height hmF2 fluctuates by up to 10 km. The TIDs above 300 km where plasma diffusion is increasingly important exhibit somewhat different propagation phases than near the F2 peak region. Below 200 km, TIDs show density fluctuations primarily due to neutral density and temperature changes. However, they have nearly opposite vertical propagation phases to those above 200 km. This simulation demonstrates GW induced effects on the ionospheric photochemistry and dynamics.

7.3.4 LSTID excitation in the auroral zone

Numerous natural and some man-made sources are available to create AGWs in the atmosphere and TIDs in the ionosphere. In this section, we focus on LSTIDs induced by geospace disturbances in the auroral and high latitudes.

Enhanced coupling processes within the magnetosphere-ionosphere-thermosphere system during geomagnetic storms and substorms result in substantial material, momentum, and energy depositions in the auroral zone. The most plausible auroral sources for AGW excitation include (1) perturbations in the auroral electrojet, (2) atmospheric response to auroral particle precipitation, and (3) dynamics associated with fast-moving plasma structures and large-scale phenomena (Hunsucker, 1982). Overall, there is a fairly well-defined one-to-one correspondence between LSTIDs and identifiable auroral disturbances, either being related to a substorm onset or to auroral streamers without a substorm that often occurs during interplanetary magnetic field (IMF) southward periods of CME storm main phases (Lyons et al., 2019). Fig. 7.32 presents a modern example with simultaneously colocated developments in the geomagnetic field disturbances, airglow emission brightness, and TID onset. On the dayside during the storm main phase, however, LSTIDs are also frequently observed that propagate into lower latitudes (Ding et al., 2008), and it appears reasonable to link them to dayside auroral events. The dayside-to-nightside trans-polar propagation of TIDs has been also reported as a result of dayside auroral/cusp disturbances during IMF southward in geomagnetic disturbances (Zhang et al., 2019c; Nishimura et al., 2020, also in Fig. 7.33).

Auroral electrojet affects the neutral atmospheric gas and launches AGW through a momentum force associated with the magnetic field **B**, the Lorentz force $\mathbf{J} \times \mathbf{B}$, and frictional heating associated with electric field **E**, the Joule heating $\mathbf{J} \cdot \mathbf{E}$. Many earlier studies showed their relative importance in exciting AGWs (Chimonas and Hines,

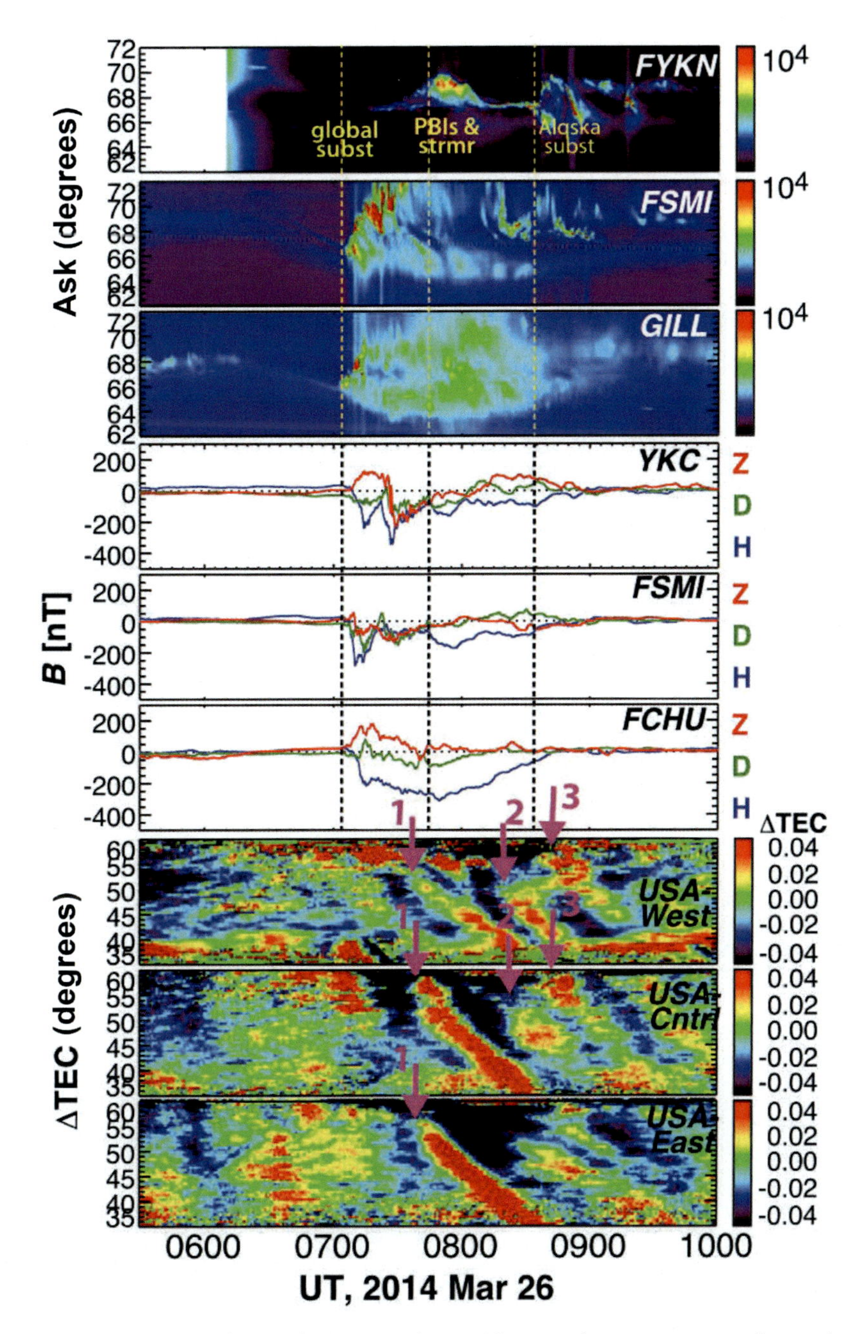

Fig. 7.32 All-sky imaging and ΔTEC keograms, along with ground magnetometer observations, for a nonstorm night of March 26, 2014. The onsets of auroral oval disturbances are shown by *vertical dashed lines* and labeled as to whether they are a substorm (subst), streamer (strmr), or poleward boundary intensification (PBI). A disturbance seen over a wide range of Canadian longitudes is labeled as "Global," and an event seen over Alaska is labeled as "Alaska." "West," "Central," and "East" refer to the Canadian longitude sectors. *Magenta arrows and numbers* identify equatorward moving large-scale traveling ionospheric disturbance. *Maroon arrows* give normals to large-scale traveling ionospheric disturbance phase front. *(From Lyons, L.R., Nishimura, Y., Zhang, S.-R., Coster, A.J., Bhatt, A., Kendall, E., Deng, Y., 2019. Identification of auroral zone activity driving large-scale traveling ionospheric disturbances. J. Geophys. Res. Space Phys.)*

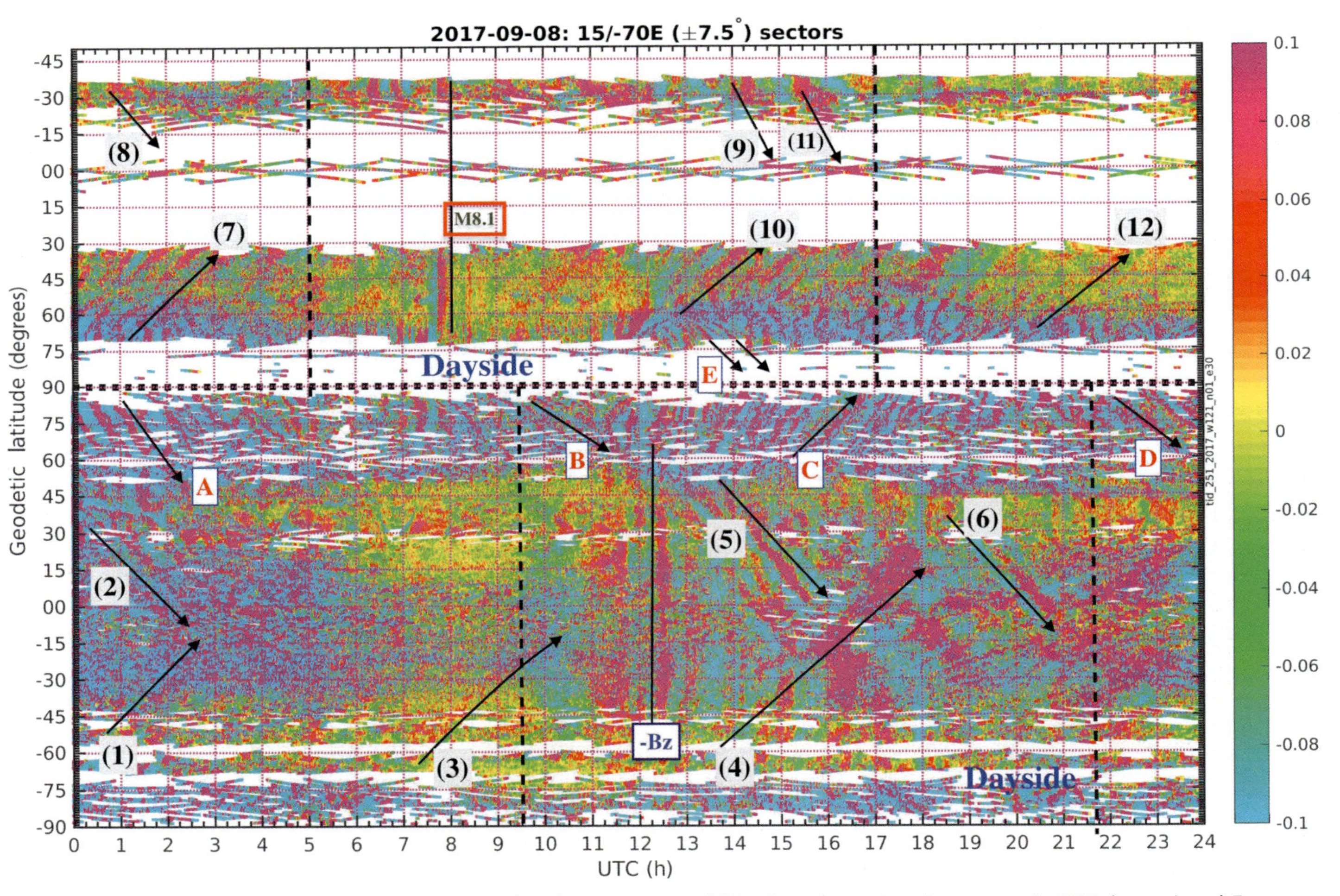

Fig. 7.33 Keograms of differential TEC showing meridional propagation of TIDs for eastern America sectors (−70°E, *bottom*) and European sectors (15°N, *top*) during September 2017 storms. TID propagation directions at mid- and low-latitudes are marked by *arrows* (1)–(4) in the America sector and (7)–(12) in the European sector. TID propagation directions in the northern polar region are marked by *arrows* (A)–(E). M8.1 solar flare is also marked, as well as the time when B_z turned to southward. *Vertical dashed lines* are terminators. *(From Zhang, S.-R., Erickson, P.J., Coster, A.J., Rideout, W., Vierinen, J., Jonah, O., Goncharenko, L.P., 2019c. Subauroral and polar traveling ionospheric disturbances during the 7–9 September 2017 storms. Space Weather17 (12), 1748–1764. 10.1029/2019SW002325.)*

1970; Testud, 1973). The Lorentz to Joule contribution L/J (Chimonas and Hines, 1970) may be estimated as $L/J \simeq gHB_z\sigma_c/(C_Lj)$ where B_z is the geomagnetic vertical component, σ_c is Cowling conductivity, j current density, and C_L is characteristic speed of propagation frequencies $= (\omega_b/\omega_c)c_0 = 350$ m s^{-1}. An accurate assessment of their relative importance appears to have faced large uncertainty, nevertheless, it was believed that the Lorentz force source function tended to generate medium-scale AGWs, while the Joule heating terms are more effective in generating large-scale AGWs (Jing and Hunsucker, 1993).

The upper atmosphere responds to auroral particle precipitation in several ways, in particular, auroral particle precipitation can significantly increase the ionospheric plasma density (by more than an order of magnitude) and therefore increase directly the ionospheric conductivity, including Cowling, Pederson, and Hall conductivities via $\sigma_c = (\sigma_P^2 + \sigma_H^2)/\sigma_P$. Thus the electric current density $\mathbf{j} = \boldsymbol{\sigma} \cdot \mathbf{E}$ increases accordingly. The effects of perturbations of the auroral electrojet via the electric field and auroral particle precipitation as source functions for AGWs are difficult to measure separately. But simulations on the relative contribution of ion convection (the auroral electric field) and particle precipitation to exciting LSTIDs show comparable contributions to the total Joule heating, although the changes of height-integrated Joule heating due to these two forcing terms may display different distributions (Sheng et al., 2020).

7.3.5 LSTID global propagation

Once excited, LSTIDs propagate away from their auroral source region, in particular, equatorward into mid- and low-latitudes, and sometimes through the equator into the conjugate hemisphere. Fig. 7.33 demonstrates these scenarios of TID global propagation using an example event of the September 2017 storm. The trans-equatorial propagation of TIDs at mid- and low-latitudes on the dayside following IMF southward turning was very significant, so was the trans-polar propagation from the dayside to nightside.

Several factors impact LSTID propagation. The first is the neutral wind filtering effect on AGWs and LSTIDs. The wave pocket of AGWs excited at auroral latitudes can propagate with an upward group velocity if the wave frequency, which is Doppler-shifted by the horizontal winds according to Eq. (7.21) or $\omega_{Ir} = \omega_r - \mathbf{k}_h \cdot \mathbf{w}$, allows that the wavenumber $\mathbf{k}_h$ is real. This can be easily satisfied if the horizontal wind $\mathbf{w}$ blows against the horizontal phase propagation $\mathbf{k}_h$. If they are in the same direction, the wind-wave critical coupling will take place as $\omega_r = \mathbf{k_h} \cdot \mathbf{w}$ ($w = \omega_r/k_h$) and GWs loss their energy fully to the background winds (Cowling et al., 1971; Yeh et al., 1972; Ding et al., 2003; Fritts and Vadas, 2008).

Other factors impacting LSTID propagation are associated with energy dissipation due to ion drag, molecular viscosity, and thermal conductivity. The viscosity caused by the vertical spatial derivative seems unlikely important for large size and low-frequency GWs in the thermosphere. Neutral wind disturbances and plasma density inhomogeneity associated with GWs can induced also electrodynamic disturbances in the electric field and current (Zhang et al., 2021). In particular, Joule forcing $\mathbf{J} \times B$, where $\mathbf{J} = \sigma_P(\mathbf{U} \times \mathbf{B})$, can damp wave energy through heating.

The AGW damping is an obvious consequence if TIDs are plasma motion fluctuation set up by the neutral fluctuations. Thus the higher the background electron density, the stronger the TAD/TID amplitude damping is Tsugawa et al. (2003). As a result, the damping is potentially more significant on the dayside than on the nightside and perhaps is weak in the nightside mid-latitude ionospheric trough and strong in the dayside EIA crest latitudes.

If the neutral fluctuations transverse the magnetic field line, since the induced plasma motion is constrained to be field-aligned only but not perpendicular, these neutral oscillations experience substantial damping effect by the ion drag. Consequently, the LSTIDs amplitude will be damped accordingly. This effect is sensitive to the angle between the wave propagation **k** and the magnetic field **B**. During the LSTID global propagation around Earth, this angle varies and so the projection of neutral wind disturbance to the field lines varies. Therefore, the LSTID amplitudes may appear larger as their propagation becomes more field-aligned (e.g., at low and equatorial latitudes, see Tsugawa et al., 2004). It may be inferred that near the magnetic equator where the field lines are horizontal, the meridional propagation with a smaller angle to the field lines will be less damped.

LSTIDs propagating through mid-latitudes are typically elongated in the zonal direction as shown in Fig. 7.34. This GNSS TEC-based observation near the dusk sector was made when IMF B_z had been at ~ -17 nT for an hour due to the arrival of a CME. These zonal wavefronts of TIDs extended across the continental United States between the west and east coasts, with a meridional wavelength up to 500 km. Further analysis shows these LSTIDs propagate equatorward at a 400–500 m s^{-1} phase speed.

Observations also show the downward phase progression of LSTID electron density variations. The case in Fig. 7.35 shows TIDs/TADs observations by the Millstone Hill Incoherent Scatter Radar (ISR) and by the Arecibo ISR and an on-site Fabry-Perot Interferometer (FPI) during an extended IMF B_z period of the March 17, 2015 great geomagnetic storm (Zhang et al., 2017a, b). These TIDs at Millstone Hill have $\sim$2.5 h period, $\sim$20% amplitude and vertical wavelength of $\sim$200 km. When the disturbances arrived at Arecibo, which is $\sim$2500 km in the south of Millstone Hill, the TADs had a nearly 4-h period as indicated in the thermospheric meridional wind disturbances. These TADs caused TIDs with a similar periodicity and clear phase progression in the electron density disturbances as shown in Fig. 7.35. It is reasonable to attribute the TADs to GWs. The wave periodicity difference between Millstone Hill and Arecibo exhibits an expected behavior of TID/AGW filtering during their long-distance propagation where longer period waves tend to survive (Richmond, 1978).

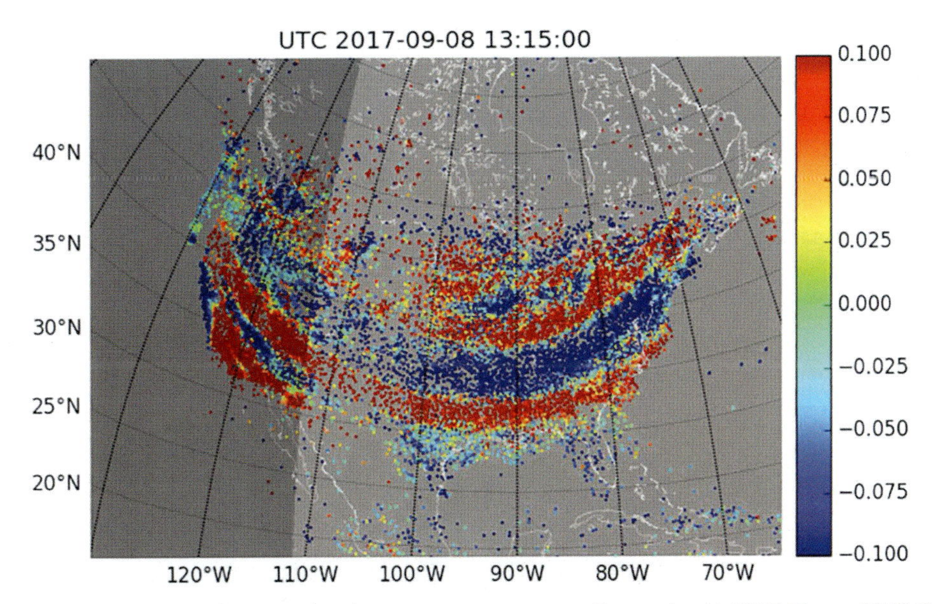

Fig. 7.34 An example of LSTIDs during a geospace storm on September 7, 2017. These GNSS TEC-based observations at 13:15 UT near the dusk sector were made when IMF B_z had been at ~17 nT for an hourly. *(From Zhang, S.-R., Coster, A.J., Erickson, P.J., Goncharenko, L.P., Rideout, W., Vierinen, J., 2019a. Traveling ionospheric disturbances and ionospheric perturbations associated with solar flares in September 2017. J. Geophys. Res. Space Phys.60 (8), 895.)*

We note that during their equatorward propagation, LSTIDs will encounter the dynamical subauroral region, where dramatic ion-neutral coupling processes associated with Subauroral Polarization Stream (SAPS) (Foster and Burke, 2002) can alter conditions for TAD/AGW propagation. These conditions include the enhanced westward thermospheric wind and the elevated neutral temperature (e.g., Wang et al., 2011; Zhang et al., 2017a). It seems possible that SAPS initiates certain plasma instability in the F-region (e.g., Perkins instability, see Perkins, 1973) (Zhang et al., 2019c) and/or similarly causes strong frictional heating as the auroral heating does, leading to TID excitation (Guo et al., 2018). The Perkins instability usually has a small (linear) growth rate because of relatively weak quiet-time electric fields (associated with the wind dynamo). During geomagnetic storms, there are abundant storm-time electric fields, some of which are intense (such as SAPS). They can effectively amplify the growth of the Perkins instability development and initiate MSTIDs. Observations, as depicted in Fig. 7.36, show indeed the LSTID rotation into south-westward propagation and the subsequent presence of MSTIDs. Although the exact physical mechanism remains an active research topic, the influence of SAPS which is characterized for its frequent occurrence during geospace storms is fairly significant.

Several thermospheric and ionospheric observations from the satellite in situ instruments and ground-based instruments, as well as numerical simulations using general

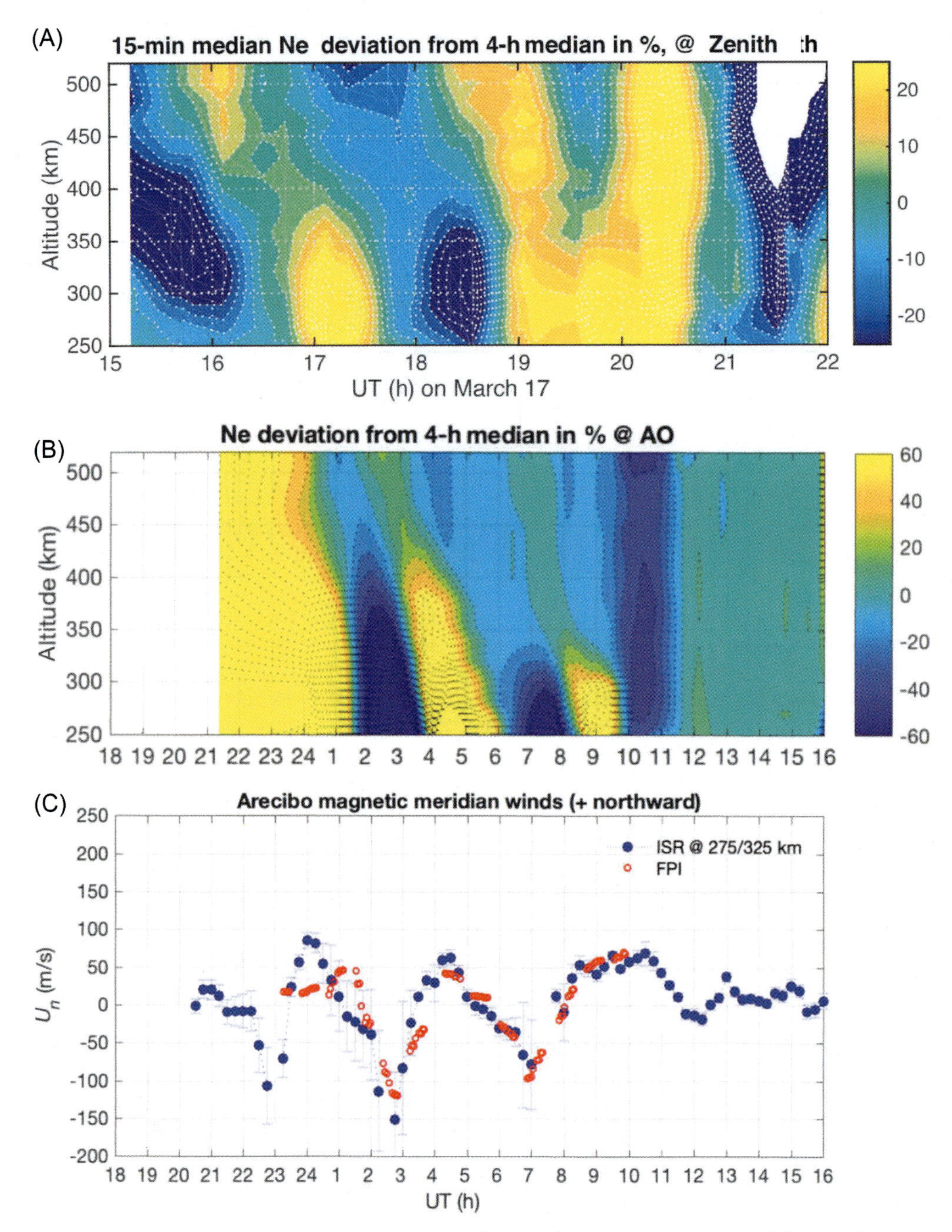

Fig. 7.35 Observations of storm-induced TIDs and TADs during the 2015 St. Patrick's Day Storm at Millstone Hill and Arecibo. (A) F-region electron density height versus universal time variations at Millstone Hill, shown as the percentage deviation from 4-h median Ne in m^{-3}. An enhancement immediately after 20 UT at indicating the local passage of storm-enhanced density (SED); (B) same as (A) but for Arecibo during a later time period; (C) meridional neutral wind observations at Arecibo based on ISR (*blue*) and FPI (*red*) data. *(From Zhang, S.-R., Erickson, P.J., Zhang, Y., Wang, W., Huang, C., Coster, A.J., Holt, J.M., Foster, J.F., Sulzer, M.P., Kerr, R., 2017a. Observations of ion-neutral coupling associated with strong electrodynamic disturbances during the 2015 St. Patrick's Day storm. J. Geophys. Res. Space Phys.122 (1), 1314–1337.)*

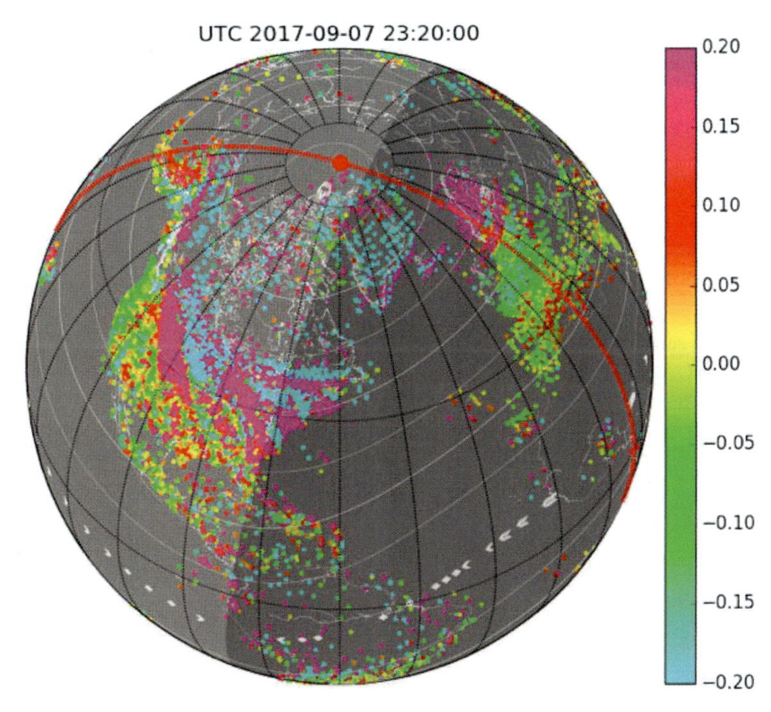

Fig. 7.36 A differential TEC map at 2330 UT on September 7 demonstrates dynamic evolution of concurrent LSTIDs and MSTIDs near the dusk sector as a bright SAPS feature occurred. *White thin lines* mark the iso-magnetic latitude at a 10 degrees interval. *(From Zhang, S.-R., Erickson, P.J., Coster, A.J., Rideout, W., Vierinen, J., Jonah, O., Goncharenko, L.P., 2019c. Subauroral and polar traveling ionospheric disturbances during the 7–9 September 2017 storms. Space Weather17 (12), 1748–1764. 10.1029/2019SW002325.)*

circulation models, have suggested that large-scale TADs/TIDs induced by severe geomagnetic storms have sufficient energy to reach equatorial latitudes, and further propagate into the conjugate hemisphere as deep as in magnetic mid-latitudes (Fuller-Rowell et al., 1994; Guo et al., 2014; Zakharenkova et al., 2016; Lu et al., 2020). Lu et al. (2020) indicated the meridional phase propagation speed for some of these LSTIDs were $\sim$640 m s^{-1}. Large propagation speeds seem necessary for the disturbances to land at low latitudes, whereas slow meridional propagation of LSTIDs normally cannot travel that far (Tsugawa et al., 2004; Ding et al., 2008).

In the polar region, observations often showed clear and persistent trans-polar transportation of plasma density disturbances from the dayside to nightside following geomagnetic disturbances with IMF southward turning. The TEC disturbances, as discussed in Zhang et al. (2019c) and shown in Fig. 7.33, can be up to 15% ($\sim$1 TECu or less) amplitudes over a very large longitudinal sector (2000+km zonal wave fronts) and propagate at a phase speed of several hundred (e.g., 500–600 m s^{-1}). These LSTIDs seem consistent with large-scale GW characteristics. GWs can be possibly excited by the dayside auroral heating near cusp during the storm main phase and travel across the pole into the nightside. However, the question is whether these trans-polar GWs can directly induce corresponding LSTID trans-polar propagation with sufficiently large amplitudes, because as discussed in Section 7.3.3.1, the nearly vertical magnetic fields prevent horizontal wind forcing on vertical plasma drift, typical for mid-latitudes. The disturbance wind-forcing across **B** associated with GWs, however, may induce disturbance electric fields embedded in the neutral waves that propagate *across* the magnetic field, and therefore it seems possible that GWs cause electrified LSTIDs.

The trans-polar LSTIDs can be also manifestation of polar cap patches that the anti-sunward convection flow entrains, not of GWs (Nishimura et al., 2020). As discussion in Section 3.3.4 (see also references therein), polar cap patches are originated from subauroral plasma and/or cusp precipitation, both of which can form the polar tongue of ionization with an elevation density from the ambient ionosphere. Particularly, enhanced energy input to the cusp initiated by IMF southward turning can trigger periodical ionospheric density enhancements by particle precipitation. Nishimura et al. (2020) demonstrated the close correlation between those LSTID observations in TEC and some of those very weak polar cap patches quasiperiodically drifting from the cusp into the polar cap via dayside reconnection.

7.3.6 LSTIDs: Challenges and future directions

Throughout this section, we assume that LSTIDs are manifestation of thermospheric neutral disturbances as TADs during storms or more generally GWs. This idea, however, has not been well examined. Due to the magnetic field control on plasma dynamics and electromagnetics, the TID and TAD/GW correlation becomes complicated at high latitudes. In regions of substantial plasma density gradients (EIA, SED, mid-latitude ionospheric trough) and dramatic dynamical processes, ionospheric influences on TADs/GWs remain some of the unresolved scientific questions.

While auroral heating is the fundamental source of TAD/GW/LSTID, there are certainly other heating processes in the upper atmosphere during storm and substorm periods. Effects of these other disturbances on LSTID excitation or on propagation of existing TIDs have been rarely explored. While heating is the essential process for exciting atmospheric waves, electrified plasma disturbance waves unrelated to heating are known to exist (e.g., as MSTIDs). The question is whether the storm-time electric field intensification can trigger certain TIDs.

To advance the LSTID science and further the understanding of TADs/GWs and LSTIDs coupling, we face some observational challenges for both ionic and neutral components in the upper atmosphere. Neutral observations for TA/GW study have much less availability than the ionospheric observations for TID study. Furthermore, the diverse ionospheric data correspond to different ionospheric properties. It remains challenging to reconcile results from these different datasets. Finally, numerical simulations using IT coupling models have achieved reasonable success in reproducing essential LSTID morphology and its relationship with TADs/GWs, however, specifying high-latitude electromagnetic forcing and particle inputs with appropriate spatiotemporal accuracy is the key. Corresponding observations of these auroral characteristics are fundamental to enable accurate model-based specification and forecast of LSTIDs.

7.4 Observational characteristics of medium-scale traveling ionospheric disturbances

Hyosub Kil[a], Woo Kyoung Lee[b], and Larry J. Paxton[a]

[a]The Johns Hopkins University Applied Physics Laboratory, Laurel, MD, United States

[b]Korea Astronomy and Space Science Institute, Daejeon, Republic of Korea

7.4.1 Introduction

Traveling ionospheric disturbances (TIDs) can be observed at any latitudes and at any time. The existence of TIDs was noticed more than a half century ago from the observations of wave-like modulations of the F-region height by radio experiments (Munro, 1948, 1958). The study of TIDs with modern instrumentation started with the incoherent scatter radar observations at Arecibo, Puerto Rico (Behnke, 1979; Harper, 1972; Thome, 1964). Two-dimensional structures of TIDs were first identified from Arecibo radar observations (Behnke, 1979). Significant advance in our understanding of TIDs began with the emergence of the all-sky imaging technique (Mendillo et al., 1997) and establishment of dense global positioning system (GPS) network (Saito et al., 1998). Since then, these techniques have been employed worldwide for the observation of TIDs. Most of our current knowledge of the characteristics of TIDs relies on the observations by these techniques.

TIDs develop from various sources such as geomagnetic storms, hurricanes, tornados, volcanos, earthquakes, eclipses, solar flares, sunset, sunrise, and rocket launches (Borries et al., 2009; Chou et al., 2017; Coster et al., 2017; Ding et al., 2007; Jonah et al., 2018; Lin et al., 2014; Nishioka et al., 2013; Ogawa et al., 2009; Tsugawa et al., 2011; Zakharenkova et al., 2016; Zhang et al., 2019a, b). For some unique events whose occurrence locations and times can be specified, associated TIDs can also be specified. For example, concentric TIDs develop around tornados (Nishioka et al., 2013), and large-scale TIDs develop at subauroral region during geomagnetic storms (Zhang et al., 2019b). Section 7.3 provides more extended discussion on those storm-related large-scale TIDs.

Frequently observed TIDs in mid-latitudes are Medium-scale TIDs (MSTIDs) (Makela and Otsuka, 2012). Because the generation of MSTIDs is known to be closely related to atmospheric gravity waves (AGWs) and because AGWs can be generated by various unknown sources, the prediction of the occurrence of MSTIDs in connection with their sources is challenging. Moreover, complex physical processes are involved in the development of MSTIDs from seed perturbations (e.g., Yokoyama, 2014).

Fig. 7.37 shows samples of MSTIDs observed (a) at night over Japan (Saito et al., 2001) and (b) on the dayside over Europe (Otsuka et al., 2013). Nighttime MSTIDs are detected simultaneously from a GPS total electron content (TEC) perturbation map and an airglow image of oxygen atom (OI, indicating in spectroscopic notation emission from neutral atomic oxygen at 630 nm). The TEC perturbation map is the map of the residual TEC after the subtraction of the background TEC. OI 630 nm emissions are produced by the dissociative recombination of molecular oxygen ions (O_2^+) with ionospheric electrons. The O_2^+ ions are created by the charge exchange reaction of ionospheric O^+ and thermospheric O_2, and therefore, the intensity is proportional to the oxygen ion and molecular oxygen densities. OI 630 nm emission is most frequently used for the detection of nighttime MSTIDs. As Fig. 7.37A shows, nighttime MSTIDs in the Northern hemisphere are typically elongated in the northwest-southeast direction and propagate toward southwest. As we show later, nighttime MSTIDs in the Southern hemisphere are mirror images of MSTIDs in the Northern hemisphere with respect to the magnetic equator. This property is closely related to conjugacy in nighttime MSTIDs. In Fig. 7.37B, daytime MSTIDs are elongated in the east-west direction and propagate equatorward. Conjugate MSTIDs have not yet been identified on the dayside. The different properties of nighttime and daytime MSTIDs are related to their generation mechanisms. Some of the characteristics of MSTIDs identified at a local region are commonly observed at other places. The elongation and propagation direction of nighttime MSTIDs are good examples. However, differences exist among the observations made at different locations and different times. The differences may represent actual variations of the characteristics, but the discrepancies caused by the use of different detection techniques or detection criteria cannot be ruled out.

This review report aims to provide a comprehensive understanding on the global characteristics of MSTIDs by synthesizing observations worldwide. The characteristics of nighttime and daytime MSTIDs are reviewed in Sections 7.4.2 and 7.4.3, respectively. In Section 7.4.4, current understanding of the generation mechanisms of daytime and nighttime MSTIDs are discussed. In Section 7.4.5, we summarize common features in MSTIDs and describe future work.

7.4.2 Nighttime MSTIDs

7.4.2.1 Wave characteristics

Modern study of MSTIDs began with the emergence of the techniques to detect two-dimensional structures of MSTIDs. The fundamental characteristics of nighttime MSTIDs that we know now have been revealed by early observations. The tilt of the ionospheric structures in the northwest-southeast direction and the propagation of the tilted structure toward the southwest was identified by the observations of the incoherent

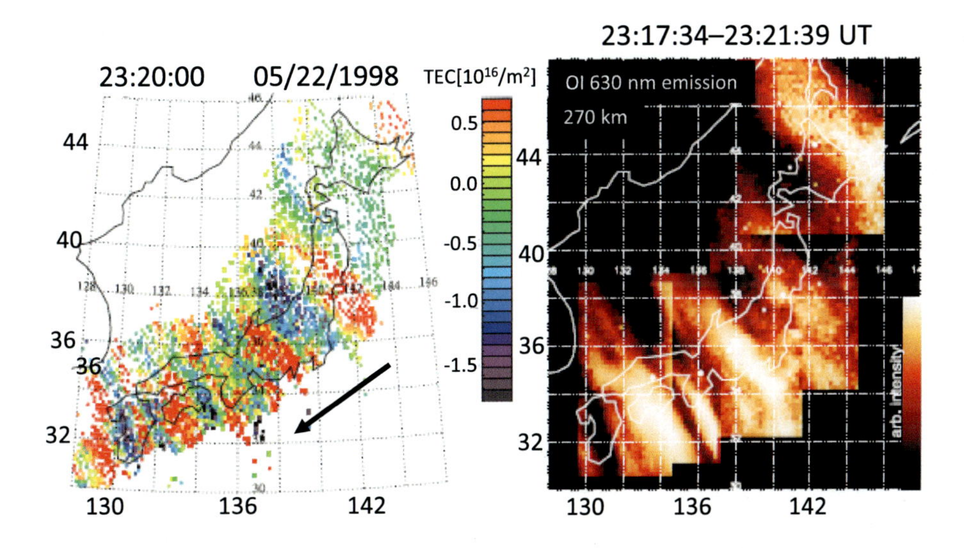

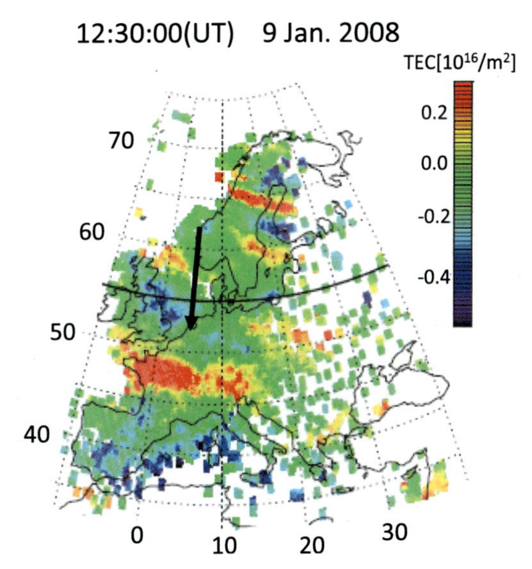

Fig. 7.37 Examples of nighttime and daytime MSTIDs. (A) Nighttime MSTIDs on a GPS TEC perturbation map and an OI 630-nm airglow image observed on the night of May 22, 1998 over Japan (Saito et al., 2001) and (B) daytime MSTIDs on a GPS TEC perturbation map observed on the night of January 9, 2008 over Europe (Otsuka et al., 2013).

scatter radar at Arecibo, Puerto Rico (Behnke, 1979). The observations of MSTIDs using All-Sky Imagers and GPS networks began in the late 1990s (Mendillo et al., 1997; Saito et al., 1998), and common characteristics of nighttime MSTIDs have been identified from these observations worldwide. Garcia et al. (2000) first reported the statistical results of the wave characteristics in nighttime MSTIDs derived from the OI 630-nm airglow observations over Arecibo, Hawaii, and Ithaca. Nighttime MSTIDs observed at those locations propagated toward the southwest with velocities of 50–180 m s^{-1}. The scale sizes and oscillation periods of nighttime MSTIDs were typically 50–500 km and periods 0.25–2.5 h, respectively. Garcia et al. (2000) also noted the inverse relationship between the solar activity and the occurrence of nighttime MSTIDs.

The wave characteristics of nighttime MSTIDs derived from the observations at different locations are compared in Fig. 7.38. Fig. 7.38A and B shows the azimuth distributions of the propagation direction over China (Ding et al., 2011) and Brazil (Amorim et al., 2011). The results over China and Brazil are derived from TEC maps in May–-August 2009 and OI 630-nm airglow observations in May 1995–July 1996, respectively. The concentric circles in Fig. 7.38A and B represent different parameters. In Fig. 7.38A, the radius of the circle represents the accumulation hours of the MSTID observation. The larger number means the more frequent occurrence. The radius in Fig. 7.38B is the phase velocity. Nighttime MSTIDs propagate toward the southwest over China and toward the northwest over Brazil. The propagation velocity over Brazil is in the range of 100–200 m s^{-1}. The propagation velocities of nighttime MSTIDs over China (Ding et al., 2011) are 50–230 m s^{-1}. Therefore, the propagations of nighttime MSTIDs in the opposite hemispheres appear to be mirrored with respect to the equator. The observations of period, velocity, amplitude, and wavelength of nighttime MSTIDs over Japan and China are shown in Fig. 7.38C and D, respectively. The results at two locations in Japan are derived from the observations of OI 630 nm airglow in 1998–2000 (Shiokawa et al., 2003). The results over China are derived from the observations of OI 630 nm airglow and measurements of GPS TEC perturbations in 2013–15 (Huang et al., 2016). The phase velocity and wavelength over Japan are similar to those over China. In the results derived from GPS TEC perturbation maps over Japan in 2002 (Otsuka et al., 2011), the phase velocity and wavelength MSTIDs are of 50–200 m s^{-1} and 150–450 km, respectively. These observations are close to those observed over China. On average, the periods are 1 h, velocities are ~100 m s^{-1}, amplitudes are 5%–10%, and wavelengths are 200–300 km. These results are consistent with those observed over Puerto Rico (Garcia et al., 2000) and Western United States (Kotake et al., 2007). Synthesis of the observations over Brazil (Amorim et al., 2011; Pimenta et al., 2008) and other places in the Southern hemisphere (Martinis et al., 2011, 2019) indicate that the wave characteristics of nighttime MSTIDs in the Southern hemisphere are comparable to those in the Northern hemisphere.

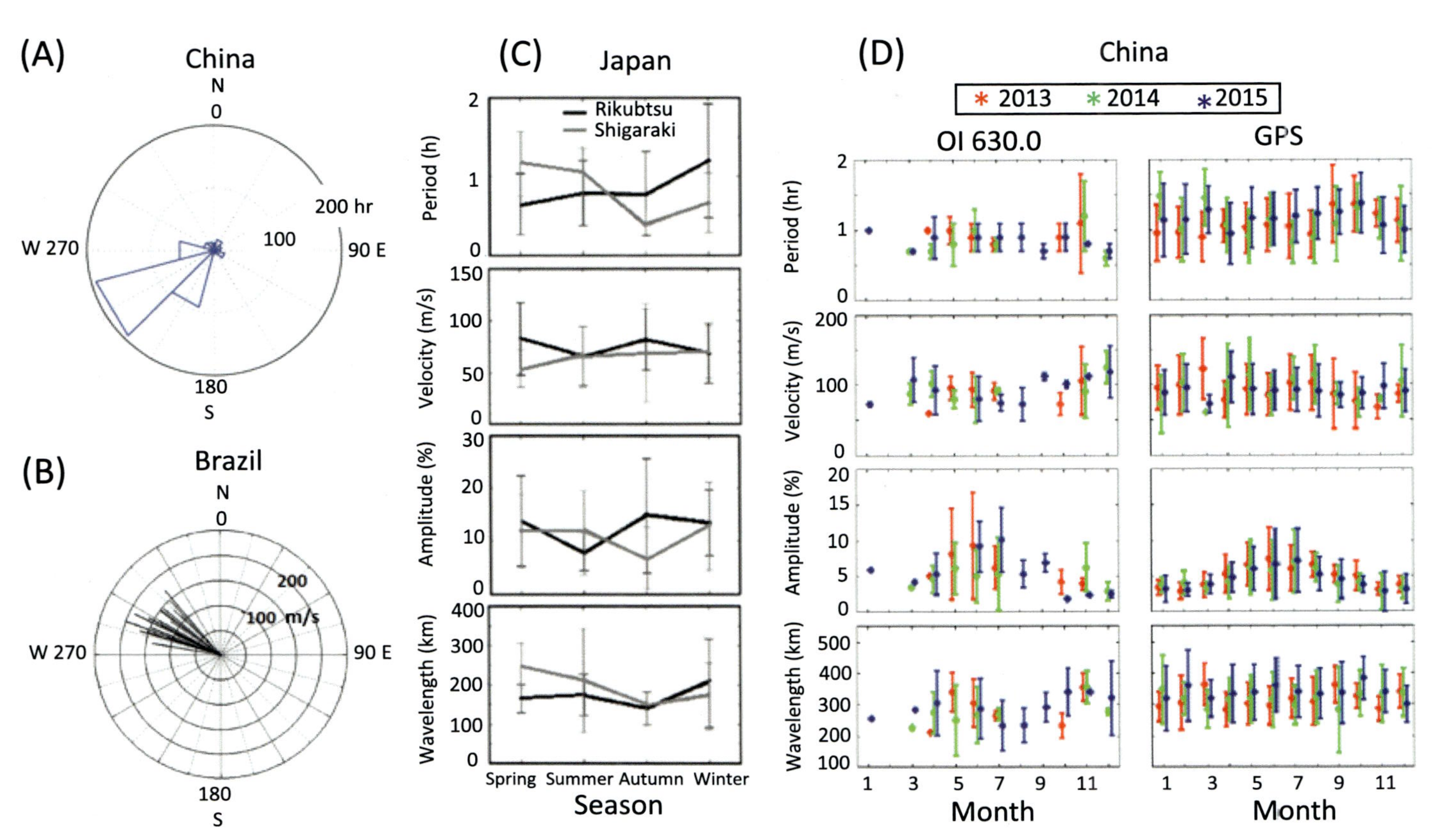

Fig. 7.38 Wave characteristics of nighttime MSTIDs. (A) Azimuth distribution of the propagation direction over China (Ding et al., 2011). The results are derived from TEC perturbation maps during summer (May–August) in 2009. The radii of circles indicate the sum of observation hours. (B) Azimuth distribution of the propagation velocities over Brazil (Amorim et al., 2011). The results are derived from OI 630-nm airglow observations in May 1995–July 1996. (C) Wave parameters at two locations in Japan (Shiokawa et al., 2003). The results are derived from OI 630-nm airglow observations in 1998–2000. (D) Wave parameters over China (Huang et al., 2016). The results are derived from OI 630 nm airglow and GPS TEC observations in 2013–15. *Vertical bars* in C and D are standard deviations.

MSTIDs also occur in the equatorial region (Fukushima et al., 2012; Makela et al., 2010; Paulino et al., 2016; Shiokawa et al., 2006). Fig. 7.39 shows the characteristics of nighttime MSTIDs in the equatorial region derived from OI 630-nm airglow observations at (a) Kototabang (geographic latitude: 0.2°S, geomagnetic latitude: 10.6°S) in Indonesia during 2002–09 (Fukushima et al., 2012) and (b) São João do Cariri (geographic coordinates: 7.4°S, geomagnetic latitude: 0.5°N) in Brazil during 2000–10 (Paulino et al., 2016). The velocity, period, and wavelength at Kototabang are of 300 m s^{-1}, 45 min, and 800 km, respectively. Shiokawa et al. (2006) also reported observations of similar wave characteristics at Kototabang. The wave parameters observed at Kototabang are about 2–4 times greater than those at São João do Cariri. The velocities and wavelengths observed at São João do Cariri are comparable to those observed in Japan and China (Fig. 7.38C and D). However, periods observed at Kototabang are close to those observed in Japan and China. The propagation directions of wave fronts at Kototabang are preferentially southward (poleward), whereas they at São João do Cariri are northward and southeastward. Thus, significant differences exist between the wave characteristics of equatorial nighttime MSTIDs in the Asian and South American sectors. The wave characteristics in the equatorial region also appear to be different from those in higher latitudes. More observations in the equatorial region at different longitudes are required for the assessment of the longitudinal and latitudinal variability of the characteristics of nighttime MSTIDs.

7.4.2.2 Occurrence climatology

Ground-based observations

The occurrence climatology of MSTIDs has been investigated using ground-based observations of airglow, GPS TEC, and spread F (Amorim et al., 2011; Candido et al., 2008; Ding et al., 2011; Duly et al., 2013; Hernández-Pajares et al., 2006; Huang et al., 2016; Kotake et al., 2006, 2007; Martinis et al., 2010; Otsuka et al., 2013; Pimenta et al., 2008; Shiokawa et al., 2003; Valladares and Sheehan, 2016). By combining ground-based observations at different locations, we can obtain partial information on the global occurrence climatology of MSTIDs. Fig. 7.40 presents the annual distributions of nighttime MSTIDs over (a) Japan (Shiokawa et al., 2003), (b) Puerto Rico (Martinis et al., 2010), (c) Brazil (Amorim et al., 2011), (d) China (Huang et al., 2016), (e) Hawaii (Duly et al., 2013), and (f) Chile (Duly et al., 2013). Observation years are indicated on each plot. The results over China are derived from GPS TEC measurements. The results over other places are derived from OI 630-nm airglow observations. In the Northern hemisphere (a, b, d, and e), the distribution of nighttime MSTIDs can be described by a semiannual variation characterized by the smaller occurrence rate in equinoxes than in solstices. However, there exists longitudinal difference in the semiannual pattern; the occurrence rate shows a pronounced peak in the months near June solstices over Japan and China, whereas the occurrence rates during June and December solstices

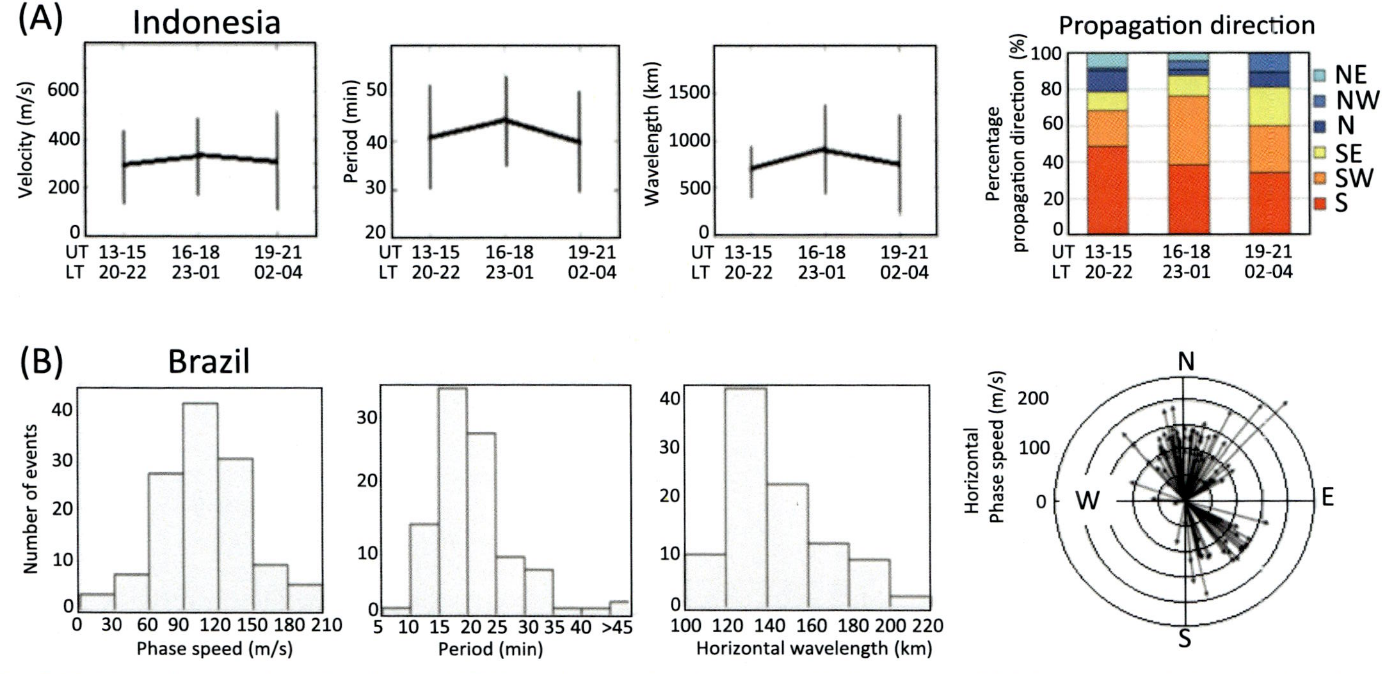

Fig. 7.39 Wave characteristics of nighttime MSTIDs in the equatorial region at (A) Kototabang (10.6°S geomagnetic latitude) in Indonesia (Fukushima et al., 2012) and (B) São João do Cariri (0.48°N geomagnetic latitude) in Brazil (Paulino et al., 2016). The results are obtained from the OI 630-nm airglow observations in 2002–09 (Kototabang) and 2000–10 (São João do Cariri).

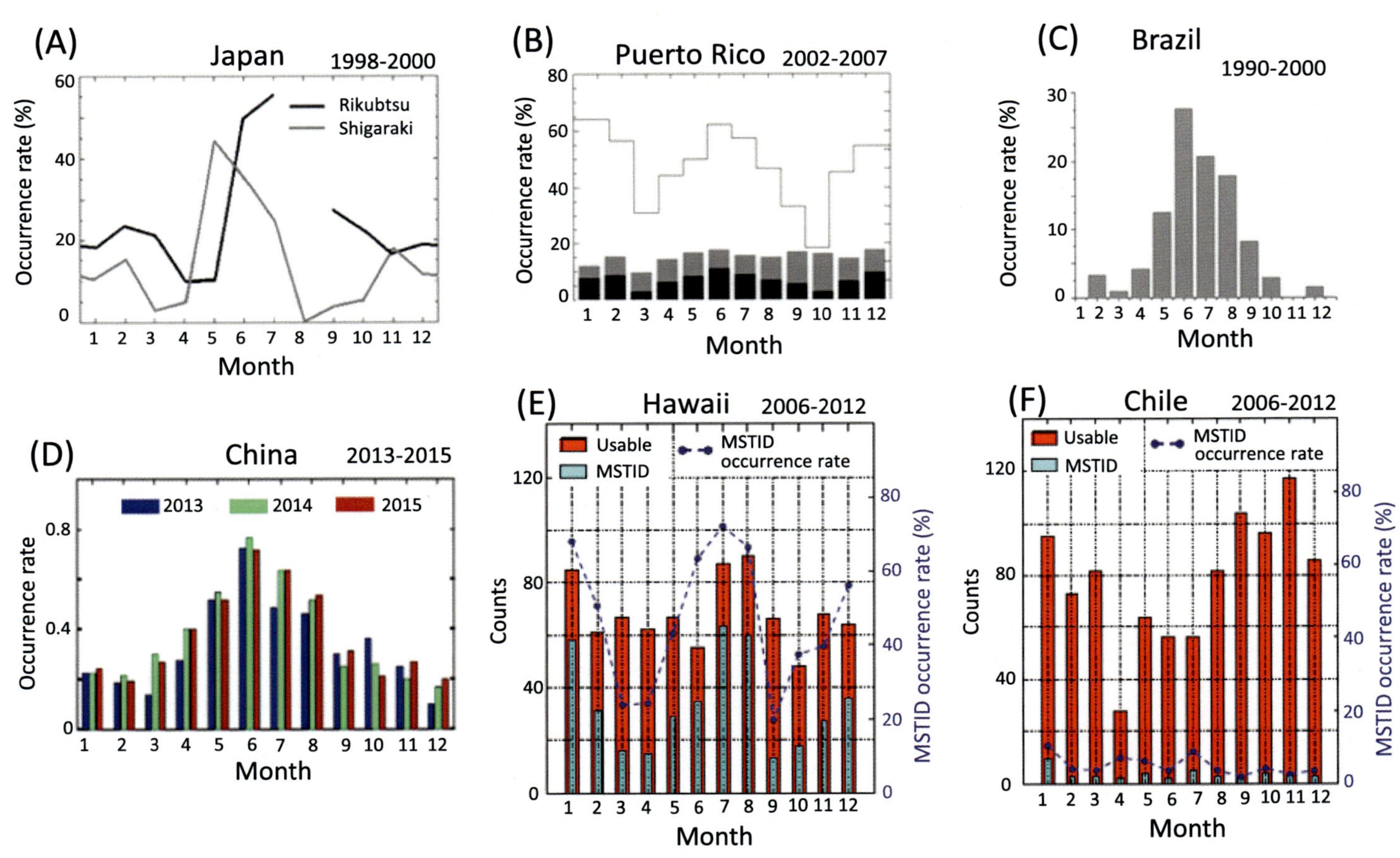

Fig. 7.40 Occurrence climatology of nighttime MSTIDs derived from (A) OI 630-nm airglow observations over Japan (Shiokawa et al., 2003), (B) OI 630-nm airglow observations over Arecibo (Martinis et al., 2010), (C) OI 630 nm airglow over Brazil (Amorim et al., 2011), (D) GPS TEC observations over China (Huang et al., 2016), and (E, F) OI 630-nm airglow observations over Hawaii and Chile (Duly et al., 2013). The observation years are indicated on the plots.

are comparable at Puerto Rico and Hawaii. This difference does not purely represent the longitudinal variation because the observations were made during different periods and the annual pattern could vary with the solar cycle. The morphology over Brazil is similar to that over Japan and China. Considering the fact that June solstice is summer in the Northern hemisphere and winter in the Southern hemisphere, the occurrence of nighttime MSTIDs in both hemispheres is not described by a seasonal behavior. The occurrence rate over Chile is too small to discern the annual pattern. The difference between the observations over Brazil and Chile may reflect the longitudinal, latitudinal, or yearly (or solar cycle) variability of the pattern, but there is an insufficient number of ground-based observations for the assessment of the effect of these factors.

Nighttime MSTIDs show a strong dependence on the solar activity. The observations in Fig. 7.41 present the variation of the nighttime MSTID activity with the solar cycle over (a) Japan (Shiokawa et al., 2003), (b) Puerto Rico (Martinis et al., 2010), and (c) Brazil (Amorim et al., 2011). The results over Japan are obtained from the analyses of ionosonde observations assuming that spread F in ionograms is produced by MSTIDs (Bowman, 2001). The observations at five ionosonde stations in Japan are distinguished by different lines, and the results on the top and bottom panels are derived from the observations during the periods of solar maximum (1979–82) and solar minimum (1973–79, 1985–87), respectively. The results over Puerto Rico and Brazil are derived from the OI 630-nm airglow observations in 2002–07 and 1990–2008, respectively. All observations demonstrate the inverse relationship between the occurrence rate of nighttime MSTIDs and solar activity. In Fig. 7.41A, the annual pattern of the occurrence rate during solar maximum is represented by a single peak during June solstices, but a semiannual pattern appears during solar minimum. As this observation shows, the annual pattern of nighttime MSTIDs can appear differently depending on the solar cycle.

Ground-based observations reveal the variation of the occurrence of nighttime MSTIDs with month, longitude, hemisphere, and solar cycle. However, the global morphology of nighttime MSTIDs is difficult to establish with ground-based observations because there are few observations in the Pacific and Atlantic and fewer in the Southern hemisphere than in the Northern hemisphere. When we infer the global morphology by synthesizing observations at different locations, we may use only the occurrence pattern. If the occurrence rate of MSTIDs derived from ionosonde observations over Japan is greater than that derived from airglow observations over Arecibo, this observation does not simply mean that MSTIDs occur more frequently over Japan than over Arecibo because the difference over two regions can be related to the detection sensitivity of the two methods. Even if MSTIDs are observed using the same method, the criteria for the detection of MSTIDs can be different from person to person. The difference in observation years (or cycle) must also be taken into account.

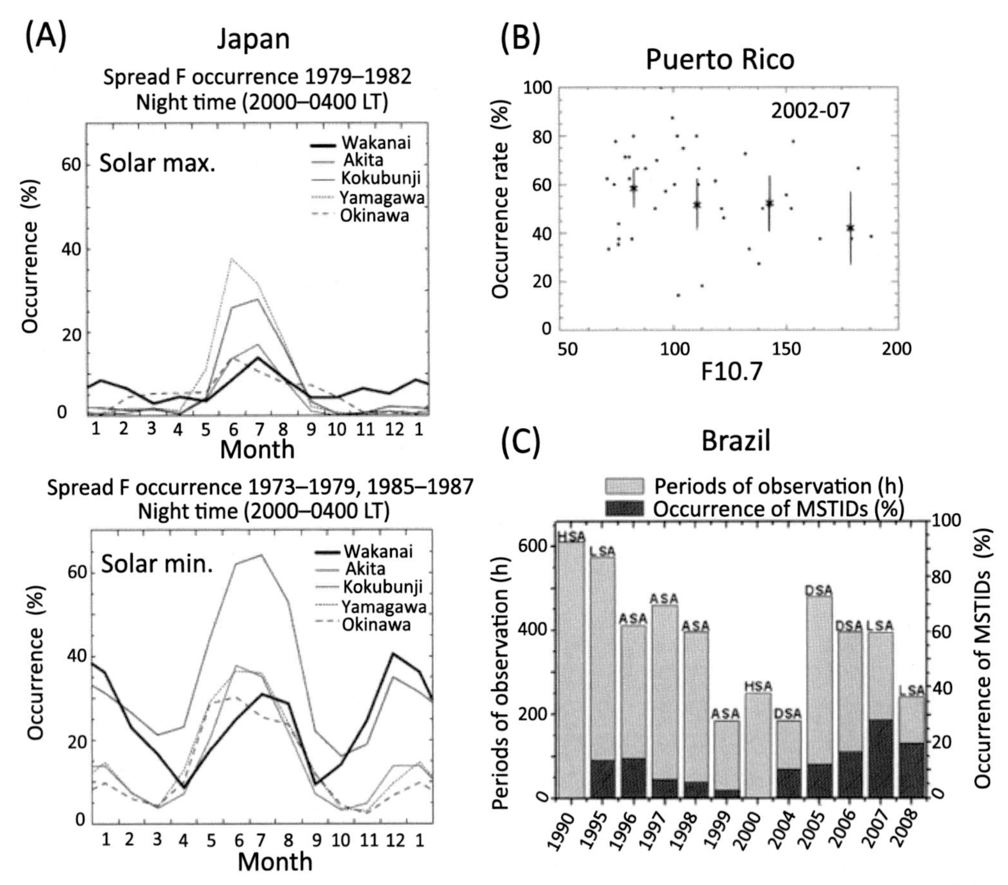

Fig. 7.41 Solar cycle dependence of the nighttime MSTID activity. (A) Seasonal variation of the occurrence rate of spread F over Japan during (*top*) solar maximum and (*bottom*) solar minimum (Shiokawa et al., 2003). (B) MSTID occurrence rate as a function of F10.7 index derived from OI 630-nm airglow observations in 2002–07 over Arecibo (Martinis et al., 2010). (C) Yearly variation of the MSTID occurrence rate over Brazil derived from the observations of OI 630 nm airglow (Amorim et al., 2011). The solar activity in each year is indicated by high solar activity (HSA), low solar activity (LSA), ascending solar activity (ASA), and descending solar activity (DSA).

Satellite observations

Satellite observations are advantageous for investigating the global behavior of MSTIDs because of the global coverage of observations and the application of consistent detection methods and criteria. For ground-based observations, the occurrence of MSTIDs can be identified by examining the morphology and propagation of ionospheric structures. Because these properties are not available from satellite observations, MSTIDs are detected from satellite observations using detection proxies. We assess the global morphology of nighttime MSTIDs by comparing the results derived using different detection proxies together with ground-based observations.

Park et al. (2010) first derived the global morphology of nighttime MSTIDs from satellite observations. Absolute electron density fluctuation (ΔN) was used as the detection proxy of nighttime MSTIDs from the Challenging Mini-Satellite Payload (CHAMP) satellite observations. Some of the results in Park et al. (2010) are presented in Fig. 7.42A and B to examine the dependence of nighttime MSTIDs on longitude, season, and solar cycle. Fig. 7.42A shows the results during solar minimum (2006–07). The pronounced peak of the occurrence rate during June solstices in the Northern Asian sector is consistent with the ground-based observations over Japan and China in Fig. 7.40. Over the locations of Hawaii and Puerto Rico, the occurrence rate during June solstices is greater than that during December solstices. This observation is somewhat different from the ground-based observations which show comparable magnitudes of the occurrence rates during June and December solstices (see Fig. 7.40B and E). In the Southern American sector, the occurrence rate during December solstices seems to be greater than that during June solstices in the satellite observations, but the results derived from the ground-based observations over Brazil (Fig. 7.40C) show a pronounced peak occurrence rate during June solstices. Overall, the occurrence pattern in Fig. 7.40A can be characterized by the higher occurrence frequency of nighttime MSTIDs in the summer hemisphere than in the winter hemisphere. Similar behavior was observed during solar maximum.

Now we examine the solar cycle dependence of the nighttime MSTIDs by comparing the observations during solar maximum (2001–02) and solar minimum (2006–07). Because the solar cycle dependences during June and December solstices are similar, Fig. 7.42B presents the observations only during December solstices. The occurrence rate during solar minimum is greater than that during solar maximum, which is consistent with the results obtained from the ground-based observations (Fig. 7.41). Magnetic field perturbations (ΔB) have also been used for the detection of nighttime MSTIDs (Park et al., 2009; Park et al., 2015). Because ΔB is induced by fluctuations in ionospheric currents which are proportional to the electron density (or conductance) (e.g., Stolle et al., 2006), ΔB has the signatures of nighttime MSTIDs. To compare the MSTID distributions derived using different parameters, the global distributions of magnetic field fluctuations derived from the CHAMP observations during the same periods are shown in Fig. 7.42C. The longitudinal distributions are similar in Fig. 7.42B and C. However, the results exhibit notable differences in the hemispheric distribution and solar cycle dependence. The hemispheric difference is pronounced in Fig. 7.42B, but it is not pronounced in Fig. 7.42C. The occurrence rate during solar minimum is greater than that during solar maximum in Fig. 7.42B, but the trend is opposite in Fig. 7.42C.

Although the proxy parameters ΔN and ΔB represent some characteristics of the morphology of nighttime MSTIDs, notable deviations exist between the morphologies of the proxy parameters and nighttime MSTIDs. ΔN is a good indicator of the strength of nighttime MSTIDs, but this parameter is subject to the background electron density; nighttime MSTIDs developed under a higher background electron density condition

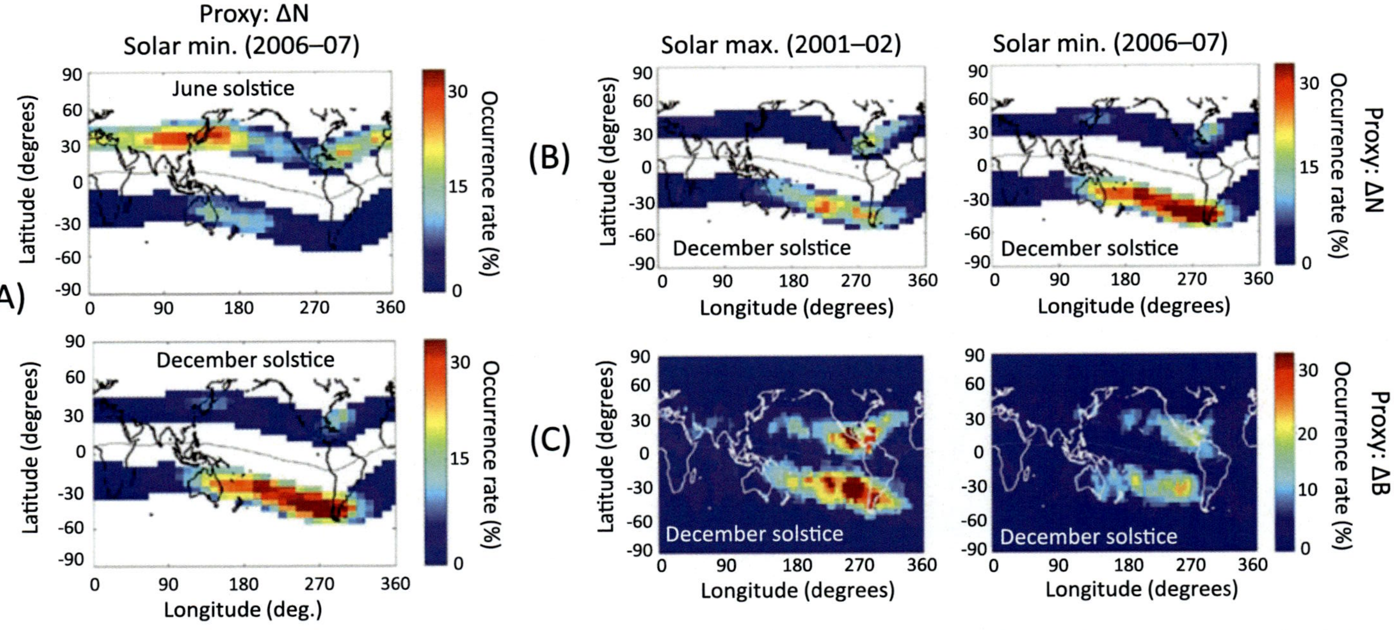

Fig. 7.42 The global distributions of nighttime MSTIDs derived from CHAMP satellite observations at 18–06 h LT during solar maximum (2001–02) and solar minimum (2006–07). (A) Comparison of the MSTID distributions in June and December solstices during the solar minimum. Absolute electron density fluctuations (ΔN) are used as the detection proxies (Park et al., 2010). (B) Solar cycle dependence of the occurrence rate of MSTIDs during December solstices. Absolute electron density fluctuations (ΔN) are used as the detection proxies (Park et al., 2010). (C) Solar cycle dependence of the occurrence rate of MSTIDs during December solstices. Absolute magnetic field perturbations (ΔB) are used as the detection proxies (Park et al., 2009).

have a greater chance of being detected. ΔB is also affected by the background electron density. Thus, the distributions of nighttime MSTIDs derived using these parameters (ΔN and ΔB) represent the distributions of the combination of the occurrence rate and strength of nighttime MSTIDs. For example, the higher occurrence rate of nighttime MSTIDs in the summer hemisphere in Fig. 7.42A can be interpreted in terms of the higher background electron density in the summer hemisphere (Kil et al., 2006; Lee et al., 2018). The higher occurrence rate of nighttime MSTIDs during solar minimum than during solar maximum in Fig. 7.42B, despite the greater electron density during solar maximum, may indicate that the occurrence rate is a more important factor than is the strength in the distribution of nighttime MSTIDs. The interpretation of the distribution of ΔB is more complex because ΔB is affected by field-aligned currents as well as by ionospheric conductance (Park et al., 2009). The higher occurrence rate of nighttime MSTIDs during the higher solar activity in Fig. 7.42C seems to be related to the proportionality of the proxy parameter (ΔB) on the solar activity.

To minimize the effect of the background electron density in the detection of nighttime MSTIDs, Kil and Paxton (2017) used the parameter S defined as

$$S = 100 \times \left[\frac{1}{n-1}\sum_{i=0}^{n-1}(\log_{10}N_i - L_i)^2\right]^{1/2} \Big/ \left[\frac{1}{n}\sum_{i=0}^{n-1}\log_{10}N_i\right]. \tag{7.25}$$

where N_i is the electron density, L_i is the linear fitting line of $\log_{10}N_i$, and n is the number of data points. With 20 data points (10 s worth of data) $S = 0.05$ was used as a threshold for irregularities. The annual and longitudinal distributions of nighttime MSTIDs in the Northern (25°–40°N magnetic latitude) and Southern (25°–40°S magnetic latitudes) hemispheres derived using this method are shown in Fig. 7.43. The results are obtained from the analyses of Swarm-A and Swarm-B observations at 21–03 h local time (LT) from December 2013 to January 2017. The morphology of nighttime MSTIDs in Fig. 7.43 can be summarized as follows. (1) The annual and longitudinal distributions of nighttime MSTIDs in the opposite hemispheres are quasisymmetric. (2) The occurrence of nighttime MSTIDs is represented by a semiannual variation with the primary and secondary peaks in June and December solstices, respectively. (3) The occurrence of nighttime MSTIDs is most pronounced over the Asian sector during June solstices in both hemispheres. These properties are consistent with the results of ground-based observations.

The CHAMP observations of about 9 years provide a unique resource to investigate the variability of the global morphology of nighttime MSTIDs with the solar cycle. Fig. 7.44 is the same format as Fig. 7.43 for the CHAMP data in 2001–03 (solar maximum), 2004–06 (intermediate), and 2007–09 (solar minimum) (Lee et al., 2021). Irregularities were detected with a parameter δ defined as

$$\delta = |\log_{10}N - \log_{10}F| \tag{7.26}$$

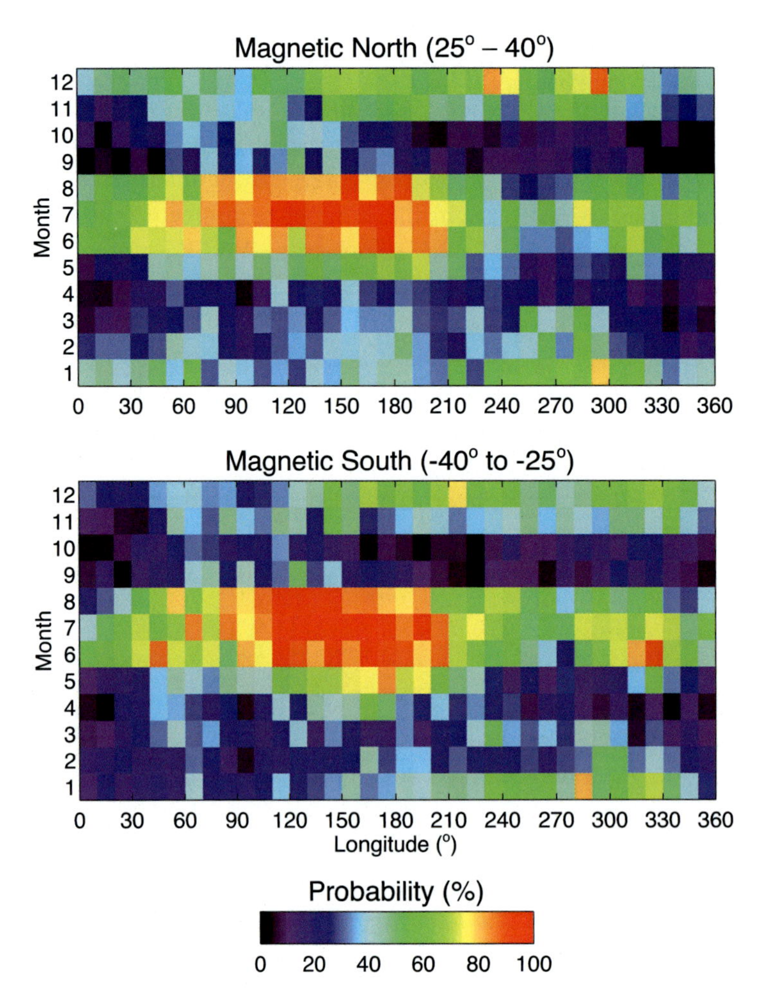

Fig. 7.43 Distributions of nighttime MSTIDs in the Northern and Southern hemispheres derived from Swarm satellite data at 21–03 h in December 2013–January 2017. Normalized logarithms of electron density fluctuations are used as the detection proxies (Kil and Paxton, 2017).

N is the electron density and F is the low-pass-filtered density obtained using the Savitzky-Golay filter (Savitzky and Golay, 1964). In Fig. 7.44, the magnitude of the occurrence rate increases with decreasing solar activity, which is consistent with ground-based observations (e.g., Fig. 7.41). The features of nighttime MSTIDs, except for the magnitude of the occurrence rate, do not vary much with the solar cycle; the quasisymmetry in the hemispheric distribution, semiannual variation, and highest occurrence rate in the Asian sector during June solstices are the same on the course of the solar cycle.

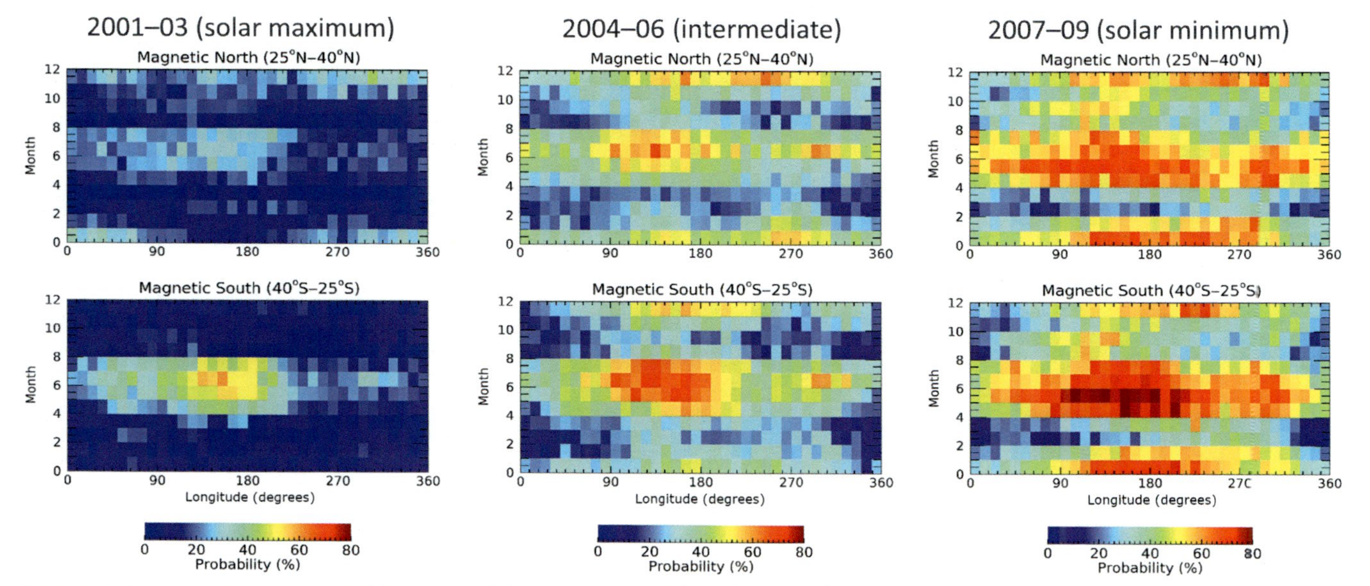

Fig. 7.44 Distributions of nighttime MSTIDs during different periods of the solar cycle derived from CHAMP data at 20–04 h LT. Logarithms of the electron density fluctuations are used as the detection proxies (Lee et al., 2021).

7.4.2.3 Conjugacy

One of the distinguishing characteristics of nighttime MSTIDs from other TIDs is conjugacy. Developments of mirror structures of MSTIDs in the opposite hemispheres are identified by simultaneous observations of OI 630 nm airglow at the magnetic conjugate locations (Martinis et al., 2011, 2019; Otsuka et al., 2004; Shiokawa et al., 2005). Fig. 7.45 shows the observations of conjugate MSTIDs using All-Sky Imagers in the (a) Asian (Otsuka et al., 2004) and (b) American (Martinis et al., 2011) sectors. Bright and dark emission bands are aligned in the northwest-southeast direction and propagate toward the southwest in the Northern hemisphere. In the Southern hemisphere, the mirror structures of northern MSTIDs develop and propagate toward the northwest. Conjugacy in nighttime MSTIDs has an important physical meaning in understanding their generation mechanisms and distributions. The simultaneous formation of conjugate structures in both hemispheres indicates the operation of electrodynamical processes (Miller and Kelley, 1997; Kelley and Fukao, 1991; Kelley and Miller, 1997). The quasihemispheric symmetry in the nighttime MSTID distributions (Figs. 7.43 and 7.44) can be explained in terms of the conjugate property in nighttime MSTIDs. As long as conjugacy is effective the sources of MSTIDs in one hemisphere control the generation of MSTIDs in both hemispheres. In that notion, the hemispheric symmetry in the distribution of nighttime MSTIDs provides supporting evidence of conjugacy. However, the hemispheric symmetry is not perfect. In Fig. 7.44, the occurrence rate in the Southern hemisphere is greater than that in the Northern hemisphere during June solstices and this difference is the largest during the solar maximum. The solar cycle dependence of the conjugacy in nighttime MSTIDs may be attributed to the solar cycle dependence of the F-region conductance, but we do not yet have a clear understanding of the reason for the higher occurrence rate of nighttime MSTIDs in the Southern hemisphere.

7.4.2.4 Anomalous features in nighttime MSTIDs

Here we introduce two phenomena which occasionally occur at night in mid-latitudes: plasma blobs and mid-latitude bubbles. These phenomena are interpreted in terms of either equatorial plasma bubbles or nighttime MSTIDs. Plasma bubbles are ionospheric structures whose plasma density is significantly reduced compared with the background density. Because severe bubbles frequently develop at night in the equatorial region, bubbles are often understood as nighttime equatorial phenomena. The bottomside of the equatorial F-region becomes unstable after sunset due to the formation of a steep vertical gradient in the plasma density and by the vertical motion of the ionosphere (Kelley, 1989). Bubbles are generated by the transport of plasma from the bottomside to the topside by the generalized Rayleigh-Taylor instability (Sultan, 1996; Woodman and La Hoz, 1976). Because this process occurs in the whole magnetic flux tubes, bubbles appear as elongated structures along magnetic field lines (Huba et al., 2008; Kil et al., 2009; Martinis and Mendillo, 2007). For further information regarding the characteristics of

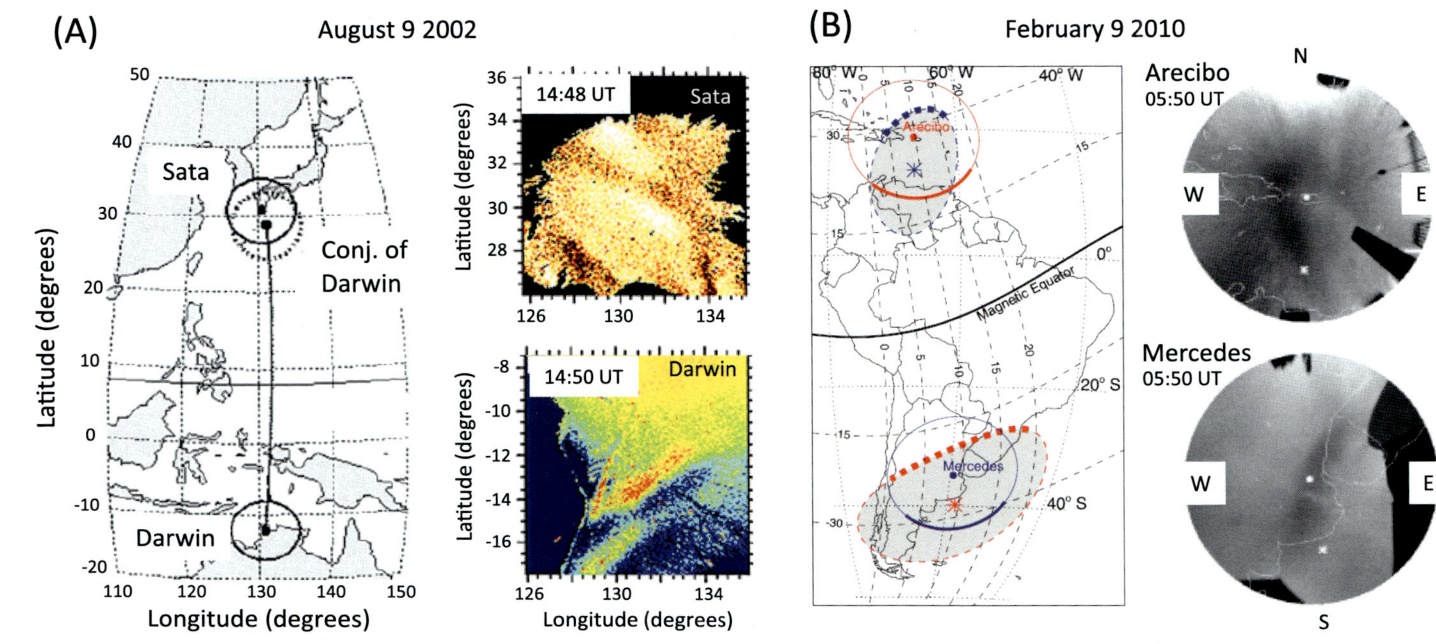

Fig. 7.45 Observations of conjugate MSTIDs in the (A) Asian sector (Otsuka et al., 2004) and (B) American sector (Martinis et al., 2011).

equatorial bubble, readers are referred to review papers (Kil et al., 2015; Makela, 2006; Woodman, 2009). In this section, we examine how well the characteristics of blobs and mid-latitude bubbles can be explained with the known characteristics of equatorial bubbles and nighttime MSTIDs.

Plasma blobs are mesoscale plasma density enhancements relative to ambient plasma. In satellite observations, the plasma density at blobs are enhanced a few times above the background density and the longitudinal widths of blobs are comparable to those of equatorial bubbles (Kil et al., 2011; Le et al., 2003; Oya et al., 1986; Park et al., 2003; Watanabe and Oya, 1986; Yokoyama et al., 2007). Unlike equatorial bubbles, blobs are most frequently detected subtropical region around ±20–30 degrees magnetic latitudes (Choi et al., 2012; Haaser et al., 2012; Watanabe and Oya, 1986).

Blobs observed at night in low- and mid-latitudes are often considered in association with equatorial plasma bubbles based on the observations of bubbles and blobs in the same magnetic meridian (Huang et al., 2014; Le et al., 2003; Martinis et al., 2009; Park et al., 2003; Yokoyama et al., 2007) and the similarity in the distributions of bubbles and blobs (Huang et al., 2014; Park et al., 2008). Model simulations also show the formation of blobs at the poleward edge of bubbles when bubbles develop in the equatorial region (Krall et al., 2010a, b). However, blobs are sometimes detected in the absence of equatorial bubbles (Kil and Paxton, 2017; Kil et al., 2011, 2015). There also exist discrepancies in the distributions of bubbles and blobs (Choi et al., 2012; Haaser et al., 2012). These observations indicate that equatorial bubbles are not prerequisites for the development of blobs. More frequent occurrence of blobs in the months near June solstices and during the periods of low solar activity (Choi et al., 2012; Haaser et al., 2012) are similar to the occurrence pattern of nighttime MSTIDs. The association of blobs with nighttime MSTIDs is further supported by the observations of blobs at the locations of MSTIDs (Kil and Paxton, 2017; Kil et al., 2019; Miller et al., 2014).

Fig. 7.46A shows the detection of blob structures at around ±25 degrees magnetic latitudes by Swarm-A satellite on the night of May 1, 2014 (Kil and Paxton, 2017). The development of MSTIDs at the time of the blob detection over Japan is verified by TEC perturbation maps. Note the detection of blobs in both hemispheres. This observation is consistent with the conjugate property of nighttime MSTIDs. Bubbles are not detected in the equatorial region on that night. The spatial structures of blobs also provide observational evidence of the association of blobs with nighttime MSTIDs. Fig. 7.46B presents the observations of blobs by Swarm satellites and the detection locations of blobs (Kil et al., 2019). Solitary blobs are detected around ±20 degrees magnetic latitudes. The aliments of blobs in the Northern and Southern hemispheres are consistent with the typical alignments of nighttime MSTIDs. Similar morphologies of blobs are observed at different locations on other days. For the blob events detected near Japan, the development of MSTIDs at the times of the blob detection are identified from GPS TEC perturbation maps. These observations and the absence of the signatures of bubbles at the magnetic

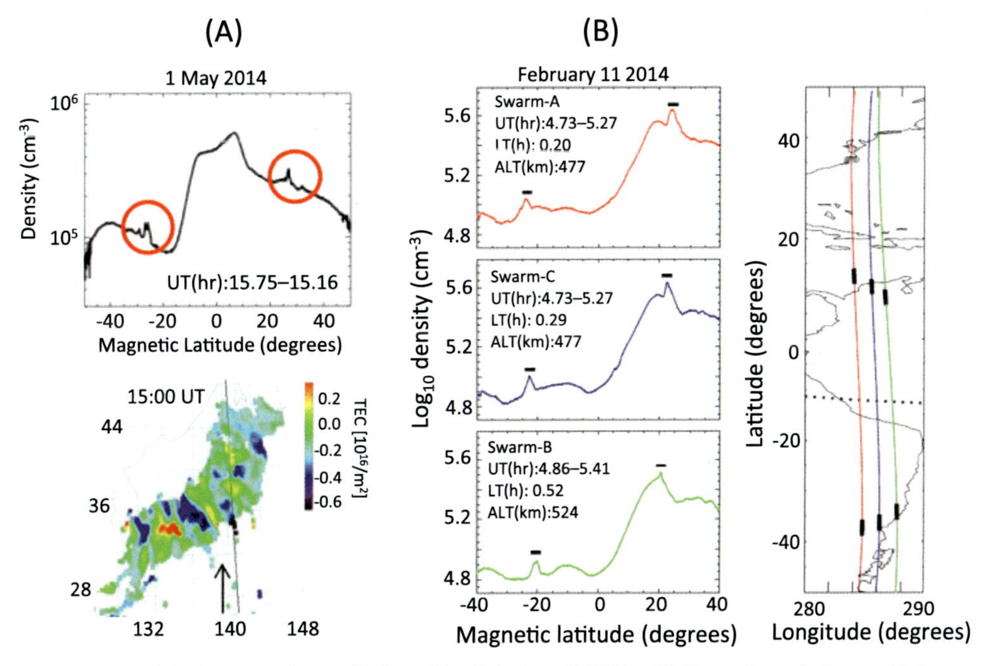

Fig. 7.46 Causal linkage of plasma blobs with nighttime MSTIDs. (A) Detection of plasma blobs at the location of MSTIDs (Kil and Paxton, 2017). (B) Spatial alignment of plasma blobs (Kil et al., 2019).

equator at the times of the blob detection corroborate the association of blobs with MSTIDs on both days.

The implication of these observations is that MSTIDs can be the sources of blobs. We do not know yet how often the generation of blobs is associated with MSTIDs or equatorial bubbles. A simple calculation shows that the uplift of the F-region of 50 km causes about 50% increase in the density relative to the background at an altitude of 500 km (Kil et al., 2019). This magnitude of modulations in the F-region height is observed at the locations of MSTIDs (Behnke, 1979; Harper, 1972). Because the uplift of the ionosphere accompanies the plasma redistribution along magnetic field lines and changes in the plasma loss rate by chemical reactions, more rigorous numerical simulations are desirable to identify the generation processes of blobs from MSTIDs.

Severe plasma depletions that look like equatorial bubbles occasionally occur at night in mid-latitudes during geomagnetic storms. These depletions (we call them "mid-latitude bubbles") are anomalous features because their occurrence latitudes are far beyond the typical boundary of equatorial bubbles. Equatorial bubbles are mostly confined within ±25 degrees magnetic latitudes (Kil and Heelis, 1998; Su et al., 2006).

The majority of studies claim that anomalously large equatorial bubbles developed under storm conditions can reach mid-latitudes (Aa et al., 2018, 2019; Cherniak and Zakharenkova, 2016; Cherniak et al., 2019; Huang et al., 2007; Ma and Maruyama, 2006; Martinis et al., 2015). However, mid-latitude bubbles are also understood in association with MSTIDs (Kil et al., 2016; Nishioka et al., 2009).

From the TEC maps over Japan during the November 10, 2004 storm (Nishioka et al., 2009) identified a TEC depletion band of about 10 TECU (1 TECU = 10^{16} electrons m^{-2}). This level of severe TEC depletion occurs in equatorial bubbles, but it is rarely observed at mid-latitudes. The TEC depletion band over Japan was aligned in the northwest-southeast direction and propagated toward the southwest. Nishioka et al. (2009) interpreted the TEC depletion band as "super MSTIDs" instead of equatorial bubbles because their characteristics (morphology and propagation) coincide with those of typical nighttime MSTIDs.

Martinis et al. (2015) reported the detection of mid-latitude bubbles over Mexico during the June 1, 2013 storm. In the OI 630-nm airglow observations, emission depletion bands emerged from the south and were tilted westward. These are the typical characteristics of equatorial bubbles. Moreover, bubbles were detected in the equatorial region in the longitudes where emission depletion bands were detected. These observations led to the interpretation that the emission depletion bands in mid-latitudes were the signatures of anomalously large equatorial bubbles. If equatorial bubbles were responsible for them, those bubbles had grown to the magnetic apex heights of ~4500 km. However, Kil et al. (2016) pointed out that the connection of the dark bands to MSTIDs could not be ruled out because MSTIDs had developed in North America on that night. The westward tilt and propagation toward the southwest are also the characteristics of nighttime MSTIDs. A similar scenario would be applicable to the mid-latitude bubble features in Aa et al. (2019) and Zhang et al. (2019b).

The origins of mid-latitude bubbles are an open question. One of the difficulties in determining the origins of mid-latitude bubbles is the occurrence of both equatorial bubbles and MSTIDs when mid-latitude bubbles are detected. As described earlier, equatorial bubbles and MSTIDs are difficult to be distinguished by their morphologies when we have partial information about them. An effective way to verify the origin of mid-latitude bubbles is tracing the evolution of bubbles from the magnetic equator to mid-latitudes. Currently, high-resolution TEC maps that cover from the equatorial region to mid-latitudes are available only at limited longitudes. The field of view of one All-Sky Imager is not sufficient for this purpose. This data gap is the major uncertainty in the interpretation of the origins of mid-latitude bubbles.

7.4.3 Daytime MSTIDs

Daytime MSTIDs have not been investigated as extensively as nighttime MSTIDs. The paucity of observations of daytime MSTIDs arises because the detection of daytime

MSTIDs is difficult. First of all, optical observations are not available on the dayside. Because ground-based observations are limited and the global morphology has not yet been established from satellite observations, our assessment of the characteristics of daytime MSTIDs is also limited.

7.4.3.1 Wave characteristics

The statistics of wave characteristics of daytime MSTIDs is derived from the analyses of TEC measurements (Ding et al., 2011; Figueiredo et al., 2018; Hernández-Pajares et al., 2006; Otsuka et al., 2013) and Super Dual Auroral Radar Network (SuperDARN) observations (Frissell et al., 2014; Ishida et al., 2008). Fig. 7.47A shows the distributions of period, wavelength, and velocity of daytime MSTIDs derived from the Blackstone (37.10°N, 77.95°W) SuperDARN observations in June 2010–May 2011 over the United States (Frissell et al., 2014). The mean values of period, wavelength, and velocity are about 40 min, 300 km, and 150 m s^{-1}, respectively. Ishida et al. (2008) derived the statistical results of daytime MSTIDs from the Hokkaido SuperDARN (43.53°N, 143.61°E) observations in November 2006–February 2007 over Northern Japan. In their results, the mean values of period, wavelength, and velocity are about 40 min, 400 km, and 150 m s^{-1}, respectively. In the observational results derived from GPS TEC perturbation maps over Japan in 2002 (Otsuka et al., 2011), the period, wavelength, and velocity are in the ranges of 20–40 min, 100–400 km, and 50–200 m s^{-1}. Over Central China, the period and phase velocity derived from GPS TEC maps in 2009–10 are in the ranges of 20–60 min and 100–400 m s^{-1}, respectively (Ding et al., 2011). Thus, the wave characteristics of daytime MSTIDs over the United States, Japan, and China are within comparable ranges. These characteristics are also comparable to the characteristics of nighttime MSTIDs.

The wave characteristics of daytime MSTIDs over Brazil in Fig. 7.47B are derived from GPS TEC maps in December 2012–February 2016 (Figueiredo et al., 2018). The periods are ~25 min, wavelengths are ~400 km, and velocities are ~320 m s^{-1}. The wavelengths and velocities over Brazil are about twice greater than those over the United States, whereas the periods over Brazil are about half of those over the United States. These differences are not attributable simply to the hemispheric difference of the characteristics because the United States and Brazil are in different latitudes and observations are made during different periods and different techniques. Currently, we do not have sufficient observations in the Southern hemisphere for the cross validation of the results obtained over Brazil. More observations at different longitudes in the Southern hemisphere are necessary to assess the characteristics of daytime MSTIDs in the global context.

The propagation directions of daytime MSTIDs over (a) Europe (Otsuka et al., 2013), (b) the United States (Frissell et al., 2014), (c) Japan (Ishida et al., 2008), (d) China (Ding et al., 2011), and (e) Brazil (Figueiredo et al., 2018) are assembled in Fig. 7.48. The observation methods and periods are referred to the figure caption. In the Northern

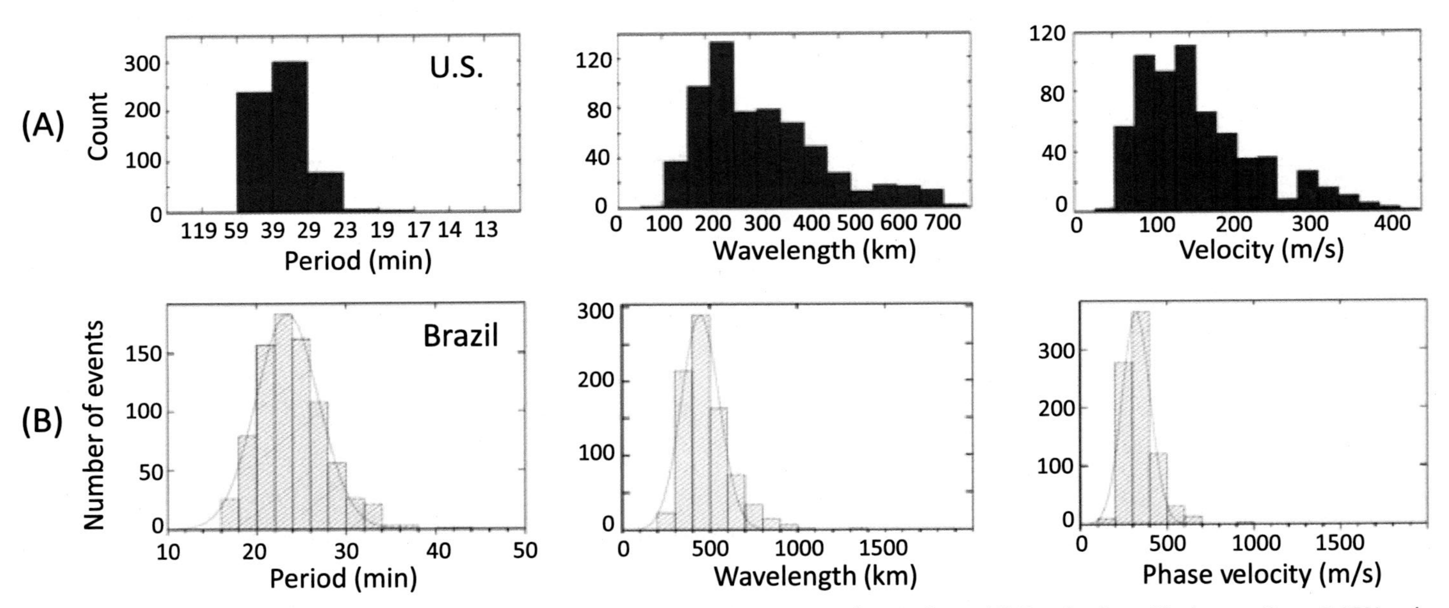

Fig. 7.47 Comparison of the wave parameters of daytime MSTIDs in the opposite hemisphere. (A) Results from Blackstone SuparDARN radar observations at 08–17 h magnetic local time in June 2010–May 2011 over the United States (Frissell et al., 2014) and (B) Results from GPS TEC perturbation maps in December 2012–February 2016 over Brazil (Figueiredo et al., 2018).

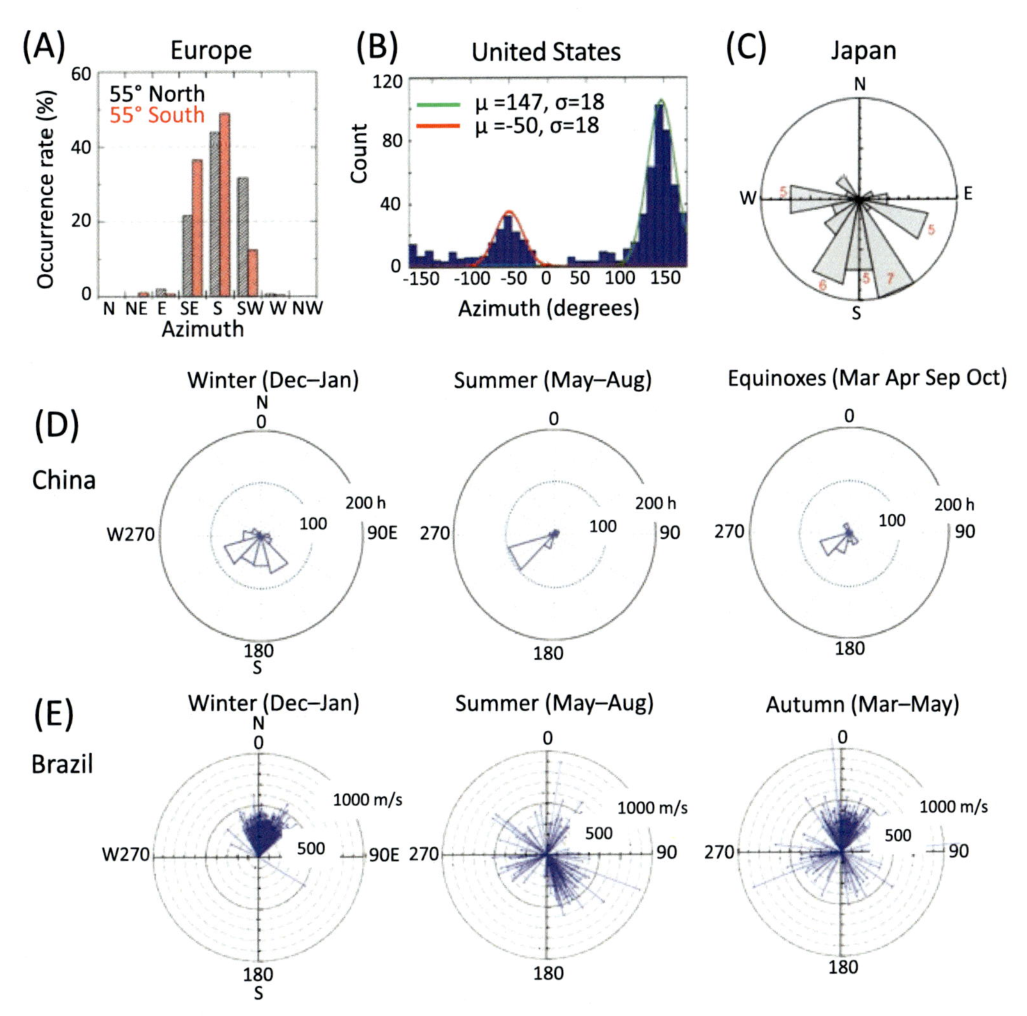

Fig. 7.48 Propagation directions of daytime MSTIDs derived from (A) GPS TEC maps at 08–16 h UT in 2008 over Northern (*gray bars*) and Southern (*red bars*) Europe (Otsuka et al., 2013), (B) Blackstone SuperDARN observations at 08–17 h magnetic local time in June 2010–May 2011 over the United States (Frissell et al., 2014), (C) Hokkaido SuperDARN observations at 01–03 UT in November 2006–February 2007 over Northern Japan (Ishida et al., 2008), (D) TEC maps in 2009 over Central China (Ding et al., 2011), and (E) GPS TEC observations in December 2012–February 2016 over Brazil (Figueiredo et al., 2018). The radii of the circles in panel D indicate the sum of observation hours.

hemisphere (a–d), the propagation direction is preferentially southward with a certain level of spread in the east-west direction. In the Southern hemisphere (e), the propagation is highly directional toward the equator during winter. The propagation direction appears to be randomly distributed in other seasons, especially during summer. Comparing Fig. 7.48D and E, daytime MSTIDs propagate toward the equator in the opposite hemispheres during winter. Because daytime MSTIDs occur most frequently during winter in both hemispheres (see Fig. 7.49), the equatorward propagation can be considered to be the representative property of daytime MSTIDs.

7.4.3.2 Occurrence climatology

Because the global morphology of daytime MSTIDs has not yet been established using satellite observations, we make a coarse assessment of the global morphology by synthesizing available ground-based observations. Fig. 7.49 presents observations over (a) the United States (Frissell et al., 2014), (b) Europe (Otsuka et al., 2013), (c) Central China (Ding et al., 2011), (d) Japan (Otsuka et al., 2011), (e) Brazil (Figueiredo et al., 2018), and (f) six locations (Kotake et al., 2006). Fig. 7.49A shows the number of daytime MSTID events derived from the Blackstone SuperDARN observations in June 2010–May 2011. Over the United States, daytime MSTIDs preferentially occur during winter. The distributions of MSTIDs over Europe (Fig. 7.49B) and Japan (Fig. 7.49D) are derived from TEC perturbation maps in 2008 and 2002, respectively. The annual occurrence pattern of daytime MSTIDs over Europe and Japan is similar to that over the United States with the peak occurrence rate during winter. The occurrence rates of daytime (red bars) and nighttime (blue bars) MSTIDs over Central China (Fig. 7.49C) are derived from GPS TEC measurements in 2009–10. The sophisticated detection methods of MSTIDs and wave characteristics are described in Ding et al. (2011). The seasonal variation of the occurrence rate of daytime MSTIDs over China is not as severe as that observed over the United States, Europe, and Japan, but daytime MSTIDs preferentially occur during winter over China. The distribution of MSTIDs over Brazil (Fig. 7.49E) is derived using TEC perturbation maps in December 2012–February 2016. The occurrence of MSTIDs is concentrated in the afternoon and in the months near July (Southern hemisphere winter).

Fig. 7.49F presents the distributions of normalized GPS TEC fluctuations over six locations. The results are derived from the observations in 2000. Focusing on the results on the dayside, daytime MSTIDs are winter phenomena in both hemispheres. This seasonal pattern is consistent with the observations in Fig. 7.49A–E. The large percentages near dawn (06 h LT) in Fig. 7.49F are seen to be caused by the rapid plasma density enhancement after sunrise (Kotake et al., 2007). This feature does not appear in Fig. 7.49B and D. MSTIDs rarely occur at night during June solstices over Europe in Fig. 7.49B and F. This daily pattern is different from that observed over Japan and China. Over China (Fig. 7.49C), the occurrence rates of MSTIDs at night are about a factor of 2 greater than those on the dayside. The occurrence rates of daytime and nighttime MSTIDs may not be directly comparable because their characteristics are different. If the amplitudes of daytime and nighttime MSTIDs are significantly different, the use of the same detection threshold would under represent smaller amplitudes of them.

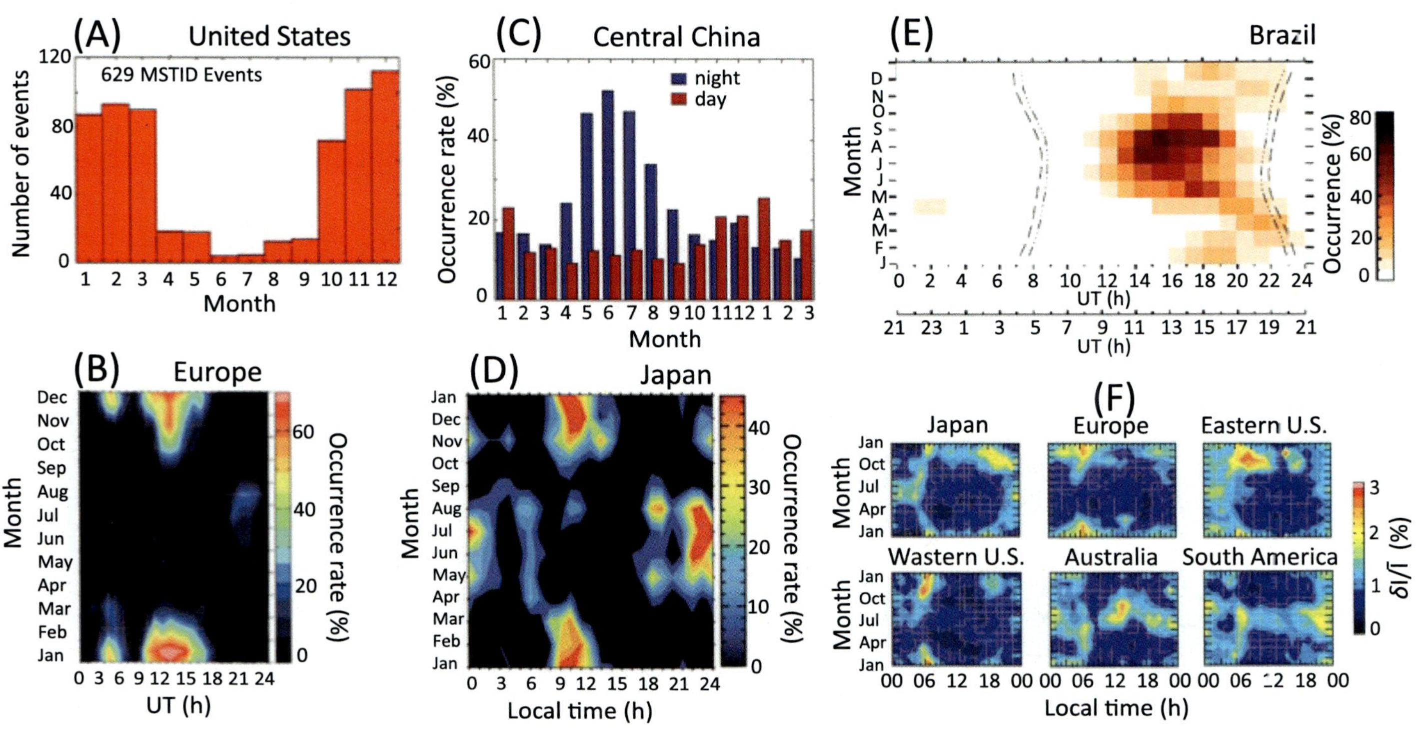

Fig. 7.49 Comparison of the distributions of daytime MSTIDs. (A) Number of daytime MSTID events derived from the Blackstone SuperDARN observations at 08–17 magnetic local time in June 2010–May 2011 over the United States (Frissell et al., 2014). (B) Occurrence rate derived from GPS TEC perturbation maps in 2008 over Europe (Otsuka et al., 2013). (C) Occurrence rates of nighttime (*blue*) and daytime (*red*) MSTIDs derived from GPS TEC measurements in 2009 over Central China (Ding et al., 2011). (D) Occurrence rate derived from GPS TEC perturbation maps over Japan in 2002 (Otsuka et al., 2011). (E) Occurrence rate derived using TEC perturbation maps in December 2012–February 2016 over Brazil (Figueiredo et al., 2018). (F) Normalized strength of TEC fluctuations derived from GPS observations in 2000 (Kotake et al., 2006).

7.4.4 Discussion

7.4.4.1 Atmospheric gravity waves

The generation of MSTIDs is understood in terms of three key factors: AGWs, Perkins instability, and E- and F-region coupling. AGWs have long been suspected as sources of wave-like modulations of the ionosphere (e.g., Behnke, 1979; Harper, 1972; Hines, 1960; Kelley and Fukao, 1991; Mendillo et al., 1997; Thome, 1964). Neutral motions induced by AGWs drive plasma motions along magnetic field lines via neutral-ion collisions and modulate the F-region height and plasma density. The generation of MSTIDs on the dayside is explained by this process. Some of the characteristics of daytime MSTIDs can be predicted based on the properties of AGWs. The seasonal behavior of daytime MSTIDs (the high occurrence frequency during winter) is consistent with the observations of the high occurrence frequency of AGWs in the thermosphere during winter (Forbes et al., 2016; Garcia et al., 2016; Park et al., 2014). Model simulations show that upward propagating AGWs generate TIDs preferentially during winter (Miyoshi et al., 2018). The preference of the equatorward propagation of daytime MSTIDs is attributed to the greater oscillation amplitudes of neutral particle motions parallel to the magnetic field lines for AGWs propagating toward the equator (Hooke, 1968). Longitudinal distributions and solar cycle dependences of daytime MSTIDs and AGWs are useful for further validation of their relationship, but observations of daytime MSTIDs are not sufficient for these investigations. As has been done for nighttime MSTIDs, the establishment of the global morphology of daytime MSTIDs using satellite observations is required for this purpose.

7.4.4.2 Perkins instability

The Perkins instability describes the amplification of ionospheric waves under certain conditions for the geometry of the magnetic field lines, the propagation direction of ionospheric waves, and the directions of background electric fields and neutral winds (Hamza, 1999; Perkins, 1973; Zhou and Mathews, 2006). Because the Perkins instability predicts the development of ionospheric perturbations aligned in the northwest-southeast direction at night, nighttime MSTIDs are often understood in association with the Perkins instability. Here we briefly review the Perkins instability and its growth conditions following the illustration in studies by Kelley et al. (2003b) and Makela and Otsuka (2012).

The dominant processes that determine the F-region height at night in mid-latitudes are the plasma motions parallel to the magnetic field lines by the neutral drag and gravitational-diffusion and the plasma motion perpendicular to the magnetic field (**B**) by electric fields (**E**) (i.e., $\mathbf{E} \times \mathbf{B}$ drift). Fig. 7.50A illustrates the vertical plasma motions driven by these three components. The equational form of the vertical plasma velocity V_z at a given height is expressed as

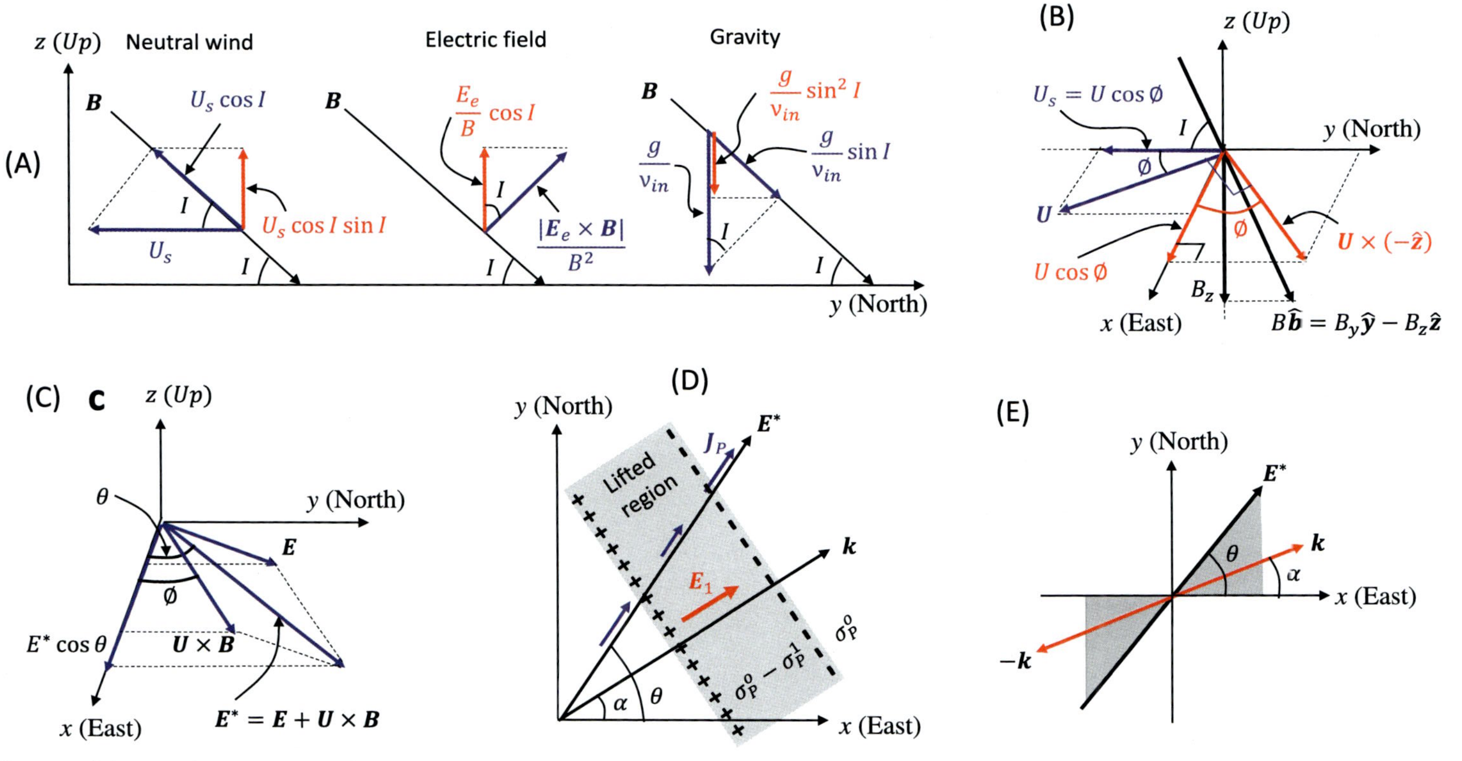

Fig. 7.50 (A) Vertical components of plasma motions driven by the southward component of neutral wind (U_s), eastward component of electric field (E_e), and gravitational force (g). (B) Representation of U_s using the eastward component of $\mathbf{U} \times (\hat{\mathbf{z}})$. (C) Representation of the summation of $\mathbf{E}$ and $\mathbf{U} \times \mathbf{B}$ using an effective electric field ($\mathbf{E}^*$). (D) Schematic illustration of the F-region uplift (*gray shading*) by a wave propagating toward $\hat{\mathbf{k}}$ and the generation of a polarization electric field ($\mathbf{E}_1$). (E) The regions (*gray shadings*) of the wave vector ($\hat{\mathbf{k}}$) where the growth rate of the Perkins instability is positive for a given $\mathbf{E}^*$).

$$V_z = U_s \cos I \sin I + \frac{E_e}{B} \cos I - \frac{g}{\nu_{in}} \sin^2 I \tag{7.27}$$

where U_s, E_e, B, and g are the magnitudes of the southward component of neutral wind (**U**), eastward component of electric field (**E**), magnetic field (**B**), and gravitational acceleration (**g**), respectively. I is the dip angle of the magnetic field and ν_{in} is the ion-neutral collision rate. The first two terms on the right-hand side of Eq. (7.27) can be represented with an effective electric field ($\mathbf{E}^*$) defined as $\mathbf{E}^* = \mathbf{E} + \mathbf{U} \times \mathbf{B}$. As Fig. 7.50B illustrates, the southward component of neutral wind ($U_s = U \cos\phi$) is equivalent to the eastward component of $\mathbf{U} \times (-\hat{\mathbf{z}})$:

$$U_s = |\mathbf{U} \times (-\hat{\mathbf{z}}) \cos\phi| = \left|\mathbf{B} \times \left(\frac{B}{B_z}\hat{\mathbf{b}} - \frac{B_y}{B_z}\hat{\mathbf{y}}\right) \cos\phi\right| = \frac{|\mathbf{U} \times \mathbf{B} \cos\phi|}{B \sin I} \tag{7.28}$$

The first two terms on the right-hand side of Eq. (7.27), as shown in Fig. 7.50C, can be written as

$$U_s \cos I \sin I + \frac{E_e}{B} \cos I = \frac{|\mathbf{E}^*| \cos\theta}{B} \cos I \tag{7.29}$$

In equilibrium, this term is balanced with the gravitational diffusion

$$\frac{|\mathbf{E}^*| \cos\theta}{B} \cos I = \frac{g}{\nu_{in}} \sin^2 I \tag{7.30}$$

The uplift of the F-region by the eastward electric field enhances the downward plasma diffusion because ν_{in} decreases with increasing altitudes. Thus, the F-region height is stabilized. For a larger dip angle, a stronger eastward electric field is required to maintain the F-region height.

The equilibrium state can be perturbed by modulations of the F-region height. Let us consider a situation where the F-layer height is enhanced by waves propagating toward the northeast direction. Fig. 7.50D illustrates the lifted F-layer (gray shading) whose wave front is aligned in the northwest-southeast direction by the waves propagating toward the northeast (**k**). The Pedersen current ($\mathbf{J}_P$) is induced by $\mathbf{E}^*$, but the current is not continuous because the electrical conductance is smaller at the lifted region. The accumulation of positive and negative charges at the boundaries develops polarization electric field ($\mathbf{E}_1$) and maintains the continuity of the current in the **k** direction:

$$\sigma_P^o E^* \cos(\theta - \alpha) = (\sigma_P^o - \sigma_P^1)[E^* \cos(\theta - \alpha) + E_1] \tag{7.31}$$

The superscripts "o" and "1" on σ_P denote the background and perturbation conductivities, respectively. In the F-region, where the gyro frequencies of electrons and ions are much greater than their collision frequencies with neutral particles, the Pedersen conductivity is proportional to ν_{in} and Eq. (7.31) can be written as

$$\nu_{in}^{o} E^{*} \cos(\theta - \alpha) = (\nu_{in}^{o} - \nu_{in}^{1})[E^{*} \cos(\theta - \alpha) + E_1] \tag{7.32}$$

Using Eqs. (7.30), (7.32) and trigonometric formulas, Eq. (7.27) at the lifted region is expressed as

$$V_z = \frac{E^{*} \cos\theta + E_1 \cos\alpha}{B} \cos I - \frac{g}{\nu_{in}^{o} - \nu_{in}^{1} \sin^2 I} = \frac{\nu_{in}^{1}}{\nu_{in}^{o} - \nu_{in}^{1}} \frac{E^{*}}{B} \cos I \sin(\theta - \alpha) \sin\alpha \tag{7.33}$$

Eq. (7.33) tells us that V_z is positive when the direction of **k** is between the direction of the electric field and y-axis (east). Under this condition, V_z increases with increasing altitude because ν_{in}^{o} decreases with increasing altitude. Because the upward motion is not balanced with the downward diffusion, the F-region height becomes unstable.

A rigorous calculation requires the use of field-line integrated conductance and currents. A simple form of the linear growth rate (γ) of the Perkins instability (Perkins, 1973) under the condition that the ion gyro frequency is much greater than the ion-neutral collision rate is similar to Eq. (7.33):

$$\gamma = \frac{E^{*}}{BH_n} \cos I \sin(\theta - \alpha) \sin\alpha = \frac{g \sin^2 I}{\langle\nu\rangle H} \frac{\sin(\theta - \alpha)}{\cos\theta} \tag{7.34}$$

where H_n is the neutral scale height and $\langle\nu\rangle$ is the plasma density-weighted ion-neutral collision rate. In Fig. 7.50E, the regions where γ has a positive value is indicated with gray shadings. Because the background electric field (**E**) in mid-latitudes is small under magnetically quiet conditions (Makela and Otsuka, 2012), the effective electric field (**E***) is dominated by **U** ×**B** which represents the southward component of neutral winds. The global wind pattern is eastward and equatorward at night in mid-latitudes. Therefore, **E*** in Fig. 7.50 is close to the actual effective electric field driven by southeastward neutral winds. Then, ionospheric perturbations aligned in the northwest-southeast direction become unstable for the Perkins instability. The morphology of MSTIDs at night is consistent with the prediction of the Perkins instability.

7.4.4.3 E- and F-region coupling

The tilted structures of nighttime MSTIDs are explained by the Perkins instability, but the growth rate of the Perkins instability is known to be too small to explain the observed magnitudes of nighttime MSTIDs (Garcia et al., 2000; Kelley and Fukao, 1991; Kelley and Miller, 1997; Miller and Kelley, 1997). This problem is addressed by appealing to E- and F-region coupling. Electric fields driven by plasma instabilities in the E-region can produce MSTIDs at night directly or by promoting the Perkins instability (Cosgrove and Tsunoda, 2002; Cosgrove et al., 2004; Haldoupis et al., 2003; Kelley and Fukao, 1991;

Kelley et al., 2003a, b; Tsunoda, 2006; Tsunoda and Cosgrove, 2001; Yokoyama, 2014; Yokoyama and Hysell, 2010; Yokoyama et al., 2009).

Here we discuss the relationship between sporadic E (Es) in the E-region and nighttime MSTIDs. Es represents patches of the electron density enhancements in the ionospheric E layer. Because perturbations in the E layer are often associated with the electron density gradient caused by Es (Haldoupis et al., 2003; Otsuka et al., 2008), Es is used as an indicator of perturbations in the E layer or vice versa. If the E- and F-region coupling plays a crucial role in the generation of nighttime MSTIDs, the occurrences of Es and nighttime MSTIDs would be correlated. The relationship between Es and nighttime MSTIDs is supported by the observations of their coincident occurrence (Helmboldt, 2012, 2016; Helmboldt and Hurley-Walker, 2020; Lakshmi Narayanan et al., 2014; Otsuka et al., 2007, 2008) and the similarity in their longitudinal distributions (Lee et al., 2021). Model simulations also show the significant role of Es in the generation of MSTIDs at night (Yokoyama, 2014; Yokoyama and Hysell, 2010; Yokoyama et al., 2009).

Lee et al. (2021) investigated the global distributions of Es and nighttime MSTIDs and their dependence on the solar cycle using the observations of the CHAMP satellite at 20–04 h LT in 2001–09. Nighttime MSTIDs are detected using the fluctuations of the logarithm of the electron density. The measurements of the signal-to-noise ratio from the GPS radio occultation experiment are used for the detection of Es. Fig. 7.51A and B shows the morphologies of nighttime MSTIDs and Es, respectively, as a function of longitude and month. The occurrence of nighttime MSTIDs shows a semiannual pattern in both hemispheres, but Es shows a seasonal pattern with the peak occurrence rate in summer. Therefore, the annual behavior of Es is different from that of MSTIDs. The morphology of nighttime MSTIDs during summer (June solstices in the Northern hemisphere and December solstices in the Southern hemisphere) can be explained in terms of Es in the local hemisphere, but the high occurrence rate of nighttime MSTIDs during winter (December solstices in the Northern hemisphere and June solstices in the Southern hemisphere) is not explained by Es in the local hemisphere because of the low occurrence rate of Es in winter. To explain the morphology of nighttime MSTIDs in both hemispheres in terms of Es, nighttime MSTIDs must be conjugate. As long as conjugacy is effective, Es in one hemisphere can generate MSTIDs in both hemispheres. Thus, asserting the role of Es (or the E- and F-region coupling) in the generation of nighttime MSTIDs is contingent on the observation of conjugacy in nighttime MSTIDs.

The dependence of nighttime MSTIDs and Es on the solar activity is shown in Fig. 7.51C. The occurrence rate of nighttime MSTIDs decreases with increasing solar activity, but the occurrence rate of Es shows only a minor variation with the solar activity. These observations indicate that factors other than Es are involved in the generation of nighttime MSTIDs. Because the growth rate of the Perkins instability, mapping of

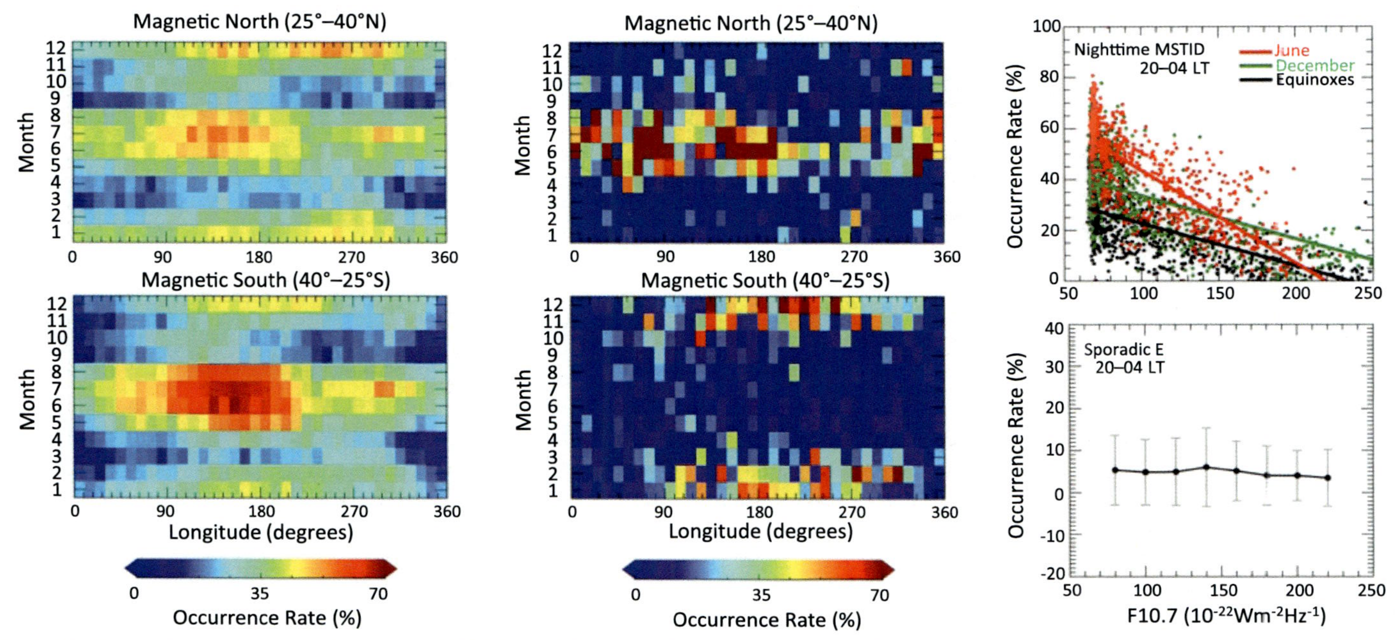

Fig. 7.51 Comparison of the morphologies of MSTIDs and Es. (A) MSTID distributions as a function of longitude and month in the magnetic North (25°–40°N) and magnetic South (25°–40°S). (B) The same format for Es. (C) Occurrence rates of (*top*) nighttime MSTIDs and (*bottom*) Es as a function of the solar F10.7 index. For nighttime MSTIDs, the observations during different months are distinguished with different colors and the lines are least square linear fitting lines. The *vertical bars* in the bottom panel are the standard deviations. The occurrence rates of MSTIDs and Es are derived using the CHAMP data at 20–04 h LT in 2001–09 and 2001–08, respectively (Lee et al., 2021).

electric fields along magnetic field lines, and propagation of atmospheric gravity waves vary with the solar cycle, these factors would determine the solar cycle dependence of nighttime MSTIDs.

7.4.5 Summary and future work

The characteristics of MSTIDs are variable by various factors, and discrepancies exist among observations. Some of the discrepancies may be related to observation techniques or data processing. Because the consistent features in all observations are difficult to identify, we summarize common features seen in majority observations.

1. Nighttime MSTIDs are elongated in the northwest-southeast direction and propagate toward the southwest in the Northern hemisphere. The elongation and propagation direction of nighttime MSTIDs in the Southern hemisphere are mirror images of those in the Northern hemisphere.
2. The periods, wavelengths, and phase velocities of nighttime MSTIDs in mid-latitude are of 0.5–1.5 h, 100–300 km, and 50–200 m s^{-1}.
3. The occurrence rate of nighttime MSTIDs is described by the semiannual variation with a primary peak during June solstices and a secondary peak during December solstices in both hemispheres. The semiannual variation is pronounced during the periods of lower solar activity. However, the amplitudes of the primary and secondary peaks are comparable or reversed in some longitude regions.
4. The annual-longitudinal patterns of nighttime MSTIDs in the opposite hemispheres are quasisymmetric.
5. The occurrence rate of nighttime MSTIDs increases with declining solar activity.
6. The occurrence rate of nighttime MSTIDs is the largest in the Asian sector during June solstices during any periods of the solar cycle.
7. The periods, wavelength, and phase velocities of daytime MSTIDs are comparable to those of nighttime MSTIDs.
8. Daytime MSTIDs preferentially develop in winter in both hemispheres.
9. Daytime MSTIDs preferentially propagate toward equator in both hemispheres. This directional behavior is pronounced in winter.
10. The longitudinal distribution of nighttime MSTIDs in the summer hemisphere is similar to the longitudinal distribution of Es in the local hemisphere. The morphology of nighttime MSTIDs in both hemispheres can be explained in terms of Es when nighttime MSTIDs have conjugacy.

Ground-based observations show that the occurrence of MSTIDs in the equatorial region is not ignorable. Characteristics of MSTIDs in the equatorial region seen to be different from those in higher latitudes, but more observations in the equatorial region are necessary for the investigation of the latitudinal variation of the characteristics. We have poor knowledge about the longitudinal and hemispheric distributions of daytime

MSTIDs. Satellite observations are a useful tool for the establishment of the global morphology of daytime MSTIDs. Numerical simulations are desirable to identify the factors that affect the conjugate property in nighttime MSTIDs and the generation of blobs and mid-latitude bubbles from MSTIDs.

Acknowledgments

JL and TD were supported by the B-type Strategic Priority Program of the Chinese Academy of Sciences (XDB41000000), the National Natural Science Foundation of China (41831070 and 41974181), and the Open Research Project of Large Research Infrastructures of CAS-"Study on the Interaction Between Low/Mid-Latitude Atmosphere and Ionosphere Based on the Chinese Meridian Project." TD was supported by the National Natural Science Foundation of China (42174198, 41904138), the National Postdoctoral Program for Innovative Talents (BX20180286), the China Postdoctoral Science Foundation (2018M642525), and the Fundamental Research Funds for the Central Universities.

HL acknowledges support from JSPS KAKENHI Grants 18H01270, 18H04446, 17KK0095, and JSPS-DFG Japan-Germany bilateral program (JPJSJRP 20181602). SE acknowledges NASA contracts 80GSFC18C0061 and NNG12FA45C.

The work at JHU/APL was supported by US Air Force MURI FA9559-16-1-0364, US Air Force International Cooperative Research and Development Program, and NSF-AGS2029840. WKL acknowledges the National Research Foundation of Korea (NRF) grant funded by the Korea Government (MSIT) (No. 2018R1C1B6006700).

S-RZ acknowledges support from the US AFOSR MURI program FA9559-16-1-036, NSF awards AGS-2033787 and AGS-1952737, NASA LWS programs 80NSSC21K1310, NNX15AB83G and 80NSSC19K0078, and US ONR Grant N00014-17-1-2186.

We also thank the support from the ISSI and ISSI-BJ for the international team on "Multi-Scale Magnetosphere-Ionosphere-Thermosphere Interaction."

References

Aa, E., Huang, W., Liu, S., Ridley, A., Zou, S., Shi, L., Chen, Y., Shen, H., Yuan, T., Li, J., et al., 2018. Midlatitude plasma bubbles over China and adjacent areas during a magnetic storm on 8 September 2017. Space Weather 16 (3), 321–331.

Aa, E., Zou, S., Ridley, A., Zhang, S.-R., Coster, A.J., Erickson, P.J., Liu, S., Ren, J., 2019. Merging of storm-time midlatitude traveling ionospheric disturbances and equatorial plasma bubbles. Space Weather 17, 285–298. 2018SW002101.

Acuña, M.H., Connerney, J.E.P., Lin, R.P., Mitchell, D., Carlson, C.W., McFadden, J., Anderson, K.A., Rème, H., Mazelle, C., Vignes, D., et al., 1999. Global distribution of crustal magnetization discovered by the Mars Global Surveyor MAG/ER experiment. Science 284 (5415), 790–793.

Alken, P., 2009. Modeling Equatorial Ionospheric Currents and Electric Fields From Satellite Magnetic Field Measurements (Ph.D. thesis). University of Colorado at Boulder.

Amorim, D.C.M., Pimenta, A.A., Bittencourt, J.A., Fagundes, P.R., 2011. Long-term study of medium-scale traveling ionospheric disturbances using OI 630 nm all-sky imaging and ionosonde over Brazilian low latitudes. J. Geophys. Res. Space Phys. 116, A06312.

Anastassiadis, M., Matsoukas, D., 1969. Electron content measurements by beacon S-66 satellite during the May 20, 1966, solar eclipse. J. Atmos. Terr. Phys. 31 (9), 1217–1222.

Appleton, E.V., 1946. Two anomalies in the ionosphere. Nature 157, 691.

Axford, W.I., 1963. The formation and vertical movement of dense ionized layers in the ionosphere due to neutral wind shears. J. Geophys. Res. 68 (3), 769–779.

Axford, W.I., Hines, C.O., 1961. A unifying theory of high-latitude geophysical phenomena and geomagnetic storms. Can. J. Phys. 39 (10), 1433–1464.

Azeem, I., Yue, J., Hoffmann, L., Miller, S.D., Straka III, W.C., Crowley, G., 2015. Multisensor profiling of a concentric gravity wave event propagating from the troposphere to the ionosphere. Geophys. Res. Lett. 42 (19), 7874–7880.

Balthazor, R.L., Moffett, R.J., 1999. Morphology of large-scale traveling atmospheric disturbances in the polar thermosphere. J. Geophys. Res. Space Phys. 104 (A1), 15–24.

Banks, P.M., 1977. Observations of joule and particle heating in the auroral zone. J. Atmos. Terr. Phys. 39 (2), 179–193.

Banks, P.M., Foster, J.C., Doupnik, J.R., 1981. Chatanika radar observations relating to the latitudinal and local time variations of joule heating. J. Geophys. Res. Space Phys. 86 (A8), 6869–6878.

Behnke, R., 1979. F layer height bands in the nocturnal ionosphere over Arecibo. J. Geophys. Res. Space Phys. 84 (A3), 974–978.

Bellchambers, W.H., Piggott, W.R., 1958. Ionospheric measurements made at Halley bay. Nature 182, 1596–1597.

Bilitza, D., 2018. IRI the international standard for the ionosphere. Adv. Radio Sci. 16, 1–11.

Biondi, M.A., Meriwether Jr., J.W., 1985. Measured response of the equatorial thermospheric temperature to geomagnetic activity and solar flux changes. Geophys. Res. Lett. 12 (5), 267–270.

Booker, H.G., Wells, H.W., 1938. Scattering of radio waves by the F-region of the ionosphere. Terr. Magn. Atmos. Electr. 43 (3), 249–256.

Borries, C., Jakowski, N., Wilken, V., 2009. Storm induced large scale TIDs observed in GPS derived TEC. Ann. Geophys. 27 (4), 1605–1612.

Bougher, S.W., Engel, S., Hinson, D.P., Murphy, J.R., 2004. Mgs radio science electron density profiles: interannual variability and implications for the Martian neutral atmosphere. J. Geophys. Res. Planets 109 (E3).

Bowman, G.G., 2001. A comparison of nighttime TID characteristics between equatorial-ionospheric-anomaly crest and midlatitude regions, related to spread F occurrence. J. Geophys. Res. Space Phys. 106 (A2), 1761–1769.

Bruinsma, S.L., Forbes, J.M., 2010. Large-scale traveling atmospheric disturbances (LSTADs) in the thermosphere inferred from CHAMP, GRACE, and SETA accelerometer data. J. Atmos. Sol. Terr. Phys. 72 (13), 1057–1066.

Burkard, O., 1965. Die halbjaehrige periode der F2-Schicht-ionisation. Arch. Meteorol. Bioklim. Wien 4, 391–402.

Burns, A.G., Wang, W., Killeen, T.L., Solomon, S.C., 2004. A “tongue” of neutral composition. J. Atmos. Sol. Terr. Phys. 66 (15–16), 1457–1468.

Burns, A.G., Zeng, Z., Wang, W., Lei, J., Solomon, S.C., Richmond, A.D., Killeen, T.L., Kuo, Y.-H., 2008. Behavior of the F_2 peak ionosphere over the South Pacific at dusk during quiet summer conditions from COSMIC data. J. Geophys. Res. Space Phys. 113 (A12).

Burns, A.G., Solomon, S.C., Wang, W., Jee, G., Lin, C.H., Rocken, C., Kuo, Y.H., 2011. The summer evening anomaly and conjugate effects. J. Geophys. Res. Space Phys. 116 (A1), A01311.

Candido, C.M.N., Pimenta, A.A., Bittencourt, J.A., Becker-Guedes, F., 2008. Statistical analysis of the occurrence of medium-scale traveling ionospheric disturbances over Brazilian low latitudes using OI 630.0 nm emission all-sky images. Geophys. Res. Lett. 35, L17105.

Chang, L.C., Thayer, J.P., Lei, J., Palo, S.E., 2009. Isolation of the global MLT thermal response to recurrent geomagnetic activity. Geophys. Res. Lett. 36, L15813.

Chang, L.C., Liu, H., Miyoshi, Y., Chen, C.-H., Chang, F.-Y., Lin, C.-H., Liu, J.-Y., Sun, Y.-Y., 2015. Structure and origins of the Weddell Sea Anomaly from tidal and planetary wave signatures in FORMOSAT-3/COSMIC observations and GAIA GCM simulations. J. Geophys. Res. Space Phys. 120 (2), 1325–1340. https://doi.org/10.1002/2014JA020752.

Chen, C.H., Huba, J.D., Saito, A., Lin, C.H., Liu, J.Y., 2011. Theoretical study of the ionospheric Weddell sea Anomaly using SAMI2. J. Geophys. Res. Space Phys. 116 (A4), A04305. https://doi.org/10.1029/2010JA015573.

Cherniak, I., Zakharenkova, I., 2016. First observations of super plasma bubbles in Europe. Geophys. Res. Lett. 43 (21), 11–137.
Cherniak, I., Zakharenkova, I., Sokolovsky, S., 2019. Multi-instrumental observation of storm-induced ionospheric plasma bubbles at equatorial and middle latitudes. J. Geophys. Res. Space Phys. 124 (3), 1491–1508.
Chimonas, G., Hines, C.O., 1970. Atmospheric gravity waves induced by a solar eclipse. J. Geophys. Res. Space Phys. (1978–2012) 75 (4), 875. 875.
Choi, H.-S., Kil, H., Kwak, Y.-S., Park, Y.-D., Cho, K.-S., 2012. Comparison of the bubble and blob distributions during the solar minimum. J. Geophys. Res. Space Phys. 117, A04314.
Chou, M.Y., Lin, C.C.H., Yue, J., Tsai, H.F., Sun, Y.Y., Liu, J.Y., Chen, C.H., 2017. Concentric traveling ionosphere disturbances triggered by super Typhoon Meranti (2016). Geophys. Res. Lett. 44 (3), 1219–1226.
Clark, R.M., Yeh, K.C., Liu, C.H., 1971. Interaction of internal gravity waves with the ionospheric F2-layer. J. Atmos. Terr. Phys. 33 (10), 1567–1576.
Collinson, G., McFadden, J., Mitchell, D., Grebowsky, J., Benna, M., Espley, J., Jakosky, B., 2019. Traveling ionospheric disturbances at Mars. Geophys. Res. Lett. 46 (9), 4554–4563. https://doi.org/10.1029/2019GL082412.
Cosgrove, R.B., Tsunoda, R.T., 2002. A direction-dependent instability of sporadic-E layers in the nighttime midlatitude ionosphere. Geophys. Res. Lett. 29 (18). https://doi.org/10.1029/2002GL014669.
Cosgrove, R.B., Tsunoda, R.T., Fukao, S., Yamamoto, M., 2004. Coupling of the Perkins instability and the sporadic E layer instability derived from physical arguments. J. Geophys. Res. Space Phys. 109 (A6), A06301. https://doi.org/10.1029/2003JA010295.
Coster, A.J., Goncharenko, L., Zhang, S.-R., Erickson, P.J., Rideout, W., Vierinen, J., 2017. GNSS observations of ionospheric variations during the 21 August 2017 solar eclipse. Geophys. Res. Lett. 44, 12. https://doi.org/10.1002/2017GL075774.
Cowling, D.H., Webb, H.D., Yeh, K.C., 1971. Group rays of internal gravity waves in a wind-stratified atmosphere. J. Geophys. Res. Space Phys. (1978–2012) 76 (1), 213–220.
Dang, T., Luan, X., Lei, J., Dou, X., Wan, W., 2016. A numerical study of the interhemispheric asymmetry of the equatorial ionization anomaly in solstice at solar minimum. J. Geophys. Res. Space Phys. 121 (9), 9099–9110.
Dang, T., Lei, J., Wang, W., Burns, A., Zhang, B., Zhang, S.R., 2018a. Suppression of the polar tongue of ionization during the 21 August 2017 solar eclipse. Geophys. Res. Lett. 123, 2918–2925. https://doi.org/10.1002/2018GL077328.
Dang, T., Lei, J., Wang, W., Zhang, B., Burns, A., Le, H., Wu, Q., Ruan, H., Dou, X., Wan, W., 2018b. Global responses of the coupled thermosphere and ionosphere system to the August 2017 great American Solar Eclipse. J. Geophys. Res. 123, 7040–7050. https://doi.org/10.1029/2018JA025566.
Dang, T., Lei, J., Wang, W., Wang, B., Zhang, B., Liu, J., Burns, A., Nishimura, Y., 2019. Formation of double tongues of ionization during the 17 March 2013 geomagnetic storm. J. Geophys. Res. Space Phys. 124 (12), 10619–10630.
Davis, C.J., Lockwood, M., Bell, S.A., Smith, J.A., Clarke, E.M., 2000. Ionospheric measurements of relative coronal brightness during the total solar eclipses of 11 August, 1999 and 9 July, 1945. Ann. Geophys. 18 (2), 182–190.
Deng, Y., Ridley, A., 2006. Role of vertical ion convection in the high-latitude ionospheric plasma distribution. J. Geophys. Res. Space Phys. 111, A09314.
Deng, W., Killeen, T.L., Burns, A.G., Johnson, R.M., Emery, B.A., Roble, R.G., Winningham, J.D., Gary, J.B., 1995. One-dimensional hybrid satellite track model for the dynamics explorer 2 (DE 2) satellite. J. Geophys. Res. Space Phys. 100 (A2), 1611–1624.
Ding, F., Wan, W., Yuan, H., 2003. The influence of background winds and attenuation on the propagation of atmospheric gravity waves. J. Atmos. Sol. Terr. Phys. 65 (7), 857–869.
Ding, F., Wan, W., Ning, B., Wang, M., 2007. Large-scale traveling ionospheric disturbances observed by GPS total electron content during the magnetic storm of 29–30 October 2003. J. Geophys. Res. Space Phys. 112, A06309.
Ding, F., Wan, W., Liu, L., Afraimovich, E.L., Voeykov, S.V., Perevalova, N.P., 2008. A statistical study of large-scale traveling ionospheric disturbances observed by GPS TEC during major magnetic storms over the years 2003–2005. J. Geophys. Res. Space Phys. (1978–2012) 113 (A3).

Ding, F., Wan, W., Ning, B., Liu, L., Le, H., Xu, G., Wang, M., Li, G., Chen, Y., Ren, Z., Xiong, B., Hu, L., Yue, X., Zhao, B., Li, F., Yang, M., 2010. GPS TEC response to the 22 July 2009 total solar eclipse in East Asia. J. Geophys. Res. Space Phys. 115, A07308. https://doi.org/10.1029/2009JA015113.

Ding, F., Wan, W., Xu, G., Yu, T., Yang, G., Wang, J.-S., 2011. Climatology of medium-scale traveling ionospheric disturbances observed by a GPS network in central China. J. Geophys. Res. Space Phys. 116, A09327.

Drob, D.P., Emmert, J.T., Meriwether, J.W., Makela, J.J., Doornbos, E., Conde, M., Hernandez, G., Noto, J., Zawdie, K.A., McDonald, S.E., Huba, J.D., Klenzing, J.H., 2015. An update to the horizontal wind model (HWM): the quiet time thermosphere. Earth Space Sci. 2, 301–319. https://doi.org/10.1002/2014EA000089.

Copernicus GmbH Duly, T.M., Chapagain, N.P., Makela, J.J., 2013. Climatology of nighttime medium-scale traveling ionospheric disturbances (MSTIDS) in the central pacific and south American sectors. Ann. Geophys. 31 (12), 2229–2237.

Duncan, R.A., 1960. The equatorial F-region of the ionosphere. J. Atmos. Terr. Phys. 18 (2–3), 89–100.

Dungey, J.W., 1961. Interplanetary magnetic field and the auroral zones. Phys. Rev. Lett. 6, 47–48. https://doi.org/10.1103/PhysRevLett.6.47.

Eccles, J.V., 1998. A simple model of low-latitude electric fields. J. Geophys. Res. 103 (A11), 26,699–26,708. https://doi.org/10.1029/98JA02657.

Emmert, J.T., 2015. Altitude and solar activity dependence of 1967–2005 thermospheric density trends derived from orbital drag. J. Geophys. Res. Space Phys. 120 (4), 2940–2950. https://doi.org/10.1002/2015JA021047.

Emmert, J.T., Picone, J.M., 2010. Climatology of globally averaged thermospheric mass density. J. Geophys. Res. Space Phys. 115 (A9). https://doi.org/10.1029/2010JA015298.

England, S.L., Immel, T.J., Sagawa, E., Henderson, S.B., Hagan, M.E., Mende, S.B., Frey, H.U., Swenson, C.M., Paxton, L.J., 2006a. Effect of atmospheric tides on the morphology of the quiet time, postsunset equatorial ionospheric anomaly. J. Geophys. Res. Space Phys. 111 (A10). https://doi.org/10.1029/2006JA011795.

England, S.L., Maus, S., Immel, T.J., Mende, S.B., 2006b. Longitudinal variation of the E-region electric fields caused by atmospheric tides. Geophys. Res. Lett. 33. https://doi.org/10.1029/2006GL027465.

England, S.L., Immel, T.J., Huba, J.D., 2008. Modeling the longitudinal variation in the post-sunset far-ultraviolet OI airglow using the SAMI2 model. J. Geophys. Res. Space Phys. 113 (A12), 1309. https://doi.org/10.1029/2007JA012536.

England, S.L., Zhang, X., Immel, T.J., Forbes, J.M., DeMajistre, R., 2009. The effect of non-migrating tides on the morphology of the equatorial ionospheric anomaly: seasonal variability. Earth Planets Space 61, 493–503.

England, S.L., Immel, T.J., Huba, J.D., Hagan, M.E., Maute, A., DeMajistre, R., 2010. Modeling of multiple effects of atmospheric tides on the ionosphere: an examination of possible coupling mechanisms responsible for the longitudinal structure of the equatorial ionosphere. J. Geophys. Res. Space Phys. 115, A05308.

England, S.L., Liu, G., Yiğit, E., Mahaffy, P.R., Elrod, M., Benna, M., Nakagawa, H., Terada, N., Jakosky, B., 2017. MAVEN NGIMS observations of atmospheric gravity waves in the Martian thermosphere. J. Geophys. Res. Space Phys. 122 (2), 2310–2335. https://doi.org/10.1002/2016JA023475.

Evans, J.V., 1965. Cause of the mid-latitude evening increase in foF2. J. Geophys. Res. 70, 1175–1185.

Fejer, B.G., De Paula, E.R., Gonzalez, S.A., Woodman, R.F., 1991. Average vertical and zonal F region plasma drifts over Jicamarca. J. Geophys. Res. 96 (A8), 13,901–13,906. https://doi.org/10.1029/91JA01171.

Figueiredo, C.A.O.B., Takahashi, H., Wrasse, C.M., Otsuka, Y., Shiokawa, K., Barros, D., 2018. Investigation of nighttime MSTIDS observed by optical thermosphere imagers at low latitudes: morphology, propagation direction, and wind filtering. J. Geophys. Res. Space Phys. 123 (9), 7843–7857.

Forbes, J.M., Marcos, F.A., Kamalabadi, F., 1995. Wave structures in lower thermosphere density from satellite electrostatic triaxial accelerometer measurements. J. Geophys. Res. 100, 14693–14701.

Forbes, J.M., Bruinsma, S.L., Doornbos, E., Zhang, X., 2016. Gravity wave-induced variability of the middle thermosphere. J. Geophys. Res. Space Phys. 121, 6914–6923. https://doi.org/10.1002/2016JA022923.

Foster, J.C., Burke, W.J., 2002. SAPS: a new categorization for sub-auroral electric fields. Eos Trans. Am. Geophys. Union 83 (36), 393–394.

Foster, J.C., Coster, A., Erickson, P.J., Holt, J.M., Lind, F.D., Rideout, W., McCready, M., Van Eyken, A., Barnes, R.J., Greenwald, R., et al., 2005. Multiradar observations of the polar tongue of ionization. J. Geophys. Res. Space Phys. 110, A09S31.

Francis, S.H., 1975. Global propagation of atmospheric gravity waves: a review. J. Atmos. Terr. Phys. 37 (6–7), 1011–1054.

Frissell, N.A., Baker, J.B.H., Ruohoniemi, J.M., Gerrard, A.J., Miller, E.S., Marini, J.P., West, M.L., Bristow, W.A., 2014. Climatology of medium-scale traveling ionospheric disturbances observed by the midlatitude blackstone SuperDARN radar. J. Geophys. Res. Space Phys. 119 (9), 7679–7697.

Fritts, D.C., Vadas, S.L., 2008. Gravity wave penetration into the thermosphere: sensitivity to solar cycle variations and mean winds. Ann. Geophys. 26 (12), 3841–3861.

Fukushima, D., Shiokawa, K., Otsuka, Y., Ogawa, T., 2012. Observation of equatorial nighttime medium-scale traveling ionospheric disturbances in 630-nm airglow images over 7 years. J. Geophys. Res. Space Phys. 117, A10324.

Fuller-Rowell, T.J., 1998. The thermospheric spoon: a mechanism for the semiannual density variation. J. Geophys. Res. 103, 3951–3956.

Fuller-Rowell, T.J., Rees, D., 1980. A three-dimensional time-dependent global model of the thermosphere. J. Atmos. Sci. 37 (11), 2545–2567.

Fuller-Rowell, T.J., Codrescu, M.V., Moffett, R.J., Quegan, S., 1994. Response of the thermosphere and ionosphere to geomagnetic storms. J. Geophys. Res. Space Phys. (1978–2012) 99 (A3), 3893–3914.

Fuller-Rowell, T.J., Codrescu, M.V., Fejer, B.G., Borer, W., Marcos, F., Anderson, D.N., 1997. Dynamics of the low-latitude thermosphere: quiet and disturbed conditions. J. Atmos. Terr. Phys. 61, 1533–1540.

Garcia, F.J., Kelley, M.C., Makela, J.J., Huang, C.-S., 2000. Airglow observations of mesoscale low-velocity traveling ionospheric disturbances at midlatitudes. J. Geophys. Res. Space Phys. 105 (A8), 18407–18415.

Garcia, R.F., Bruinsma, S., Massarweh, L., Doornbos, E., 2016. Medium-scale gravity wave activity in the thermosphere inferred from GOCE data. J. Geophys. Res. Space Phys. 121 (8), 8089–8102.

Guo, J., Liu, H., Feng, X., Wan, W., Deng, Y., Liu, C., 2014. Constructive interference of large-scale gravity waves excited by interplanetary shock on 29 October 2003: CHAMP observation. J. Geophys. Res. Space Phys. 119 (8), 6846–6851.

Guo, J.P., Deng, Y., Zhang, D.H., Lu, Y., Sheng, C., Zhang, S.-R., 2018. The effect of subauroral polarization streams on ionosphere and thermosphere during the 2015 St. Patrick's day storm: global ionosphere-thermosphere model simulations. J. Geophys. Res. Space Phys. 123 (3), 2241–2256.

Haaser, R.A., Earle, G.D., Heelis, R.A., Klenzing, J., Stoneback, R., Coley, W.R., Burrell, A.G., 2012. Characteristics of low-latitude ionospheric depletions and enhancements during solar minimum. J. Geophys. Res. Space Phys. 117, A10305.

Hagan, M.E., Maute, A., Roble, R.G., 2009. Tropospheric tidal effects on the middle and upper atmosphere. J. Geophys. Res. Space Phys. 114, A01302.

Haldoupis, C., 2011. A tutorial review on sporadic E layers. In: Abdu, M., Pancheva, D. (Eds.), Aeronomy of the Earth's Atmosphere and Ionosphere. IAGA Special Sopron Book Series, vol 2. Springer, Dordrecht. https://doi.org/10.1007/978-94-007-0326-1_29.

Haldoupis, C., Kelley, M.C., Hussey, G.C., Shalimov, S., 2003. Role of unstable sporadic-E layers in the generation of midlatitude spread F. J. Geophys. Res. Space Phys. 108 (A12), 1446.

Hamza, A.M., 1999. Perkins instability revisited. J. Geophys. Res. Space Phys. 104 (A10), 22567–22575.

Hanson, W.B., Moffett, R.J., 1966. Ionization transport effects in the equatorial F region. J. Geophys. Res. (1896–1977) 71 (23), 5559–5572. https://doi.org/10.1029/JZ071i023p05559.

Harper, R.M., 1972. Observation of a large nighttime gravity wave at Arecibo. J. Geophys. Res. 77 (7), 1311–1315.

Hartman, W.A., Heelis, R.A., 2007. Longitudinal variations in the equatorial vertical drift in the topside ionosphere. J. Geophys. Res. Space Phys. 112, A03305.

Hays, P.B., Jones, R.A., Rees, M.H., 1973. Auroral heating and the composition of the neutral atmosphere. Planet. Space Sci. 21 (4), 559–573.

He, M., Liu, L., Wan, W., Ning, B., Zhao, B., Wen, J., Yue, X., Le, H., 2009. A study of the Weddell Sea Anomaly observed by FORMOSAT-3/COSMIC. J. Geophys. Res. 1114. https://doi.org/10.1029/2009JA014175.

Hedin, A.E., Mayr, H.G., 1973. Magnetic control of the near equatorial neutral thermosphere. J. Geophys. Res. 78, 1688–1691.

Hedin, A.E., Reber, C.A., Newton, G.P., Spencer, N.W., Brinton, H.C., Mayr, H.G., Potter, W.E., 1977. A global thermospheric model based on mass spectrometer and incoherent scatter data MSIS, 2. Composition. J. Geophys. Res. (1896–1977) 82 (16), 2148–2156. https://doi.org/10.1029/JA082i016p02148.

Heelis, R.A., 2004. Electrodynamics in the low and middle latitude ionosphere: a tutorial. J. Atmos. Sol. Terr. Phys. 66, 825–838. https://doi.org/10.1016/j.jastp.2004.01.034.

Heelis, R.A., Kendall, P.C., Moffett, R.J., Windle, D.W., Rishbeth, H., 1974. Electrical coupling of the E- and F-regions and its effect on F-region drifts and winds. Planet. Space Sci. 22 (5), 743–756. https://doi.org/10.1016/0032-0633(74)90144-5.

Helmboldt, J.F., 2012. Insights into the nature of Northwest-to-Southeast aligned ionospheric wavefronts from contemporaneous very large array and ionosonde observations. J. Geophys. Res. Space Phys. 117, A07310.

Helmboldt, J., 2016. A multi-platform investigation of midlatitude sporadic E and its ties to E-F coupling and meteor activity. Ann. Geophys. 34 (5), 529–541.

Helmboldt, J.F., Hurley-Walker, N., 2020. Ionospheric irregularities observed during the GLEAM survey. Radio Sci. 55 (10). https://doi.org/10.1029/2020RS007106. e2020RS007106.

Henderson, S.B., Swenson, C.M., Gunther, J.H., Christensen, A.B., Paxton, L.J., 2005. Method for characterization of the equatorial anomaly using image subspace analysis of Global Ultraviolet Imager data. J. Geophys. Res. Space Phys. 110 (A9). https://doi.org/10.1029/2004JA010830.

Heppner, J.P., Maynard, N.C., 1987. Empirical high-latitude electric field models. J. Geophys. Res. Space Phys. 92 (A5), 4467–4489.

Hernández-Pajares, M., Juan, J.M., Sanz, J., 2006. Medium-scale traveling ionospheric disturbances affecting GPS measurements: spatial and temporal analysis. J. Geophys. Res. Space Phys. 111 (A7), A07S11.

Hines, C.O., 1959. An interpretation of certain ionospheric motions in terms of atmospheric waves. J. Geophys. Res. Space Phys. (1978–2012) 64 (12), 2210–2211.

Hines, C.O., 1960. Internal atmospheric gravity waves at ionospheric heights. Can. J. Phys. 38 (11), 1441–1481.

Hocke, K., Schlegel, K., 1996. A review of atmospheric gravity waves and travelling ionospheric disturbances: 1982–1995. Ann. Geophys. 14 (9), 917–940.

Hooke, W.H., 1968. Ionospheric irregularities produced by internal atmospheric gravity waves. J. Atmos. Terr. Phys. 30 (5), 795–823.

Horvath, I., Lovell, B.C., 2009. Investigating the relationships among the South Atlantic magnetic Anomaly, southern nighttime mid-latitude trough, and nighttime Weddell Sea Anomaly during southern summer. J. Geophys. Res. 114, 693–706. https://doi.org/10.1029/2008JA013719.

Hsu, M.L., Lin, C.H., Hsu, R.R., Liu, J.Y., Paxton, L.J., Su, H.T., Tsai, H.F., Rajesh, P.K., Chen, C.H., 2011. The OI 135.6 nm airglow observations of the midlatitude summer nighttime anomaly by TIMED/GUVI. J. Geophys. Res. Space Phys. 116 (A7), A07313. https://doi.org/10.1029/2010JA016150.

Huang, C.-S., Kelley, M.C., 1996. Nonlinear evolution of equatorial spread F: 2. Gravity wave seeding of Rayleigh-Taylor instability. J. Geophys. Res. Space Phys. 101 (A1), 293–302.

Huang, C.-S., Foster, J.C., Sahai, Y., 2007. Significant depletions of the ionospheric plasma density at middle latitudes: a possible signature of equatorial spread F bubbles near the plasmapause. J. Geophys. Res. Space Phys. 112, A05315.

Huang, C.-S., Le, G., de La Beaujardiere, O., Roddy, P.A., Hunton, D.E., Pfaff, R.F., Hairston, M.R., 2014. Relationship between plasma bubbles and density enhancements: observations and interpretation. J. Geophys. Res. Space Phys. 119 (2), 1325–1336.

Huang, F., Dou, X., Lei, J., Lin, J., Ding, F., Zhong, J., 2016. Statistical analysis of nighttime medium-scale traveling ionospheric disturbances using airglow images and GPS observations over central China. J. Geophys. Res. Space Phys. 121 (9), 8887–8899.

Huba, J.D., Joyce, G., 2007. Equatorial spread F modeling: multiple bifurcated structures, secondary instabilities, large density 'bite-outs,' and supersonic flows. Geophys. Res. Lett. 34, L07105.

Huba, J.D., Joyce, G., 2010. Global modeling of equatorial plasma bubbles. Geophys. Res. Lett. 37, L17104.

Huba, J.D., Joyce, G., Krall, J., 2008. Three-dimensional equatorial spread F modeling. Geophys. Res. Lett. 35, L10102.

Huba, J.D., Ossakow, S.L., Joyce, G., Krall, J., England, S.L., 2009. Three-dimensional equatorial spread F modeling: zonal neutral wind effects. Geophys. Res. Lett. 36, L19106.

Hunsucker, R.D., 1982. Atmospheric gravity waves generated in the high-latitude ionosphere: a review. Rev. Geophys. 20 (2), 293–315.

Immel, T.J., Sagawa, E., England, S.L., Henderson, S.B., Hagan, M.E., Mende, S.B., Frey, H.U., Swenson, C.M., Paxton, L.J., 2006. Control of equatorial ionospheric morphology by atmospheric tides. Geophys. Res. Lett. 33. https://doi.org/10.1029/2006GL026161.

Immel, T.J., England, S.L., Mende, S.B., Heelis, R.A., Englert, C.R., Edelstein, J., Frey, H.U., Korpela, E. J., Taylor, E.R., Craig, W.W., Harris, S.E., Bester, M., Bust, G.S., Crowley, G., Forbes, J.M., Gérard, J.-C., Harlander, J.M., Huba, J.D., Hubert, B., Kamalabadi, F., Makela, J.J., Maute, A.I., Meier, R.R., Raftery, C., Rochus, P., Siegmund, O.H.W., Stephan, A.W., Swenson, G.R., Frey, S., Hysell, D.L., Saito, A., Rider, K.A., Sirk, M.M., 2017. The ionospheric connection explorer mission: mission goals and design. Space Sci. Rev. 214 (1), 13. https://doi.org/10.1007/s11214-017-0449-2.

Ishida, T., Hosokawa, K., Shibata, T., Suzuki, S., Nishitani, N., Ogawa, T., 2008. SuperDARN observations of daytime MSTIDs in the auroral and mid-latitudes: possibility of long-distance propagation. Geophys. Res. Lett. 35, L13102.

Jacchia, L.G., 1965. Static diffusion models of the upper atmosphere with empirical temperature profiles. Smithson. Contrib. Astrophys. 8, 215–257.

Jee, G., Burns, A.G., Kim, Y.-H., Wang, W., 2009. Seasonal and solar activity variations of the Weddell Sea Anomaly observed in the TOPEX total electron content measurements. J. Geophys. Res., A04307. https://doi.org/10.1029/2008JA01380.

Jiang, J., Wan, W., Ren, Z., Yue, X., 2018. Asymmetric DE3 causes WN3 in the ionosphere. J. Atmos. Sol. Terr. Phys. 173, 14–22. https://doi.org/10.1016/j.jastp.2018.04.006.

Jin, H., Miyoshi, Y., Fujiwara, H., Shinagawa, H., 2008. Electrodynamics of the formation of ionospheric wave number 4 longitudinal structure. J. Geophys. Res. 113. https://doi.org/10.1029/2008JA013301.

Jing, N., Hunsucker, R.D., 1993. A theoretical investigation of sources of large and medium scale atmospheric gravity waves in the auroral oval. J. Atmos. Terr. Phys. 55 (13), 1667–1679.

Johnson, C.Y., 1966. Ionospheric composition and density from 90 to 1200 kilometers at solar minimum. J. Geophys. Res. (1896–1977) 71 (1), 330–332. https://doi.org/10.1029/JZ071i001p00330.

Jonah, O.F., Coster, A., Zhang, S., Goncharenko, L., Erickson, P.J., de Paula, E.R., Kherani, E.A., 2018. TID observations and source analysis during the 2017 memorial day weekend geomagnetic storm over North America. J. Geophys. Res. Space Phys. 123, 8749–8765. https://doi.org/10.1029/2018JA025367.

Jones Jr., M., Forbes, J.M., Hagan, M.E., 2014. Tidal-induced net transport effects on the oxygen distribution in the thermosphere. Geophys. Res. Lett. 41 (14), 5272–5279. https://doi.org/10.1002/2014GL060698.

Jones Jr., M., Emmert, J.T., Drob, D.P., Picone, J.M., Meier, R.R., 2018. Origins of the thermosphere-ionosphere semiannual oscillation: reformulating the "thermospheric spoon" mechanism. J. Geophys. Res. Space Phys. 123 (1), 931–954. https://doi.org/10.1002/2017JA024861.

Kato, S., 1957. Horizontal wind-systems in the ionospheric E region deduced from the dynamo theory of the geomagnetic S_q variation. Part IV. J. Geomagn. Geolectr. 9, 107–115.

Kelley, M.C., 1989. The Earth's Ionosphere. Academic Press, New York.

Kelley, M.C., 2009. The Earth's Ionosphere: Plasma Physics and Electrodynamics. Academic Press.

Kelley, M.C., Fukao, S., 1991. Turbulent upwelling of the mid-latitude ionosphere: 2. Theoretical framework. J. Geophys. Res. Space Phys. 96 (A3), 3747–3753.

Kelley, M.C., Miller, C.A., 1997. Electrodynamics of midlatitude spread F 3. Electrohydrodynamic waves? A new look at the role of electric fields in thermospheric wave dynamics. J. Geophys. Res. Space Phys. 102 (A6), 11539–11547.

Kelley, M.C., Haldoupis, C., Nicolls, M.J., Makela, J.J., Belehaki, A., Shalimov, S., Wong, V.K., 2003a. Case studies of coupling between the E and F regions during unstable sporadic-E conditions. J. Geophys. Res. Space Phys. 108 (A12), 1447.

Kelley, M.C., Makela, J.J., Vlasov, M.N., Sur, A., 2003b. Further studies of the Perkins stability during Space Weather Month. J. Atmos. Sol. Terr. Phys. 65 (10), 1071–1075.

Kil, H., Heelis, R.A., 1998. Global distribution of density irregularities in the equatorial ionosphere. J. Geophys. Res. Space Phys. 103 (A1), 407–417.

Kil, H., Paxton, L.J., 2017. Global distribution of nighttime medium-scale traveling ionospheric disturbances seen by Swarm satellites. Geophys. Res. Lett. 44 (18), 9176–9182.

Kil, H., DeMajistre, R., Paxton, L.J., Zhang, Y., 2006. Nighttime F-region morphology in the low and middle latitudes seen from DMSP F15 and TIMED/GUVI. J. Atmos. Sol. Terr. Phys. 68 (14), 1672–1681.

Kil, H., Oh, S.-J., Kelley, M.C., Paxton, L.J., England, S.L., Talaat, E., Min, K.-W., Su, S.-Y., 2007. Longitudinal structure of the vertical E× B drift and ion density seen from ROCSAT-1. Geophys. Res. Lett. 34, L14110.

Kil, H., Heelis, R.A., Paxton, L.J., Oh, S.-J., 2009. Formation of a plasma depletion shell in the equatorial ionosphere. J. Geophys. Res. 114, A11302. https://doi.org/10.1029/2009JA014369.

Kil, H., Choi, H.-S., Heelis, R.A., Paxton, L.J., Coley, W.R., Miller, E.S., 2011. Onset conditions of bubbles and blobs: a case study on 2 March 2009. Geophys. Res. Lett. 38, L06101.

Kil, H., Kwak, Y.-S., Lee, W.K., Miller, E.S., Oh, S.-J., Choi, H.-S., 2015. The causal relationship between plasma bubbles and blobs in the low-latitude F region during a solar minimum. J. Geophys. Res. Space Phys. 120 (5), 3961–3969.

Kil, H., Miller, E.S., Jee, G., Kwak, Y.-S., Zhang, Y., Nishioka, M., 2016. Comment on "The night when the auroral and equatorial ionospheres converged" by Martinis, C., J. Baumgardner, M. Mendillo, J. Wroten, A. Coster, and L. Paxton. J. Geophys. Res. Space Phys. 121 (10), 10599.

Kil, H., Paxton, L.J., Lee, W.K., Jee, G., 2019. Daytime evolution of equatorial plasma bubbles observed by the first Republic of China satellite. Geophys. Res. Lett. 46 (10), 5021–5027.

Kirchengast, G., Hocke, K., Schlegel, K., 1996. The gravity wave-TID relationship: insight via theoretical model—EISCAT data comparison. J. Atmos. Terr. Phys. 58 (1–4), 233–243.

Komjathy, A., Yang, Y.M., Meng, X., Verkhoglyadova, O., Mannucci, A.J., Langley, R.B., 2016. Review and perspectives: understanding natural-hazards-generated ionospheric perturbations using GPS measurements and coupled modeling. Radio Sci. 51 (7), 951–961.

Kondo, T., Richmond, A.D., Liu, H., Lei, J., Watanabe, S., 2011. On the formation of a fast thermospheric zonal wind at the magnetic dip equator. Geophys. Res. Lett. 38, L10101.

Kosch, M.J., Cierpka, K., Rietveld, M.T., Hagfors, T., Schlegel, K., 2001. High-latitude ground-based observations of the thermospheric ion-drag time constant. Geophys. Res. Lett. 28 (7), 1395–1398.

Kotake, N., Otsuka, Y., Tsugawa, T., Ogawa, T., Saito, A., 2006. Climatological study of GPS total electron content variations caused by medium-scale traveling ionospheric disturbances. J. Geophys. Res. Space Phys. 111, A04306.

Kotake, N., Otsuka, Y., Ogawa, T., Tsugawa, T., Saito, A., 2007. Statistical study of medium-scale traveling ionospheric disturbances observed with the GPS networks in Southern California. Earth Planets Space 59 (2), 95–102.

Krall, J., Huba, J.D., Joyce, G., Yokoyama, T., 2010. Density enhancements associated with equatorial spread F. Ann. Geophys. 28 (2), 327–337.

Krall, J., Huba, J.D., Ossakow, S.L., Joyce, G., 2010b. Equatorial spread F fossil plumes. Ann. Geophys. 28, 2059–2069.

Kwak, Y.-S., Richmond, A., 2007. An analysis of the momentum forcing in the high-latitude lower thermosphere. J. Geophys. Res. Space Phys. 112, A01306.

Kwak, Y.S., Richmond, A.D., Roble, R.G., 2007. Dependence of the high-latitude lower thermospheric momentum forcing on the interplanetary magnetic field. J. Geophys. Res. Space Phys. 112, A06316.

Lakshmi Narayanan, V., Shiokawa, K., Otsuka, Y., Saito, S., 2014. Airglow observations of nighttime medium-scale traveling ionospheric disturbances from Yonaguni: statistical characteristics and low-latitude limit. J. Geophys. Res. Space Phys. 119 (11), 9268–9282. https://doi.org/10.1002/2014JA020368.

Le, G., Huang, C.-S., Pfaff, R.F., Su, S.-Y., Yeh, H.-C., Heelis, R.A., Rich, F.J., Hairston, M., 2003. Plasma density enhancements associated with equatorial spread F: ROCSAT-1 and DMSP observations. J. Geophys. Res. Space Phys. 108 (A8), 1318.

Le, H., Liu, L., Yue, X., Wan, W., Ning, B., 2009. Latitudinal dependence of the ionospheric response to solar eclipses. J. Geophys. Res. Space Phys. 114, A07308.

Lee, W.K., Kil, H., Paxton, L.J., 2018. Tropical ionization trough in the ionosphere seen by swarm—a satellite. Geophys. Res. Lett. 45 (22), 12135.

Lee, W.K., Kil, H., Paxton, L.J., 2021. Global distribution of nighttime MSTIDs and its association with E region irregularities seen by CHAMP satellite. J. Geophys. Res. Space Phys. 126, e2020JA028836.

Lei, J., Roble, R.G., Kawamura, S., Fukao, S., 2007. A simulation study of thermospheric neutral winds over the MU radar. J. Geophys. Res. Space Phys. 112, A04303.

Lei, J., Thayer, J.P., Wang, W., Luan, X., Dou, X., Roble, R., 2012a. Simulations of the equatorial thermosphere anomaly: physical mechanisms for crest formation. J. Geophys. Res. Space Phys. 117, A06318. https://doi.org/10.1029/2012JA017613.

Lei, J., Thayer, J.P., Wang, W., Richmond, A.D., Roble, R., Luan, X., Dou, X., Xue, X., Li, T., 2012. Simulations of the equatorial thermosphere anomaly: field-aligned ion drag effect. J. Geophys. Res. Space Phys. 117, A01304.

Lei, J., Dang, T., Wang, W., Burns, A., Zhang, B., Le, H., 2018a. Long-lasting response of the global thermosphere and ionosphere to the 21 August 2017 solar eclipse. J. Geophys. Res. Space Phys. 123 (5), 4309–4316.

Lei, J., Huang, F., Chen, X., Zhong, J., Ren, D., Wang, W., Yue, X., Luan, X., Jia, M., Dou, X., et al., 2018b. Was magnetic storm the only driver of the long-duration enhancements of daytime total electron content in the Asian-Australian sector between 7 and 12 September 2017? J. Geophys. Res. Space Phys. 123 (4), 3217–3232.

Li, J., Wang, W., Lu, J., Yuan, T., Yue, J., Liu, X., Zhang, K., Burns, A.G., Zhang, Y., Li, Z., 2018. On the responses of mesosphere and lower thermosphere temperatures to geomagnetic storms at low and middle latitudes. Geophys. Res. Lett. 45 (19), 10128.

Lillis, R.J., Fillingim, M.O., Ma, Y., Gonzalez-Galindo, F., Forget, F., Johnson, C.L., Mittelholz, A., Russell, C.T., Andersson, L., Fowler, C.M., 2019. Modeling wind-driven ionospheric dynamo currents at Mars: expectations for insight magnetic field measurements. Geophys. Res. Lett. 46 (10), 5083–5091.

Lin, C.H., Richmond, A.D., Heelis, R.A., Bailey, G.J., Lu, G., Liu, J.-Y., Yeh, H.C., Su, S.-Y., 2005. Theoretical study of the low-and midlatitude ionospheric electron density enhancement during the October 2003 superstorm: relative importance of the neutral wind and the electric field. J. Geophys. Res. Space Phys. 110, A12312.

Lin, C.S., Immel, T.J., Yeh, H.-C., Mende, S.B., Burch, J.L., 2005b. Simultaneous observations of equatorial plasma depletion by IMAGE and ROCSAT-1 satellites. J. Geophys. Res. Space Phys. 110 (A6).

Lin, C.-H., Hsiao, C.C., Liu, J.Y., Liu, C.H., 2007. Longitudinal structure of the equatorial ionosphere: time evolution of the four-peaked EIA structure. J. Geophys. Res. Space Phys. 112 (A12).

Lin, C.H., Liu, J.Y., Cheng, C.Z., Chen, C.H., Liu, C.H., Wang, W., Burns, A.G., Lei, J., 2009. Three-dimensional ionospheric electron density structure of the Weddell Sea Anomaly. J. Geophys. Res. 114. https://doi.org/10.1029/2008JA013455.

Lin, C.-H., Lin, J.-T., Chen, C.H., Liu, J.Y., Sun, Y.Y., Kakinami, Y., Matsumura, M., Chen, W.H., Liu, H., Rau, R.-J., et al., 2014. Ionospheric shock waves triggered by rockets. Ann. Geophys. 32, 1145–1152.

Lin, C.C.H., Shen, M.-H., Chou, M.-Y., Chen, C.-H., Yue, J., Chen, P.-C., Matsumura, M., 2017. Concentric traveling ionospheric disturbances triggered by the launch of a Spacex Falcon 9 rocket. Geophys. Res. Lett. 44 (15), 7578–7586. https://doi.org/10.1002/2017GL074192.

Liu, H., Watanabe, S., 2008. Seasonal variation of the longitudinal structure of the equatorial ionosphere: does it reflect tidal influences from below? J. Geophys. Res. 113. https://doi.org/10.1029/2008JA013027.

Liu, J.Y., Chen, Y.I., Pulinets, S.A., Tsai, Y.B., Chuo, Y.J., 2000. Seismo-ionospheric signatures prior to M $\geq$ 6.0 Taiwan earthquakes. Geophys. Res. Lett. 27 (19), 3113–3116.

Liu, H., Lühr, H., Henize, V., Köhler, W., 2005. Global distribution of the thermospheric total mass density derived from CHAMP. J. Geophys. Res. 110.

Liu, H., Lühr, H., Watanabe, S., Köhler, W., Henize, V., Visser, P., 2006. Zonal winds in the equatorial upper thermosphere: decomposing the solar flux, geomagnetic activity, and seasonal dependencies. J. Geophys. Res. 111.

Liu, H., Lühr, H., Watanabe, S., 2007a. Climatology of the equatorial Mass Density Anomaly. J. Geophys. Res. 112.

Liu, H., Stolle, C., Watanabe, S., Abe, T., Rother, M., Cooke, D.L., 2007b. Evaluation of the IRI model using CHAMP observations in polar and equatorial regions. Adv. Space Res.

Liu, H., Watanabe, S., Kondo, T., 2009a. Fast thermospheric wind jet at the Earth's dip equator. Geophys. Res. Lett. 36.

Liu, H., Yamamoto, M., Lühr, H., 2009b. Wave-4 pattern of the equatorial mass density anomaly—a thermospheric signature of tropical deep convection. Geophys. Res. Lett. 36.

Liu, H., Thampi, S., Yamamoto, M., 2010. Phase reversal of the diurnal cycle in the midlatitude ionosphere. J. Geophys. Res. 115.

Liu, H., Doornbos, E., Nakashima, J., 2016. Thermospheric wind observed by GOCE: wind jets and seasonal variations. J. Geophys. Res. 121, 6901–6913.

Liu, H., Thayer, J., Zhang, Y., Lee, W., 2017. The non-storm time corrugated upper thermosphere: what is beyond MSIS? Space Weather 15, 746–760.

Liu, H., Lei, J., Yamazaki, Y., 2020. Day-to-day variability of the thermosphere and ionosphere. In: Wang, W., Zhang, Y. (Eds.), Geophysical Monograph Series: Upper Atmosphere Dynamics and Energetics. American Geophysical Union.

Lu, G., Goncharenko, L., Nicolls, M.J., Maute, A., Coster, A., Paxton, L.J., 2012. Ionospheric and thermospheric variations associated with prompt penetration electric fields. J. Geophys. Res. Space Phys. 117, A08312.

Lu, G., Zakharenkova, I., Cherniak, I., Dang, T., 2020. Large-scale ionospheric disturbances during the 17 March 2015 storm: a model-data comparative study. J. Geophys. Res. Space Phys. 125 (5), A08312.

Lühr, H., Rother, M., Häusler, K., Alken, P., Maus, S., 2008. The influence of nonmigrating tides on the longitudinal variation of the equatorial electrojet. J. Geophys. Res. Space Phys. 113, A08313.

Lyons, L.R., Nishimura, Y., Zhang, S.-R., Coster, A.J., Bhatt, A., Kendall, E., Deng, Y., 2019. Identification of auroral zone activity driving large-scale traveling ionospheric disturbances. J. Geophys. Res. Space Phys. 124 (1), 700–714.

Ma, G., Maruyama, T., 2006. A super bubble detected by dense GPS network at East Asian longitudes. Geophys. Res. Lett. 33, L21103.

Makela, J.J., 2006. A review of imaging low-latitude ionospheric irregularity processes. J. Atmos. Sol. Terr. Phys. 68 (13), 1441–1458.

Makela, J.J., Otsuka, Y., 2012. Overview of nighttime ionospheric instabilities at low-and mid-latitudes: coupling aspects resulting in structuring at the mesoscale. Space Sci. Rev. 168 (1–4), 419–440.

Makela, J.J., Miller, E.S., Talaat, E.R., 2010. Nighttime medium-scale traveling ionospheric disturbances at low geomagnetic latitudes. Geophys. Res. Lett. 37, L24104.

Martinis, C., Mendillo, M., 2007. Equatorial spread F-related airglow depletions at Arecibo and conjugate observations. J. Geophys. Res. Space Phys. 112, A10310.

Martinis, C., Baumgardner, J., Mendillo, M., Su, S.-Y., Aponte, N., 2009. Brightening of 630.0 nm equatorial spread-F airglow depletions. J. Geophys. Res. Space Phys. 114, A06318.

Martinis, C., Baumgardner, J., Wroten, J., Mendillo, M., 2010. Seasonal dependence of MSTIDs obtained from 630.0 nm airglow imaging at Arecibo. Geophys. Res. Lett. 37, L11103.

Martinis, C., Baumgardner, J., Wroten, J., Mendillo, M., 2011. All-sky imaging observations of conjugate medium-scale traveling ionospheric disturbances in the American sector. J. Geophys. Res. Space Phys. 116, A07310.

Martinis, C., Baumgardner, J., Mendillo, M., Wroten, J., Coster, A., Paxton, L., 2015. The night when the auroral and equatorial ionospheres converged. J. Geophys. Res. Space Phys. 120 (9), 8085–8095.

Martinis, C., Baumgardner, J., Mendillo, M., Wroten, J., MacDonald, T., Kosch, M., Lazzarin, M., Umbriaco, G., 2019. First conjugate observations of medium-scale traveling ionospheric disturbances (MSTIDs) in the Europe-Africa longitude sector. J. Geophys. Res. Space Phys. 124 (3), 2213–2222.

Maruyama, N., Watanabe, S., Fuller-Rowell, T.J., 2003. Dynamic and energetic coupling in the equatorial ionosphere and thermosphere. J. Geophys. Res. 108.

Matuura, N., 1966. Reaction rates in the F region. Rep. Ionos. Space Res. Jpn. 20, 289–303.

Mayr, H.G., Hedin, A.E., Reber, C.A., Carignan, G.R., 1974. Global characteristics in the diurnal variations of the thermospheric temperature and composition. J. Geophys. Res. 79, 619–628.

Mayr, H.G., Harris, I., Herrero, F.A., Spencer, N.W., Varosi, F., Pesnell, W.D., 1990. Thermospheric gravity waves: observations and interpretation using the transfer function model (TFM). Space Sci. Rev. 54 (3–4), 297–375.

McDonald, S.E., Dymond, K.F., Summers, M.E., 2008. Hemispheric asymmetries in the longitudinal structure of the low-latitude nighttime ionosphere. J. Geophys. Res. Space Phys. 113, A08308.
Mendillo, M., Baumgardner, J., Nottingham, D., Aarons, J., Reinisch, B., Scali, J., Kelley, M., 1997. Investigations of thermospheric-ionospheric dynamics with 6300-Å images from the Arecibo observatory. J. Geophys. Res. Space Phys. 102 (A4), 7331–7343.
Mendillo, M., Erickson, P.J., Zhang, S.-R., Mayyasi, M., Narvaez, C., Thiemann, E., Chamberlain, P., Andersson, L., Peterson, W., 2018. Flares at Earth and Mars: an ionospheric escape mechanism? Space Weather 315 (5811), 501.
Miller, C.A., Kelley, M.C., 1997. Horizontal plasma flow at midlatitudes: more mechanisms and the interpretation of observations. J. Geophys. Res. Space Phys. 102 (A6), 11549–11555.
Miller, E.S., Kil, H., Makela, J.J., Heelis, R.A., Talaat, E.R., Gross, A., 2014. Topside signature of medium-scale traveling ionospheric disturbances. Ann. Geophys. 32 (8), 959–965.
Miyoshi, Y., Fujiwara, H., Jin, H., Shinagawa, H., Liu, H., Terada, K., 2011. Model study on the formation of the equatorial mass density anomaly in the thermosphere. J. Geophys. Res. 116.
Miyoshi, Y., Fujiwara, H., Jin, H., Shinagawa, H., Liu, H., 2012. Numerical simulation of the equatorial wind jet in the thermosphere. J. Geophys. Res. 117.
Miyoshi, Y., Jin, H., Fujiwara, H., Shinagawa, H., 2018. Numerical study of traveling ionospheric disturbances generated by an upward propagating gravity wave. J. Geophys. Res. Space Phys. 123 (3), 2141–2155.
Moffett, R.J., Hanson, W.B., 1965. Effect of ionization transport on the equatorial F-region. Nature 206 (4985), 705–706.
Munro, G.H., 1948. Short-period changes in the F region of the ionosphere. Nature 162 (4127), 886–887.
Munro, G.H., 1958. Travelling ionospheric disturbances in the F region. Aust. J. Phys. 11 (1), 91–112.
Namba, S., Maeda, K.I., 1939. Radio Wave Propagation. Corona Publishing, Tokyo, p. 86.
Nesse Tyssøy, H., Stadsnes, J., Sørbø, M., Mertens, C.J., Evans, D.S., 2010. Changes in upper mesospheric and lower thermospheric temperatures caused by energetic particle precipitation. J. Geophys. Res. Space Phys. 115, A10323.
Nishimura, Y., Zhang, S.-R., Lyons, L.R., Deng, Y., Coster, A.J., Moen, J.I., Clausen, L.B., Bristow, W. A., Nishitani, N., 2020. Source region and propagation of dayside large-scale traveling ionospheric disturbances. Geophys. Res. Lett. 47 (19), 619.
Nishioka, M., Saito, A., Tsugawa, T., 2009. Super-medium-scale traveling ionospheric disturbance observed at midlatitude during the geomagnetic storm on 10 November 2004. J. Geophys. Res. Space Phys. 114, A05326.
Nishioka, M., Tsugawa, T., Kubota, M., Ishii, M., 2013. Concentric waves and short-period oscillations observed in the ionosphere after the 2013 Moore EF5 tornado. Geophys. Res. Lett. 40 (21), 5581–5586.
Oberheide, J., Pedatella, N.M., Gan, Q., Kumari, K., Burns, A.G., Eastes, R.W., 2020. Thermospheric composition O/N response to an altered meridional mean circulation during sudden stratospheric warmings observed by GOLD. Geophys. Res. Lett. 47 (1).
Ogawa, T., Nishitani, N., Otsuka, Y., Shiokawa, K., Tsugawa, T., Hosokawa, K., 2009. Medium-scale traveling ionospheric disturbances observed with the SuperDARN Hokkaido radar, all-sky imager, and GPS network and their relation to concurrent sporadic E irregularities. J. Geophys. Res. Space Phys. 114, A03316.
Otsuka, Y., Shiokawa, K., Ogawa, T., Wilkinson, P., 2004. Geomagnetic conjugate observations of medium-scale traveling ionospheric disturbances at midlatitude using all-sky airglow imagers. Geophys. Res. Lett. 31, L15803.
Otsuka, Y., Onoma, F., Shiokawa, K., Ogawa, T., Yamamoto, M., Fukao, S., 2007. Simultaneous observations of nighttime medium-scale traveling ionospheric disturbances and E region field-aligned irregularities at midlatitude. J. Geophys. Res. Space Phys. 112 (A6).
Otsuka, Y., Tani, T., Tsugawa, T., Ogawa, T., Saito, A., 2008. Statistical study of relationship between medium-scale traveling ionospheric disturbance and sporadic E layer activities in summer night over Japan. J. Atmos. Sol. Terr. Phys. 70 (17), 2196–2202.
Otsuka, Y., Kotake, N., Shiokawa, K., Ogawa, T., Tsugawa, T., Saito, A., 2011. Statistical study of medium-scale traveling ionospheric disturbances observed with a GPS receiver network in Japan. In: Aeronomy of the Earth's Atmosphere and Ionosphere, Springer, pp. 291–299.
Otsuka, Y., Suzuki, K., Nakagawa, S., Nishioka, M., Shiokawa, K., Tsugawa, A., 2013. GPS observations of medium-scale traveling ionospheric disturbances over Europe. Ann. Geophys. 31 (2), 163–172.

Oya, H., Takahashi, T., Watanabe, S., 1986. Observation of low latitude ionosphere by the impedance probe on board the Hinotori satellite. J. Geomagn. Geoelectr. 38 (2), 111–123.
Park, J., Min, K.W., Lee, J.-J., Kil, H., Kim, V.P., Kim, H.-J., Lee, E., Lee, D.Y., 2003. Plasma blob events observed by KOMPSAT-1 and DMSP F15 in the low latitude nighttime upper ionosphere. Geophys. Res. Lett. 30, 2114.
Park, J., Min, K., Kim, V., Kil, H., Kim, H., Lee, J.J., Lee, E., Kim, S.J., Lee, D.Y., Hairston, M., 2008. Statistical description of low-latitude plasma blobs as observed by DMSP F15 and KOMPSAT-1. Adv. Space Res. 41 (4), 650–654.
Park, J., Lühr, H., Stolle, C., Rother, M., Min, K.W., Chung, J.-K., Kim, Y.H., Michaelis, I., Noja, M., 2009. Magnetic signatures of medium-scale traveling ionospheric disturbances as observed by CHAMP. J. Geophys. Res. Space Phys. 114, A03307.
Park, J., Lühr, H., Min, K.W., 2010. Neutral density depletions associated with equatorial plasma bubbles as observed by the CHAMP satellite. J. Atmos. Sol. Terr. Phys. 72, 157–163.
Park, J., Lühr, H., Lee, C., Kim, Y.H., Jee, G., Kim, J.-H., 2014. A climatology of medium-scale gravity wave activity in the midlatitude/low-latitude daytime upper thermosphere as observed by CHAMP. J. Geophys. Res. Space Phys. 119, 2187–2196.
Park, J., Lühr, H., Nishioka, M., Kwak, Y.-S., 2015. Plasma density undulations correlated with thermospheric neutral mass density in the daytime low-latitude to midlatitude topside ionosphere. J. Geophys. Res. 120, 6669–6678.
Paulino, I., Medeiros, A.F., Vadas, S.L., Wrasse, C.M., Takahashi, H., Buriti, R., Leite, D., Filgueira, S., Bageston, J.V., Sobral, J.H.A., et al., 2016. Periodic waves in the lower thermosphere observed by OI630 nm airglow images. Ann. Geophys. 34 (2), 293–301.
Pawlowski, D.J., Ridley, A.J., 2008. Modeling the thermospheric response to solar flares. J. Geophys. Res. Space Phys. 113, A10309.
Pedatella, N.M., Liu, H.L., 2018. The influence of internal atmospheric variability on the ionosphere response to a geomagnetic storm. Geophys. Res. Lett. 45 (10), 4578–4585.
Perkins, F., 1973. Spread F and ionospheric currents. J. Geophys. Res. 78 (1), 218–226.
Philbrick, C.R., McIsaac, J.P., 1972. Measurements of atmospheric composition near 400 km. Space Res. 12, 743–750.
Picone, J.M., Hedin, A.E., Drob, D.P., Aikin, A.C., 2002. NRLMSISE-00 empirical model of the atmosphere: statistical comparisons and scientific issues. J. Geophys. Res. 107. https://doi.org/10.1029/2002JA009430.
Pimenta, A.A., Amorim, D.C.M., Candido, C.M.N., 2008. Thermospheric dark band structures at low latitudes in the Southern hemisphere under different solar activity conditions: a study using OI 630 nm emission all-sky images. Geophys. Res. Lett. 35, L16103.
Ponthieu, J.J., Killeen, T.L., Lee, K.M., Carignan, G.R., Hoegy, W.R., Brace, L.H., 1988. Ionosphere-thermosphere momentum coupling at solar maximum and solar minimum from DE-2 and AE-C data. Phys. Scripts 37 (3), 447.
Qian, L., Solomon, S.C., Kane, T.J., 2009. Seasonal variation of thermospheric density and composition. J. Geophys. Res. Space Phys. 114 (A1).
Qian, L., Burns, A.G., Wang, W., Solomon, S.C., Zhang, Y., Hsu, V., 2016. Effects of the equatorial ionosphere anomaly on the interhemispheric circulation in the thermosphere. J. Geophys. Res. Space Phys. 121 (3), 2522–2530.
Qian, L., Wang, W., Burns, A.G., Chamberlin, P.C., Coster, A., Zhang, S.-R., Solomon, S.C., 2019. Solar flare and geomagnetic storm effects on the thermosphere and ionosphere during 6–11 September 2017. J. Geophys. Res. Space Phys. 124 (3), 2298–2311.
Raghavarao, R., Wharton, L.E., Spencer, N.W., Mayr, H.G., Brace, L.H., 1991. An equatorial temperature and wind anomaly (ETWA). Geophys. Res. Lett. 18, 1193–1196.
Rastogi, R.G., 1960. Abnormal features of the region of the ionosphere at some southern high-latitude stations. J. Geophys. Res. 65, 585–592.
Rees, D., Fuller-Rowell, T.J., 1989. The response of the thermosphere and ionosphere to magnetospheric forcing. Philos. Trans. R. Soc. Lond. A Math. Phys. Sci. 328 (1598), 139–171.
Ren, D., Lei, J., 2017. A simulation study of the equatorial ionospheric response to the October 2013 geomagnetic storm. J. Geophys. Res. Space Phys. 122 (9), 9696–9704.

Ren, Z., Wan, W., Liu, L., Xiong, J., 2009. Intra-annual variation of wave number 4 structure of vertical E× B drifts in the equatorial ionosphere seen from ROCSAT-1. J. Geophys. Res. Space Phys. 114, A05308.

Ren, Z., Wan, W., Xiong, J., Liu, L., 2010. Simulated wave number 4 structure in equatorial F-region vertical plasma drifts. J. Geophys. Res. Space Phys. 115, A05301.

Richards, P.G., Meier, R.R., Chen, S.-P., Drob, D.P., Dandenault, P., 2017. Investigation of the causes of the longitudinal variation of the electron density in the Weddell Sea Anomaly. J. Geophys. Res. Space Phys. 122 (6), 6562–6583.

Richmond, A.D., 1978. Gravity wave generation, propagation, and dissipation in the thermosphere. J. Geophys. Res. Space Phys. 83 (A9), 4131–4145.

Richmond, A.D., 1979. Ionospheric wind dynamo theory: a review. J. Geomagn. Geoelectr. 31 (3), 287–310.

Richmond, A.D., Lathuillere, C., Vennerstrøm, S., 2003. Winds in the high-latitude lower thermosphere: dependence on the interplanetary magnetic field. J. Geophys. Res. Space Phys. 108, 1066.

Riousset, J.A., Paty, C.S., Lillis, R.J., Fillingim, M.O., England, S.L., Withers, P.G., Hale, J.P.M., 2014. Electrodynamics of the Martian dynamo region near magnetic cusps and loops. Geophys. Res. Lett. 41 (4), 1119–1125.

Rishbeth, H., 1971. The F-layer dynamo. Planet. Space Sci. 19, 263–267.

Rishbeth, H., 1972. Thermospheric winds and the F-region: a review. J. Atmos. Terr. Phys. 34 (1), 1–47.

Rishbeth, H., 2002. Whatever happened to superrotation. J. Atmos. Sol. Terr. Phys. 64, 1351–1360.

Rishbeth, H., Müller-Wodarg, I.C.F., Zou, L., Fuller-Rowell, T.J., Millward, G.H., Moffett, R.J., Idenden, D.W., Aylward, A.D., 2000a. Annual and semiannual variations in the ionospheric F2-layer: II. Physical discussion. Ann. Geophys. 18 (8), 945–956.

Rishbeth, H., Zou, L., Muller-Wodarg, I.C.F., Fuller-Rowell, T.J., Millward, G.H., Moffett, R.J., Idenden, D.W., Aylward, A.D., 2000b. Annual and semiannual variations in the ionospheric F2-layer: II. Physical discussion. Ann. Geophys. 18, 945–956.

Ruohoniemi, J.M., Greenwald, R.A., 1996. Statistical patterns of high-latitude convection obtained from Goose Bay HF radar observations. J. Geophys. Res. Space Phys. 101 (A10), 21743–21763.

Sagawa, E., Immel, T.J., Frey, H.U., Mende, S.B., 2005. Longitudinal structure of the equatorial anomaly in the nighttime ionosphere observed by IMAGE/FUV. J. Geophys. Res. Space Phys. 110 (A9).

Saito, A., Fukao, S., Miyazaki, S., 1998. High resolution mapping of TEC perturbations with the GSI GPS network over Japan. Geophys. Res. Lett. 25 (16), 3079–3082.

Saito, A., Nishimura, M., Yamamoto, M., Fukao, S., Kubota, M., Shiokawa, K., Otsuka, Y., Tsugawa, T., Ogawa, T., Ishii, M., et al., 2001. Traveling ionospheric disturbances detected in the FRONT campaign. Geophys. Res. Lett. 28 (4), 689–692.

Salah, J.E., Oliver, W.L., Foster, J.C., Holt, J.M., Emery, B.A., Roble, R.G., 1986. Observations of the May 30, 1984, annular solar eclipse at Millstone Hill. J. Geophys. Res. 91, 1651–1660.

Savitzky, A., Golay, M.J.E., 1964. Smoothing and differentiation of data by simplified least squares procedures. Anal. Chem. 36, 1627–1639.

Scherliess, L., Thompson, D.C., Schunk, R.W., 2008. Longitudinal variability of low-latitude total electron content: tidal influences. J. Geophys. Res. Space Phys. 113, A01311.

Schunk, R.W., 1983. The terrestrial ionosphere. In: Carovillano, R.L., Forbes, J.M. (Eds.), Solar-Terrestrial Physics. 609. Reidel Publ. Co., Dordrecht. vol.

Schunk, R.W., 2014. Ionosphere-thermosphere physics: current status and problems. In: Huba, J., Schunk, R., Khazanov, G. (Eds.), Geophysical Monograph Series: Modeling the Ionosphere-Thermosphere System. American Geophysical Union, Washington, DC, pp. 3–12.

Schunk, R., Nagy, A., 2009. Ionospheres: Physics, Plasma Physics, and Chemistry. Cambridge University Press.

Sheng, C., Deng, Y., Zhang, S.-R., Nishimura, Y., Lyons, L.R., 2020. Relative contributions of ion convection and particle precipitation to exciting large-scale traveling atmospheric and ionospheric disturbances. J. Geophys. Res. Space Phys. 125 (2), 1667.

Shiokawa, K., Ihara, C., Otsuka, Y., Ogawa, T., 2003. Statistical study of nighttime medium-scale traveling ionospheric disturbances using midlatitude airglow images. J. Geophys. Res. Space Phys. 108 (A1), 1052.

Shiokawa, K., Otsuka, Y., Tsugawa, T., Ogawa, T., Saito, A., Ohshima, K., Kubota, M., Maruyama, T., Nakamura, T., Yamamoto, M., et al., 2005. Geomagnetic conjugate observation of nighttime medium-scale and large-scale traveling ionospheric disturbances: FRONT3 campaign. J. Geophys. Res. Space Phys. 110, A05303.

Shiokawa, K., Otsuka, Y., Ogawa, T., 2006. Quasiperiodic southward moving waves in 630-nm airglow images in the equatorial thermosphere. J. Geophys. Res. Space Phys. 111, A06301.

Stolle, C., Liu, H., 2014. Low-latitude ionosphere and thermosphere: decadal observations from the CHAMP mission. In: Huba, J., Schunk, R., Khazanov, G. (Eds.), Geophysical Monograph Series: Modeling the Ionosphere-Thermosphere System. American Geophysical Union, Washington, DC, pp. 259–272.

Stolle, C., Lühr, H., Rother, M., Balasis, G., 2006. Magnetic signatures of equatorial spread F as observed by the CHAMP satellite. J. Geophys. Res. 111.

Su, S.Y., Liu, C.H., Ho, H.H., Chao, C.K., 2006. Distribution characteristics of topside ionospheric density irregularities: equatorial versus midlatitude regions. J. Geophys. Res. Space Phys. 111, A06305.

Sultan, P.J., 1996. Linear theory and modeling of the Rayleigh-Taylor instability leading to the occurrence of equatorial spread F. J. Geophys. Res. Space Phys. 101 (A12), 26875–26891.

Tarpley, J.D., 1970. The ionospheric wind dynamo—I. Lunar tide. Planet. Space Sci. 18, 1075–1090.

Testud, J., 1973. Ondes atmosphériques de grande échelle et sous-orages magnétiques (Ph.D. thesis). Université de Paris.

Thampi, S., Lin, C.H., Liu, H., Yamamoto, M., 2009. First tomographic observations of the mid-latitudes summer night anomaly (MSNA) over Japan. J. Geophys. Res. 114.

Thayer, J.P., Killeen, T.L., 1993. A kinematic analysis of the high-latitude thermospheric neutral circulation pattern. J. Geophys. Res. Space Phys. 98 (A7), 11549–11565.

Thayer, J.P., Killeen, T.L., McCormac, F.G., Tschan, C.R., Ponthieu, J.J., Spencer, N.W., 1987. Thermospheric neutral wind signatures dependent on the East-West component of the interplanetary magnetic field for Northern and southern hemispheres as measured from dynamics explorer-2. AnGeo 5, 363–368.

Thome, G.D., 1964. Incoherent scatter observations of traveling ionospheric disturbances. J. Geophys. Res. 69 (19), 4047–4049.

Torr, M.R., Torr, D.G., 1985. Ionization frequencies for solar cycle 21: revised. J. Geophys. Res. Space Phys. 90 (A7), 6675–6678.

Tsugawa, T., Saito, A., Otsuka, Y., Yamamoto, M., 2003. Damping of large-scale traveling ionospheric disturbances detected with GPS networks during the geomagnetic storm. J. Geophys. Res. Space Phys. (1978–2012) 108 (A3), 669.

Tsugawa, T., Saito, A., Otsuka, Y., 2004. A statistical study of large-scale traveling ionospheric disturbances using the GPS network in Japan. J. Geophys. Res. Space Phys. (1978–2012) 109 (A6), 669.

Tsugawa, T., Saito, A., Otsuka, Y., Nishioka, M., Maruyama, T., Kato, H., Nagatsuma, T., Murata, K.T., 2011. Ionospheric disturbances detected by GPS total electron content observation after the 2011 off the Pacific coast of Tohoku Earthquake. Earth Planets space 63 (7), 875–879.

Tsunoda, R.T., 2006. On the coupling of layer instabilities in the nighttime midlatitude ionosphere. J. Geophys. Res. Space Phys. 111, A11304.

Tsunoda, R.T., Cosgrove, R.B., 2001. Coupled electrodynamics in the nighttime midlatitude ionosphere. Geophys. Res. Lett. 28 (22), 4171–4174.

Vadas, S.L., 2007. Horizontal and vertical propagation and dissipation of gravity waves in the thermosphere from lower atmospheric and thermospheric sources. J. Geophys. Res. Space Phys. 112, A06305.

Vadas, S.L., Fritts, D.C., 2002. The importance of spatial variability in the generation of secondary gravity waves from local body forces. Geophys. Res. Lett. 29 (20).

Valladares, C., Sheehan, R., 2016. Observations of conjugate MSTIDs using networks of GPS receivers in the American sector. Radio Sci. 51 (9), 1470–1488.

Vickrey, J.F., Vondrak, R.R., Matthews, S.J., 1982. Energy deposition by precipitating particles and joule dissipation in the auroral ionosphere. J. Geophys. Res. Space Phys. 87 (A7), 5184–5196.

Wan, W., Liu, L., Pi, X., Zhang, M.L., Ning, B., Xiong, J., Ding, F., 2008. Wavenumber-4 patterns of the total electron content over the low latitude ionosphere. Geophys. Res. Lett. L12104.

Wan, W., Ren, Z., Ding, F., Xiong, J., Liu, L., Ning, B., Zhao, B., Li, G., Zhang, M.L., 2012. A simulation study for the couplings between DE3 tide and longitudinal WN4 structure in the thermosphere and ionosphere. J. Atmos. Sol. Terr. Phys. 90–91, 52–60.

Wang, W., Lei, J., Burns, A.G., Solomon, S.C., Wiltberger, M., Xu, J., Zhang, Y., Paxton, L., Coster, A., 2010. Ionospheric response to the initial phase of geomagnetic storms: common features. J. Geophys. Res. Space Phys. 115, A07321.

Wang, W., Lei, J., Burns, A.G., Qian, L., Solomon, S.C., Wiltberger, M., Xu, J., 2011. Ionospheric day-to-day variability around the whole heliosphere interval in 2008. Sol. Phys. 274, 457–472.

Watanabe, S., Oya, H., 1986. Occurrence characteristics of low latitude ionosphere irregularities observed by impedance probe on board the Hinotori satellite. J. Geomagn. Geoelectr. 38 (2), 125–149.

Whitehead, J.D., 1961. The formation of the sporadic-E layer in the temperate zones. J. Atmos. Terr. Phys. 20 (1), 49–58.

Wilson, G.R., Weimer, D.R., Wise, J.O., Marcos, F.A., 2006. Response of the thermosphere to Joule heating and particle precipitation. J. Geophys. Res. Space Phys. 111, A10314.

Copernicus GmbH Woodman, R.F., 2009. Spread F-an old equatorial aeronomy problem finally resolved? Ann. Geophys. 27 (5), 1915–1934.

Woodman, R.F., La Hoz, C., 1976. Radar observations of F region equatorial irregularities. J. Geophys. Res. (1896–1977) 81 (31), 5447–5466.

Wu, Q., Schreiner, W.S., Ho, S.-P., Liu, H.-L., Qian, L., 2017. Observations and simulations of eddy diffusion and tidal effects on the semiannual oscillation in the ionosphere. J. Geophys. Res. Space Phys. 122 (10), 10502–10510.

Xiong, C., Lühr, H., 2014. The midlatitude summer night anomaly as observed by CHAMP and GRACE: interpreted as tidal features. J. Geophys. Res. Space Phys. 119 (6), 4905–4915.

Yeh, K.C., Liu, C.H., 1974. Acoustic-gravity waves in the upper atmosphere. Rev. Geophys. 12 (2), 193–216.

Yeh, K.C., Webb, H.D., Cowling, D.H., 1972. Evidence of directional filtering of travelling ionospheric disturbances. Nat. Phys. Sci. 235 (59), 131–132.

Yokoyama, T., 2014. Hemisphere-coupled modeling of nighttime medium-scale traveling ionospheric disturbances. Adv. Space Res. 54 (3), 481–488.

Yokoyama, T., 2017. A review on the numerical simulation of equatorial plasma bubbles toward scintillation evaluation and forecasting. Prog. Earth Planet. Sci. 4 (1), 1–13.

Yokoyama, T., Hysell, D.L., 2010. A new midlatitude ionosphere electrodynamics coupling model (MIECO): latitudinal dependence and propagation of medium-scale traveling ionospheric disturbances. Geophys. Res. Lett. 37, L08105.

Yokoyama, T., Su, S.-Y., Fukao, S., 2007. Plasma blobs and irregularities concurrently observed by ROCSAT-1 and Equatorial Atmosphere Radar. J. Geophys. Res. Space Phys. 112, A05311.

Yokoyama, T., Hysell, D.L., Otsuka, Y., Yamamoto, M., 2009. Three-dimensional simulation of the coupled Perkins and Es-layer instabilities in the nighttime midlatitude ionosphere. J. Geophys. Res. Space Phys. 114, A03308.

Yokoyama, T., Shinagawa, H., Jin, H., 2014. Nonlinear growth, bifurcation, and pinching of equatorial plasma bubble simulated by three-dimensional high-resolution bubble model. J. Geophys. Res. Space Phys. 119 (12), 10474–10482.

Yuan, T., Zhang, Y., Cai, X., She, C.-Y., Paxton, L., 2015. Impacts of CME-induced geomagnetic storms on the midlatitude mesosphere and lower thermosphere observed by a sodium LiDAR and TIMED/GUVI. Geophys. Res. Lett. 42 (18), 7295–7302.

Yue, J., Jian, Y., Wang, W., Meier, R.R., Burns, A., Qian, L., Jones Jr., M., Wu, D.L., Mlynczak, M., 2019. Annual and semiannual oscillations of thermospheric composition in TIMED/GUVI limb measurements. J. Geophys. Res. Space Phys. 124 (4), 3067–3082.

Zakharenkova, I., Astafyeva, E., Cherniak, I., 2016. GPS and GLONASS observations of large-scale traveling ionospheric disturbances during the 2015 St. Patrick's Day storm. J. Geophys. Res. Space Phys. 121 (12), 12138.

Zhang, S.-R., 2021. Ionosphere and thermosphere coupling at mid- and subauroral latitudes. In: Huang, C., Lu, G. (Eds.), Space Physics and Aeronomy, vol. 3: Advances in Ionospheric Research: Current Understanding and Challenges. American Geophysical Union, Washington, DC, pp. 339–368.

Zhang, S.-R., Huang, X.-Y., 1995. A numerical study of ionospheric profiles for mid-latitudes. Ann. Geophys. 13 (5), 551–557.

Zhang, S.-R., Foster, J.C., Coster, A.J., Erickson, P.J., 2011. East-West Coast differences in total electron content over the continental US. Geophys. Res. Lett. 38, L19101.
Zhang, S.-R., Foster, J.C., Holt, J.M., Erickson, P.J., Coster, A.J., 2012. Magnetic declination and zonal wind effects on longitudinal differences of ionospheric electron density at midlatitudes. J. Geophys. Res. Space Phys. (1978–2012) 117 (A8), A08329.
Zhang, S.-R., Erickson, P.J., Foster, J.C., Holt, J.M., Coster, A.J., Makela, J.J., Noto, J., Meriwether, J.W., Harding, B.J., Riccobono, J., et al., 2015. Thermospheric poleward wind surge at midlatitudes during great storm intervals. Geophys. Res. Lett. 42 (13), 5132–5140.
Zhang, S.-R., Erickson, P.J., Zhang, Y., Wang, W., Huang, C., Coster, A.J., Holt, J.M., Foster, J.F., Sulzer, M.P., Kerr, R., 2017a. Observations of ion-neutral coupling associated with strong electrodynamic disturbances during the 2015 St. Patrick's Day storm. J. Geophys. Res. Space Phys. 122 (1), 1314–1337.
Zhang, S.-R., Zhang, Y., Wang, W., Verkhoglyadova, O.P., 2017b. Geospace system responses to the St. Patrick's Day storms in 2013 and 2015. J. Geophys. Res. Space Phys. 122, 6901–6906.
Zhang, S.-R., Coster, A.J., Erickson, P.J., Goncharenko, L.P., Rideout, W., Vierinen, J., 2019a. Traveling ionospheric disturbances and ionospheric perturbations associated with solar flares in September 2017. J. Geophys. Res. Space Phys. 60 (8), 895.
Zhang, S.-R., Erickson, P.J., Coster, A.J., Rideout, W., Vierinen, J., Jonah, O., Goncharenko, L.P., 2019b. Subauroral and polar traveling ionospheric disturbances during the 7–9 September 2017 storms. Space Weather 17 (1), 1748–1764.
Zhang, S.-R., Erickson, P.J., Gasque, L.C., Aa, E., Rideout, W., Vierinen, J., et al., 2021. Electrified post-sunrise ionospheric perturbations at Millstone Hill. Geophys. Res. Lett., 48 (18), e2021GL095151.
Zhou, Q., Mathews, J.D., 2006. On the physical explanation of the Perkins instability. J. Geophys. Res. Space Phys. 111 (A12).

Index

Note: Page numbers followed by *f* indicate figures and *t* indicate tables.

G

H

I

J

K

Printed in the United States
by Baker & Taylor Publisher Services